Earthquake Hazards and Mitigation

Earthquake Hazards and Mitigation

Editors
R. Ayothiraman
Hemanta Hazarika

I.K. International Publishing House Pvt. Ltd.
NEW DELHI

Published by

I.K. International Publishing House Pvt. Ltd.
S-25, Green Park Extension
Uphaar Cinema Market
New Delhi–110 016 (India)
E-mail:info@ikinternational.com
Website: www.ikbooks.com

ISBN 978-81-89866-77-8

Reprint 2015

Published by Krishan Makhijani for I.K. International Publishing House Pvt. Ltd., S-25, Green Park Extension, Uphaar Cinema Market, New Delhi–110 016 and Printed by Rekha Printers Pvt. Ltd., Okhla Industrial Area, Phase II, New Delhi–110 020.

Preface

We have the great pleasure in welcoming all the participants to the *International Workshop on Earthquake Hazards and Mitigations (EHAM-2007)* on *December 7 and 8, 2007* in Guwahati, India.

It is needless to emphasize again the importance of carrying out investigation and research on prediction of earthquake hazards and finding out appropriate mitigation measures. The extent of damage caused by the various earthquake related hazards occurred in historical time to the recent time are self explanatory. Extensive research is being carried out on prediction of earthquake hazards and its mitigations at global/national/regional levels in the last few decades. Several seminars, symposiums, workshops and conferences are being organized at national and international levels regularly to share recent experiences and research findings among the academicians, design consultants and practicing engineers. However, every earthquake teaches a new lesson; causes severe problems and imposes new challenge, thus driving us to pursue the research on finding the solutions to these problems and challenges. Organization of workshops, symposiums, seminars and conferences on a regular basis on the field of earthquake engineering and allied areas is essential to exchange the present state-of-the-art and to promote recent and advanced technologies in mitigating earthquake hazards. Thus, we have the theme of the workshop as "*Earthquake Hazards and Mitigations*".

The objective of this workshop is to: (1) bring together people from all spheres to exchange knowledge and discuss the new developments in the field of earthquake engineering and allied areas, (2) exchange information about present state-of-the-art and current practice adopted globally in prediction and mitigation of earthquake hazards, and (3) explore novel and innovative methods for prediction and mitigation of hazards considering the future earthquakes for building sustainable/ safe infrastructures and ensuring safety of community. It is expected that this workshop will contribute towards achieving these objectives through the exchange of information and ideas among the international scientific and research community.
The themes covered broadly in this workshop are:

(1) Engineering Seismology and Seismic Hazards

(3) Geotechnical Earthquake Engineering

(4) Structural Earthquake Engineering

(5) Other related issues

Under these broad themes, several sub-topics related to estimation of structural and geotechnical hazards caused by earthquake and their mitigation methods are covered in the workshop.

There has been an overwhelming response to our call for papers from all over the world. The workshop committee has been delighted at the number of interesting and stimulating papers that have been received. In this workshop, five keynote lectures and twelve invited lectures will be delivered by distinguished scholars and experts in this field. Another major attraction will be a panel discussion comprising of experts in different allied areas from various countries. Apart from keynote/ invited lectures; a total of 44 papers submitted from various countries will be presented in total of six technical sessions conducted in parallel sessions of two.

This workshop is held under the auspices of the Indian Institute of Technology Guwahati in association with Asian Technical Committee (ATC-3) on Geotechnology for Natural Hazards and Technical Committee (TC4) on Earthquake Geotechnical Engineering and Associated Problems of International Society of Soil Mechanics and Geotechnical Engineering. We would like to express sincere gratitude to both the organizations for their strong technical support.

The committee is looking forward to a very successful and convivial event for all participants. It is hoped that this workshop will provide an excellent opportunity for fruitful exchange of ideas among researchers, designers, consultants, manufactures, government officials, contractors, academicians and students and help disseminate the latest information and technology.

The publication of this volume has been possible through the sustained efforts of the core members of the organizing committee of EHAM 2007. On behalf of the organizing committee, we express our sincere thanks to all the members, who have worked tirelessly behind the scene for the success of the EHAM 2007. We also express our thanks to *Mr. Rimil Besra, Mr. Sanjoy Bhowmik (M.Tech) and Ms. Ajanta Kalita (Ph.D)* students of IIT Guwahati, who have helped in formatting of papers and for the excellent cover page design.

Finally, as the editors of this volume and secretaries of the organizing committee, we would like to express our sincere thanks for all the cooperation voluntarily given by those, who have served as international/national advisory committee members, keynote and invited speakers, discussion leaders, panelists, chairpersons, and the organizing committee members. We wish that with the active engagement from all of you during these two days, we will have a very successful workshop.

Hemanta Hazarika
Akita Prefectural University, Japan

R. Ayothiraman
IIT Guwahati, India

Advisory Committee

International Advisory Committee

Prof. A. Akbar, *Pakistan*
Prof. A. Ansal, *Turkey*
Prof. A. Huang, *Taiwan*
Prof. A. Khasanov, *Uzbekistan*
Prof. Anand Puppala, *USA*
Prof. A. K. Panah, *Iran*
Prof. A.M. Kaynia, *Norway*
Prof. A. Peiris, *Sri Lanka*
Prof. A. Zhuspbekov, *Kazakhstan*
Prof. D. T. Bergado, *Thailand*
Prof. F. H. Lee, *Singapore*
Prof. F. Tatsuoka, *Japan*
Prof. H. Rahardjo, *Singapore*
Prof. I. Towhata, *Japan*
Prof. K. Ishihara, *Japan*
Prof. K. Rollins, *USA*
Prof. K. Tokimatsu, *Japan*
Prof. K. Wakamatsu, *Japan*
Prof. K. Watanabe, *Japan*
Prof. L. Wang, *China*
Prof. M. Kim, *Korea*
Prof. M. A. Sakr, *Egypt*
Prof. M. B. Karkee, *Canada*
Prof. M. Kazama, *Japan*
Prof. M. Lin, *Taiwan*
Prof. M. Okamura, *Japan*
Prof. M. Yoshimine, *Japan*
Prof. N. Yoshida, *Japan*
Prof. Pedro Seco e Pinto, *President (ISSMGE)*
Prof. R. Orense, *Philippines*
Prof. S. Kim, *Korea*
Prof. S. Prakash, *USA*
Prof. S. Valliappan, *Australia*
Prof. S. Yasuda, *Japan*
Prof. T. Kokusuo, *Japan*
Prof. W. Sengara, *Indonesia*
Prof. W. F. Lee, *Taiwan*
Dr. Y. Iwasaki, *Japan*
Dr. Y. Sasaki, *Japan*
Dr. Y. Tsukamoto, *Japan*

National Advisory Committee

Prof. A. Boominathan, *IIT Madras*
Prof. A. Goyal, *IIT Bombay*
Prof. A. Meher Prasad, *IIT Madras*
Prof. BVS. Viswanadham, *IIT Bombay*
Prof. C.S. Manohar, *IISc Bangalore*
Prof. C.V.R. Murty, *IIT Kanpur*
Prof. D. K. Paul, *IIT Roorkee*
Prof. G. R. Reddy, *BARC, Mumbai*
Prof. K. Rajagopal, *IIT Madras*
Prof. K. S. Rao, *IIT Delhi*
Prof. M. R. Madhav, *Vice-Prest. ISSMGE*
Prof. N. Som, *Jadhavpur University*
Prof. S. K. Deb, *IIT Guwahati*
Prof. S.K. Jain, *IIT Kanpur*
Prof. S. K. Nath, *IIT Kharagpur*
Prof. S. K. Prasad, *SJCE, Mysore*
Prof. T.G. Sitharam, *IISc Bangalore*

Organizing Committee

PATRON

Prof. Gautam Barua, *Director, IIT Guwahati, India.*

CHAIRMAN

Prof. Anjan Dutta, *Head, Civil Engg Dept, IIT Guwahati, India.*

ORGANIZING SECRETARIES

Dr. R. Ayothiraman, *IIT Guwahati, India.*
Dr. Hemanta Hazarika, *Akita Prefectural University, Japan.*

LOCAL ORGANIZING COMMITTEE (IIT GUWAHATI)

Dr. Arup K. Sarma
Dr. Ashish Verma
Dr. Baleshwar Singh
Dr. B. Pradhan
Dr. Chandan Mahanta
Dr. C. Mallikarjuna
Dr. G. Barua
Dr. G. Saravana Kumar
Dr. K. D. Singh
Dr. P. Muthukumar
Dr. S. Talukdar
Dr. S.K. Deb
Dr. Saswati Chakraborty
Dr. S. Dutta
Dr. S.K. Dash
Dr. Sharad Gokhale
Dr. S. Sreedeep
Dr. P. Sreeja
Dr. STG. Raghu Kanth
Dr. Suresh A. Kartha
Dr. Teiborlong R. Lyngdoh
Dr. V. Prabhu
Mr. Arun Borsaikia
Mr. Kumar Pallav

Contents

Preface	***v***
Advisory Committee	***vii***
Organizing Committee	***viii***
Part 1 : Keynote Lectures	**1**
1. Slope Failures in 2004 Niigataken-Chuetsu Earthquake in Japan and their Evaluation by Energy *Takaji Kokusho and Tomohiro Ishizawa*	3
2. Granular Piles for Liquefaction Mitigation - Effect of Drainage, Densification and Dilation *M. R. Madhav and A. Murali Krishna*	18
3. Estimating Seismic Hazard in Indian Cities *R.N. Iyengar*	31
4. Structural Control for Seismic Disaster Mitigation *T.K. Datta*	45
5. Qinghai-Tibet Railway, China and its Earthquake Damage Mitigation *Lanmin Wang, Zhijian Wu, Junjie Sun and Luxin Zhang*	57
Part 2 : Invited Lectures	**71**
6. Countermeasures Against Seismic Failure of Existing Reclaimed Areas by Banking *S. Yasuda*	73
7. Seismic Site Characterization Using Geotechnical and Geophysical Techniques *T.G. Sitharam and P. Anbazhagan*	85
8. National Seismic Network and Earthquake Activities in Egypt *Kamal A. El-Sayed, Mohamed A. Sakr and Enayat A. Awad*	88
9. Seismic Hazard Estimation at Bed Rock and Ground Surface Levels for Chennai City *A. Boominathan and A. Suganthi*	102
10. New Strategy and Tools for Mitigation of Landslide Disasters *Ikuo Towhata*	116
11. Verification of Push-Over Analysis Method with Static Load Testing *N. Lakshmanan, K. Muthumani, G.V. Rama Rao, N. Gopalakrishnan and G.R. Reddy*	125
12. Research on Earthquake Resistant Structural Design at IIT Madras *S. R. Satish Kumar*	138
13. Studies on System Identification of Multistoreyed Buildings based on Strong Motion Data Recorded in Guwahati City Region *K. Suresh, Sajal Kanti Deb and Anjan Dutta*	150
14. Possible Coupling between Seismic Activity and Groundwater Chemistry around the Shillong Plateau, North-Eastern India *C. Mahanta, A.D.L. Skelton, L. Claesson, G. Chakrapani, J. Routh, M. Mörth and P.P. Khanna*	155
15. Seismic Design of Nuclear Facilities in India - Issues and R&D Efforts *G.R. Reddy*	165
16. Seismotectonics of Northeastern India with Special Reference to Earthquake Hazard *D.R. Nandy*	191

Part 3 : Engineering Seismology and Seismic Hazard **201**

17. Siting of LNG Terminal in Chile
Y. Zaczek and R. Boroschek 203
18. Seismic Scenario of Guwahati City
S.K. Nath, K.K.S. Thingbaijam, A. Raj, K. Shukla, I. Pal, D.R. Nandy, M.K. Yadav, B.K. Bansal, S. Dasgupta, K. Majumdar, J.R. Kayal, A.K. Shukla, S.K. Deb, J. Pathak, P.J. Hazarika and D.K. Paul 210
19. Maximum Credible Earthquake Prediction - A Neural Network Approach
K.K.S. Thingbaijam and S.K. Nath 219
20. Seismic Hazard and Microzonation Methodology For Flat Terrain
P. Anbazhagan 228
21. Seismic Hazard Assessment and Soil Effect Studies in the City of Tabriz
M. Hosseinpour, M. Hajialilue-Bonab and M. Zare 235
22. Effects of Shillong Topography on Ground Motion
Kumar Pallav, STG Raghukanth and Konjengbam Darunkumar Singh 244

Part 4 : Geotechnical Earthquake Engineering **247**

23. Effects of Embankment Rigidity on Behavior of Naturally Deposited Soils During/After Earthquakes
T. Noda, A. Asaoka, M. Nakano K. Nakai, and H. Takeuchi 249
24. Undrained Cyclic Torsional Shear Tests on Sand up to Extremely Large Strain Levels
J. Koseki, T. Kiyota, T. Sato and A.M. Mohammad 257
25. Liquefaction Susceptibility of Some Selected Sites in Haryana - A Case Study
Ashwani Jain, Manish Kumar and D.K. Soni 264
26. Stress Distributions in Earth and Rockfill Dams
B.K. Maheshwari and P. Anuradha 272
27. Testimony from Eyewitnesses of Showa Bridge Collapse in the 1964 Niigata Earthquake
K. Wakamatsu, T. Tazoh, N. Yoshida, H. Nakazawa, H. Kiku, S. Yasuda and I. Towhata 280
28. Causes of Showa Bridge Collapse in the 1964 Niigata Earthquake
N. Yoshida, T. Tazoh, K. Wakamatsu, S. Yasuda, I. Towhata, H. Kiku and H. Nakazawa 288
29. Liquefaction Resistance and Dynamic Shear Moduli of In-Situ Frozen and Reconstituted Sandy Soils
T. Kiyota, J. Koseki and T. Sato 296
30. Shaking Table Model Tests on Residual Displacements of Earth Dams for Their Performance-Based Design
S. Sendir, J. Sato and I. Towhata 303
31. Potential Flaws in Computing Liquefaction Potential Index, P_L
Chihping Kuo and Muhsiung Chang 310
32. Model Tests on Seismic Performance of Reinforced Soil Retaining Walls by Using Different Geo-Grids
S. Nakajima, K. Hong, S. Mulmui, J. Koseki, K. Watanabe and M. Tateyama 319
33. Experimental Study on Bore-type Tsunami Wave Acting on a Coastal Dike
Kazuo Tominaga, Susumu Nakano, Takeshi Okabe and Seiji Amou 326
34. Seismic Performance of Earth Embankments
S.K. Prasad, G.P. Chandradhara and P. Nanjundaswamy 335

35. Study of the Effect of Excess Pore Pressure Dissipation in Liquefaction Using Shaking Table Test
K. Kaneda, H. Yamazaki, K. Nagano and H. Hazarika 343
36. On the Choice of Input Motion for Site-Specifiic Earthquake Response Analysis
L. Govindarajju and Subhamoy Bhattacharya 349
37. Effect of Lateral Load on Piles Embedded in Sandy Slope
N. Almas Begum and K. Muthukkumaran 359
38. Empirical Correlation of Liquefaction Induced Settlement of Coromandel Coastal Sand Deposits
S. Senthamilkumar, C. Natarajan and K. Muthukumaran 370
39. Assessment of Seismic Site Characteristics of Nagapattinam Coromandel Coast Sand Deposit
S. Senthamilkumar, C. Natarajan and K. Muthukumaran 378
40. Evalaution of Liquefaction Potential of Guwahati City
R. Ayothiraman, S.T.G. Raghu Kanth and S. Sreelatha 387
41. Effect of Pile Diameter on Deflection Behavior of Piles Under Seismic Loads
R. Ayothiraman and G. Chandra Prakash 399
42. Behavior of Kaolinite Clays under Cyclic Loading
J. Rakesh Pillai, R.G. Robinson and A. Boominathan 409
43. Low Strain Shear Modulus from Field and Laboratory Tests
R. Uma Maheswari, A. Boominathan and G. R. Dodagoudar 415
44. Grain and Specimen Size Effects on Liquefaction of Fine Sands to Undrained Cyclic Triaxial Loading
K. Rangaswamy, A. Boominathan, and K. Rajagopal 423
45. Steady State of Sand in Triaxial Extension Test
M. Yoshimine and M. Kataoka 431
46. Multi Channel Analysis of Surface Wave (MASW) Testing for Site Characterization of Delhi Region
D. Neelima Satyam and K.S. Rao 439
47. Soil-Structure Interaction in the Analysis of Bridge with Deep Well Foundation
Supratic Gupta, Sumant Gupta, Pallavi Agarwal, Shailesh Chandra and K.S. Rao 447
48. Earthquake Hazard Mitigation Measures Using Tire Derived Recycled Geomaterials
H. Hazarika 453

Part 5 : Structural Earthquake Engineering **461**

49. Elasto-Plastic Energy Dissipation Device for Passive Seismic Response Control
K. Muthumani, K. Sathish kumar, N. Gopalakrishnan, R. Sreekala, G.R. Reddy and Y.M. Parulekar 463
50. Modelling Post Peak Behaviour of Concrete for Seismic Applications
R. Sreekala, N. Lakshmanan, K. Muthumani, N. Gopalakrishnan and K. Sathish Kumar 471
51. Vibration Characteristics of Base Isolated Low Rise Building using Microtremor Measurements
T. Sato, K. Tokeshi, C. Cuadra and H. Hazarika 480
52. Experimental Investigation on Retrofitting and Rehabilitation of Reinforced Concrete Opening Corners
A. Patel and M.R. Champatrao 487
53. Research on the Evaluation of the Resistance to Earthquake for the Wooden House Structure by using Microtremor Measurement
N. Koushige, K. Hayakawa, S. Honda and K. Edane 495

54. Effect of Change in Founding Depth on the Structural Response of Safety Related Nuclear Power Plant Structure
G. R. Patil, Rajiv Ranjan, K. Giridhar, G. Prabhakar, D.K Jain and U. S. P. Verma 502
55. Seismic Response Control of Complex Piping Systems Using Elasto-Plastic Dampers-Experiments and Analysis
Y.M. Parulekar, G.R. Reddy, K.K. Vaze, A.K. Ghosh, H.S. Kushwaha and Ramesh Babu 511
56. Performance of Base Isolated RCC Framed Building Under Actual Earthquake
P.N. Dubey, G.R. Reddy, S.K. Deb, K.K. Vaze, A.K. Ghosh and H. S. Kushwaha 521
57. Estimation of Hysteretic Energy Demand including P-Delta Effect Using Equivalent Systems
A. Roy Chowdhury and S. Ghosh 530
58. Non Linear Analysis of Shear Wall—Slab Interface Under Lateral Loading
S. Greeshma, K.P. Jaya, S. Asadhullah, S. Balakumar, V. Palanisamy and C. Prakash 538
59. Performance Evaluation of Structural Engineering Opensees Software on HPC and Grid-Computing Systems
M.S. Shah and P.G. Gavali 548

Part 6 : Other Related Issues **555**

60. Diaster Management Strategies in Transportation System—A Reasearch Review
E. Shalparni and A. Verma 557

Part-1

Keynote Lectures

SLOPE FAILURES IN 2004 NIIGATAKEN-CHUETSU EQ. IN JAPAN AND THEIR EVALUATION BY ENERGY

***Takaji Kokusho*[1] *and Tomohiro Ishizawa*[2]**
[1]*Professor, Civil Engineering Dept., Chuo University, Tokyo, kokusho@civil.chuo-u.ac.jp*
[2]*Assistant Professor, ditto, ishizawa@civil.chuo-u.ac.jp*

ABSTRACT

The Niigata-ken Chuetsu earthquake caused more than 4000 slope failures in the middle part of the main island of Japan. Bedding planes had a strong effect on the slope failures. Slope failures due to this particular earthquake are classified into 3 types and their mechanisms are discussed. Then, an energy approach for run-out distance of failed soil mass in view of earthquake energy together with potential energy is applied to representative slope failures during the earthquake. It is found that equivalent friction coefficient back-calculated by the energy approach is strongly dependent on the initial slope inclination though the absolute value of the former is smaller than the latter.

Keywords: Energy balance, Run-out distance, Seismic wave energy, Friction coefficient

1 INTRODUCTION

The Niigata-ken Chuetsu earthquake (M_J=6.8) occurred on October 23, 2004, which caused more than 4000 slope failures in the middle part of the main island of Japan due to the main shock and also several strong aftershocks. The damaged area (Fig.1) is known as a landslide-prone area of green-tuff, with geological structures of active folding which cover active faults underneath. Synclines and anticlines are running parallel in the north-south direction, among which rivers are flowing in the same direction. Mountains are about 400 m at the highest, and the slopes are composed of weak sedimentation rock of Neogene, alternative layers of strongly weathered mud stones and sand stones. Bedding planes had a strong effect on the slope failures. Some of the earthquake-induced slides were obviously influenced by previous landslides. Rainfalls in three days prior to the quake were 120 mm, which may have influenced the slope instabilities. It was disclosed that similar disasters accompanying countless landslides in the green-tuff soft rock areas had occurred once in every 25-30 years on average in the north and central Main Island of Japan, the lessons of which had not been learned before.

In this paper, slope failures during the earthquake are first addressed to discuss various factors influencing failure modes. Then, the baic idea of an energy approach for run-out distance of slope failure is explained by comparing a simple block model with innovative shake table tests of a model slope of dry sand and showing a flow-chart for evaluation. This approach is then applied to typical slope failures during the Niigata-ken Chuetsu earthquake to backcalculate equivalent friction coefficients to demonstrate its applicability.

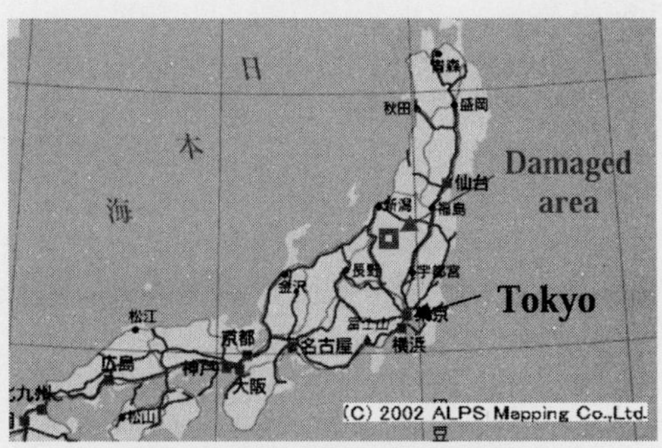

Fig.1 Area damaged during 2004 Niigataken Chuetsu earthquake.

2 CASE HISTORY DURING 2004 NIIGATAKEN CHUETSU EARTHQUAKE

2.1 Classifications of Slope Failures

The slope failures due to this particular earthquake are classified into 3 types as illustrated in Fig. 2;

Type-A: Deep slips parallel to sedimentation planes (dip slip), in gentle slopes of around 20 degrees. In many cases, displaced soil mass had originally been destabilized by river erosions or road constructions, and glided as a rigid body along the slip plane. The displaced soil volumes were very large, translating ground surface with little disturbance.

Type-B: Shallow slips of 1~2 m deep not parallel to sedimentation planes at steep slopes (>30 degrees). This type far outnumbered Type-A, but the individual soil volume was not so large. Soils normally fell down as pieces, sometimes leaving trees with deep roots at original places.

Type-C: Slips in strongly weathered colluvial soils in places where koi-ponds and terraced paddy fields were located. This type seems very peculiar in this earthquake because countless koi-ponds were located in the damaged area. This failure type obviously involved ponds, which seems to have provided water for piping and caused delayed flow-type failure of the colluvial soils. Soil liquefaction or cyclic softening may have contributed to large ground deformation including cracks because sand boils were actually witnessed at some sites.

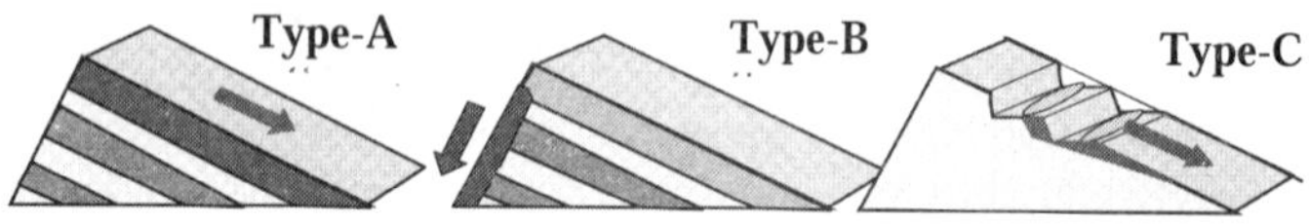

Fig.2 3 types of slope failures, A, B, and C, occured 2004 Niigataken Chuetsu Earthquake.

A number of slope failures stopped streams and made more than 80 natural reservoirs. Some of the largest ones belong to Type-A. The most typical one in Higashi-Takezawa, shown in Fig.3, where highly weathered sandstone (actually dense sandy soil) of about 20m thick glided about 100 meters horizontally along a underlying mudstone slip plane of 20 degrees. On the natural dams stopping a river, emergency dewatering and spillway-constructions were implemented to prepare for melting snow flooding in the next spring as shown in Fig.4. This treatment made the natural dam sustainable as long as engineered embankment dams.

Fig.3 Photograph of Higashi-Takezawa slide (Type-A) seen from top of scarp

Fig.4 Photograph of Higashi-Takezawa slide (Type-A) after completion of spillway.

The other typical case of Type-A is shown in Fig.5, where the upper rock mass slid down literally as a rigid body along weathered silty sand seam (inclined by 20-23 degrees) sandwiched in mudstone. In most Type-A failures, slope stabilities were deteriorated even before the earthquake because the slope toes had been cut by river flows (in Higashi-Takezawa) or road constructions (in Yokowatashi).

The Dainichi-Yama slide shown in Fig.6 is the largest slope failure occurred during the earthquake, which may be classified into Type-A. A block of 500 m by 500 m in horizontal dimension and more than 30 m thick slid as a block along a dip plane of 16 degrees, exposing an up-slope scarp of 36 degrees, and a top part of the displaced block rotated by 25 degrees along a slip circle, leaving a very unusual landscape after the event.

In type-B, the soil volume in each failure was much smaller than Type-A, though the number of failures was extremely greater. Among them, two bigger failures are shown in Fig.7 (Haguro Tunnel Entrance) and in Fig.8 (Naranoki). Their slopes were steeper than 30 degrees, slid by crossing dip planes and their volumes were exceptionally large as Type-B. Unlike Type-A failure, the failed soil mass was disintegrated into small pieces, though the run-out distance was not so long generally despite presumably high water content due to heavy rainfall prior to the earthquake.

In Figs.9 and 10, slope failures of Type-C are exemplified. Type-C failures were essentially similar to the dip slip of Type-A, but the displaced soil mass had been very much weathered and utilized as koi ponds or terraced paddy fields. The koi ponds seem to have played an important role in triggering the failures; they kept the soil in a condition of high water content and prone to seismic instability, developing ground fissures and internal erosions by pond water which eventually caused large-volume failures. Some of the failures probably occurred no sooner than the earthquake shaking while the others seem to have occurred a few days or

Fig.5 Photograph of Yokowatashi slide where mirror-like slip plane of mudstone appeared (Type-A)

Fig.6 Photograph of Dainchi-Yama slide where top of a huge slidebelonging to Type-A rotated by 25 degrees.

Fig.7 Photograph of Haguro Tunnel entrance; Large failure of Type-B.

Fig.8 Photograph of Naranoki, Large failure of Type-B.

weeks later. However, there are still a lot remained before the exact mechanism of Type-C failures is understood in detail.

3 ENERGY APPROACH

Seismically induced slope failures have normally been evaluated based on equilibrium of forces acting on a potentially sliding soil mass. This force approach can evaluate the safety factor against the slope failure, but cannot predict slide deformations, once failure occurs. From the viewpoint of the performance based design or the risk evaluation of slope failures, it is very important to know not only the safety factor but also how far the effect reaches down-slope.

Fig.9 Photograph of Kajigane failure involving koi-ponds in the upper slope (Type-C)

Fig.10 Photograph of Musigame failure involving koi-ponds (Type-C) (http://www:ajiko.co.jp)

Newmark method (Newmark 1965) or its modifications by using FEM analyses (e.g. Makdisi and Seed 1978) can evaluate the displacement of a rigid soil block along a assumed fixed slip surface based on a double integration of acceleration exceeding some threshold acting on it. In actual slope failures, however, sliding soil may not necessarily behave as a rigid body but deforms continuously without fixed slip surfaces. It sometimes tends to become destructive due to a shift from slow rigid-block slide to fast debris flow. The Newmark method normally gives displacement not exceeding about 1 m, indicating a significant limitation for evaluating flow-type failures with a long run-out distance. There exists no simple method available for performance-based design in which different levels of slope performance including flow-type failure can be evaluated at present.

In order to evaluate slope failures from their initiation to termination, an energy approach was proposed by Kokusho and Kabasawa (2003) and actually developed by Kokusho and Ishizawa (2006). In that method, four energies; potential energy change by the gravity $-\delta E_p$, earthquake energy contributing to the slope failure E_{EQ}, dissipated energy in the sliding soil mass E_{DP} and kinetic energy E_k of the sliding soil mass are correlated by the following equation;

$$-\delta E_p + E_{EQ} = E_{DP} + E_k \qquad (1)$$

or in an incremental form as;

$$-\Delta\delta E_p + \Delta E_{EQ} = \Delta E_{DP} + \Delta E_k \qquad (2)$$

Note that the potential energy change before and after failure δE_p in Eq.(1) or $\Delta\delta E_p$ in Eq.(2) is normally negative.

Once failure starts, the amount of the dissipated energy is critical to decide if it develops as a flow-type failure and how far it flows. If ΔE_{DP} is smaller than $-\Delta\delta E_p$ in a time increment when earthquake shaking has already ended; ΔE_{EQ}=0, then in Eq.(2) ΔE_k increases and the soil mass accelerates. It is also inferred from Eq.(2) that a shift from slow slide to fast flow may occur not only due to an increase in $-\Delta\delta E_p$ but also due to a drastic decrease of ΔE_{DP} caused by pore-pressure buildup in liquefiable soil, strength loss in high-sensitivity clay, etc. In fast flow failures, soil mass will keep flowing unless kinetic energy plus subsequent potential energy change ($-\delta E_p$) is all dissipated in the sliding soil mass. If $-\Delta\delta E_p$ is smaller than ΔE_{DP}, then ΔE_k is negative, hence the soil mass decelerates and stops when reserved kinetic energy E_k is all consumed. Thus, provided that failure mode and energy dissipation mechanism in flowing soil mass are known, it is possible to evaluate run-out distance in flow-type slides by the energy approach.

In the following, shake table tests carried out to investigate the energy balance in a model slope made from dry sand are addressed. The test results are compared with an energy balance in a simple rigid block model to develop an evaluation method for slope deformation based on the energy concept.

3.1 Shake Table Model Tests

A spring-supported shaking table shown in Fig. 11(a) was utilized to test a model slope made from sand, called Model-A (Fig.11(b)) here, in a rectangular lucite box of 80 cm length, 50 cm height and 40 cm in width. The model slope was made by air-pluviated dry clean Toyoura sand (total mass 30 kg) of relative density $Dr \approx 40\%$. The slope angle was parametrically changed as 29, 20, 15 and 10 degrees. The table was initially pulled to several different horizontal displacements and then released to generate decayed free vibration. The frequency of the vibration was changed in 4 steps, 2.7, 2.5, 2.2 and 2.0 Hz by attaching additional steel plates to the table.

Dissipated energy, which can be calculated from displacement amplitudes in the decay vibration depends not only on the energy dissipation due to slope deformation but also on other energy loss mechanisms such as radiation into shake table foundation and frictions in the springs and their joints. In order to evaluate the dissipated energy exclusively due to slope deformation in Model-A,

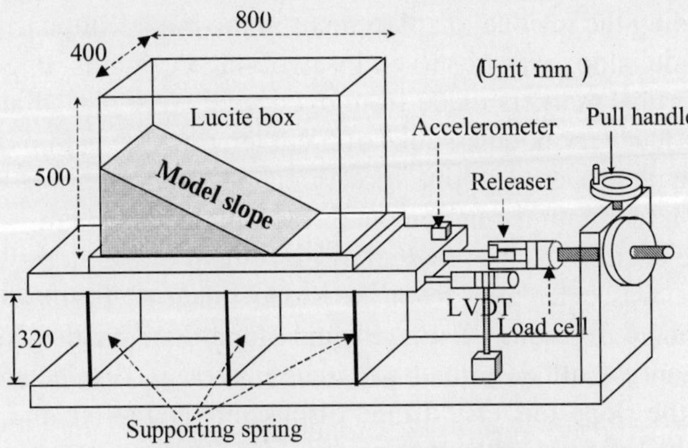

(a) Spring support shake table

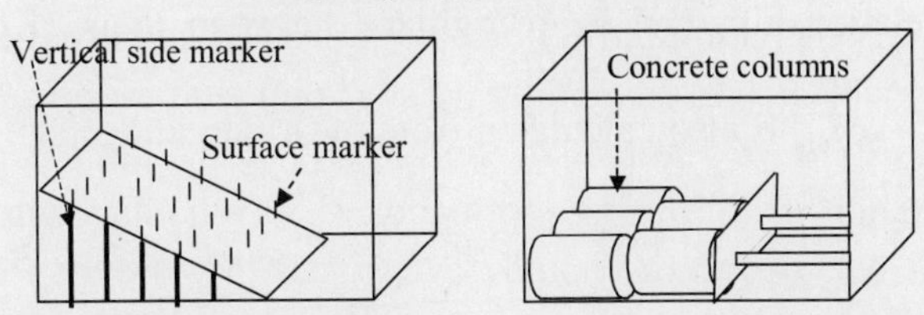

(b) 2 models compared; Model-A (left) & Model-B (right)

Fig.11 Shake table test apparatus for model slopes (a) and 2 models compared (b).

a dummy model, called Model-B consisting of a pile of rigid concrete blocks, was tested in the same lucite box in the same way (See Fig.11(b)). The total weight and the center of gravity were adjusted to be almost identical in the two models.

The decay in amplitudes, measured by a LVDT displacement gauge in both Model-A and B are exemplified in Fig.12. Note that the difference in amplitudes grows larger with the number of cycles, though the initial table displacement and the vibration period of the table are almost the same between the two models. It may be reasonable to assume that this difference reflects a greater energy dissipated in Model-A (the model slope) due to its deformations, since almost negligible energy dissipation is expected in the rigid concrete blocks in Model-B. The loss energy per cycle ΔW can be calculated as

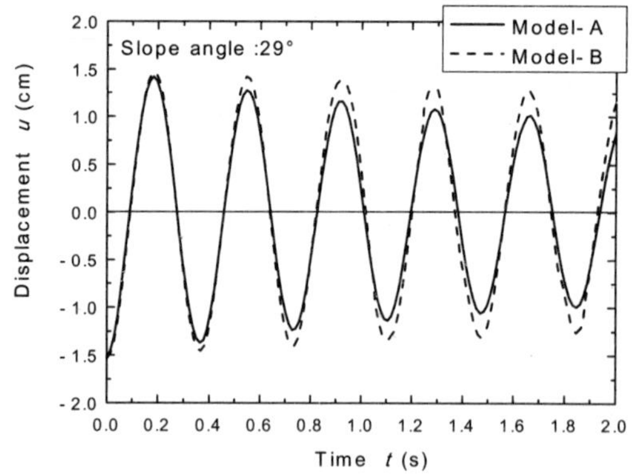

Fig.12 Decay vibrations measured by displacement gauge in Model-A and B.

$$\Delta W = 4\pi WD \tag{3}$$

in which $W = \kappa u^2/2$, representing the strain energy in the corresponding cycle, can be evaluated from a spring constant κ and a displacement amplitude u of the shaking table. Earthquake energy increment in the model slope ΔE_{EQ} in Eq.(2) can then be evaluated from a loss energy per cycle in Model-A ΔW_A ,and that in Model-B, ΔW_B as;

$$\Delta E_{EQ} = \Delta W_A - \Delta W_B \tag{4}$$

Total energy E_{EQ} calculated as a sum of ΔE_{EQ} in each cycle represents the amount of earthquake energy involved in producing the residual displacement in the model slope.

Deformation of the model slope was observed by two video cameras, one from the side and the other from the above. Vertical markers made from dyed sand were placed at the side of the model (Fig.11(b)). On the slope face, dry noodle sticks of 5 cm length were set up in line. The interval of these markers was 10 cm in the down-slope direction. The deformation was measured in each cycle of the input vibration to obtain the incremental residual displacement.

In order to correlate the energies to the residual displacement of the slope, the horizontal residual displacement of the slope was calculated from the video images. From an engineering point of view, there may be various definitions of the residual slope displacement of the deformable soil mass; average displacement of all deformed soil mass, average displacement of slope surface, average displacement of the slope toe, etc. In the discussions hereafter, the average displacement of slope surface (δ_{rs}) will be dealt with, though the other displacements could be used instead because very stable interrelationships can be recognized between them (Kokusho and Ishizawa 2007).

Potential energy change $-\Delta\delta E_p$ is also calculated from the video image as;

$$\Delta\delta E_p = \Delta\left(\rho g B \int z dx dz\right) \tag{5}$$

where z is the vertical coordinate of the slope surface and ρ is the soil density (assumed constant), and the integration is carried out over the cross-sectional area of the slope. The incremental energies, ΔE_{EQ} and $-\Delta\delta E_p$, calculated in each cycle are summed up to evaluate the corresponding total energies, E_{EQ} and $-\delta E_p$. Then, the dissipated energy E_{DP} can be readily evaluated from Eq.(1) in which E_k=0 if the energy balance after the end of slope failure is concerned. The total residual displacement δ_{rs} is also calculated by summing up all incremental displacements $\Delta\delta_{rs}$.

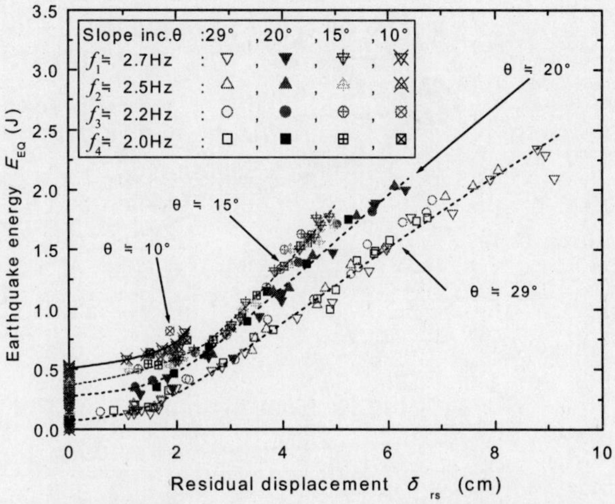

Fig.13 Earthquake energy versus residual displacement for 4 slope angles under different input frequencies

In Fig.13 the residual displacements δ_{rs} are plotted versus the vibration energy E_{EQ} contributed to slope deformations for 4 different slope angles of 29, 20, 15 and 10 degrees under 4 different input frequencies. It is remarkable that, for each slope angle, all plots can be approximated as a single curve despite the difference in the input frequency, indicating that the energy can serve as a unique determinant for slope displacement even under different shaking frequencies. Obviously, the gentler the slope is, the greater is the energy E_{EQ} to attain the same residual displacement δ_{rs}. Also noted in Fig.13 is that there seems to exist a threshold energy, corresponding to each slope angle, below which no residual displacement occurs, indicating that the energy determines not only residual displacements but also the initiation of slope displacement.

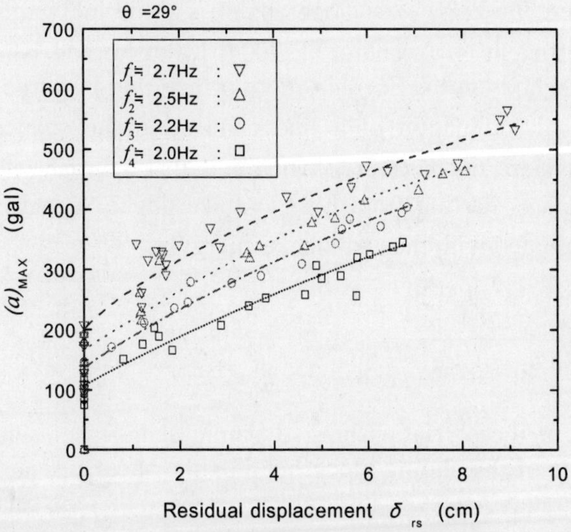

Fig.14 Maximum acceleration versus residual displacement under different input frequencies.

In order to emphasize the uniqueness of the displacement versus energy relationship, the same displacement data of the 29 degrees slope are plotted versus maximum accelerations $A_{\max}$ in the first cycle of the decayed free vibration in place of the energy in Fig.14. Obviously, the same acceleration results in different residual displacements under different input frequencies despite some data scatters, indicating that acceleration cannot be a unique determinant for slope failure not only for the residual slope displacement but even for the initiation of failure.

3.2 Test Data Interpretation by Rigid Block Model

The application of the energy approach to a rigid block model in Fig.15(a) gives the potential energy change $-\delta E_p$ and the dissipated energy due to the block slippage E_{DP} to be correlated with horizontal residual displacement δ_r as;

$$-\delta E_p = Mg\beta\delta_r \qquad (6)$$

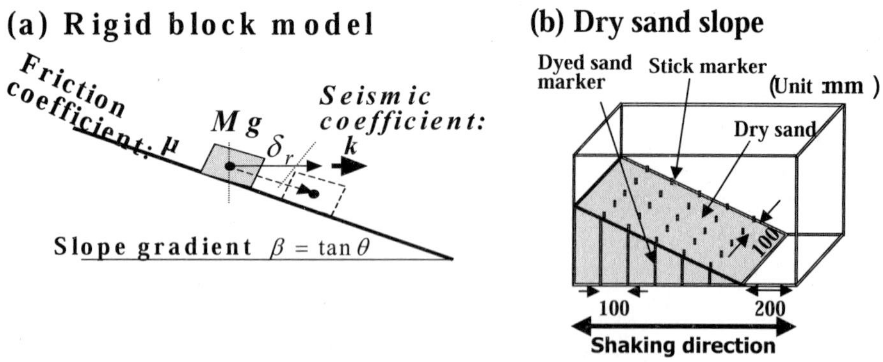

Fig.15 Comparison of models of rigid block and dry sand slope.

$$E_{DP} = Mg\frac{\mu\left(1+\beta^2\right)}{1+\mu\beta}\delta_r \tag{7}$$

where $\beta = \tan\theta$ (θ= slope angle) is slope inclination and $\mu = \tan\phi$ (ϕ= friction angle) is friction coefficient (Kokusho et al. 2004; Kokusho and Ishizawa 2006). Even for slopes not necessarily straight as in Fig.8(a) but more or less winding, the same equation also holds if β is taken as the global inclination of a line connecting the start and the end of the sliding block, and μ is considered as the average friction in that interval.

Then, starting from Eq. (1) and using $E_k = 0$ if compared before and after slope failure, δ_r is correlated with the earthquake energy as;

$$\delta_r = \frac{1+\mu\beta}{\mu-\beta}\frac{E_{EQ}}{Mg} \tag{8}$$

In these relationships, dynamic changes of seismic inertia force affect not only the driving force of the sliding block but also the shear resistance along the slip surface. If the slip plane is saturated, however, it should be assumed that the seismic inertia force is carried by temporary pore-water pressure and does not change the effective stress normal to the plane and hence the shear resistance. Consequently, for saturated slip plane, Eqs. (6) and (7) are replaced by Eqs.(6') and (7'), in which $\sigma_{n0} = Mg\Big/\left[\left(1+\beta^2\right)A\right]$ is total stress normal to the slip plane, σ'_{n0} is the corresponding effective stress and A is the horizontal area of the sliding soil mass (Kokusho and Ishizawa 2007).

$$-\delta E_p = \beta\sigma_{n0}\delta_r A\left(1+\beta^2\right) \tag{6'}$$

$$E_{DP} = \mu\sigma'_{n0}\delta_r A\left(1+\beta^2\right) \tag{7'}$$

Then, Eq. (9') is obtained in place of Eq.(9).

$$\delta_r = \frac{1}{\mu\sigma'_{n0} - \beta\sigma_{n0}}\frac{E_{EQ}}{\left(1+\beta^2\right)A} \tag{8'}$$

It is needless to say that the rigid block model cannot exactly reproduce the failure of the deformable sand slope. One of the most significant differences is that, in actual slope failures, soil mass may not slide as a rigid body along a fixed slip plane but deforms continuously with or without movable slip planes. It is interesting therefore to examine how much the failure mechanism in the sand slope can be captured by the rigid block model.

In Fig.16, the residual displacements δ_{rs}, which is considered here to be equivalent to δ_r in the rigid block model, obtained by a number of tests for different input frequencies and different slope angles are plotted versus the normalized earthquake energies E_{EQ}/Mg. The weight of the displaced soil mass Mg was evaluated from Eq. (6) using the measured potential energy $-\delta E_p$ and the measured displacement δ_{rs} to comply with the rigid block theory.

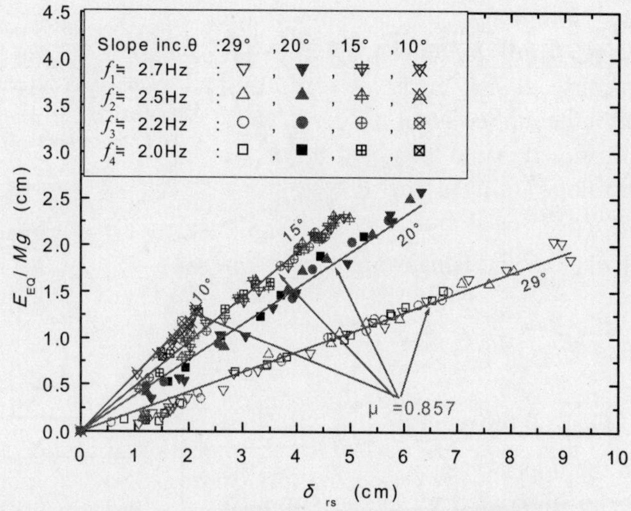

Fig.16 Normalized earthquake energy versus residual displacement for 4 slope angles under different input frequencies.

It is remarkable that the straight lines in Fig.16 corresponding to Eq. (8) for μ=0.86 can predict the residual slope displacement almost perfectly for all slope angles and all input frequencies. This indicates that if an appropriate friction coefficient is known in advance, the simple rigid block model, which apparently involves the failure mechanism quite different from reality, can successfully simulate realistic sand slopes.

3.3 Energy-Based Slope Failure Evaluation Method

Based on the model test results and their interpretation by the rigid block theory, an energy-based design method for earthquake-induced residual displacement can be proposed as shown in Fig.17. First, a sloping ground is idealized by an equivalent horizontal 2-layers system consisting of an upper layer including the slope and a base layer as shown in Fig.18. The input energy E_{IP} transmitting upward is designated at the base layer. In general, assuming 1D wave propagation of the SH wave, the wave energy E passing through a unit horizontal area A can be formulated (Kokusho et al. 2007) as;

$$E/A = \rho Vs \int (\dot{u})^2 dt \tag{9}$$

where $\dot{u}$ is the particle velocity of a wave propagating in one direction in a soil layer and ρVs is its seismic impedance (ρ = soil density and Vs = S-wave velocity). By introducing the energy radiating downward in the base layer, E_d, the earthquake energy, E_{EQ}, which is consumed in the upper ground including the slope can be obtained as;

$$E_{EQ} = E_{IP} - E_d \tag{10}$$

By assuming that no downward wave from the upper layer exists because all the energy transmitting into the upper layer is completely consumed inside the upper layer due to slope failure, the energy ratio E_{EQ}/E_{IP} can be formulated as (Kokusho and Ishizawa 2007);

$$E_{EQ}/E_{IP} = 4\alpha/(1+\alpha)^2 \quad (11)$$

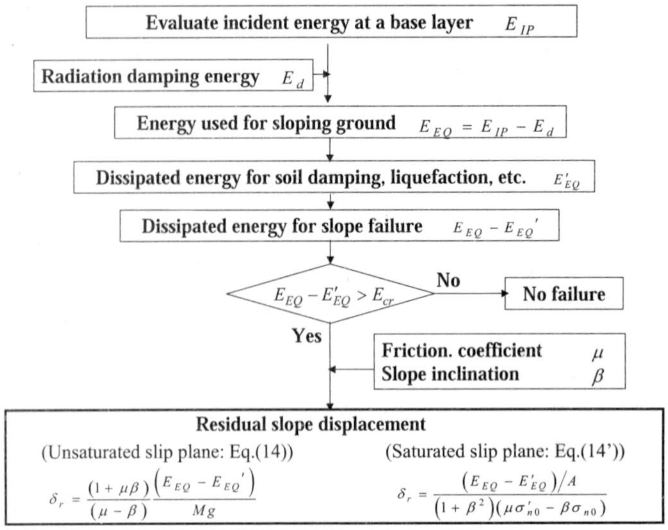

Fig.17 Flow chart for evaluating residual displacement based on input earthquake energy.

Out of the energy E_{EQ} supplied in the upper layer, a part of it seems to be dissipated by cyclic straining of soil or internal soil damping, If this cyclically dissipating energy is denoted by E'_{EQ}, then the energy used exclusively for slope failure becomes $(E_{EQ} - E'_{EQ})$. Consequently, E_{EQ} in Eq.(8) or (8') should be replaced by $(E_{EQ} - E'_{EQ})$, if E'_{EQ} is not negligibly small.

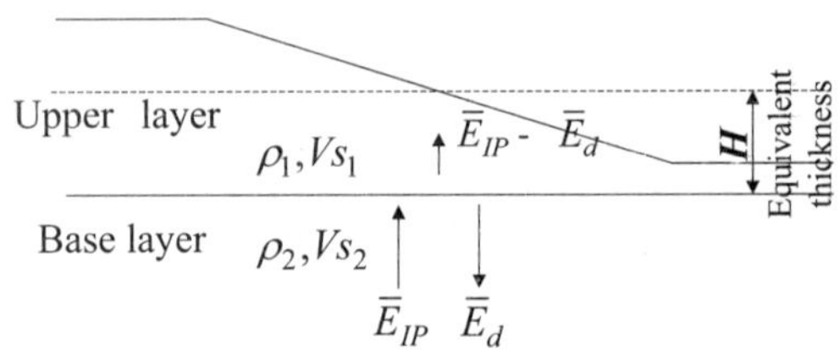

Fig.18 Idealization of sloping ground by equivalent horizontal 2-layers system

In general, the thickness or the mass of sliding soil necessary in Eq.(8) or (8') may be determined by conventional slip surface analyses to look for a potential slip surface with a minimum factor of safety. In some cases, the potential slip surface may be reasonably assumed to coincide with a bedding plane or a weaker seam.

In the above, the slope was idealized to be straight. However, even for slopes not necessarily straight, the same energy balance also hold if β is taken as global inclination and μ as averaged friction coefficient. If such a slope as illustrated in Fig.19 is given for example, a very simple procedure is possible to evaluate the run-out distance for seismically induced slope failure, which may be convenient for developing a hazard map considering the effect of run-out distance. In Fig.19, a center of gravity of a soil mass is supposed to move from P to Q during the failure. The drop height PO can be expressed as $-\delta E_p/Mg$ and the horizontal displacement OQ is δ_r. The line PQ corresponds to the global inclination β, while the inclination of the local slope is β_0. For the case of saturated slip surface, the energy ratio can be expressed as

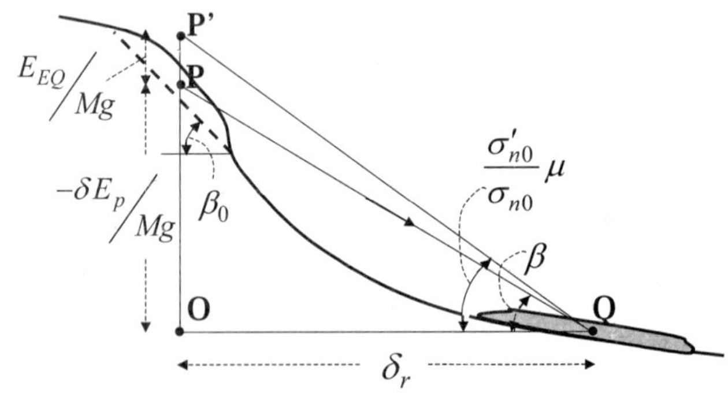

Fig.19 Simplified evaluation method for run-out distance of seismically induced slope failure.

$$\frac{-\delta E_p}{E_{EQ}} = \frac{\beta\sigma_{n0}}{(\mu\sigma'_{n0} - \beta\sigma_{n0})} \quad (12)$$

Hence, due to the earthquake energy, the centroid P is considered to climb up to P' by E_{EQ}/Mg. From Eqs.(8') and (12), the inclination of the line P'Q can be obtained as $(\sigma'_{n0}/\sigma_{n0})\mu$.

Consequently, the procedure for run-out distance evaluation is as follows;

1) Determine the dimension of a potential sliding soil mass and its centroid P.
2) Evaluate the earthquake energy E_{EQ} from Eqs.(9) and (11).
3) Locate the point P' higher than P by E_{EQ}/Mg (or $(E_{EQ}-E'_{EQ})/Mg$ if E'_{EQ} is not negligible small).
4) Draw the line P'Q with the inclination of $(\sigma'_{n0}/\sigma_{n0})\mu$ starting from the point P' and then the point Q and δ_r can be readily obtained.

4 APPLICATION OF ENERGY APPROACH TO 2004 CHUETSU EARTHQUAKE.

The most important point of the energy approach is that an equivalent friction coefficient can properly be evaluated in advance. It may be possible in some cases to evaluate it directly from soil tests. Because of complexity involved in actual slope failures in the field, however, more robust is to accumulate as many case studies as possible and back-calculating them to obtain proper friction coefficients depending on various site conditions such as topography, geology, mechanical properties, water content, etc. The 2004 Niigata-ken Chuetsu earthquake provides us with a valuable opportunity for that.

Evaluation of input energy

The input earthquake energy E_{IP} defined at base layers of slopes was extrapolated from several KIK-net vertical array records around the area (Ishizawa 2007). In Fig.20, the values of calculated input energy per unit area E_{IP}/A at several sites are plotted versus corresponding focal distance R on a log-log chart. The solid line indicates the energy per unit area calculated as:

$$E_{IP}/A = E_0/\left(4\pi R^2\right) \qquad (13)$$

The total wave energy E_0 (unit: erg=10^{-10} kJ) in Eq. (13) released from a point source is determined using an empirical equation by Gutenberg (1955);

$$\log E_0 = 1.5M + 11.8 \qquad (14)$$

in which M is the Richter's Magnitude, although M_J (Japanese Earthquake Magnitude almost equivalent to the Richter's Magnitude) is actually used in the computation. The plots for the calculated energy at bedrocks may be approximated by the dashed line parallel to the solid line corresponding to M=6.69. Thus, input energies E_{IP} at bedrocks during the earthquake may be readily computed by Eqs.(13) and (14) by earthquake magnitude and focal distances

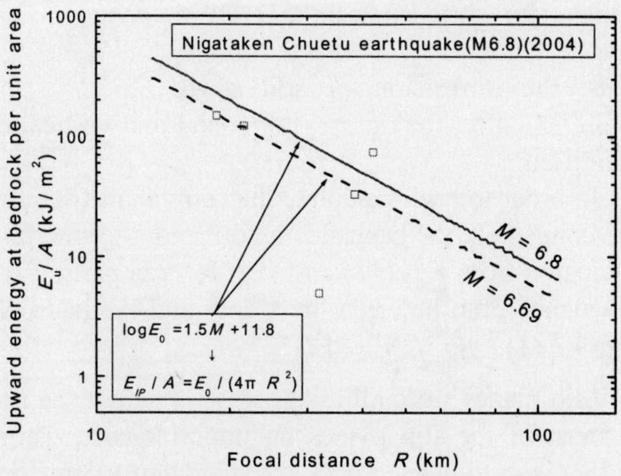

Fig.20 Input energy at bedrock, E_{IP}, versus focal distance evaluated from KIK-net records during 2004 Niigataken Chuetsu EQ.

for seismic slope hazard study in general. Fig.21 shows the spatial distribution of input energy at the base layer E_{IP} by the main shock for all slope failures tabulated as a database in this research.

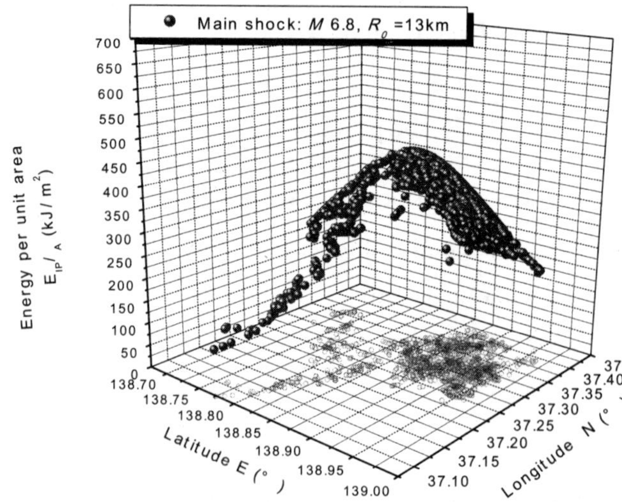

Fig.21 Distribution of input energy E_{IP} for the main shock of the 2004 Niigataken Chuetsu earthquake

Back-calculation of friction coefficient

Seven slope failures addressed earlier in the case history are chosen here again for the back-calculation; 3 for Type-A, 2 each for Types-B and C. Fig.22 depicts horizontal displacements (run-out distance) of those slopes, δ_{rn} for centroids and δ_{rt} for toes of displaced soil masses, versus pre-failure slope inclinations evaluated from 3D DEM data before and after the earthquake (Ishizawa 2007). A clear difference is obvious in pre-failure slope inclinations of failed slopes β_0 due to the difference in the failure types; the most gentle in Type-A, .the steepest in Type-B and in between in Type-C. Also obvious is larger run-out distance in Type-C presumably due to the involvement of pond water.

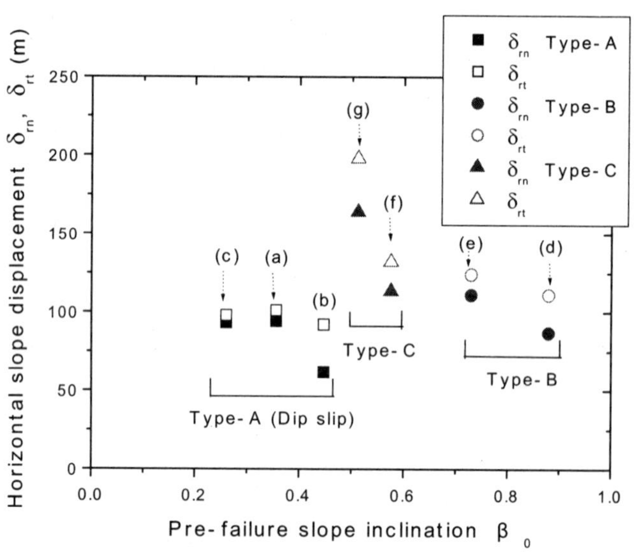

Fig.22 Horizontal slope displacement (δ_{rn}; centroid, ; δ_{rt} tip) versus pre-failure slope inclination for 3 failure types.

Fig.23 illustrates rigid block models for the 7 slope failures. The alphabets (a)–(g) in Fig.22 correspond to those in this figure. In each model, the displaced soil was idealized by a rectangular block of equivalent horizontal area and average thickness. In most of the slope failures, it was not so difficult to estimate the depth of hidden slip surfaces from 3D DEM images before and after the failure. Thus, the dimension of soil block before and after the failure, horizontal and vertical displacements of centroid and its gradient were quantified.

In order to back-calculate the equivalent friction coefficients, the average slope inclination of a line connecting the centroids before and after the failure was chosen as β=tanθ. The input energies calculated from Eqs.(13) and (14) for this particular earthquake were E_{IP}/A=244 - 301 kJ/m^2. The impedance ratio between the slopes and the bedrock was assumed $\alpha = 0.3$ for all the slopes, then from Eq.(11) $E_{EQ}/E_{IP} = 0.71$.

Slip planes were all assumed saturated at the time of earthquake. Water was actually running on most of the slip planes on impervious mudstone several days after the earthquake. Hence, Eq.(8') is used here to back calculate equivalent friction coefficient μ or equivalent friction angle from the

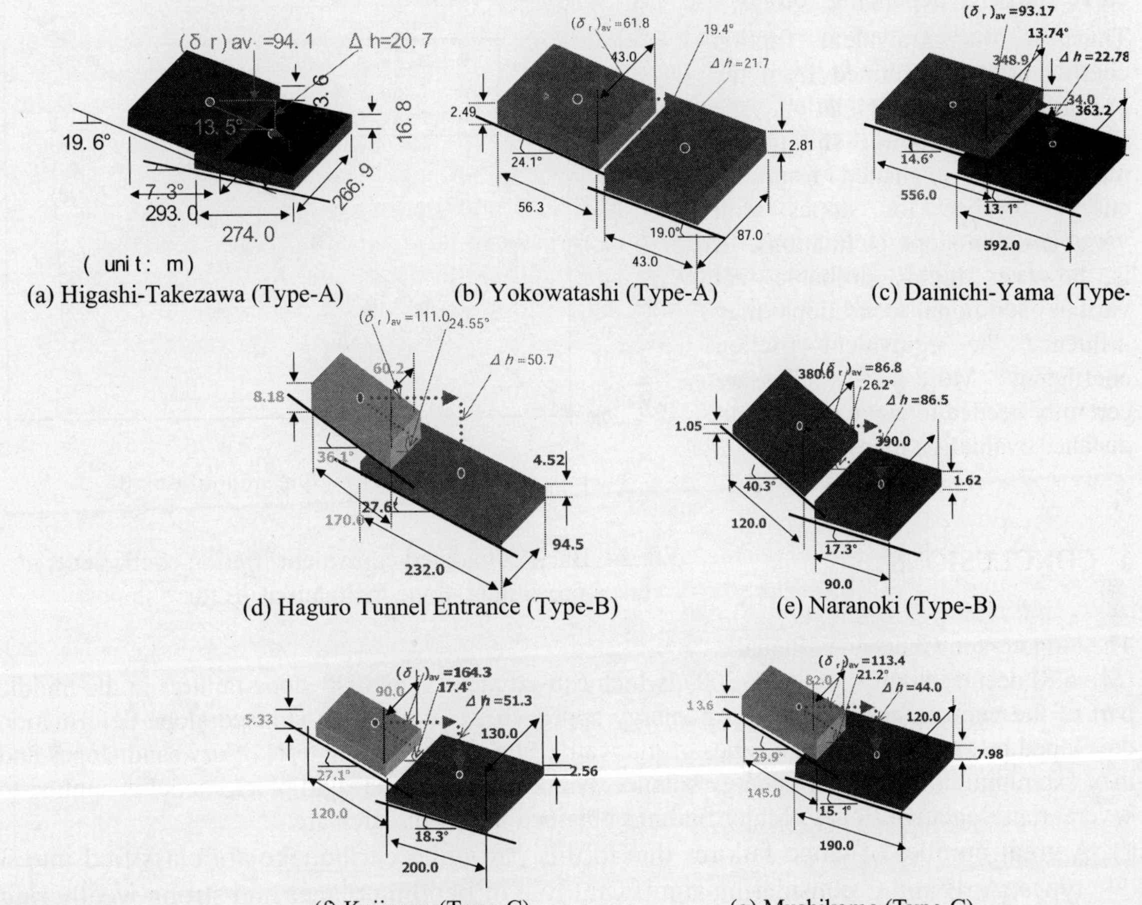

(a) Higashi-Takezawa (Type-A) (b) Yokowatashi (Type-A) (c) Dainichi-Yama (Type-

(d) Haguro Tunnel Entrance (Type-B) (e) Naranoki (Type-B)

(f) Kajigane (Type-C) (g) Mushikame (Type-C)

Fig.23 Simplified rigid block models for 7 representative slope failures; (a)-(g), during the earthquake.

models in Fig.23. The internal dissipating energy E'_{EQ} by liquefaction or soil damping was assumed negligibly small compared to other energies and ρ_t =1.8 t/m^3, assuming a realistic site condition that the major portion of the slid soil block was unsaturated except along the slip plane.

Fig.24 indicates friction coefficients μ versus pre-failure slope inclinations β_0 for the 7 slopes. The symbols (a)–(g) in this figure correspond to those in Fig.23. Note that all back-calculated μ-values are smaller than β_0 indicating that friction coefficients μ, originally larger than β_0 needless to say, decreased drastically due to the earthquake shaking. This implies that the failed soil mass accelerated first and then decelerated due to gentler or reverse slope angles in down-slope sections. In Higashi-Takezawa for example, the large difference between μ and β_0 allowed the failed soil mass to accelerate and push up a front portion of displaced soil on the opposite side of the valley as indicated in Fig.3.

The exact mechanism why this low friction coefficient could be realized is yet to be clarified. In some site conditions, seismically induced pore-pressure build-up or liquefaction may have been involved. The sandstone overlying the slip plane of cemented mudstone in Higashi-Takezawa was highly weathered and almost equivalent to decomposed medium dense sand. In Type-C, weathered soils of high water content may have shifted to high-speed mudflow due to pore-pressure build-up.

It is interesting to note that the μ versus β relationship may be approximated by a dotted curve in Fig.24 which appears to be uniquely dependent on pre-failure slope inclinations β_0 irrespective of the failure types. This may suggest that the friction coefficient which used to be larger than the pre-failure slope inclination (represented by the dashed line) tends to drop to a level of the dotted curve largely depending on β_0. Thus, if the equivalent friction coefficient is determined from the pre-failure slope inclination, the run-out distance of the sliding soil mass can be evaluated from the energy balance for slopes with variable down-slope inclinations. It is, however, highly probable that various additional conditions may influence the equivalent friction coefficient. More case studies are certainly needed to make the run-out distance evaluation more reliable.

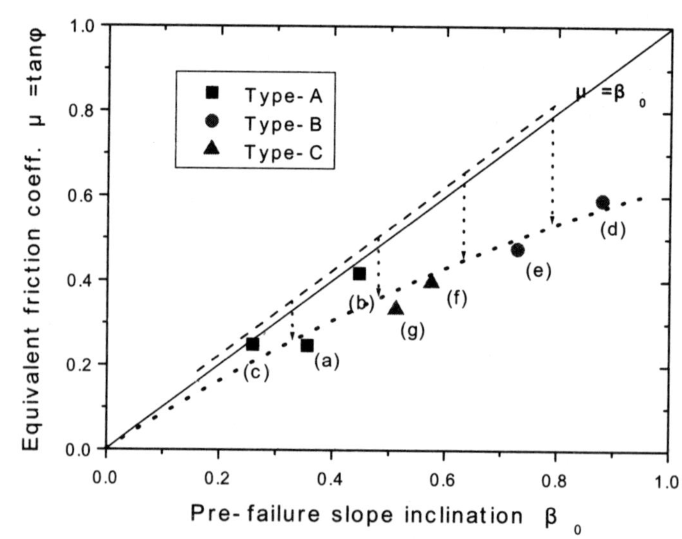

Fig.24 Back-calculated equivalent friction coefficients μ versus pre-failure slope inclinations β_0 for 7 slopes.

5 CONCLUSIONS

The Niigata-ken Chuetsu earthquake (M_J=6.8) occurred on October 23, 2004, which caused more than 4000 slope failures in the middle part of the main island of Japan. An energy approach for earthquake-induced slope deformation developed by first conducting a series of innovative shake table model tests of dry sand slopes and then examining the associated energy balance by comparing with a rigid block model is applied to several representative slopes. Major findings obtained in this research are:

1) A great number of slope failures due to this particular earthquake are classified into 3 types, A, B and C considering significant role of bedding planes and strong weathering effect.
2) Among the 3 types, there are still a lot remained before the exact mechanism of Type-C failures is understood in detail.
3) Shake table model tests yielded a unique relationship for each slope inclination between residual slope displacement δ_r and the energy E_{EQ}, which is independent of input frequency or acceleration, indicating that energy can serve as a proper parameter for evaluating residual slope displacement.
4) Comparison of the test results with energy balance for a Newmark-type rigid block indicated that the rigid block model can almost perfectly emulate continuously deforming sand slope provided that an appropriate friction coefficient μ can be estimated.
5) In order to back-calculate the friction coefficient μ, the energy-based method was applied to representative slope failures during the 2004 Niigataken Chuetsu earthquake. A simplified procedure of run-out distance evaluation for arbitrarily changing slope inclinations is indicated.
6) The back-calculated friction coefficients μ were all lower than the initial slope inclinations β_0, indicating that the failed soil mass first accelerated and then decelerated due to gentler or reverse slopes in down-slope sections.
7) The back-calculated μ versus β_0 curve seems to be almost unique irrespective of different failure types, suggesting that the run-out distance of sliding soil mass of failed slopes during this particular earthquake may be evaluated from the initial slope inclination.

ACKNOWLEDGEMENTS

A part of this research was supported by Special Coordination Funds for Promoting Science and Technology, "Earthquake damage in active-folding areas - Creation of a comprehensive data archive and suggestions for its application to remedial measures for civil-infrastructure systems -"

of Japan Science & Technology Agency. KIK-net earthquake data was provided by the internet of NIED (National Research Institute for Earth Science and Disaster Prevention) in Tsukuba, Japan.

REFERENCES

1. Gutenberg, B. (1955). The energy of earthquakes, *Quarterly Journal of the Geological Society of London*, Vol.CXII, No.455, 1-14.
2. Ishizawa, T. (2007). Evaluation of seismically induced slope displacement in terms of energy, PhD thesis, Graduate School of Science and Engineering, Chuo University, Tokyo, Japan (in Japanese).
3. Kokusho, T. and Kabasawa, K. (2003). Energy approach to flow failure and its application to flow due to water film in liquefied deposits, *Proc. of International Conference on Fast Slope Movement*, Naples, Prediction and Prevention for Risk Mitigation, Naples, 297-302.
4. Kokusho, T., Ishizawa, T. and Harada, T. (2004). Energy approach for earthquake induced slope failure evaluation, *Proc. 11th International Conference on Soil Dynamics & Earthquake Engineering and 3rd International Conference on Earthquake Geotechnical Engineering*, Berkeley, California, Vol.2, 260-267.
5. Kokusho, T. and Ishizawa, T. (2006). Energy approach for earthquake induced slope failure evaluation, *Soil Dynamics and Earthquake Engineering* 26, Elsevier, 221-230.
6. Kokusho, T. and Ishizawa, T. (2007). Energy approach to earthquake-induced slope failures and its implications. *Journal of Geotechnical and Geoenvironmental Engineering*, ASCE, Vol.133, No.7, 828-840.
7. Kokusho, T., Motoyama, R. and Motoyama, H.(2007). Wave energy in surface layers for energy-based damage evaluation, *Soil Dynamics & Earthquake Engineerng* 27, 354-366.
8. Makdisi, F. I. and Seed H. B. (1978). Simplified procedure for estimating dam and embankment earthquake –induced deformations, *Journal of Geotechnical Engineering*, Div. ASCE, Vol.104, No.GT7, 849-867.
9. Newmark, N. M. (1965). Effects of earthquakes on dams and embankments, Fifth Rankine Lecture, *Geotechnique* Vol.15, 139-159.

GRANULAR PILES FOR LIQUEFACTION MITIGATION - EFFECT OF DRAINAGAE, DENSIFICATION AND DILATION

M. R. Madhav[1] and A. Murali Krishna[2]
[1]*Professor Emeritus, Jawaharlal Nehru Technological University, Hyderabad, India.*
Email: madhavmr@gmail.com
[2]*Research Scholar, Indian Institute of Science, Bangalore, India.*
Email: amurali@civil.iisc.ernet.in

ABSTRACT

Soil liquefaction and associated ground failures have been a major source of damage during the past earthquakes. The risk of liquefaction and associated ground deformation can be reduced by various ground-improvement methods including the stone column (gravel drain) technique. The paper presents and summarises comprehensively the drainage, densification and dilation effects of granular piles in liquefaction mitigation. Liquefaction mitigation of the ground treated by granular piles (GP) is assessed considering the pore pressure generation and dissipation taking into account the densification and dilation effects along with drainage effect. Model of pore-water pressure generation and dissipation developed by Seed and Booker (1977) is applied, with some modifications for the densification and dilation effects of GPs, in terms of modifications in coefficients of volume change and permeability and generation of negative pore pressures due to dilatancy, to the analysis of columnar gravel drains under a variety of earthquake conditions. The combined effects of densification and dilation of the granular piles are proved to be more effective in mitigating the liquefaction potential of sand deposits.

Keywords: Granular piles, Ground improvement, Liquefaction mitigation, Earthquake,

1 INTRODUCTION

Seismic hazards date from many centuries. Earthquakes are constantly posing risk to life and infrastructure facilities. Seismic hazards can be described by ground shaking, structural hazards, liquefaction, landslides, failure of earth structures, lifeline, tsunami and seiche hazards. One of the most dramatic causes of damage to structures during earthquakes has been the development of liquefaction in loose saturated sand deposits, manifested either by the formation of boils and mud-spouts at the ground surface, by seepage of water through ground cracks or in some cases by the development of quick sand conditions over substantial areas (Seed and Idriss 1982). Soil liquefaction is the state at which the soil deposit looses its strength and flows as a fluid A qualitative understanding of the mechanism underlying the liquefaction of saturated sands, subjected to cyclic loading, such as that induced by earthquakes, had been recognized widely since first being examined by Casagrande in 1936 (Martin et al. 1975).

Saturated granular material, when subjected to cyclic loading involving the reversal of shear stress, tends to compact or densify. Under undrained conditions, in cases where the soil consists of loose granular materials and high water table, the tendency to get densified may result in the development of excess hydrostatic pore water pressure during each cycle of loading. If the magnitude of pore-water pressure generated equals the confining pressure, the effective stress becomes zero, the soil is said to have liquefied. Pore pressure buildup leading to liquefaction may be due to static or cyclic stress applications and the possibility of its occurrence depends on the void ratio or relative density of sand and the confining pressure (Seed 1979). Resistance to liquefaction can be improved by increasing the density (densification), modifying the grain size distribution (grouting/mixing), stabilizing the soil fabric (reinforcing), reducing the degree of saturation, dissipation of the excess pore pressures generated and interception of the propagation of excess pore pressures (drainage), etc. (Madhav and Arlekar, 2000).

A possible method of stabilizing a soil deposit susceptible to liquefaction is to install a system of gravel or rock drains (Seed and Booker 1977). Provision of sand drains/granular piles/stone columns is the most commonly adopted ground treatment methodology even for liquefaction mitigation and it has proved its effectiveness in many instances (Mitchell and Wentz 1991). Adalier and Elgamal (2004) reviewed the current state of the stone column technologies as a liquefaction countermeasure. Due to installation of gravel drains, the generated pore water pressures due to repeated loading may get dissipated almost as fast as they are generated. Thus granular piles are effective in mitigating liquefaction damage due to the drainage facility. In addition to the drainage effect, densification and reinforcement of the ambient soil around the rammed granular pile (RGP) should also be considered for total and better evaluation of the improvement.

Granular piles are of the displacement type and hence densify in situ ground during the process of installation (Madhav 2001). The effect of densification is manifested through an increase in the coefficient of earth pressure at rest and in the values of modulus of deformation of the soil (Ohbayashi et al. 1999). Densification by RGP causes increase in deformation moduli and decreases in the coefficients of permeability and volume change. The densification effect, however, decreases with distance from the center of the compaction point and may become negligible at the periphery of the unit cell. Thus, the coefficients of horizontal permeability, $k_h(r)$, and volume change, $m_v(r)$, of the soil around the granular can be considered to vary with distance, r, from the center of the granular pile. Moreover densified soils experience volume increase during shearing. Generation of very high negative pore pressures can be observed due to suppression of the tendency for dilation, which further enhances the liquefaction mitigation (Madhav and Arlekar 2000).

In this paper the pore pressure generation and dissipation of the treated ground under earthquake conditions is analyzed to quantify the densification effect of ground treatment with RGP and due to dilation of RGP. Theory of pore-water pressure generation and dissipation developed by Seed and Booker (1977) is applied, with some modifications for evaluating the densification and dilation effects of RGPs, to the analysis of columnar gravel drains under earthquake conditions.
The papers presented at the International Workshop on Earthquake Hazards and Mitigations will be printed in form of the workshop proceedings in hard copy along with a CD and be distributed during the workshop. The following is the description, and is formatted in, of the desired format for the papers. Use 11 TNR size for complete text of the paper.

2 RAMMED GRANULAR PILES

Ground improvement by means of granular piles/stone columns/geopiers, which is associated with partial substitution of the in–situ soil, originated in sixties. Granular piles ground provide increased bearing capacity, significant reduction in settlement, free drainage, increase of liquefaction resistance, etc. Granular piles are installed by vibro-compaction, vibro-replacement, cased bore hole (rammed stone columns/RGP) or by simple auger boring methods. RGP are installed into the ground by partial or full displacement methods and by ramming in stages, using a heavy falling weight, within a 'pre-bored casing' or 'driven closed end casing', retracting the casing pipe stepwise. In the latter case, driving of closed end tube itself densifies the surrounding soil.

2.1 Densification effect

Ramming action of the falling weight tends to ram the stone into the sides of the hole, which densifies more and reinforces the ground. Densification and reinforcement effects cause modifications in the properties of the in situ soil. This densification and modification in the soil parameters of the ground are not uniform over the entire zone of surrounding ground but is a function of the distance from the point of densification. During the process of installation of RGP, the soil adjacent to and in the vicinity of the point of treatment gets densified most. This densification effect decreases with the distance from the point of densification. Thus, the densification effect of the ground, due to the installation of granular pile, is maximum near the periphery of the granular pile and decreases with the distance from the granular pile. Densification by RGP causes increase in deformation moduli and decrease in the coefficients of permeability and of volume change.

Sujatha (1998) analyzed several field test data and postulated linear and exponential variations of the stiffness of soil with distance as limits for the densification effect. Similar linear variations are considered for the reductions of flow and deformation parameters of the in situ soil (coefficients of permeability and volume change). This reduction will be most at the point of densification and decays with distance towards the periphery of the unit cell reaching the original in-situ values at the farthest end of the unit cell.

2.1.1 Linear variation of the flow parameters

A linear variation of the coefficient of horizontal permeability, k_h, with distance from the center of RGP is considered with k_h value becoming minimum at the edge of RGP (r = a) and maximum at the periphery of the unit cell (r = b) (Fig. 1). The general expression for the variation of the coefficient of permeability at a distance 'r' from the centre of the granular pile considering linear variation can be expressed as

$$k_h(r)=\left(\frac{k_{sb}-k_{sa}}{b-a}\right).(r-a)+k_{sa} \tag{1}$$

where k_{sa} and k_{sb} are the coefficients of permeability at the edge of RGB (r=a) and at the edge of the unit cell (r=b) respectively. Non-dimensionalising with k_{hi}, initial in-situ value of coefficient of permeability

$$R_k(r)=\left(\frac{R_{kb}-R_{ka}}{b-a}\right).(r-a)+R_{ka} \tag{2}$$

where $R_k=k_h/k_{hi}$, $R_{kb}=k_{sb}/k_{hi}$ and $R_{ka}=k_{sa}/k_{hi}$

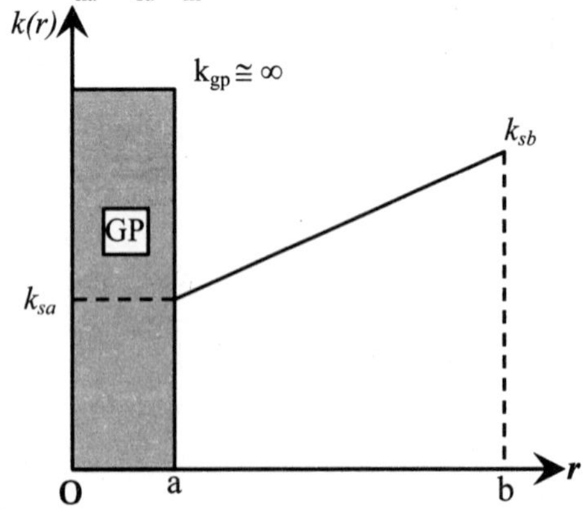

Fig. 1. Variation of coefficient of permeability of soil with distance

Similarly, the expression for linear variation of coefficient of volume change, m_v, is

$$m_v(r)=\left(\frac{m_{vb}-m_{va}}{b-a}\right).(r-a)+m_{va} \tag{3}$$

where m_{va} and m_{vb} are respectively the coefficients at distances 'a' and 'b' from the centre. Non-dimensionalizing with 'm_{vi}', initial in situ value of coefficient of volume change, one gets

$$R_{mv}(r)=\left(\frac{R_{mb}-R_{ma}}{b-a}\right).(r-a)+R_{ma} \tag{4}$$

2.2 Dilation effect

Fig. 2(a) (Vaid et al. 1981) is a typical example of stress-strain behaviour of granular material under drained conditions in a simple shear test at different vertical stress conditions. While initially loose samples undergo volume decrease, dense samples experience volume increase (dilation) during shearing. The rate of dilation increases with relative density. The response of saturated sand under undrained triaxial conditions (Leonards 1962) can be seen in Fig. 2(b). While positive pore pressures are generated in loose sands, generation of very high negative pore pressures can be observed due to suppression of the tendency for dilation.

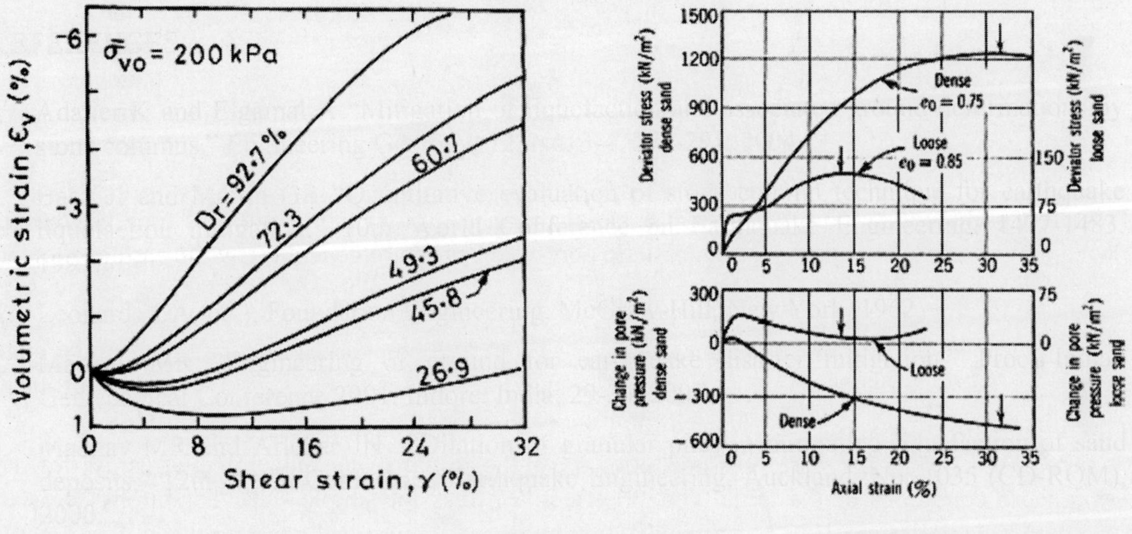

Fig. 2. (a) Stress-strain behaviour for granular material under drained conditions (Vaid et al. 1981)
(b) Response of saturated sand under undrained triaxial conditions (after Leonards 1962)

The dilation angle is one single parameter which can be readily obtained from both laboratory (drained triaxial or simple shear) and in situ (self-boring pressuremeter) tests, which can give a measure of the liquefaction resistance. Baez and Martin (1991) present an evaluation of the relative effectiveness of stone columns for the mitigation of liquefaction of soil. They also describe tests on footings on soil reinforced with stone columns, which have been used to calibrate a finite element program. The most interesting result obtained is that the stone columns experience an increase in effective stress simultaneously with the development of negative pore pressures. The effect of dilation of granular pile material on settlement response of granular pile treated ground has been investigated by Poorooshasb and Madhav (1985) and Van Impe and Madhav (1992). In the former, the response of a granular pile reinforced soil subjected to uniform loading through a relatively rigid raft is studied considering the granular pile material to follow the rigid plastic dilatant strain hardening postulates of Poorooshasb et al. (1966). The tendency for dilation is resisted by the soil, which offers larger interaction (confining) stresses. As a result, both the pile and the soil become stiffer with increasing applied stress. As a consequence of dilatant nature of the granular material, the settlement versus the intensity of loading curve exhibits a non-linear relation. The mechanical effect of dilatancy of granular material in increasing the stiffness of soft or loose soil deposits has been quantified in the above two works. The effect of dilatancy during undrained state that exists within a gravel drain/granular pile during a seismic event is also considered in the analysis.

3 SEED & BOOKER (1977) MODEL

In most practical cases, the horizontal permeability of sand/gravel deposit will be several times greater than its vertical permeability and usually the spacing between vertical drains is closer than the distance required for pore water to drain vertically towards the free surface. Furthermore, many natural deposits of sand are interspersed with narrow horizontal layers of relatively impermeable silt, which may severely inhibit vertical drainage. For these reasons, the dominant mechanism in the operation of a gravel drain system often is one of pure horizontal drainage (Seed and Booker 1977).

For flow into a gravel drain, assuming pure radial flow, and constant coefficients of permeability (k_h) and volume compressibility (m_v), the governing equation for the phenomenon can be written as (Seed and Booker 1977):

$$\frac{k_h}{\gamma_w . m_v}\left(\frac{1}{r}\frac{\partial u}{\partial r}+\frac{\partial^2 u}{\partial r^2}\right)=\frac{\partial u}{\partial t}-\frac{\partial u_g}{\partial N}.\frac{\partial N}{\partial t} \tag{5}$$

where u is the excess pore pressure at a radial distance r, from the centre, t is time, γ_w the unit weight of water, u_g = peak excess hydrostatic pore-water pressure generated by the earthquake; and

$$\frac{\partial u_g}{\partial N}=\frac{\sigma'_o}{\alpha \pi N_l}\frac{1}{\sin^{2\alpha-1}\left(\frac{\pi}{2}r_u\right)\cos\left(\frac{\pi}{2}r_u\right)} \tag{6}$$

where $r_u = u/\sigma'_o$ = the pore pressure ratio, σ'_o = the initial mean bulk effective stress for axi-symmetric conditions or the initial vertical effective stress for simple shear conditions; N_l is the number of cycles required to cause liquefaction and α = an empirical constant which is a function of the soil properties with typical average value of 0.7. The irregular cyclic loading induced by an earthquake is converted to an equivalent number, N_{eq}, of uniform cycles at an amplitude of 65% of the peak cyclic shear stress, i.e. $\tau_{cyc} = 0.65.\tau_{max}$, occurring over a duration of time, t_d, and

$$\frac{\partial N}{\partial t}=\frac{N_{eq}}{t_d} \tag{7}$$

4 NEW MODEL CONSIDERING DENSIFICATION and dilation OF GROUND

The governing equation of Seed and Booker (1977) (Eq. 5) is modified to include the densification effect of RGP in dissipating the excess pore-water pressures. In the ground treated with RGP, coefficient of permeability (k_h) and coefficient of volume compressibility (m_v) are considered to depend on the distance, r, instead being constant values. Considering an element of soil, in polar coordinates as shown in Fig. 3, laminar flow and using Darcy's law, the expression for the flow in radial direction is obtained as,

$$\frac{k(r)}{r}\frac{\partial h}{\partial r}+k(r)\frac{\partial^2 h}{\partial r^2}+\frac{\partial k(r)}{\partial r}\frac{\partial h}{\partial r}=\frac{S}{1+e}.\frac{\partial e}{\partial t} \tag{8}$$

From the stress-strain relations and definitions of coefficient of compressibility (a_v) and coefficient of volume change (m_v) and for S =1(fully saturated),

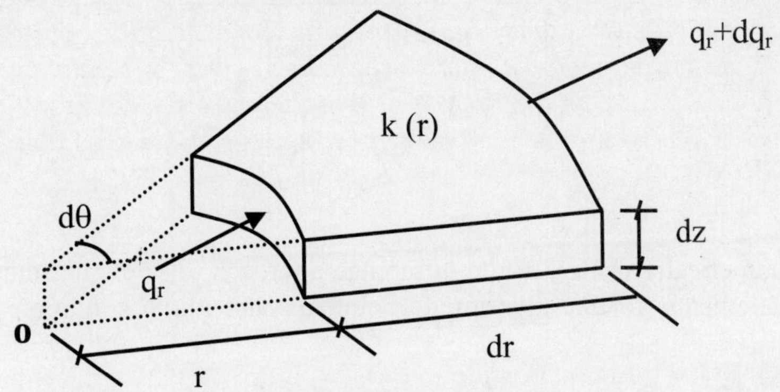

Fig. 3. Flow through soil element.

$$\frac{1}{1+e}.\frac{\partial e}{\partial t}=\frac{m_v}{1+e}.\left(\frac{\partial u}{\partial t}-\frac{\partial \sigma}{\partial t}\right) \tag{9}$$

From Eq. 9 and with h = u/γ_w; (u = excess pore pressure; γ_w = unit weight of water) Eq. 9 is rewritten as:

$$\frac{k_h(r)}{\gamma_w.m_v(r)}\left(\frac{1}{r}\frac{\partial u}{\partial r}+\frac{\partial^2 u}{\partial r^2}\right)+\frac{1}{\gamma_w.m_v(r)}.\frac{\partial(k_h(r))}{\partial r}.\frac{\partial u}{\partial r}=\frac{\partial u}{\partial t}-\frac{\partial \sigma}{\partial t} \tag{10}$$

where $k_h(r)$ and $m_v(r)$ are defined in Eqs. (1) and (3) respectively.
Considering the rate of change of total stress ($\partial\sigma/\partial t$) as the rate of pore pressure generation

$$\frac{\partial u_g}{\partial t}=\frac{\partial u_g}{\partial N}.\frac{\partial N}{\partial t} \tag{11}$$

The final form of Eq. 10 is,

$$\frac{k_h(r)}{\gamma_w.m_v(r)}\left(\frac{1}{r}\frac{\partial u}{\partial r}+\frac{\partial^2 u}{\partial r^2}\right)+\frac{1}{\gamma_w.m_v(r)}.\frac{\partial(k_h(r))}{\partial r}.\frac{\partial u}{\partial r}=\frac{\partial u}{\partial t}-\frac{\partial u_g}{\partial N}.\frac{\partial N}{\partial t} \tag{12}$$

Eq. 12 is a more generalized equation considering all forms of non-homogeneity. In non-dimensional form, with normalized pore pressure $W = u/\sigma'_o$, which is the same as the pore pressure ratio r_u (Seed and Booker 1977) and $W_g = u_g/\sigma'_o$, Eq. 12 becomes

$$\frac{k_h(r)}{\gamma_w.m_v(r)}\left(\frac{1}{r}\frac{\partial W}{\partial r}+\frac{\partial^2 W}{\partial r^2}\right)+\frac{1}{\gamma_w.m_v(r)}.\frac{\partial(k_h(r))}{\partial r}.\frac{\partial W}{\partial r}=\frac{\partial W}{\partial t}-\frac{\partial W_g}{\partial N}.\frac{\partial N}{\partial t} \tag{13}$$

Non-dimensionalising the terms 'r' with 'R' (=r/b) and 't' with 'T' (=t/t_d), Eq. 13 becomes

$$T_{bd} \cdot \frac{R_k(R)}{R_{mv}(R)} \left(\frac{\partial^2 W}{\partial R^2} + \frac{1}{R} \frac{\partial W}{\partial R} \right) + \frac{T_{bd} \cdot}{R_{mv}(R)} \frac{\partial (R_k(R))}{\partial R} \cdot \frac{\partial W}{\partial R} = \frac{\partial W}{\partial T} - \frac{\partial W_g}{\partial N} N_{eq} \tag{14}$$

where

$$\frac{\partial W_g}{\partial N} = \frac{1}{\alpha \pi N_l} \cdot \frac{1}{\sin^{2\alpha\text{-}1}\left(\frac{\pi}{2} r_u \right) \cos\left(\frac{\pi}{2} r_u \right)} \tag{15}$$

and

$$T_{bd} = \left(\frac{k_{hi}}{\gamma_w} \frac{t_d}{m_{vi}} \right) \cdot \frac{1}{b^2} \tag{16}$$

Eq. 14 is solved numerically using a finite difference approach, discretizing the unit cell radially into a number of elements, for the appropriate boundary and initial conditions (Murali Krishna 2003).

4.1 Boundary Conditions

The material in the drains is far more permeable than the surrounding sand layer (Gravel drain is infinitely permeable). If the effect of dilatancy of gravel drain is not considered the excess pore-water pressure in the drain is effectively zero i.e., at r = a or R = a/b, u = 0 or W= 0. Considering the dilation effect, since the granular piles tend to dilate under undrained conditions, they develop negative pore pressure which is estimated in a manner very similar to the estimation of positive pore pressure in loose sand deposits (Madhav and Arlekar 2000). Then the pore pressure at r = a is,

$$W_{g\ at\, r=a} = \left(\frac{u_g}{\sigma'_o} \right)_{at\, r=a} = -d_c \left(\frac{2}{\pi} \text{arc} \sin \left(\frac{N}{N_l} \right)^{1/2\alpha} \right) \tag{17}$$

where d_c is a constant that depends on the degree of dilatancy of the granular pile material and depends on the densification achieved during the installation of granular piles. At the outer boundary of the unit cell, due to symmetry, rate of pore-water pressure in the radial direction is zero, i.e.,

$$\text{at } r = b \text{ or } R = 1, \quad \frac{\partial u}{\partial r} = 0 \text{ or } \frac{\partial W}{\partial R} = 0 \tag{18}$$

4.2 Initial Condition

At t = 0 or T = 0, pore pressures at all the nodes in the soil are equal to the average of pore water pressure generated over the initial time period of 'dt' (or 'dT'). i.e., average of pore-water pressure generated over an initial cycle interval, dN.

$$W_{g\ at\, T=0} = \left(\frac{u_g}{\sigma'_o} \right)_{at\, t=0} = \frac{1}{2} \left(\frac{2}{\pi} \text{arc} \sin \left(\frac{dN}{N_l} \right)^{1/2\alpha} \right) \tag{19}$$

5 RESULTS AND DISCUSSION

Under the assumption of purely radial flow, the pore pressure ratio, $w= u/\sigma'_o$, throughout the sand depends on the dimensionless parameters: a/b = a ratio representing the geometric configuration of the RGPs; N_{eq}/N_l = cyclic ratio characterizing the severity of the earthquake shaking in relation to the liquefaction characteristics of the sand; T_{bd}, relating the duration of the earthquake to the consolidation properties of the sand; α, a parameter characterizing the shape of the pore pressure generation curve. It is found that for many materials $\alpha = 0.7$ fits the experimental data well, and this value is adopted (Seed and Booker, 1977); $R_k(r)$ & $R_{mv}(r)$, functions of radial distance, r, define the variations of coefficients of horizontal permeability and volume compressibility of the in situ soil respectively and d_c, a constant that depends on the degree of dilatancy of the granular material and depends on the densification achieved during the installation of granular piles.

If coefficients of permeability and volume change of the surrounding soil are unchanged and there is no dilation effect due to installation of RGP, the condition represents the 'No Densification & No Dilation' case. In this case the ambient soil will have the initial k_h and m_v (i.e., $R_{ka}= R_{kb}= R_{ma}= R_{mb}= 1$) and $d_c = 0$. The effect of a/b ratio on maximum pore water pressure is shown in Fig. 4 for cyclic ratio (N_{eq}/N_l) of 2 and a range of a/b (0.1 to 0.4) with T_{bd} value of 1. Maximum pore pressure ratio $W_{max}(T)$, i.e. maximum value of u/σ'_o throughout the layer is plotted against $T(t/t_d)$. These results are for no densification effect and agree closely with the results of Seed and Booker (1977). For cyclic ratio, (N_{eq}/N_l) equals to 2, if no drains are present (a/b = 0) the soil liquefies at $T_l = 1/(N_{eq}/N_l)$ i.e., at T = 0.5. For a/b = 0.1, initial liquefaction is deferred for a period of 'T' but eventually occurs at about T=0.6, the liquefied state, $W_{max} = 1$, continues until the end of the period of strong shaking. Thereafter pore-water pressure dissipates. It can be observed that at higher a/b values, initial liquefaction may be prevented as the maximum pore pressure ratio decreases with increase in a/b value.

Effect of cyclic ratio on maximum pore pressure ratio, without densification and dilation effects, is shown in Fig. 5. A range of cyclic ratios, N_{eq}/N_l, 1 to 5 for a/b of 0.3 and T_{bd} of 1 is considered. As cyclic ratio increases, pore water pressure increases and leads to liquefaction depending on the generation and dissipation of pore pressures. The time at which the ground liquefies decreases with increase of cyclic ratio as per T_l (=$1/(N_{eq}/N_l)$). If attained, the liquefied state continues until the end of the period of strong shaking. i.e., T=1. Thereafter pore pressure ratio decreases gradually. It can be observed from Fig. 5 that the peak value of W_{max} increased from about 0.27 to 0.52 for an increase in N_{eq}/N_l from 1 to 2. For N_{eq}/N_l values 3, 4 and 5, W_{max} reaches a value of one, at T of about 0.62, 0.35 and 0.24 respectively.

Densification effect, but with no dilation effect ($d_c = 0$), with respect to coefficient of volume change only (i.e., no change in coefficient of permeability ($R_{ka}=R_{kb}= 1$)) on maximum pore pressure ratio, W_{max}, is shown in Figs. 6 and 7. Fig. 6 shows the effect of densification at the near end, R_{ma}, on W_{max}, for no modification in the coefficient of volume change at the farthest end (R_{mb}=1). A range of R_{ma}(1-0.1) for T_{bd} = 1, a/b = 0.3 and N_{eq}/N_l = 2 is considered. It can be observed that W_{max} decreases with decrease in R_{ma}. Maximum pore pressure ratio, W_{max}, decreased from 0.525 to 0.375 for a decrease in R_{ma} from 1 to 0.1. The effect of densification at the farthest end (R_{mb}) is shown in Fig. 7, considering the densification at the near end as R_{ma}=0.3. A range of R_{mb} from 1 to 0.5 for T_{bd} = 1, a/b = 0.3 and N_{eq}/N_l = 2 is considered. It can be observed that W_{max} decreases from 0.41 to 0.25 for a decrease in R_{mb} from 1 to 0.5. It can be concluded from Figs. 6 and 7 that densification effects with respect to coefficient of volume change reduces the pore pressures and mitigates the liquefaction potential.

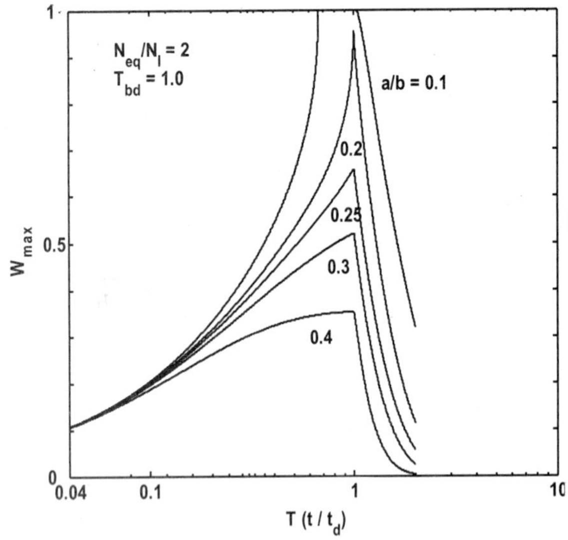

Fig. 4. Effect of area ratio on W_{max} for 'No Densification & No Dilation' Case

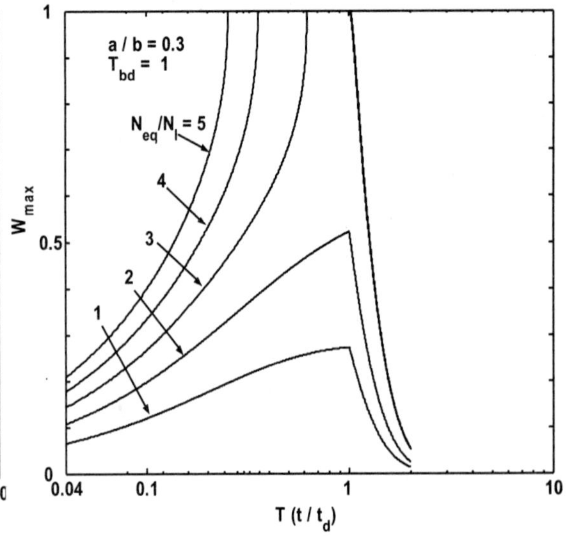

Fig. 5. Effect of N_{eq}/N_l ratio on W_{max} for 'No Densification & No Dilation' Case

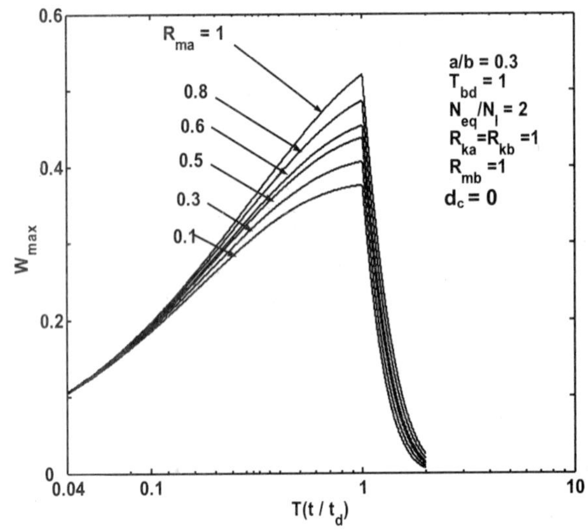

Fig. 6. Effect of R_{ma} on W_{max}

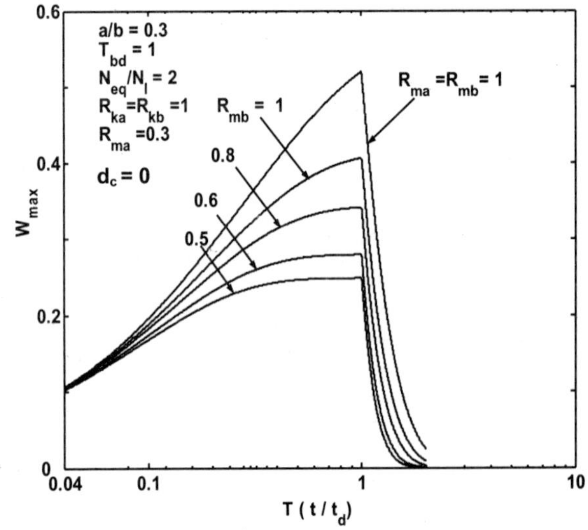

Fig. 7. Effect of R_{mb} on W_{max}

Densification effect in terms of coefficient of permeability only, i.e., no change in coefficient of volume change (R_{ma}=R_{mb}= 1), but with no dilation effect (d_c = 0), on maximum pore pressure ratio is shown in Figs. 8 and 9. The effect of densification at the near end, R_{ka}, on W_{max} for no modification in the coefficient of permeability at the farthest end (R_{kb}=1) is shown in Fig. 8. A range of R_{ka} (1-0.5) for T_{bd} = 1, a/b = 0.3 and N_{eq}/N_l = 2 is considered. W_{max} increases from 0.525 to 0.675 for a decrease in R_{ka} from 1 to 0.8. For R_{ka}=0.6 the maximum pore pressure ratio reaches 1, i.e. the liquefied state. Further decrease in R_{ka} will reduce the time to attain the liquefaction state. Densification effect at the farthest end (R_{kb}) on maximum pore pressure ratio is shown in Fig. 9 considering the densification at the near end as R_{ka}=0.6. A range of R_{kb} from 1 to 0.6 for T_{bd} = 1, a/b = 0.3 and N_{eq}/N_l = 2 is considered. It can be observed that the maximum pore pressure ratio is slightly affected by the densification at the farthest end, R_{kb}. It can be concluded from Figs. 8 and 9 that densification effects with respect to reduction in coefficient of permeability increase the pore pressures. This observation is revealing in that the densification effect with respect to coefficient of permeability alone results in a negative effect.

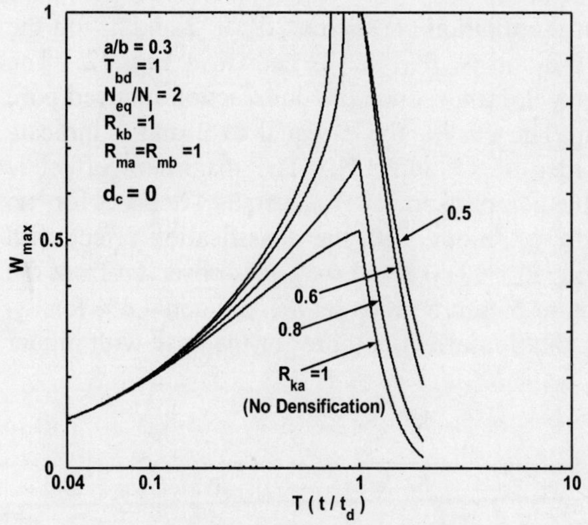

Fig. 8. Effect of R_{ka} on W_{max}

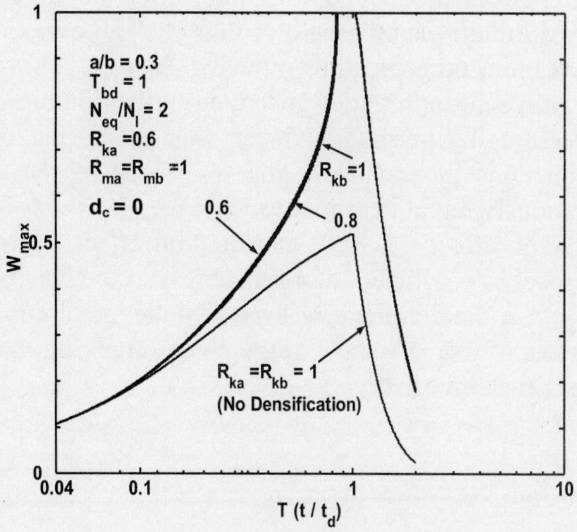

Fig. 9. Effect of R_{kb} on W_{max}

Densification effect with respect to both the coefficients of permeability and volume change, but without dilation effect (d_c = 0), on W_{max} is shown in Figs. 10 and 11. Fig. 10 shows the densification effect on both the coefficients of permeability and volume change at near end on maximum pore pressure ratio, W_{max}. In this case the coefficients of permeability and volume change at the farthest end remain unaffected (R_{kb}= R_{mb}=1). A range of R_{ka} and R_{ma} (1-0.3) for T_{bd} = 1, a/b = 0.3 and (N_{eq}/N_l) = 2 is considered. The densification effect considered is the same for both R_{ka} and R_{ma}. Maximum pore pressure ratio increases from 0.525 to 0.62 for a decrease in R_{ka} and R_{ma} from 1 to 0.8. W_{max} almost reaches 1, i.e. attains liquefaction state, at T =1 for R_{ka}=R_{ma}=0.6. Further decrease in R_{ka} and R_{ma} reduces the time corresponding to W_{max}=1. The effect of densification at the farthest end with respect to both coefficients of permeability (R_{kb}) and volume change (R_{mb}) on maximum pore pressure ratio, W_{max}, is shown in Fig. 11, considering densification at the near end as R_{ka}= R_{ma}=0.6. The densification effect considered is the same for both R_{kb} and R_{mb} (1 – 0.7). W_{max} reaches 1 indicating liquefaction state, at T =1 for R_{kb}= R_{mb}=1 i.e. for no densification at the farthest end. Densification at the farthest end, i.e. reduction in R_{kb} and R_{mb} (from 1 to 0.7) reduces W_{max} (from about 1 to about 0.57), and thus establishes that liquefaction of ground is prevented by overall densification of the ground.

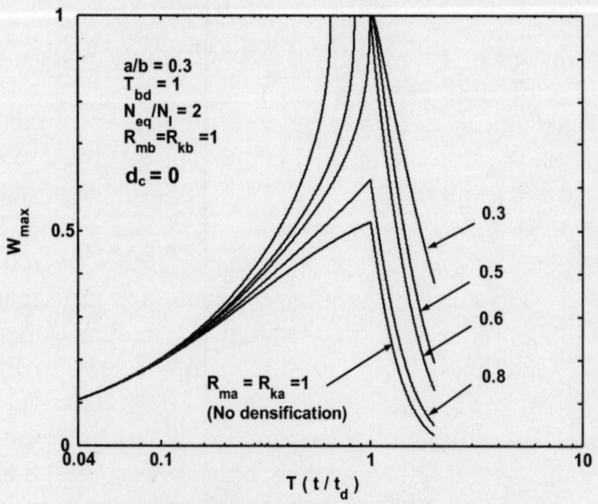

Fig. 10. Effect of R_{ma} and R_{ka} on W_{max}

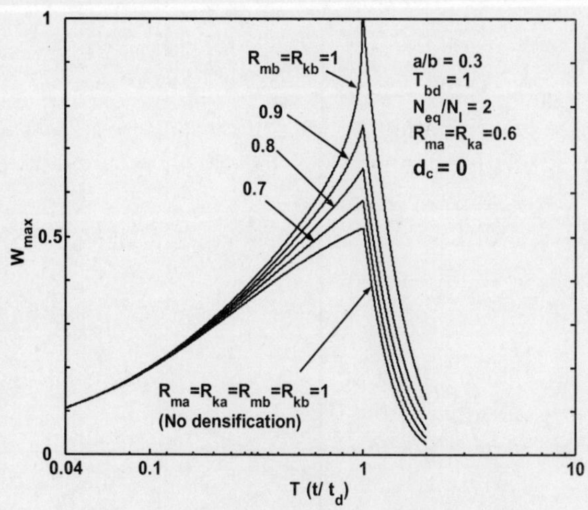

Fig. 11. Effect of R_{mb} and R_{kb} on W_{max}

Effect of dilation alone, for no densification effect (i.e., $R_{ka}= R_{kb}= R_{ma}= R_{mb}= 1$), on maximum pore pressure ratio is presented in Fig. 12. The effect of the dilation coefficient, d_c (0, 2 and 5), on the maximum pore pressure ratio for $T_{bd} = 1$, $a/b = 0.2$ and $N_{eq}/N_l = 2$, is shown in Fig. 12. The negative pore pressures generated in the dilating gravel drain reduce the liquefaction induced pore pressures by permitting faster rates of dissipation. The curves for d_c equal to 2 and 5 indicate reductions in maximum pore pressures of the order of 13 and 17%. The maximum effect is obviously felt at t/t_d equal to 1.0. Fig. 13 provides the comparison of 'W_{max} versus T' curves for 'no dilation effect' ($d_c = 0$) and dilation effect with $d_c = 2$, along with the densification effect with respect to coefficient of permeability at the near end ($R_{ka} = 1$, 0.8 and 0.6). It is observed from the Fig. that the dilation effect reduced the maximum pore pressure ratios by 8.2, 5.2 and 4.3% for R_{ka} values of 0.6, 0.8 and 1 respectively implying that dilation effect is more for the case with higher densification effect.

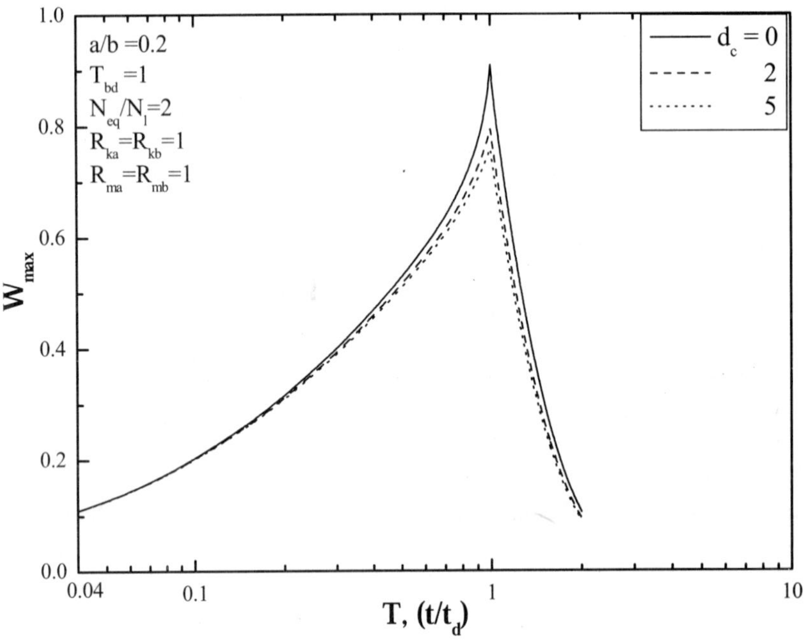

Fig. 12. Maximum pore pressure ratio for different values of dilation coefficient

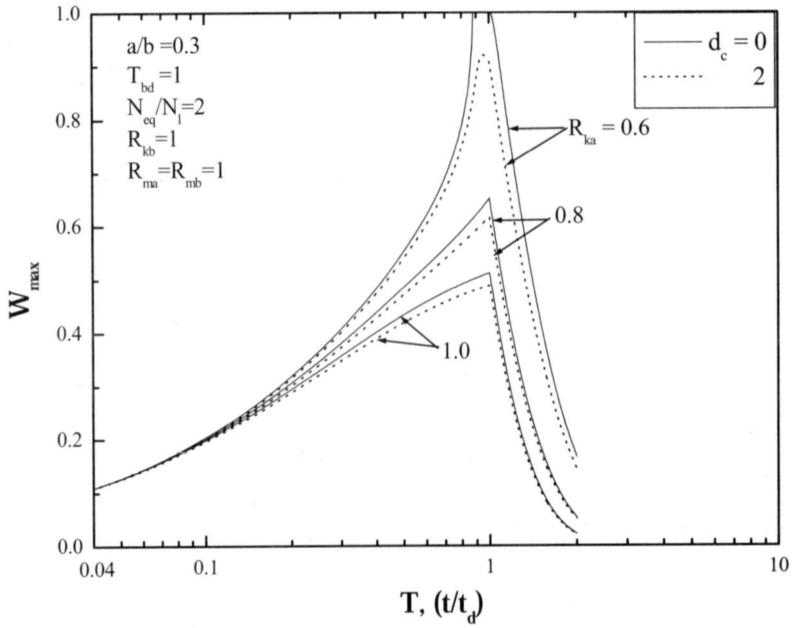

Fig. 13. Effect of dilation and densification on W_{max}

6 CONCLUSIONS

An analysis of ground treated with gravel drains (stone columns) considering the effect of installation in densifying the ground and its dilation effect is proposed. Both the coefficients of volume change and permeability are considered to be affected due to densification. In fact they decrease because of densification. Dilation effect generates negative pore water pressures in the granular piles. Maximum pore pressure ratio decreases with closer spacing of gravel drains and increases with increase in cyclic ratio (N_{eq}/N_l). Densification effect on the coefficient of volume change alone is positive in that the maximum induced pore water pressure ratios get reduced. Densification effect on the coefficient of permeability alone increases the maximum pore water pressure ratios giving a negative effect.Densification effect on both coefficients of permeability and volume change gives a slightly negative or positive effect depending on the extent of densification. The negative pore pressures generated in dilating gravel drain reduce liquefaction induced pore pressures by permitting faster rates of dissipation, hence enhance liquefaction mitigation. The negative effect of the densification is balanced by the dilation effect thus proving the effectiveness of granular piles in the liquefaction mitigation.

REFERENCES

1. Adalier K and Elgamal A "Mitigation of liquefaction and associated ground deformations by stone columns," Engineering Geology, 72, No. 3-4, 275-291, 2004.
2. Baez JI and Martin GR "Quantitative evaluation of stone column technique for earthquake liquefaction mitigation," 10th World Conference on Earthquake Engineering, 1477-1483, 1991.
3. Leonards GA (ed.), Foundation Engineering, McGraw-Hill, New York, 1962,
4. Madhav MR "Engineering of ground for earthquake disaster mitigation," Proc. Indian Geotechnical Conference 2001, Indore, India, 29-34, 2001.
5. Madhav MR and Arlekar JN "Dilation of granular piles in mitigating liquefaction of sand deposits," 12th World Conference Earthquake Engineering, Auckland. No: 1035 (CD-ROM), 2000.
6. Martin WDL Finn and Seed HB "Fundamentals of liquefaction under cyclic loading," J. of the Geotechnical Engineering Division, ASCE, 101(GT5), 425-438, 1975.
7. Mitchell JK and Wentz FK "Performance of improved ground during Loma Prieta earthquake," Report No. EERC91/12, Earthquake Engineering Research Center, University of California, Berkeley, 1991.
8. Murali Krishna A "Liquefaction mitigation of loose sand deposits by rammed granular piles," M.Tech Thesis, Indian Institute of Technology, Kanpur, India, 2003.
9. Ohbayashi J, Harda , and Yamamoto M "Resistance against liquefaction of ground improved by sand compaction pile method," Earthquake Geotechnical Engineering, Seco e Pinto (ed.), Balkama, Rotterdam, 549-554, 1999.
10. Poorooshasb HB and Madhav MR "Application of rigid plastic dilatancy model for prediction of granular pile settlements," 4th International Conference on Numerical Methods in Geomechanics, Nagoya, Japan, 1805-1808, 1985.
11. Poorooshasb HB, Holubec I and Sherbourne AW "Yielding and flow of sand in triaxial compression," Canadian Geotechnical Journal, 3, 179-190, 1966.
12. Seed HB "Soil liquefaction and cyclic mobility evaluation for level ground during earthquakes," J. of the Geotechnical Enigineering Division, ASCE, 105, No. 2, 201-255, 1979.

13. Seed HB and Booker JR "Stabilization of potentially liquefiable sand deposits using gravel drains," J. of the Geotechnical Enigineering Division, ASCE, 103, No. 7, 757-768, 1977.

14. Seed HB and Idriss IM. "Ground Motions and Soil Liquefaction During Earthquakes," Earthquake Engineering Research Institute, Oakland, California, 1982.

15. Sujatha KP "Analysis of reinforcement and densification effects in improved ground," M.Tech Thesis, Indian Institute of Technology, Kanpur, India, 1998.

16. Vaid Y, Byrne PM and Hughes JMO "Dilation angle and liquefaction potential," Journal of Geotechnical Engineering, ASCE, 103(7), 1003-1008, 1981.

17. Van Impe WF and Madhav MR "Analysis and settlement of dilating stone column reinforced soil," Austrian Geotechnical Journal, 137:114-121, 1992.

ESTIMATING SEISMIC HAZARD IN INDIAN CITIES

R. N. Iyengar
Raja Ramanna Fellow, Dept. of Civil Engineering,
Indian Institute of Science, Bangalore 560012.
rni@civil.iisc.ernet.in

ABSTRACT

Success of disaster mitigation strategies depends on how well the stakeholders are sensitized to the necessity of reducing the vulnerability of built up structures. Since earthquakes are rare and unpredictable, there is a tendency among the less informed persons, to deny the importance of city level hazard estimation as a pro-active step in the right direction. In recent years, it is well recognized that the IS-1893 division of the country into four seismic zones is a preliminary exercise. They are useful as guides but cannot be substitutes for accurate estimation of hazard to cities existing near active faults. The much-discussed seismic microzonation maps the hazard in a city, on a local scale. It aims at educating building owners, local administrators and elected representatives responsible for the welfare of a ward of the size of one square kilometer, or less. The challenge one faces here, is to scientifically compute the basic hazard without sacrificing geological, seismological and geotechnical details. A simple approach for finding seismic hazard is discussed with Delhi and Mumbai as examples.

Keywords: Hazard, Attenuation, Spectrum, Probabilistic Analysis.

1 INTRODUCTION

The recent spate of earthquakes in the country has highlighted the importance of having a well-planned reduction strategy built into the larger policy framework of sustainable development. The vulnerability of Indian habitat was dramatically demonstrated by the Kutch earthquake of 26th January 2001, which did not belong to the more frequent Himalayan sources. The looming seismic risk to Guwahati, Shillong, Dehradun, Delhi, Agra, Chandigarh and other cities can be perceived in the backdrop of active faults in the Himalayas and in the Indo-Gangetic plains. Time and again earthquakes have occurred without warning. As population increases and our cities expand, the potential for earthquake loss increases. It has to be remembered that the principal cause of death and damage is from collapse of buildings, the possibility of which even otherwise increases with time. While the financial loss in absolute terms is a function of the state of development, initial investment and cost of living indices, the vulnerability to natural disasters is dependent on the social and economic condition of the population. It is estimated that loss to GNP due to a strong earthquake can be twenty times greater in developing countries than in developed countries. Thus, it is prudent to plan, develop and protect built environment from possible future destruction. In this context seismic hazard estimation plays a key role in planning and protecting the built environment in the earthquake prone cities of India. The challenge, is to scientifically compute the ground vibration in the cities due to earthquakes oroginating on faults that are hundreds of kilometers long at unknown depths in the contributing seismo-tectonic region of about 300,000 Sq. Km. Municipal engineers who like to evaluate the vulnerability or safety of existing buildings, need to know how the ground will vibrate in future at their sites within the city. This ground motion lasting for a few seconds is due to the release of energy stored for centuries at the fault level, nonetheless strongly influenced by the local soil conditions at the site.This article highlights how city level hazard can be estimated including uncertainties of seismic data, available fault maps and local site conditions. Delhi and Mumbai are taken as typical examples to present some preliminary results.

2 CONTROL REGION AROUND A CITY

Seismic hazard at a site can be defined as the propensity of the ground to move and/or vibrate at the surface level due to earthquakes originating in a region of about 300 Km radius around the site. This represents a large catchment that may include several known faults of different seismic potential. Preparation of the fault map relevant to the city under investigation can be carried out using the Seismotectonic Atlas of India. As the first step in hazard analysis, it is essential to prepare a catalogue of all past earthquakes in the control region. This helps us to characterize the region, for its seismicity, in terms of the Gutenberg-Richter recurrence relation. Since historical data is of varied quality, statistical analysis following one or more standard methods described in the literature is essential.

2.1 Database For Delhi

A general review of the seismic status of Delhi, has been previously presented by the author (Iyengar 2000). Varāha Mihira, who lived in 5th century A.D., has mentioned northern India including Delhi and its surroundings as the felt region of severe earthquakes. However, the first historically reported earthquake in Delhi city was on 17th July 1720. The MMI value of this event at Delhi has been estimated to be IX. Even though the focus of this event has not been identified, from historical accounts it is surmised that the source should have been near Delhi city. It is known that Agra was affected severely by an earthquake on 6th July 1505. Another event of high intensity felt in the Agra-Mathura region on 25th August 1803 has been described by Nazir Akbarabadi (1740-1830) in his Urdu poem *bhūcāl-nāma.* This event caused damage to Kutub Minar, one of the famous landmarks of Delhi. It would appear that Taj Mahal withstood this event with no recognizable distress. In recent times an event of magnitude M_L=6 occurred on 27thAugust 1960, having its epicentral tract in the Delhi-Gurgaon region. Distant events originating in the Himalayan plate boundary, such as Uttarakashi (M_L=6.4, 20th October 1991) and Chamoli (M_L=6.8, 29th March 1999) earthquakes have also caused nonstructural damage in parts of Delhi. Thus, the hazard at Delhi is controlled broadly by two different seismogenic regimes. The seismo-tectonic region affecting Delhi city is shown in Fig.1. A detailed catalogue of past earthquakes with their epicenter in the above region has been prepared (Iyengar and Ghosh 2004). In Fig.1, twenty important faults are shown along with the known 278 past epicenters. This figure highlights Delhi's susceptibility to earthquakes originating either from the Himalayas or from local sources. The faults are numbered for further reference and some of their details are presented in Table 1. The magnitude M_u given in this table refers to the estimated maximum magnitude a particular fault may be able to produce. This value has been assigned based on past information, without excessive conservatism. The hazard at any point in Delhi is a function of the seismic activity of the region, which in turn is directly related to the individual recurrence property of the listed twenty faults. This amounts to stating that the quantified regional seismicity has to be accounted by the twenty faults in differing proportions. The control region has been divided into two parts denoted as the Himalayan region (HR) and the Delhi region (DR) as shown in Fig.1. The maximum likelihood method of Kijko and Sellvol (1989, 1992) has been used for each of the two parts, with an upper bound on the maximum magnitude, to obtain the regional (*a, b*) values. This leads to the values b=0.78±0.04 for HR and b=0.87±0.04 for DR.

2.2 Mumbai City (19^0 N, 72.8^0 E)

Mumbai (Bombay) the financial capital of India is a mega city with a population exceeding ten million. The city is located in Peninsular India (PI), which has experienced devastating Koyna (1967, M_w=6.3), Khillari (1993,M_w=6.1), Jabalpur (1999, M_w=5.8) and Bhuj (2001, M_w=7.7) earthquakes. A detailed hazard study as applicable to Mumbai has been presented by Raghukanth and Iyengar (2006,2007). As per this study there are some twentythree faults that may induce ground motion in Mumbai. The fault map of this region prepared from the Seismo-tectonic Atlas of India is shown in Fig.2 superposed with known epicenters. It is observed that the seismic situation around Mumbai is much different from that around Delhi. Mumbai is in the Panvel zone which is seismically active. This zone strikes in the North-Northwest direction along the west coast. Its extent is from 16^0N to 21^0N approximately, within which Mumbai is located. A large vertical fault with the same strike is exposed on the western slopes of the Western Ghats in the Belgaum-Vengurla region. Hence a system of N-NW striking faults is surmised to exist below the Deccan Traps in this zone. Nandy (1995) in his review of *Neotectonism and Seismic Hazards in India* has identified the N-S trending Panvel flexure as an active fault. There is evidence, as per Wadia (1957), to conclude that the old land surface has subsided by some ten meters along the coast. A study by Misra shows that there are basically four active fault zones surrounding Mumbai. These are the Cambay-structure, Narmada-Tapti tectonic zone, Kukdi-Ghod lineament zone and the Sahyadri mountain ranges. It is also observed that parallel to the Panvel fault, a number of faults exist in the area. A total of 1274 events covering the period 1594-2002 A.D., have been collected from various sources to prepare a database specific to the region around Mumbai. This catalogue has been evaluated for its degree of completeness, following the method of Stepp (1972). Statistical analysis shows that data is complete, in the following fashion: ($3 \leq m < 4$) is complete in 1952-2002; ($4 \leq m < 5$) is complete in 1942-2002; ($5 \leq m < 6$) is complete in 1932-2002; and ($6 \leq m < 7$) is complete in 1902-2002. following the maximum likelihood method of Kijko and Sellevoll, the parameters of the Gutenberg and Richter equation are a=0.77±0.04 and b = 0.86±0.02.

3 FAULT DEAGGREGATION

The recurrence relation derived above is for the entire region and is not specific to any particular fault. Fault level recurrence is necessary to discriminate nearby sources from far off sources and to differentiate activity rate among faults. Such, fault level recurrence relation is rarely known due to paucity of large scale data accruing in historical times. An alternative is to empirically fix the b-value by measuring the slip rate of a fault. For the faults under consideration no slip values are available Hence, one has to proceed on a heuristic basis invoking the principle of conservation of seismic activity. As per this, the regional seismicity measured in terms of the number of earthquakes per year with $m \geq m_o$, should be equal to the sum of such events occurring on individual faults. This amounts to, $N(m_o) = \Sigma N_i(m_o)$ (i=1,,2…23) which is a fact. However, in finding $N_i(m_o)$ some assumptions are necessary. First, it is argued that a longer fault can produce more number of small events of magnitude m_o than a shorter fault. Hence, $N_i(m_o)$ may be taken as being proportional to the length of the fault, leading to a simple weight factor $\alpha_i = L_i/\Sigma L_i$. On the other hand one may also argue that future activity will continue, at least in the short run, similar to past activity. Hence, irrespective of its length seismic activity of a fault should be related to the number of past events associated with it in the catalogue. This way, one can arrive at another weight factor χ_i as the ratio of the past events associated with fault i to the total number of events in the region. Here, the average of α_i and χ_i is taken as the final weight to get

$$N_i(m_o) = 0.5(\alpha_i + \chi_i)\, N(m_o). \quad (1)$$

As an example, the above weight factors for Mumbai are shown in Table 1. For individual faults it would be appropriate to use the regional b-value. This gives

$$N_i(m) = N_i(m_0)\left[1 - \frac{1 - e^{-\beta(m-m_0)}}{1 - e^{-\beta(m_u - m_0)}}\right] \quad (2)$$

where, m_u is the maximum potential magnitude of the i-th fault and $\beta = 2.303b$. The above argument provides a basis for decomposing the regional hazard into fault level recurrence relations. Such final recurrence results are shown in Fig. 3 for Delhi and in Fig.4 for Mumbai.

4 PROBABILISTIC HAZARD ANALYSIS

In the past, Cornell and Merz (1975), Kiureghian and Ang (1977), Bullen and Bolt (1985) have discussed PSHA. Outlines of the method are also given by Kramer (1996). Presently, international engineering standards and manuals such as USDOE-1024, IBC-2003, USNRG-1.165 and EM-1110 of US Army Corp of Engineers, specify hazard in terms of a given value being exceeded at a particular probability level in a time window of one or more years. The steps involved are:

(i) Calculation of the probability distribution of distance r_j from site to source, given a magnitude m_i on the source, $P\lfloor r_j \mid M = m_i \rfloor$.This caters to the source to site distance uncertainty.

(ii) Computation of the frequency of occurrence $\lambda(m_i)$ of a particular event of magnitude m_i. This takes care of the magnitude uncertainty. The magnitude M is taken as an exponential distributed random variable between a minimum and maximum value.

(iii) Computation of the probability $P\lfloor Y > y^* \mid r_j, m_i \rfloor$ that a specified ground motion level y* is exceeded for a source to site distance r_j and event m_i. In this step the uncertainties in ground motion prediction in the attenuation law is handled. This requires a suitable attenuation law connecting Y with magnitude and distance, in a statstical fashion. The total annual probability of exceedance of y* is then obtained by summing over all distances for a given event magnitude, and then over all magnitudes for each fault. Eventually, the total probability of exceedance at the site is the sum of the individual probabilities due to all faults. If the number of earthquakes randomly occurring on a fault is taken to be a Poisson process, the probability of the random variable (*Y*=PGA/*g*) exceeding the level *a* in *T* years at the site can be expressed in terms of the annual rate of exceedance μ_a as,

$$P(Y > a) = 1 - \exp(-\mu_a T) \tag{3}$$

The basic expression for μ_a is

$$\mu_a = \sum_{s=1}^{N_s} N_s(m_0) \int_m \int_r P(Y > a \mid m, r) p_{R|M}(r \mid m) p_M(m) dr dm \tag{4}$$

It is emphasized here that the above expression sums up the individual contributions of all faults (s=1,2,…N_s) to obtain the annual probability of exceedance of PGA. As an example, for the central point (India Gate) of Delhi, the individual contribution of all the twenty faults to the mean annual hazard is presented in Fig.5, along with the total value (Iyengar and Ghosh 2004). Once the value of μ_a is found at a site, the probability of PGA exceeding a design value *a* in a life period of *T* years is given by equation (3). By repeating the above computations at various points representing Delhi and its environs, one can construct a micro-zonation map to any required level of detail. Here, a rectangular region of 40 Km x 30 Km, with India Gate as the central point is taken as the target region. This region is divided into 1320 grid points at 1km x 1km distance apart. PSHA incorporates all known faults, epicenters, past data and local characteristics in a judicious manner to arrive at the site-level hazard. Here, this result is shown in Fig. 6, for nearly B-type soil condition, for PGA with 2% exceedance probability in 50 years. This corresponds to a return period of 2500 years as suggested by the International Building Code IBC-2003. It is seen that the hazard peaks up near the local Delhi fault at the center of the city. This effect is also seen from Fig.5, where the contribution of individual faults to hazard at the center point is shown. As one moves away from the center towards the corners, the distance from the Delhi fault increases and its contribution to the hazard decreases. The pattern of the contours in Fig.6 indicates that the hazard is strongly influenced by the southwest-northeast trending Mahendragarh-Dehradun fault. It is noted that the result of Fig.6 does not include the effect of soft soil deposits that are widely spread over Delhi. Further refinement of the microzonation map of Fig.6 incorporating the effect of soft soil deposits is yet to be done. On similar lines, PSHA for response spectrum can be carried out with the help of frequency dependent attenuation laws.

5 PSHA FOR MUMBAI

It has been pointed out that a suitable attenuation relation is needed for PSHA. For the case of Delhi a relation based on recorded data was used. However, it is well known that we do not have strong motion records available for all magnitude values and distances. Hence empirical relations based on low and medium magnitude earthquake records are not credible for forecasting ground motion, which is demanded in PSHA. This difficulty is circumvented by large-scale synthetic simulation of accelerograms using a seismological model. Previously (Raghukanth and Iyengar 2006, 2007) have studied attenuation of PGA and S_a in Peninsular India from considerations of engineering seismology. The form of the attenuation equation proposed for bedrock (*br*) condition is

$$\ln (y_{br})=\mathbf{c_1}+\mathbf{c_2}\,(m\text{-}6)+\mathbf{c_3}\,(m\text{-}6)^2 -\mathbf{c_4}\,r -\ln r +\ln \varepsilon_{br} \quad (5)$$

This form is similar to the one used in other stable continental regions of the world. In this equation, y_{br} stands for the spectral acceleration (Sa/g); *m* and *r* refer to moment magnitude and hypocentral distance respectively. Fault data relevant to the present study of Mumbai are reported in Table 2. Coefficients of Eq. 5 are presented in Table 3. The above relation is initially obtained for bedrock sites with a shear wave velocity of 3.6 km/s. It is known that the surface level spectral acceleration values can be different from bedrock values, depending on local soil conditions. To account for this effect, the average shear wave velocity in the top 30m of a site (V_{30}) is universally adopted as the classifying parameter. As per the NEHRP sites are classified as; A: $V_{30} > 1.5$ Km/s; B: $0.76 < V_{30} \leq 1.5$; C: $0.36 < V_{30} \leq 0.76$; D: $0.18 < V_{30} \leq 0.36$ and E,F: $V_{30} \leq 0.18$ Km/s. Results of equation 5 are modified for local site condition further as,

$$y_s=y_{br}F_s \quad \text{and} \quad lnF_s=a_1 y_{br}+a_2+ln\delta_s \quad (6,7)$$

Here a_1 and a_2 are regression coefficients and δ_s is the error term, corresponding to site classification s = A, B, C, D. For E and F-type sites, rigorous nonlinear site response analysis would be needed with specific local soil parameters. This work is yet to be done. The above coefficients along with the standard deviation of the error $\sigma(ln\ \delta_s)$ are presented in Table 4. With the help of these correction factors, the average 5% response spectrum can be found for any A, B, C and D-type site in Peninsular India including Mumbai. The variation from the mean is characterized by the standard deviation expressed as

$$\sigma(\ln \varepsilon_s) = \sqrt{\sigma(\ln \varepsilon_{br})^2 + \sigma(\ln \delta_s)^2} \quad (8)$$

Seismic hazard curves for PGA computed separately for these sites in Mumbai are shown in Fig.7. It is observed that at shorter return periods, PGA value of C and D-type sites is higher than that at A and B-type sites. On the other hand, for longer return periods, PGA obtained for C and D-type sites is less than the B-type site value. This is attributable to the disposition of faults *vis-à-vis* Mumbai, the probability of occurrence of large magnitude events on the faults and finally the nonlinear behavior of C and D-type sites. The response spectra (5% damping) for Mumbai with 500-year return period is shown in Fig. 8.

6 SUMMARY

A brief discussion on how seismic hazard in Indian cities can be found is described in this article with the help of two examples. In summary the important steps in hazard estimation are as follows:

(i) All active faults and past epicenters in a region of about 300 Km around the city are identified. The city is taken as a point for this exercise. The Seismotectonic Atlas of India of GIS and other data base may be consulted for this step. The historical catalogue of earthquakes of magnitude greater than 3-4 is prepared to find the Gutenberg-Richter recurrence relation for the region using a well attested statistical procedure. The (*a, b*) values characterize the region affecting the city.

(ii) Fault deaggregation is carried out to arrive at the G-R relation for individual faults in the region of influence. This may be done by heuristic approaches depending on the potential of faults to generate earthquakes or using measured slip rates, if available. In the Indian context the heuristic approach is unavoidable.

(iii) A suitable attenuation relation valid for the region at the bedrock/basement level (V_s~3.5 Km/s) is selected. Past strong motion data are too few to arrive at an equation valid for all magnitudes and distances. Hence this has to be generated by large scale computer simulation (at least 10,000 samples) of a stochastic seismological model catering to the stress drop, quality factor, magnitude, and hypocentral distance variations possible in the region. The derived empirical spectral attenuation relations have to be validated with respect to available instrumental records. Such an investigation has been carried out for Peninsular India.

(iv) At the bed rock level it is generally found that ground acceleration is not too sensitive to distance variations of a few kilometers over which a city would be extending. However, at the surface level there can be considerable variation depending on the local soil properties which sometimes vary within 10-20 meters. Investigating the variation of soil properties with respect to depth going up to the bed rock level is a very challenging task. This work has to be broken up in to at least two stages; the deep section and the shallow section. The deep section is usually investigated through seismological signature analysis. The shallow section is amenable for borehole studies. Sites are usually classified as A, B, C, D, E and F depending on the average shear wave velocity in the top 30 meters of the soil. For the first four types of sites connecting the deep section with the shallow section is possible through detailed numerical simulation to arrive at results applicable over wide ranges of land mass that are nearly horizontally layered. For other cases, city specific analysis would be necessary to understand how basement ground motion gets modified as it arrives at the surface level. Microtremor measurements help us to estimate the first natural frequency of the ground, when the layering is horizontal. If a valid correlation between the first frequency and the site classification can be established for Indian conditions, microtremor measurements can provide a quick method of site classification with in a city. However, their use in quantifying site amplification is not valid except under some unrealistically ideal conditions.

(v) Probabilistic seismic hazard analysis is carried out for a benchmark uniform site condition to find the cumulative ground motion probability due to all faults in various time windows. Depending on the economic design period of the structures hazard maps are prepared for return periods of 500, 1000, and 2500 years. These are not the intervals at which an earthquake could occur in the region. For example, 500-year PGA refers to the value of peak ground acceleration that is expected to occur in a 50 year design period with a probability of exceedance of 10%.

(vi) A table of frequency dependent amplification values connecting the benchmark results to the actual site condition is needed to map the surface level ground motion on a fine scale of say 1:5000, leading to seismic microzonation of the target city. Eqs. (5,6,7,8) in association with Tables 3 and 4 can be used to compute site specific seismic hazard of cities in Penisular India.

REFERENCES

1. Bullen, K.E. and Bolt, B.A., (1985) *An introduction to the theory of seismology*. 4th edition. Cambridge University Press. Cambridge.
2. Cornell, C.A. and Merz, H.A., (1975) "Seismic risk analysis of Boston" *J. of Struct. Div. ASCE*, **101**, ST10, pp. 2027-2043.

3. Der Kiureghian, A. and Ang, A. H.-S., (1977) "A fault rupture model for seismic risk analysis" *Bull. Seism. Soc. Am.*, **67**, 4, pp. 1173-1194.
4. Iyengar, R. N., (2000) "Seismic Status of Delhi Megacity" *Current Science.*, **78,** 5, pp. 568-574.
5. Iyengar, R. N. and Ghosh, S.(2004) "Microzonation of earthquake hazard in greater Delhi area" *Curent Science*, 87(9), 1193-1202
6. Iyengar R. N. and Raghu Kanth S. T. G. (2004), "Attenuation of ground motion in Peninsular India" *Seismological Research Letters*, 79(5), 530-540
7. Kijko, A., and Sellevoll, (1989) "Estimation of Seismic Hazard Parameters from Incomplete Data Files. Part I: Utilization of extreme and complete catalogues with different threshold Magnitudes" *Bull. Seism. Soc. Am.*, **79**, pp. 645-654.
8. Kijko, A., and Sellevoll, M. A., (1992) "Estimation of Earthquake Hazard Parameters from Incomplete Data Files. Part II: Incorporation of Magnitude Heterogeneity" *Bull. Seism. Soc. Am.*, **82**, pp. 120-134.
9. Kramer S. L. (1996), *Geotechnical Earthquake Engineering,* Prentice Hall, Inc., New Jersey, USA.
10. Nandy D. R. (1995), "Neotectonism and Seismic Hazards in India" *Ind. J. Geology*, 67, 34-48.
11. Raghukanth S.T.G and Iyengar R.N. (2006) "Seismic Hazard estimatioin for Mumbai City" *Current Science,* 91(11), 1486-94.
12. Raghukanth S.T.G and Iyengar R.N. (2007) "Estimation of seismic spectral acceleration in Peninsular India" *J. Earth Syst. Sci.* 116, No. 3, 199–214.
13. Stepp J. C. (1972) "Analysis of Completeness of the Earthquake Sample in the Puget sound area and its effect on statistical estimates of earthquake hazard", *Int. Conf. on Microzonation* II, 897-909.
14. Wadia D. N. (1957), *Geology of India,* Macmillan & Co. Ltd., New York.

Figure 1. Seismo-tectonic map of Delhi Region (Iyengar and Ghosh 2004).

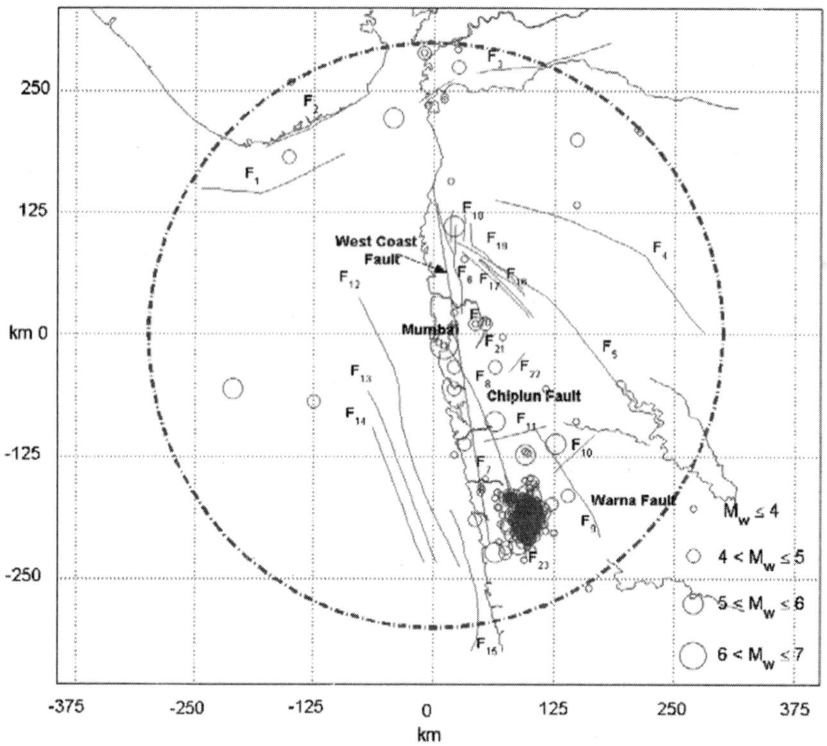

Figure 2. Seismotectonic map of Mumbai region (Raghukanth and Iyengar 2006).

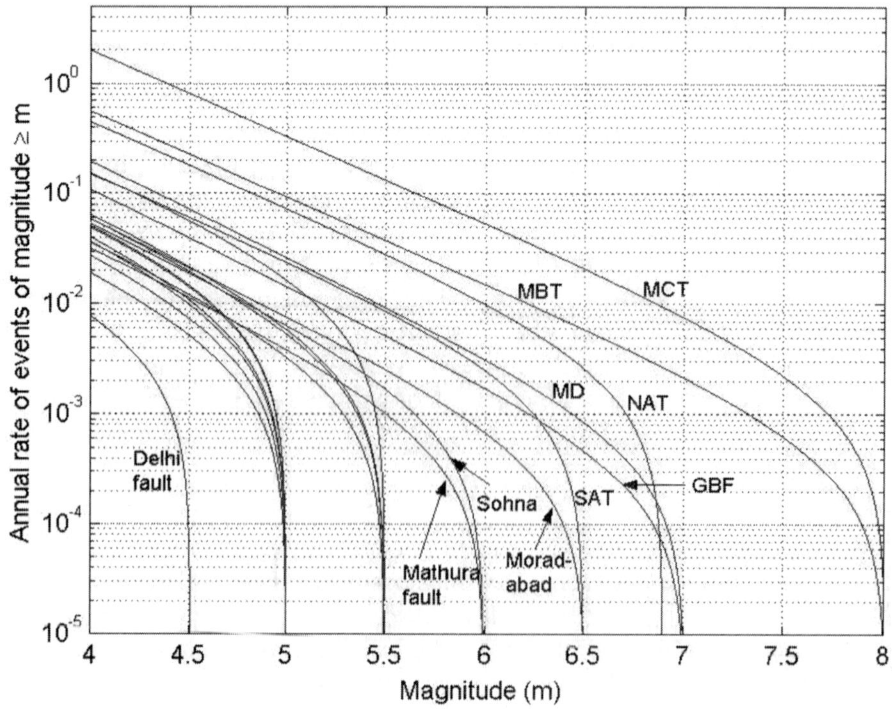

Fig.3. Fault Level Recurrence Relation for Delhi Region.

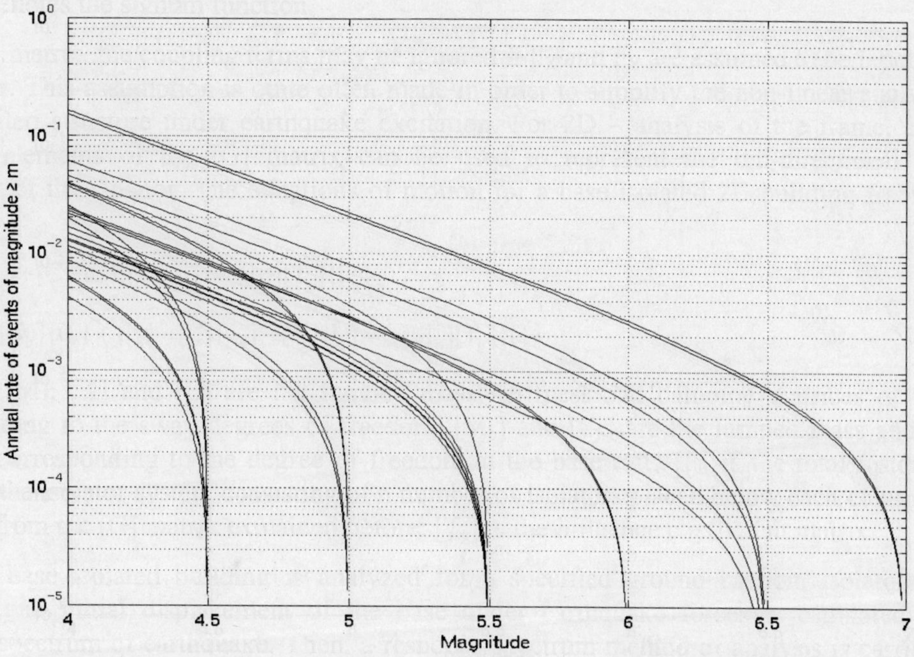

Fig.4 Fault Level Recurrence Relation for Mumbai Region.

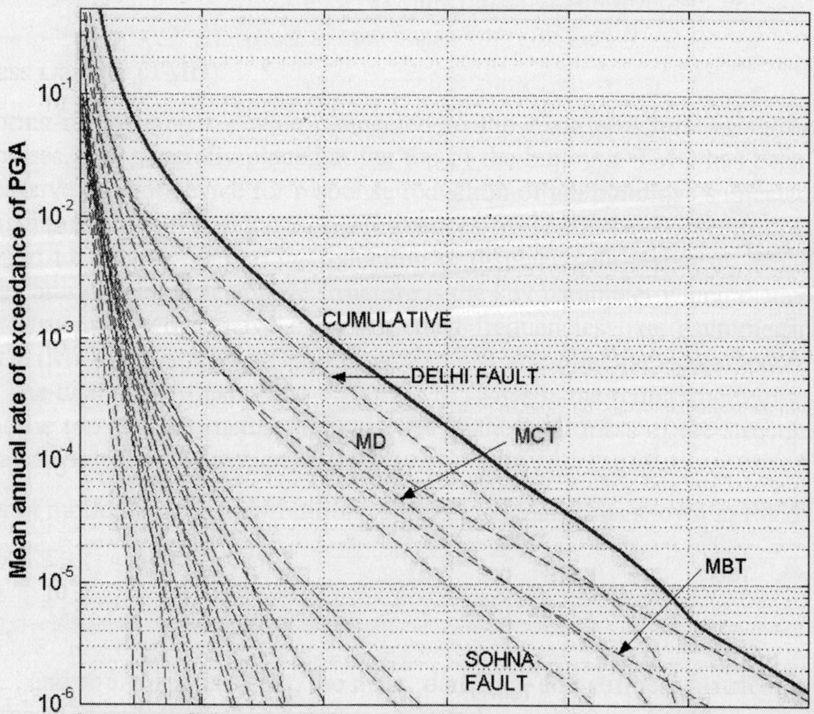

Figure 5. Hazard Deaggregation at India Gate, Delhi City due to Individual Faults.

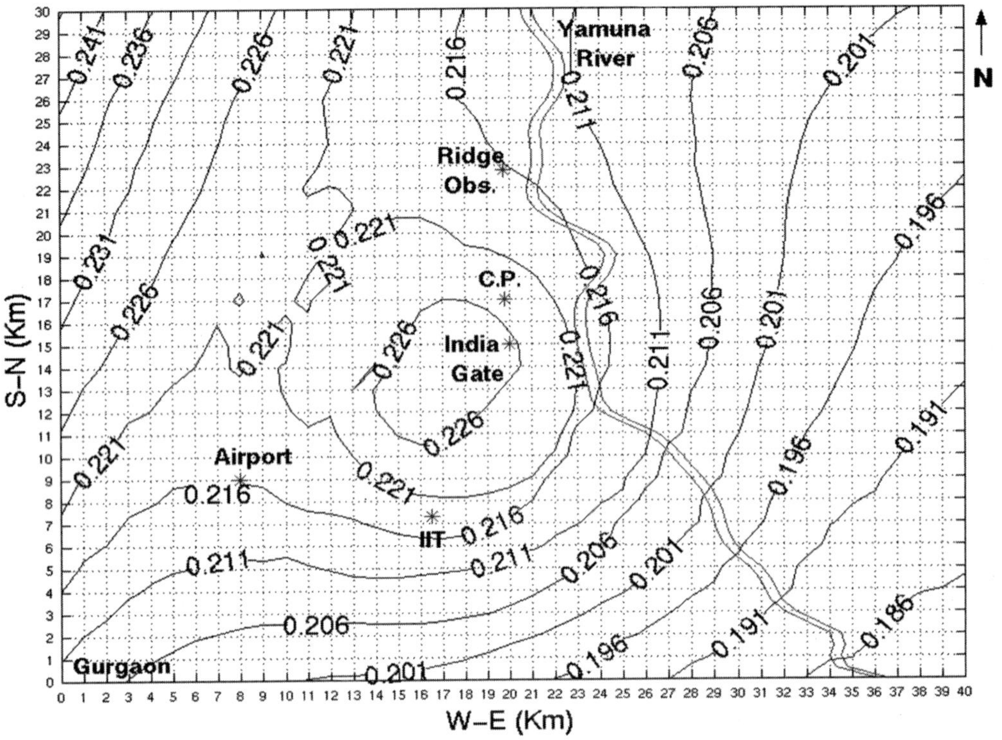

Fig.6. Microzonation of Seismic Hazard in Delhi. (PGA/g) at Hard soil (B-Type) Sites with 2% Probability of Exceedance in 50 Years.

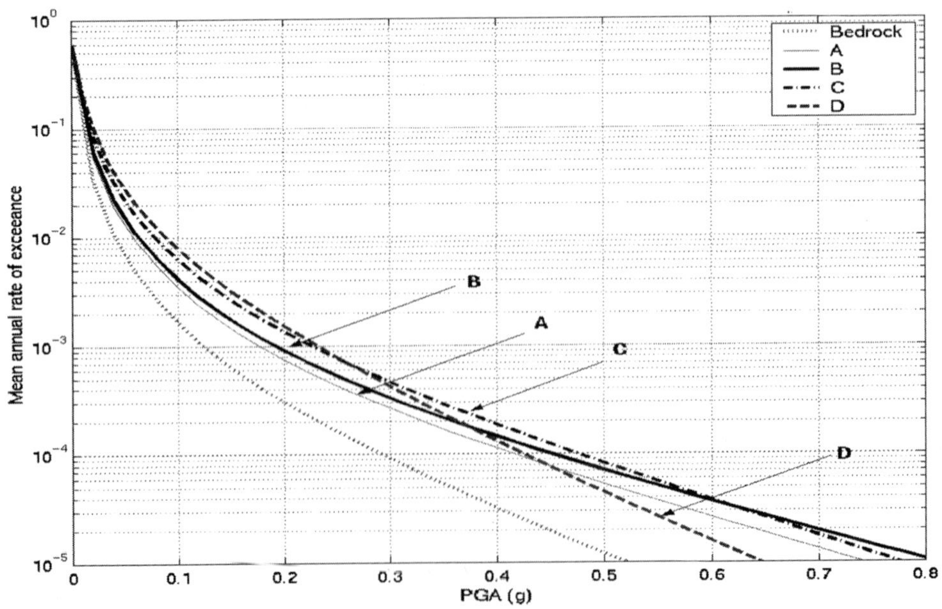

Fig. 7. Effect of Local Site Condition on Seismic Hazard at Mumbai.

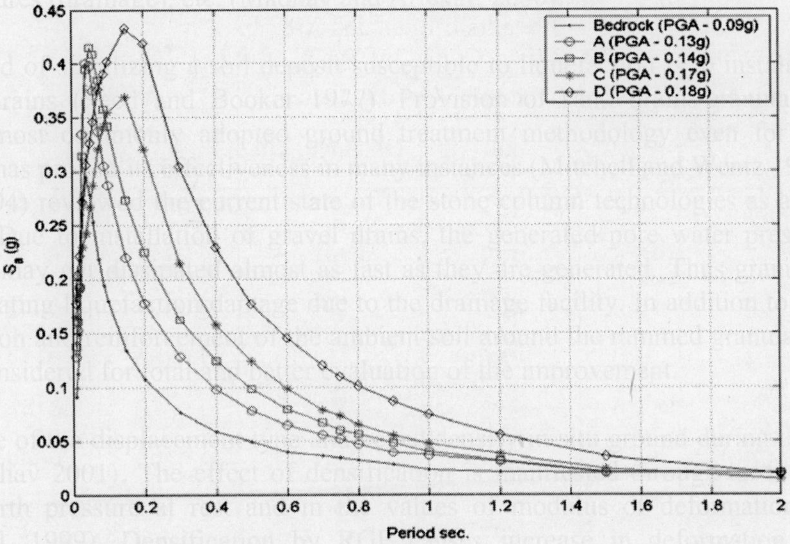

Fig.8. Mumbai Response Spectra.10 % probability of Exceedance in 50 years (5 % damping).

Table 1: Faults Influencing Delhi City

No.	Fault Name	M_u	Length Km.	w_s
1	Great boundary fault (GBF)	7	320	0.1462
2	Mahendraghar-Dehradhun	7	300	0.2657
3	Moradabad	6.5	165	0.0790
4	Chahapoli	5.5	215	0.0872
5	Sabi Fracture	5.5	195	0.0796
6	Near Mathura	5	84	0.0371
7	Parallel fault to no. 6	5.5	115	0.0490
8	Fault left of Alwar	5	130	0.0547
9	Fault near Alwar	5	55	0.0260
10	Fault near Jaipur	5	117	0.0497
11	Mathura	6	100	0.0432
12	Sohna	6	105	0.0719
13	Delhi	4.5	7	0.0106
14	Main Central Thrust (MCT)	8	350	0.5847
15	North Almora thrust (NAT)	6.9	280	0.1315
16	Main Boundary Thrust (MBT)	8	450	0.1647
17	Alaknanda	5.5	51	0.0454
18	Ropar	5	35	0.0144
19	Near Ramgarh	5	37	0.0149
20	South Almora Thrust (SAT)	6.5	130	0.0444

Table 2. Faults Influencing Mumbai City

Fault No.	m_u	Length km	Hypocent. Distance, km	Weights	
				α_i	χ_i
1	6.0	161	207	0.0518	0.0041
2	5.5	109	248	0.0352	0.0028
3	5.0	72	272	0.0231	0.0042
4	5.0	266	151	0.0856	0.0154
5	5.0	252	93	0.0810	0.0146
6	6.5	93	36	0.0299	0.0044
7	6.5	372	23	0.1195	0.0400
8	6.5	240	16	0.0773	0.0400
9	7.0	127	144	0.0408	0.3557
10	6.0	58	188	0.0182	0.0414
11	6.0	66	121	0.0212	0.0413
12	5.5	297	68	0.0955	0.0076
13	4.5	190	89	0.0612	0.0048
14	4.5	149	114	0.0480	0.0038
15	5.5	125	180	0.0403	0.0032
16	5.5	84	49	0.0269	0.0030
17	5.5	96	34	0.0310	0.0030
18	5.5	94	60	0.0303	0.0030
19	5.0	60	62	0.0193	0.0030
20	4.5	40	45	0.0129	0.0030
21	4.5	25	46	0.0081	0.0030
22	4.5	42	76	0.0135	0.0030
23	7.0	91	157	0.0293	0.3957

Table 3. Coefficients of Attenuation Equation (5)

Period (s)	c_1	c_2	c_3	c_4	σ (ε_{br})
0.000	1.6858	0.9241	-0.0760	0.0057	0.4648
0.010	1.7510	0.9203	-0.0748	0.0056	0.4636
0.015	1.8602	0.9184	-0.0666	0.0053	0.4230
0.020	2.0999	0.9098	-0.0630	0.0056	0.4758
0.030	2.6310	0.8999	-0.0582	0.0060	0.5189
0.040	2.8084	0.9022	-0.0583	0.0059	0.4567
0.050	2.7800	0.9090	-0.0605	0.0055	0.4130
0.060	2.6986	0.9173	-0.0634	0.0052	0.4201
0.075	2.5703	0.9308	-0.0687	0.0049	0.4305
0.090	2.4565	0.9450	-0.0748	0.0046	0.4572
0.100	2.3890	0.9548	-0.0791	0.0044	0.4503
0.150	2.1200	1.0070	-0.1034	0.0038	0.4268
0.200	1.9192	1.0619	-0.1296	0.0034	0.3932
0.300	1.6138	1.1708	-0.1799	0.0028	0.3984
0.400	1.3720	1.2716	-0.2219	0.0024	0.3894
0.500	1.1638	1.3615	-0.2546	0.0021	0.3817
0.600	0.9770	1.4409	-0.2791	0.0019	0.3744
0.700	0.8061	1.5111	-0.2970	0.0017	0.3676
0.750	0.7254	1.5432	-0.3040	0.0016	0.3645
0.800	0.6476	1.5734	-0.3099	0.0016	0.3616
0.900	0.4996	1.6291	-0.3188	0.0015	0.3568
1.000	0.3604	1.6791	-0.3248	0.0014	0.3531
1.200	0.2904	1.7464	-0.3300	0.0013	0.3748
1.500	-0.2339	1.8695	-0.3290	0.0011	0.3479
2.000	-0.7096	1.9983	-0.3144	0.0011	0.3140
2.500	-1.1064	2.0919	-0.2945	0.0010	0.3222
3.000	-1.4468	2.1632	-0.2737	0.0011	0.3493
4.000	-2.0090	2.2644	-0.2350	0.0011	0.3182

Table 4. Site Coefficients in Equations (6, 7, 8)

Period	F_A (a_1=0)		F_B (a_1=0)		F_C			F_D		
(sec)	a_2	$\sigma(\ln\delta_s)$	a_2	$\sigma(\ln\delta_s)$	a_1	a_2	$\sigma(\ln\delta_s)$	a_1	a_2	$\sigma(\ln\delta_s)$
0.000	0.36	0.03	0.49	0.08	-0.89	0.66	0.23	-2.61	0.80	0.36
0.010	0.35	0.04	0.43	0.11	-0.89	0.66	0.23	-2.62	0.80	0.37
0.015	0.31	0.06	0.36	0.16	-0.89	0.54	0.23	-2.62	0.69	0.37
0.020	0.26	0.08	0.24	0.09	-0.91	0.32	0.19	-2.61	0.55	0.34
0.030	0.25	0.04	0.18	0.03	-0.94	-0.01	0.21	-2.54	0.42	0.31
0.040	0.31	0.01	0.29	0.01	-0.87	-0.05	0.21	-2.44	0.58	0.31
0.050	0.36	0.01	0.40	0.02	-0.83	0.11	0.18	-2.34	0.65	0.29
0.060	0.39	0.01	0.48	0.02	-0.83	0.27	0.18	-2.78	0.83	0.29
0.075	0.43	0.01	0.56	0.03	-0.81	0.50	0.19	-2.32	0.93	0.19
0.090	0.46	0.01	0.62	0.02	-0.83	0.68	0.18	-2.27	1.04	0.29
0.100	0.47	0.01	0.71	0.01	-0.84	0.79	0.15	-2.25	1.12	0.19
0.150	0.50	0.02	0.74	0.01	-0.93	1.11	0.16	-2.38	1.40	0.28
0.200	0.51	0.02	0.76	0.02	-0.78	1.16	0.18	-2.32	1.57	0.19
0.300	0.53	0.03	0.76	0.02	0.06	1.03	0.13	-1.86	1.51	0.16
0.400	0.52	0.03	0.74	0.01	-0.06	0.99	0.13	-1.28	1.43	0.16
0.500	0.51	0.06	0.72	0.02	-0.17	0.97	0.12	-0.69	1.34	0.21
0.600	0.49	0.01	0.69	0.02	-0.04	0.93	0.12	-0.56	1.32	0.21
0.700	0.49	0.01	0.68	0.02	-0.25	0.88	0.12	-0.42	1.29	0.21
0.750	0.48	0.02	0.66	0.02	0.36	0.86	0.09	-0.36	1.28	0.19
0.800	0.47	0.01	0.63	0.01	-0.34	0.84	0.12	-0.18	1.27	0.21
0.900	0.46	0.01	0.61	0.02	-0.29	0.81	0.12	0.17	1.25	0.21
1.000	0.45	0.02	0.62	0.11	0.24	0.78	0.10	0.53	1.23	0.15
1.200	0.43	0.01	0.57	0.03	-0.11	0.67	0.09	0.77	1.14	0.17
1.500	0.39	0.02	0.51	0.04	-0.10	0.62	0.09	1.13	1.01	0.17
2.000	0.36	0.03	0.44	0.06	-0.13	0.47	0.08	0.61	0.79	0.15
2.500	0.34	0.04	0.40	0.08	-0.15	0.39	0.08	0.37	0.68	0.15
3.000	0.32	0.04	0.38	0.10	-0.17	0.32	0.09	0.13	0.60	0.13
4.000	0.31	0.05	0.36	0.11	-0.19	0.35	0.08	0.12	0.44	0.15

STRUCTURAL CONTROL FOR SEISMIC DISASTER MITIGATION

T. K. Datta
Dogra Chair Emeritus Professor
Department of Civil Engineering, IIT-Delhi, tkdatta@civil.iitd.ac.in

ABSTRACT

The paper provides an overview of the theory of structural control as applied to civil engineering structures. To begin with, a brief outline of various passive control strategies like, the use of base isolation, TMDs, and VEDs in protecting structures for earthquake forces is given. Modeling and dynamic theory based on which they operate are discussed. The relative merits and demerits of the various passive control devices are outlined. Second part of the paper deals with active control of structures. Basic principles of the active strategies are briefly explained with special reference to some of the control algorithms that have been developed. The control algorithms include linear quadratic control (LQR), pole placement technique and sliding mode control using Liapunov stability conditions. A brief outline of more promising semi active control of structures using hydraulic dampers is also presented. Finally, practical implementation of a few active control strategies is shown.

Keywords: Structural Control, Passive Control, Active Control, Semi Active Control, Base Isolation, Visco-elastic Dampers, Tuned Mass Dampers, Linear Quadratic Control.

1. INTRODUCTION

It is gradually being recognized that the most effective means of mitigating earthquake forces in structures is by way of structural control. Therefore, research in the area of structural control in civil engineering applications is going to be one of the challenges for the next millennium. Structural control is not only a good strategy for mitigating earthquake forces, but also it is equally effective in controlling undesirable responses of structures produced due to wind and other dynamic excitations. Broadly speaking, structural control methods can be classified as passive and active control methods. The passive control method is activated by structural motion. No external force or energy is applied to effect the control. On the other hand, active control method is effected by externally activated device to change the response. The activation of external force is based on the measurement of external disturbance and (or) structural response. Sensors are employed for measurement purposes and with the help of computers, the digital signal activates (converted to analog signal) the required external force. Combination of these two methods of structural control has been used to evolve hybrid control methods for realizing stringent control requirements. In recent years a class of active control systems, for which the external energy requirements are orders of magnitude smaller than typical active control system, have been evolved. They are known as semi-active control system, in which control action is produced by the movement of the structure but is regulated by external source of energy.

There are a number of mechanisms by which the structural control action is achieved. For example, in passive control base isolation devices, visco-elastic dampers, tuned mass dampers, liquid-column dampers, liquid-mass dampers, metallic yield dampers and friction dampers have been proposed. Out of these, base isolation devices have been widely implemented in practice. In active control, active mass/tuned mass dampers (ATMD), active tendon systems, actuators/ controllers have been used. Out of these, active mass/ tuned mass dampers have been implemented for response reduction of tall buildings for controlling wind-induced vibration. In hybrid control, the concepts using base isolation and ATMD, and visco-elastic dampers and ATMD have been proposed. In semi active control, electro rheological/ magneto-rheological dampers, fluid viscous dampers and tuned mass dampers are being investigated. Some of the control mechanisms which are widely reported in the literature are shown in Figs. 1-8.

There are a few state of the art papers on the structure control, which provide a good over-view of different structural control strategies developed and implemented in practice. They include state of the art review on base isolation of structures by Kelly (1986) and Jangid and Datta (1995). On active structural control two state of the art papers by Soong (1988) provide an excellent overview of different active structural control systems developed till 1990. A relatively recent comprehensive review on structural control is presented by Housener et al. (1997) and a state of the art review on semi active control is recently presented by Symans and Constantinou (1999).

Herein, a brief outline of a few important and widely used control strategies for seismic protection of structures is presented. The control strategies include (i) Base isolation, Tuned mass dampers and Visco-elastic dampers amongst passive control of structures; (ii) Linear Quadratic Control, Sliding mode control and pole placement technique amongst active control of structures; and (iii) Semi active control using semi hydraulic dampers.

2. PASSIVE CONTROL OF STRUCTURES

In the passive control of structures, the control devices are actuated because of the motion of the structure. No external source of energy is required to actuate them. Further, the control of response depends upon the properties of the control devices and the mechanism of control. A base isolation device controls the response of the structure by way of both deflecting and absorbing the seismic energy. Deflection of seismic energy is caused by making the base of the structure flexible and hence by increasing the time period of the isolated structure. Absorption of seismic energy takes place due to the hysteretic behavior of the isolator TMD controls the response by dissipating a part of seismic energy i.e., there is a transfer of energy from main structure to the TMD. Visco elastic dampers control the response by absorbing a part of seismic energy through a damping mechanism.

2.1 Base Isolation

A number of base isolation devices have been developed, tested and implemented in practice. They include laminated rubber bearing (LRB), New Zealand rubber bearing (NZB), friction pendulum system (FPS), pure friction system, resilient friction base isolator (RFRB) etc. These blocks like devices shown in Fig.1 are placed between the found action and the structure. The load deformation behaviour of the isolator is obtained from experimental test and is modelled for theoretical analysis. The laboratory test is carried out under cyclic loading and the hysteretic loop obtained from experimental test is used to characterize the isolator stiffness and damping. One of the popular models representing the restoring forces Fx and Fy of an isolator in X and Y directions are given by the relation

$$\begin{Bmatrix} F_X \\ F_Y \end{Bmatrix} = \alpha K_0 \begin{Bmatrix} U_X \\ Y_Y \end{Bmatrix} + (1-\alpha) K_0 \begin{Bmatrix} Z_X \\ Z_Y \end{Bmatrix} \tag{1}$$

in which U_X and U_Y are the lateral displacements, Z_X and Z_Y are the hysteretic displacement component of restoring forces in X and Y directions respectively; α is the post to yielding stiffness ratio; $K_0 = Q / q$, is the initial stiffness; Q and q are the yield force and displacement respectively; Z_X and Z_Y satisfy the following coupled non-linear first order differential equations:

$$q\{\dot{Z}\} = [G]\{\dot{U}\} \tag{2}$$

$$[G] = \begin{bmatrix} A - \beta\,\mathrm{sgn}(\dot{U}_X)\,|Z_X|\,Z_X - \tau Z_X^2 & -\beta\,\mathrm{sgn}(\dot{U}_Y)\,|Z_Y|\,Z_X - \tau Z_X Z_Y \\ -\beta\,\mathrm{sgn}(\dot{U}_X)\,|Z_X|\,Z_Y - \tau Z_X Z_Y & A - \beta\,\mathrm{sgn}(\dot{U}_Y)\,|Z_Y|\,Z_Y - \tau Z_Y^2 \end{bmatrix} \tag{3}$$

in which, β, τ and A are the parameters which control the shape and size of the hysteretic loop; and sgn denotes the signum function.

In the [G] matrix, the coupling terms may be ignored if F_X and F_Y are assumed to be independent of each other. This assumption is quite often made in order to simplify the non-linear analysis of the base isolated structure under earthquake excitation. For 2D – analysis of the frame, any of the diagonal elements of the [G] matrix can be used to represent the uni-directional hysteretic behaviour of the isolator. The equations of motion for a base isolated 2D building frame may be written as

$$[M](\ddot{u}+\ddot{u}_b)+[C]\{\dot{u}\}+[K]\{u\}=-[M][T_g]\{\ddot{u}_g\} \quad (4)$$

$$[M](\ddot{u}_b)+[C_b]\{\dot{u}_b\}+\{F_b\}-[C]\{\dot{u}_b\}-[K]\{U\}=-[M_b][T_g]\{\ddot{u}_g\} \quad (5)$$

in which [M], [K] and [C] are the lumped mass, stiffness and damping matrices of the frame corresponding to the sway degrees of freedom; $[M_b]$ and $[C_b]$ are the lumped mass and damping matrices corresponding to the degree of freedom at the base raft; $\{F_b\}$is the total restoring force vector of the isolator system consisting of n number of isolators placed below each column; $\{F_b\}$ is obtained from the [G] matrix explained before; $[T_g]$ is the influence coefficient matrix.

Before a base isolated building is analyzed for a specified ground motion, isolators are first designed and initial displacement of the base under earthquake forces is estimated from the response spectrum of earthquake. Then, a response spectrum method of analysis is carried out for the base isolated structure. The details of the initial design and the response spectrum method of analysis are given in the text book on dynamics by Hart (2001). The time history analysis of the base isolated frame with specified base isolators can be performed using incremental time integration procedure. In standard softwares like SAP2000, this analysis can be performed once the backbone curves for the hysteretic behaviour of the isolators are specified.

2.2 Tuned Mass Damper (TMD)

TMD is a spring-mass-damper device connected to the main structure as shown in Fig.5. For building structures, it is generally placed at the top of the building. TMD has been widely used as an effective passive control device for response reduction of tall buildings subjected to gustiness of wind. For seismic control of structures, performances of TMDs have also been investigated and have been found to provide 30 to 35% reduction in displacement response. Proper tuning of the TMD with the natural frequencies of the structure is the key parameter which dictates the control of response. For structures which have closely spaced frequencies like, asymmetric tall buildings, multiple TMDs (MTMDs) are found to be more robust and effective as compared to providing a single TMD. The difficulty of using the TMDs or MTMDs is the requirement of the large mass to be installed at the top of the structure. About 10% of the total mass of the structure is required to provide desirable control of the response.

The equations of motion a multistory building frame with a TMD as shown in Fig.5 takes the form

$$M\ddot{X}+C\dot{X}+KX=MI\ddot{X}_g+G \quad (6)$$

$$m_t\ddot{y}+C_t(\dot{y}-\dot{x}_n)+K_t(y-x_n)=-m_t\ddot{X}_g \quad (7)$$

in which M, C, and K are, respectively, the mass, damping, and stiffness matrices corresponding to the sway degrees of freedom of the frame; m_t, C_t and K_t are, respectively, the mass, damping, and stiffness of the AMD; X is a vector of displacement of the frame with respect to ground; y is the displacement of the mass of the AMD; over dots indicate time derivatives; $\ddot{X}_g$ is the ground acceleration; I is unit vector of size n; and G is vector of size n having all elements as zero except the nth elements, which is given by

$$C_t(\dot{y}-\dot{x}_n)+K_t(y-x_n) \quad (8)$$

For a specified time history of ground motion, the above equation of motion can be solved in modal coordinates by decoupling Eqn.6 and solving n+1 coupled equation where n is the number of modes considered. Alternatively, [C] matrix for the frame can be generated using Rayleigh damping and then, writing Eqns.6 and 7 together in state space form. The state space equation can be solved in Matlab.

2.3 Visco-Elastic Dampers

Another type of passive control of building frames subjected to earthquake excitation can be achieved by way of putting visco-elastic dampers (VEDs) as shown in Figs.3 and 4. VEDs have both linear springs and dashpots provided in parallel and these are modeled in different ways. The linear model uses a linear spring which models a potential energy, quadratic in deformation and linear dashpot which models a dissipative force proportional to deformation rate. Three types of linear model have been used in the literature (Shukla, 2002) namely, Kelvin model, Linear hysteretic model and Maxwell model. Maxwell model is described here.

A Maxwell element consists of a linear spring with constant β in series with a linear viscous dashpot with constant $\tau\beta$. This model satisfies the following differential equation

$$\dot{f}(t)+\frac{1}{\tau}f(t)=\beta\dot{\delta}(t) \tag{9}$$

in which f(t) is the force in the dampers, $\delta(t)$ is the displacement and τ is the relaxation time. By defining the relaxation time τ and β, the Maxwell model of VED can be defined. For a multi-storey building frame with VEDs, the equation of motion takes the form

$$M\ddot{y}(t)+C\dot{y}(t)+Ky(t)+\sum_{i=1}^{N_e}B_i{}^{t}f_i(t)=I_0\ddot{x}_g(t)\,; \qquad y(0)=y_0; \qquad \dot{y}(0)=\dot{y}_0 \tag{10}$$

where y(t) is vector of displacements; M, K, C represent mass, stiffness and damping matrices, respectively; $\ddot{x}_g(t)$ represents the ground acceleration; I_0 is the influence coefficient vector; $f_i(t)$ is the force in the ith energy dissipation device, and N_e is the number of such devices; B_i is the coefficient matrix which relates y(t) and $\delta_i(t)$ as

$$\delta_i(t)=B_i y(t) \tag{11}$$

$f_i(t)$ follows the relationship given by Eqn.9. The equation of motion can be solved in frequency domain or time domain. The problem is one of non-classical damping and can be handled by different methods i.e. by constructing [C] matrix or by using modal analysis techniques using iterative pseudo force method or modal strain energy method Hart, 2001.

3. ACTIVE CONTROL OF STRUCTURES

Active control of structure has been topic of intensive research in the past. Many active control algorithms, their implementations, advantages and disadvantages have been discussed in the literature. The state of the art research papers on structural control summarize the research work carried out in this area (Soong, 1988; Housner et al. 1997). It is not possible to describe all different types of control strategies here. Two active control strategies namely, linear quadratic control, and sliding mode control are discussed in the following.

Most of the control concepts have feed back structure. Actuators are powered by external energy sources to produce the required control force. Based on the control strategy, active control can be classified as open loop control system; close loop control and open–close loop control. When the control force signal given to the actuator is computed based on the measured ground motion, then it is called open loop control. When signal is computed based on only measured responses, then it is called closed loop control. Open – close loop control is a combination of both. An adaptive control system is a variation of open close loop control with a controller which can adjust parameters of the system. The adaptive systems are generally used to control structures whose parameters are unknown and are based on tracking error between the measured response and the observed response. A learning control system is one which has an ability to learn and can switch over from open loop control system to close loop control system depending upon the requirements.

Based on mathematical modeling, the active control of structures can be classified as (i) linear control in which all mathematical equations and operations are linear; (ii) non-linear control in which these operations are non-linear; (iii) time varying control in which the control functions and parameters change with time; (iv) discrete time control system in which the control parameters vary at discrete time interval; (v) lumped parameter control system in which the control parameters are lumped; (vi) distributed control parameter system in which the parameters are function of space; (vii) deterministic control system in which all mathematical operations are deterministic and finally, (viii) stochastic control system in which mathematical equations and operations are stochastic.

Most of the control theories reported in the literature are based on deterministic control system with lumped parameters with time varying control operations. The mathematical equation of motion for all control methods has the following structure:

$$M\ddot{x}(t)+C\dot{x}(t)+Kx(t)=Du(t)+Ef(t) \tag{12}$$

where M, C, K are mass, damping and stiffness matrices; X(t) is the displacement vector; f(t) is the excitation due to ground motion; D is the location matrix of control force; E is the location matrix for the excitation forces; u(t) is the control force vector. The control force vector has the following approximate form.

$$u(t)=K_1x(t)+C_1\dot{x}(t)+E_1f(t) \tag{13}$$

where K_1, C_1, and E_1 are respectively control gains which can be time-dependant. From Eq (12) and (13), it follows that,

$$M\ddot{x}(t)+(C-DC_1)\dot{x}(t)+(K-DK_1)x(t)=(E+DE_1)f(t) \tag{14}$$

It is seen from Eq.(14) that the effect of structural control is to mathematically modify the damping, the stiffness and the excitation in such a way that the response of the system is controlled. The matrices K_1, C_1, and E_1 are called gain matrices and can be obtained in such a way that the response in principle can be totally eliminated. However, in real practice it is not possible to totally eliminate the response. Different degrees of control of response are achieved by deriving the control gain matrices K_1, C_1, and E_1. Derivation of these matrices depends on the control algorithms selected. In general, the control algorithms have some objective functions to be minimized. Accordingly, the gain matrices are derived. Different control algorithms differ in respect of finding these gain matrices or finding the control force vector keeping in view an objective function that reduces the structural response. The solution of the control problem and the development of control algorithm are obtained by writing the control equation of motion in state space of the form:

$$\dot{z}(t)=Az(t)+Bu(t)+Hf(t) \tag{15}$$

in which z(t) is the state vector.

3.1 LQR Control of Building Frames with ATMD

A multistory shear frame with AMD placed on the top storey (nth storey) is shown in Fig.7. The equation of motion of the system under ground excitation is given by:

$$M\ddot{X}+C\dot{X}+KX=-MI\ddot{X}_g+G' \tag{16}$$

$$m_t\ddot{y}+c_t(\dot{y}-\dot{x}_n)+k_t(y-x_n)=u-m_t\ddot{x}_g \tag{17}$$

in which M, C, and K are, respectively, the mass, damping, and stiffness matrices corresponding to the sway degrees of freedom of the frame; mt, ct, and k_t are, respectively, the mass, damping, and stiffness of the AMD; X is the vector of sway displacement of the frame; y is the displacement of the mass of the AMD; over dotes indicate time derivatives; $\ddot{X}_g$ is ground acceleration; u is control force; I is unit vector of size n; and G' is vector of size n having all elements as zero except the nth element, which is given by

$$c_t(\dot{y}-\dot{x}_n)+k_t(y-x_n)-u \tag{18}$$

In order to reduce the number of equations to be solved, the equations of motion are written in terms of the modal coordinates of the building and the displacement of the AMD as

$$\bar{M}\ddot{Z}+\bar{C}\dot{Z}+\bar{K}Z=-\phi^T MI\ddot{X}_g+\phi^T G' \tag{19}$$

$$m_t\ddot{y}+c_t(\dot{y}-\dot{x}_n)+k_t(y-x_n)=u-m_t\ddot{x}_g \tag{20}$$

in which $\bar{M}$, $\bar{C}$, and $\bar{K}$ are m x m diagonal matrices; $\bar{M}=\phi^T M\phi$; $\bar{C}=\phi^T C\phi$; $\bar{K}=\phi^T K\phi$; ϕ is undamped mode shape matrix of size n x m; Z is vector of generalized coordinates; and m is number of mode shapes considered.

For the formulation of the control problem, Eqns. 19 and 20 are written in the state space as

$$\dot{V}=AV+Bu+F\ddot{X}_g \tag{21}$$

in which $V^T=[Z_y \dot{Z}_{\dot{y}}]$A is matrix of size 2(m+1) x 2(m +1) whose elements are function of $\phi_i, \eta_i, \omega_i, c_i, m_t$ (ω_i is ith natural frequency of the building; η_i is modal damping ratio); B is control force coefficient vector; and F is excitation force vector.

The control force is determined by minimizing a quadratic performance index given by

$$\bar{J}=\int_0^t \frac{1}{2}(V^T QV+ru^2)dt \tag{22}$$

in which Q, a positive semi definite matrix, is a weighting matrix associated with the components of V; and r is positive weighting factor.

The minimization of Eqn.22, with t assumed to be large, provides the control force as [Chang and Soong (1980)]

$$u=K^t V \tag{23}$$

where $$K^T=\frac{B^T P}{r} \tag{24}$$

in which P is determined by solving the following Riccati equation

$$PA+A^T P-PBr^{-1}B^T P+Q=0 \tag{25}$$

in which Q is the size 2 (m+1) x 2 (m +1). The matrices A and Q and the vectors B and F for the closed loop system are given by Sarbjeet and Datta (1998). The controlled equation of motion in state space can be solved in Matlab.

3.2 Sliding Mode Control of Building Frame using ATMD

The equations of motion in the modal coordinates remain the same as LQR control. In the state space form, the equation of motion remains also the same. The design approach for nonlinear sliding mode control consists of two stages. The first stage is concerned with the design of a switching manifold (a set of discontinuous surfaces constituting the sliding surface) S_S upon which the desired dynamic behavior can be generated for the system. The second stage is concerned with the non-linear control strategy that guarantees bounded motion about the desired dynamic behavior for the first stage and to take into account the different uncertainties present in the system.

In the first stage, the sliding surface is designed with the help of k sliding variables such that

$$S_S = \lambda_S V = 0 \tag{26}$$

where S_S is a vector consisting of k sliding variables, i.e., $S_S = [S1, S2, S3, \ldots Sk]^T$; and λ_S is a k x 2 (n + 3) matrix. The sliding surface is designed such that the motion on the sliding surface is stable.

The sliding surface is achieved by employing a transformation of the state vector U and using the classical feedback design technique to obtain λ_S. The procedure involves the minimization of a quadratic performance function like linear quadratic control.

In the second stage, the controlled actions are designed to force the system state to reach the sliding surface and maintain it there. Generally, this requires the derivation of a discontinuous control law since the system state is altered each time the state tends to violate the condition S = 0. One of the methods to derive the control law is to use Lyapunov's direct method to guarantee that S = 0 is an asymptotically stable equilibrium point. The details of the control algorithm are given by Sarbjeet and Datta (2000).

3.3 Pole Placement Technique

For any system with an active controller, the equation of motion in the state space can be written in the form

$$\{\dot{Z}\} = [A]\{Z\} + [B]w \tag{27}$$

where w is a scalar control force. According to the second method of Liapunov stability analysis, for a system represented by Eq. 27, the eigen-values of the system need to be investigated. If all eigen-values of matrix [A] have negative real part, the system is asymptotically stable. Further, a necessary and sufficient condition for all eigen -values to have negative real part is that they have determinant with positive coefficient of the leading terms of their characteristic polynomial. The eigen-values of the matrix [A] are called the poles of the original system. The stability of a linear closed-loop system can be determined from the location of the closed-loop poles in the complex s plane, in which s are the poles of the system. If any of these poles lie on the right half of the s plane, then the system is unstable. Therefore, closed-loop poles on the right half of s plane are not permissible in the usual linear control system. If all closed-loop poles lie to the left half of s plane, then the system is stable. This concept is utilized to control the flutter of suspension bridge.

In suspension bridges at flutter condition ($U = U_f$, $\omega = \omega_f$, and $k = k_f$, where subscript f denotes the flutter condition), the damping of the dominant mode (mostly the torsional mode) is either zero or negative. Therefore, according to Liapunov's second method, the real part of dominant pole becomes either zero or positive, and makes the system unstable. If the system considered is completely state controllable, then poles of the closed-loop system may be placed at any desired location by means of state feedback through an appropriate state-feedback gain matrix. If K1 is the gain matrix, than the eigen-values of the matrix ([A] – [B] x [K1] are the desired poles of the controlled system, which would replace the poles of unstable system. The necessary and sufficient condition for arbitrary pole placement to make all eigen-values to have negative real part is that the rank of the controllability matrix [H], defined in Eq. 28, to be equal to the dimension of the matrix itself, i.e.,

$$[H] = [B \mid AB \mid \ldots \mid A^{n-1}B] \quad (28)$$

where n is dimension of the matrices. The details of the control of suspension bridge flutter using pole placement technique are given by Pouzeynali and Datta (2005).

4. SEMI-ACTIVE CONTROL

It originates from the passive control system modified to allow the adjustment of mechanical properties based on feed back from excitation or measured response. As an active control system, it monitors the feed back measurement and generates appropriate command signal. As a passive control system, control forces are developed as a result of the motion of the structure. Control forces primarily act to oppose the motion and are developed through appropriate control algorithms. Semi active control generally requires small amount of external power for the operation (of the order of few watts in comparison to tens of Kilo-watts). Different types of semi active control devices include (Dyke et al., 1996). (i) Stiffness control devices (ii) Electro rheological dampers/magneto rheological dampers; (iii) Friction control devices; (iv) Fluid viscous devices; (v) TMDS and TLDS.

Here, the semi-active control of building frame using hydraulic damper and Fuzzy control theory is described briefly.

For the building frame shown in Fig.9, the equation of motion corresponding to the structural sway degrees of freedom of the frame under seismic excitation $\ddot{X}_g(t)$ and the control actions F_d can be written as

$$M\ddot{X} + C\dot{X} + KX + HF_d = -MI\ddot{X}_g \quad (29)$$

in which F_d is force vector containing forces due to the r number of semi-active hydraulic dampers (SHD in series with braces); and H is (n x r) matrix denoting the location of r dampers.

The semi-active hydraulic dampers are installed between the bracing and the beam in the structural plane as shown in Fig.9. The SHD has been modeled as a Maxwell element consisting of a damping element with a variable damping coefficient (C_t) in series with the stiffness of the damper, K_d , and the stiffness of the bracing, K_b . The equation of motion considering the hydraulic damper characteristics can be written in state space form and the control of the frame under seismic excitation can be obtained using Simulink and Fuzzy Logic tool box of Matlab. The control commands as obtained from the fuzzy algorithm are modified to take into account the damper capacity and development of the actual force in the damper. The details are given by Bhardwaj and Datta (2006).

5. ACTUAL APPLICATION OF ACTIVE CONTROL

Some actual applications of active control schemes for the reduction of wind-induced vibration of tall buildings have been reported. These control schemes have also been found to control the seismic response of the building to moderate earthquakes. Salient features of three such applications on tall buildings are presented below:

5.1 AMD on Kyobashi Seiwa Building

The building is located at Chuo-Ku, Tokyo with a frontage of 4m and a total height of 33m. The building has 11 floors and is made of rigid steel frames. Two AMDs are located at the top floor spaced apart. First AMD has a mass of 4 tons and second one has a mass of 1 ton. The idea of providing two AMDs was to control the torsional response of the structure also. The schematic diagram of the AMD is shown in Fig. 10. The hydraulic pump of 22 KW is used to actuate the masses. Sensors are placed at basement, at 6th floor and at 11th floor. The computer is provided on the top floor itself. This is the world's first AMD installed on building

5.2 Duox on ANDO Nighikicho

It is located in Chiyoda – Ku, Tokyo. It has 14 stories and two basement levels, and is made of rigid steel frames. Above the ground weight of the building is 2600 tonnes-f. Two directional simultaneous AMD is placed on the top of a TMD which is placed on the top floor. The damping system of the TMDs consists of oil dampers. The TMD weight is 18 t-f, while the AMDs have 2 t-f weight each. The schematic diagram of the Duox system is shown in Fig. 11. The Duox system operates on the principle that if the active control system fails, the TMD will provide at least the minimum control of response.

5.3 Trigon on Shinjuku tower

It is located in Shinjuku – Ku, Tokyo with a floor area of 264,100 m2. The structure is made of steel and partially reinforced concrete frames. It is a 52 storied plus 5-basement structure with an above ground weight of 130,000 t-f. Three control devices are installed in the form roller pendulum mass on 36th floor. Control masses are 110 tonnes each. Maximum stroke of the pendulum is 110 cm. Period adjusted is 3.7258 seconds and the motor capacity to drive the pendulum is 75KW. The schematic diagram of the roller pendulum is shown in Fig. 12.

REFERENCES

1. Bhardwaj, M. K., and Datta, T. K. (2006). "Semiactive Fuzzy Control of the Seismic Response of Building Frames." J. Struct. Eng., 132(5), 791-799.

2. Chang, J. C. H., and Soong, T. T. (1980). "Structural Control using Active Tuned Mass Damper." *J. Eng. Mech.*, 106(6), 1091-1098.

3. Garry C. Hart (2000), *Structural Dynamics for Structural Engineers*, John Wiely & Sons, New York.

4. Housner, G. W., Bergman, L. A., Caughey, T. K., Chassiakos, A. G., Claus, R. O., Marsi, S. F., Skelton, R. E., Soong, T. T., Spencer Jr., B. F., and Yao, J. T. P. (1997). "Structural Control: Past, Present and Future." *J. Eng. Mech.*, 123(9), 897-971.

5. Jangid, R. S., and Datta, T. K. (1995). "Seismic Behaviour of Base Isolated Building – A State of the Art Review." *J. Struct. And Buildings*, *Proc. Inst. Civil Engrs. (London)*, 110, 186-203.

6. Kelly, J. M., (1986). "A Seismic Base Isolation: A Review and Bibliography." *Soil Dyn. and Earthquake Eng.*, 202-216.

7. Pourzeynali, S., and Datta, T. K. (2005). "Semiactive Fuzzy Logic Control of Suspension Bridge Flutter." J. Struct. Eng., 131(6), 900-912.

8. Shukla, A. K. (2002). *Passive and Hybrid Control of Structures under Seismic Excitation*, Ph.D thesis, submitted to IIT-Delhi.

9. Singh, S., and Datta, T. K. (1998). "Open-Closed Linear Control of Building Frames Under Seismic Excitation." *J. Struct. Eng.*, 124(1), 43-51.

10. Singh, S., and Datta, T. K. (2000). "Nonlinear Sliding Mode Control of Seismic Response of Building Frames." *J. Eng. Mech.*, 126(4), 340-347.

11. Soong, T. T. (1988). "State-of-the-Art Review: Active Structure Control in Civil Engineering." *J. Eng. Struct.*, 10(2), 73-84.

12. Symans, M. D., and Constartinon, M. C. (1999). "Semiactive Control Systems for Seismic Protection of Structures: a State of the Art Review." *Eng. Struct.*, 21, 469-487.

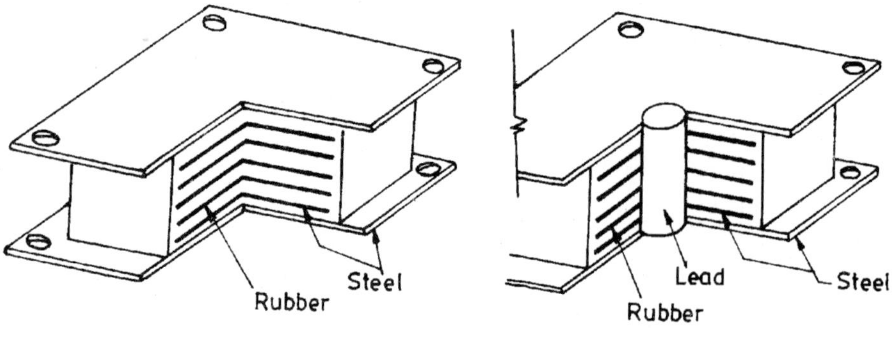

Figure 1: Laminated rubber bearing base isolators

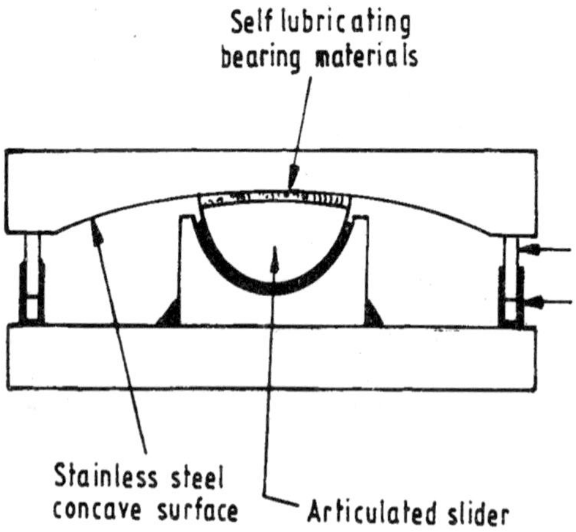

Figure 2: Friction pendulum base isolators

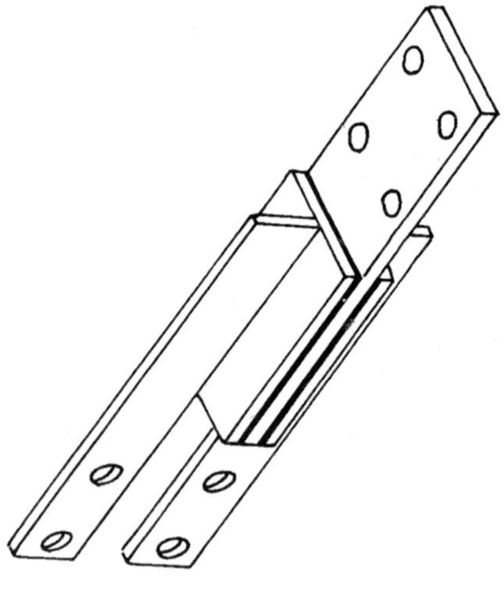

Figure 3: Visco-elastic damping device

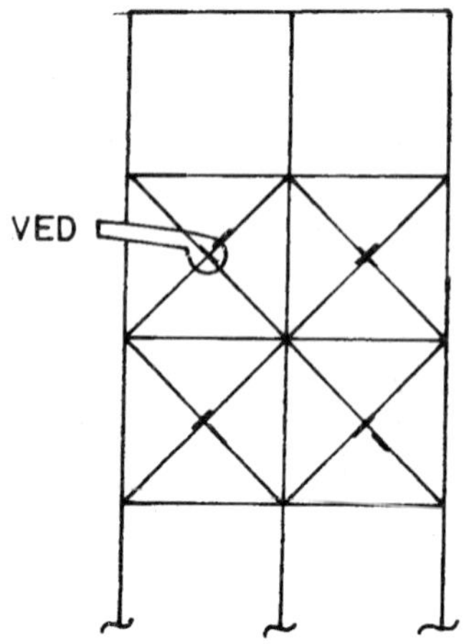

Figure 4: Use of VED in Frame

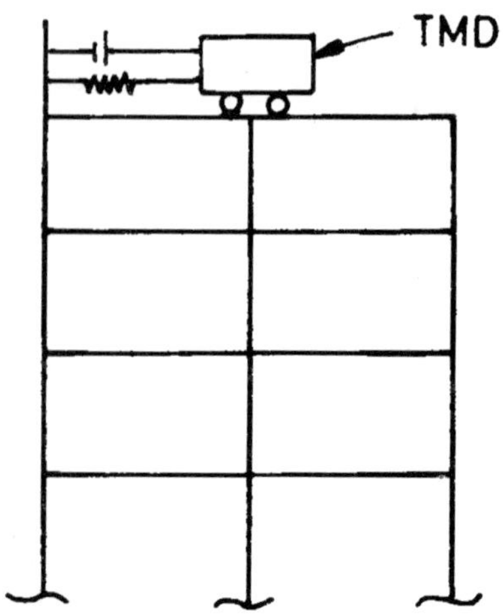

Figure 5: Use of TMD in Frame

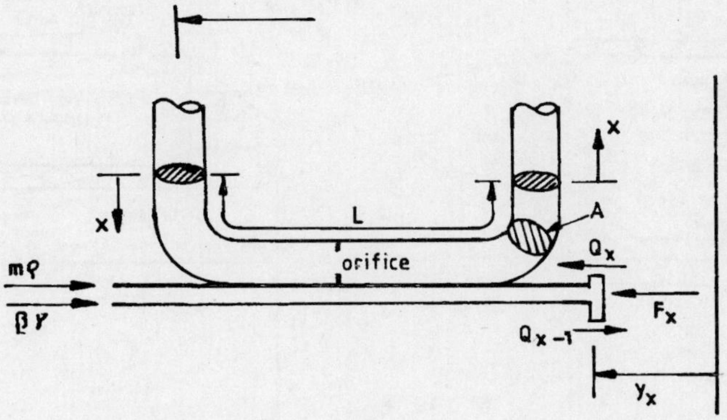

Figure 6: Tuned liquid column damper

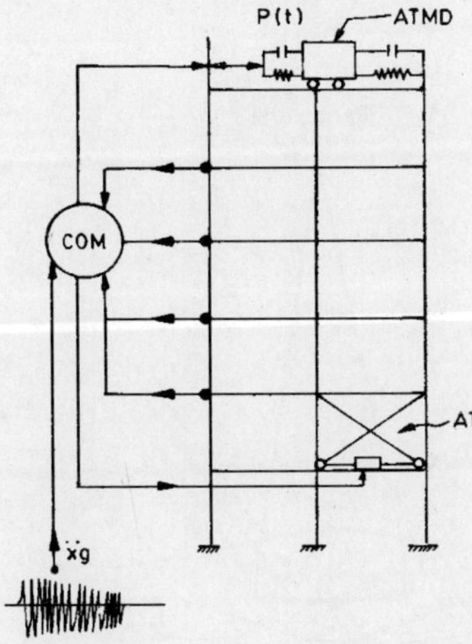

Figure 7: Active control strategies

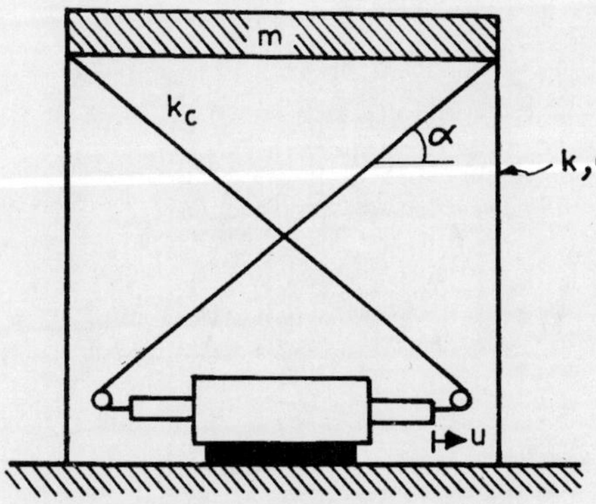

Figure 8: Active tendon scheme

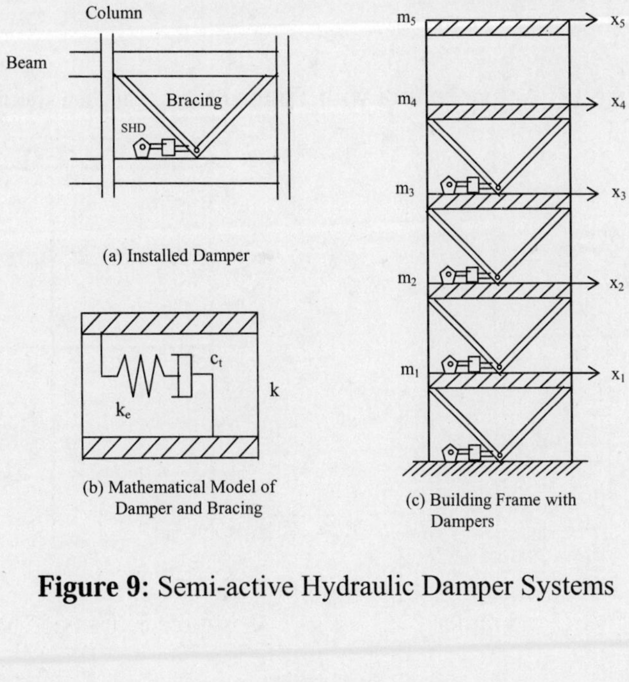

Figure 9: Semi-active Hydraulic Damper Systems

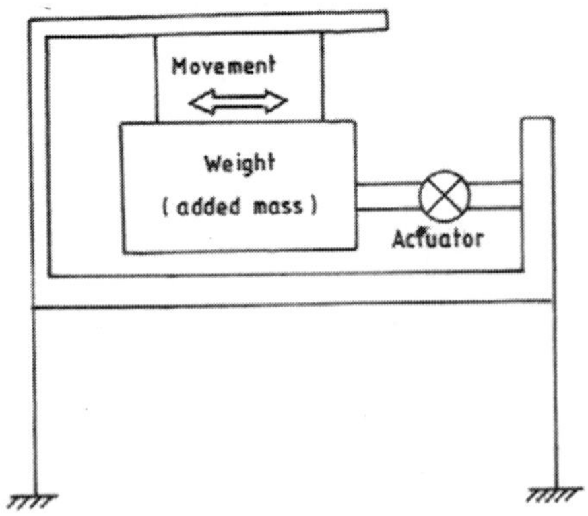

Figure 10: Active control with AMD system

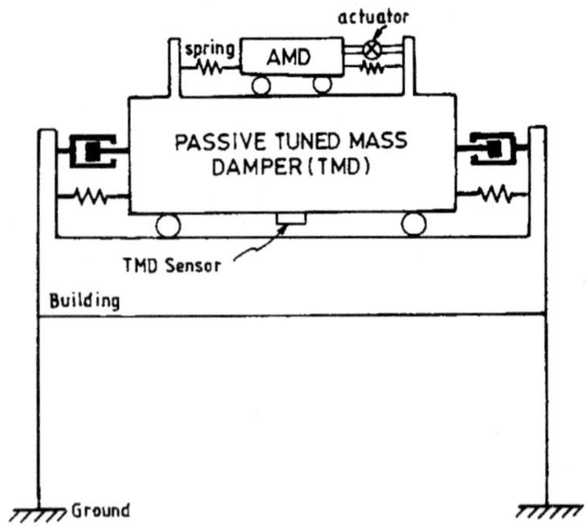

Figure 11: Active control with Duox system

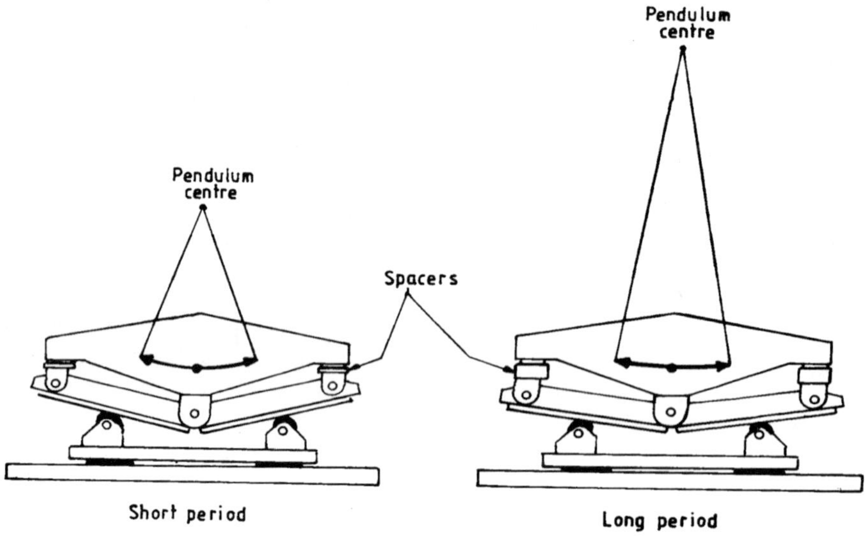

Figure 12: Active control with Trigon (Roller pendulum system)

QINGHAI-TIBET RAILWAY, CHINA AND ITS EARTHQUAKE DAMAGE MITIGATION

Lanmin Wang[1], Zhijian Wu[1,2], Junjie Sun[1] and Luxin Zhang[3]

[1]*Lanzhou 730000, China, Lanzhou Institute of Seismology, China Earthquake Administration (CEA), e-mail* wanglm@gssb.gov.cn

[2] *Lanzhou 730000, China, Cold and Arid Regions Environmental and Engineering Institute, Chinese Academy of Sciences (CAS), e-mail* zhijianlz@163.com

[3] *Xining 810007, China, Qinghai-Tibet Railway Construction Headquarters, e-mail* zlx22@vip.sina.com

ABSTRACT

The Qinghai-Tibet Railway (QTR) is the highest-elevation passenger train in the world and the first to connect central China with Tibet. Construction of this railway had to contend with major geotechnical challenges, such as permafrost, environmental protection and geologic hazards, e.g. seismic activity, steep and unstable slope and variable hydrologic conditions. Its completion is a remarkable feat in world's railway building history and crystallization of wisdom of Chinese workers. In this paper, the planning and preparing history and the construction project of QTR are introduced. The major three thorny problems, permafrost, lack of oxygen and environmental frangibility for the construction and their solutions are presented, which are active methods of riprap roadbeds, heat pipe roadbed and bridges over land for permafrost, health care system for lack of oxygen and environmental protection measures for construction and operation. Sesimic safety assessment and study on the influence of earth temperature on dynamic charactristics of permafrost and ground motion were carried out for earthquake damage mitigation of the railway. The laboratory test, field test and observation and testing operation have shown that design, construction, measures for earthquake hazards mitigation and environmental protection of the Qinghai-Tibet Railway are successful.

Keywords: Qinghai-Tibet Railway, Hazard mitigation, Permafrost, Engineering measures

1 INTRODUCTION

China made history on July 1, 2006, when the Qinghai-Tibet Railway (QTR) was put into test operation. QTR is the highest-elevation railway for passenger train in the world and the first railway to connect central China with Tibet, providing a direct tie between Tibet and other partitions of China with more economic, cultural and political significances. Stretching about 1,142 km, it runs from Golmud in Qinghai province to Lhasa, the capital city of the Tibet Autonomous Region, China (Fig.1).

Early in 1956, China government prepared to construct QTR, which begins at Xining, the capital of Qinghai province, ends at Lhasa. However the great project had to be laid aside because of the financial difficulty in 1961. And then, China aspired after putting it into practice, the second time in 1974, but it was stopped in 1978 due to three insoluble problems at that time, i.e. permafrost, lack of oxygen and environmental frangibility. It was the third time that China government put the issue of QTR construction on the table in 2000 while the country's economy had grown dramatically and Chinese scientists and engineers had accumulated enough data both in field and laboratory for 40 years before 2000 to design and construct QTR across the permafrost areas on the Qinghai-Tibet Plateau (QTP). After more than one year preparing, the project started in June 29, 2001 and completed in June 30, 2006.

Completion of QTR project is a remarkable feat in world's railway construction history and crystallization of wisdom of Chinese scientists, technicians, workers and managers. It would benefit a lot to all ethnic groups in Qinghai province and the Tibet and speed the economic growth and social development in this area significantly.

In this paper, the great project of QTR, the three major thorny problems and their solutions and earthquake hazards mitigation for the railway are introduced in details.

2 THE PROJECT OF QTR

QTR traverses the spectacular topography of QTP, cutting across four mountain chains – Kunlun, Fenghuo, Tangula and Nianqintanggula – where elevations of the trackbed are all above 4,600m. It also crosses five major rivers, the Yellow River, Yangtse River, Mekong (Lancang) River, Nujiang River and Lhasa-Brahmaputra River, and passes through the Three River National Natural Protection Region. And most of its length is above 4,000m in elevation, and 50km is above 5,000m. (Fig.2)

The whole QTR consists two sections extensively.The first section of QTR from Xining to Golmud is 814km long, which has been operated since 1984. The second section from Golmud to Lhasa is about 1142km long, which was started to construct in June 29, 2001 and opened to transportation in July 1, 2006. Construction of the railway had to contend with major geotechnical challenges, permafrost, environmental protection and geologic hazards, e.g. seismic activity, steep and unstable slope and variable hydrologic conditions.
The major reasons causing the engineering geology problems, are the special harsh climate and the change of microtopography, sand blown by wind, and hydrologic conditions under engineering construction. So it is important to pay more attention to construction geological conditions, apply the dynamical design and optimizing enhancing intension design to constructing QTR.

The railway crosses the largest area of low-latitude permafrost in the world, in which there are 546.41km locating at continuous permafrost areas. There are total 10 tunnels, 9518m long, in QTR, of which 2 tunnels locate at permafrost areas, the elevations of 7 tunnels are above 4000m, the Fenghuo Mountain tunnel is the highest elevation tunnel of the world, 4905m above sea level, the Yangbajing tunnel is the longest one along QTR, 3345m. Bridges with a total length of 159.88km are constructed along QTR, in which 451 bridges, a 120.28km length, locate at permafrost regions and the highest one is Sancha River Bridge, which is 54.1m high, and the longest one is Qingshuihe Bridge, which is 11.7km long (Fig.3). The elevation of the highest place of QTR is 5072m, which is also the highest point of railroad all over the world. The Tanggula station, 5068m, is the highest station in the world.

There are total 24 institutes and colleges, more than 100 permafrost scientists and 200 geotechnical engineers, 4 design institutes, 25 construction companies and 13 supervision companies, around 36,000 peoples took part in construction of the QTR project.

Passengers can begin this train trip in Beijing, Shanghai, Guangzhou, Chongqing, Chengdu, Lanzhou and Xining and end in Lhasa. From Beijing, the capital of China, to Lhasa, it's a journey of 4,064Km, and the tour time is only 46 hours and 30 minutes. QTR trains are grouped by 14 cars and 2 locomotives, and the seating capacity is 936 persons. Every passenger on the train is given an oxygen mask to alleviate symptoms of altitude sickness. The oxygen level in each car is monitored, and cars are also equipped with a digital display that gives continuous updates of elevation, train speed, outside temperature and distance to the next station.

3 THE MAJOR THREE THORNY PROBLEMS AND THEIR SOLUTIONS

The permafrost, lack of oxygen and environmental frangibility are three thorny problems for construction of QTR across the Tibetan Plateau. How to keep the stability of embankment in permafrost regions is the biggest scientific and technical challenge for scientists and engineers, because of the influences of thermal variations due to the artificial disturbance during construction of QTR and the increase of environmental temperature in the future.

3.1 Permafrost

QTR crosses 546.41km continuous permafrost regions, 82 km discontinuous permafrost regions, and the regions which annual average ground temperature above -1.0°C occupy 275km, high-contained ice regions 221km, high temperature and high-contained ice section about 134km (Fig.4). The permafrost is a very special soil, and it is quite sensitive to temperature, and other characteristics, including physical, chemical, and engineering features, are unstable and nearly correlative with temperature (Fig.5). Moreover, the characteristics are also influenced by the ice content, which is directly correlated to temperature and decreases with the rising of temperature (Xu, 2001).

As a result of the influences of global climate warming and railway engineering on the permafrost degradation, stabilizing permafrost is the most pervasive challenge for both construction and maintenance of the railway. Protecting the permafrost is a key to preventing embankment and roadbed failure from frost heaving and thaw collapse, two main roadbed diseases from the degradation of permafrost. If the permafrost absorbs thermal energy that is not dissipated, then the permafrost will melt and the roadbed will be damaged.

3.1.1 Methods

Several active methods are developed and taken to protect permafrost along QTR, which include 3 three kinds of riprap roadbeds (namely ripped-stone embankment (116.07km), ripped-stone protect slope roadbed (156.21km), and U-shaped combination roadbed (or the multi-roadbed)), heat pipe roadbed (30.38km), bridge on land (bridge instead of roadbed, 451 bridges, 120.28km), and other assistant methods, such as, the ventilated-pipe roadbed, the awning roadbed, the heat preservation materials and heat pipe multi-roadbed etc. They have been applied in engineering design and construction. (Fig.6)

(1) Riprap Roadbed: Much of the embankments and roadbeds are protected with "riprap", which are coarse, angular pieces of rock used to stabilize a surface. The riprap-engineering measures shield the roadbeds, embankments or both of them from solar heating and cool them by convection to maintain a higher thaw depth, which thus protects the permafrost. The larger rock sizes, 10 to 15cm and 40-50cm, and the large spaces between rocks are the key to increasing cooling energy and heat exchange under the riprap. The penetration of the cold energy is two to four times greater with the riprap than through normal graded roadbeds.

(2) Heat Pipe Roadbed: Heat pipes stabilize permafrost by removing heat from permafrost and dissipating it into the ambient cold air. They employ evaporative cooling to transfer thermal energy from one point to another by evaporation and condensation of a coolant. Heat pipes rely on a temperature difference between the ends of the pipe and cannot lower temperatures at either end beyond the ambient temperature.

(3) Bridge on land (Bridge Instead of Roadbed): If the annual average earth temperature of the permafrost is very high (0~-0.5°C), it's not sufficient for proteceting permafrost under the embankment. To minimize the thermal disturbance caused by construction, we adopted bridges instead of roadbeds.

(4) Ventilated-pipe Roadbed: The ventilated-pipe roadbed has an open pipe installed beneath the roadbed through the embankment that cools the ground, as denser cold air flows into the pipes and displaces less dense, warmer air. Any accumulation of heat will be carried away through the pipe, thus maintaining the frozen permafrost.

(5) Awning Roadbed: Because of the strength of solar radiation on QTP, awning roadbed is constructed to shade the roadbeds and embankment from solar heating. The awnings also prevent rain and snow from directly impacting the roadbed.

3.1.2 Testing Sections

As the beginning of the design phase, some kinds of special roadbed structures and engineering methods, such as the insulation material embankment (as dominant measure), the riprap roadbed, ventilated-pipe roadbed, and thermopile embankment etc, were adopted to keep the embankment deformation within an acceptable range in the aspect of temperature variation. In order to verify the initial design idea and the chosen measures applicability, four experimental sections regarded as pioneer engineering had been conducted before construction in 2001. Experimental sections consist of three roadbed engineering sections and two tunnel engineering sections, which are, the very-warm-fine-particle experimental engineering along Qingshui river, the thick ground ice experimental engineering along Beilu river, the thawing area and permafrost zone transition section experimental engineering along Tuotuo river, and the permafrost tunnel experimental engineering sections on Kunlun mountains and Fenghuo mountains, respectively. The settlement and deformation of the embankment, the stress and deformation of kinds of structure, as well as air temperature and meteorologic observation were carried through. At the same time, the laboratory tests were proceeded as well.

3.1.3 Observation in Situ

The deformation of the permafrost is a relatively long-term and slow process, which is further complicated by the fluctuations in ground temperatures. Therefore, the monitoring of soil temperatures and deformation are quite important during the construction and operations of QTR. The temperature profiles, the permafrost table and the variations of mean annual ground temperatures can be obtained from inspecting the soil temperatures, which directly reflect the thermal regimes of roadbed. The frost heave and thaw collapse are unique phenomena in permafrost zone, and the deformation directly reflects the stability of the permafrost roadbed.

(1) Temperature Monitoring: Considering of the distribution of the geomorphic unit along QTR, the sub-zone of the mean annual ground temperature in the permafrost region and the setting principle of the different engineering measures, the system of 6 temperature monitoring holes is set at every transect, of which, one is in the centerline of the railway embankment, one each at both sides of the railway shoulder (3~3.5m away from the centerline), and the slope foot, respectively, the last one at the natural area (about 20m away from the slope foot line). The temperature monitoring hole should extend to a depth about 5~10m below the natural permafrost table (Fig.7). In the typical measurement sections, the measurements also include the parameters in meteorology, such as air temperatures, wind velocity, wind direction, solar radiation, precipitation and evaporation.

(2) Deformation Monitoring: Taking the post-constructing deformation of settlement and the active layer thawing deformation into consideration, there should be set 6 spots for deformation monitoring at each transect, of which, three are on the roadbed surface (one at the centerline and two at each shoulder), and the other three are set below the roadbed (originally natural surface), which are located directly blow the former three sites (Fig.7).

In order to investigate the deformation under the influence of the frost heave and thaw settlement, The monitoring system has been constructed since 2001, which includes the deformation monitor, 6 spots at each transect every 100m along the 546.71Km of permafrost regions, and the temperature monitor, a total of 100 monitoring section with 6 temperature monitoring holes at each transect. The degradations are measured every 15 days manually; however, the temperature data are collected automatically every 1 hour and transferred wirelessly from the field in situ to the lab in Golmud. The data collection covers a wide range of major geomorphic units such as Kunlun mountainous area, Chumare River plain, Kekexili mountainous region, Fenghuo Mountains and Wuli mountainous region, Tuotuo river basin and Yanshi plain. These regions include the very warm and very unstable permafrost zone (section I: $-0.5°C < T_a < 0$), the warm and unstable permafrost zone (section II: $-1°C < T_a < -0.5°C$), the low temperature and almost stable permafrost zone (section III: $-2°C < T_a < -1°C$), the low temperature and stable permafrost zone (section IV: $T_a < -2°C$). The temperature observation points were distributed mainly at section I, section II and section III. The profiles also covered different engineering structures in permafrost region.

3.1.4 Brief Summary

Based on laboratory and field tests, the insulation material embankment is proved to be a passive protecting permafrost measure and it is not effective for protecting permafrost from deformation at the warm permafrost region, which belongs to section I and section II (Sheng, 2003). However, the other measures can cool ground temperature and protect permafrost from deformation effectively (Ma, 2002; Niu, 2003; Lai, 2003; Wu, 2005; Ma, 2005).

In the second design phase, the active measures protecting permafrost were adopted, such as the riprap roadbed, heat pipe roadbed and dry bridge. And these engineering measures applied proved to be capable of diminishing deformation of roadbed and reducing ground temperature of the permafrost along QTR. So, these active engineering measures can be used to keep the stability of the embankment in the permafrost.

Chinese scientists have gained better understanding of permafrost and developed new techniques to stabilize railway embankment on permafrost through observations and studies since 1960's. These knowledge and techniques make it successful to construct the QTR.

The speed of the train and the damage ratio are two major criteria for evaluating the railroad in permafrost areas. The speed of trains running on the QTR is around 120km/h and 100km/h through unfrozen soil areas and permafrost regions, respectively, which is the highest speed in permafrost regions in the world. The damage ratio is only less than 1% by the end of 2006.

3.2 Lack of Oxygen

At 4,500m elevation in the Tibetan Plateau, with atmospheric pressure and oxygen 45 percent lower than at sea level, an annual average air temperature of 5 degrees bellow 0□, and extremes including low temperatures of -47.8°C and wind speeds above 30m/s. Add in solar and ultraviolet radiation 1.5 to 2.5 times what it is at sea level, and not only is preconstruction research and fieldwork a challenge, but so is the construction itself.

During the period of construction for QTR, the three-class hospital system was set up. There were total 115 hospitals and more than 600 doctors and nurses working on the Tibetan Plateau, and 17 oxygen producing stations and 25 hyperbaric oxygen chambers were equipped for the constructors along QTR. As a result of successful operating of the health care system, the sick constructors could get effective treatment in 30 minutes, that the success ratio of emergency treatment for the acute altitude disease, such as hydrocephalus and pulmonary edema, was 100%, and there was nobody died for altitude disease when constructing for QTR.

3.3 Environmental Frangibility

The Tibetan Plateau has a unique and fragile ecosystem. During the period of designing for QTR, the route of the railway was designed to avoid nature protection area as far as possible.

In order to protect the wild animals, 33 animal underpasses were constructed along QTR. Additionally, to minimize environmental disturbance during construction, staging areas were planned to minimize surface disturbance and avoid sensitive areas, and were revegetated at the completion of the construction. (Fig.8)

In order to protect the environment during the life of the railway's operation, the trains are designed as a sealed system to protect the environment it passes through. The train cars are completely sealed, so no waste, litter or discharges are thrown out from the cars. All refuse is contained on board the train for disposal at the stations.

The fund put into the project of environmental protection is about 150 million US dollars, and the environmental protection supervision was firstly operated during construction of QTR.

4 EARTHQUAKE DAMAGE MITIGATION

The Tibetan Plateau is one of the most tectonically active areas of the world and also one of the most seismically active regions, as earthquake can be both of frequent and strong. There are a total of 33 Ms 6.0-6.9 earthquakes and 3 Ms 7.0-8.5 earthquakes occurring in the QTP between Jan.1, 1980 to Jul.31, 2006 (Fig.9). Especially, the west of Kunlun Mountain pass Ms 8.1 earthquake, which created a surface rupture 430km long in the permafrost regions, on November 14, 2001, caused a certain extent of damage and potential hazard to the infrastructure and construction projects on the Qinghai-Tibet Plateau which relates to the strategy of the developing western China, such as the QTR and the Qinghai-Tibet Highway (Fig.10).

Faced with the problem of how to construct a stable railway on the Tibetan Plateau, the construction of the railway has been paid much attention to earthquakes and active faults. The railway traverses the active faults at several places, which were the main boundaries of active plates and large earthquakes took place there. Earthquakes will disturb the stabilities of the railway. Therefore, it is very important and pressing to study earthquake damager mitigation on the Tibetan Plateau.

4.1 Identification of Active Faults and Seismic Zonation at the Key Sections along QTR

For the entrustment of China Railway First Survey and Design Institute, we a group of experts from CEA undertook the project of "Identification of active faults and seismic zonation at the key sections along the QTR" from 2000 to 2002. The work zone was divided into two classes, normal section and key section. There are two key sections along the railway, which are the South Mt. Pass-Kunlun section and Dangxiong-Yangbajin section. The South Mt. Pass-Kunlun Mt. section is about 130km in length, which includes three large size active faults, namely, the Middle Kunlun Mt. Fault, where a earthquake of magnitude 6.9 occurred in 1902, the East Kunlun Mt. Fault, where a earthquake of magnitude 7.5 occurred in 1937, and the Kunlun Mt. Pass Fault, where a earthquake of magnitude 8.1 occurred in 2001. The Dangxiong-Yangbajing section is about 180 km in length, including the Bengco Fault, where an earthquake of magnitude 8 occurred in 1951, the East Nianqingtanggula Mt. Fault, where an earthquake of magnitude 7.5 occurred at Dangxiong in 1952, and an earthquake of magnitude 8 occurred at the north of Dangxiong in 1411. Faults of these key sections are more active than other sections along QTR. Since the sections, Kunlun Mt. Pass to Sangxiong and Yangbajing to Lhasa along the railway, belongs to normal sections, there are many active faults. Such as the active faults from Kunlun Mt. Pass to Tanggula Mt. Pass, the north of Anduo to Naqu, and so on, and which may cause moderate or strong earthquakes.

The harsh climate on the Tibetan Plateau is a challenge to investigate those active faults. However, in a limited preparing time of QTR Project, geologists took proper methods to get the following major achievements: (1) The report on fault activity and seismic zoning of key sections along QTR; (2) The map (1:250,000) of active faults, which covers 25km wide, both sides of the railway; (3) The map (1:250,000) of seismic intensity zoning at the key sections, which covers 25km wide, both sides of the railway; (4) The map (1:1,000,000) of seismic zonation at the common sections, which covers 25km wide, both sides of the railway; (5) The report of evaluation for activity of major faults locates at of the sections of Kunlun Mt. Pass-Sangxiong and Yangbajing-Lhasa along the railway.

The results shown that there are 44 actives faults identified at the sections of Kunlun Mt. Pass-Sangxiong and Yangbajing-Lhasa, which has a length of 790km and a width of 50km along the railway. Hereinto 24 recent active faults were assessed as the Holocene Epoch crossing QTR, in which 18 active faults could move and dislocate in the future 100 years. During designing and constructing QTR, the appropriate engineering measures should be taken at these sites, where the 18 Holocene Epoch active faults cross the railway. The Wenquan Fault, the Sangxiong section and the Yangbajing section of the east Niantanggula Mt. Fault, are all the Holocene Epoch active ruptures, which are parallel with close to the railway. The railway should depart from these two faults as far as possible.

Seismic intensity zoning map is shown as Fig.11 with the exceedance probabilities of 10% in 50 years. The proportion of length for each intensity zone is given in table 1. The largest intensity along the railway exceeds IX degree, and the smallest is VII degree.
Compared with "Map of China seismic intensity zoning (1990)", it shows that a higher seismic intensity region extends along the QTR. Northern boundary of VIII zone, which locates at the north of Budongquan, enlarges about 20 km northwards. However, the southern boundary retracts about 10km norwards. The section, from Wudaoliang to Sidaoban, belonged to VI zone, but it is VII in the new zonation map. Northern boundary of IX zone in the section, Dangxiong-Yangbajing, enlarges about 80km northwards. The VIII zone near it, reduces accordingly, and its northern boundary moves at south about 20 km.

4.2 Seismic Safety Assessment of Large Bridges

Seismic safety assessments for four significant bridges sites of QTR were carried out. This work play an important role for assessing the stability of four grand bridges in QTR, which are Dunlang grand bridge, Jiuzila grand bridge, Dangquhe grand bridge, and Langbuqu grand bridge respectively from north to south, The objective of this assessment effort was to offer parameters for design, by which the bridges would have an enough capability to withstand a strong earthquake event.

In seismic safety assessments of the four sites, the analysis on seismo-geological features and assessment of seismic activity in regional and near fields, attenuation relations of ground motion, seismic risk analysis, in-situ test related to velocity of shear wave were involved. And finally both the parameters of design ground motion and seismic intensities with the exceedance probabilities of 63.5%, 10% and 2% in 50 years at the four bridges' sites are provided (Table 2).
The results also shown that there are several faults in the range of near four sites, which do not across sites of the four bridges and have no destroying effect on the bridges while the faults take a movement. However, Dunlang and Jiuzila grand bridges are too near the cross of faults belt at the southeast foot of Nyainqntanglha Mt. and Jiuzila Faults to avoid being influenced by some seismic secondary disasters, e.g. ground fissure. Fortunately, there is no liquefaction potential at the four sites.

4. 3 Effect of Earth Temperature on Ground Motion of Permafrost Sites along QTR

The variation of temperature is one of the most important factors that influence the mechanical properties of frozen soil and also the one that affects the carrying capacity of the structure foundation at permafrost areas. Without disturbing of the outside factors and human being engineering activities, permafrost is beneficial to anti-earthquake. Unfortunately, global climate becoming warmer in the past decades, the average air temperature per year on the Tibetan Plateau has increased by 0.2-0.4°C (IPCC, 2001). The temperature increase is more predominant in winter and the annual variation range of air temperature is decreasing gradually. Thus, the degradation of permafrost appears regionally (Wang S.L.,1996), and consequently the natural geography, geological environment, and engineering geological environment of permafrost area along Qinghai-Tibet Highway, running parallel to QTR at the most places, have changed greatly due to the environmental change (Wang G.S.,1996). In order to estimate the effect of climate warming on anti-seismic performance of permafrost sites, the influences of earth temperature on dynamic characteristics of permafrost and ground motion of permafrost sites were studied.

The research showed that the soil velocities at the permafrost regions along QTR are greater than the ones at the non-frozen soil areas (Table 3). Ice content and earth temperature are important factors influencing velocity at permafrost region. Dynamic modulus and strength of permafrost decrease with increasing of earth temperature. Compared with the spectral accelerations for typical unfrozen soil areas, the earthquake influence coefficient and characteristic period are less, and there are much more abundant components of spectral acceleration in moderate and short periods in the permafrost region along QTR. However, earth temperature remarkably influences the parameters of seismic ground motion in the permafrost region along QTR (Wang L.M.,2005). Under the certain inputting of seismic motion, the ground displacement and velocity increase predominantly with going up of earth temperature. The change of acceleration with the increase of temperature follows a pattern of decrease-increase-decease and then tends to be stable. Predominate period of ground motion acceleration decreases along with the falling of temperature (Fig. 12, Fig.13 and Fig.14).

5 CONCLUSIONS

1. Successful construction of QTR is a remarkable feat in railway construction history of permafrost region, which makes all the provincial capital cities of China to be connected by railways. It would benefit all ethnic groups in both Qinghai province and Tibet Autonomous Region of China, where economic growth and social development will be speed up.

2. All the engineering and management measures to solve three major problems, permafrost, lack of oxygen and environmental protection, have been proved to be efficiency and practical, which may be useful to the other railway construction projects in high elevation and permafrost areas.

3. The research achievements in earthquake damage mitigation have been applied into the design, construction, and maintenance of Qinghai-Tibet Railway, which have partly examined by the West Kunlun Mountain Pass 8.1 Earthquake in 2001 and will continue to show its effects on seismic safety of the railway operation.

3 ACKNOWLEDGEMENT

This research is financially supported by the scientific research fund of QTR Construction Headquarters and basic science research fund of the Institute of Earthquake Prediction, CEA. Prof. Ming Lu, Prof. Li Xiaojun, Dr. Dongli Zhang, Mr. James Whitt, Mr. Qiang Wang, and Mr. Lijun He provided some data, advice and assistance for our research and paper preparing. The authors deeply appreciate all above-mentioned institutions and persons.

REFREENCES

1. China Earthquake Administration and Ministry of Construction of the People Republic China (1990). "Seismic intensity zonation of China."
2. General Administration of Quality and Technical Supervision (1999). "Code for seismic safety evaluation of engineering sites (GB 17741-1999)." Standards Press of China.
3. General Administration of Quality and Technical Supervision (2001). "Seismic shock parameter zonation of China (GB 18306-2001)," Standards Press of China.
4. IPCC (2001). "Climate change 2001: the scientific basis, summary for policymakers and technical summary of working group I report," Eds. By Houghton J T, Ding Yihui, D. Griggs et al., Cambridge University Press, 98.
5. Lai Yuanming, Zhang Luxin, Zhang Shujuan, et al., (2003). "The cooling effect of ripped-stone embankments on Qing-Tibet railway under climatic warming," Chinese Science Bulletin, 48(6), 598-604.
6. Wei Ma, Guodong CHENG, Qingbai WU et al., (2005). "Application on idea of dynamic design in Qinghai-Tibet railway construction," Cold Regions Science and Technology, 41(3), 165-173.
7. Ma Wei, Cheng Guodong and Wu Qingbai (2002). "Preliminary Study on Technology of Cooling Foundation in Permafrost Regions," Journal of Glaciology and Geocryology, 24(5), 579-587.
8. Niu Fu-jun, Yu Qi-hao and Lai Yuan-ming (2003). "Analysis of ground temperature change and thermal process in the testing duct-ventilated embankment of the Qinghai-Tibet Railway," Journal of Glaciology and Geocryology, 25(6), 614-260.
9. Sheng Yu, Wen Zhi and Ma Wei (2003). "Preliminary analysis on insulation treatment of embankment at Beiluhe Test section of Qinghai-Tibet Railway," Chinese Journal of Rock Mechanics and Engineering, 22(supp. 2), 2659-2664.
10. Wang Guoshang, Jin Huijun and Lin Qing (1996). "Changes of natural and engineering environments in permafrost regions along the Qinghai-Tibet Highway," Proceeding of the 5th Chinese Conference on Glaciology and Geocryology, Lanzhou, Vol. 1.
11. Wang Lanmin, Zhang Dongli and Wu Zhijian (2005). "The influence of earth temperature on dynamic characteristics of frozen soil and the parameters of ground motion on the sites of permafrost," Proceedings of the 16th International Conference on Soil Mechanics and Geotechnical Engineering, Osaka Janpan, 2733-2736.
12. Wang Shaoling (1996). "An approach on permafrost degradation and environmental problems in the Tibetan plateau," Proceeding of the 5th Chinese Conference on Glaciology and Geocryology, Lanzhou, Vol. 1.
13. Wu Zhijian, Ma Wei, Sheng Yu, et al. (2005). "Cooling effectiveness analysis of the vent-pipe, cast-detritus and heat preservation material on protecting embankment in permafrost regions," Rock and Soil Mechanics, 26(8), 1288-1293.
14. Xu Xuezu, Wang Jiacheng and Zhang Lixin (2001). "Physics of Frozen Soil," Science Press, Beijing.

Figures (Fig.1~14):

Fig. 1 Schetching line of QTR
(From http://www.yahoo.com, edited)

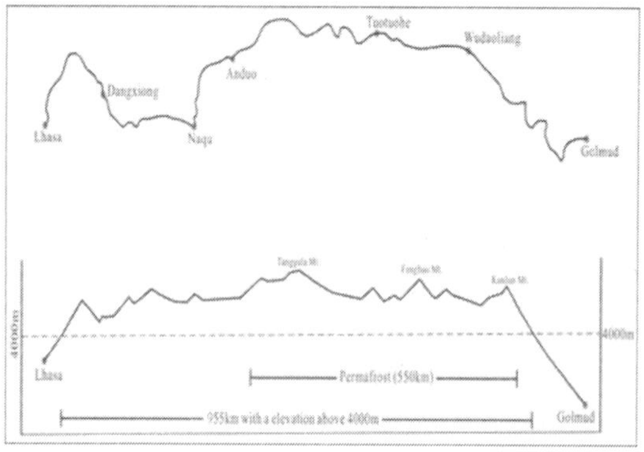

Fig. 2 Elevation and route of QTR

Fig.3 The Sancha River Bridge (left) and the Qingshui River Bridge (right)

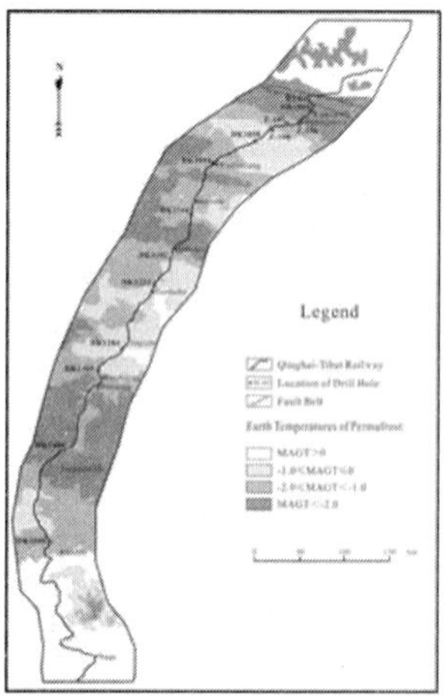

Fig. 4 The distribution of ground temperature of the permafrost and drill holes along QTR

Fig. 5 The ice in the permafrost
(by prof. Ma Wei)

Fig. 6 Examples of Engineering Measures Applied in QTR

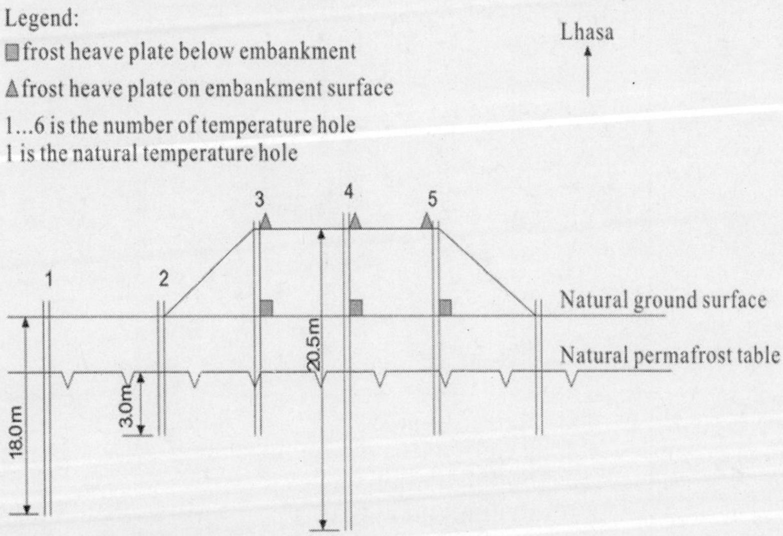

Fig. 7 A typical distribution of monitoring transect for temperatures and deformation (by prof. Ma Wei)

Fig. 8 Staging area (left) and animal path (right, from Xinhua Press)

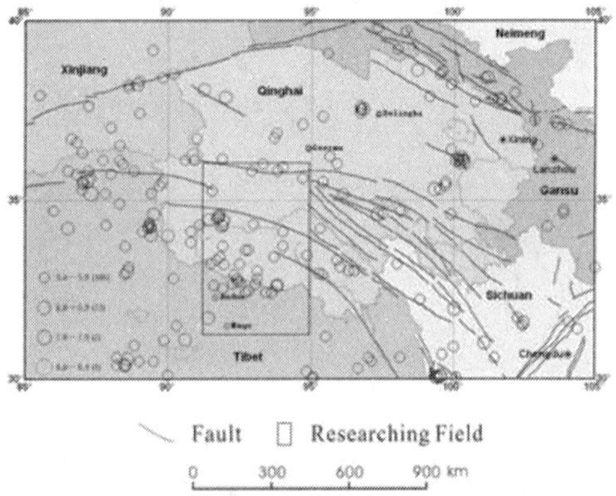

Fig. 9 Distribution of epicenters at and around the Qinghai-Tibet Plateau between Jan. 1, 1980 and Jul. 31, 2006

Fig. 10 A train was running across the fracture belt caused by the 8.1 earthquake in 2001

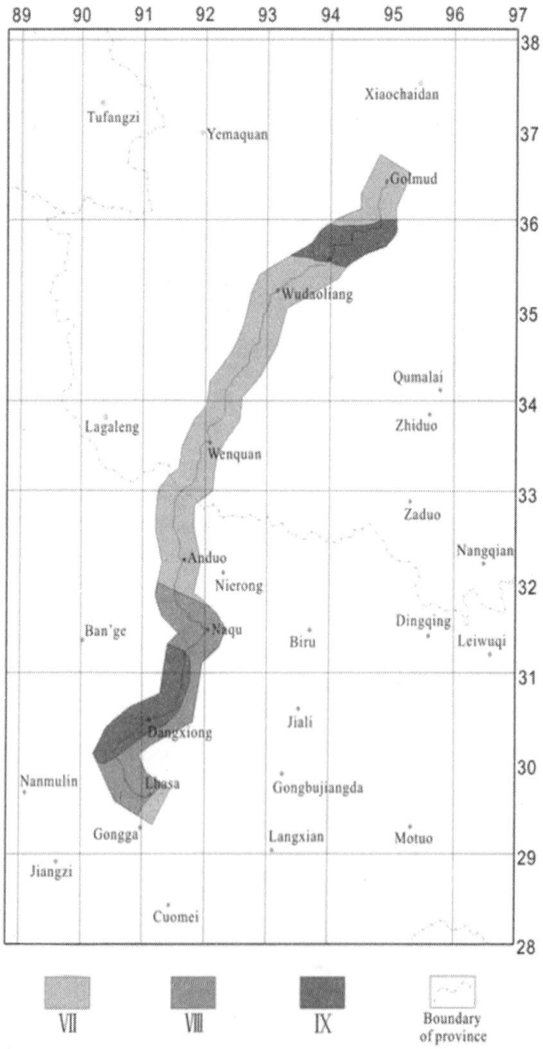

Fig. 11 Seismic intensity zonation for QTR (1:1,000,000)

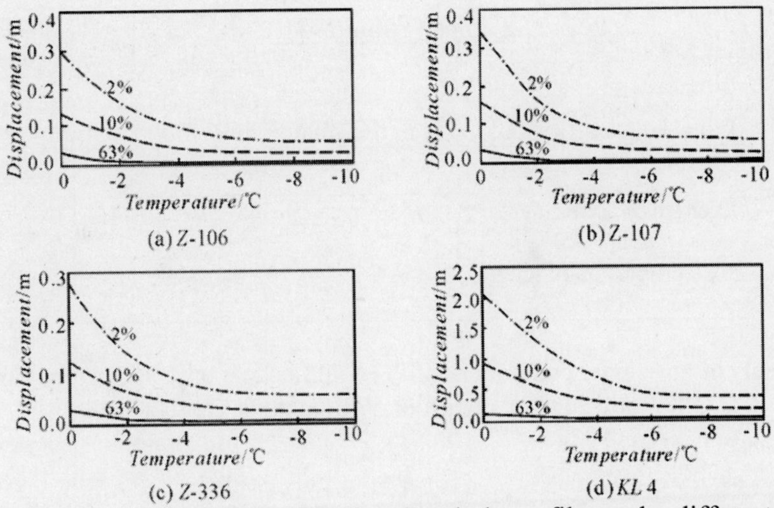

Fig. 12 Curves of displacement at the four borehole profiles under different temperatures

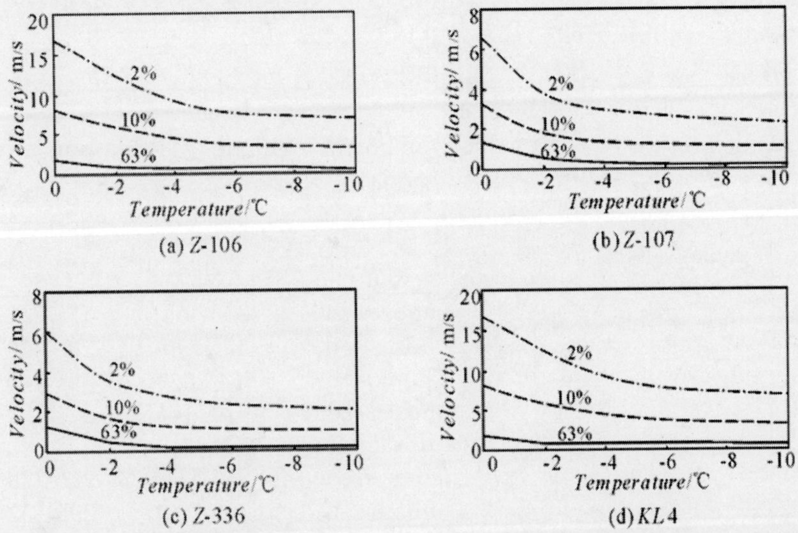

Fig. 13 Curves of velocity at the four borehole profiles under different temperatures

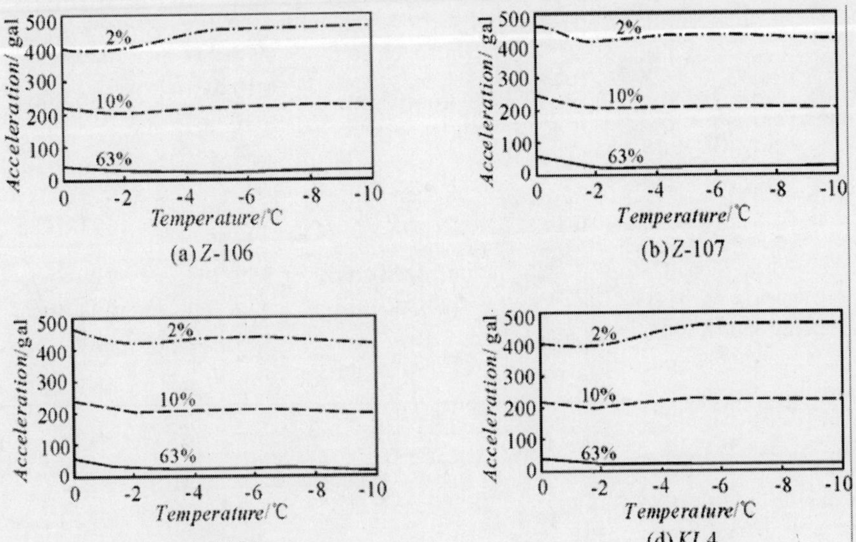

Fig. 14 Curves of acceleration at the four borehole profiles under different temperatures

Tables (Table 1~3):

Table 1 Proportions of QTR in the seismic intensity zonation

Intensity zonation	VII	VIII	≥IX
Length of accumulative total (km)	612	365	141
The proportion of length (%)	54.8	32.6	12.6

Table 2 Parameters of horizontal component of PGA and its response spectrum (damping ratio of 5%) for design at site ground of Dunlang great bridge

Exceedance probabilities in 50 years	A_{max} (Gal)	T_1 (sec)	T_2 (sec)	β_m	α
10%	450	0.18	0.50	2.70	1.0
5%	600	0.18	0.55	2.70	1.0
1%	990	0.20	0.80	2.75	1.1

Table 3 Observation results of wave velocity in boreholes along QTR (data acquired in summer, 2004)

Lithology	Depth (m)	Velocity (m/s)		
		Soil	Vp	Vs
silty clay	0-4	unfrozen soil	230-368	150-242
		frozen soil	305-590	206-420
	4-10		564-1050	395-740
	10-15	frozen soil	1026-1368	780-943
mudstone	0-4	unfrozen soil	330-541	216-328
		frozen soil	395-730	296-532
	4-10		770-1350	545-869
	10-15	frozen soil	1489-1620	910-1043
marl	0-4	unfrozen soil	812-904	526-644
		frozen soil	1020-1332	750-890
	4-10	frozen soil	1377-1592	878-1050
	10-15		1579-1653	1121-1184
fine sandy soil	0-4	unfrozen soil	189-367	120-263
		frozen soil	278-562	204-400
	4-10	frozen soil	382-643	251-456
	10-15		632-765	465-550

Part-2

Invited Lectures

COUNTERMEASURES AGAINST SISMIC FAILURE OF EXISTING RECLAIMED AREAS BY BANKING

S. Yasuda
Tokyo Denki University, yasuda@g.dendai.ac.jp

ABSTRACT

Slope failures in housing lots cause sever damage to houses and human lives. In this paper, slope failures during past three earthquakes in Japan are reviewed at first. In these cases filled slopes failed due to liquefaction of filled soil, weathering of filled materials and strong seismic motion. Based on these case histories, patterns of unstable fill slopes during earthquake were summarized. Then, as many reclaimed lands have been developed without considering seismic stability, appropriate procedure to evaluate seismic stability of existing reclaimed areas is discussed. In Japan, a technical committee chaired by the author was organized to deal this problem and published a guideline in this year. Contents of the guideline are introduced in the second part of this paper. In the guideline, it is recommended to evaluate the seismic stability by five steps; a) finding put of filled zone, b) field investigation, c) soil investigation and laboratory tests, d) analyses on slope stability and deformation, and e) selection of appropriate countermeasures.

Keywords: Seismic stability, Slope failure, Existing fill, Countermeasures

1 INTRODUCTION

Recently urban areas of many big cities have been expanded due to the increase of population. In Japan, almost big cities are located in lowlands. However, new expanding areas are located in hills or mountains. Therefore, reclamation works must be done by cutting the hills and filled in valleys. In general, the fill is not fully compacted and water level in the bank is high because the bank is located on former valley. Then fill slopes are susceptible to failure during earthquakes. Once the failure of the slope occurred, severe damages to human lives and housings occur. Therefore, slopes of banks must be treated not to cause failure.

In this paper, damages to reclaimed areas by banking for housing lots during past Japanese earthquakes are described. Then methods to find out existing filled zones, prediction methods on slope failure and countermeasures against slope failure are introduced.

2 DAMAGES TO RECLAIMED AREA BY BANKING DURING PAST EARTHQUAKES IN JAPAN

(1) Liquefaction-induced settlements of houses on filled grounds

An artificially filled housing lot at Kiyota district in Sapporo City were severely damaged during the 1968 Tokachi-oki earthquake. This might be the first experience of the damage to artificially filled housing lots due to earthquakes, in Japan. The maximum surface acceleration at Kiyota was estimated as about 0.08 g. The housing lot was constructed by cutting soils of hills and filled in valleys as shown in Figure 1. There were 279 houses in the housing lot. Of them, 27.5 % of houses were damaged during the earthquake. As shown in Figure 1, no house constructed on the cut ground was damaged. On the contrary, 56 % of houses constructed on the filled ground were settled and tilted. Ground water table in the fill was shallow and filled soil is volcanic ash sand.

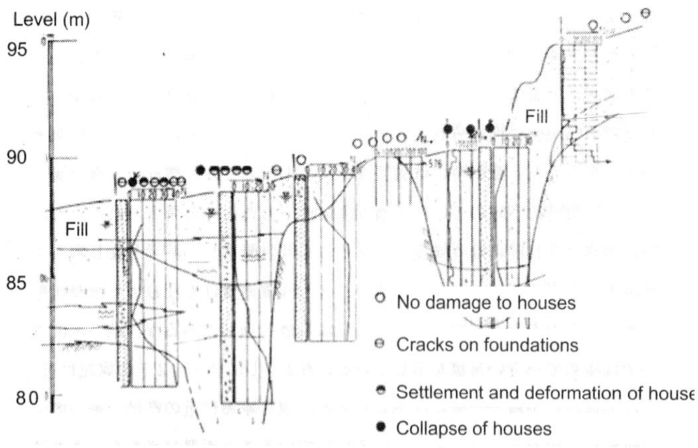

Figure 1 Soil cross section and damaged houses during the 1968 Tokachi-oki earthquake (Technical Committee on the 1968 Tokachi-oki earthquake, 1969)

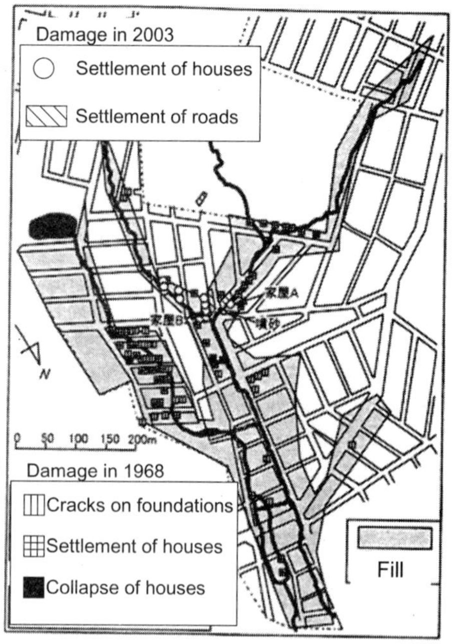

Figure 2 Comparison of damaged houses during two earthquakes (JGS, 2004)

Photo 1 Settled houses due to liquefaction during the 2003 Tokachi-oki earthquake at Utsukushigaoka district

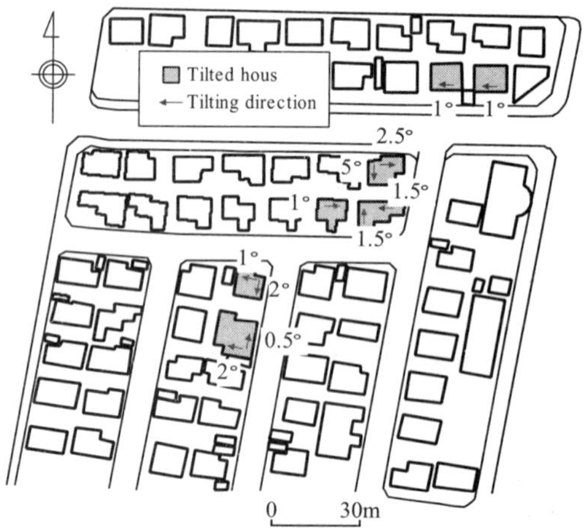

Figure 3 Rough measurement of inclination of settled houses (Yasuda et al., 2004)

Therefore it is estimated the fill soil liquefied during the earthquake and caused settlement and tilt of houses.

In 2003, 35 years after the 1968 Tokachi-oki earthquake, new Tokachi-oki earthquake occurred and caused damage to houses in this district again. Figure 2 compares damaged houses during the two earthquakes. Both damaged houses were located on fill ground. However, damaged houses during the previous earthquake were survived during the new earthquake. After the previous earthquake underground conduits were constructed in the damaged zone. It is estimated that ground water level decreased due to dewatering effect. This may be one of the reasons why the damaged houses survived during the next earthquake.

At Utsukushigaoka district near Kiyota district, several houses settled as shown in Photo 1 during the new Tokachi-oki earthquake. It can be judged that liquefaction occurred in this site because traces of sand volcanoes were observed around the damaged houses. The boiled soil was volcanic sandy silt. The damaged ground had been constructed recently by filling a channel in a hill zone. Therefore, it is estimated that the filled soil liquefied and caused the settlement of the houses. The authors measured the angle of inclination of the damaged houses roughly, and indicated on Figure 3. As shown in this figure, roughly several houses settled and tilted more than 1 degree. According to the experience of the damage during the 2000 Tottoriken-seibu earthquake, inhabitants felt giddy and nausea, and could not live after the earthquake in the houses which tilted more than 1/100 (Yasuda, Hitomi and Hashimoto, 2004). The angles of inclination of the tilted houses shown in Figure 3 were more than 1/100.

(2) Collapse of houses due to failure of fill slopes

Several housing lots were damaged during the 1978 Miyagiken-oki earthquake in Sendai City in Japan. Of them severe slope failure of a bank occurred at Midorigaoka Ichi-chome housing lot. As shown in Photo 2, many houses were collapsed due to the slope failure. In 1957 to 1958, the bank was constructed by filling a vally. During the reclamation work, trees on surrounding hills were felled and threw down into the valley, rocks cut from the hill also threw into the valley, then soil cutting form the hills were filled as schematically shown in Figure 4. Figure 5 shows a soil cross section along the slope. Thickness of the fill was 5 to 15 m. Density of the fill was very low with 0 to 10 of SPT *N*-values. It is estimated that burried trees became humus during 30 years. Moreover, as big voids existed between buried rocks, groundwater flew in the big voids and caused

Photo 2 Collapsed houses due to slope failure at Midorigaoka in Sendai City during the 1978 Miyagiken-oki earthquake

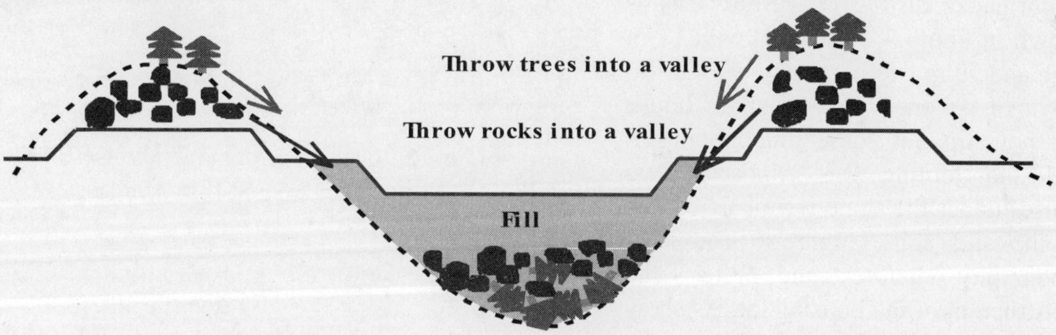

Figure 4 Procedure of reclamation work conducted at Midorigaoka

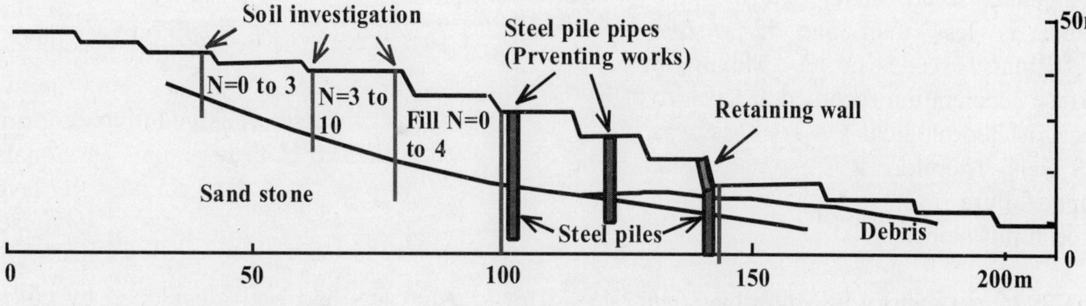

Figure 5 Soil cross section along failed slope (Tohoku Branch, JSCE)

weathering of rocks. Then bottom of the fill became very loose and caused slope failure during the Miyagiken-oki earthquake.

During restroration, steel pipe piles with 318.5 mm in diameter were installed to increase resistance against slide, as shown in Figure 5. The steel pipes was strengthend by filling with H bars and concrete inside the pipes. The steel pipe piles were installed in two rows with 2 m pitch. A concrete retainig wall and a drainage well were also constructed. Safety factor against sliding increased up to 1.2.

Photo 3 Failed slope and damaged houses at Midorigakoka in Kushiro City during the 1993 Kushiro-oki earthquake

(3) Damage to houses due to slide or deformation of slopes

Many houses built on terraces were damaged due to slide or deformation of fill slopes during the 1993 Kushio-oki earthquakes in Japan. Small valleys existed at the terraces before reclamation. Peat and soft clay were deposited at the valleys. The top of terraces were cut and filled in the valleys. One slope failure at Midorigaoka district in Kushiro City is shown in Photo 3. Soil mass with 11 m high and 90 m width slid and caused slip down of a house. One neignbor house did not slip but some foundation soil slid and footing was exposed. Two houses under the slope were barried due to the slid soils. Figure 6 compares topography before and after the construction of the housing lot. As show in this figure, collapsed zone coincides with filled zone. SPT *N*-value of the fill was 0 to 4. According to the slope failure analyses conducted after the earthquake, safety factor against slope failure is less than one if seismic coefficient exceeds 0.12. Maximum surface acceleration recorded at Kushiro Local Meteorological Observatory was 919 Gals. Therefore it is estimated that slopc failure occurred due to inertia force at this slope.

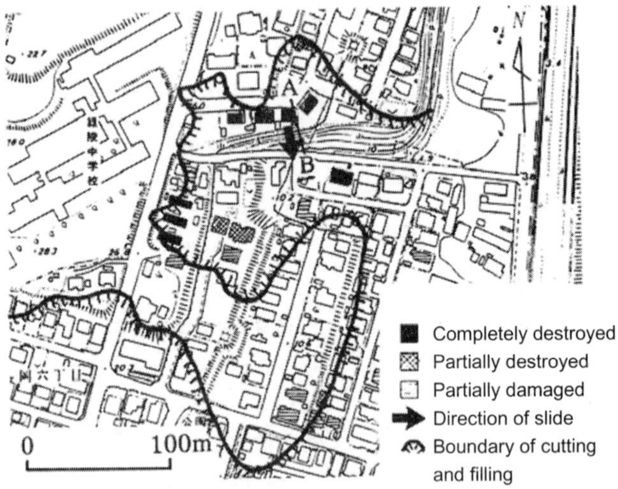

Figure 6 Location of damaged houses at Midorigaoka (JSSMFE, 1994)

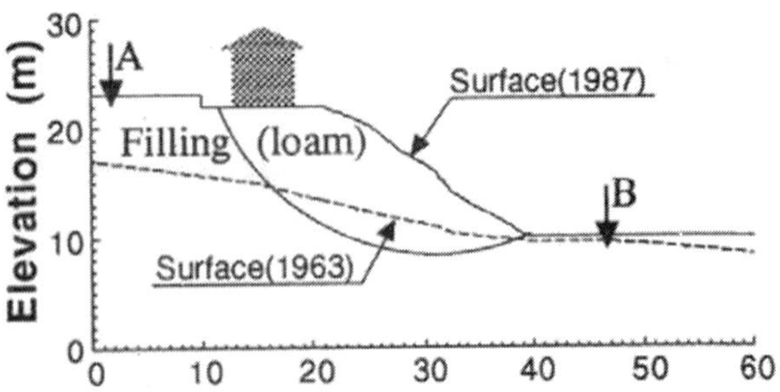

Figure 7 Mechanism of slope failure (JSSMFE, 1994)

At Kayanuma district in Shibecha Town, a land for housing lots had been developed by cutting north slopes and filling in south zones as shown in Figure 8. Four valleys were filled with the thickness of 10 m in maximum. No special dainage conduites were put on the bottom of the valleys.

Filled soils slid and nine houses were severely damaged during the 1993 Kushiro-oki earthquake as shown in Photo 4. Soil investigations by Swedish Weight Sounding and other tests were carried out after the earthquake to study the mechanism of the damage. The banking soil was volcanic silty sand. Figure 9 shows soil cross section along A-A' and B-B' lines which cross heavily damaged zones. The filled soil was loose with SPT *N*-vales of around 10, and lower part of the fill was saturated. It was difficult to recognize the occurrence of liquefaction during the site survey conducted just few days after the earthquake, because the land was covered with snow and sand volcanoes could not be observed. However, detailed seismic ad liquefaction analyses prevailed the occurrence of liquefaction of filled sand. In the slope stability analysis, safety factor agains sliding was less than 1.0 if sesmic coefficient exceeds 0.2. And, deformation analysis by Tara 3 showed large deformation of about 1 m, as shown in Figure 10.

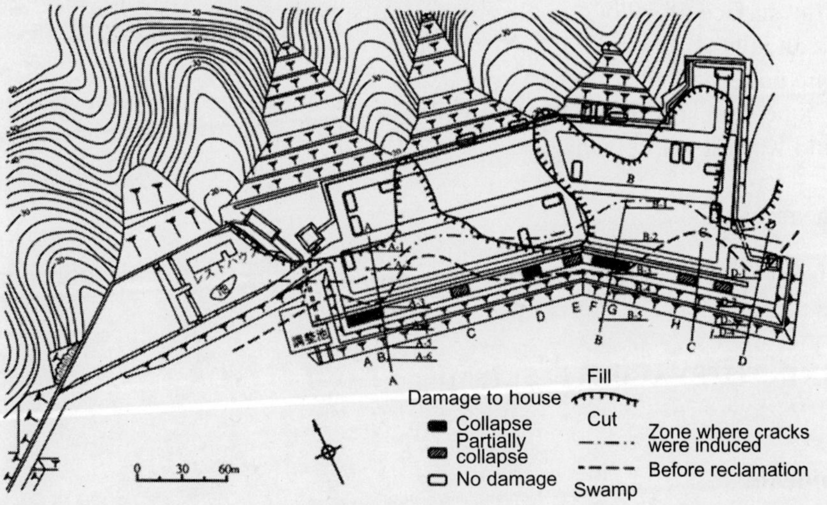

Figure 8 Filled zones and damaged houses at Kayanuma (JSSFE, 1994)

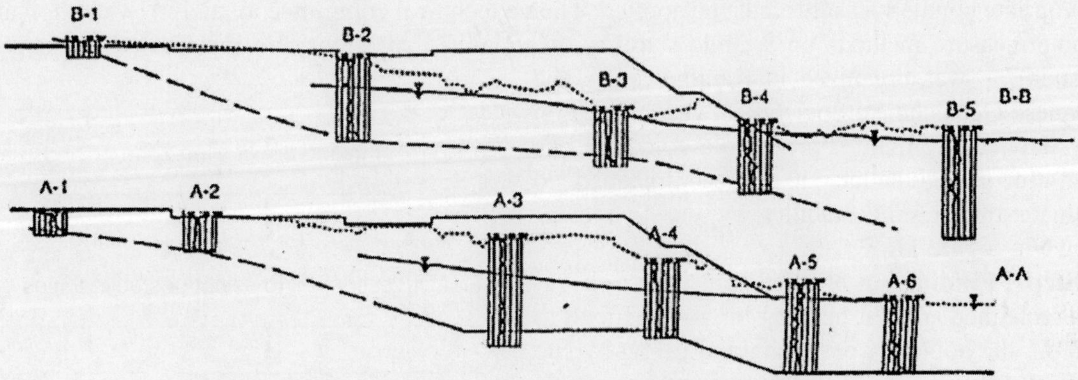

Figure 9 Soil cross section along the slope (JSSMFE, 1994)

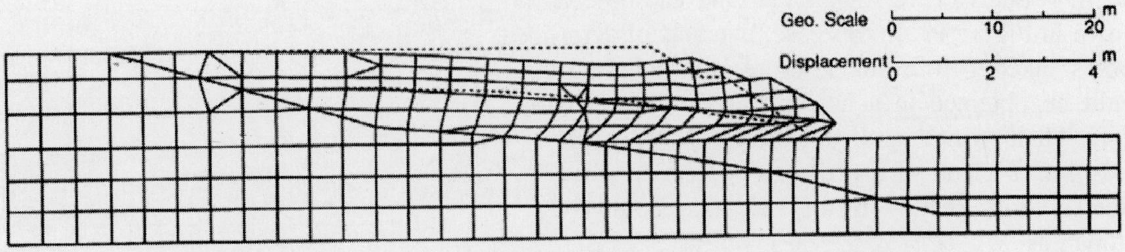

Figure 10 Deformation of bank analyzed by Tara 3 (JSSMFE, 1994)

3 BRIEF SUMMALY OF UNSTABLE FILL SLOPES DURING EARTHQUAKES

Case histories above mentioned suggest that the following reclaimed areas by banking are unstable during earthquakes:

(1) Fills are not compacted.
(2) Soft clayey layers or weathered materials exist at the bottom of fills or at the surface of valleys.
(3) Drainage conduits at the bottom of banks are not installed.
(4) Fills are liquefiable because of high ground water level.
(5) Slopes are steep.
(6) Embankments are high.

Photo 4 Deformed bank and damaged houses at Kayanuma during the 1993 Kushiro-oki earthquake

Then, for existing reclaimed area by banking, unstable slopes must be estimated and applied appropriate countermeasures by considering these items.

4 GUIDELINE TO EVALUATE SEISMIC STABILITY OF EXISTING RECLAIMED AREAS

(1) Aim of guideline

In Japan, many reclaimed areas for housing lots been constructed around big cities from around 1970. Of them only few reclaimed areas were constructed by considering stability during future earthquakes. Seismic stability of many existing reclaimed areas by banking must be evaluated and appropriate countermeasures must be applied. Then a technical committee to study "Evaluation and countermeasure methods on seismic stability of reclaimed areas for housing lots (Chairman: S. Yasuda)" was organized in Kanto Branch, the Japanese Geotechnical Society last year. Based on the study, the committee prepared a guideline this year. In the guideline, it is recommended to evaluate the seismic stability by the five steps shown in Figure 11.

(2)Step1: Finding out of filled zone

In a reclaimed area for housing lot, it is difficult to find out the boundary of cutting and filling by site survey, if there is no record on reclamation work. Then a special technique to find filled zoned is recommended in the guideline by comparing old and new maps or aerial photos. One example is shown in Figure 12. In this case, two sets of aerial photos taken before and after reclamation work could be obtained in a housing lot in Kawasaki City. Then zones and thickness of fill were measured by comparing the two sets of aerial photos. Cross section of the fill also could be estimated.

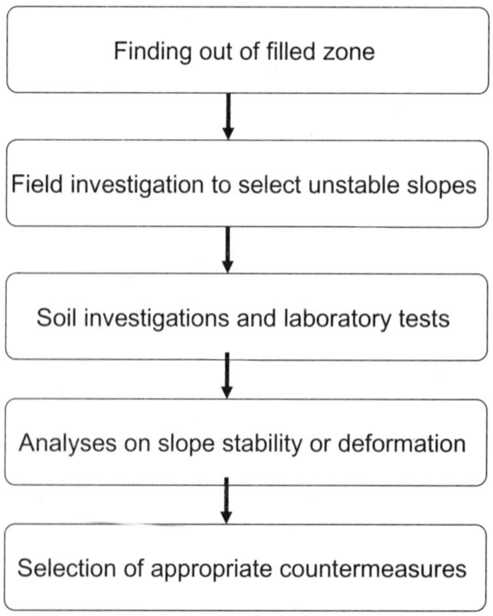

Figure 11 Recommended steps in the guideline

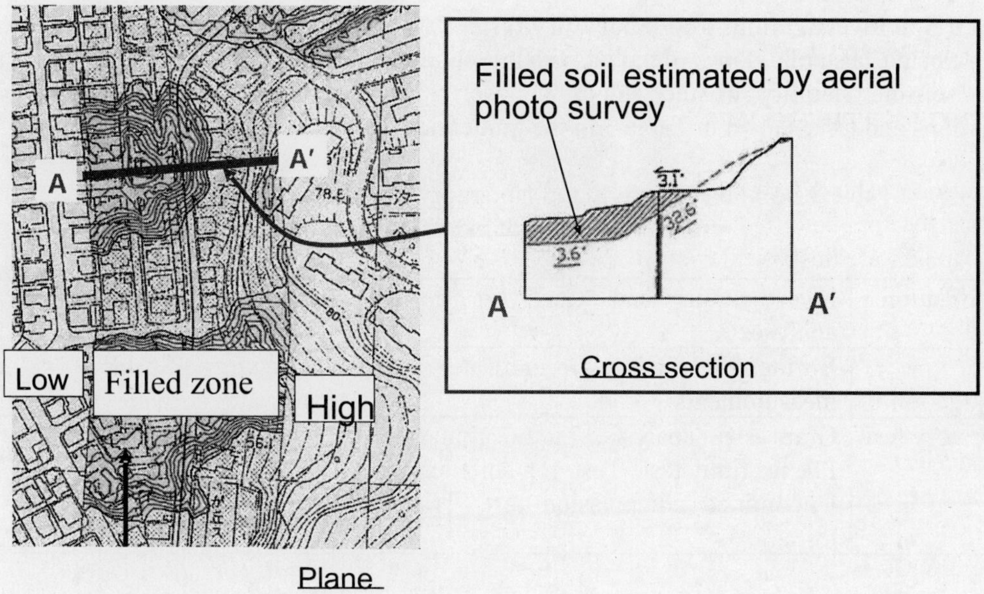

Figure 12 Filled zone measured by two sets of aerial photos before and after reclamation in Kawasaki City

Filled zone can be estimated also by comparing two sets of topographical maps drawn before and after the reclamation work. However accuracy of estimated zone is lower than that measured by aerial photos. Table 1 compares accuracy of estimated fill zone.

Table 1 Accuracy of estimated filled zone

Grade	Scale of aerial photos or topographical maps before and after reclamation		Error of estimated fill zone
	Aerial photos	Topographical maps	
A	1/10,000	1/2,500 to 1/3,000	Less than 1 m
B	1/10,000 (old)	1/3,000 to 1/5,000	1 m to 2 m
C	1/10,000 (old)	1/10,000	2 m to 3 m
D		1/20,000	More than 4 m

(3) Step2: Field investigation to select unstable slopes

The next step is the selection of existing unstable fill slopes during earthquakes. Stability of existing embankment slopes during earthquakes can be evaluated roughly by visual inspection. Then, in the guideline, field investigation of the found fills by visual inspection based on several factors shown in Table 2, is recommended. Several evaluation methods based on these factors have been developed in Japan. These methods are introduced in the guideline.

Table 2 Factors introduced in visual inspection to evaluate seismic stability of existing fill slope

Item	Factors introduced in visual inspection
Fill	Height of fill, inclination of slope, Banking material, Type of slope protection work, Deformation of slope, Springwater, Drainage well
Retaining wall, house	Type of retaining wall, Type of weeping, Deformation of retaining wall, Springwater, Deformation of foundation of house
Topography	Catchment, Land use on the slope

(4) Step3: Soil investigations and laboratory tests

For the selected unstable slopes, detailed soil investigation and laboratory tests are necessary to evaluate seismic stability of the slopes precisely. It is recommended to conduct several investigations and tests shown in Table 3 in the guideline.

Table 3 Soil investigations and laboratory tests necessary to evaluate seismic stability of existing fill slope

Classification	Investigations and tests for normal analyses	Additional investigations and tests for precise analyses
Soil investigation	Boring, SPT, CPT, Groundwater level measurements	PS logging, Surface exploration, Pore water pressure measurements
Laboratory test	Grain size analysis, Liquid limit test, Plastic limit test, Test for soil density, Unconfined compression test, Triaxial test	Cyclic triaxial test, Cyclic torsional shear test

Several attentions in the soil investigations for existing fills are paid in the guideline. On of them is the judgement of groundwater level in existing fills. If there is no drainage conduit under a fill, pore water pressure is hydrostatic pressure as shown in Figure 13(a). Then, groundwater level in bore holes is not affected by the depth of bore holes. However, if there are drainage conduits at the bottom of the fill, pore water pressure near the drainage conduits is lower than hydrostatic pressure. Therefore, groundwater level decreases with increase of the depth of bore hole as shown in Figure 13 (b). So, in the guideline, it is recommended to pay special attention in the judgement of groundwater level.

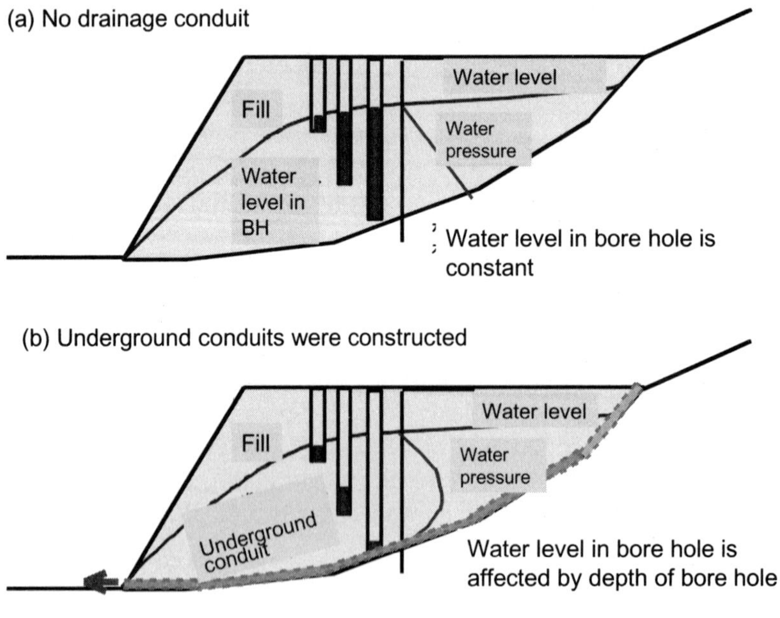

Figure 13 Relationships between distribution of pore water pressure and groundwater level

(5) Step 4: Analyses on slope stability or deformation

Based on soil investigations and laboratory tests, analyses for slope stability or deformation are conducted. In the slope stability analyses during earthquake, seismic force and excess pore water pressure must be considered as shown in Figure 14. Then, in the guideline, methods to evaluate excess pore water pressure are introduced at first.

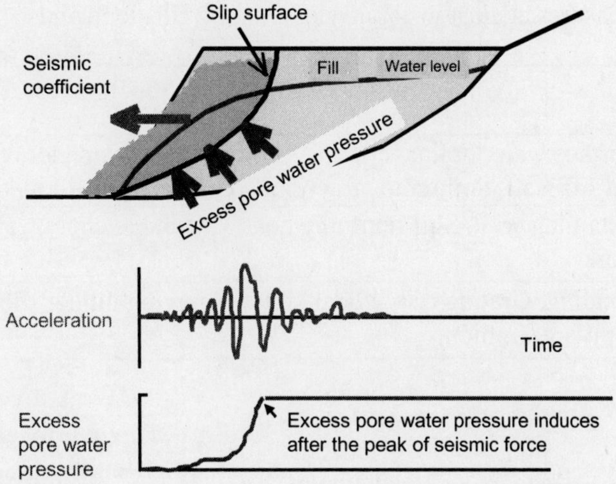

Figure 14 Time of the occurrence of seismic force and excess pore water pressure

As shown in Figure 14, only few excess pore water pressures induce before the peak of seismic motion. Therefore two sets of analyses; a) to consider seismic coefficient only, and b) to consider excess pore water pressure only, are recommended for the evaluation of seismic slope stability as shown in Table 4.

Table 4 Two factors to be considered in stability analysis

Type of filled soil		Factors to be considered	
		Excess pore water pressure	Seismic coefficient
Sand	Loose and saturated	O	X
	Dense and/or unsaturated	X	O
Clay		X	O

In performance-based design, it is necessary to estimate not only the stability but also deformation of banks. In the performance-based design for river dikes and road embankments, deformation analyses shown in Figure 15 become popular. Then, several methods to estimate the deformation of banks are introduced in the guideline.

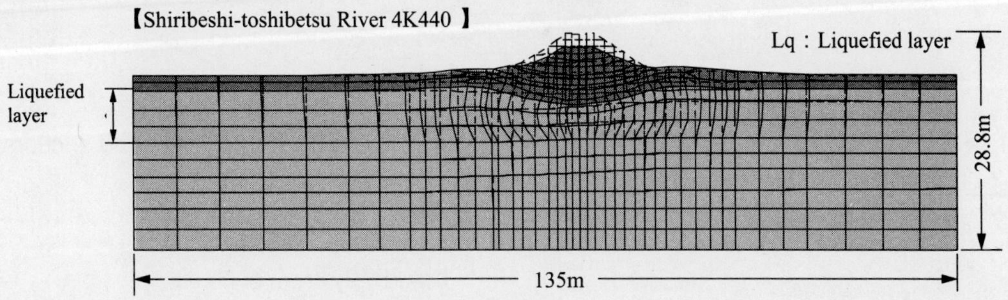

Figure 15 An example of analyzed liquefaction-induced deformation of a river dike

(6) Step5: Selection of appropriate countermeasures
Final step is the selection of appropriate countermeasures against slope stability during earthquakes. Three types of countermeasures against instability of existing fill slopes; control works, preventing works and slope protection works, are introduced in the guideline. Name of the countermeasures were listed in Table 5. And, examples of arrangement plans are shown in Figure 16 and 17.

Table 5 Countermeasures against instability of existing fill slope and arrangement plans

Type of countermeasure	Countermeasure methods
Control works	Surface water drainage work, Groundwater drainage work, Ground water cut-off wall, Soil removal work, Counterweight fill method
Preventing works	Retaining wall, Soil reinforcement, Ground anchorage, Pile works, Shaft work
Slope protection works	Sodding, Grating crib work, Mortal spraying gunite shooting, Concrete spraying, Gabion

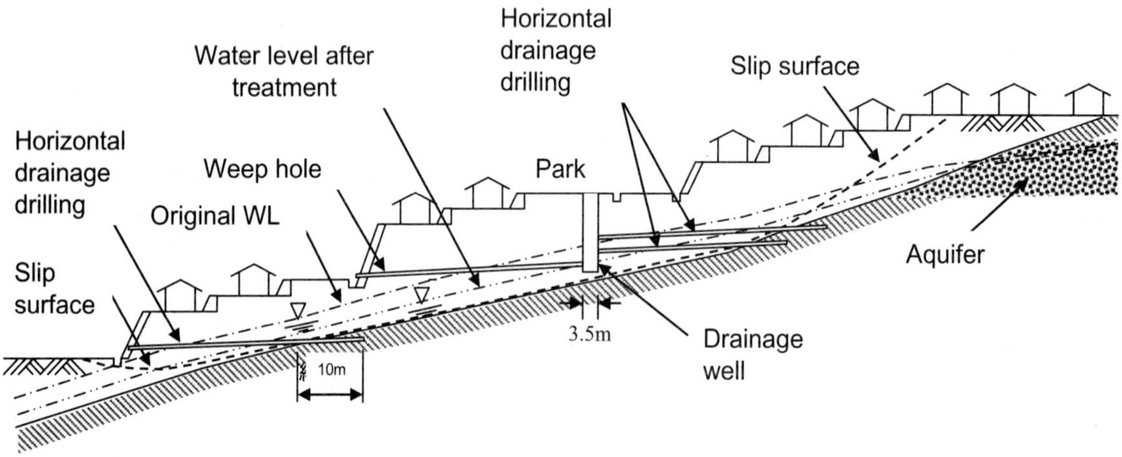

Figure 16 Countermeasure by horizontal drainage drilling and drainage well (Kanto Branch, JGS, 2007)

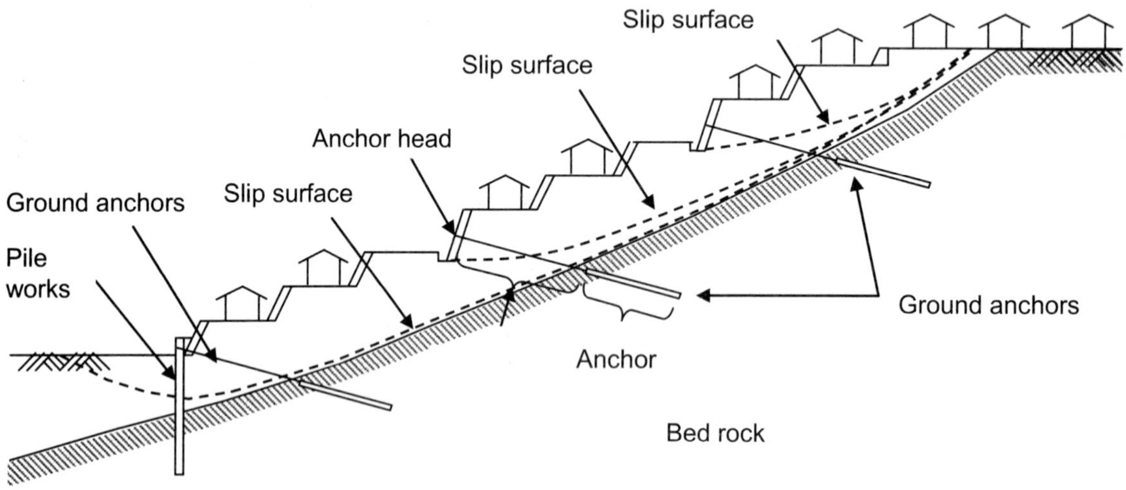

Figure 17 Countermeasure by ground anchors and pile work (Kanto Branch, JGS, 2007)

In addition, countermeasure methods for existing retaining walls shown in Figure 18, and countermeasures against liquefaction-induced settlement of houses shown in Figure 19 are introduced in the guideline.

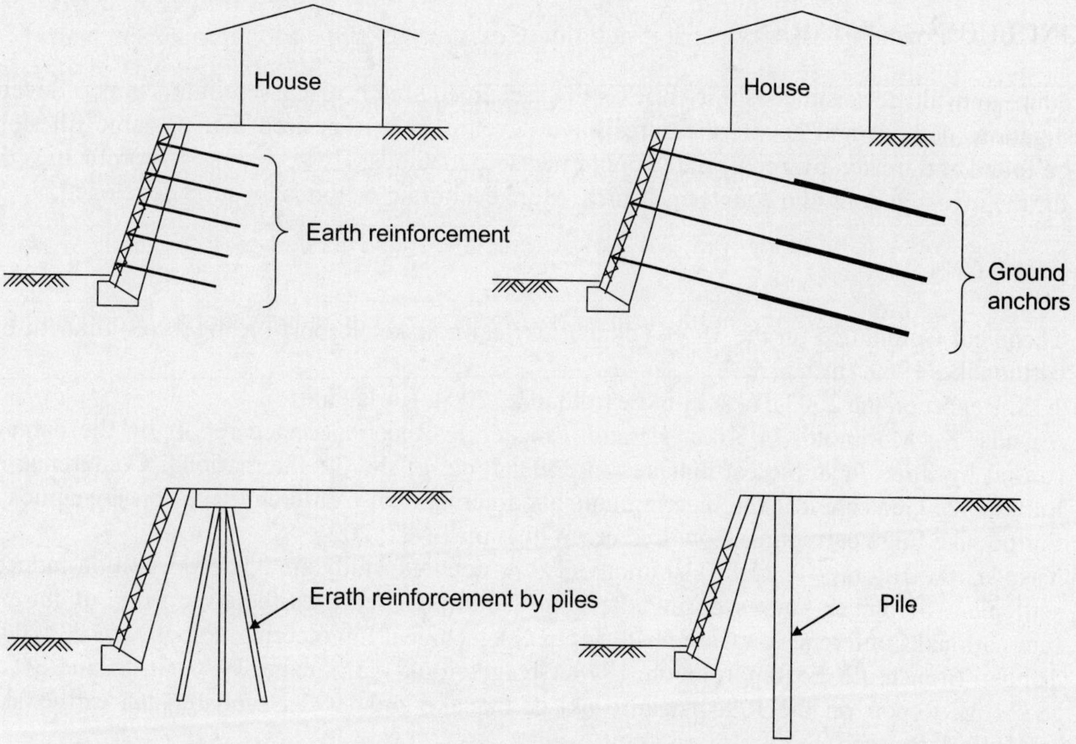

Figure 18 Countermeasures for existing retaining walls (Kanto Branch, JGS, 2007)

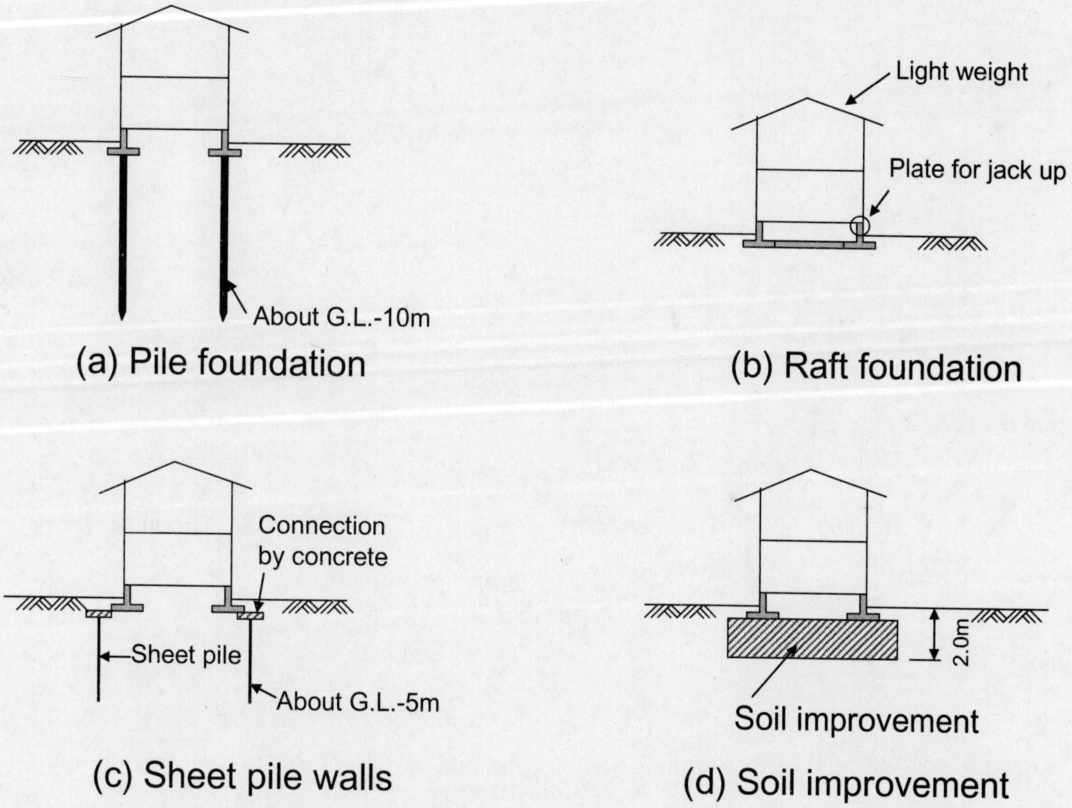

Figure 19 Countermeasures for existing houses (Kanto Branch, JGS, 2007)

7 CONCLUDING REMARKS

Procedure to evaluate seismic stability of existing reclaimed area is discussed in this paper. Several investigation methods and countermeasures have developed. It is desired that unstable fill slope will be found and treated by appropriate countermeasure methods. However, serious problem is the cost for the investigations and countermeasures. More economic methods must be developed.

REFERENCES

1. Technical Committee on the 1968 Tokachi-oki earthquake: Report on the 1968 Tokachi-oki earthquake, 1969. (in Japanese)
2. JGS: Report on the 2003 Tokachi-oki earthquake, 2004. (in Japanese)
3. Yasuda, S., Morimoto, I., Kiku, H. and Tanaka, T: Reconnaissance report on the damage caused by three Japanese earthquakes in 2003, Proc. of the 3rd International Conference on Earthquake Geotechnical Engineering and 11th International Conference on Soil Dynamics & Earthquake Engineering, Keynote Lecture, Vol.1, pp.14-21, 2004.
4. Yasuda, S., Hitomi, T. and Hashimoto, T.: A detailed study on the liquefaction-induced settlement of timber houses during the 2000 Tottoriken-seibu earthquake, Proc. of the 5th International Conference on Case Histories in Geotechnical Engineering, Paper No.3.33, 2004.
5. Tohoku Branch, JSCE: Report on the 1978 Miyagiken-oki earthquake, 1980. (in Japanese)
6. JSSMFE: Report on the 1993 Kushiro-oki earthquake and 1993 Notohanto-oki earthquake, 1994. (in Japanese)
7. Kanto Branch, JGS: Guideline on evaluation of seismic stability of existing reclaimed area by banking, 2007. (in Japanese)

SEISMIC SITE CHARACTERIZATION USING GEOTECHNICAL AND GEOPHYSICAL TECHNIQUES

Sitharam T.G. and Anbazhagan.P
Department of Civil Engineering,
Indian Institute of Science, Bangalore, Karnataka, India, 560012.
sitharam@civil.iisc.ernet.in

A complete site characterization is essential for the seismic site classification and site response studies, both of them can be used together for seismic microzonation. Seismic microzonation is the initial phase of earthquake risk mitigation and requires multidisciplinary approach with major contributions from geology, seismology, geotechnical and structural engineering (Ansal, 2004). Site Characterization should include an evaluation of subsurface features, sub surface material types, subsurface material properties and buried/hollow structures to determine whether the site is safe against earthquake effects. Site characterization should provide data on the following:

- Site description and location
- Geotechnical data
- Soil conditions
- Geological data
- Hydrogeology/ ground water data
- Aquifer or permeable characteristics
- Presence and distribution of contaminants (if any)
- Climatic conditions (if needed)

Assessment of available data should include an analysis of the sufficiency and validity of the data in relation to the proposed application/ study.As part of the site characterization, experimental data should be collected, interpolated and represented in the form of maps. The representation maps can be further used for the site classification and seismic studies. Here an attempt has been made to characterize the Bangalore site using Geotechnical and Geophysical experimental data. About 850 collected geotechnical borehole information with standard penetration "N" values data and 58 geophysical data of multichannel analysis of surface wave field test results are used for Bangalore site characterization. The detailed discussion about the SPT data, interpretations, mappings and corrections applied for SPT 'N' are presented in this presntation. About geophysical method of MASW, field testing, results, evaluation of dynamic properties of soil, mapping of shear wave velocity, site classification and correlation between SPT corrected "N" values and shear wave velocity is also discussed in disscused in this presentation.

The 3-D subsurface model with geotechnical data has been generated for site characterization of Bangalore. The base map of Bangalore city (220sq.km) with several layers of information (such as Outer and Administrative boundaries, Contours, Highways, Major roads, Minor roads, Streets, Rail roads, Water bodies, Drains, Landmarks and Borehole locations) has been generated. GIS database for collating and synthesizing geotechnical data available with different sources and 3-dimensional view of soil stratum presenting various geotechnical parameters with depth in appropriate format has been developed. In the context of prediction of reduced level of rock (called as "engineering rock depth" corresponding to about Vs > 700 m/sec) in the subsurface of Bangalore and their spatial variability evaluated using Artificial Neural Network (ANN). Observed SPT 'N' values are corrected by applying necessary corrections, which can be used for engineering studies such as site response and liquefaction analysis.
Site characterization has also been carried out using measured shear wave velocity with the help of

shear wave velocity survey using MASW. MASW (Multichannel Analysis of Surface Wave) is a geophysical method, which generates a shear-wave velocity (Vs) profile (i.e., Vs versus depth) by analyzing Raleigh-type surface waves on a multichannel record. MASW system consisting of 24 channels Geode seismograph with 24 geophones of 4.5 Hz capacity was used in this investigation. The shear wave velocity of Bangalore subsurface soil has been measured and correlation has been developed for shear wave velocity (Vs) with the standard penetration tests (SPT) corrected 'N' values. About 58 one-dimensional (1-D) MASW surveys and 20 two-dimensional (2-D) MASW surveys has been carried out with in 220 sq.km Bangalore urban area. Dispersion curves and shear velocity 1-D and 2-D have been evaluated using SurfSeis software. Using 1-dimensional shear wave velocity, the average shear wave velocity of Bangalore soil has been evaluated for depths of 5m, 10m, 15m, 20m, 25m and 30m (Vs30) depths. The sub soil classification has been carried out for local site effect evaluation based on average shear wave velocity of 30m depth (Vs30) of sites using NEHRP (National Earthquake Hazard Research Programme) and IBC 2000 (International Building Code) classification. Bangalore falls into site class D type of soil. Mapping clearly indicates that the depth of soil obtained from MASW is closely matching with the soil layers in the bore logs. The measured shear wave velocity at 38 locations close to SPT boreholes, which are used to generate the correlation between the shear wave velocity and corrected 'N' values using a power fit. Figure 1 shows the actual data along with the fitted equation along with upper bound and lower bound curves. Also, developed relationship between shear wave velocity and corrected 'N' values corresponds well with the published relationships of Japan Road Association (1980). To fulfil G_{max} equation for site response study an attempt has been made to correlate the measured G_{max} (calculated from measured shear wave velocity and densities of each layer) of each borehole to the corrected SPT-N values. At about 38 locations, MASW test locations were very close to the SPT locations. From these 38 locations, about 190 data pairs of Vs and SPT corrected blow count have been used for the regression analysis. Figure 2 also shows the actual data along with the fitted equation and in addition upper bound and lower bound curves are also shown. Fitted equation matches up to a corrected SPT-N [(N1)60cs] value of 40 with Imai and Tonouchi (1982) equation. Beyond the "(N1)60cs" values of 40 fitted equation G_{max} is heigher than the Imai and Tonouchi (1982) equation. This is attributed to the soil type in Bangalore. The soil type is residual in nature and it is classified as silty sand, sandy silt with clay content.

ACKNOWLEDGEMENTS

Authors thank the Department of Science and Technology (DST), Seismology Division, Govt. of India for funding the project "Seismic microzonation of Bangalore" (Ref no.DST/23(315)/SU/2002 Dated October 2003).

REFERENCES

1. Ansal, A (ed) (2004), "Recent Advances in Earthquake Geotechnical Engineering and Microzonation," Kluwer Academic Publishers. Printed in Netherlands.
2. IBC, (2000), "International Building Code," International Code Council: Inc. 5th Edition, Falls Church, VA.
3. Imai, T., and Tonouchi, K. (1982), "Correlation of N-value with S-wave velocity and shear modulus", Proceedings, 2nd European Symposium on Penetration Testing, Amsterdam, pp 57-72.
4. Japan Road Association (JRA), (1980), "Specification for highway bridges" Part V, Earthquake Resistant Design.
5. TC4-ISSMGE (1999), "Manual for Zonation on Seismic Geotechnical Hazard," Revised edition, Technical Committee for Earthquake Geotechnical Engineering (TC4) of the International Society of Soil Mechanics and Geotechnical Engineering (ISSMGE), 209p.

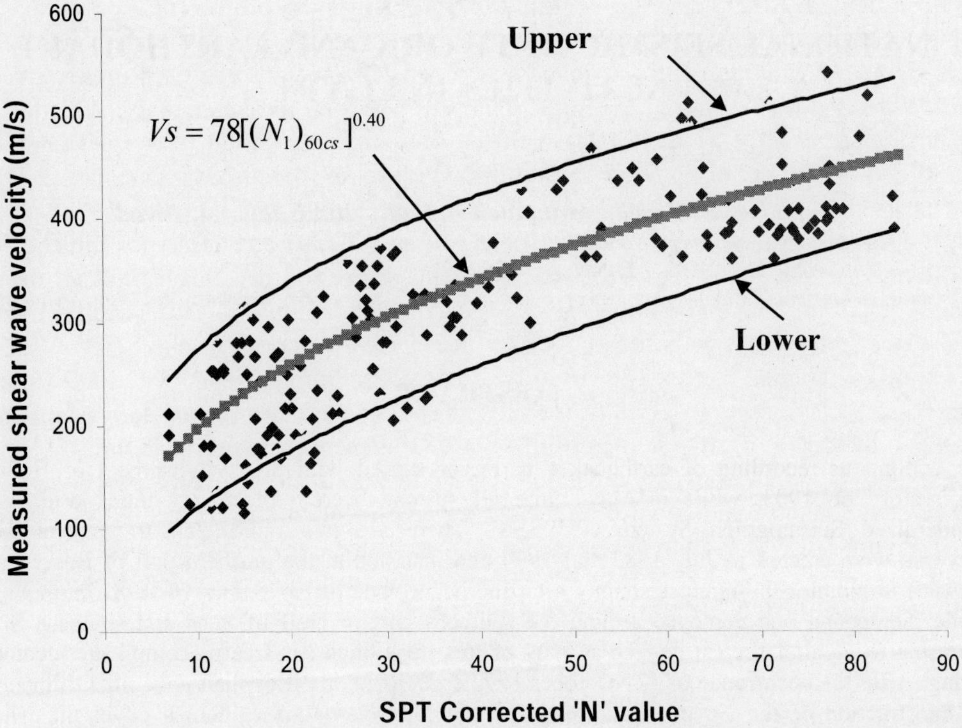

Figure 1: Shear wave velocity versus corrected SPT 'N' values

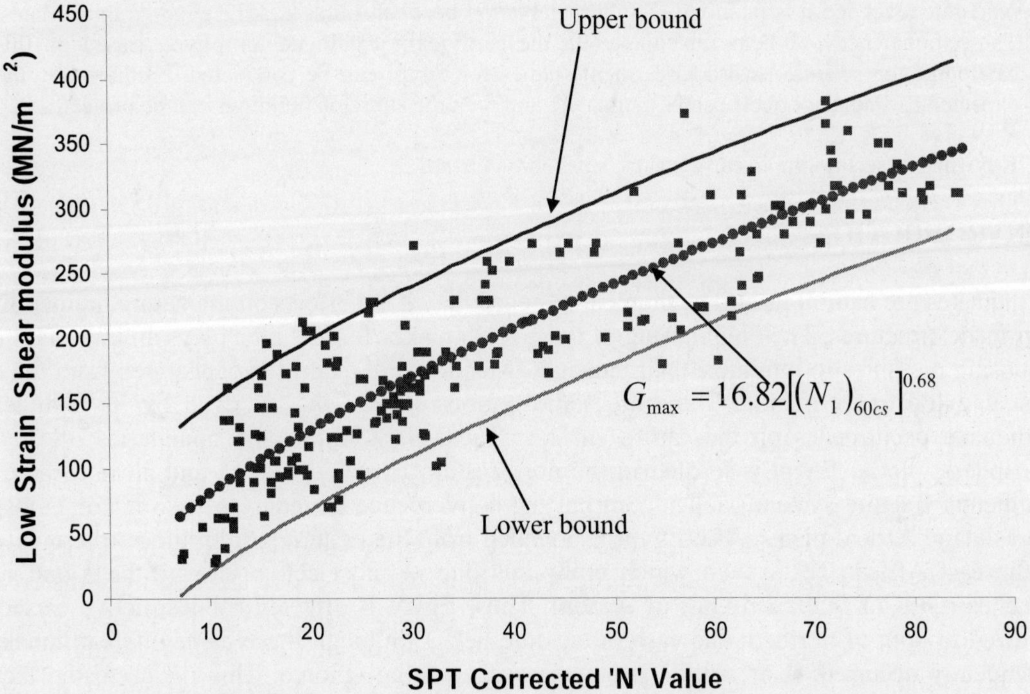

Figure 2: Shear modulus versus corrected SPT "N" values

NATIONAL SEISMIC NETWORK AND EARTHQUAKE ACTIVITIES IN EGYPT

Kamal A. El-Sayed[1], Mohamed A. Sakr[2] and Enayat A. Awad[1]
[1] Seismology Department, National Research Institute of Astronomy & Geophysics, Egypt.
E-mail: ka_rahmaneg@yahoo.com
[2] Director, Geotechnical Engineering Research Laboratory, Faculty of Engineering, Tanta University, Egypt
E-mail: mamsakr@yahoo.com .

ABSTRACT

The continuous recording of earthquakes in Egypt started with the establishment of Helwan Observatory on 1903, while in May 1962; all systems were replaced by the World Wide Standardized Seismograph System (WWSSS). There are two radio-telemetric seismograph networks were erected in July 1982 and 1990 and installed at the northern part of Naser Lake (Aswan) to monitor the microseismicity after the occurrence of November 14, 1981 earthquake. While the second one deployed around the southern part of Gulf of Suez and southern Sinai Peninsula to monitor the earthquake activity of this triple-junction area of complicated tectonic setting. After the occurrence of 12th October 1992 earthquake, the Egyptian government financed the construction of the Egyptian National Seismic Network (ENSN), which cover the whole Egyptian territory and can detect and record the majority of local, regional and teleseismic events. This network consists of 63 seismic stations equipped with short-period, broadband and borehole seismometers and supported by advanced and more sophisticated systems for recording, transmission, receiving and analysis of digital data. Most of the recorded data are transmitted to the main center via Satellite technology. It is concluded that, the seismicity of Egypt is of moderate level and it is mainly due to the interaction between Eurasia, Africa and Arabian plates. The national network helps in completing the earthquake catalogue for Egypt. Based on this catalogue, the seismic hazard assessment studies for Egypt can be conducted. Furthermore, the expected earthquake effects can be estimated, and then the strategic buildings can be protected.

Keywords: seismograph, earthquakes, seismicity, hazard.

1 INTRODUCTION

Earthquakes are natural hazard phenomenon; hence they have effects on the nature, human life and man-made structures. Then monitoring of these earthquakes for any area plays an important role in proposing measures to minimize their damages. Most of earthquakes are generated from the earth's crust and the upper mantle. There are some regions (belts) characterized by its high level of earthquake occurrences on the earth's surface. These belts define the boundaries of the major lithospheric plates. Egypt is located in the northeastern corner of Africa and close to one of the continental fracture system (Hellenic arc) at the convergence boundary of two major lithospheric (Eurasia and Africa) plates. Also Egypt is bounded from the eastern part by the northern extension of the east African rift system which represents one of most active tectonic belts and its two branches (Gulf of Suez and Gulf of Aqaba). Thus, Egypt is affected by destructive earthquakes occurred in both of northern and eastern tectonic belts. Although these damaging earthquakes are infrequently occurred, their risky consequences can not be ignored. This is due to the fact that, most of earthquakes occur near overpopulated cities and villages, coupled with the poor methods and materials of construction. Soil characteristics in different localities and their amplification impact on seismic waves are important parameters that control earthquake risk. Monitoring of earthquake activities represents the basic data for assessment of seismic hazard which is greatly needed to identify areas with different degrees of vulnerability. They will serve for further risk studies, construction codes and also for land-use planning.

2 HISTORICAL EARTHQUAKE ACTIVITIES

By reviewing early earthquakes reported occurring in Egypt, many searches in Arabic and Persian writings revealed a considerable amount of information concerning early earthquakes in Egypt and surrounding areas. The historical activity indicates that, Egypt was seismically active from as long back as 2200 BC. There are several publications comprising earthquakes that happened in and around Egypt since 2200 BC and 1903 (Sieberg, 1932; Ambraseys, 1961, 1978, 1983; Maamoun, 1979; Maamoun et al, 1984; Ambraseys et al, 1994; and El-Sayed and Wahlstrom, 1995). There are many historical earthquakes affected Egypt (Fig.1), the following is the description of the major and destructive historical events and their effects;

2.1 2200 BC: Sharkia Province earthquake

Sharp shock with a felt area of approximately 45000 km^2 was reported at Sharkia province. The estimated maximum intensity was VII near the village Tell Basta and Abu Hammad, about 16 km southeast of Zagazig. This earthquake was a severe one and caused deep fissures and soil cracks at these places.

2.2 May 26, 1111: East Cairo earthquake

A damaging earthquake in Lower Egypt affected Cairo and Fustat (Old Cairo area) around 9:00 am. It was felt over an area of approximately 25000 Km^2 and ruined a number of places. Maximum intensity VI was reported from Re-Hachope temple where slight damage was reported.

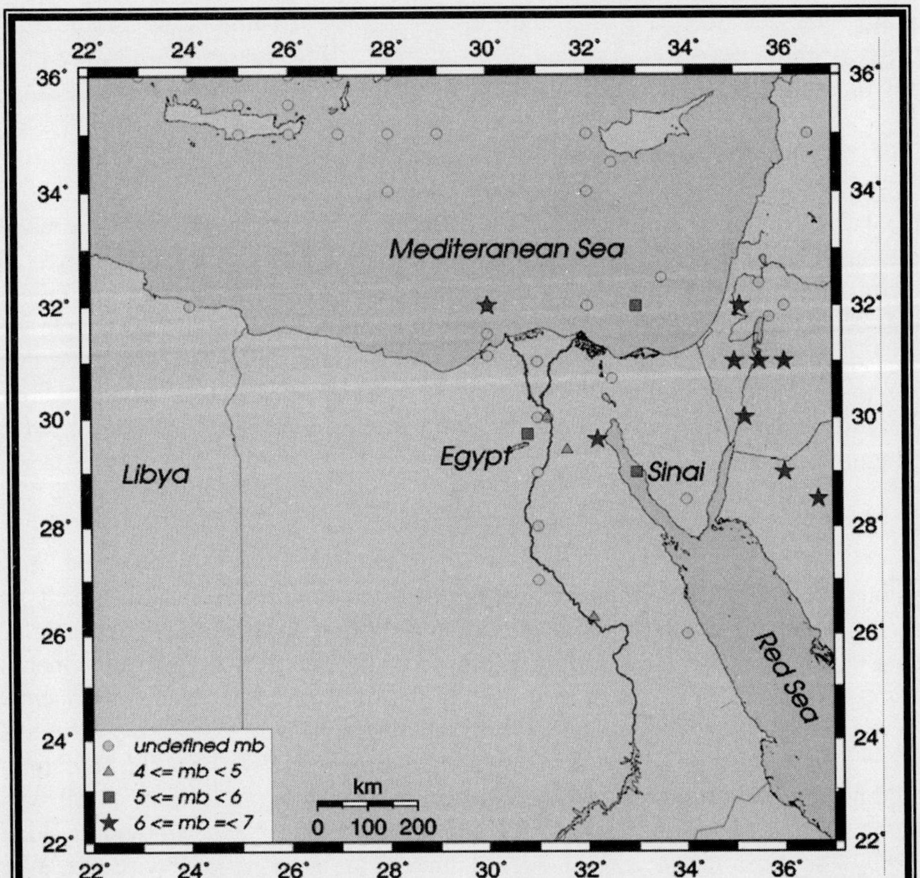

Fig. (1) Distribution of the historical earthquakes.

2.3 August 8, 1303: Mediterranean Earthquake

Many researchers (Ambraseys, 1961; Maamoun et al., (1984) and Ambraseys et al., (1994) stated that, a violent earthquake affected Syria and Egypt, particularly in Alexandria, where many houses were ruined and much of the city walls were destroyed and killing a number of people. This event occurred in the Mediterranean Sea off shore of Egypt and triggered a seismic sea waves (tsunami) that submerged half of the city of Alexandria. The lighthouse at Alexandria was shattered and its top collapsed. The maximum intensity was IX at Alexandria. In the Nile Delta there was a widespread damage. Two villages in Sharkia province were destroyed. Great number of houses, mosques and streets were collapsed within Delta area. Heavy damage was observed in Fayoum, Cairo, Beni-Suef, slight damages were reported as far as Assiut in the south of Nile River. This shock was felt also in Greece and especially in Rhodes area. Several sequences were reported through three weeks.

2.4 October 2, 1698: Rosetta earthquake

This earthquake was felt over an area bordering Bihira, Alexandria, Kafr-Elsheikh and Gharbia provinces. Maximum intensity VI was observed in Rossetta, where no serious damage was reported.

2.5 September 1754: Tanta area (Nile Delta) earthquake

A locally and destructive earthquake was felt over an area of approximately 150,000 km^2. Maximum intensity VIII was estimated in Gharbia province and VII in the Nile Delta and Cairo. Kebeasy et al., (1981) located the epicenter of this earthquake at Tanta in the middle of Nile Delta, where the maximum damage was reported. Two thirties of old houses were heavy damaged and consequently thousands of people were killed and injured.

2.6 August 7, 1847: Fayoum earthquake

This earthquake is the largest in the Nile Valley where it was felt over nearly the whole area of Egypt. Maximum intensity VIII was assigned in Fayoum and its neighboring area. The heavy damages were reported in the Nile Valley till Assiut. In Fayoum 85 people were killed, 62 were injured, 3000 houses and many mosques were destroyed. In middle Egypt, 27 people were killed. In Cairo, 100 people were killed and thousands of houses were destroyed. Moreover, thousands of people were injured and thousands of houses were differently damaged in the whole country. Another shock strongly felt in Cairo and Alexandria in the morning of 10 August, which caused damage and injuries in Alexandria.

3 INSTRUMENTAL EARTHQUAKE ACTIVITIES

The continuous recording of earthquakes in Egypt started with the establishment of Helwan Observatory on 1903. Helwan Seismological Station (HLW) is located at 29° 51' N, 31° 20' E, and its elevation is 115 meters above S.L. on limestone bedrock (Fig. 2). An E-W component of Milne-Show seismograph was the only instrument used at that time. Another N-S component of Milne-Show and a vertical component of Galitzin-Willip seismographs were respectively initiated in 1922 and 1938. There are different episodes of modernization for Helwan Station; the first one was started by adding another set of short-period Sprengnether Seismographs in 1951. All systems were replaced by the standardized set of Benioff- S.P and Sprengnether- L.P seismographs when Helwan was chosen by the U.S. Coast and Geodetic Survey to be equipped with the World Wide Standardized Seismograph System since May 1962. A Japanese S.P. Seismograph system with visual recording and frequency analyzer was added and operated since December 1972.

In 1975 three stations were initiated at Aswan (south Egypt), Abu-Simbel (south Egypt) and Matrouh (northwest Egypt) as a part of the national network. These stations which were equipped with Russian standard S.P. Seismometers. National program of updating these stations and also adding new seismological stations at selected sites has been started since April, 1984. Aswan station was provided by 3 S.P- and 3 L.P seismometers and visual/digital recording system (Table 1). The Nasser (Aswan) Lake event occurred in November 14, 1981. In July 1982, a radio-telemetered seismograph network was installed around the northern part of the Lake (reservoir) to monitor the microseismicity (Simpson et al., 1984).

Table (1) Parameters of seismological stations from 1899 to 1984.

Station Code	Location	Instrument type	Comp.	T_o	T_g	Record speed	Magnific.	Period of operation
HLW	Helwan	Milne-Show	E-W	12	11.1	10 & 15	250	1899 to 1962
HLW	Helwan	Milne-Show	N-S	12	-	10 & 15	250	1922 to 1962
HLW	Helwan	Galitzin-Willip	V	11.2	11.1	15	1000	1938-1962
HLW	Helwan	Sprengnether	3-comp.	1.5	1.05	60	3000	1951-1962
HLW	Helwan	Benioff	3-comp.	1.0	0.74	60	30000	1962 to 2000
HLW	Helwan	Sprengnether	3-comp.	15	100	30	3000	1962 to 1995
HLW	Helwan	Japanese	3-comp.	S.P.	-	120	30000	1972- 2003
MTH	Matrouh	SKM-3	3-comp.	S.P.	-	60	10000	1975-1990
ASW	Aswan	SKM-3	3-comp.	S.P.	-	60	10000	1975-1996
ASW	Aswan	SKM-3	3-comp.	L.P.	-	30	2000	1972-1996
ABS	A.Simbel	SKM-3	3-comp.	S.P.	-	60	20000	1975-1991
Aswan telemeter Network	Northern Nasser Reservoir		13 V 2 E-W 2 N-S	1.0	-	Analog & Digital	30000 to 100000	1982 to present

Using Helwan seismograms, Ismail (1960) had reported all recorded earthquakes which have epicentral distance within 10 degrees from Helwan, in the period between 1903 and 1950. The majority of these events were originating from the known active zones of the eastern Mediterranean basin. Some of these were reported felt at Cairo. The epicenters of the other earthquakes were found in some surrounding Helwan station and Cairo (Fig. 3). The low sensitivity and the slow recording speed (low time resolution) of the seismograph system could not help in determining the location of these events. Ismail's work was extended by Gergawi and El-Khashab (1968) for the time period between 1950 and 1962 to local and nearby earthquakes within distance 4.5 degree from Helwan. In spite of the short time interval, the sensitivity of the seismograph system had enabled them to investigate some local seismic activity. The instrumental Helwan magnitude scale (Maamoun, 1978) was brought into the study of seismicity of Egypt (Maamoun et al., 1981).

Maamoun et al., (1984) collected available information on earthquakes occurring in the period between 2200 B. C. and 1981 A. D. in the region between 22.0° – 33.5° N and 25.0° – 34.5° E. They investigated the data to construct a homogeneous catalog to be used on the basis of qualitative and quantitative studies. They distinguished the data in relation to time into two main periods; from 2200 B. C. to 1903 (historical observations) and from 1900 to 1981 (instrumental earthquakes) that located depending upon the number of the recording stations, their distribution around the epicenter, and the size of the event. Kebeasy (1990) showed that the seismic activities are concentrated in four narrow belts and these belts are: 1) Levant-Aqaba trend, 2) Northern Red Sea-Gulf of Suez-Cairo-Alexandria trend; 3) Eastern Mediterranean-Cairo-Fayum trend, and 4) Mediterranean Coastal Dislocation trend.

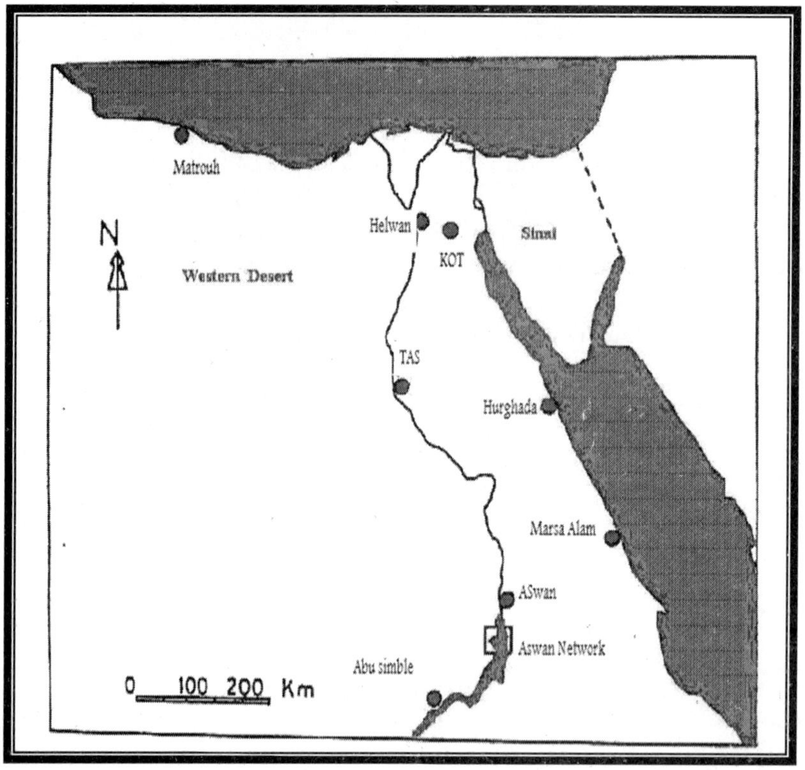

Fig. (2) Geographical distribution of seismic station before 1997.

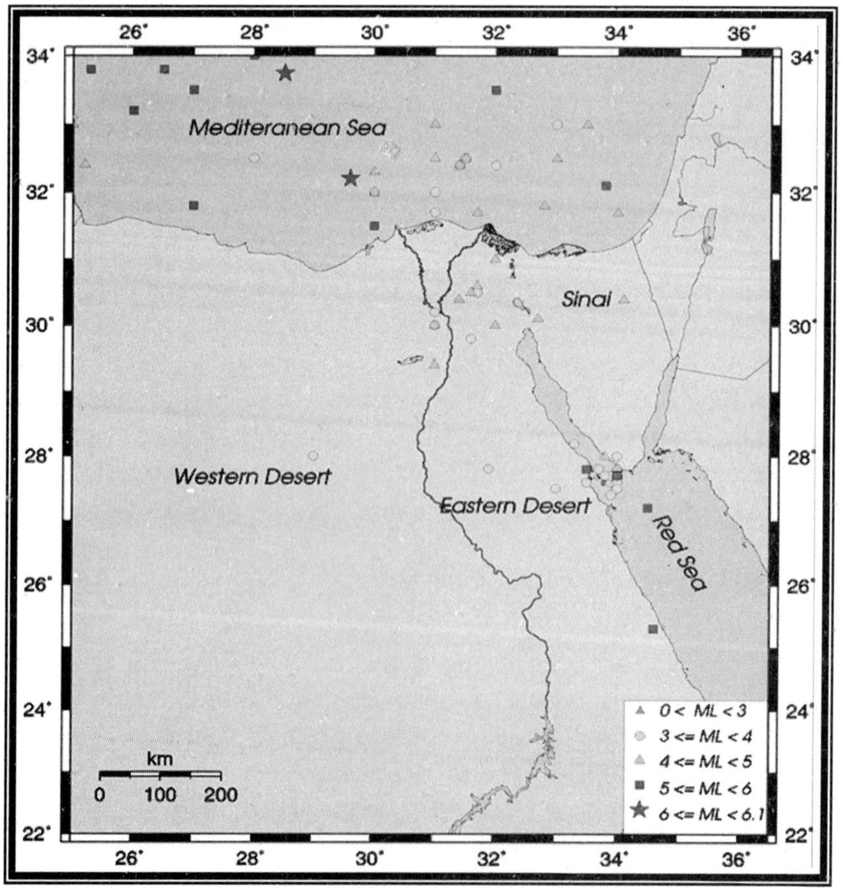

Fig. (3) Seismicity of Egypt from 1900 – 1962.

All the available data for earthquakes that occurred in an area between 22° – 34° N and 25° – 37° E in the period of time from 1900 to July 1997 have been collected in this study (Fig.4). There are major and destructive instrumental earthquakes have been occurred and cause panic damage and they are;

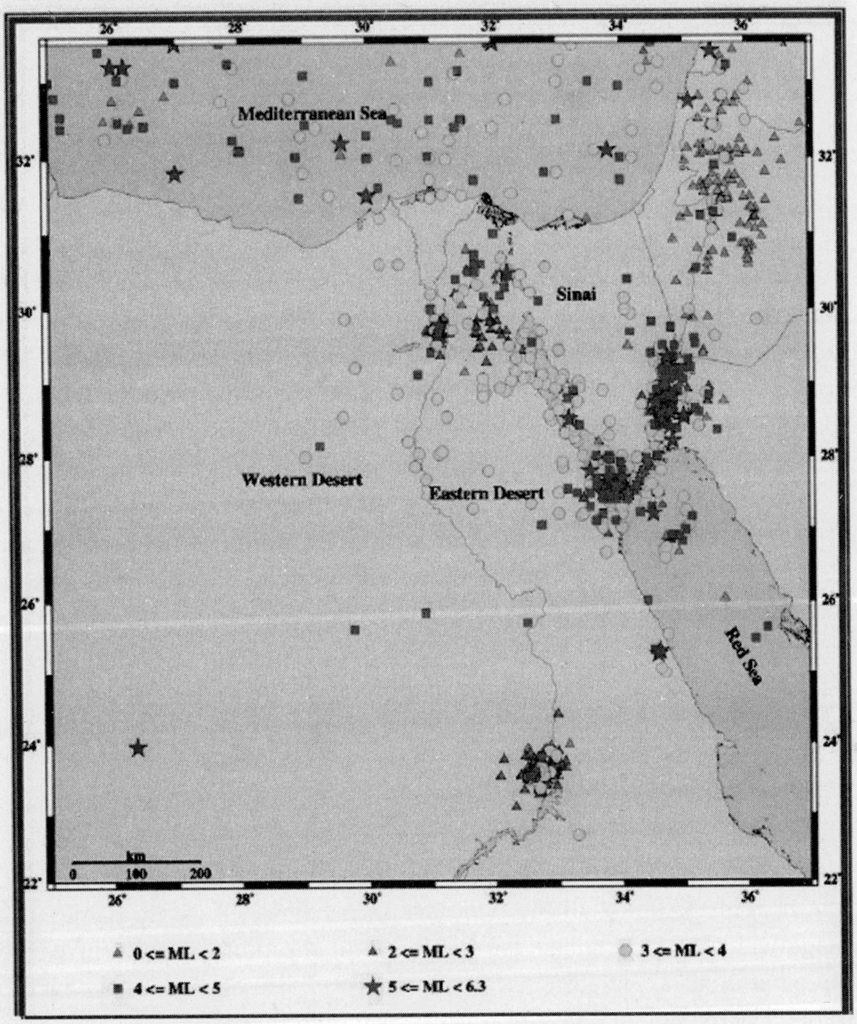

Fig. (4) Distribution of earthquakes from 1900 – July 1997.

3.1 September 12, 1955: Alexandria earthquake

This earthquake caused catastrophic damage in many localities of Egypt; 20 killed, 106 injured and hundreds of buildings destroyed. The epicenter of this earthquake located at 32.2° N and29.6° E with normal focal depth. Rothe (1969) assigned 6.7 as a surface wave magnitude. The main shock was widely felt in an area represented by a circle with radius more than 300 km. in Egypt and neighboring area. A maximum intensity of VIII was estimated in a limit area in Bihira province, where 5 persons were killed and 41 or more injured. In Rashid and Idku, the water pipes were damaged, furthermore destruction of more than 300 buildings of older brick constructions were reported in Rashid, Idku, Damanhur, Mahmodia and Abu Hommos. The Damanhur police office and two schools of Abu Hommos were also suffered further damage. A maximum intensity VII occurred at different localities of Bihira, Kafr-El Sheikh, Tanta, Monofia and Alexandria provinces where 3 persons were killed and 26 or more injured. Many of poor buildings suffer heavy damage. The police offices of Kafr-El Dawar, Kom Hamada and Quesna are slightly damaged. The high electric power cables of Dessouk were fallen.

3.2 March 31, 1969: Shadwan earthquake

The epicenter of this earthquake is located to the northwest of Shadwan Island (27.6o N and 32.9° E). A maximum intensity of IX was assigned in Shadwan Islands, Tawila and Jubal, and a body-wave magnitude of 6.3 (Maamoun and El-Khashab, 1978). On the Shadwan Island itself, landslides, earthslumps and rock falls were common. Fissures and cracks in soil extend parallel to the Red Sea –Gulf of Suez direction. Ten Kilometers west of the fractured area in the sea a submerged coral reef was raised above sea level. Heavy damage was reported in south Sinai and Hurghada and extends to the middle and southern Egypt (Qus, Luxor and Esna).

3.3 October 12, 1992: Cairo earthquake

A relatively moderate-magnitude earthquake (M_s 5.9) with epicenter about 40 Km south of Cairo centre, in Dahshour (Fig. 5), caused a disproportional amount of damage (estimated at more than LE 500 million) and the loss of many lives. The area mostly affected was Cairo, in particular the old section of Cairo, Bulaq and the region to the south, along the west bank of the Nile to Gerza and El-Rouda. The shock was also strongly felt and caused damage and life loss in the southern Delta, around Zagazig. Damage extended throughout Fayoum and as far as south as Beni-Suef and Minia. In all, 350 buildings collapsed completely and 9000 were irreparably damaged, killing 561 persons and injuring 9832. Many of the causalities in Cairo were victims of panic-stricken stampedes of pupils rushing out of schools. About 350 school and 216 mosques were ruined and 50000people were made homeless. The maximum assigned intensity was VIII at the epicentral area (Fig. 6).

3.4 November 22, 1995: Gulf of Aqaba earthquake

A strong earthquake with magnitude 7.2 with epicenter located in the southern part of the Gulf of Aqaba, 450 km. southeast of Cairo. Most of reported damage was concentrated in Sinai where a number of hotels were totally collapsed leading to the loss of three lives and the injury of ten people in Nuweiba city. Almost all the buildings in Nuweiba city were damaged on the Gulf of Aqaba. Damage was also reported to the platforms of the port facilities in Nuweiba city. This damage recorded at cities along eastern side of the Gulf of Aqaba in Jordan and Saudi Arabia as well as Elat Port. The maximum intensity was IX for this earthquake.

4 EGYPTIAN NATIONAL SEISMIC NETWORK (ENSN)

As a result of this damage, the Egyptian government supports the National Research Institute of Astronomy and Geophysics (NRIAG) to install the Egyptian National Seismic Network (ENSN) and the strong motion network (Figs. 7 and 8). ENSN consist of main center at Helwan and five sub-centers at Hurghada, Burg El-Arab, Marsa Alam, Aswan and Kharga. The main center receives the seismic data from the near distance stations through a telemetry communication and from the remote stations and the sub-centers via the telephone lines and the satellite communications. The received data is analyzed for determining the earthquake parameters. The sub-centers collect the data from some close distance stations and send it to Helwan via Satellite communications. The distribution of the seismic stations and the strong motion units are chosen to cover the known seismic sources as well as possible. Also, this distribution includes the regions of historical earthquakes.

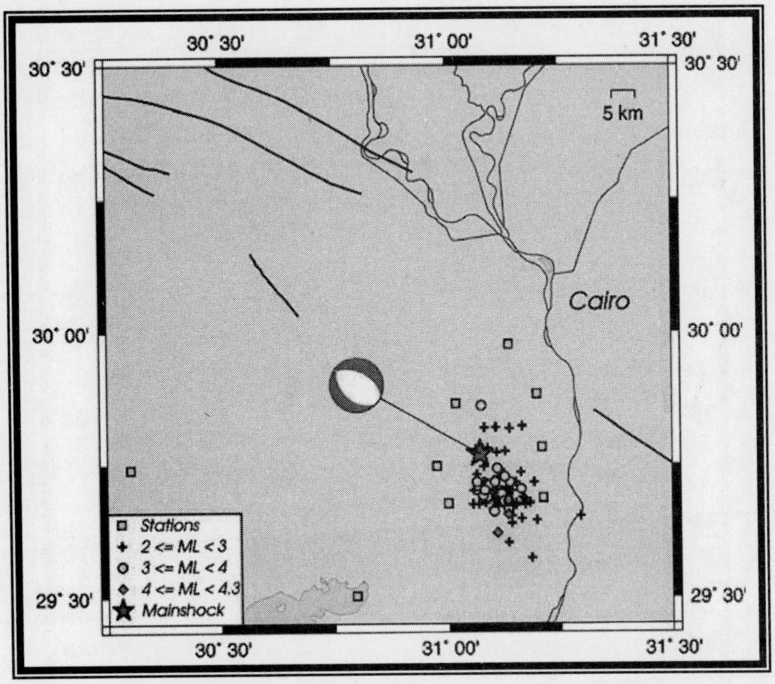

Fig. (5) Location of 12th Oct. 1992 earthquake.

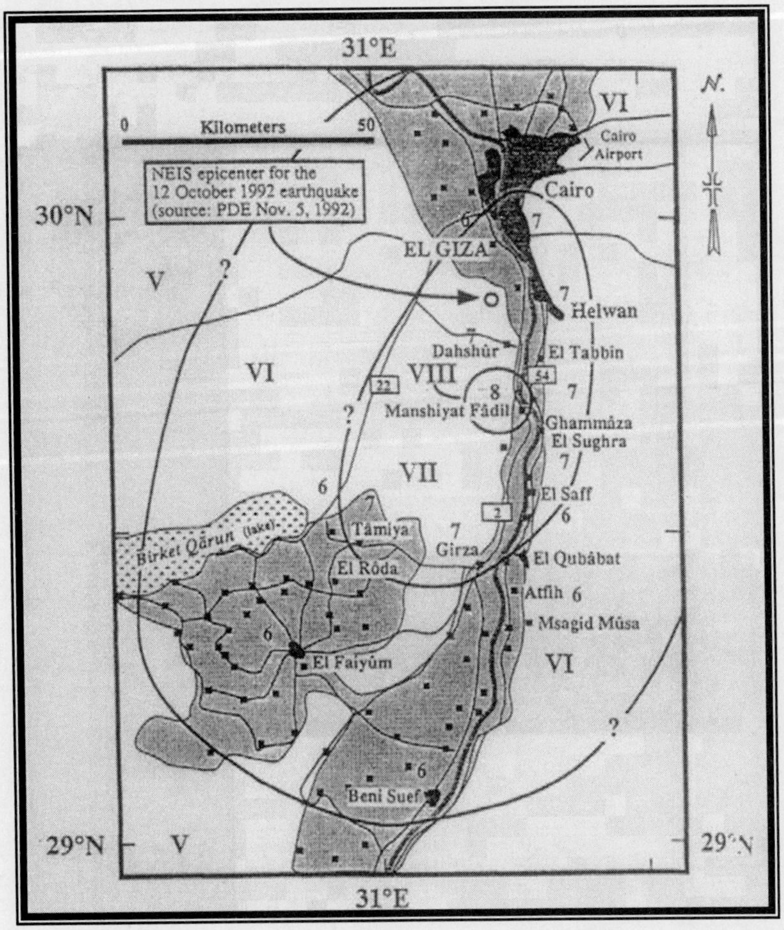

Fig. (6) Intensity map for Oct. 12, 1992 earthquake.

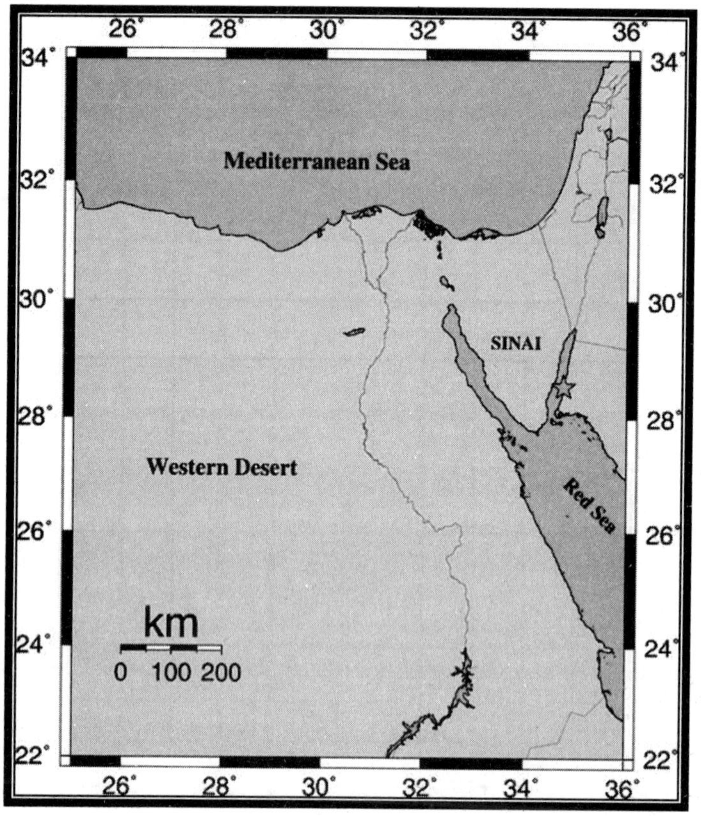

Fig. (7) Location of 25th Nov. 1995 earthquake.

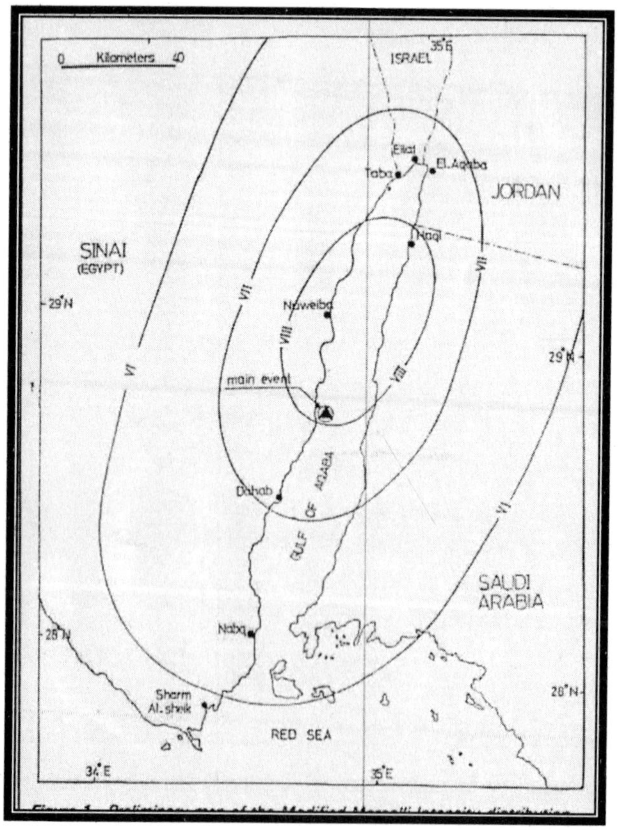

Fig. (8) Intensity map for Nov. 25, 1995 earthquake.

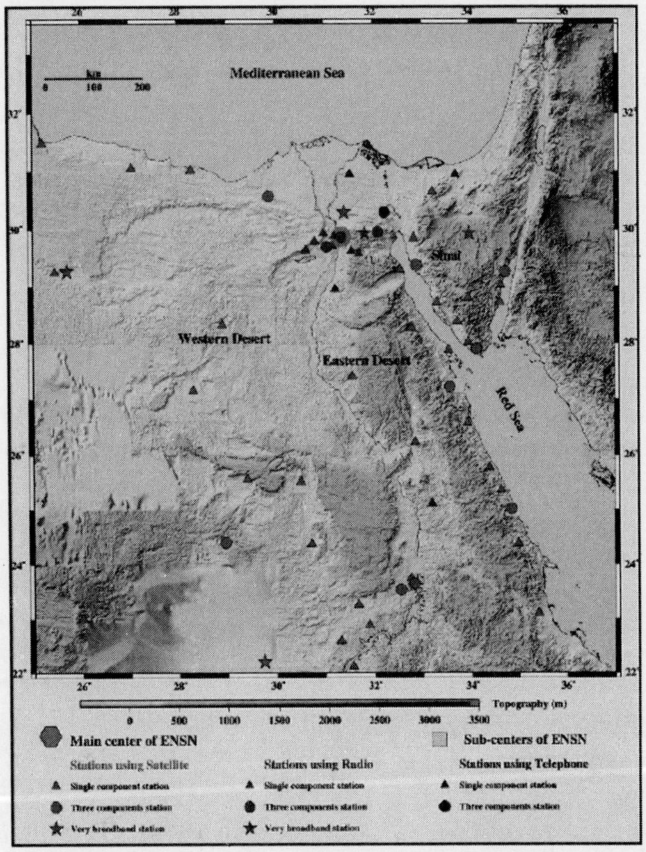

Fig. (7) Geographical distribution of ENSN stations.

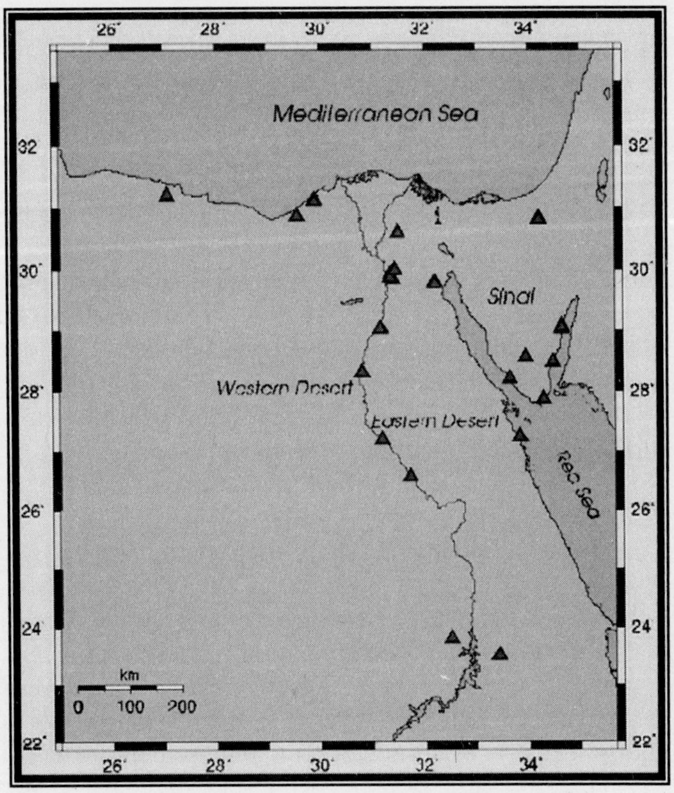

Fig. (8) Geographical strong motion stations.

4.1 Data Processing

The seismic Data Analysis software (DAN ver.2.56) provided by Nanometerics Inc. (Canada) extracts the digital data of remote site online from the ring buffer. The arrival times of P- and S-waves from the available stations are picked. The location program assign a numeric factor to qualitative quality factors assigned to phase reading during picking. The location of seismic events could be done by two implemented programs, Hypo71 PC) and LOC (Geological Survey of Canada). LOC program is preferable to locate regional events, while it is preferable to use Hypo71 when well azimuth coverage of stations is available.

Two magnitude scales are used to calculate the relative size of each earthquake (M_n and M_l). The response is the value read from the response file for given frequency and phase. Amplitude is the zero to peak counts and period as measured between the peaks of the peak to peak measurements. DAN firstly converts the amplitude to counts into actual ground motion in nm/sec suing the system response file. The program then converts this ground motion to ground displacement and then to ground motion that would be recorded on Wood-Anderson Seismograph. Finally, the maximum amplitude is selected and Richter empirical formula for southern California is used to calculate the Ml value (Eq. 1).

$$\mathbf{Ml = Log_{10} A + 2.56 \log_{10}\Delta - 1.67} \qquad (1)$$

Where A; is the ground motion in μm; Δ in km $10 < \Delta < 600$ km. While, M_n is calculated using Nuttli (1973) for distance $\Delta > 4$ degrees (Eq. 2).

$$\mathbf{M_n = -0.10 + 1.66 \log_{10} (\Delta\ km) + \log_{10} (amp/magnification) - 1.0 * \log_{10} (period)} \qquad (2)$$

4.2 Seismic activity

The local seismic activities (Fig.9) recorded by ENSN from Aug. 1997 (starting time of ENSN recording) to May 2007 are located at specific seismic zones that reflect their tectonic activity. These zones are; Cairo-Suez District; Beni Suef zone; Southwest Cairo zone; Central Gulf of Suez; Northern Red Sea; Gulf of Aqba; Northern Naser's Lake zone; Abu Dabbab zone; El-Minya-Sohag zone and Mediterranean Sea Zone.

Cairo – Suez shear zone extends from the northern Gulf of Suez toward Cairo city and is characterized by moderate activity. The northern part of eastern Desert which reflects a tendency of ENE –WSW active faults (these faults transverse the Gulf of Suez main faults). One of these faults is shown by a cluster of seismic activity from the Gulf of Suez to Beni-Suef city. Southwest Cairo zone (Dahshour area) is characterized by moderate seismic activity. A cluster of seismic activity comes clearly from the central part of Gulf of Suez, which indicates that the entire Gulf is active. A cluster of earthquakes is shown at the northern part of Red Sea at the triple junction area. The activity tends to lie at the Aqaba fault extension to the south. There are high seismic activity is also related to the Gulf of Aqaba and its extension towards the north.

A cluster of earthquakes is clearly shown along the central part of the Red Sea after the operation of ENSN stations that located along the western Red Sea Coast. This activity is related to the active rifting of the northern Red Sea. Moderate seismic activity is occurred along the northern part of Naser's Lake to the southwest of Aswan High Dam. A new cluster of seismic activity is recorded along the southern part of Naser's Lake. This cluster tends to constitute in E-W direction in agreement with the major fault trends in this area. Another seismic zone is identified with increasing the delectability of ENSN by installation of some stations in the central part of Egypt. This zone extends from Abu Dabbab area along the Red Sea coast to the western direction until the Nile River. ENSN recorded a scattered activity from the western Desert to the west of El-Minya and Sohag cities. Scattered events lie along the Mediterranean coast. The regional seismic activity (Fig.10) is located at the southern part of Hellenic and Cyprean arcs. The regional events lie as far as to the northern part of Turkey and Greece. Few numbers of earthquakes are located along southern Jordan, southern Red Sea and northern Sudan.

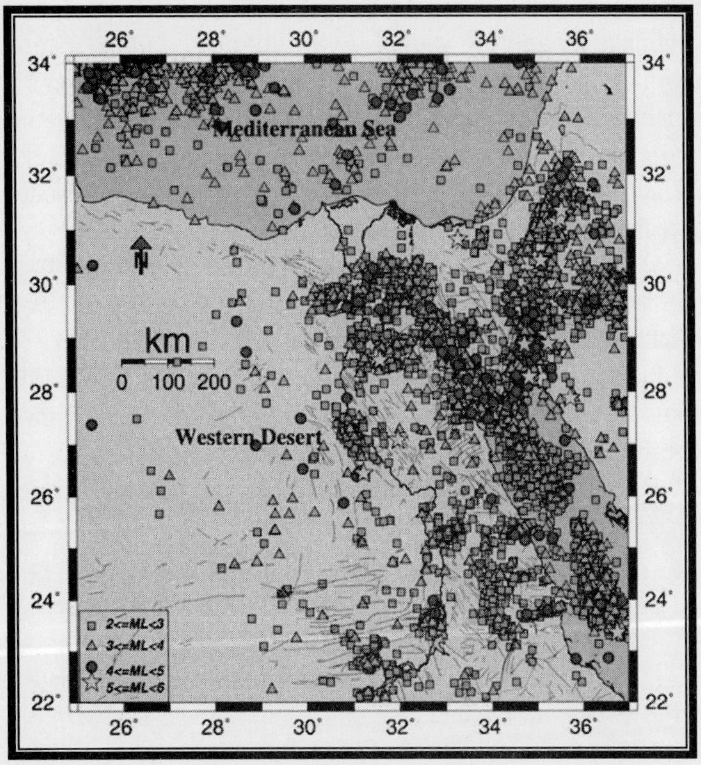

Fig. (9) Local earthquakes recorded by ENSN from Aug. 1997–May 2007.

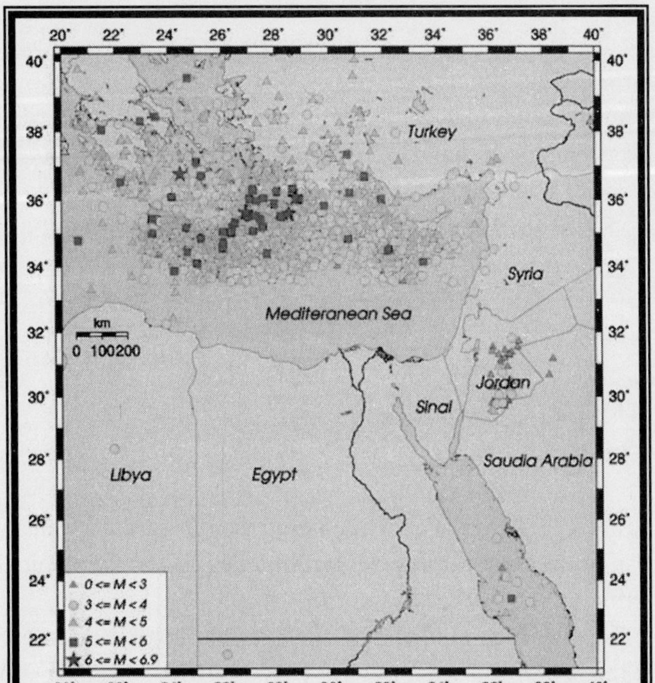

Fig. (10) Regional earthquakes recorded by ENSN from Aug. 1997–May 2007.

CONCLUSIONS

Egypt is affected by many destructive and damaging earthquakes through its long history and many people were killed, injured and a lot of structures were destroyed. The recording of earthquakes started with one station since 1898. Then there are consequences of modernization have been carried out to the earthquake monitoring stations by installing additional stations in different sites. Besides the local networks that deployed around the northern part of Naser's Lake in Aswan and around the southern part of Gulf of Suez. In July 1997 the National Egyptian Seismic Network (ENSN) started to monitor the seismic activity that distributed allover Egypt. This network is designed to cover the active areas. This network consists of 63 seismic stations and transmitted their data via satellite technology of communications.

By installing the ENSN, the seismicity map of Egypt is completely changed where; the local seismic activities reflect the incredible large number of smaller earthquakes. This large number of events is related to the increasing of the detectability of the network. New earthquake source zones are discovered and some of these zones characterized by its moderate to high earthquake activity. These zones should be taken into consideration in the hazard assessment studies for Egypt. The continuous upgrading of ENSN is highly recommended because this will increase the efficiency of the ENSN for the detection of very small earthquakes.

REFERENCES

1. Ambraseys, N. N., (1961). On the seismicity of the south-west Asia (date from a XV century Arabic manuscript), Revue pour L'etude calamites, **37**, Geneva, 18-30.
2. Ambraseys, N. N., (1978). Middle East a reappraisal of seismicity, Q.Jr. Engr. Geol. 11, 19-32. Ambraseys, N. N., (1983): A note on historical seismicity. Bull. Seismological Society of America, **73**, 1917-1920.
3. Ambraseys, N. N., Melville, C. P., and Adams, R. D. (1994): The seismicity of Egypt, Arabia and the Red Sea, a historical review. King Abdul-Aziz City for Science and Technology, Cambridge Univ. Press, Amsterdam, 1-137.
4. El-Sayed, A. and Wahlstrom, R. (1995): Distribution of the energy release, b-value and seismic hazard in Egypt. Proceedings of the XXI General Assembly of IUGG, July 2-14, Colorado, USA.
5. Gwegawi, A. and El-Khashab, H., (1968): Seismicity of U. A. R., Helwan Observatory Bull., **76**.
6. Ismail, A. (1960): Near and local earthquakes of Helwan (1903-1950). Helwan Observatory Bull. ,**49**, 12-54.
7. Kebeasy, R. M., Maamoun, M., Ibrahim, E. (1981): Aswan Lake induced earthquakes. Japan Bull. of IISEE, **19**, 155-160.
8. Kebeasy, R. M. (1990): Seismicity in Geology of Egypt (R. Said, Ed.) A.A. Balkerma, Rotterdam, 51-59.
9. Lee, W. H.and Lahr, J. (1972): HYPO71: A computer program for determining hypocenter, magnitude and first motion pattern of local earthquakes, U. S. Geological Survey, Open File Report, 72-224.
10. Maamoun, M. and El Khashab, H. M. (1978): Seismic studies of the Shadwan (Red Sea) earthquake. Helwan Inst. Astr. and Geophysics, **171**, 22-46.
11. Maamoun, M. (1979): Macroseismic observations of principal earthquakes in Egypt. Bull. Helwan. Inst. Astron. Geophys. (HIAG), **183**, 81-103.

12. Maamoun, M, and Allam, A. (1981): A study of the seismicity of the east Mediterranean region, Japan, Bull. of IISEE, **19**, 57-80.
13. Maamoun, M. , Megahed, A., and Allam, A. (1984): Seismicity of Egypt. Helwan Institute of Astronomy and Geophysics (HIAG) Bull., **4**, ser. B, 109-162.
14. Nuttli, O. W. (1973): Seismic wave attenuation and magnitude relations for eastern North America. J. Geophys. Res., **78**, 876-885.
15. Rothe, J. P. (1969): Seismicity of the earth 1954-1965, Earth Sciences. UNESCO 1.
16. Simpson, D. W., Kebeasy, R. M., Nicholson, C., Maamoun, M., Albert, R. N., Ibrahim, E. M., Megahed, A., Gharib, A., Hussain, A. (1984): Aswan telemetered seismograph network. Bull. of Egyptian Geophysical Society, **2**, 184-213.

SEISMIC HAZARD ESTIMATION AT BED ROCK AND GROUND SURFACE LEVELS FOR CHENNAI CITY

***Boominathan A*[1] *and Suganthi A*[2]**

[1]*Professor, Indian Institute of Technology Madras, boomi@iitm.ac.in*

[2]*Research Scholar, Indian Institute of Technology Madras, suganthi.iitm@gmail.com*

ABSTRACT

Chennai city suffered moderate tremors during 2001 Bhuj and Pondicherry earthquakes and 2004 Sumatra Earthquake. After Bhuj earthquake, Indian Standards (IS: 1893–2002) has been revised and the Chennai city has been upgraded from Zone II to Zone III which leads to substantial increase of the design ground motion parameters. Therefore, a comprehensive study is carried out to assess the seismic hazard of the Chennai city based on probabilistic and deterministic approaches. A fault map has been developed containing 36 faults within 300km radial distance from the city. The one dimensional ground response analysis was carried out for large number of sites by equivalent linear method using SHAKE91 program to estimate the ground motion parameters considering the local site effects. The shear wave velocity profile was obtained from the corrected SPT N- values and it was verified with the Multichannel Analysis of Surface Wave (MASW) test. At bed rock level, probabilistic based PGA hazard contours are presented for 500 years return period. The seismic hazard at ground level is represented in terms of deterministic based PGA, predominant period contours and Spectral Acceleration Ratio (SAR) contours for the entire city.

Keywords: Seismic Hazard, PSHA, DSHA, Site effects

1 INTRODUCTION

Peninsular India has been considered as a stable continental region since many years. It is primarily the damages caused during the 2001 Bhuj earthquake (M_w 7.6) demanded the immediate study of the Peninsular region. Earthquakes of Koyna (1967; M_w 7.6), Latur (1993; M_w 6.1) and Jabalpur (1997; M_w 5.8) also occurred in the "stable" Indian shield. A review of the historical as well as recent earthquake activity in peninsular India indicated that different parts of the peninsular region are characterized by a low to moderate level of seismic activity. But it is only in recent decades that the occurrence of some large and damaging earthquakes has caused concern, which led to the study of peninsular seismicity in greater detail (Chandra, 1977). Moderate to large magnitude earthquakes in these seismic zones perhaps occur with extremely long recurrence intervals. Such long recurrence intervals strengthened the false belief about the stability of the Peninsular India. With the revised seismic zoning map pegged Chennai at a higher activity zone (Zone III), there is an urgent need to study the seismic hazard potential of Chennai city and the effect of local soil conditions on the ground motion parameters which will enable urban planners to design the earthquake resistant structures and strengthen existing unstable structures. This paper presents a comprehensive study carried out for estimation of seismic hazard for Chennai city by both deterministic and probabilistic approaches. The ground motion parameters are estimated by considering site effects.

2 SEISMICITY AND SEISMOTECTONICS OF THE REGION

The data on the occurrence of past earthquakes and the knowledge of various tectonic features with which the earthquakes can be associated are the two basic information components required for all seismic hazard analyses studies. These are technically known as the seismicity and the seismotectonic details of area that are used for the better understanding of the seismic susceptibility of the area. Indian seismicity is characterized by a relatively high frequency of great earthquakes and a relatively low frequency of moderate earthquakes. Typical seismicity of Peninsular India based mainly on Gauribidanur seismic array (GBA) detections of regional earthquakes spanning two decades (1978–1997) is shown in Fig. 1.

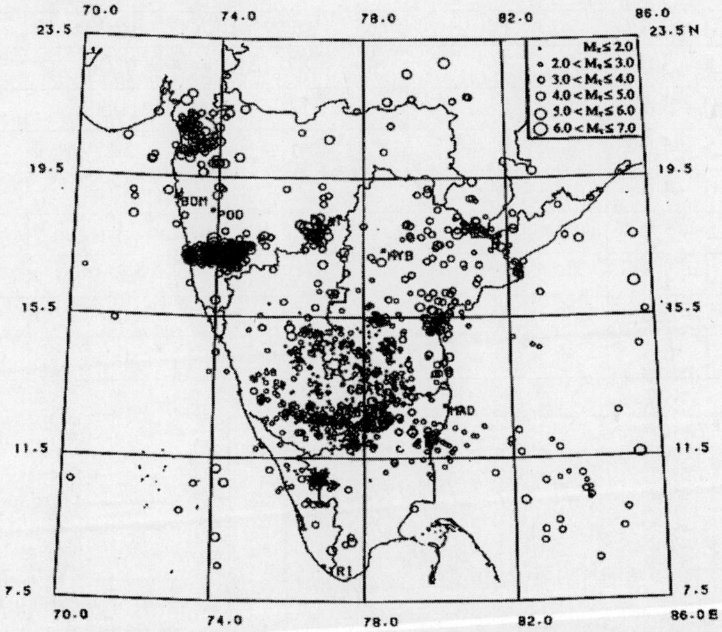

Figure 1. Seismicity of Peninsular India (Gangrade and Arora, 2000)

Seismotectonic Atlas of India (2000) provides useful information on the seismicity details of the environment. Regions with pronounced variation in thickness show higher seismicity as compared to the parts with more or less uniform thickness. Historical earthquake information within 300 km radial distance from Chennai was obtained from NEIC, USA since 1800 A.D. onwards. A total of 65 earthquake data was obtained from this catalogue. The seismological details of the region reported in Ghosh (1994) and NEIC, USA are also used in this study. The summary of the earthquake events is presented in Table 1. Repeated events were removed and finally a new catalogue of 638 earthquake data was prepared. Seismic events with magnitude greater than 2 are only considered in the preparation of earthquake catalogue.

Seismicity data alone are usually not sufficient for the purpose of identifying and characterizing earthquake sources. Seismotectonic data provide a valuable and necessary supplement to historical and instrumental seismicity data. The first step in modeling the seismicity of a region is to define the seismic source zones. It is a general practice to consider the seismic and seismotectonic information around 300 km radial distance from the site for the best representation of the seismic status of the region. Seismological information and seismotectonic features of the region were collected from the latest Seismotectonic Atlas of India (2000) and Ghosh (1994). Considering the limitations on the knowledge of the behaviour and slip rate properties of the faults, all the seismic sources in the study are considered as line sources. The Seismotectonic map combines the fault map with geological features of the area under consideration. A total of 36 faults were identified around the city and a fault map has been developed as shown in Fig. 2. The fault map has encircled the faults lying within 300 km radius from the city. It can be observed that the faults 24 and Palar fault are found to be the longest faults within this distance. In addition to that, there are two principal deep-seated faults, i.e., fault 17 and 26. Right now, the activities of the faults are not properly known. Due to the lack of sufficient information about the seismic activity details of the faults, all the faults are considered to be seismically active and have been given equal weightage in the Seismic Hazard Analysis.

TABLE 1 Summary of Earthquake Events

Sr. No.	Magnitude range	Global data		GBA Data	
		No. of events in the range	No. of events With M > M	No. of events in the range	No. of events with M > M
1	2.0 – 2.5	1	39	109	417
2	2.5 – 3.0	3	38	110	308
3	3.0 – 3.5	1	35	103	198
4	3.5 – 4.0	7	34	54	95
5	4.0 – 4.5	19	27	35	41
6	4.5 – 5.0	1	8	4	6
7	5.0 – 5.5	6	7	2	2
8	5.5 – 6.0	0	1	0	0
9	6.0 – 6.5	1	1	0	0

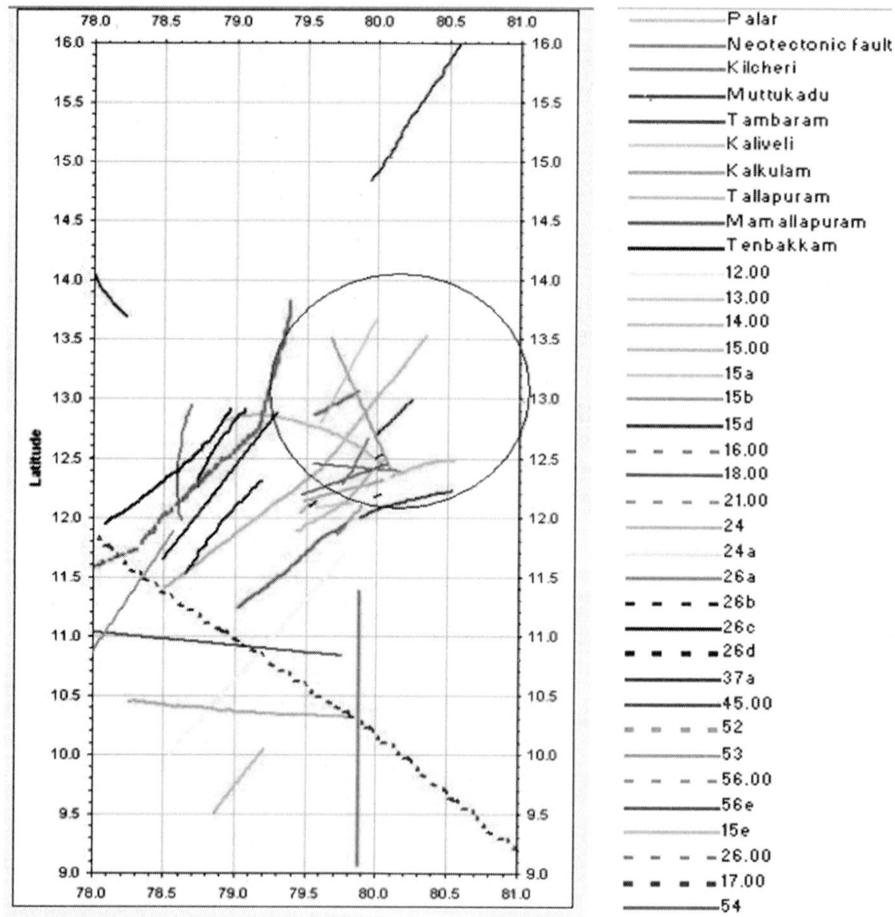

Figure 2. Fault Map

3 BASE MAP PREPARATION

A base map is the one of the important ingredients of the seismic microzonation studies; a preparation of which requires a special consideration. Over the last four decades Geographical Information Systems (GIS) have emerged as the predominant medium for graphic representation of geospatial data, including geotechnical, geologic and hydrologic information. Toposheets of scale

1:50,000 obtained from Survey of India were used for creating the base map. The GIS software, Arc Info was used for creation of base map of the city (Fig. 3). The scanned toposheets were digitized onscreen with several layers including administrative boundaries, highways, railroads, water bodies, and land marks as shown in Fig. 3. A total of 400 borehole data were collected from the reputed geotechnical agencies and marked on the map. Typical borehole locations for some of the regions of the Chennai city are shown in the figure.

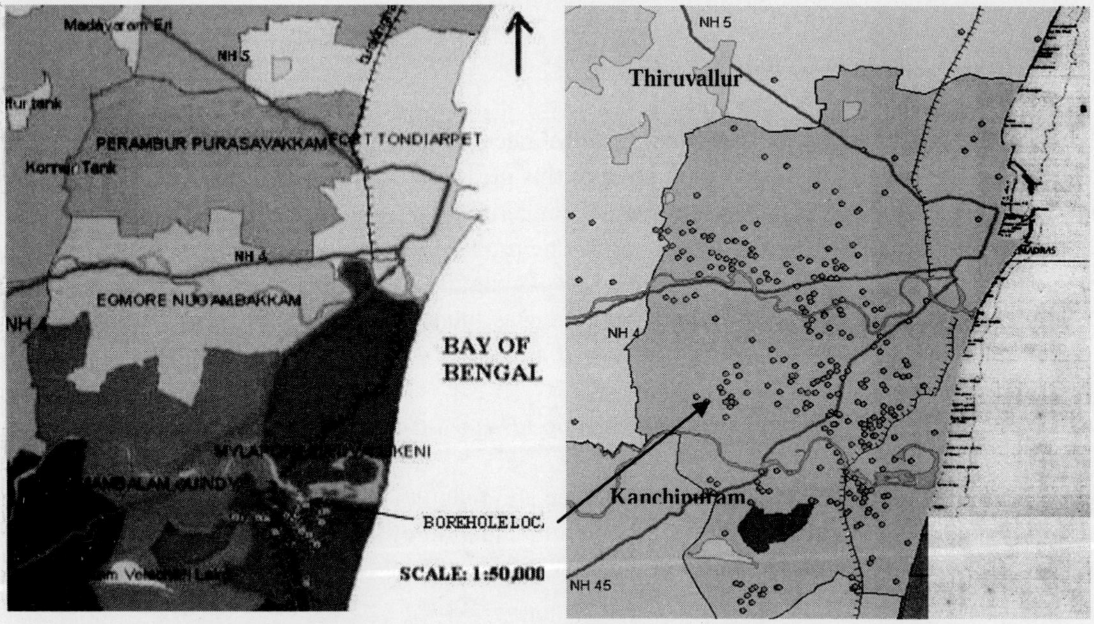

Figure 3. Base Map

4 SEISMIC HAZARD ASSESSMENT

Seismic hazard analysis is the primary tool to support both building codes and preparedness. It involves the quantitative evaluation of ground shaking hazards at a specified location. Approaches to seismic hazard assessment can be grouped into two broad categories: Deterministic and Probabilistic. The deterministic approach uses an accepted earthquake scenario, whereas the probabilistic approach quantifies the rate (or probability) of exceeding various ground motion levels at a site, given all possible earthquakes. In the present study, seismic hazard analysis has been carried out by both probabilistic and deterministic approaches.

4.1 Probabilistic Seismic hazard Analysis (PSHA)

The foundations of probabilistic seismic hazard analysis was established by Cornell (1968) and further refined by many researchers in subsequent years. In this approach, the distances to potential seismic sources and the magnitudes of earthquakes generated by the sources are treated as random variables (Reiter, 1990). The occurrences of earthquakes are considered probabilistically.

4.1.1 Earthquake Source Characterization

The three principal components of seismic source characterization are source location and geometry, maximum earthquake magnitude, and earthquake recurrence. Within these sources, earthquakes are assumed to occur randomly, in terms of their epicentral locations, as well as in terms of their occurrence times. The maximum earthquake magnitude that a seismic source is capable of generating defines the upper bound to the earthquake recurrence relationship. In the present scenario, the properties and the rupture characteristics of the faults are not well established. Hence, the present study considers the method of addition of an increment to the largest historical earthquake. Thus for each seismic source, the largest observed past magnitude is increased by 0.5 units to arrive at the maximum possible magnitude (Kijko and Graham 1998; Sokolov et al. 2001). Completeness analysis has been carried out using Stepp method as explained in (Shanker and Sharma, 1997). It was observed that for 3-4, 4-6 and 5-6 magnitude groups the data is complete for 30, 60 and 100 years respectively. Gutenberg and Richter recurrence relationship was used to characterize the distribution of earthquake sizes in a given period of time. The a and b parameters were found to be 1.53 and 0.65 respectively.

4.1.2 Spatial and size uncertainty

The entire Chennai city and its surroundings have been divided into 1 km x 1 km grid size, and the hazard level has been assessed for each node of this grid. The source zones are divided in to a large number of segments of equal length and the uncertainty in the source to site distance can be described using a probability density function. The probability density function of the magnitude is also found in the similar manner. A lower level magnitude of 3 was assumed. The maximum possible magnitude on a particular seismic source was found using the popular method, which is to increase the largest observed past magnitude by 0.5.

4.1.3 Probability of ground motion exceedance and Hazard Curves

For each combination of the hypocentral distance and magnitude, peak ground acceleration (PGA) at bed rock level is estimated using the attenuation equation of strong ground motion proposed for the Peninsular India by Iyengar and Raghukanth (2004)

$$\text{Ln}\,(PGA/g) = C_1 + C_2\,(M\text{-}6) + C_3\,(M\text{-}6)^2 - \ln(R) - C_4\,R + \ln\varepsilon \qquad (1)$$

where $C_1 = 1.7816$; $C_2 = 0.9205$; $C_3 = -0.0673$; $C_4 = 0.0035$ and $\sigma\,(\ln\varepsilon) = 0.3136$. M, y and R refer to moment magnitude, PGA (g) and hypocentral distance respectively. Since the peninsular shields in general are more susceptible to the shallow focus earthquake, a focal depth of 10 km is assumed and the hypocentral distance is calculated for all fault sources.

The probabilities that various target peak acceleration levels y^* that will be exceeded are calculated as

$$\mathrm{P}\,[Y > y^* \mid m,\ r] = 1 - F_Y(y^*) \qquad (2)$$

where $F_Y(y)$ is the value of CDF of Y at m and r.

Seismic hazard curves can be obtained by computing the mean annual rate of exceedance λ_{y^*}, for different specified ground motion values y^*. The annual rate of exceedance λ_{y^*} of each peak acceleration level for each magnitude and distance interval for a particular source zone can be computed as

$$\lambda_{y^*} = \nu_i\ \mathrm{P}[\,Y > y^* \mid M = m,\ R = r\,]\ \ \mathrm{P}[\,M = m\,]\ \ \mathrm{P}[\,R = r\,] \qquad (3)$$

where $\nu_i = \exp(\alpha\text{-}\beta m)$, $\alpha = 2.303$ a; $\beta = 2.303$b

These curves are first obtained individually for all the thirty-six faults and combined to estimate the aggregate hazard at the site. Consequently, the probability a particular target acceleration will be exceeded by an earthquake of magnitude of $M>m_o$ on any of these sources can be interpreted from the cumulative seismic hazard curve. The reciprocal of the annual probability of exceedance gives the return period for the corresponding PGA value.

4.1.4 Temporal Uncertainty

Temporally, the earthquakes are assumed to follow a simple Poisson process. The seismic hazard curve combined with the Poisson model to estimate the probabilities of exceedance of a particular value y^* in finite time period, T can be expressed as

$$P[\, Y_T > y^* \,] = 1 - e^{\lambda_y * T} \qquad (4)$$

The procedure is repeated for all grid points to obtain PGA at bed rock level. A MATLAB code was written to carryout the PSHA analysis by the above procedure and finally the seismic hazard contour map was obtained. Figs. 4 shows the contours of rock level PGA for 10% probability of exceedance in 100 years which to a return period of 500 years. It can be observed from the figure that the PGA values are in similar pattern with that of the fault map. The rock level PGA varies from 0.10g to 0.25g.

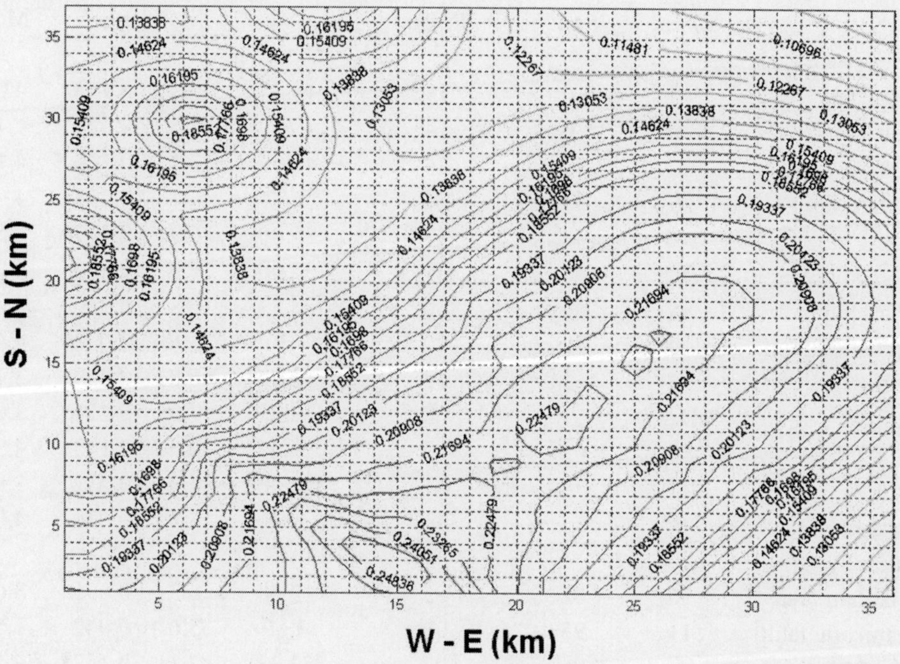

Figure 4. Contours of rock level PGA with 10% probability of exceedance in 100 years

4.2 Deterministic Seismic hazard Analysis

The deterministic approach basically requires the determination of the maximum magnitude of a causative fault or seismotectonic source. Generally, a radial distance of 100 km from the site is considered for carrying out the deterministic seismic hazard analysis. Hence out of 36 faults, 16 faults that lie within 100 km radial distance are selected for this analysis. The maximum magnitude for each fault source can be assigned based on the empirical relationships or from the observed seismicity in the near vicinity of the source. Assuming the faults to be normal faults and that a fault would create a surface rupture equivalent to $1/3$ of its total length (Mark, 1977), the maximum magnitude can be computed using the Wells and Coppersmith relationship (1994) as given below:

$$M_w = 4.86 + 1.32 \log L \qquad (5)$$

where M_w is the Moment magnitude, which corresponds to the maximum magnitude for a fault source and L is the fault surface rupture length in km. Maximum magnitude of the faults estimated by Mark (1977) method gives very high values due to the dependency of the method on the fault length alone.

The earthquake data collected from various sources gives an overall idea about the seismicity details associated with each fault source. This information is used for assigning the maximum magnitude for each fault source by taking into consideration the seismicity around that particular fault source. The maximum magnitude on a particular seismic source was taken as the largest observed past magnitude plus 0.5 (Kijko and Graham 1998; Sokolov et al. 2001). The maximum magnitude for all the faults that were obtained from the Wells and Coppersmith relationship (1994) and observed seismicity details are presented in Table 2 (Boominathan et al. 2007).

Table 2. The details of faults and seismicity

S.No.	*Name of fault*	*Fault length L, km*	*Closest distance km*	*Hypocentral distance, R, km*	*Using Wells and Coppersmith relation* L/3	M_w	PGA (g)	*Using observed seismicity* M_w	PGA (g)
1	Fault 15d	40	10	14	13.33	6.3	0.745	4.0	0.066
2	Fault 24	365	10	14	121.67	7.6	**2.025**	4.4	**0.106**
3	Fault 53	137	32	34	45.67	7.1	0.526	4.1	0.029
4	Kilcheri fault	26	33	34	8.67	6.1	0.228	4.0	0.025
5	Fault 15a	105	46	47	35.00	6.9	0.317	4.5	0.032
6	Neotectonic fault	105	48	49	35.00	6.9	0.302	3.8	0.013
7	Palar Fault	85	59	60	28.33	6.8	0.216	4.0	0.013
8	Tambaram fault	10	59	60	3.33	5.6	0.072	4.4	0.021
9	Fault 15	96	61	62	32.00	6.8	0.220	3.7	0.009
10	Fault 52	115	67	68	38.33	7.0	0.214	3.6	0.007
11	Fault 15e	50	68	69	16.67	6.5	0.142	4.5	0.020
12	Fault 54	129	70	71	43.00	7.0	0.212	3.8	0.009
13	Mahapalipuram fault.	5	75	76	1.63	5.1	0.036	4.0	0.010
14	Kalkulam fault	36	82	83	12.08	6.3	0.096	3.6	0.005
15	Muttukadu fault	11	95	96	3.67	5.6	0.042	3.5	0.004
16	Fault 26d	160	96	97	53.33	7.1	0.157	4.5	0.013
17	Fault 56e	75	97	98	25.00	6.7	0.110	4.5	0.013
18	Fault 26	1000	98	99	333.33	8.2	0.318	4.5	0.013

The peak ground acceleration (PGA) at bedrock level is estimated using the attenuation equation as presented in section 4.1.3 and is shown in Table 2. It can be observed from the table that the maximum magnitude computed using the empirical relationship show very high values ranging from 5.6 to 8.2. These calculations further lead to the unrealistic PGA values that are irrelevant to the observed seismicity of the study area. On the other hand, maximum magnitudes obtained from the past earthquake details are quite smaller with a range from 3.5 to 4.5 and the computed PGA value ranges from 0.004g to 0.106g. However, the controlling earthquake is found to be the one resulting from the fault 24 in both the methods. It is not logical to use the unrealistic PGA values obtained from the empirical method; therefore, PGA values calculated using the observed seismicity is selected for further analysis. Thus it can be established from DSHA that a hazard of 0.106 g (PGA) results from the controlling earthquake of M_W 4.4 and epicentral distance of 11 km from fault 24. This PGA has been used as an input acceleration after suitable scaling of input acceleration time history and the same is used for the ground response analysis.

5 LOCAL SITE EFFECTS

Local site conditions profoundly influence most of the important characteristics mainly the acceleration amplitude and frequency characteristics of ground motion during an earthquake. In the present study one dimensional ground response analysis (GRA) was carried out to estimate the ground motion parameters at the surface. Ground response analysis can be carried out using linear, equivalent linear or non-linear methods. The equivalent linear approach is adopted to perform the site-specific ground response analysis at various locations of the Chennai city. A computer program SHAKE91, which uses equivalent linear approximation for layered soils, is used for computing the seismic response of horizontally layered soil deposits of the study area. The input data includes the acceleration time history at bedrock, the shear wave velocity or maximum shear modulus, modulus reduction and damping curves of soil layers.

5.1 Selection of input acceleration time history

The important characteristics of the bed rock motion include acceleration amplitude, frequency and duration content. From DSHA, the peak ground acceleration was estimated as 0.106 g. The frequency content and bracketed duration for the study area was estimated as 5 Hz and 3 s based on the charts proposed by Seed et al (1969) and Chang and Krinitzsky (1977) respectively. In the present study, 1952 Kern County (Taft) earthquake recorded ground motion time history with M_w 7.7 and PGA of 0.185g was selected and scaled to the PGA, frequency content and bracketed duration of the study region considered. The developed scaled time history of bed rock motion is shown in Fig. 5.

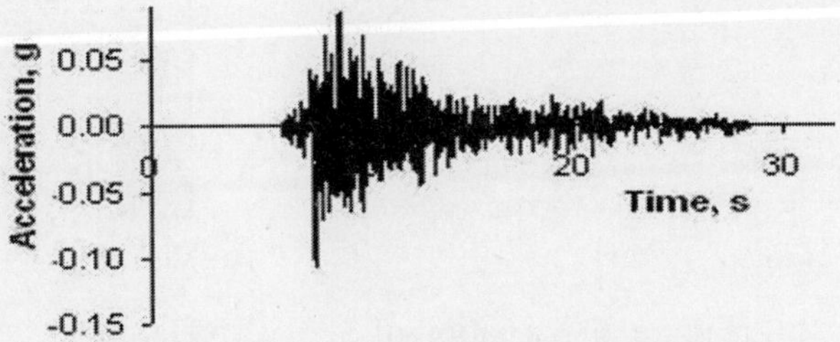

Figure 5. Input Acceleration Time History

5.2 Determination of shear wave velocity

A large and reliable borehole data for a number of sites in and around the Chennai city has been obtained from the reputed geotechnical agencies. The details of the soil layers and their engineering properties were assessed from the compiled data. The SPT N-values obtained in the field were corrected for various factors: overburden pressure, hammer energy, borehole diameter, rod length and fines content. The shear wave velocity, V_s was estimated from the corrected SPT-N values ($(N_1)_{60}$) using the following empirical equations (Japanese Road Association 1980):

$$V_s\ (\text{m/sec}) = 100\ (N_1)_{60}^{1/3} \quad \text{(for clay)} \tag{6}$$

$$V_s\ (\text{m/sec}) = 80\ (N_1)_{60}^{1/3} \quad \text{(for sand)} \tag{7}$$

In order to verify the shear wave velocity obtained from the empirical correlations, MASW test has been conducted at selected sites. The result of MASW tests carried out at a typical site near West Mogappair area located in the North West region of Chennai city is described below. The subsoil at the site consists of four layers. The top layer of about 6 m thickness consists of medium stiff to stiff silty clay with SPT N values varying from 6 to 18. This layer is followed by loose to medium dense sand deposit of 7 m thickness with N values increase from 4 to 16 with depth. The subsequent layer consists of soft to hard clay deposit of 14 m thickness with N value varies from 5 to 78. This layer is followed by dense sand deposit of 3 m thickness with N value >100. The soil profile with the variation of SPT-N values with depth for the site is shown in Fig. 6.

Geometrics make 24 channels Geode seismic recorder with SGOS operating software is used to carry out MASW tests. The vertical geophones of 4.5 Hz (24 Nos.) with 2 m interval are used to receive the wavefield generated by the active source of 5 kg sledgehammer. The sledgehammer was discharged at an offset of 5, 10 and 15 m from the first geophone. The acquired surface wave data were processed using Surfseis software to develop 1-D shear wave velocity profile which is shown in Fig. 7. The measured shear wave velocity from MASW test and the calculated shear wave velocity using empirical relation are presented in Table 3. It can be observed from the table that the estimated shear wave velocity is comparable with the measured shear wave velocity for various soil layers.

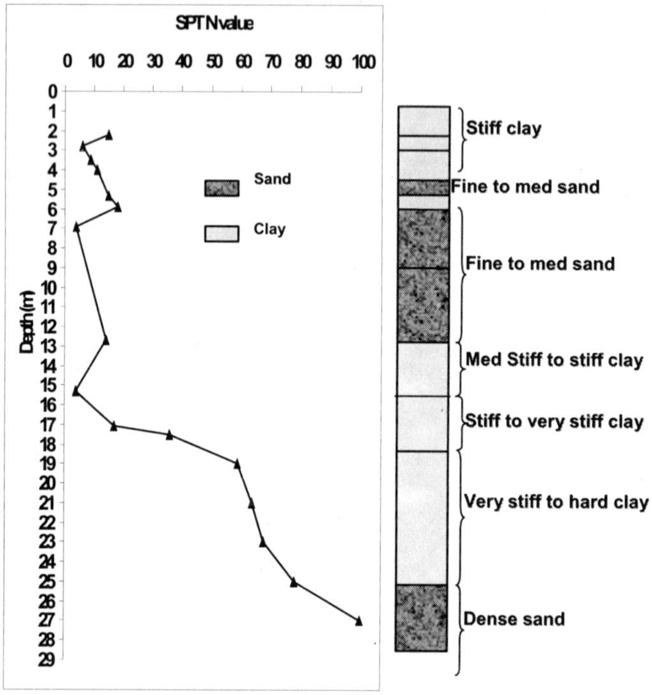

Figure 6. Variation of SPT with Depth

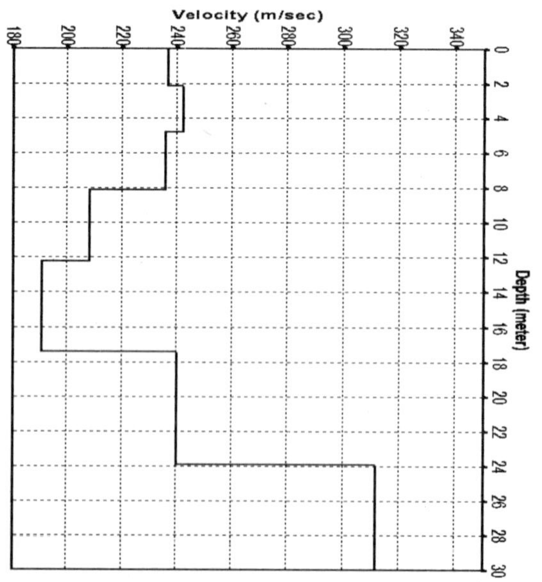

Figure 7. Variation of Shear Wave Velocity with Depth

Table 3. Shear wave velocity of soil obtained from MASW and the corrected SPT-N value

Soil description	*V_S from MASW (m/s)*	*V_S from SPT-N (m/s)*
Stiff clay	230	190
Fine to med sand	240	180
Stiff clay	230	200
Fine to med sand	200	180
Med Stiff to stiff clay	190	170
Stiff to very stiff clay	240	330
Very stiff to hard clay	290	400
Dense sand	320	460

6 GROUND RESPONSE ANALYSIS

One Dimensional Ground Response Analysis (GRA) was carried out at various locations of the Chennai city using SHAKE91 program which is based on the equivalent linear approach. Fig. 8 shows the various locations in Chennai where the ground response analysis has been carried out. The output of the ground response analysis includes the soil response and structural response in terms of surface acceleration and spectral acceleration. The amplification factor i.e. ratio of the Fourier amplitude at the surface to the Fourier amplitude at bedrock level at different frequencies obtained from GRA carried out for three typical sites, namely, the rocky, sandy and clayey sites located in the area of Perungudi, Vysarpadi and Adyar respectively are given in Fig. 9. It can be observed from the figure that the amplification occurs at low frequency (about 1.2 Hz) for the clayey site, mid-range frequency (about 4 Hz) for the sandy site and higher frequency (about 6.5 Hz) for the rocky site. Unlike other sites, the second peak for the rocky site occurs at much higher frequencies. The amplification of the clayey site is about 1.5 times higher than the rocky and sandy site due to the presence of thick deposits of low shear wave velocity. The predominant period, Tp, defined as the period for which the maximum soil amplifications occur is obtained from GRA for various sites. The predominant period is found to be 0.25s, 0.83s and 0.15s for sandy, clayey and rocky site respectively. The predominant period contour for the entire city is shown in Fig. 10. The fundamental or natural period, T_1, corresponding to the first mode of vibration of the soil system, is used by several authors as microzoning parameter. However, it should be considered that Tp is more significant parameter than T_1 as far as their relation to damage in most standard dwelling buildings (Roca et al. 2006).

The response spectra obtained from the ground response analysis carried out for the three typical sites above considered are given in Fig. 11. It can be noticed from the figure that the spectral acceleration values for the clayey site are high at wide range of periods up to 0.8s in contrast to the rocky sites where high spectral acceleration values occurs at narrow range of low periods.

Response spectra obtained for various sites are compared with the response spectra obtained for the input bedrock. The Spectral Acceleration Ratio (SAR) i.e., ratio of spectral acceleration at the surface to the bedrock contours for a typical period of 1.0 s (flexible structures) is plotted (Fig. 12). It can be observed from Fig. 12 that there is a low level of amplification for all sites except the clayey sites where significant amplification occurs due to period lengthening effect.

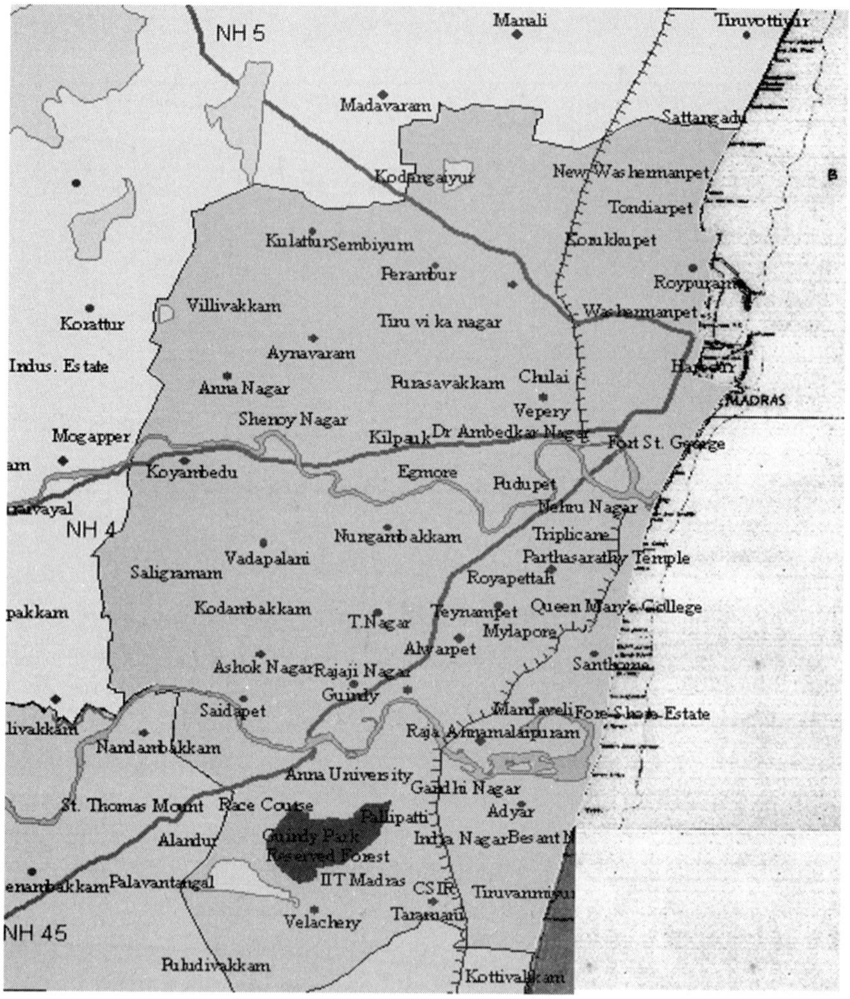

Figure 8. Sites where Ground Response Analysis done

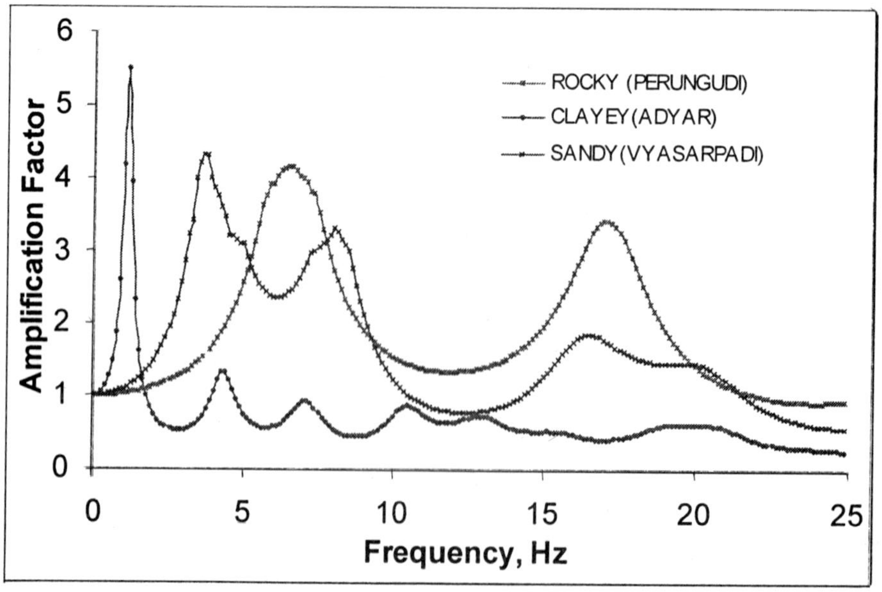

Figure 9. Frequency Response Function of Selected Regions of Chennai

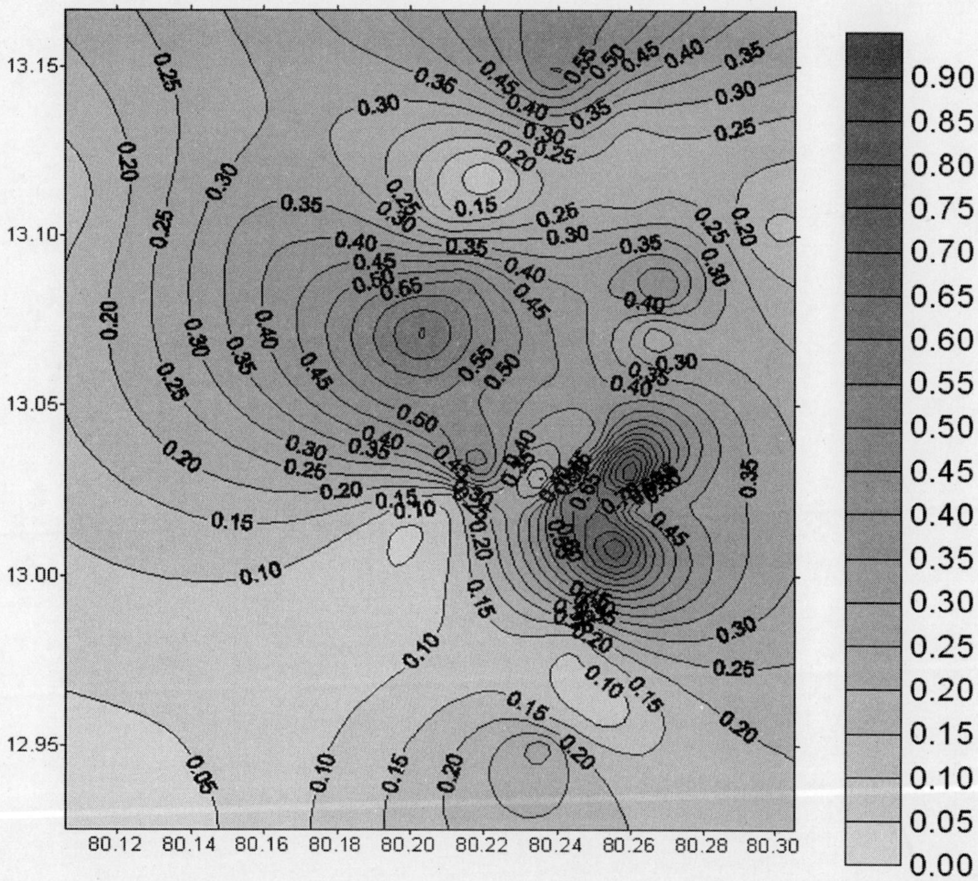

Figure 10. Predominant period contours

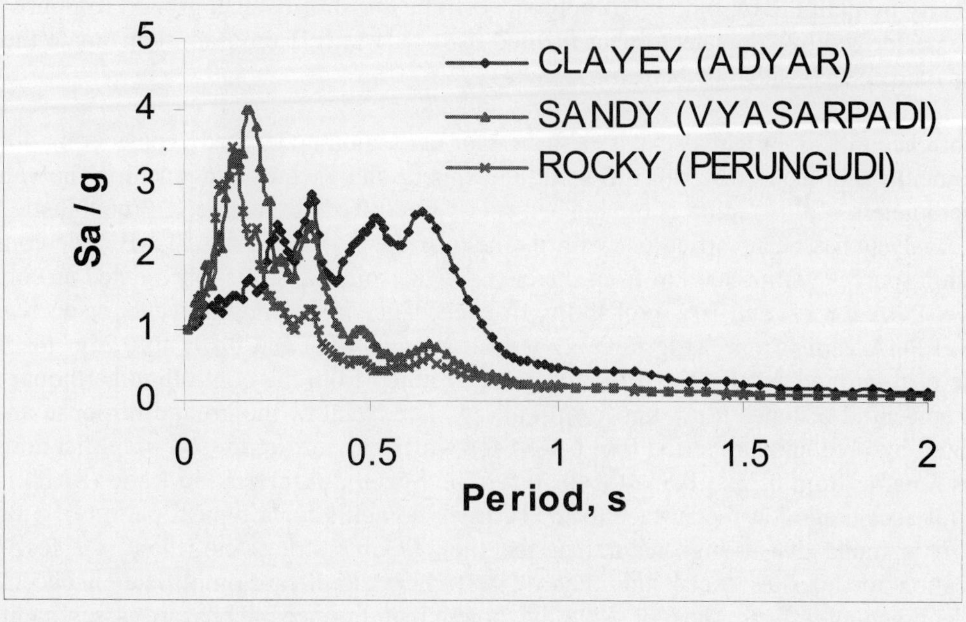

Figure 11. Response Spectra for the Selected Regions of Chennai

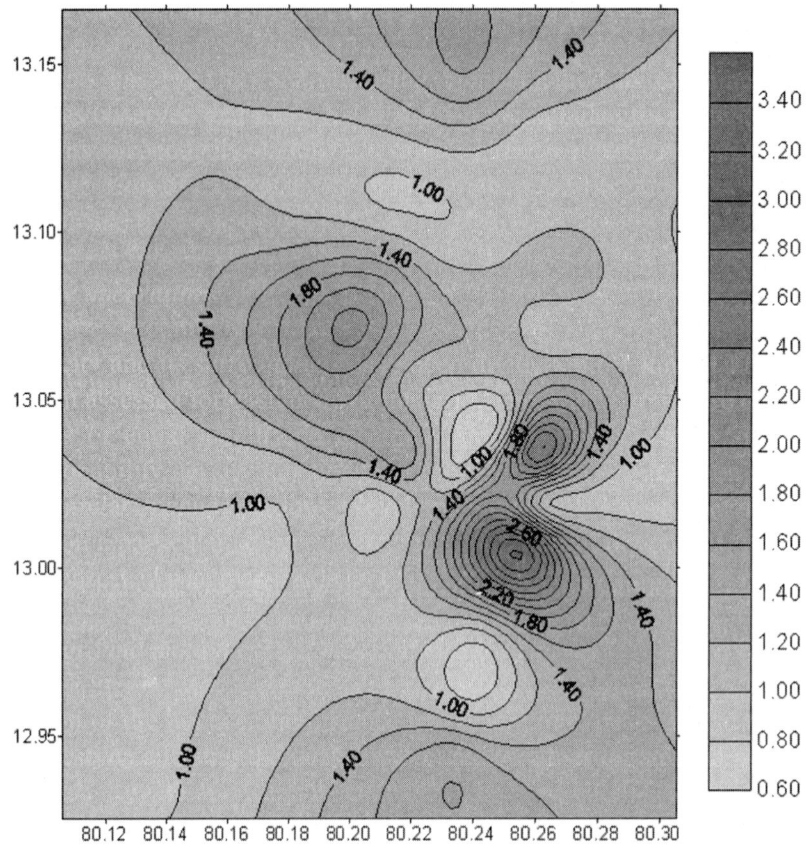

Figure 12. SAR contours for 1.0s

7 SUMMARY AND CONCLUSIONS

The seismic hazard analysis was carried out for the establishment of PGA at bedrock level for the Chennai city by both PSHA and DSHA approaches. The one dimensional ground response analysis was also performed by equivalent linear method using SHAKE91 program to estimate the ground motion parameters considering *local site effects.*

The recurrence rate of earthquakes is established for the region under study considering 400 events with moment magnitude greater than 3 using Gutenberg and Richter relationship. The values of a and b parameters of this region is established as 1.53 and 0.65 respectively. Probabilistic Seismic Hazard Analysis has been carried out with the help of the developed MATLAB code considering the seismic sources within 300 km from the city. The seismic hazard is represented as contours of rock level PGA for 2% and 10% probability of exceedance in 50 and 100 years respectively. The rock level PGA varies from 0.14g to 0.32g (for 2500 years RP) and 0.10g to 0.25g (for 500 years RP). The peak ground acceleration of 0.106 g has resulted from the controlling earthquake of M_w 4.4 and epicentral distance of 11 km from fault 24. The result of the ground response analysis is represented by predominant period (the period for which the maximum soil amplifications occur) contours ranging from 0.25s - 0.83s for the city. The Spectral Acceleration Ratio (SAR) i.e., ratio of spectral acceleration at the surface to the bedrock contours for a typical period of 1.0s is also plotted. It is found that at high natural period (i.e., flexible structures), there is a low level of amplification for all sites except the clayey sites where significant amplification occurs due to period lengthening effect. Thus, it is to be noted that the seismic hazard assessment has to incorporate the local site effects as realistically as possible in the analysis procedure in order to place reliability on the estimated peak ground acceleration.

ACKNOWLEDGEMENTS

The author likes to thank Dr. G.R. Dodagoudar, Assistant Professor, Ms. R. Uma Maheswari, Research Scholar, Department of Civil Engineering, IIT Madras for their valuable input for the preparation of this paper. The author gratefully acknowledge the Department of Science and Technology, Government of India for funding the sponsored research project entitled "Seismic Site Characterization and Site Amplification Studies for Chennai City" (DST No: 23(497)/ SU/2004 Dt. 09/08/2005). The author extends his thanks to M/s. Geotechnical Solutions, Chennai for providing the borehole data of various parts of the city.

REFERENCES

1. Boominathan, A., G.R. Dodagoudar, A. Suganthi, and R. Uma Maheswari (2007) Seismic Hazard Assesment Considering Local Site Effects for Microzonation Studies for Chennai City", Proceedings of Workshop on Microzonation, IISC Bangalore, Interlie Publishing, pp. 94-104.
2. Chandra, U. (1977) Earthquakes of Peninsular India – A Seismotectonic Study, Bulletin of Seismological Society of America, **67**, 1387-1413.
3. Chang, F.K. and E.L. Krinitzsky (1977) Duration, spectral content and predominant period of strong motion earthquake records from Western United States, Miscellaneous paper 5-**73**-1, U.S. Army Corps Engineers Waterways Experiment Station Vicksburg, Missipppi.
4. Cornell, C.A. (1968) Engineering Seismic Risk Analysis, Bulletin of Seismological Society of America, **58**, 1, 583–1,606.
5. Gangrade, B.K. and S.K. Arora (2000) Seismicity of the Indian peninsular shield from regional earthquake data; Pure and Applied Geophysics, **157**, 683 – 1705.
6. Ghosh, A.K. (1994) Design basis ground motion parameters for PFBR site, Kalpakkam, BARC, Mumbai.
7. Iyengar, R.N. and S.T.G. Raghukanth (2004) Attenuation of Strong Ground Motion in Peninsular India, Seismological Research Letters, **75**, 530-540.
8. Japan Road Association (1980) Specification and Interpretation of Bridge Design for Highway – Part V: Resilient Design, 14-15.
9. Kijko, A. and G. Graham (1998) Parametric-historic Procedure for Probabilistic Seismic Hazard Analysis, Part I: Estimation of Maximum Regional Magnitude Mmax, Pure and Applied Geophysics, **152**, 413–442.
10. Mark, R.K. (1977) Application of linear statistical model of earthquake magnitude versus fault length in estimating maximum expectable earthquakes, Geology, **5**, 464–466.
11. Reiter, L. (1990), Earthquake Hazard Analysis: Issues and Insights, Columbia Univ. Press, New York.
12. Roca, A., C.S. Oliveira, A. Ansal and Figueras (2006) Local site effects and Microzonation, In Assessing and Managing Earthquake Risk (Eds. Oliveria, C.S., A. Roca and X. Goula), Springer publishers, Netherlands, Chapter 4.
13. Seed, H.B, I.M. Idriss, and F.W. Kiefer (1969) Characteristics of rock motions during earthquakes; Journal of the Soil Mechanics and Foundations Divisions, ASCE, 95, 5, 1199 – 1218.
14. Seismotectonic Atlas of India (2000) Geological Survey of India, New Delhi.
15. Shanker, D. and M.L. Sharma (1997) Statistical analysis of completeness of seismicity data of the Himalayas and its effect on earthquake hazard determination", Bulletin of Indian Society of Earthquake Technology, **34**, 3, Paper No.368, 159-170.
16. Sokolov, Y,V., C.H. Loh, and K.L. Wen (2001) Empirical Models for Site- and Region-Dependent Ground Motion Parameters in the Taepei Area: A Unified Approach, Earthquake Spectra, **17**, 2, 313-332.
17. Wells, D.L. and K.J. Coppersmith (1994) New Empirical Relationships among Magnitude, Rupture Length, Rupture Width, Rupture Area, and Surface Displacement, Bulletin of Seismological Society of America, **84**, 974 - 1002.

NEW STRATEGY AND TOOLS FOR MITIGATION OF LANDSLIDE DISASTERS

Ikuo Towhata

University of Tokyo, Japan, E-mail: towhata@geot.t.u-tokyo.ac.jp

1. INTRODUCTION

Slope failure due to heavy rainfall is one of the severest natural disasters which occur in many parts of the world. Because a landslide occurs without an evident warning, it could kill a number of residents if it occurred in a densely populated region. Although the safest countermeasure against landslide disasters is "not to live near an unstable slope", this idea scarcely makes sense. This is because people are living in their present place in order to cultivate land, run business near clients' place and maintain their life. Farmers cannot survive if they are forced to live at far distance from their field. Merchants cannot maintain their business if they have to live without customers. Thus, it is important that disaster mitigation policy should take care of the way for people to maintain their life.

It is of the primary importance to identify potentially unstable slopes. This procedure is possible for slopes which moved in the past and have landslide topography today (Fig.1). It is however important that many slope failures occur at places where no failure occurred in the past (Fig.2). Engineers therefore somehow identify potentially dangerous slopes, with some safety margin due to the aforementioned uncertainty, and determine the range of debris runoff where the risk of disaster is high. It has been pointed out that the current practice of landslide hazard assessment (Tables 1 and 2) relies on the topography, which supplies very important information on slope instability, but that little attention is paid to local soil conditions.

Fig.1. Cililin slope in Indonesia which moved in the past.

Fig. 2. Another slope failure in Cililin without previous preceding failure.

Table 1. Assessment of slope instability (MLIT method; translated by the author).

Issues	Remarks	Points	Weight
Symptom of landslide	Major distortion or cracks Minor distortion None	20 10 0	W_1
Landslide topography	Very clear Clear Not clear	10 6 2	
Geological structure	Fault Volcanic morphology Strata parallel to slope Strata normal to slope Intrusion of volcanic rock Cap rock structure Other	5 5 4 2 1 1 0	W_2
Age and Rock type	Paleozoic and Mesozoic rocks (crystalline schist and sedimentary rock) Tertiary sedimentary rock Greenstone, serpentinite Quaternary sedimentary rock Others (igneous rock etc.)	2 2 2 1 0	
Water spring	Yes No	3 0	
Any legend on landslide history	Yes No	20 0	W_3
Total	$W_1+W_2+W_3$	**≧40** A; **39 – 20** B; **19≧** C	

Table 2. Risk assessment of steep slope (version until 1987, MLIT; translated by the author).

Issue	Category	Point
Height of slope	> 10m	W_1=7
	< 10m	3
Slope angle	> 45 deg.	W_2=1
	< 45 deg.	0
Overhang	Yes	W_3=3
	No	0
Thickness of surface soil	> 0.5 m	W_4=1
	< 0.5 m	0
Water spring	Yes	W_5=1
	No	0

Issue	Category	Point	
Any instability nearby	Yes	W_6=3	
	No	0	
Good slope reinforcement	No	W_7=3	
	Yes	0	
Distortion in slope reinforcement	Yes	W_8=3	
	No	0	
Total point			
$W_1+W_2+W_3+W_4+W_5+W_6+W_7+W_8$			
	≧9	**8 – 6**	**≦5**
Risk	**A**	**B**	**C**

Note: the target of assessment is those slopes which is higher than 5 m and inclined by more than 30 degrees.

2. ON RECENT SLOPE FAILURES

The 2004 Niigata-Chuetsu earthquake triggered many slope failures. Fig.3 illustrates one of them where two failures occurred beside each other. It may be seen that, in the smaller one which involved only the surface weathered soil, the soil mass simply fell down to the bottom of a valley and stayed there. In contrast, the bigger one on the left-hand side involved soil mass of a greater depth, and the debris traveled in the horizontal direction after fall and stopped the river flow, forming a natural dam. Thus, the effects of size on the extent of damage are clear. It is therefore reasonable to state that risk assessment should pay attention to the size of unstable soil mass. Since this earthquake was preceded by heavy rainfall three days in advance, it has been supposed that water content in the failed soil mass was high.

An attempt was made (Mizuhashi et al., 2006) to improve the methods in Tables 1 and 2 by adding more information on local soils. Soils were collected from sites of rainfall-induced landslides which occurred in Japan in the years of 2004 and 2005. Fig.4 illustrates that current methods can identify slope failure risk to some extent. Particularly, the method in Table 1 may be able to assess the size of slope failure (volume of soil).

Fig. 3. Slope failures at Naranoki during 2004 Niigata-Chuetsu earthquake.

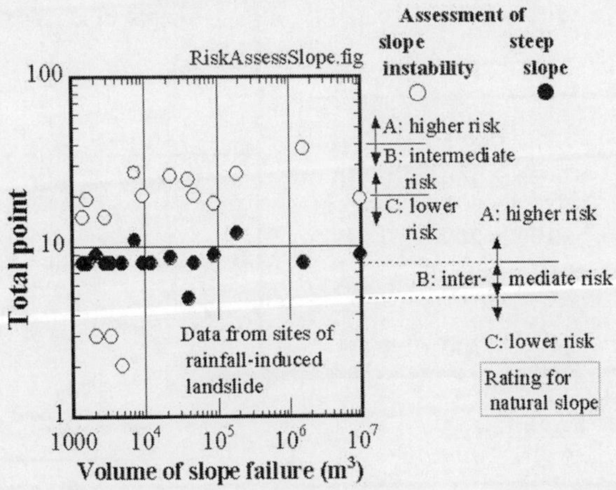

Fig. 4. Risk assessment by existing hazard assessment methods (Mizuhashi et al., 2006).

Another example of risk assessment, which concerns seismically-induced landslide, is shown in Table 3 which was proposed for practice by the Kanagawa Prefectural Governments. Apparently, this method does not take into local soil conditions. To improve this method, soil samples were collected from sites of seismic slope failures in the Niigata-Chuetsu region. Note that detailed soil investigation such as undisturbed soil sampling and triaxial undrained shear was not conducted because practice of risk assessment cannot afford these expensive measures. Only physical properties of soils were measured.

Fig. 5 compares the assessed risk point by the Kanagawa method with the volume of collapsed soil. Although the Joganji site is very exceptional, there is some correlation between the point and the volume. To improve this correlation such physical properties as swelling strain of reconstituted soil samples and the plasticity index were added to the point calculation. The swelling was taken into consideration because of the preceding heavy rainfall in the Niigata-Chuetsu earthquake. On the other hand, the plasticity index is certainly an easy parameter to measure. Accordingly, Figs. 6 and 7 give some improvement in the correlation, although the Joganji site remains exceptional.

Table 3. Risk assessment of slope failure during earthquakes (Kanagawa Prefectural Government, 1986).

Factor	Category	Weight	Factor	Category	Weight
Maximum surface	0 to 200	0.0	Hardness of	Soil	0.0
acceleration (Gal);	200 to 300	1.004	rock, W_4	Soft rock	0.169
W_1	300 to 400	2.306		Hard rock	0.191
	>400	2.754	Length of	No fault	0.0
Length of a	0 to 1000	0.0	faults (m)	0 to 200	0.238
contour line at mean	1000 to 1500	0.071	W_5	>200	0.710
elevation (m) W_2	1500 to 2000	0.320	Length of artificial	0 to 100	0.0
	>2000	0.696	slopes (m), W_6	100 to 200	0.539
Difference between	0 to 50	0.0		>200	0.845
highest site and	50 to 100	0.550	Vertical section,	(1)	0.0
lowest site (m), W_3	100 to 200	0.591	of slope, W_7	(2)	0.151
	200 to 300	0.814		(3)	0.184
	>300	1.431		(4)	0.207a

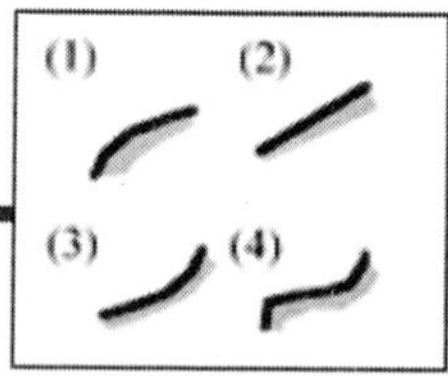

$W=W_1+W_2+W_3+W_4+W_5+W_6+W_7$	2.93	3.53	3.68	
Rank	A	B	C	D
Number of slope failures within 500m×500m grid	0	1 to 3	4 to 8	>9

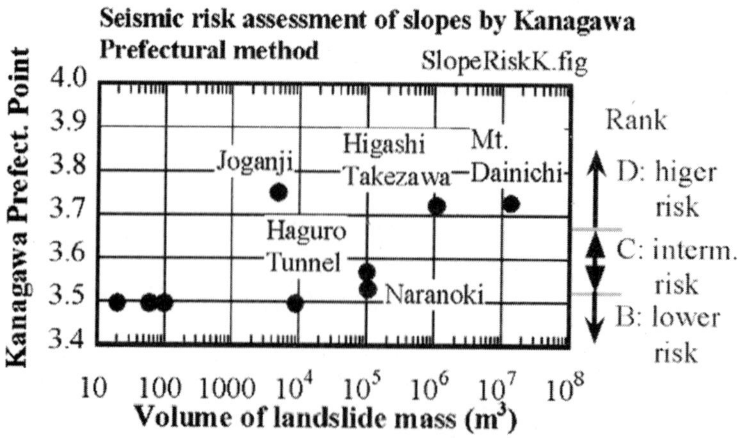

Fig. 5. Seismic hazard assessment of slopes by using Kanagawa Prefectural method.

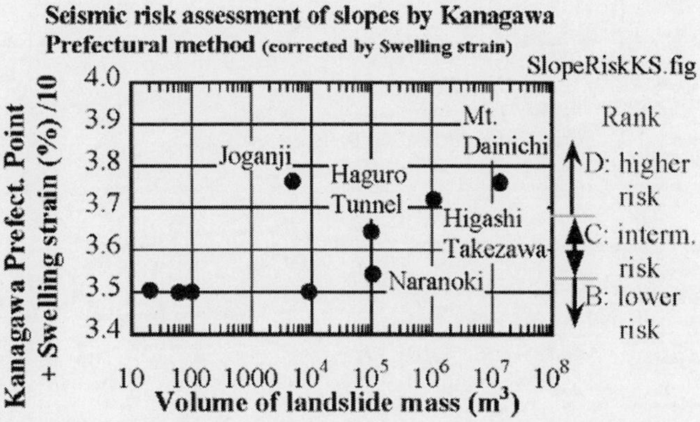

Fig. 6. Correction of Kanagawa Prefectural method by swelling strain.

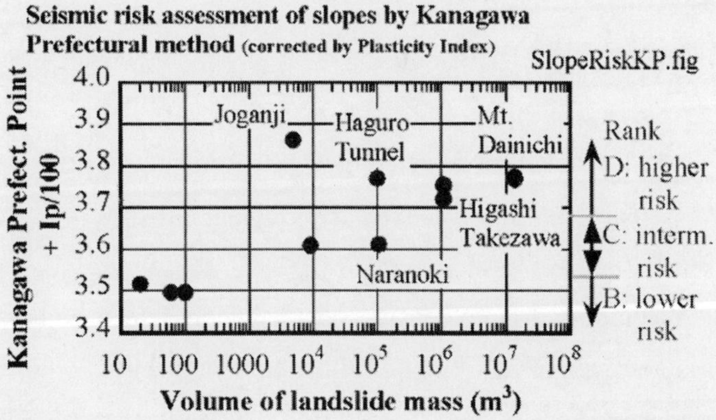

Fig. 7. Correction of Kanagawa Prefectural method by plasticity index (Ip).

3. EARLY WARNING AND QUICK EVACUATION

It was mentioned that people scarcely want to move to safer places, even if slope failure is considered likely in their present places. This is because they have jobs in the present place, while not in the place of relocation. In this regard, it is practical to develop tools with which an imminent slope failure is detected and early warning is issued for quick evacuation.

Fig. 8. Slope failure test in natural slope of Kaba-san Mountain.

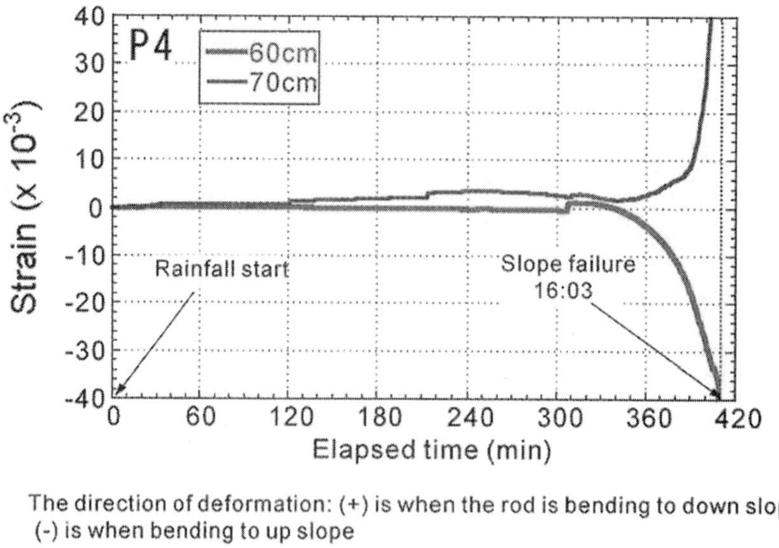

Fig. 9. Minor soil displacement detected by pipe strain gauges prior to final failure (Kaba-san Mountain test by Ochiai et al. 2004).

The author's group is developing an inexpensive sensor which detects minor displacement of soil during rainfall which is a precursor of ultimate failure of s slope. Fig.8 shows a site of controlled slope failure which was subjected to artificial rainfall. The soil strain (Fig.9), measured by pipe strain gauges, shows that minor displacement started approximately two hours before failure. Therefore, it is supposed that displacement of soil is a good index of failure.

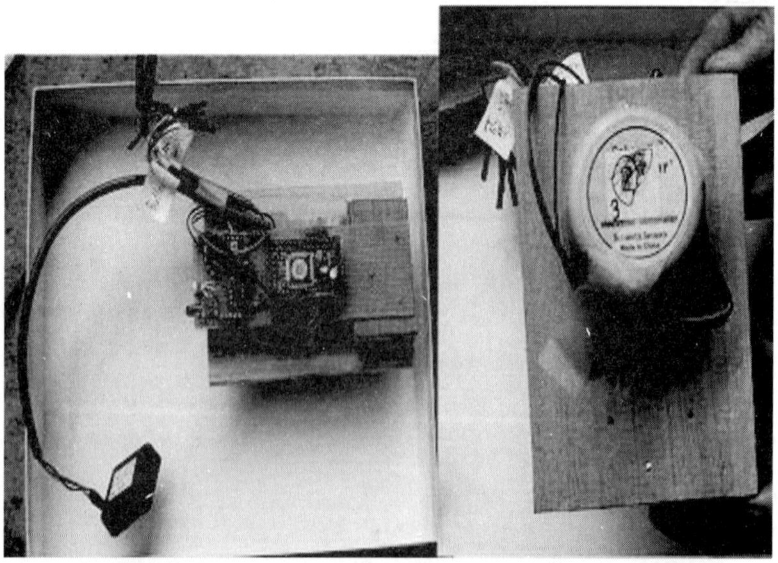

Fig. 10. First generation of displacement sensor (hand-made).

Fig. 11. Model embankment after rainfall-induced failure (Public Works Research Institute).

Fig. 10 illustrates a hand-made transducer which is embedded in a slope and detects minor displacement. The data is sent by wireless to an internet system. Four of them were embedded in a 3.6-meter model embankment which was subjected to artificial rain fall (Fig. 11). Similarly to Fig.9, minor displacement was recorded to start at 30 minutes before the failure. It is important that moisture content in soil was recorded by the same instruments and that the some of the moisture records decreased due to open cracks and drainage into them. Therefore, moisture content is not an appropriate target of monitoring, although soil failure is certainly induced by increased moisture content.

4. NEED FOR FURTHER TECHNICAL DEVELOPMENT

The equipment of the first generation (Fig.10) had a problem of energy efficiency. Because the instrument is operated by battery in the field, low energy efficiency leads to loss of power within a short time. More efficient electronic circuit was, therefore, designed and fabricated in which the power is activated only once a minute. This made four batteries sufficient for more than one year of field operation. Fig.12 illustrates the second generation of the equipment. The fabrication cost of this device is approximately 300 US $ or less. In future, it is interesting to install acceleration transducer, thermal transducer, and even a video camera for monitoring different kinds of natural disasters etc. The new circuit is now going to be installed in water-proof container and will be installed at a variety of sites so that its long-term durability and operability will be validated.

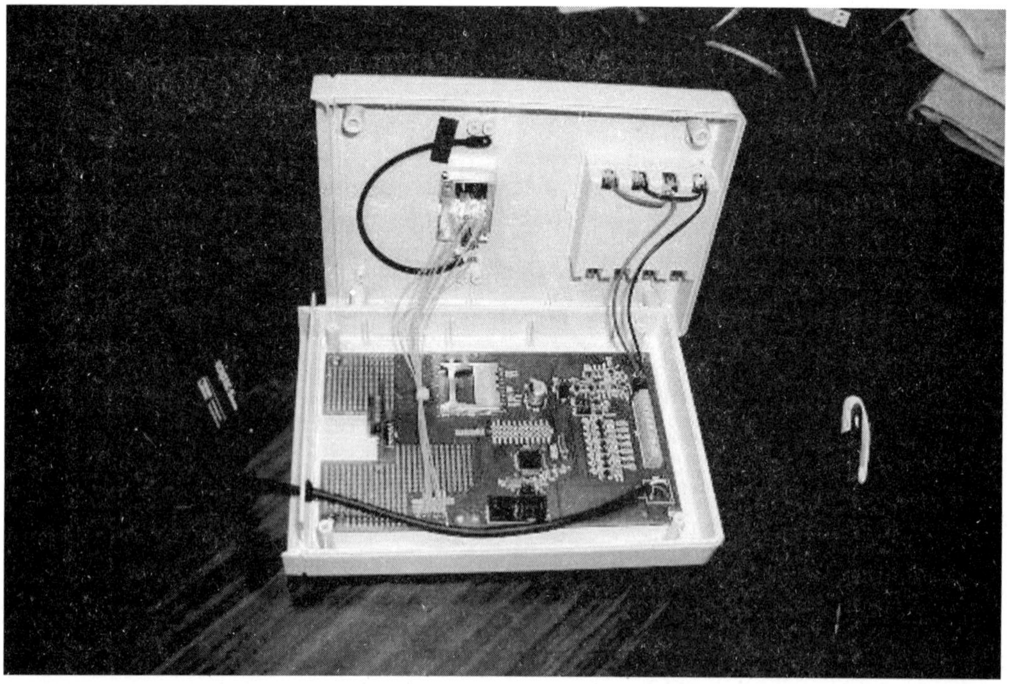

Fig.12 Second generation of field displacement transducer.

5. CONCLUDING REMARKS

Efforts have been made to mitigate landslide-induced risk. However, more attention should be paid to the size of the unstable soil mass and the effect of local soil characteristics. Although many risk specialists talk about relocation of local people to safer places, this idea is not welcome by the people. As an alternative and more practical measure, the author is investigating an inexpensive tool which monitors minor soil displacement which is a precursor of the ultimate slope failure. It is expected that this tool will be installed in many potentially unstable slopes and make early warning which prompts evacuation of endangered people.

LIST OF REFERENCES

1. Kanagawa Prefectural Government (1986) Prediction of seismic damage in Kanagawa Prefecture, pp.13-63 (in Japanese).
2. Mizuhashi, M.., Towhata, I., Sato, J., and Tsujimura, T. (2006) Examination of slope hazard assessment by using case studies of earthquake- and rainfall-induced landslides, Soils and Foundations, Vol.46, No.6, pp.843-853.
3. Ochiai, H., Okada, Y., Furuya, G., Okura, Y., Matsui, T., Sammori, T., Terajima, T., and Sassa, K. (2004) A fluidized landslide on a natural slope by artificial rainfall, Landslides, Vol.1, No.3, pp.211-219.

VERIFICATION OF PUSH-OVER ANALYSIS METHOD WITH STATIC LOAD TESTING

N. Lakshmanan[1], K.Muthumani[2], G. V. Rama Rao[2], N. Gopalakrishnan[2] and G. R. Reddy[3]
[1]- Director, [2]-Scientist, Structural Engineering Research Centre, CSIR Campus, Taramani, Chennai-600113.
[3]-Scientist, BARC, Mumbai.

ABSTRACT

Push-over analysis, a step ahead of linear response spectrum analysis, is a handy tool for engineers to predict the seismic performance of existing RC frames and for performance based design of new structures. Push-over analysis involves monotonic application of lateral load or displacement in steps, with progressive refining of the stiffness matrix of the system, such that ultimate load and ultimate lateral displacement of the structure can be captured with reasonable accuracy. Paper presents a brief description of the seismic evaluation procedure through push-over analysis. A brief description of an experimental investigation on the push-over test of a three-storied reinforced concrete model frame is presented. Test results are compared with analysis results of SAP-2000, with default hinge properties and also with modifications. Suitable future directions are indicated.

1. INTRODUCTION

Push-over analysis towards performance based seismic design (PBSD) is an evolving methodology with a goal to design and construct structures, with predictable seismic performance. Shortly, performance based earthquake engineering and design will become the standard method for design and construction of earthquake resistant structures. The manual on "Seismic evaluation and retrofit of concrete buildings", published by American Technology Council, ATC-40 is a popular design tool for engineers all over the world for PBSD. This document provides a practical guide to the entire evaluation and retrofit process using performance based objectives. However, the document requires adaptations and modifications to suit Indian conditions.

2. METHODOLOGY OF PUSH-OVER ANALYSIS

Push-over analysis is nothing but a nonlinear static analysis for a reinforced concrete framed structures subjected to lateral loading. The gravity loads are applied and then lateral loading is applied first in x-direction starting at the end of gravity push and next in y-direction again starting at the end of gravity push (IDARC, SAP 2000). The concept of plastic hinge is extremely important in the nonlinear analysis. While a concrete element undergoes large deformation in the post-yield stage, it is assumed that all the deformation is assumed to take place at a point called "plastic hinge", which has approximately a length of the order of effective depth (also called as plastic hinge length, l_d). The rotation capacity of a plastic hinge 'θ' is taken as $l_d(\phi_u - \phi_y)$. A similar approach can be used for obtaining rotation capacity of columns under axial force and bending moment in two directions. Similar plastic hinges with limit capacities on deformation can be defined for all six degrees of freedom, namely, axial force, transverse shear forces in X- and Y-direction, moments about Y- and Z-axis, and torsion (moment about X-axis). More details on evaluation of ductility, energy absorption, damage modeling, and detailing are available elsewhere (Lakshmanan, 2003, 2005). A typical response at a plastic hinge may be as given in Fig.1. Point A is the origin. B is the point of yielding. BC represents strain hardening region. C is the point corresponding to maximum force. DE is the post-failure capacity region. On the frame structure the analyst identifies possible locations for plastic hinge formation from his experience.

Mathematically nonlinear static analysis does not lead to a unique solution. Small changes in properties or sequence of loading can lead to large variation in nonlinear response.

The pushover analysis may be carried out using force control or deformation control. In the first option, the structure is subjected to an incremental distribution of lateral force and incremental displacements are calculated. In the second option, the structure is subjected to a deformation profile, and lateral forces needed to generate the displacement are computed. Since deformation profile is unknown, the first option is commonly used. For the displacement control the user specifies the target maximum displacement at the control point. In certain software, displacement control is not the same as applying displacement loading on the structure. Displacement control is simply used to measure the displacement that results from the applied loads and to adjust the magnitude of the loading in an attempt to reach a certain measured displacement value. The so-called displacement control in this case is essentially a modified form of force control. The force control strategy can have the following options:

i) Uniform distribution; ii) triangular distribution; iii) generalised power distribution; and iv) modal adaptive distribution with single or multiple mode participation.

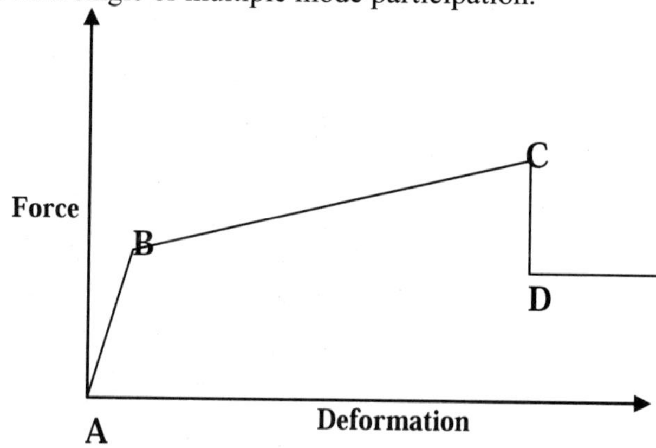

Fig-1. Idealised Force-Deformation Curve.

3. IDEALIZED SYSTEM PARAMETERS

The multi-linear force-displacement curve obtained after push-over analysis is idealized as bilinear as shown in Fig. 2 with positive or negative post-yield slope.

First let us consider the equivalent linearization method, which is the basis for capacity spectrum approach. The maximum response of the equivalent system is computed using

$$\ddot{x}_{eq} + 2\xi_{eq}.\omega_{eq}.\dot{x}_{eq} + \omega_{eq}^2.x_{eq} = -\ddot{x}_g \qquad [1]$$

where ξ_{eq} and ω_{eq} are the viscous damping ratio, and natural circular frequency of the equivalent linear system. For a bilinear system, the time period of the equivalent system T_e as compared to that of the original system T_0 is given by

$$T_{eq} = T_o\sqrt{\frac{\mu}{(1-\alpha)+\alpha\mu}} \qquad [2]$$

where μ = ductility ratio

and α = ratio of positive post-yield stiffness to original stiffness

The equivalent damping is given by

$$\xi_e = \xi_o + \frac{2}{\pi}\left[\frac{(1-\alpha)(\mu-1)}{\mu - \alpha\mu + \alpha\mu^2}\right] \qquad [3]$$

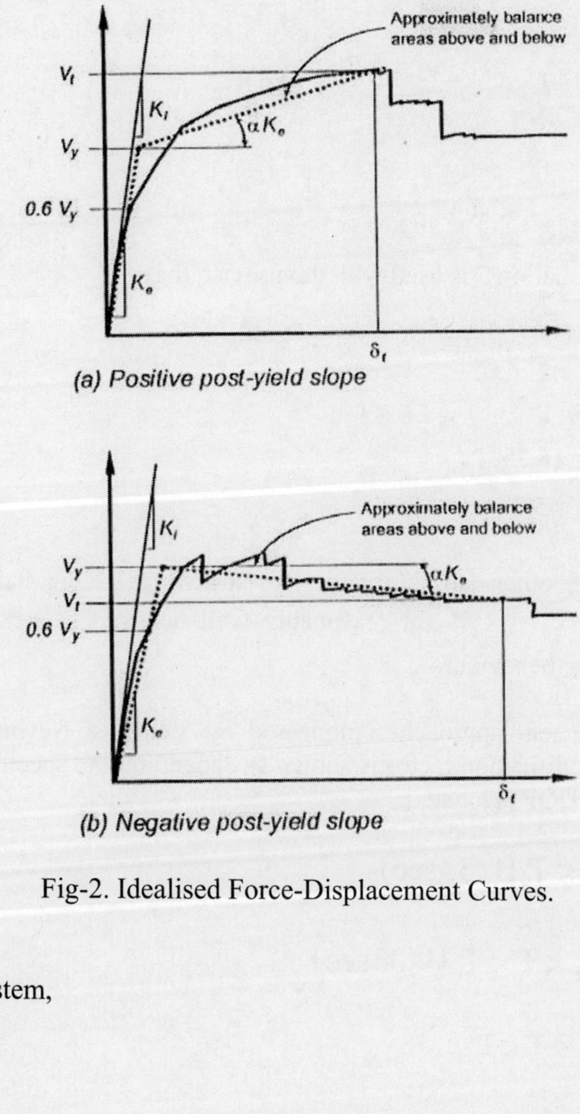

Fig-2. Idealised Force-Displacement Curves.

For an elasto-plastic system,

$$T_{eq} = \sqrt{\mu}\, T_o \qquad [4]$$

and $$\xi_{eq} = \xi_o + \frac{2}{\pi}\left(1 - \frac{1}{\mu}\right) \qquad [5]$$

Using Takeda hysteretic model and experimental investigations on model reinforced concrete frames, an empirical equation for computing equivalent damping was given by Gulkan and Sozen (1974).

$$\xi_{eq} = \xi_o + 0.2\left(1 - \frac{1}{\sqrt{\mu}}\right) \qquad [6]$$

Kowalsky (1994) used the secant stiffness at maximum deformation for defining the period together with hysteretic model given by Takeda and an unloading stiffness factor of 0.5.

$$\xi_{eq} = \xi_o + \frac{1}{\pi}\left[1 - \mu^n\left(\frac{1-\alpha}{\mu} + \alpha\right)\right] \qquad [7]$$

where n = 0 for steel structures
= 0.5 for concrete structures

For an elasto plastic system

$$\xi_{eq} = \xi_o + \frac{1}{\pi}\left(1 - \frac{1}{\sqrt{\mu}}\right) \qquad [8]$$

In ATC-40, ξ_{eq} given in Equation 3 is used with the proviso that

$$\xi_{eq} = \xi_o + \chi\xi_h \qquad [9]$$

where ξ_h is limited to 45%.

χ = 1.0 for ξ_h = 16.25%
= 0.77 for ξ_h = 45%

with linear interpolation for other damping values for structures having stable and full hysteretic behaviour. χ is 0.67 for ξ_h = 25% for reasonably well behaved systems and χ = 0.33 for systems with poor hysteretic behaviour.

In the displacement coefficient approaches pioneered by Velestos, Newmark, and Hall (1960, 1982), the displacement modification factor is shown to depend on the spectral region in which the period of vibration of the SDOF is located.

$$\begin{aligned} C &= \mu && \mathrm{T} < \mathrm{T_a}(1/33\,\mathrm{sec}) \\ &= \frac{\mu}{(2\mu-1)^{\beta}} && \mathrm{T_a} < \mathrm{T} < \mathrm{T_b}(1/33\,\mathrm{sec}) \\ &= \frac{\mu}{\sqrt{(2\mu-1)}} && \mathrm{T_b} < \mathrm{T} < \mathrm{T_c'} \\ &= \frac{T_c}{T} && \mathrm{T_c'} < \mathrm{T} < \mathrm{T_c}(0.57\,\mathrm{sec}) \\ &= 1.0 && \mathrm{T} > \mathrm{T_c} \end{aligned} \qquad [10]$$

where $T_c' = \dfrac{\sqrt{2\mu-1}}{\mu}.T_c$

and $\beta = \dfrac{\log(T/T_a)}{2\log(T_b/T_a)}$

Miranda (2001) has recently suggested that

$$C=\left[1+\left(\frac{1}{\mu}-1\right)e^{-12T\mu'}\right]^{-1} \quad [11]$$

where $\mu' = \mu^{-0.8}$.

More recently, Garcia and Miranda (2003) and Chopra (2003) have proposed expressions for C_R and $C\mu$ based on extensive investigation of field data and using regression analysis.

4. CAPACITY SPECTRUM APPROACH

The spectral coefficient given in IS:1893-2002 is multiplied by the zonation factor Z to obtain plot of spectral acceleration versus time period. The Acceleration Displacement Response Spectrum (ADRS) called also as Demand spectrum is obtained by plotting S_a Vs. S_d where $S_d = (T_1^2/4\pi^2)(Z)(S_a/g)$. The capacity spectrum is derived from the inelastic shear-roof displacement curve using:

$$S_a = \{(V/W).g\}/\alpha_1, \quad [12]$$

and $$S_d = \Delta_{roof}/(PF_{R,1}) \quad [13]$$

where α_1 = modal mass coefficient for first made

and $PF_{R,1}$ = modal participation factor for first mode.

Based on the multiplying factors given in IS:1893-2002 for obtaining seismic response acceleration coefficients and the variations of (S_a/g) with S_d for various damping ratios can be plotted (Fig.3). Performance point will be a point on the capacity spectrum where T_e and ξ_e would be satisfied. To satisfy life safety, the roof displacement shall be less than 1.2% of height for ordinary frames and 2% for special moment resisting frames. Life safety condition can be considered as design basis earthquake. For collapse prevention, the performance point should exist. In any case, the total drift shall not exceed 0.33 V/P, where V is the base shear and P is the gravity load for collapse prevention. Also, in a degrading nonlinear response, the limit of degradation on strength compared to the peak is 20%.

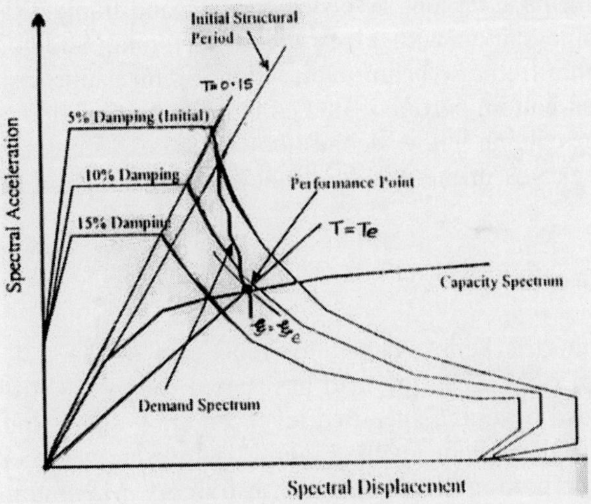

Fig-3. Demand and Capacity Spectra.

The above discussion clearly reveals that push-over analysis needs to be used with lot of caution and expert judgment. In fact, it could become a dangerous tool in the hands of practitioners who have less exposure to nonlinear/dynamic behaviour of concrete structures.

5. DESCRIPTION OF THE STRUCTURE TESTED

The structure tested is a two bay three storied reinforced concrete framed structure consisting of three frames (Frame-1 to Frame-3). The structure corresponds to a half-scale physical model. The centre to centre distance between the columns is 1500 mm and the height of each storey is 1800mm. The column section is of 150 X 200 mm size consisting of two 16 mm bars at top and bottom. The beam section is also of 150 X 200 mm size consisting of two 12 mm bars on either side. Column has 8 mm diameter ties spaced at 150 mm centres and beam has 8 mm diameter two legged stirrups at 100 mm centres. The slab is 100 mm thick with 12 mm bars at 200 mm centres with reinforcement for support moment provided separately. Figure 4 shows the elevation and typical plan view of the frame tested for verification of push-over analysis methodology. Figure 5 shows the photographic view of the tested structure. Suitable weights to simulate the effects of live loads are added at each floor level.

5.1 Loading Arrangement

Static loads are applied through hydraulic Enerpac jacks at three levels of the structure. The hydraulic jacks are connected at one end to the heavy-duty reaction wall and the other end is connected to the structure, through a calibrated load-cell and a heavy distribution beam. The ratio of the loading between each storey is kept in the ratio of squares of the heights, namely 1:4:9 for the first, second and the third storey levels. As loading cannot be applied simultaneously, the drop in loading is compensated by finer adjustments at each increment. Each increment of loading is kept as 5000 N at the top floor. This increment is made finer in the final stages. For the third floor, the load is applied through the two corner columns, whereas for the second and first floors, the load is applied on the middle column. Through this arrangement and with the loading ratio of 1:4:9, each frame is loaded equally, preventing flexing of diaphragm slabs.

5.2 Instrumentation for the Test

The instrumentation consist of displacement (using LVDT and theodolite) at three levels, strain gauges pasted on the reinforcement bars at potential hinge locations, surface mounted strain gauges at critical locations and tilt meters at floor levels. The model is elaborately instrumented and measurements were taken at every load step. LVDTs are kept against a stationery steel frame for measuring the displacements at various levels on frame-1 and frame-2. Deformation of frame-3 is monitored by a theodolite station with a resolution of 1 mm. Strain gauges are pasted on the column re-bars at 100 mm from the beam intersection. Strain gauge instrumentation on re-bars is carried out at the top and bottom portion of the column in the first floor for frames 1 and 2 and in the bottom portion of the column in the second floor. Inter-connecting beams at these locations are also pasted with strain gauges in the pre-dominant bending direction at 100 mm away from the column intersection.

5.3 Discussion of Test Results

Even though the load ratio is kept constant for most part of the test, the frame could not be controlled under ultimate stage when the load has started dropping. Stroke of the testing actuator also could not be extended beyond a specified level. Figure 6 shows the lateral load deformation behaviour of the frame at three floor levels. Figure 7 shows the variation of strain values against base shear, for the bottom portion of the columns in frame-1. Maximum displacement attained at each floor level is 95, 190 and 252 mm. Corresponding un-recovered displacement values after removal of load are 55, 110 and 150 mm. It may be seen that a near straight line deflected profile at ultimate stage/ maximum displacement stage indicates the bottom column hinges at foundation level. Maximum strain values are recorded at the tension side of the end-columns and middle columns with strain values reaching up to 42000 με at the last load step. This corresponds roughly to a curvature of $3X10^{-4}$ rad/mm and for a plastic hinge length of d/2 (100 mm), the maximum plastic rotation is roughly $3X10^{-2}$ radians or 1.7^0.

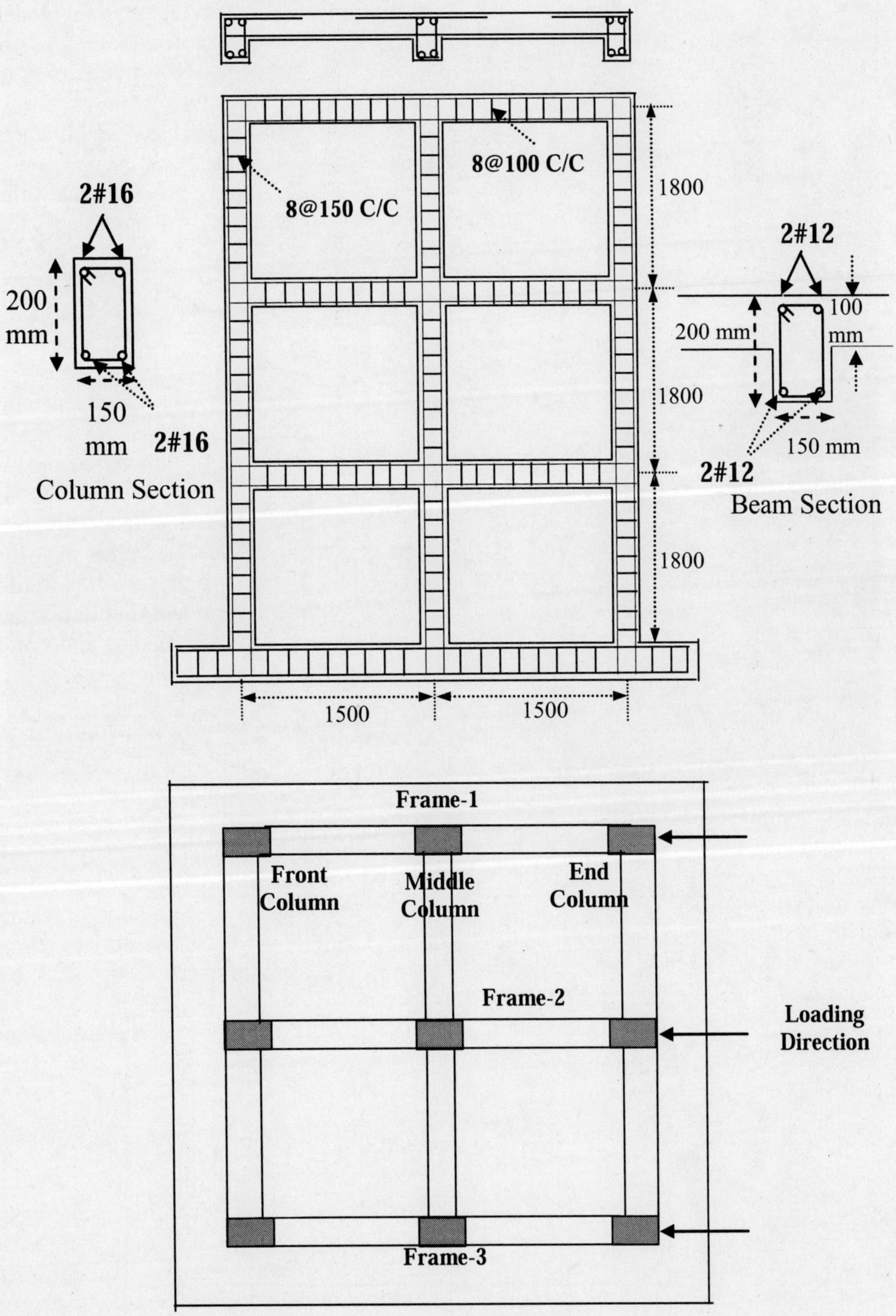

Fig-4. Elevation and Typical Plan View of the Frame Tested for Verification of Push-over Analysis.

Fig-5. Typical View of the Frame.

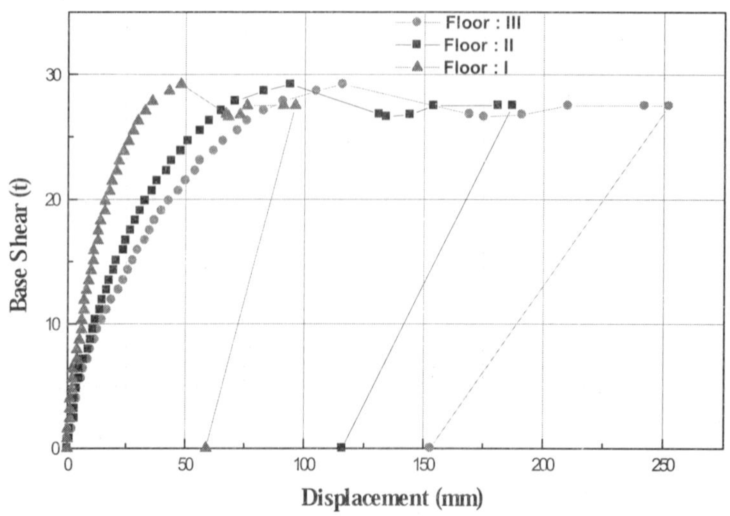

Fig-6. Displacement at Three Floor Levels of the Frame.

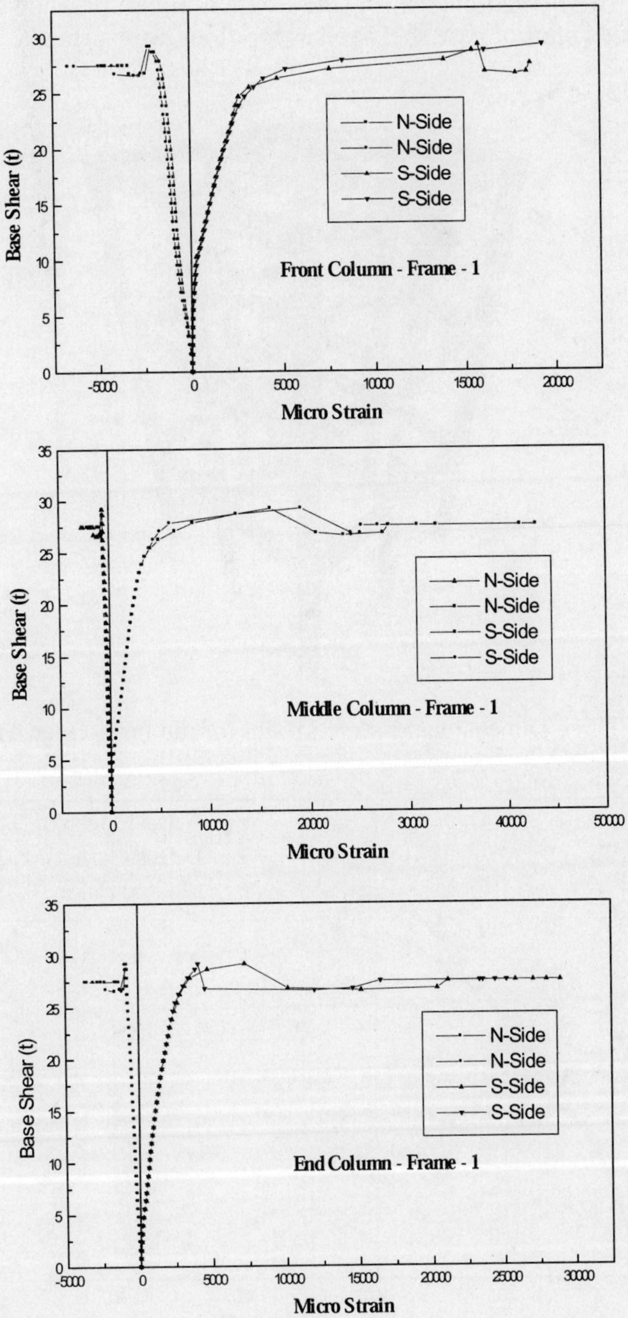

Fig-7. Strain Variation on the Reinforcement Bars at the Bottom Portion of Column for Frame-1.

6. PUSH-OVER ANALYSIS

Push-over analysis of the frame is carried out using SAP-2000 software. Initially a gravity load analysis is carried out for the applied loads as well as for the self-weight of the system. Stress resultants arrived during gravity analysis is super-imposed on the results of the lateral load analysis at the push-over stage. Push-over analysis is carried out for two conditions

(a) Using default hinge properties available in SAP-2000
(b) Modified hinge properties with externally computed ultimate load and rotation values suggested by ATC-40.

Figure 8 shows the SAP-2000 model of the frame. Figure 9 shows the analytically and experimentally evaluated plots of base shear versus top floor displacement.

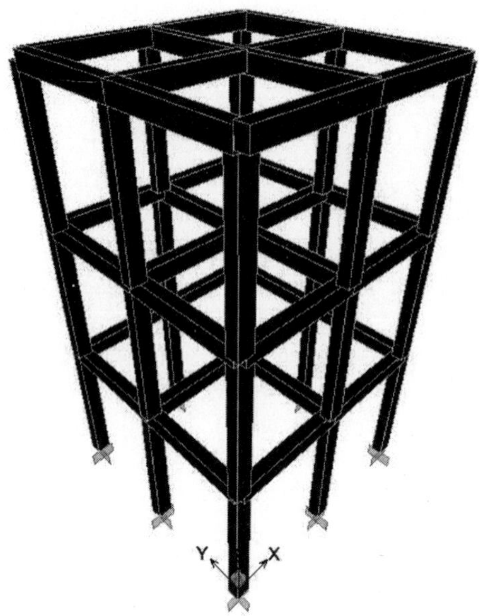

Fig-8. Three Dimensional Frame Model for the Push-Over Analysis.

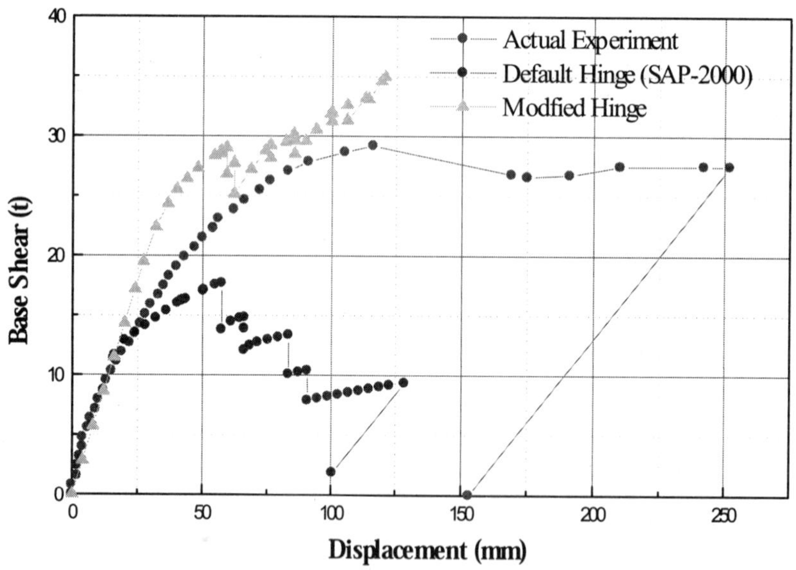

Fig-9 Comparison of Experimental results with Analysis.

An interesting method proposed by Midori Karva et al. has been used to convert the lateral force deformation characteristics of the frame into ADRS format using the following set of expressions:

$$S_a = \frac{\delta^T M \delta}{\left[\delta^T M 1\right]^2} Q_b$$

$$\omega_1^2 = \frac{\{\delta\}^T \{f\}}{\{\delta\}^T M \{\delta\}} \qquad [14]$$

$$S_d = \frac{S_a}{\omega_1^2}$$

Above equations are used to translate the force deformation characteristics to ADRS format and is shown in Figure 12.

It is seen from Figure 9 that default hinge properties cannot be adopted, even though section details, areas of main steel reinforcement, concrete cover and steel yield stress are input to the program. Neither the load nor the ultimate displacement values are correctly predicted. Even though, there is some agreement in the prediction of the maximum load, when default hinges are modified with user-defined properties, further work needs to be done to back-work the hinge properties, that would have given rise to this lateral load characteristics.

7. MISMATCH OF APPLIED LATERAL LOAD PROFILE WITH RESULTING DISPLACEMENT

The essential pre-requisite for simulating the dynamic action of a seismic event through equivalent static lateral loads, is through application of a static force profile, which should match with the inertial load profile. This means that for a uniform mass distribution, displacement and applied force profiles should match. Towards observing this effect, both for experimental and analytical curves, applied force and displacement profiles, normalised with reference to the third floor force and displacements are plotted in Figures 10 and 11 (for experiment and analysis respectively). Whereas the force profile is 1: 0.444: 0.111, displacement profile varies as 1: 0.77: 0.4. Hence more force had been applied in the experiment at the top, than what is actually required. However, this ratio remains nearly constant throughout the experiment as seen from Figure 10. The trend for the analysis is such that, for lower displacement values (< 60 mm), the ratio is 1: 0.67: 0.33. For Displacement values at 80 mm, the ratio changes to 1: 0.76: 0.62 and at 100 mm, the values are 1:0.87:0.72. In essence, analysis predicts the displacement as nearly constant values, along the height at lateral stages, from an initial linear displacement profile. However, in experiment, the linear profile nearly remains the same throughout.

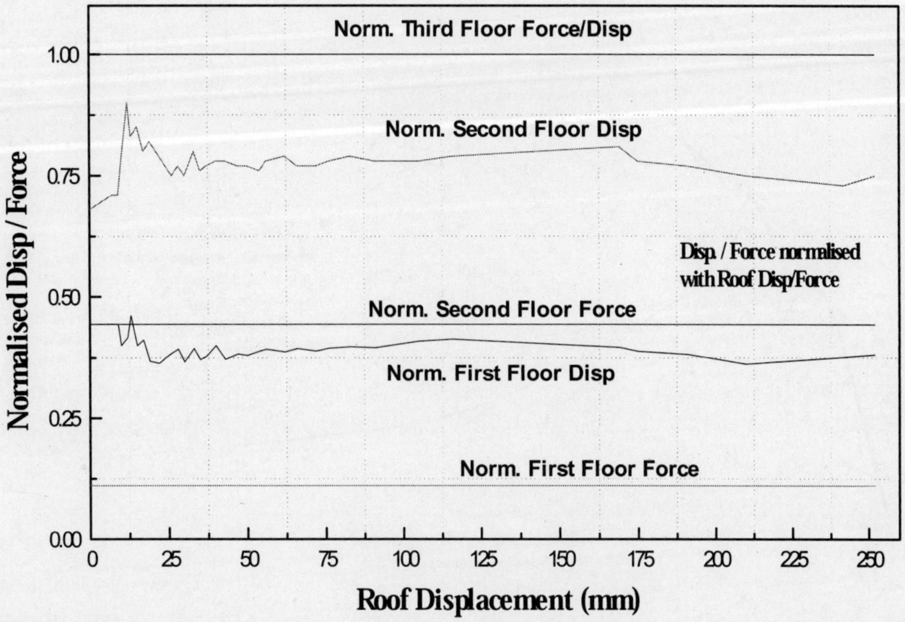

Fig-10. Comparison of Experimental Displacement Profile with Applied Force Profile.

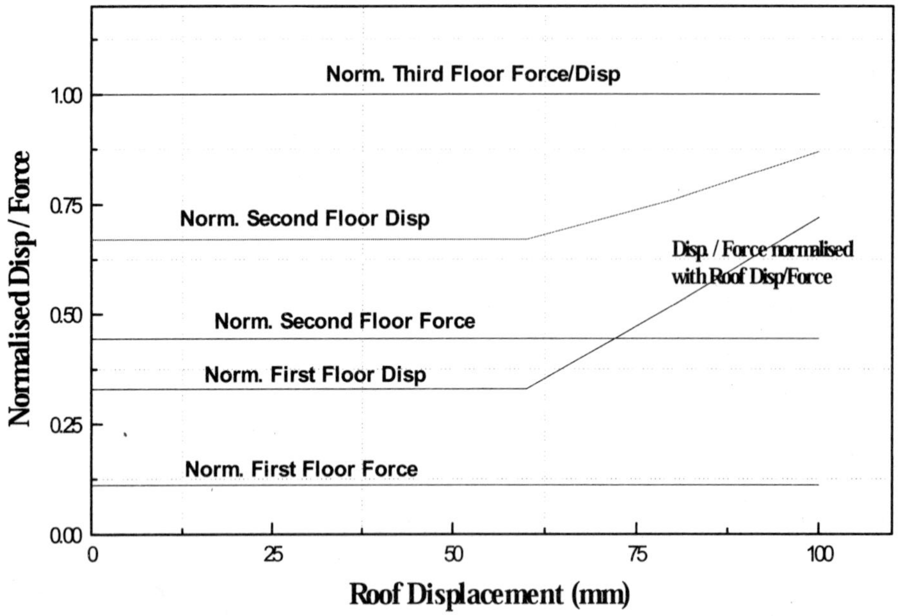

Fig-11 Comparison of Analytical Displacement Profile with Applied Force Profile.

Hence suitable corrections are carried out in the $\delta^T F$ term of Equation 14 and the curves are re-plotted. The results of analytical values are increased by a factor of 1.33 for displacement values more than 60 mm. Similarly, ω^2 term of the experimental curve is corrected and re-plotted in Figure 12. Now a good matching is seen between the experiment and analysis.

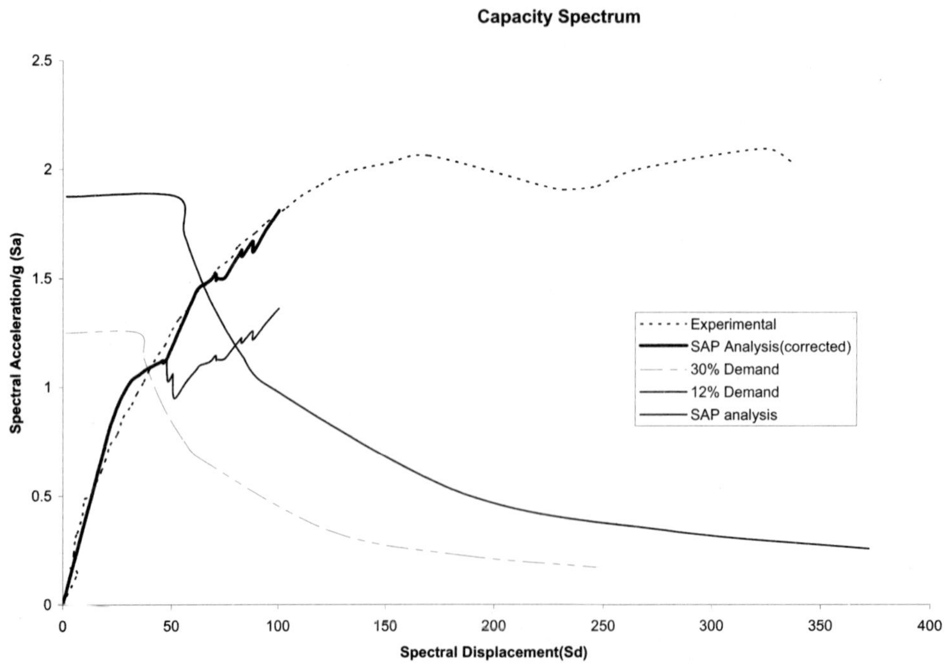

Fig-12 Experimental and Analytical Results in ADRS format and Corrected Analytical and Experimental Values.

8. CONCLUSIONS

Important conclusions are:

(a) The capacity curve needs to be corrected for displacement profile not-matching with inertial load profile.

(b) Stiffness reduction beyond yield point is continuous for the experiment and discrete and jagged in the analysis.

Push-over analysis is an alternative simple methodology available to engineers for seismic evaluation of existing structures and for performance based design of new structures. However, the methodology has to be judiciously handled considering the vagaries of the predicted output parameters with reference to input hinge properties. The tested structure at SERC could be a good bench-marking tool for refining analysis results.

REFERENCES

1. ATC-40 (1996), Seismic Evaluation and Retrofit of Concrete Buildings, Vol.1 and 2, Report No.SSC 96-01, California Seismic Safety Commission

2. Chopra, A.K., and Chintanapakdee, C. (2003), "Inelastic Deformation Ratios for Design and Evaluation of Structures: Single Degree of Freedom Bi-linear Systems", Report No. EERC 2003-09, College of Engineering, University of California, Berkeley

3 Garcia, R.J., and Miranda, E. (2003), "Inelastic Displacement Ratios for Evaluation of Existing Structures", Earthquake Engineering and Structural Dynamics, 1237-1258

4 Gulkan, P., and Sozen, M. (1974), "Inelastic Response of Reinforced Concrete Structures to Earthquake Motions", Journal of American Concrete Institute, pp 604-610

5 IS: 13920-1993, "Ductile Detailing of Reinforced Concrete Structures subjected to Seismic Forces", Bureau of Indian Standards, New Delhi

6. IS:1893-2002, (2002), Indian Standard – Criteria for Earthquake Resistant Design of Structures, Part-1, General Provisions and Buildings, BIS, New Delhi.

7. Kowalsky, M.J., Priestly, M.J.N., and Machae, G.A. (1994), "Displacement-based Design, a Methodology for Seismic Design Applied to SDOF Reinforced Concrete Structures", Report No. SSRP-94/16, Structural Systems Research Project, Univ. of California, California

8. Lakshmanan, N. (2003), "Ductility, Energy Absorption and Damage Modeling of Concrete Structural Elements" Proc. International Conference on Recent Trends in Concrete Technology and Structures", Coimbatore, India, INCONTEST – 2003, Vol.1, September, pp.69-85.

9. Lakshmanan, N., Krishnamoorthy, T.S., et al. (Oct. 2004), "Flexural Behaviour of R.C. Beam Retrofitted with Steel Plates and CFRP Wrapping", Indian Concrete Journal, pp 45-51

10. Lakshmanan, N. (2005), "Ductility Based Design and Detailing for Lateral Loads", NPCBEERM-SERC, pp 3.137 - 3.176

11. Midori Karva, M., Okawa, I., Liba, M. and Teshigarvara, M. (2003), "Performance Based Seismic Design Code for Buildings in Japan", Earthquake Eng. and Engineering seismology, Vol:4, No:1, 15-23

12. Miranda, E. (2001), "Estimation of Inelastic Deformation demands of SDOF Systems", Journal of Structural Engineering, ASCE, pp 1005-1012

13. Newmark, N.M., and Hall, W.J. (1982), "Earthquake Spectra and Design", EERI Monograph Series, Earthquake Engineering Research Institute, Oakland, California.

14. Valestos, A.S., and Newmark, N.M. (1960), "Effect of Inelastic Behaviour on the Response of Simple Systems to Earthquake Motions", Proc. World Conf. on EQ Engineering, Japan, Vol.2, 895-912.

RESEARCH ON EARTHQUAKE RESISTANT STRUCTURAL DESIGN AT IIT MADRAS

S R Satish Kumar
Department of Civil Engineering, IIT Madras, Chennai 600 036. E-mail kim@iitm.ac.in

ABSTRACT

Earthquake Resistant Design was not given its rightful importance in India However, a series of earthquakes since 1993 have underscored the need to design and construct structures to withstand seismic loading. Keeping this trend in view, considerable research work is being carried out at IIT Madras. In this paper, some of the work carried out recently is briefly reviewed. These analytical works include research on the response of frames with semi-rigid connections, frames with buckling restrained braces and performance based design of reinforced concrete frames. The experimental works include tests on rectangular hollow sections and their connections, beam-column joints/connections in reinforced concrete and steel-concrete composite construction and retrofitting of RC beams. In addition, parametric studies have also been carried out on connection details using finite element methods. The results obtained can be used to improve the seismic performance of building structures and thus mitigate possible future disasters.

Keywords: Semi-rigid connections, performance-based design, non-buckling brace, rectangular hollow sections

1 INTRODUCTION

Earthquake Resistant Design was not given its rightful importance in India probably due to lack of records on historical earthquakes in the peninsula. However, a series of earthquakes, starting with the Latur earthquake in 1993, the Jabalpur earthquake in 1997 and the Gujarat earthquake in 2001, underscored the need to design and construct structures to withstand seismic loading. The rapid development in the economy and consequent increase in building stock as well as the rising stakes in business have led to greater awareness about the need to verify the safety of existing building stock and to retrofit them if required and also to improve the resistance of new structures by better design and detailing. Keeping this trend in view, considerable research work is being carried out at the Structural Engineering Laboratory of the Department of Civil Engineering, IIT Madras. In this paper, some of the work carried out recently is briefly reviewed. These analytical works include research on the response of frames with semi-rigid connections, frames with buckling restrained braces and performance based design of reinforced concrete frames. The experimental works include tests on rectangular hollow sections and their connections, beam-column joints/connections in reinforced concrete and steel-concrete composite construction and retrofitting of RC beams. In addition, parametric studies have also been carried out on connection details using finite element methods. The results obtained can be used to improve the seismic performance of building structures and thus mitigate possible future disasters.

2 RESEARCH ON ANALYSIS AND DESIGN METHODS

It has been recognized that not all beam-to-column connections can be treated as either fixed or pinned. Some types of connections behave in between and need to be classified as semi-rigid. Most codes, including the revised IS 800 (2007), allow for semi-rigid connections but stipulate that their analysis be carried out considering the moment-rotation behaviour of the connections.

Semi-rigid frames can offer several advantages over rigid frames in resisting seismic loads. However, they are likely to drift more and the ductility demands can also be higher particularly when partial strength connections are used. At present, very few guide lines are available for the design of semi-rigid frames subjected to seismic loads. Regular plane frames of two, four and six storeys were analyzed under equivalent static and seismic loads from five different earthquakes. Beams in the semi-rigid frames have been idealised as beam elements having rotational springs at either ends. Semi-rigid frames are characterized by two parameters J and α. The ratio of connection stiffness K_O to the beam flexural stiffness is called the 'joint factor (J)'.

Thus,

$$J = \frac{K_O}{(EI/L)_{beam}} \tag{1}$$

The other parameter α is defined as the ratio of the flexural stiffness of beam to the flexural stiffness of the column. Thus,

$$\alpha = \frac{(EI/L)_{beam}}{(EI/h)_{column}} \tag{2}$$

where 'h' is the height of the column.

The response of the semi-rigid frames was characterized in the form of ratios of static and dynamic roof drift factors (SDF & DDF) and time period factor (TF) taken with respect to their equivalent rigid frames. The variation of roof drifts and time periods with respect to connection stiffness is obtained. As a result of static analyses, it was first established that although the roof displacement depends on the absolute values of frame member geometric properties, the static drift factor was independent of the absolute values. Therefore the variation of static drift factor SDF with joint factor J is important since it enables the calculation of semi-rigid frame drifts from the corresponding rigid frame drifts. The variation static drift factor (SDF) with J for the two storey frame with different α values is shown in Fig. 1. It can be observed from Fig. 3 that SDF decreases with increase in J. The SDF values are about 1.2 for J equal to 10, 1.4 for J equal to 5 and shoot up as J decreases below 5. From Fig. 1, it can also be observed that the variation of SDF with α is relatively small compared to its variation with J.

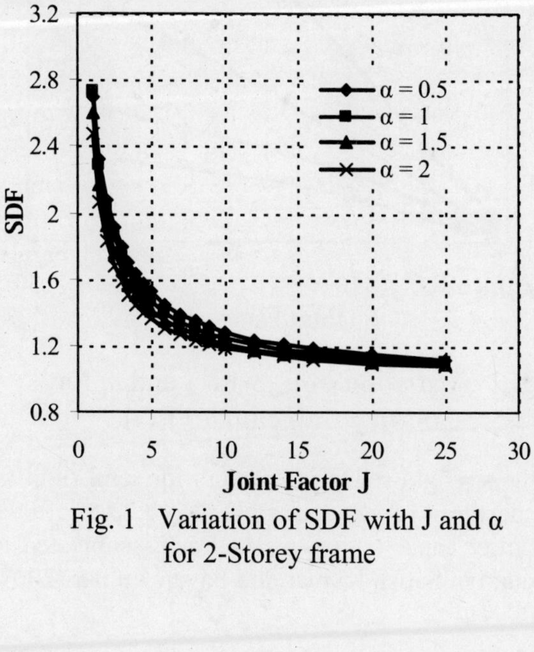

Fig. 1 Variation of SDF with J and α
for 2-Storey frame

The variation of SDF with J for α equal to 0.5 for the four and six storey frames was similar to the case of the two storey frame. The variation of time period factor TF was also similar to that of SDF. The corresponding time period factors were ten and twenty percent higher in semi-rigid frames. Knowing the static drift factor and time period factor as a function of joint factor will enable the evaluation of drifts and time periods of semi-rigid frames from the values corresponding to rigid frames.

Nonlinear time-history analyses were then carried out on the frames assuming a tri-linear moment rotation curve for the semi-rigid connection. Five different earthquake accelerograms, scaled to a peak ground acceleration of 0.36g were used. The ratios of roof drifts in semi-rigid frames to that in rigid frames under seismic loading (DDF) was found to vary in a manner analogous to the variation under equivalent static loading. The design base shears for semi-rigid frames were also lower than the base shears in rigid frames.

The variation of Connection ductility demand μ_C and global ductility demand μ_G with joint factor J and strength factor *m* were also studied. The variation of connection ductility demands with joint factor for various strength factors, for the two storey frame under a particular earthquake EQ1, is shown in Fig. 2. Connection ductility demand (μ_C) increases almost linearly with increase in J for a given strength factor. Frames with stiffer connections (higher J values) normally attract more seismic force and yield at relatively smaller rotations compared to frames with flexible connections for a given connection strength. So, the connection ductility demands for stiffer connections are significantly higher compared to that for flexible connections as shown in Fig. 2. This explains the reason for failures of such connections in past earthquakes and also underscores the advantage in using semi-rigid connections. It can also be observed from the Fig. 8 that decreasing the connection strength, further increases the ductility demands on the connection. Connection ductility demand (μ_C) for an *m* of 0.4 and J of 25, is 29 which is very high and rather difficult to obtain for semi-rigid connections. This imposes a restriction on minimum strength of semi-rigid connections.

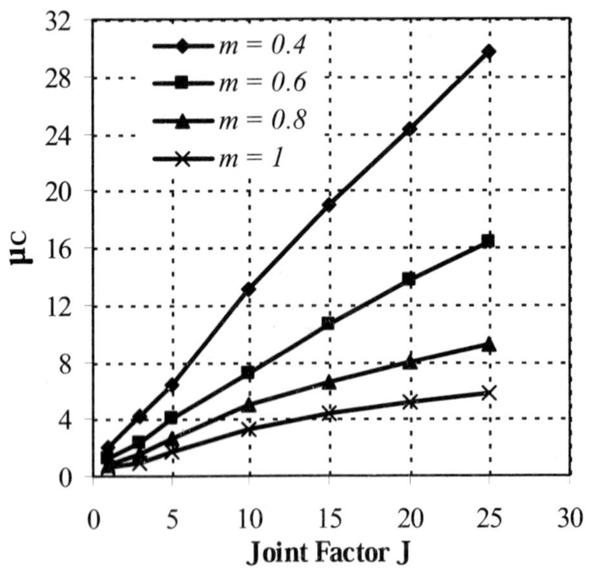

Fig. 2 Variation of μ_C with J and *m* for 2-Storey frame under EQ1

The demands also help evaluate the strength reduction factors for semi-rigid frames. The strength reduction factors not only help in designing connections but also help to evaluate the suitability of connections with a particular ductility capacity for use in frames subjected to seismic loadings. More details of the study can be found in Satish Kumar and Pavan Kumar (2007).

The bracing in steel frames also develop problems because of their high slenderness and consequent tendency to buckle at low compressive loads. To overcome these problems, a new type of brace known by various names as core-loaded, non-buckling and buckling-restrained brace was developed (Kalyanaraman *et al* 1994). The brace or core is enclosed in an outer sleeve (see cross-section in Fig. 3) which effectively restrains the buckling of the brace and improves its performance under compressive loads thereby giving stable hysteretic loops. The strength and stiffness of Non-Buckling bracing can be tailored to obtain the desired level of storey drift limit and energy absorption by using cores with non-prismatic cross sections as shown in Fig. 3. The yield load of the core is governed by the reduced area of the core (α = Areduced / Atotal), whereas the axial stiffness of the bracing is governed by the ratio of the length of core with reduced area to the total length of the core denoted by β. Thus by modifying parameters α and β, both storey shear and the drift of the building can be controlled.

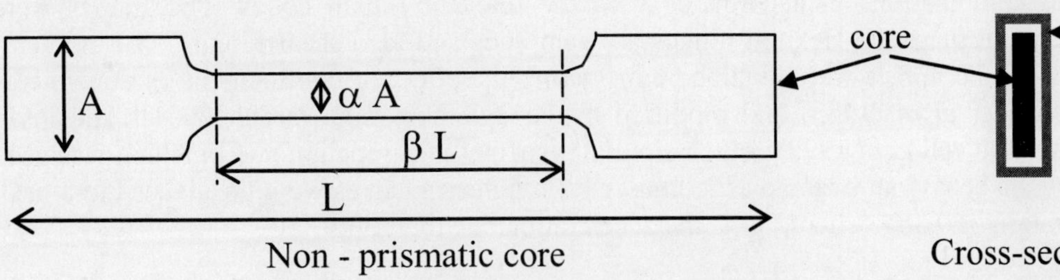

Fig. 3: Core-loaded brace details

A single bay ten-storey concentric braced frame with diagonal Non-Buckling bracing is chosen for studying the effect of brace parameters during earthquake loadings. Two types of frames were used in the analysis. Moment Resisting Braced Frame (MBRF) with concentric Non-Buckling diagonal bracings in all stories, wherein both the frame and the bracings, acting together resist lateral loads. A non-moment resisting braced Frame (BRF) with concentric Non-Buckling bracings in all stories, wherein the lateral load is resisted by bracings alone. Non-linear time history analysis was carried out using SAP2000 to study the variation of maximum base shear, roof displacement, inter-storey drift, Damage Index and stresses in various members due to El-Centro earthquake accelerogram. Based on the study, the following conclusions were arrived at: 1) Values of α between 0.25 and 0.50 give effective control of base shear for MBRF. 2) The values of α between 0.50 and 1.00 are desirable for BRF. Lower values of α ($\alpha<0.5$) lead to higher inter-storey drifts and hence are not desirable. 3) Any value of β can be used in the MBRF, if the frame is adequately stiff and does not lead to higher deformations in bracing. 4) The values of β between 0.50 and 1.00 can be used for BRF. Lower values of β ($\beta<0.5$) lead to higher damage index and may lead to the failure of bracing system. More details of the study can be obtained from Ravikumar *et al* (2007).

Seismic design codes are traditionally based on the force-based approach wherein structures are designed with a certain minimum lateral strength. It is well known that seismic response is a function of not only the strength but also the initial stiffness and the ductility and hysteretic energy dissipation capacity. The initial stiffness is taken into account by calculating the seismic force as a function of time period using a response spectrum while the ductility is accounted for by considering a response reduction factor. However, it has been observed that such an approach which relates to the elastic response does not produce consistent inelastic response in terms of the amount and distribution of damage in structural elements. When the performance of non-structural elements is also considered, the force-based approach gives widely different results (Priestley 2000).

In view of the above, the displacement based approach also known as performance based design approach has been explored in recent years and found to give better results. In this approach, the prime response quantities of interest are the inter-storey drifts and the design process directly attempts to limit these drifts to an acceptable value. However, the adoption of PBD requires the understanding of the influence of various parameters, such as the strength, stiffness, energy dissipation capacity and detailing parameters such as the ratios of beam-to-column strength and stiffness, percentage of reinforcement and amount of confining steel, on the local and overall damage. While the effect of each individual parameter on the response can be studied in isolation, understanding the effect of their combinations will require considerable study. In this study, a small portion of the overall problem is studied extensively to arrive at meaningful conclusions which can be used to develop a suitable performance based design approach.

Nonlinear time-history analyses have been carried out on regular plane frames of two, four and eight stories using an evolutionary degrading (pivot) hysteretic model and six earthquake accelerograms. The dead, live and earthquake loads were calculated by assuming suitable spacing of frames, thickness of slab and wall etc. The frames were then designed and detailed for the governing load combinations in accordance with the relevant Indian codes. The frames were modeled using point plasticity flexural hinges at beam ends and also shear hinges. The moment-rotation curves for the hinges were obtained considering the effect of confinement on concrete as proposed by Mander *et al* (1988) and modified by Panagiotakos and Fardis (2001). The Pivot hysteretic model (Dowell *et al* 1998), which simulates stiffness degradation and pinching was used to model the cyclic behaviour of the plastic hinges. All the earthquakes were normalised to a peak ground acceleration of 0.36g, where, 'g' is the acceleration due to gravity.

The force-based method considers the seismic performance of frames as a function of the strength (or response reduction factor R) and stiffness (or time period) and the properties of the ground motion characterized by the spectral displacement. In addition to these, the percentage of reinforcement has been recognized recently as an important parameter in controlling the performance. This parameter not only controls the stiffness but also determines the yield displacement and hence affects the ductility. In view of this, there have been suggestions to consider absolute drifts and deformations in place of ductility (Priestley 2000). However, since ductility has been given prime importance in traditional design, it is better to study the influence of the percentage of reinforcement on the ductility demand and this is one of the prime objectives of the present study. In view of these considerations, the design parameters considered are limited to strength (yield and ultimate), stiffness (or time period) and percentage of reinforcement while the response parameters considered are local and global ductility, roof and inter-storey drifts and damage indices. For parametric study, different values of response reduction factor R, (3, 5 and 7) and various percentage of reinforcement p_t (0.8%, 1.6% and 2.4%) were considered. In all 118 frames were analyzed for six different earthquakes giving a total of 702 analyses. However, in this paper only the results obtained for the four-storey frames are presented.

Global displacement ductility demand (μ_G) is defined as the ratio of maximum roof displacement to yield roof displacement at first yield anywhere in the structure. The variation of global ductility demands (μ_G) for the four storeyed frame with fundamental natural period is shown in Fig. 4 for various R and p_t values. The graphs are drawn for mean plus one standard deviation values obtained for the six earthquakes. It can be seen that the global ductility values increase almost linearly with time period for given R and p_t values. This is because frames with higher time periods are likely to experience higher displacement demands and the yield displacement practically remains constant with change in stiffness or time period. It can also be seen from the figures that for a given time period and pt value, global ductility demands increase with increase in R value. Similarly, for a given R and time period, global ductility demands decrease with increase in pt values.

The inter storey drift index (IDI) is defined as the difference in displacement of two consecutive floors divided by the storey height and is expressed as a percentage. The performance based design procedure stresses on the importance of limiting inter-storey drifts as a means of limiting damage to structural and non-structural elements. The maximum inter-storey drift indices, obtained for the four storeyed frames from the non-linear time history analyses, are plotted as functions of frame fundamental time periods in Fig. 5 for various R and p_t values.

Thus it is clear that global ductility demand as well as inter-storey drift index, vary with time period, response reduction factor and the percentage of reinforcement for code designed frames. Thus it can be concluded that code designed structures do not give consistent performance and it is necessary to develop procedures which will yield consistent performance. One way of implementing performance based design to get consistent performance is to adjust the response reduction factors. However, there are limitations to this procedure and the desired performance may not be always feasible to achieve. The response reduction factors, for the target IDIs of 1.5%, 2.0%, 2.5% and 3.0% at time periods 0.8, 1.3, 1.8 and 2.3 seconds, were obtained by back calculations and are as shown in Fig. 6. Thus by selecting suitable response reduction factors, the current design procedures can be easily modified to get the desired performance (Satish Kumar and Venkateswarlu 2007).

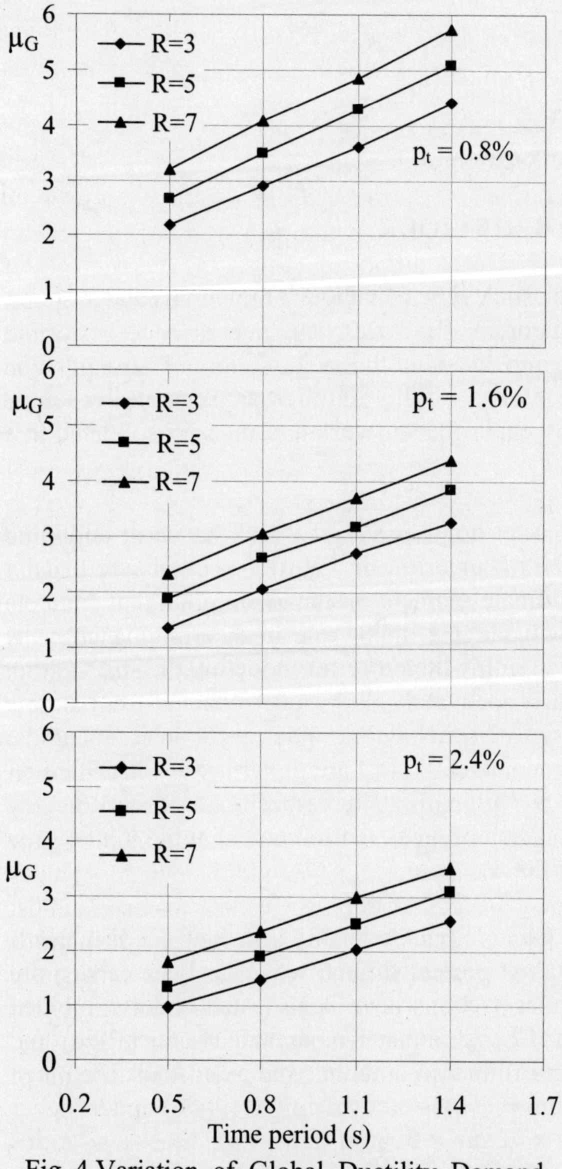

Fig. 4 Variation of Global Ductility Demand with Time Period for the 4-Storey Frame

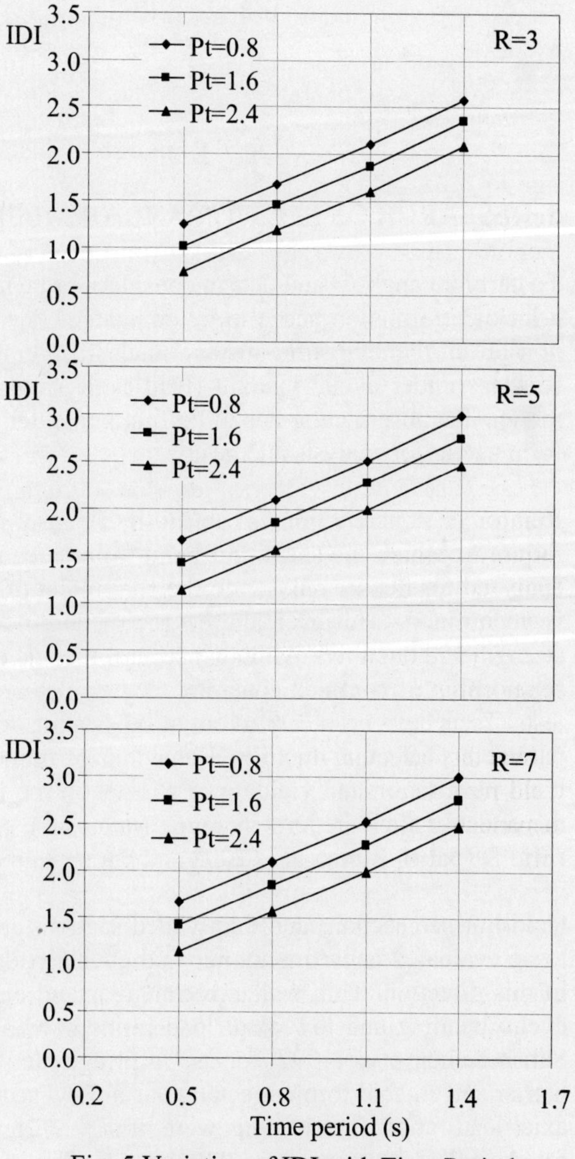

Fig. 5 Variation of IDI with Time Period for the 4-Storey Frame

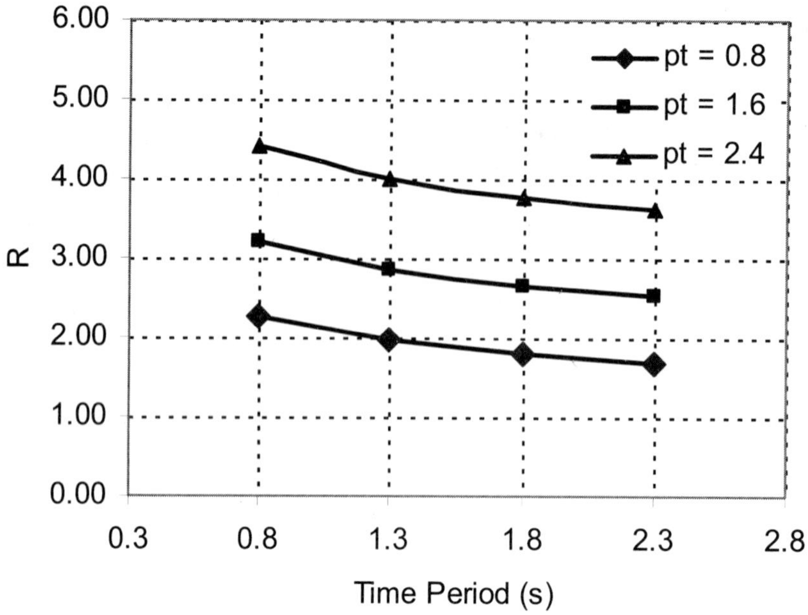

Fig. 6 Response Reduction Factors for desired IDI

3 RESEARCH ON SECTION AND MEMBER BEHAVIOUR

To carryout analysis and design considering the inelastic cyclic behaviour of materials and systems, a lot of information needs to be generated. For important structures, it is necessary to perform a non-linear dynamic time-history analysis taking into account the degradation of strength and stiffness under cyclic loading (load-cycle fatigue effects). This requires an accurate hysteretic model. The information has to be first generated by experimental work and then consolidated in a form useful for analysis and design.

Reinforced concrete frames can fail by flexure, shear or bond failure of which the shear and bond failure of which the shear and bond failures can be rather brittle in nature. Sections are usually analyzed for flexure only up to the attainment of ultimate moment on the assumption that concrete is unconfined. However, the degree of confinement plays a major role in determining both the strength and ductility. Although several models exist in the literature for modeling the stress-strain relationship of confined concrete, they do not predict adequately the results obtained from simple tests. Tests have been carried out at IIT Madras to study the behaviour under cyclic loading and the parameters affecting ductility. Behaviour of frames with weak joints are governed by shear failure, bond penetration and yielding of stirrups in the joint region. Tests have been carried out to study the effect of alternate joint detailing such as the provision of cross reinforcement at the joint region (Fig. 7) (Satish Kumar *et al* 2002 and Satish Kumar 2004).

Cold-Formed sections and thin-walled sections are liable to buckle locally and their ductility needs to be evaluated under monotonic and cyclic loading. At present limited work has been carried out in this direction. Thin walled rectangular and circular sections have been tested under combined cyclic bending and low axial load ratios of about 0.2, by Satish Kumar and Usami (1996) and Satish Kumar et al (1997) for use in bridge piers. For building columns, the axial loads are much higher and so cold formed rectangular hollow sections were tested by Srinivas (2006) under higher axial load ratios. The sections were of size 172mm x 92mm x 5.4mm (b/t = 26). Axial load ratios for the two specimens were 0.8 and 0.6. The specimens had cyclic ductilities of four and eight, respectively. A two storeyed frame with beams of hollow rectangular sections was tested by Abdul Sathar (2006) to study the effect of strength and stiffness degradation of hinging regions on the seismic performance of frames.

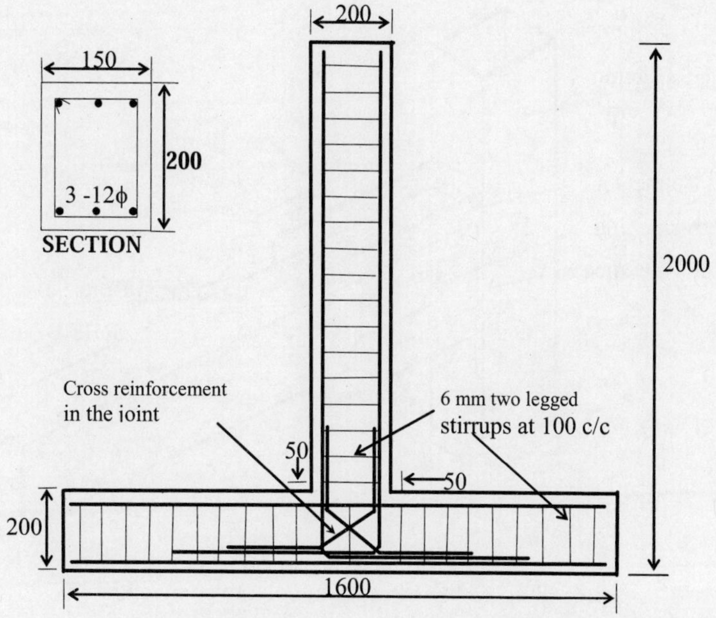

Fig. 7 Cross-reinforcement in the Joint Region

Concrete-filled tubes (CFTs), particularly of square or rectangular shape, can be extremely efficient systems for columns. In addition to the advantages of confined concrete, the delay in local buckling leads to higher strength and ductility. Both high strength and self compacting concrete can be used to fill the tubes. Limited tests have been conducted to study their behaviour under cyclic loading (Bergmann *et al* 1995). Testing of such sections filled with self-compacting concrete will be undertaken soon.

Rectangular Hollow Sections (RHS) have superior structural performance compared to conventional steel sections. However, the application of RHS in structural steel frame works is limited because suitable connection configurations have not been developed between such members. Also adequate information on the moment-rotation characteristics of connections between RHS members is not available for design. To overcome these problems, a new and efficient connection was proposed which is easy to fabricate and convenient for erection (Fig. 8). The connection employs channel connectors welded to the column flange and bolted to the beam to transfer beam flange forces into the column webs thereby avoiding the need to provide internal diaphragms in the column. An opening in the web facilitates the installation of bolts and can be used to pass service lines. By choosing suitable dimensions for the channel connectors, the strength and stiffness of the connection can be varied. The behaviour of the connection was evaluated by cyclic tests and non-linear finite element analysis. In the case of channel connectors of high strength, failure occurs at the beam net-section away from the face of the column, similar to beams with Reduced Beam Sections (RBS). It was found that the connection has adequate ductility and hysteretic energy dissipation capacity for use in severe seismic areas (Satish Kumar and Prasada Rao 2006).

The moment-rotation characteristics of the connection can be expressed in terms of the three-parameter power model, for semi-rigid frame analysis. The depth and thickness of the channel connectors are the primary parameters affecting the connection characteristics. The width-to-thickness ratio of column web also has a significant effect on the initial connection stiffness. Other effects such as shear lag and stress intensity have a profound influence on the ultimate moment capacity of the connection. Empirical equations were obtained to evaluate modal parameters based on a parametric study using finite element analysis. Guidelines for the evaluation of ductility and energy dissipation capacity were also arrived at. More details may be obtained from Prasada Rao and Satish Kumar (2006).

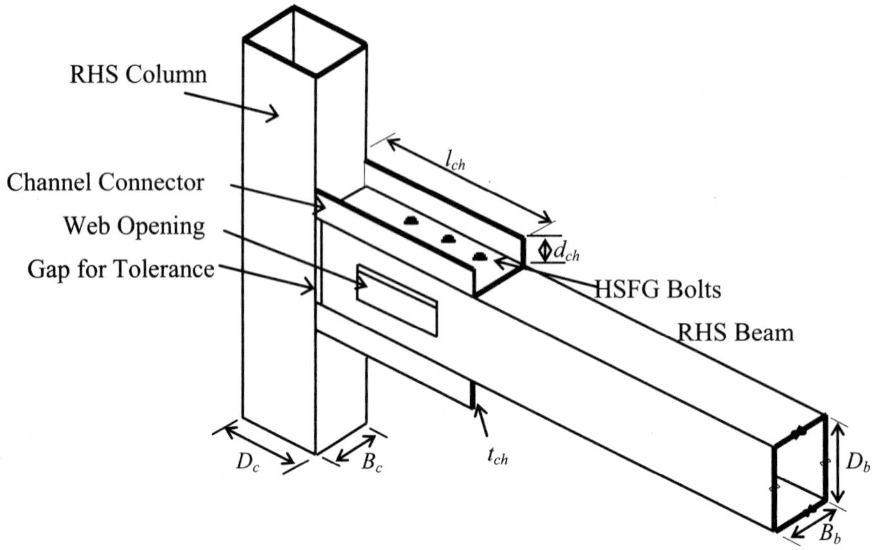

Fig. 8 Proposed RHS beam-to-column connection

Steel-concrete composite construction combines the advantages of steel and concrete. Composite beams have increased moment capacities compared to bare steel sections due to the compression taken by the concrete slab. Their connections can be easily strengthened by adding extra reinforcement in the slab at the support regions (Fig. 9). However, they will still be semi-rigid connections. At present very few guidelines exist to evaluate their moment-rotation behaviour as well as the parameters controlling their inelastic seismic performance such as ductility.

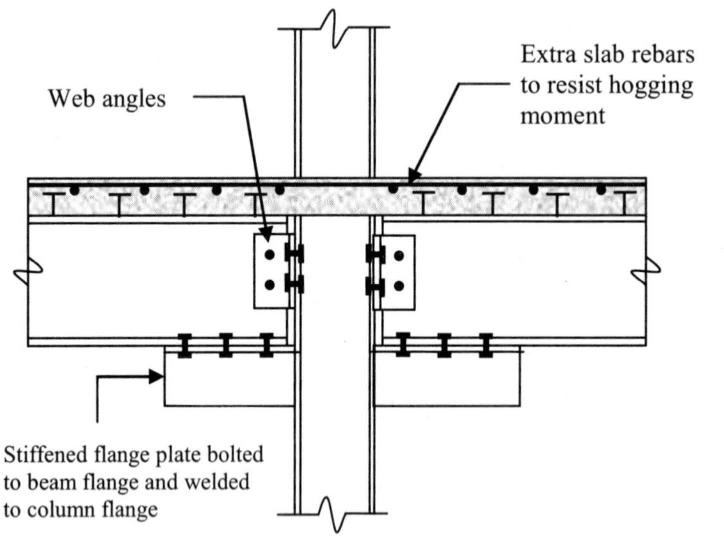

Fig. 9 Steel-concrete composite beam-to-column connection

To evaluate the performance of flange-plate and stiffened flange-plate (T-stub) connections, both experimental work and Finite element analyses have been carried out. Modelling concrete behaviour is a major issue and so three different methods were tried. It was observed that the smeared crack model gives convergence problems when analysis nears ultimate load, hence post-peak response can be obtained only if concrete is modeled using damaged plasticity model or elastic-plastic material model. The results of the FE analysis indicate that the stiffened flange plate connection will perform better than unstiffened flange plate connection under reversal of loads and hence is more suitable under large seismic loads. More details of the analyses issues can be found in Smitha and Satish Kumar (2007).

The Gujarat earthquake of 2001 highlighted the extreme vulnerability of buildings in India. In order to reduce the risk of a similar disaster a major project on retrofitting was undertaken at IIT Madras with the sponsorship of the Department of Science and Technology. More recently a handbook was also compiled for the benefit of all stakeholders (CPWD 2007). In continuation of this, further research is being carried out to develop and evaluate suitable retrofitting methods. One such work focuses on retrofitting of beams deficient in shear or flexure by concrete jacketing.

To ensure adequate confinement of jacketed concrete near the joint, closely spaced stirrups will be required. However, proper placement of stirrups is a problem in existing structures due to the presence of the slab. Therefore a scheme was devised using threaded J-hooks at larger intervals which support longitudinal anchor bars below the slab (see Fig. 10). The stirrups can then be attached to these anchor bars at closer spacing. The jacketing was done using self compacting concrete. Beam-column sub-assemblages with and without retrofitting were tested under monotonic and cyclic loading to evaluate the adequacy of this retrofit scheme (Badri Narayanan *et al* 2007).

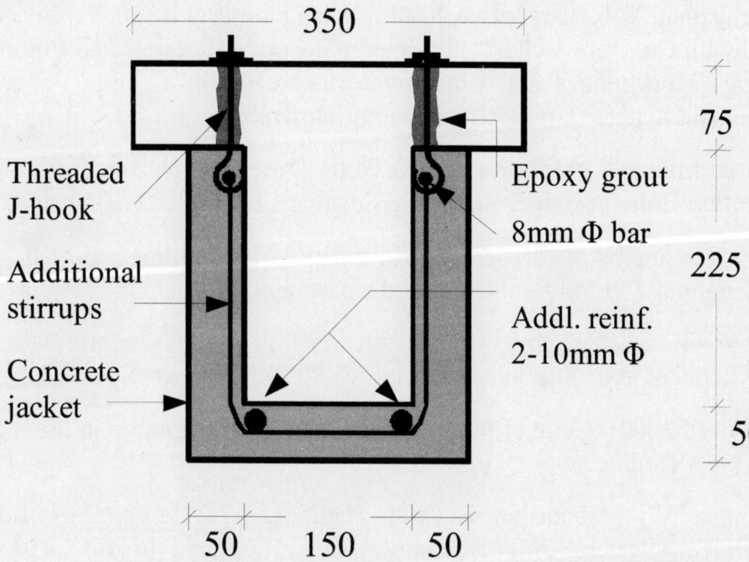

Fig. 10 Beam retrofitting by concrete jacketing

Masonry walls in load bearing and framed construction perform poorly because of lack of integrity leading to disintegration of the walls. To improve the situation, hollow interlocking blocks were developed at IIT Madras (Fig. 11). The masonry can be easily reinforced in horizontal and vertical directions. The seismic performance of such masonry is expected to be superior to conventional masonry. The out-of-plane collapse can also be minimized. Tests are being planned to study their performance under cyclic loading.

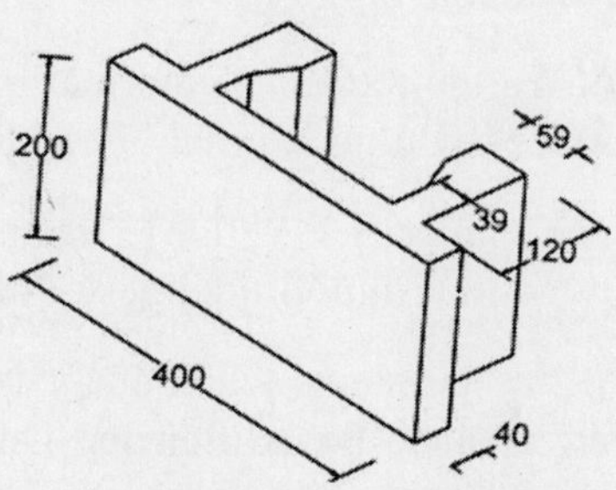

Fig. 11 A stretcher unit of Hollow Interlocking Block Masonry

Some work on fatigue crack growth behaviour under cyclic loading is also under progress but is not described here. The reader may refer to Nagapadmaja et al (2007a and 2007b) for details.

4 CONCLUSIONS

Considerable analytical and experimental research is being carried out at IIT Madras to contribute towards earthquake resistant design of structures. A brief review of various topics addressed was presented. Additional information may be obtained from the references sighted or by paying a visit to the laboratory. Although the primary focus is to improve the technology in India, the results can be applied across the world to design structures with improved seismic performance.

REFERENCES

1. Abdul Sathar (2007) 'Seismic Performance of Steel Frame with Cold-formed Rectangular Hollow Sections', Mitch Thesis, IIT Madras.
2. Badari Narayanan, V T, Sengupta A K and Satish Kumar S R (2007) 'Investigation of Beams Retrofitted with Concrete Jackets', International Conference on Civil Engineering in the New Millennium: Opportunities and Challenges (CENeM-2007), 11-14 January, 2007, Bengal Engineering and Science University, Shibpur, Howrah, India.
3. Bergmann R, Matsui C, Meinsma C and Dutta D (1995) ,Design Guide for Concrete-filled Hollow Section Columns under Static and Seismic Loading', TUV-Verlag, Koln, Germany.
4. CPWD (2007) Handbook for Seismic Retrofitting of Buildings, Edited by Devdas Menon and A K Sengupta, Central Public Works Department, New Delhi.
5. Dowell, R.K., F. Seisble. and E.L. Wilson (1998) Pivot Hysteretic Model for Reinforced Concrete Members, ACI Structural Journal, 95, 607-617.
6. IS 800 (Revised 2007) 'Code of Practice for General Construction in Steel'; Bureau of Indian Standards, New Delhi.
7. Kalyanaraman, V., Mahadevan K. and Thairani, V. (1998) 'Core Loaded Earthquake Resistant Bracing System', Jnl. of Construct. Steel Research, 46 (1-3), 437-439.
8. Mander J.B., M.J.N. Priestley and R. Park, (1988) 'Theoretical Stress-Strain Model for Confined Concrete', Jnl. of Struct. Eng., ASCE, 114, 1804-1823.
9. Nagapadmaja P, Kalyanaraman V, Satish Kumar S R and Chellapandi P (2007a) 'Experimental Study on LBB Behaviour of LMFBR Pipe Elbows', International Jnl. of Fatigue (in press, accepted manuscript, available online).
10. Nagapadmaja P, Kalyanaraman V, Satish Kumar S R and Rao B N (2007b) 'Fatigue Crack Propagation Analysis of Surface Cracks in Pipe Elbows', International Jnl. of Fatigue, (in press, accepted manuscript, available online).
11. Panagiotakos T. B. and M.N. Fardis, (2001) 'Deformation of Reinforced Concrete Members at Yielding and Ultimate', ACI Structural Journal, 98, 135-148.
12. Prasada Rao D V and Satish Kumar S R (2006) 'Beam-to-Column Connection with Web Opening – Parametric Study and Design Guidelines', Jnl. of Construct. Steel Research, 62, 747-756.
13. Priestley M.J.N, (2000) "Performance Based Seismic Design", Paper No 2831, 12th World Conference on Earthquake Engineering, New Zealand.
14. Ravi Kumar G, Kalyanaraman V and Satish Kumar S R (2007) 'Behaviour of Frames with Non-buckling Bracing under Earthquakes', Jnl. of Construct. Steel Research, 63, 254-262.

15. Satish Kumar S R and Usami T. (1996) 'Damage Evaluation in Steel Box Columns by Cyclic Loading Tests', Jnl of Struct. Eng., ASCE, 122 (6), 626-634.

16. Satish Kumar S R, Mizutani S and Okamoto T (1998) 'Nonlinear Dynamic Response of Thin-Circular Steel Tubes', in Stability and Ductility of Steel Structures, Ed. By Usami T and Ito Y., Elsevier Science, U.K.

17. Satish Kumar S R, Vijaya Raju B and Rajaram GSBVS (2002) 'Hysteretic Behaviour of Lightly Reinforced Concrete Exterior Beam-to-column Joint Sub-assemblages', Jnl of Struct. Eng., SERC, 29(1), 31-37.

18. Satish Kumar S R (2004) 'Cyclic Behaviour of Lightly Reinforced Concrete Beam-To-Column Joints', Paper No. 1907, 13 World Conference on Earthquake Engineering, Vancouver, Canada.

19. Satish Kumar S R and Prasada Rao D. V. (2006) 'RHS Beam-to-Column Connection with Web Opening – Experimental Study and Finite Element Modeling', Jnl. of Construct. Steel Research, 62, 739-746.

20. Satish Kumar S. R. and Pavan Kumar G. V. R. (2007) 'Seismic Ductility Demands on Semi-rigid Steel Frames', Structural Engineers World Congress (SEWC), 2-7 Nov, Bangalore.

21. Satish Kumar S. R. and Venkateswarlu G. (2007) 'Performance Based Design of Reinforced Concrete Frames', Structural Engineers World Congress (SEWC), 2-7 Nov, Bangalore.

22. Smitha, M.S. and Satish Kumar S R (2007) 'Performance of Composite Beam to Column Semi-rigid Connections', Structural Engineers World Congress (SEWC), 2-7 Nov, Bangalore.

23. Srinivas C (2006) 'Cyclic Testing of Light-gauge Steel Sections', MTech Thesis, IIT Madras.

STUDIES ON SYSTEM IDENTIFICATION OF MULTISTOREYED BUILDINGS BASED ON STRONG MOTION DATA RECORDED IN GUWAHATI CITY REGION

K. Suresh, Sajal K. Deb and Anjan Dutta
Department of Civil Engineering, IIT Guwahati, Guwahati-781039, Assam

ABSTRACT

Structural health monitoring of multi-storeyed shear buildings based on system identification by parametric state space modeling has been presented in this paper. The earthquake ground acceleration and the acceleration response histories of various floors of buildings are required as input. The method evaluates the modal parameters like the eigen frequencies and mode shapes using a sub-space iteration technique (N4SID) after appropriate re-sampling and filtering of the recorded response. Using the Least Square Technique, the stiffness of the structure is evaluated and localized damage can be identified, if present. The method of system identification proposed in this study is capable of identification of system parameters of buildings with non-uniform mass distribution. The method is applied for identification of a nine storeyed reinforced concrete building of BSNL Guwahati using structural response data recorded during earthquakes on 11th and 14th August, 2006.

IDENTIFICATION TECHNIQUES

Application of system identification techniques for health monitoring of structures has attracted attention of civil engineers in recent years. In terms of how input output data are translated into a mathematical relation, the field of identification can be divided broadly into two classes: parametric identification and nonparametric identification. In parametric identification, the structure of the mathematical relation is fixed a priori and parameters of the structure are fitted to the data. In nonparametric identification, no (or very little) assumption is made with respect to the model structure. So the above mentioned non-parametric identification approaches do not require knowledge of the excitation and hence it is applicable to the general problem where the excitation is from ambient sources, which is generally immeasurable. On the other hand, there are different types of parametric models in system identification like Auto Regressive eXogenous model (ARX), Auto Regressive Moving Average eXogenous model (ARMAX), Box-Jenkins model, State Space Model etc. Most of the researches carried out using parametric models were confined to the evaluation of damping and frequency of damaged / undamaged structures as an indicative of damaged state of the structure. However, in this study the stiffness of the structure is calculated to detect the location and extent of the damage in the structure. Parametric modeling can be used as a faster way to detect the state of a structure soon after it undergoes an earthquake. Researches on parametric modeling in recent years are more towards the use of these models in damage detection by using active sensing (Lynch, 2004). Here, parametric models have been used for damage detection from the data collected using passive sensors. Lynch (2004) introduced the concept of linear classifications of poles obtained from the parametric identification of a structure. The poles of ARX time-series models describing modal frequencies and damping ratios were plotted upon the discrete-time complex plane and perception linear classifiers were employed to determine if poles of the structural element in an unknown state (damaged or undamaged) could be distinguished from those of the undamaged structure. But, it is observed the damage cannot be localized by this method. Chakraverty (2005) proposed a simple and concise identification procedure using Holzer criteria along with other numerical technique.

In this study, the method of detection of location and evaluation of the extent of the damage in a structure using parametric State Space model has been carried out. Further, an iterative procedure has been developed to determine the eigenvectors and the stiffness coefficients of multi-storeyed shear buildings with non-uniform mass distribution in vertical direction.

DAMAGE LOCALIZATION

In the State Space form, the relationship between the input, noise and output signals is written as a system of first order differential or difference equations using an auxiliary state vector $x(t)$ (Ljung, 1987). Unlike other parametric models, the State Space model has the advantage wherein the state vectors provide more insight to the physical state of the system. The state vector $x(t)$ has the physical significance like positions, velocities etc. and the outputs can be expressed as the known combinations of the states. This is the reason for adopting State Space Modeling as the parameterized method in damage detection of the structure in the present work.

DETAILS OF SAMPLE BUILDING AND INSTRUMENTATION

A nine storey building of BSNL Guwahati (Fig. 1) has been instrumented to record the acceleration response histories of different floors during earthquakes. A tri-axial accelerometer has been mounted at the ground level to collect the ground acceleration. Uni-axial accelerometers are installed only at some limited floors in both X and Y directions of the building. Four uni-axial accelerometers are installed at first, third, seventh and top floor of the building along the X (shorter) direction while three uni-axial accelerometers are installed at first, fifth and top floor of the building along the Y (longer) direction of the building. The data collected from the mounted sensors in the building during earthquake on 11th and 14th of August, 2006 have been utilized for the current study.

Fig.1 *BSNL building in Guwahati*

Determination of the Modal Parameters of BSNL Building, Guwahati

First the various floor responses obtained from the accelerometers are filtered using the sixth order Butterworth filter of MATLAB tool box (version 7.0) and then the responses obtained are re-sampled at frequency of 0.06 sec which corresponds to the Nyquest frequency. Nyquest frequency is twice the highest frequency of the structure and by re-sampling to that frequency any external noise present beyond this frequency will be filtered without affecting the modal properties. The filtered re-sampled floor accelerations of various floors and the ground accelerations are given as input for N4SID model and the frequencies and the corresponding mode shapes are evaluated. Further as the even mode shapes and frequencies are corresponding to conjugate modes, they have been removed. The frequencies corresponding to a mode with 20% damping or more are ignored as they correspond to non-physical modes. The poles as extracted corresponding to acceleration data along the X-direction for the earthquake on 11th August, 2006 are plotted in Fig.2.

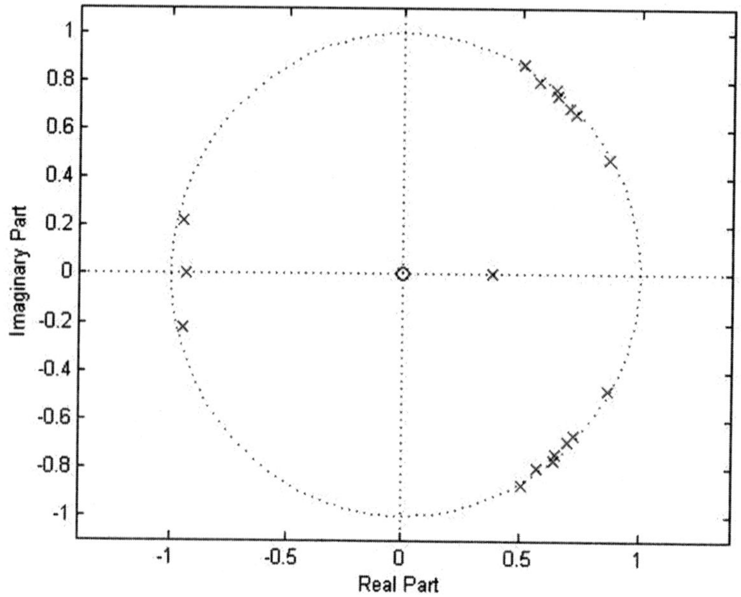

Fig. 2 Plotted Poles of BSNL building for Earthquake data(X-Direction) on 11th August, 2006

It is observed that the system is a stable system as the poles are within the unit circle. The obtained frequency and damping ratios corresponding to the acceleration along X and Y directions have been shown in Table 1 and 2.

Table 1 Frequencies and damping corresponding to the acceleration along X direction

Mode No.	Evaluated Frequency (rad/sec)	Damping Ratio (%)
1	8.357	4.16
2	12.43	2.46
3	13.03	2.35
4	14.21	1.52
5	15.87	1.77
6	48.54	0.92

Table 2 Frequencies and damping corresponding to the acceleration along Y direction

Mode No.	Evaluated Frequency (rad/sec)	Damping Ratio (%)
1	7.49	5.88
2	10.15	5.46
3	11.62	1.17
4	15.26	1.66
5	16.90	0.96
6	48.82	0.53

Similarly, modal parameters have also been obtained from the earthquake data on14th August, 2006. The poles have been plotted in Fig.3, while the frequencies and damping have been shown in Table 3 and 4.

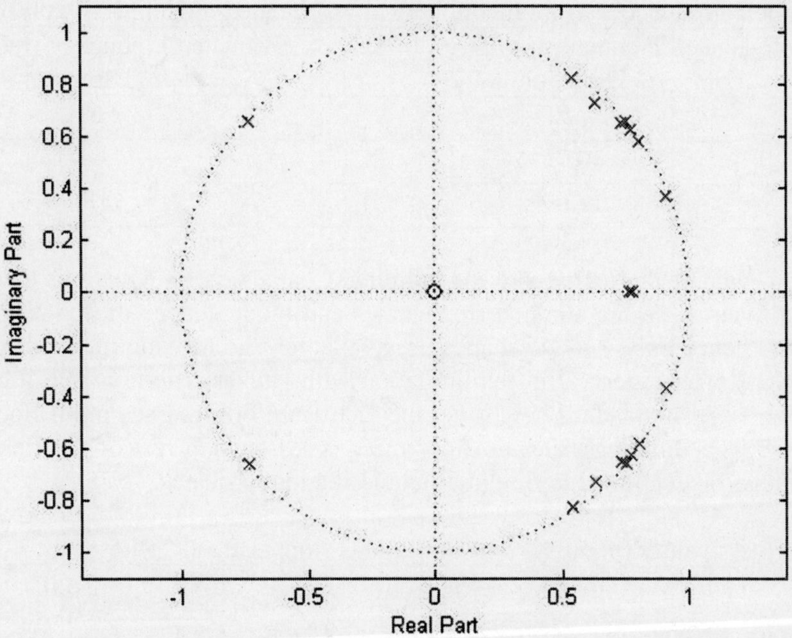

Fig.3 Plotted Poles of BSNL building for Earthquake data(X-Direction) on 14th August, 2006

Table 3 Frequencies and damping corresponding to the acceleration along X direction

Mode No.	Evaluated Frequency (rad/sec)	Damping Ratio (%)
1	8.19	4.41
2	12.57	1.40
3	13.63	1.35
4	14.53	1.82
5	17.10	1.39
6	48.18	0.71

Table 4 Frequencies and damping corresponding to the acceleration along Y direction

Mode No.	Evaluated Frequency (rad/sec)	Damping Ratio (%)
1	7.13	5.56
2	10.32	5.51
3	11.52	2.98
4	15.57	1.45
5	16.74	0.75
6	45.02	0.69

It is observed that there is not much of change in the modal parameters of the building as the magnitude of earthquakes have been very small and structure has not been undergone any major damage.

The bare frame model of the sample building has been simulated using SAP 2000 program and the predominant frequencies along X direction and Y direction are shown below in the Table 5 which shows close agreement with the identified modal parameters. The difference in the frequencies may be attributed to the stiffness of infill wall, which has not been considered in the numerically simulated model.

Table 5 Frequencies from the numerically simulated model (SAP 2000)

Mode No.	Evaluated Frequency (rad/sec) In X-Direction	Evaluated Frequency (rad/sec) In Y-Direction
1	7.20	6.82
2	10.71	8.41
3	19.09	18.48

The next step is to evaluate the stiffness of the building. Characteristics equations have been solved to evaluate the stiffness from the attained frequencies and mode shapes. In the X-direction, there were 4 sensors and hence only 4 mode shapes are available. The missing mode shapes have been evaluated using the Holzer criteria. But the limitation with Holzer criteria is that it can be applied only when the floors are of standard height i.e. they have uniform masses in all floors. But in the sample building (BSNL Building), the top floor mass is lesser than rest of the floors. Hence, an iterative approach has been utilized to find out the missing mode shapes.

The identified stiffness values (in 10^6 KN/m) for various storeys are as follows:

K_1	K_2	K_3	K_4	K_5	K_6	K_7	K_8	K_9
1.2443	2.0368	2.9244	2.4149	2.8266	2.3048	2.4772	2.7475	2.5555

The computed stiffness values (in 10^6 KN/m) of the corresponding floors are as follows:

K_1	K_2	K_3	K_4	K_5	K_6	K_7	K_8	K_9
1.0117	2.0645	2.0645	2.0645	2.0645	2.0645	2.0645	2.0645	2.0645

The calculated values are on the lower side as compared to identified stiffness. The additional stiffness is probably due to the contribution of infill walls etc.

Conclusions

The system identification using sub-space iteration (N4SID) and Least Square Method is an effective method of structural health monitoring for a multi-storeyed shear building. The data must be properly re-sampled and filtered to remove noise from recorded signals. Re-sampling at Nyquest frequency eliminates noise of higher range without affecting the modal properties. If the building is standard, then the missing mode shapes and the stiffness can be obtained even with limited number of sensors. The iterative procedure developed in this study is found to be effective for the evaluation of structural stiffness for buildings with non-uniform mass distribution. The method can be very effectively used for structural health monitoring even with limited number of sensors are available for structural response recording.

References

1. Chakravorty, S. (2005), "Identification of structural parameters of multistorey shear buildings from modal data", Earthquake Engineering and Structural Dynamics, 34, 543-554.
2. Ljung L. (1987), System Identification: Theory for the User, Prentice-Hall Inc.
3. Lynch J.P. (2004) "Linear Classification of System Poles for Structural Damage Detection using Piezoelectric Active Sensors", *SPIE 11th Annual International Symposium on Smart Structures and Materials*, San Diego, CA, USA, March.

POSSIBLE COUPLING BETWEEN SEISMIC ACTIVITY AND GROUNDWATER CHEMISTRY AROUND THE SHILLONG PLATEAU, NORTH EASTERN INDIA

Mahanta, C.[1] *Skelton, A.D.L.*[2]*, Claesson, L.*[2]*, Chakrapani, G.*[3]*, Routh, J.*[2]*, Mörth, M.*[2] *and Khanna, P.P.*[4]

1. Department of Civil Engineering, Indian Institute of Technology, Guwahati 781039, India
2. Department of Geology and Geochemistry, Stockholm University, Stockholm, Sweden
3. Department of Earth Sciences, Indian Institute of Technology, Roorkee, 247667, Uttaranchal, India
4. Wadia Institute of Himalayan Geology, DehraDun, India

1 INTRODUCTION

Transient changes in groundwater chemistry can be caused by a number of factors. The chemistry of near-surface groundwater can be influenced by rainfall, snowmelt, temperature variability and other near-surface atmospheric or hydrological processes. Changes in groundwater chemistry can also reflect contamination events, such as leakage from landfills and other waste disposal sites. Deeper sourced groundwater is less strongly influenced by these relatively short timescale processes which occur at the Earth's surface and may thus provide a more sensitive monitor of permeability changes in the Earth's crust.

The permeability of any portion of the Earth's crust is a function of its porosity and the degree to which individual pores and/or cracks are interconnected. Enhanced permeability can arise, either due to the opening of additional porosity or crack space, or if originally isolate pores or cracks become interconnected. These types of physical changes are inferred to occur in conjunction with seismic activity. Changes to the permeability of a portion of the Earth's crust can influence the flow path and source of groundwater. This may then influence chemical transport by groundwater flow. This has led to a range of hypotheses which suggest that earthquakes could be spatially and/or temporally coupled with chemical changes in groundwater. Several authors (Ulomov and Mavashev, 1971; Wakita, 1988; Thomas, 1988; Tsunogai and Wakita, 1995 and Claesson et al., 2004) report chemical changes in groundwater which occurred weeks to months before an earthquake. The magnitude of the earthquakes which were studies by these authors ranges from M 5 to M 8. These authors relate the observed changes in groundwater chemistry to changes in crustal permeability. This interpretation concurs with interpretations based on investigation of pre-seismic phenomena, such as changes in V_P/V_S, electrical conductivity and shear wave splitting in that the observed phenomena is thought to reflect redistribution of water in the crust. Claesson et al. (2004) report chemical changes in groundwater which occurred several days after an M 5.8 earthquake, which occurred north of Iceland. Claesson et al. (in prep.) relate these changes to the opening of a fracture pathway to a source region that was not tapped before the earthquake. They monitor the progressive sealing off of this source region over a period of approximately 2 years. This could be caused by mineral growth along fracture surfaces progressively clogging the fracture pathway.

In this study, we present and interpret hydrogeochemical data obtained from December 2003 to March 2005. During the period October 2004 to March 2005, 60 groundwater samples were collected on average every 2-3 days at a commercial bottling plant (Silver Drop), which is located at latitude 26° 12' 24.66" and longitude 91° 41' 27.28". This location is on the northern flank of the Shillong Plateau and close to the inferred subterranean culmination of the inferred Oldham Fault (Bilham and England, 2001; Mitra et al., 2005). We were able to extend our time series to include the period December 2003 to September 2004, by analyzing 10 samples of groundwater, which has been bottled at the plant for commercial purposes.

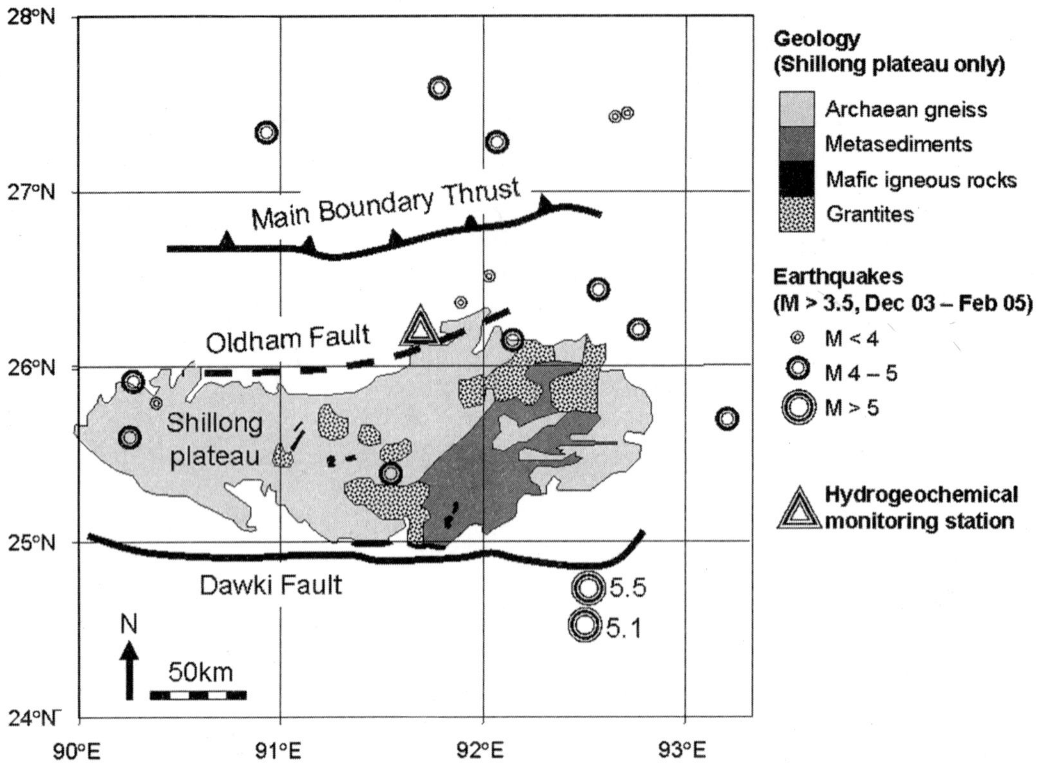

Fig. 1. Location map showing major tectonic features, regional geology, the location of the hydrogeochemical monitoring station, north of the Shillong plateau, and the epicenters of all M > 3.5 earthquakes which occurred within ca. 200 km of the monitoring station between December 2003 and February 2005. The epicenters of the two M > 5 earthquakes which occurred on December 9, 2004 (M 5.5) and February 15, 2005 (M 5.1), south of the Shillong plateau, are highlighted.

2 GEOLOGICAL BACKGROUND

The Brahmaputra valley and northern Bengal Basin, which flank the 1.6 – 2 km elevated Shillong Plateau in northeastern India are both densely populated and strongly affected by earthquakes (Fig. 1). The region, which is subjected to an average of 20-30 M ≥ 5 earthquakes per decade (1973-2005, http://neic.usgs.gov/), suffered major loss of life and infrastructural damage caused by the great 1987 Assam earthquake. The Shillong Plateau is interpreted as a 'pop-up' structure, which is bounded to the south, by a northward-dipping reverse fault (the Dauki fault), and to the north, by a buried southward-dipping reverse fault (the Oldham fault). The existence of the Oldham fault was inferred by Bilham and England (2001) based on their reappraisal of surface distortion caused by the 1987 earthquake, which indicated an 18 ± 7 m slip on this buried fault. Their interpretation of the Shillong Plateau as a 'pop-up' structure between the Dauki and Oldham reverse faults is supported by the teleseismic profile published by Mitra et al. (2005). The groundwater used in this study was sourced at a depth of 110 m within the Archaean gneiss which forms much of the uplifted Shillong Plateau.

3 METHODOLOGY

In this study, we present and interpret hydrogeochemical data obtained from December 2003 to March 2005. During the period October 2004 to March 2005, 60 groundwater samples were collected on average every 2-3 days at a commercial bottling plant (Silver Drop), which is located at latitude 26° 12' 24.66" and longitude 91° 41' 27.28". This location is on the northern flank of the

Shillong Plateau and close to the inferred subterranean culmination of the Oldham Fault. The groundwater was sourced at a depth of 110 m within the Archaean gneiss which form the uplifted Shillong Plateau. We were able to extend our time series to include the period December 2003 to September 2004, by analyzing 10 samples of groundwater, which has been bottled at the plant for commercial purposes. Groundwater samples were analyzed for Al, Mn, Ni, Cu, Pb, Cr, Fe, Co, Ba and Sr using a ICP-MS at IIT-Roorkee, and for Al, As, B, Ba, Be, Ca, Cd, Ce, Co, Cr, Cu, Eu, Fe, K , La, Li, Mg, Mn, Mo, Na, Ni, P, Pb, Rb, S, Sc, Si, Sr, Ti, V, Y, Yb, Zn and Zr using a ICP-OES at Stockholm University.

4 RESULTS

The concentrations of As, B, Be, Cd, Co, Cr, Cu, Eu, Fe, La, Mo, Ni, P, Pb, Rb, Sc, Ti, V, Y, Yb and Zr in the analyzed groundwater samples were below detection limit, and there were no significant variations in the concentrations of Al, Ce, K, Li, Mg, Mn, S and Si. The concentrations of Ca and Sr increased from baseline values of 18.6 ± 0.6 ppm and 91 ± 3 ppb to maxima of 28.8 ppm and 138 ppb between February 9 and 12 (Fig. 2a). For both Ca and Sr, the 6 samples collected over the next 12 days exhibited a power law return to the baseline value, with $R^2 > 0.99$. The behavior of Fe was shown to be similar to Ca and Sr in the ITT-Roorkee data. However, because this result could not be replicated at Stockholm University, Fe is excluded from our analysis. The concentration of Na increased from a baseline value of 17.2 ± 1.3 ppm in December 2004, reached a maximum of 25-35 ppm in January 2005 and decreased to 17.3 ± 0.7 ppm in February 2005 (Fig. 1b). The concentration of Ba decreased from a baseline value of 19.9 ± 1.5 ppb to a minimum of 7.4 ppb between February 9 and 17 (Fig. 2b). The concentration of Ba increased steadily after February 17, but had not returned to the basement value on February 28. The concentration of Zn increased from a baseline of 21 ± 6 ppb to a maximum of 153 ppb between March 15 and April 10, 2004 (Fig. 2c). The 60 samples collected over the next 10 months exhibited a power law return to the baseline value, with $R^2 = 0.78$.

(a)

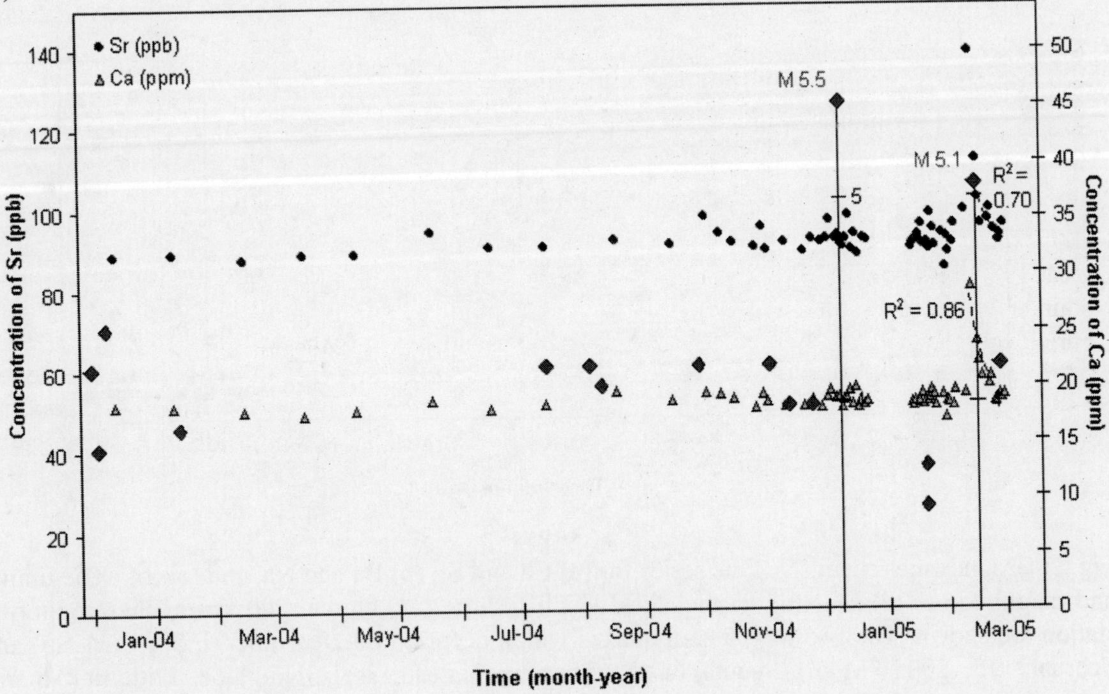

(b)

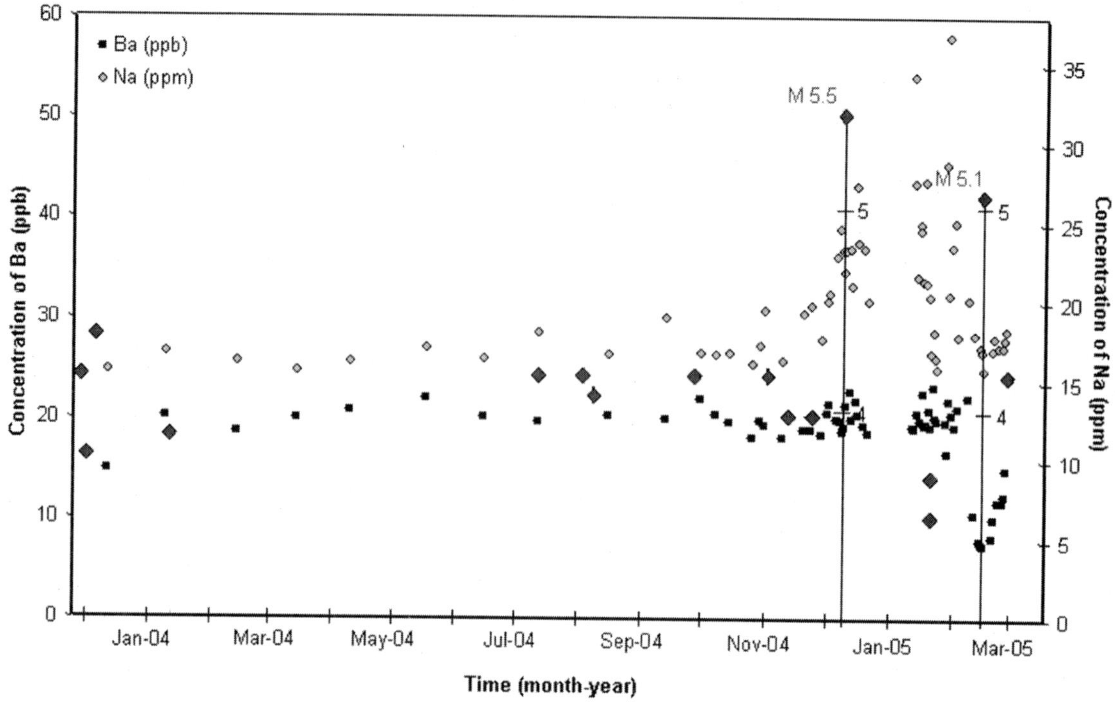

(c)

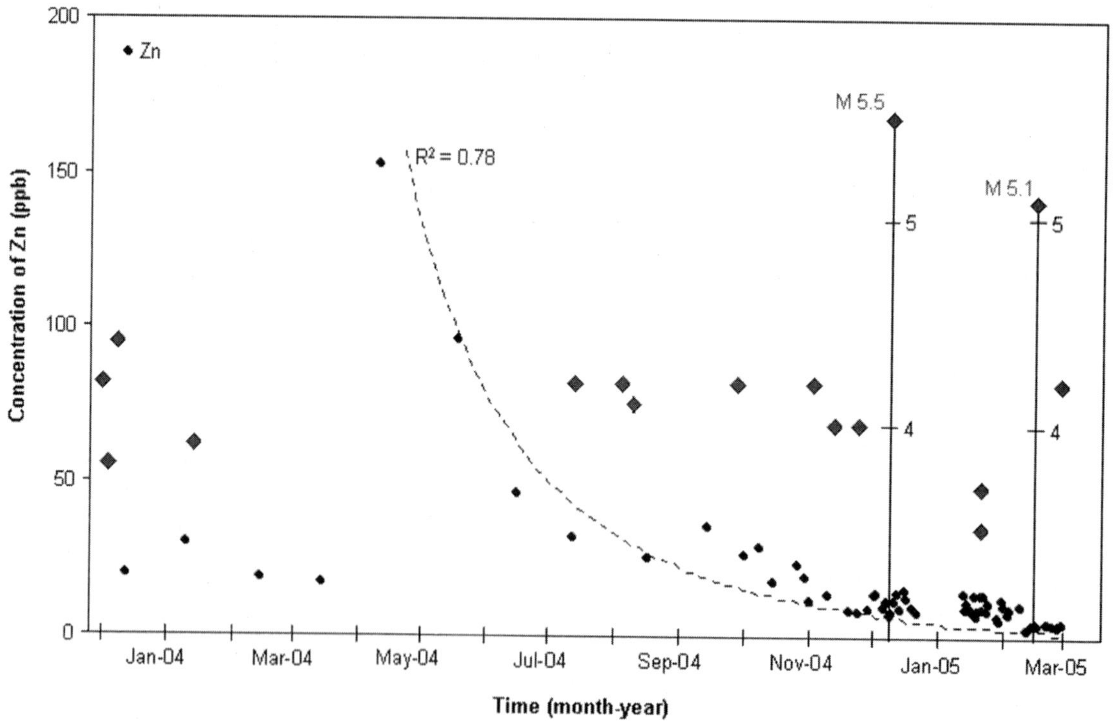

Fig. 2. Groundwater chemistry time series for (a) Ca and Sr, (b) Ba and Na, and (c) Zn. The timing and magnitude of all M > 3.5 earthquakes which occurred within ca. 200 km of the monitoring station are shown. The two M > 5 earthquakes which occurred on December 9, 2004 (M 5.5) and February 15, 2005 (M 5.1), south of the Shillong plateau, are highlighted, with tick marks indicating magnitude.

5 DISCUSSION

To help assess whether or not the transient chemical shifts shown in Figs. 2a-c could be related to seismic activity, we have plotted the epicenters of the sixteen M > 3.5 earthquakes which occurred within about 200 km of the hydrogeochemical monitoring station between December 2003 and March 2005 on Fig. 1. The magnitude and timing of these events are shown on the time series plots where they can be compared with the hydrogeochemical data (Figs. 2a-c). During the study, only two earthquakes were stronger than M 5. Both of these earthquakes occurred south of the Shillong Plateau. On December 9, 2004, an M 5.5 earthquake occurred at latitude 24° 33' 0" and longitude 92° 31' 12", and on February 15, 2005, an M 5.1 earthquake occurred at latitude 24° 45' 36" and longitude 92° 32' 24". The respective distances between the epicenters of these earthquakes and the site where samples were collected are 206.5 km and 187 km. The other fourteen earthquakes were considerably weaker, ranging from M 3.5 to M 4.4.

There is no correlation between the fourteen M 3.5 – 4.4 earthquakes and the hydrogeochemical data (Figs. 2a-c). The Na anomaly spans the time interval December 2004 to mid-February 2005. The two M > 5 earthquakes occurred within this time interval. The Ca and Sr maxima occurred three days before the M 5.1 earthquake on February 15, 2005. However, no similar Ca or Sr maximum coincided with the larger M5.5 earthquake on December 9, 2004. The concentration of Ba in groundwater dropped significantly three days before the M 5.1 earthquake on February 15, 2005, reaching its minimum two days after this earthquake. We conclude that (i) changes in Ca, Sr and Ba concentrations could be related to the M 5.1 earthquake on February 15, 2005, and (ii) the Na anomaly could be related to the increased seismic activity from December to mid-February. However, we note the absence of concentrations changes for Ca, Sr and Ba in conjunction with the M 5.5 earthquake which occurred on December 9, 2004, close to the epicenter of the M 5.1 earthquake. We further note that the Zn anomaly occurred during a period of seismic quiescence.

Ba/Sr

The Ba/Sr ratio in groundwater has been used to compare the inputs from K- and Ca-bearing minerals in the source region (Land et al., 2000). This is possible because Ba, which generally occurs in trace quantities, substitutes for K in potassic rock-forming minerals, whereas Sr, which also occurs in trace quantities, substitutes for Ca in calcic rock-forming minerals. The K-bearing minerals which commonly occur in a felsic gneiss, which is the main rock type in the source region at Silver Drop, include biotite and K-feldspar. These minerals are generally more resistant to solution by water-rock interaction processes, compared with plagioclase, which is generally the most common Ca-bearing mineral in a felsic gneiss. Thus Ba/Sr can be used as a proxy for enhanced water-rock interaction processes.

Fig. 3 shows the time series for the Ba/Sr ratio at Silver Drop. This ratio drops from 0.22 ± 0.01 to 0.08 three days before the M 5.1 earthquake on February 15, 2005. After this earthquake, the Ba/Sr ratio increased steadily, but baseline values were not yet reached on February 28, 2005.

This sudden decrease of the Ba/Sr ratio is consistent with the opening of additional porosity and/or crack space. This would increase the rock surface area exposed to fluid-rock interaction processes and the more readily dissolved plagioclase surfaces would react more quickly. The subsequent increase of the Ba/Sr ratio could either reflect the slower onset of biotite and/or K-feldspar dissolution or the progressive clogging of the newly opened pore space by clay minerals which are formed as a by-product of the dissolution process.

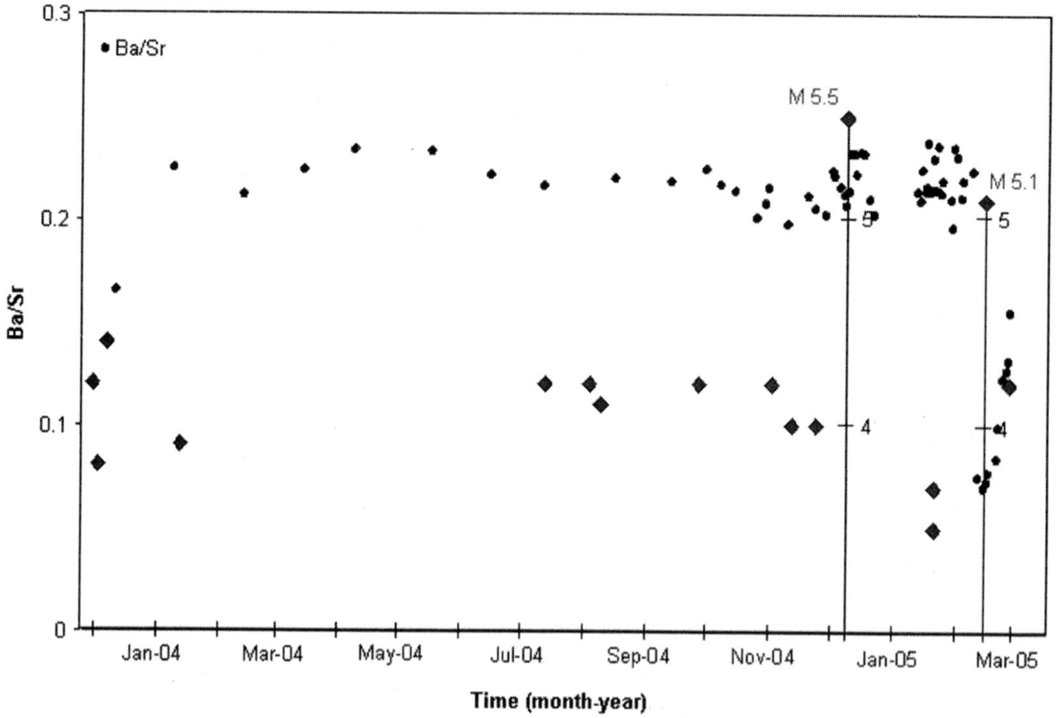

Fig. 3. Groundwater Ba/Sr ratio. The timing and magnitude of all M > 3.5 earthquakes which occurred within ca. 200 km of the monitoring station are shown. The two M > 5 earthquakes which occurred on December 9, 2004 (M 5.5) and February 15, 2005 (M 5.1), south of the Shillong plateau, are highlighted, with tick marks indicating magnitude.

This behavior of the Ba/Sr ratio thus provides strong evidence that the changes in groundwater chemistry, which occurred between February 9 and 12, 2005, were seismically-induced and could therefore be related to the M 5.1 earthquake on February 15. This conclusion raises a number of important questions. Firstly, why did the chemical shift preceded the earthquake by 3 to 6 days? Secondly, what mechanism could explain coupling between a change of groundwater chemistry, north of the Shillong plateau 'pop-up' structure, and an earthquake, ca. 200 km away, south of the Shillong plateau? Thirdly, if the chemical changes were seismically induced, why was there no similar chemical response to the earlier, stronger M 5.5 earthquake on December 9, 2004?

The first question has been tackled in other studies of pre-seismic phenomena. These studies provide evidence that crustal permeability can respond to the build up of stress before an earthquake. This might occur locally, perhaps by the opening of new crack surfaces within a few centimeters of an aquifer. The reported timescale of this process ranges from a few days to a few months.

Regarding the second question, this is not the first study to infer that pre-seismic phenomena occurred at a considerable distance from the earthquake's epicenter. The two M > 5 earthquakes on December 9, 2004 and February 15, 2005 occurred at depths of 34 and 35 km, respectively. In the extended crustal section, close to the Shillong plateau, these epicenters are within 5 km of the base of the crust, where its plastic nature may result in pre-seismic strain affecting a considerable region of the crust. The existence of a structural 'pathway' comprised of the Dauki and inferred Oldham Faults, between the epicenter and the monitoring station might also extend the size and influence the geometry of the region affected by pre-seismic stress (Fig. 4). Indeed, the inferred seismic-hydrogeochemical coupling would support the existence of the inferred Oldham Fault (Bilham and England, 2001; Mitra et al., 2005).

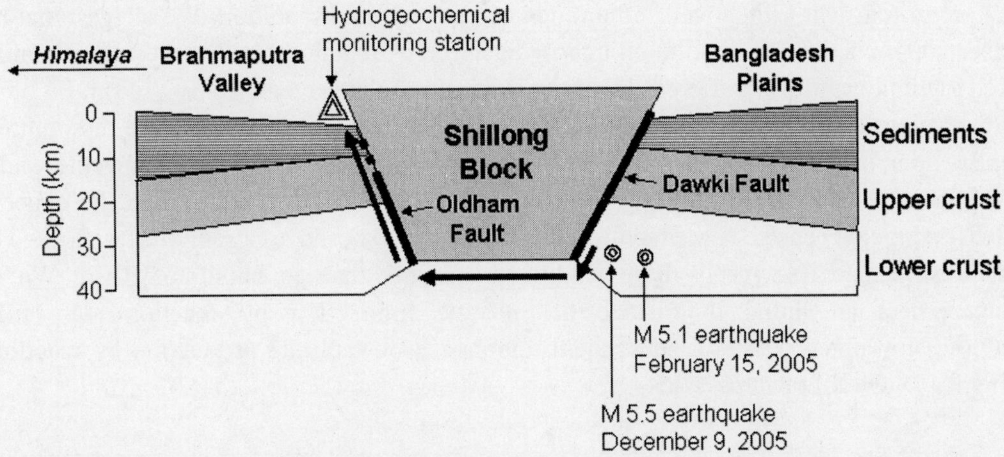

Fig. 4. Schematic cross-section of the Shillong 'pop-up' structure (modified after Mitra et al., 2005). The structural 'pathway' between the epicenter of the M 5.1 earthquake which occurred on February 15, 2005 (M 5.1), south of the Shillong plateau and the hydrogeochemical monitoring station, north of the Shillong plateau, is highlighted.

The third question is problematic and may simply point to the randomness of (or highlight our limited understanding of) the inferred seismic-hydrogeochemical coupling process. The lack of a similar chemical shift associated with the M 5.5 earthquake provides a clear warning that any attempt to use hydrogeochemical as a predictive tool must proceed cautiously.

Na

In gneiss, Na is largely concentrated in plagioclase. Na is released from the Na-component of the plagioclase solid solution (albite) by the reactions (Garrels and Mackenzie, 1967):

$$2\ NaAlSi_3O_8 + 2\ CO_2 + 3\ H_2O = Al_2Si_2O_5(OH)_4 + 2\ Na^+ + 2\ HCO_3^- + 4\ SiO_2(aq) \quad [1]$$
(albite) (kaolinite)

$$2\ NaAlSi_3O_8 + 2\ CO_2 + 2\ H_2O = Al_2Si_4O_{10}(OH)_2 + 2\ Na^+ + 2\ HCO_3^- + 2\ SiO_2(aq) \quad [2]$$
(albite) (smectite)

Reaction [1] releases Na^+ and $SiO_2(aq)$ in the proportion 1:2 ($Na^+/SiO_2(aq) = 0.5$), whereas reaction [2] releases Na^+ and $SiO_2(aq)$ in the proportion 1:1 ($Na^+/SiO_2(aq) = 1$). Fig. 5 shows the time series for the $Na^+/SiO_2(aq)$ ratio at Silver Drop. This ratio increased from 0.56 ± 0.04 to a maximum of 0.9 – 1.2 during the time period December 2004 to mid-February 2005. This time period is also characterized by increased seismic activity. The 'accumulated earthquake magnitude' within ca. 200 km of Silver Drop for the 3-month period December 2004 to February 2005 is 7.6, compared with 6.6, 0.0, 6.7 and 6.7 for the previous four 3-month periods (Table 1).

Table 1. The accumulated earthquake magnitude (**ΣM**) was calculated by summing the magnitudes (**M**) of the M > 3.5 earthquakes within ca. 200 km of Silver Drop using the expression: $\Sigma\mathbf{M} = \mathbf{Log}_{10}\left(\sum \mathbf{M}^{10}\right)$.

Time interval	M > 3.5 earthquakes	Accumulated earthquake magnitude
Dec 03 – Feb 04	3.8, 4.4, 3.9	6.6
Mar 04 – May 04		0.0
Jun 04 – Aug 04	4.2, 4.2, 4.1	6.7
Sep 04 – Nov 04	4.2, 4.0, 4.0	6.7
Dec 04 – Feb 05	5.5, 3.5, 3.7, 5.1, 4.2	7.6

We interpret a transient switchover in the mode of plagioclase dissolution. We suggest that before December 2004, plagioclase dissolution mainly occurred by reaction [1]. This reaction would have produced kaolinite and thus buffered the groundwater to the observed $Na^+/SiO_2(aq)$ ratio of ~ 0.5. We further suggest that from December 2004 to mid-February 2005, plagioclase dissolution was transiently dominated by reaction [2]. This reaction would have produced smectite and thus increased the groundwater $Na^+/SiO_2(aq)$ ratio towards ~ 1. This behavior might be expected if increased seismicity caused new fracture surfaces to be exposed to groundwater flow. This is because fresh plagioclase reacts more readily to smectite than to kaolinite (REF). We might therefore expect an initial dominance of smectite production by reaction [2], buffering $Na^+/SiO_2(aq)$ towards ~ 1, and a subsequent dominance of kaolinite production by reaction [1], buffering $Na^+/SiO_2(aq)$ towards ~ 0.5.

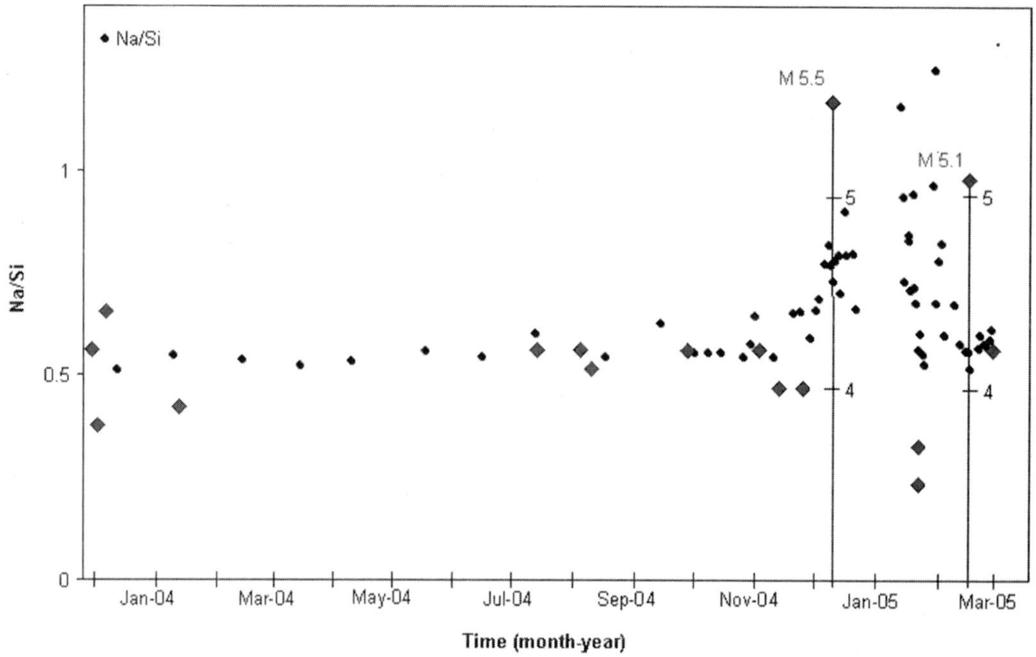

Fig. 5. Groundwater $Na^+/SiO_2(aq)$ ratio. The timing and magnitude of all M > 3.5 earthquakes which occurred within ca. 200 km of the monitoring station are shown. The two M > 5 earthquakes which occurred on December 9, 2004 (M 5.5) and February 15, 2005 (M 5.1), south of the Shillong plateau, are highlighted, with tick marks indicating magnitude.

Zn

The concentration of Zn in groundwater is not only sensitive to natural factors, e.g. the geological characteristics of the source region, but also to anthropogenic factors, e.g. transient inputs associated with mining (acid mine runoff), industry (steel production), coal burning and waste disposal. We remain uncertain as to the cause of the Zn anomaly in April 2004. However, we can infer that the likelihood that it relates to seismic coupling is very low, because it occurs during a period of seismic quiescence (Table 1).

6 CONCLUSIONS

Transient, but significant, shifts in the concentrations of Ca and Sr, and the Ba/Sr ratio in groundwater samples collected at the Silver Drop commercial bottling plant, north of the Shillong Plateau occurred 3-6 days before an M 5.1 earthquake, which occurred on February 15, 2005, ca.

200 km from the bottling plant and south of the Shillong Plateau. We tentatively suggest seismic-hydrogeochemical coupling, with local permeability changes induced by pre-seismic stress, perhaps 'transported' along the Dauki – Oldham fault system. However, we note the lack of any similar hydrogeochemical anomaly associated with the larger M 5.5 earthquake, which occurred on December 9, 2004 in the same region. We further conclude that the $Na^+/SiO_2(aq)$ ratio of groundwater was transiently increased during the time interval December 2004 to mid-February 2005. This could also result from hydrogeochemical-seismic coupling if newly-opened fractures exposed fresh plagioclase to groundwater. Initially, plagioclase would preferentially alter to smectite and $Na^+/SiO_2(aq)$ would increase. Subsequently, plagioclase would preferentially alter to kaolinite and $Na^+/SiO_2(aq)$ would decrease. We remain uncertain as to whether a Zn anomaly in April 2004 was the result of a natural or anthropogenic change. We conclude that it is likely that hydrochemical shifts detected north of the Shillong Plateau from November 2004 to February 2005 were caused by pre-seismic stress prior to two earthquakes which occurred on December 9, 2004 and February 15, 2005, south of the Shillong Plateau. Two types of hydrochemical shift were detected, which based on petrographic observation, probably record switchovers in the mode of feldspar alteration, represented by the reaction sequence [1] to [3].

REFERENCES

1 Ulomov, V.I. & Mavashev, B.Z. The Tashkent Earthquake of 26 April. *Tashkent: Acad. Nauk. Uzbeck.* SSR, FAN (1971).

2 Igarashi, G. *et al.* Ground-water radon anomaly before the Kobe earthquake in Japan. *Science* **269,** 60-61 (1995).

3 Tsunogai, U. & Wakita, H. Precursory chemical changes in groundwater: Kobe earthquake, Japan. *Science* **269,** 61-63 (1995).

4 Claesson, L., Skelton, A., Graham, G., Dietl, C., Mörth, M., Torssander, P. & Kockum, I. Hydrogeochemical changes before and after a major earthquake. *Geology* **32,** 641-644 (2004).

5 Johansen, A., Saleur, H. & Sornette, D. New evidence of earthquake precursory phenomena in the 17 January 1995 Kobe earthquake, Japan. *Eur. Phys. J. B* **15,** 551-555 (2000).

6 Scholz, C.H., Sykes, L.R. & Agarwal, Y.P. Earthquake prediction: A Physical Basis. *Science* **181,** 803-810.

7 Wyss, M. Second round evaluation of proposed earthquake precursors. *Pure Appl. Geophys.* **149,** 3-16 (1997).

8 Geller, R.J., Jacksom, D.D., Kagan, Y.Y. & Mulargia, F. Earthquakes cannot be predicted. *Science* **275,** 1616 (1997).

9 Knopoff, L., Levshina, T., Keilis-Borok, V. I. & Mattoni, C. Increased long-range intermediate-magnitude earthquake activity prior to strong earthquakes in California. *J. Geophys. Res.* **101**, 5779-5796.

10 Allègre, C.J., Le Mouel, J. L. & Provost, A. Scaling rules in rock fracture and possible implications for earthquake prediction. *Nature* **297,** 47-49 (1982).

11 USGS/NEIC Preliminary Determinations of Epicenters, Monthly Listing.

12 Bilham, R. & England, P. Plateau 'pop-up' in the great 1897 Assam earthquake. *Nature* **410,** 806-809.

13 Mitra, S., Priestley, K., Bhattacharyya, A. K. & Gaur, V.R. Crustal structure and earthquake focal depths beneath northeastern India and southern Tibet. *Geophys J. Int.* **160,** 227-248 (2005).

14 Garrels, R.M. & Mackenzie, F.T. Origin of the chemical composition of some springs and lakes. *Adv. Chem. Ser.* **67,** 222-242 (1967).

15 Land, M., Ingri, J., Andersson, P.S. & Öhlander, B. Ba/Sr, Ca/Sr and $^{87}Sr/^{86}Sr$ ratios in soil water and groundwater: implications for relative contributions to stream water discharge. *Appl. Geochem.* **15,** 311-325 (2000).

16 Busenberg, E. & Clemency, C.V. The dissolution kinetics of feldspars at 25°C and 1 atm CO_2 partial pressure. *Geochim. Cosmochim. Acta.* **40,** 41-49 (1976).

17 Stillings, L.L. & Brantley, S.L. Feldspar dissolution at 25°C and pH 3: Reaction stoichiometry and the effect of cations. *Geochim. Cosmochim. Acta.* **59,** 1483-1496 (1995).

18 Aleur, H., Sammis, C.G. & Sornette, D. Discrete scale invariance, complex fractal dimensions, and logperiodic fluctuations in seismicity. *J. Geophys. Res.* **101,** 17661-17677 (1996).

SEISMIC DESIGN OF NUCLEAR FACILITIES IN INDIA - ISSUES AND R&D EFFORTS

G.R. Reddy
Reactor Safety Division, Bhabha Atomic Research Centre
Mumbai-400085, India. email: rssred@barc.gov.in

ABSTRACT

Safety of the nuclear facilities and the environment around the plant should be ensured against all natural hazards including the earthquakes. Keeping the public safety as the main goal, nuclear facilities are designed to withstand low probability high magnitude earthquakes. Nuclear facilities comprise Structures, Systems and Components (SSCs). Integrity and serviceability of the SSCs of nuclear facility should be ensured against normal and accidental loading such as earthquakes. In this paper, brief seismic design aspects of the SSCs of the nuclear facilities are discussed. Some of the seismic design issues are identified. In addition the Research and Development work being carried out along with various research centers and Institutions to resolve some of the seismic design issues are also discussed.

Keywords: Earthquakes, Nuclear Facilities, Seismic Issues, Spectra, Time History, Damping, Ratcheting, Passive control, Re-qualification, Retrofitting

1 INTRODUCTION

Structures, support systems and components in Nuclear facilities are referred to civil structures, piping systems, electrical systems and control & instrumentation. Components are generally referred to mechanical and electrical equipment. Safety of Structures, Systems and Components (SSCs) as shown in Fig.1, implies structural integrity and leak tightness of pressure boundaries & radiation boundaries.

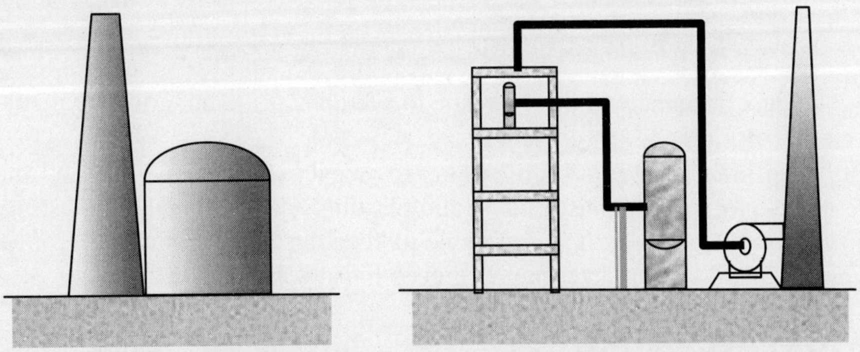

Fig.1 Schematic of Structures, Systems and components

To meet this requirement, the SSCs need to be designed for normal operating loads such as dead loads, supporting systems and component weights, pressure, temperature, normal operating vibratory loads etc and accidental loads caused due to both external and internal events such as Loss Of Coolant Accident (LOCA), natural calamities, and man made events. The severity of the external events depends on site, where the facility is going to come up. To ensure better safety of the facility, proper site need to be selected and shall be examined with respect to the frequency and severity of natural phenomenon and man induced events. The various aspects that need to be considered in site selection are as follows [1].

a. Seismology and geology
b. Hydro-Geology
c. Accessibility by road, rail etc.
d. Flooding potential
e. Potential for external man-induced events

Detailed Seismological and geological studies need to be performed considering the aspects of capable faults, frequency of different magnitude earthquakes (i.e. seismic activity), slope instability, liquefaction potential etc. The area of 300 km radius for power plants and 100 km radius for facilities handling limited radioactivity need to be considered. Based on the comprehensive investigations, design basis earthquake motion need to be evolved.

Hydrogeology studies need to be performed to evaluate the fluctuations of ground water levels, which are required for structural design in view of leak tightness and foundation stability. To evaluate realistic design water levels data of nearest ground water source, principal ground water discharge points, ground water flow directions, ground water related human activities etc. need to be considered.

The elevation of sites shall be fixed at the design stage, above the design basis flood level resulting either from probable maximum precipitation with mean recurrence interval of 1000 years or otherwise as applicable. Due considerations shall be given to aspects such as the site located down stream or up stream of the dam (Fig.2), tropical cyclones, wind waves as appropriate to inland and coastal sites.

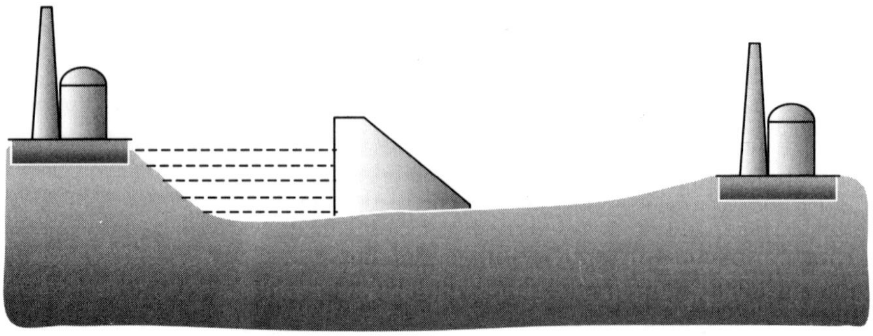

Fig.2 Schematic of the plant site located down stream or up stream of dam

Probability of man induced events such as aircraft impact (Fig. 3), storage and manufacture of inflammable and explosive materials, mining and blasting activities need to be estimated and need to be with in 10^{-6} events per year. It is preferable to keep the sites 10 km away from airports, 5 km from mines and industries manufacturing inflammable materials.

2 REGULATORY REQUIREMENTS

The general regulatory safety requirement for all nuclear installations is to protect individuals, society and the environment from harm by establishing and maintaining in the nuclear installation, effective defence safeguard against radiological hazards. All operational radiation exposure within the installation or exposure due to planned release of radioactive material from the installation is to be kept below prescribed limits and as low as reasonably achievable [2,3]. All measures must be taken to prevent accidents in nuclear installations and to mitigate their consequences if they occur. The structures designed should provide safety functions, which provides controlled environment for safe operation of SSCs such that no fault can occur due to effects of site/plant specific Postulated Initiating Events (PIEs), which might other wise have caused release of radioactivity.

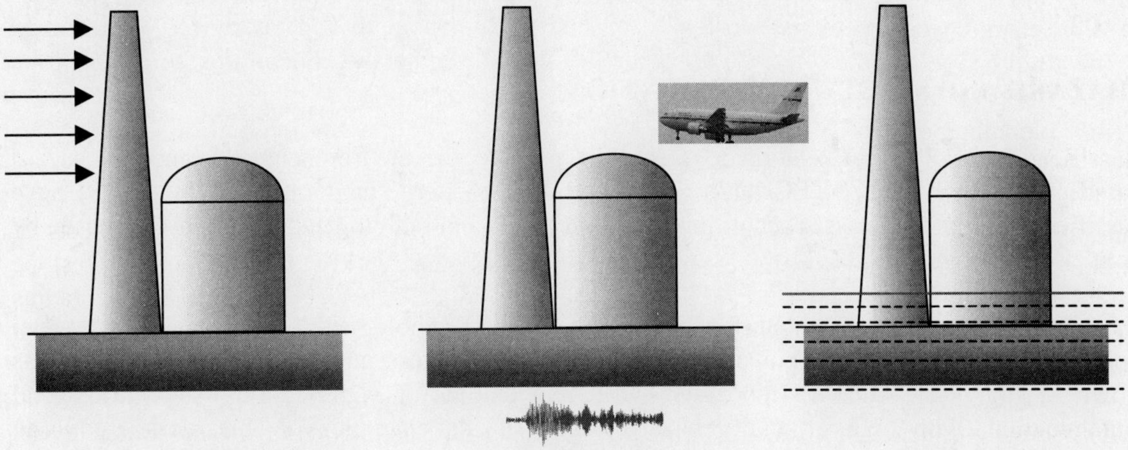

a. Wind Loads

b. Earthquake loads and air craft Impact

c. Flood Loads

Fig.3 Facility subjected to some of the external loads

To meet the above regulatory requirements the aspects to be considered for design of SSCs [2-10] are summarized as follows

(i) Structural strength
(ii) Confinement or isolation (ventilation systems).
(iii) Leak tightness serviceability [Leak tightness of containments, pools, liquid retaining structures etc. through liners or prestressing]
(iv) Avoidance of drop loads.
(v) Functionality of safety related items (e.g. heat removal, ventilation, fire control, etc.)
(vi) Appropriate anchoring of items.
(vii) Avoidance of interaction problems: mechanical, chemical or through impairment of operator action.
(viii) Stability of support ground during earthquakes, winds and floods.
(ix) Radiological shielding.

Defense in depth [11, 12] approach should be applied to the nuclear power plant facilities. Application of the concept of defense in depth in the design of the nuclear facilities provides a series of levels of defense (inherent features, equipment and procedures), which are aimed at preventing accidents and ensuring appropriate protection in the event that prevention fails. The aim of the first level of defense is to prevent deviations from normal operation and to prevent system failures by applying redundancy, independence and diversity in design. The aim of the second level of defense is to control (by detection and intervention), deviations from operational states so as to prevent anticipated operational occurrences from escalating to accident conditions. The third level of defense leads to the requirement that engineered safety features shall be provided, which are capable of transferring the nuclear facilities, first to a controlled state and subsequently to a safe shutdown state, and of maintaining at least one barrier for the confinement of radioactive material. The aim of the fourth level of defense is to address beyond design basis accidents (BDBAs) in which the design basis may be exceeded and to ensure that radioactive releases are kept as low as practicable. The most important objective for this level is the protection of the confinement function. The fifth and final level of defense is aimed at mitigation of the radiological consequences of potential releases of radioactive material that may result from accident conditions using emergency control center. In case of nuclear facilities other than NPPs, the number of levels could be maintained, but the number and role of the barriers could be different from that of a power plant, depending on the accident scenarios and the reliability that may be associated with any defence level.

3 HAZARD AND SAFETY CLASSIFICATION

Classification of SSCs in relation to risk, they pose to the environment is termed as hazard classification. Only IAEA TECDOCs [2, 3] and IAEA safety standard series [4, 5, 13] have classified SSCs in four classes according to hazard levels and radiological contamination posed by them.

Class 1 (high hazard): potential for significant off-site radiological contamination e.g. Fuel reprocessing plants and high power (> 2MW) research reactors with high criticality hazard as shown in Fig. 4(a); Class 2 (moderate hazard): potential for significant on-site radiological contamination, with high criticality hazard; some of the examples of this class are waste management plants and small size research reactors as shown in Figs. 4(b) & (c) Class 3 (low hazard): potential for significant on-site radiological contamination; Fuel conversion facility can be considered as one of the example into this class. Class 4 (conventional hazard): 'industrial risk', conventional industrial buildings.

Fig.4 (a) Large size research Reactors and Reprocessing Plants

Fig. 4(b) Waste Management Plant

Fig. 4(c) Small size research reactors

As mentioned above, the nuclear facilities need to be designed for various natural and man induced events in addition to normal operating loads. Various natural events and their annual exceedance probability are given in Table 1. The variation of Peak Ground Acceleration (PGA) with annual probability of exceedance is shown in Fig. 5.

Table 1 Annual Hazard Exceedance Probability

Hazard	High	Medium	Low	Conventional
Hazard Class	1	2	3	4
Earthquake	1 E-4	5 E-4	1 E-3	2 E-3
Wind	1 E-4	1 E-3	2 E-2	2 E-2
Flood	1 E-5	1 E-4	5 E-4	2 E-3
Aircraft Impact	1 E-5	1 E-4	NA	NA

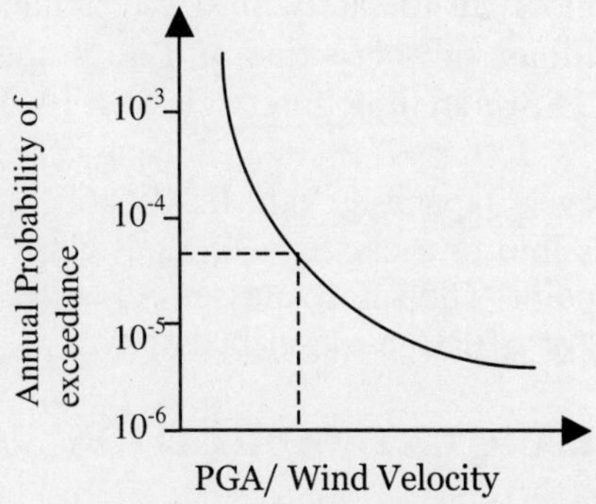

Fig 5 Annual Hazard Exceedance

The safety classification determines the relevant design criteria, which are measures of quality, by which the adequacy of each SSC in relation to its importance to plant safety is ensured. According to USNRC [9], and AERB [7], the procedure followed for this purpose is to identify various safety functions required to be performed in an NPP to achieve safety, while the criteria of classification for all civil, mechanical, electrical and I&C SSCs are same (i.e., radiological consequences on probable failure of SSC). Appropriate design requirements for each safety class are established with the most stringent requirements specified for the highest class and so on. AERB safety classification for civil and mechanical SSCs in NPPs is as follows:

Safety Class 1: The SSCs required to perform the safety functions necessary to prevent the release of a substantial fraction of core fission product inventory to the containment/environment are classified as safety class 1, e.g. those components that comprise the reactor coolant system pressure boundary, those that are necessary to shut down the reactor following a loss of coolant accident etc.

Safety class 2: The SSCs that perform the safety function necessary to *mitigate* the consequences of an accident which would otherwise lead to release of substantial fraction of the core fission product inventory or activation product inventory into the environment are classified as safety class 2. It also includes those safety functions necessary to prevent anticipated operational occurrences from leading to accident conditions; and those safety functions whose failure under certain plant conditions may result in severe consequences, e.g., failure of decay heat removal system, containment building, containment isolation features etc.

Safety Class 3: SSCs required performing a support role to safety functions in safety classes 1 and 2 and performing safety functions necessary to prevent radiation exposure to the public or site personnel from exceeding the relevant acceptable limits from sources outside the reactor coolant system are classified as safety class 3.

Safety Class 4: The SSCs, which incorporate safety functions that do not fall within safety classes 1, 2 or 3. Safety class 4 includes those components that are necessary to limit the discharge or release of radioactive waste and airborne radioactive material below prescribed limits during all operational states and would not result in the exposure of the public or site personnel in excess of prescribed limit, even if they failed. (e.g., PHT and moderator deuteration and de-deuteration system, D_2O clean up system etc).

Not Important to Nuclear Safety (NINS) Class: This class includes all other systems, which are not associated with any of the safety functions listed above and designed as per industrial standards.

AERB [7] has also classified electrical power systems, electrical power system equipment and electrical components of process equipment into two main safety classes. All systems and equipment, important to safety, are classified as safety class EA. Equipment/components, which are not important to safety but have to meet special requirements relating to radiation and/or seismic conditions, are classified as safety class EB. AERB has classified Instrumentation and Control (I&C) SSCs in three safety classes: IA, IB and IC.

However, Japanese Code JEAG 4601-1987 [10] has classified various facilities of Nuclear power plants into four classes named as class As, A, B and C based on their importance from the safety viewpoint. The safety classes As, A, B and C correspond respectively to the safety classes 1, 2, 3 and 4 of AERB, as described above.

4 SEISMIC CATEGORIZATION

The SSCs of NPPs are broadly classified into structures, equipment and piping with unique characteristics of their own. The various structures of an NPP site are reactor building, control building, reactor auxiliary building, service building, waste management building, turbine building, spent fuel storage bay building etc. These structures are directly founded on the ground and support the various equipment and piping systems. SSCs necessary to assure capability for shut down, decay heat removal and confinement of radioactive material have to be designed to remain functional throughout the plant life in the event of natural phenomenon such as earthquakes, cyclones and floods hence they are seismically categorized as follows.

USNRC [9] and AERB [7] have categorized SSCs into three categories namely:

Category I SSCs: Those whose failure could directly or indirectly cause accident conditions or which are required for shutting down the reactor, monitoring critical parameters, maintaining it in a safe shutdown condition and removing decay heat on a long term basis, (e.g. reactor containment structure, reactor core etc). These SSCs are designed for S1 and S2 level earthquakes. S1 is referred to as Operating Basis Earthquake (OBE) and S2 is referred to as Safe Shutdown Earthquake (SSE).

Category II SSCs: Those, which prevent the escape of radioactivity beyond the limits prescribed for normal operation and not covered in category-1 or mitigate those accident conditions which last for such long periods that there is a reasonable likelihood of an earthquake of the defined severity occurring during this period and not covered in category-I (e.g. waste management building). These SSCs are designed for S1 level earthquake.

Category III SSCs: Those, which are not important to safety and not covered in category-I or II (e.g. turbine building, turbo-generator etc). These SSCs are designed for earthquake resistance as per national practice [14].

Safety Classification and Seismic Categorization of Nuclear Facilities need to be performed as per the above definitions and details provided [15] and further design of SSCs as follows.

5 DESIGN OF STRUCTURES SYSTEMS AND COMPONENTS

For the design of SSCs, the following data need to be collected.

Structures

a. Foundation details including soil data
b. Structural geometry
c. Material properties including damping
d. Equipment loading
e. Live loads
f. Wind loads
g. Flood levels
h. Design response spectra and design earthquake time history

Systems & Components

a. Details of supports
b. Weights of supporting systems such as valves
c. Weight of process media
d. System geometry
e. Material property including damping
f. Pressure loading
g. Thermal loading
h. Broadened floor response spectra
i. Base or floor design earthquake time history

Once the data is collected, the system can be idealized as a single degree of freedom or multi degree of freedom system. However, for preliminary design purpose, most of the SSCs can be approximated to a single degree of freedom system and estimate their responses for earthquake excitation. At later stage, detailed analysis can be performed considering multi degree of freedom system as explained below.

6 MATHEMATICAL MODELING OF STRUCTURAL SYSTEMS

A mathematical model of the system shall adequately represent the dynamic characteristics of the physical system e.g. mass, stiffness and damping. For evaluating the accurate seismic responses, the model should represent the structure as accurately as possible. Finite element method has been most popular in modeling the structure for seismic analysis. Structure (Fig. 6(a)) can be modeled accurately either by plate and shell (Fig. 6(b)) or 3D brick elements. Otherwise, for ease, beam models (Fig. 6(c)) can be used for obtaining the global seismic responses. These are finally applied on the 3D finite element model for evaluating the design stresses.

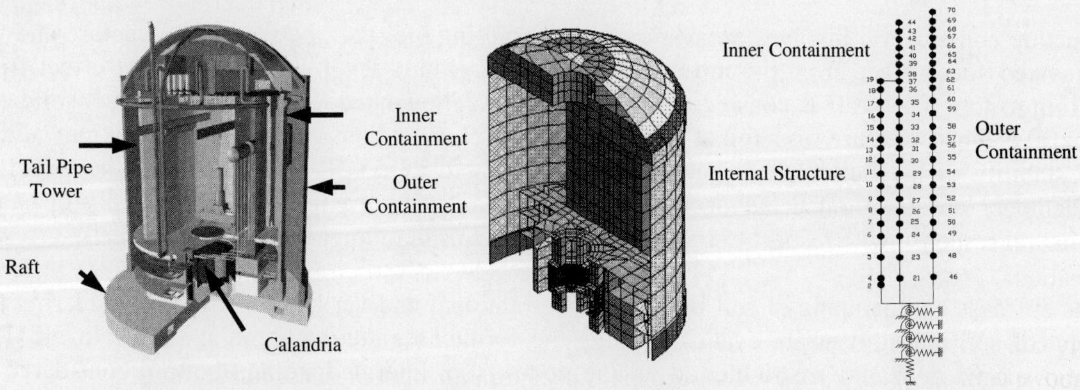

a. Typical reactor Building b. Plate and Shell element model c. Beam Element Model

Fig.6 Mathematical models of typical Reactor Building

While modeling the structures due importance should be given to soil-structure-interaction, soil-structure-equipment-interaction, damping. Few details of the same are explained below.

6.1 Soil Structure Interaction

The response of a structure during dynamic loading such as earthquake, wind, aircraft impact etc. depends on the characteristics of the excitation, the surrounding soil, water table and the structure itself. In the mathematical model of the superstructure, it is necessary to include the effect of the foundation conditions. The foundation conditions may have a considerable effect on the response of structure, depending on the nature of the soil, structure, embedment and the extent of interaction as shown in Fig. 7.

There are two acceptable methods of SSI analysis. One is the direct method and the other is impedance function method (Indirect method). In the former, soil is modeled along with the structure and the motion is applied at the bottom of the soil. This approach is generally recommended for soft and medium soils. In the second method, soil is replaced with impedance functions and the ground motion derived at free field is used. There are two types of impedance functions; frequency dependent and frequency independent. The former is preferred for non-uniform soil layers and the latter is suitable for uniform soils. In our country most of the reactors built/being built are on rock sites, which have relatively uniform rock to a depth equal to the largest foundation dimension. Hence frequency independent impedance functions are used. Frequency independent functions such as spring constants and damping ratios for different directions of motion are evaluated using ASCE code [16]. However, soil damping is limited to 7%, 20% and 30% of critical damping for rocking, horizontal translation and vertical translation respectively.

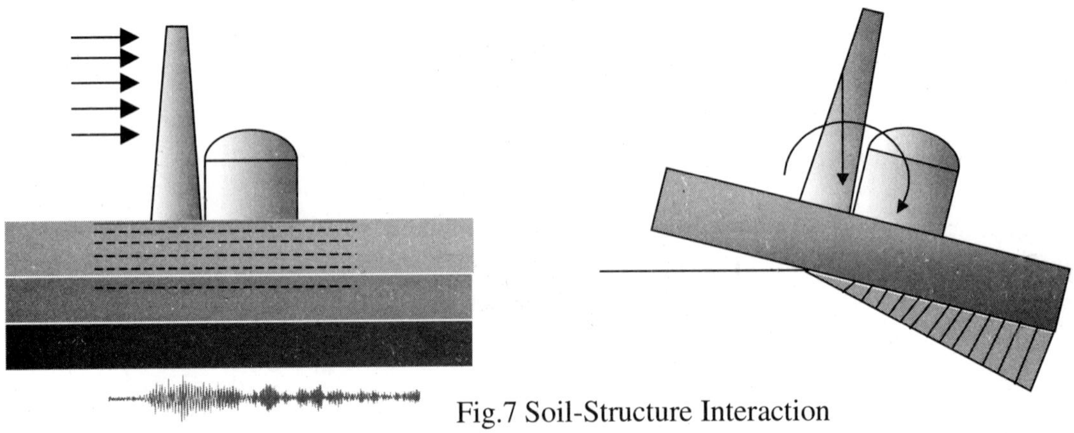

Fig.7 Soil-Structure Interaction

Structure construction involves excavation and backfilling the space between the structure and the excavated soil surface from the top of the raft to the ground level. The embedment effect from bottom to top of the raft is considered in the analysis. The embedment effect of the consolidated backfill against structure from top of raft to ground level is not considered. However, it may add to the stability of the building. It has also been indicated in ASCE code [16] that the effect of embedment on structural behavior is to increase its resonant frequencies and usually decrease structural response with respect to a structure without consideration of embedment effects.

Side stiffness and damping of soil in horizontal, rotational and vertical directions considering the effect of soil embedment are evaluated using the formulae given by Beredugo and Novak [17]. These spring values are distributed at all the nodes (top, intermediate and bottom considered to represent the raft, in proportion to the height of the element.)

6.2 Structure- Equipment Interaction

Large size equipment and structure may interact with each other during an earthquake. This results in variation of the uncoupled response. The best way of accounting for the interaction effects of the structure and equipment is by coupling together and analyzing for the given earthquake load. However, due to practical reasons, it is not preferred. In fact, for the equipment that affects significantly the structure/equipment, uncoupled natural frequencies are identified using decoupling criteria [16, 18] as given below.

Let, R_f be the ratio of frequency or modal frequency of uncoupled equipment to the uncoupled structure and R_m is the ratio of mass or modal mass of the uncoupled equipment to the uncoupled structure. Dominant modes (>20% mass participation) are only considered for calculating the frequency and mass ratios.

a. Decoupling can be done for any R_f, if $R_m < 0.01$
b. If $0.01 \leq R_m \leq 0.1$, decoupling can be done provided $0.8 \geq R_f \geq 1.25$
c. If $R_m > 0.1$, an approximate model of the secondary system (e.g. equipment) should be included in the primary system (Supporting Structure)
d. For rigid equipment whose frequency is more than 33 Hz, only mass of the equipment should be included.

The above criteria are not straightforward for applying to the complex structure such as containment structure. Reddy et al [19] have suggested, certain modifications that make the criteria applicable to the complex structure such as reactor containment structure. The above criteria are also not applicable for multi-connected equipment. A new criterion that Reddy et al [20] have developed can be used for checking the decoupling requirement of multi-connected equipment.

6.3 Damping of Structures

Damping is the common designation of all kinds of energy absorption of a vibratory system due to mainly its hysteretic, friction and viscous characteristics. In structural analysis, the representation of energy dissipation through equivalent viscous damping is very popular because it leads to linear differential equations of motion, which can be readily solved. Damping ratios for structural materials are generally less than 10% and for different materials the damping values are different. In order to obtain modal response, in case the structure contains material with different damping, the percentage of critical damping in each mode has to be evaluated using the weighted strain energy principle. The lower bound damping values for the various materials constituting the different type of structures are tabulated in Table 2.

Table 2 Material Damping

Material	Damping	
	SSE	OBE
Prestressed Concrete	5%	2%
Reinforced Concrete	7%	4%

In order to obtain modal damping from different damping values of the materials, criteria of weighted damping based on strain energy stored in various materials as given in ASCE code [16] is adopted. However, the values of computed modal damping should be restricted to a maximum of 20% as per ASCE code [16].

7 MODELING TECHNIQUES- EQUIPMENT AND PIPING SYSTEMS

For the seismic analysis of piping and equipment, generally finite element methods are used. In this method, the stiffness, mass and damping are modeled appropriately and one of the following techniques is adopted.

Simple models can be developed for valves, pumps, motors, fans and some heat exchangers as shown in Fig. 8. Valves, motors, fans are generally rigid whose frequency is more than 30 Hz. However, its supports are generally flexible which can be modeled as springs or beam elements as shown in Fig 8.

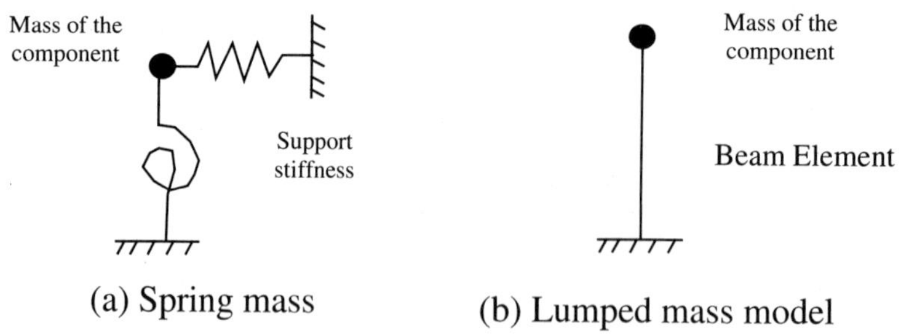

Fig. 8 Simple models

7.1 Beam Model or One-Dimensional Finite Element

This modeling is typically applied to beams, columns, frames, ducts, cable trays, conduits, tanks, cabinets, storage racks, pressure vessels and heat exchangers and are expressed as a continuous or one dimensional finite element in a two or three- dimensional space. Masses are represented by lumped parameters, which develop a diagonal elemental mass matrix, or by the means of consistent mass matrices.

7.2 Plate/Shell or Two-Dimensional Finite Element

This type of modeling is adopted for items whose primary mode of failure is by biaxial bending stress, plane stress or plane strain. Typically included in this category are: cabinets, tanks, pressure vessels and heat exchangers whose shell support significant eccentric loads which would tend to excite shell or local modes of vibration.

7.3 Beam Model Based on the Energy of the Plate or Shell Finite Element

For the equipment having sections of irregular shapes, a beam model of this equipment is generated using energy equivalence between plate or shell finite element model and beam model is developed.

7.4 Three-Dimensional Finite Element

This type of modeling is expensive and is preferable for local analysis to obtain correct stress picture at openings, nozzle junctions etc.

7.5 Piping Models

Straight pipe elements, pipe bends are used for modeling piping. The supports are modeled as beam element and valves are modeled as rigid element. A typical piping model of PHT system of PHWR is generated using combination of straight pipe element, pipe elbow element, beam element and spring element.

8 DESIGN OF STRUCTURES

8.1 Design Checks

The forces and moments calculated due to external events shall be combined with the forces and moments due to other loadings. The structure is designed as per standard civil engineering design codes [6, 21, 22, 23]. Some of the important aspects need to be considered while designing structures for seismic load are as follows.

8.2 Layout selection

Seismic effects should be minimized by

a. Locating the centre of gravity of all structures as low as possible;
b. Selecting a plan and elevation that is as simple and regular as possible;
c. Avoiding protruding sections (lack of symmetry) as far as possible;
d. Making the centre of rigidity at the various elevations as close to the centre of gravity as practicable.
e. To reduce undesirable differential movements between structures, consideration should be given to locate the structures, to the extent practicable, on a common foundation structure, or at least, avoiding different embedment depths.

8.3 Design Guidelines

a. Brittle failure in shear and/or bond, or in the compressive zones of concrete should be prevented in RC structures
b. An appropriate minimum compressive strength of concrete should be fixed to ensure the ultimate strength of the structural members is governed by the reinforcement;
c. For rebars, an appropriate minimum ratio between UTS and YS (~1.25) should be defined, to guarantee a minimum ductility (>10 %);
d. Appropriate design of structural joints, particularly in RC structures should provide a high ductility and a capability to accommodate high displacements and rotations. Guidelines of Indian Standard on "Ductile Detailing of Reinforcement" [23] may be followed in this regard.

Typically the structure is designed for Limit state of Serviceability and Limit state of strength with various load combinations. The various loads considered are permanent actions such as dead Load (D), prestress load (F), Variable actions such as test pressure (P_t), test temperature (T_t), Operating Basis Earthquake (OBE) (E_o), Live load on IC Dome (L), Live load during construction (L_C), Operating temperature (T_o), Wind loads [24] during construction (W_c) and accidental actions such as accident pressure (P_a), Peak temperature after accident (T_a), Safe Shutdown Earthquake (SSE) (E_S).

9 DESIGN OF COMPONENTS AND PIPING SYSTEMS

Stress analysis needs to be performed using appropriate finite element models as explained above for various loads viz. pressure, thermal, mechanical, impact loads, weights, wind, vibrations and earthquake load.

9.1 Loading Conditions

The loadings that shall be taken into account in designing a component include, but are not limited to those in a. through g. below:

a. Internal and external pressure
b. Impact loads, including rapidly fluctuating pressures
c. Weight of components and normal contents under operating and test conditions, including additional pressure due to static and dynamic heads of liquids.
d. Superimposed loads such as other components, operating equipment, insulation, corrosion resistant or erosion resistant linings.
e. Wind loads, snow loads, vibrations and earthquake loads where specified.
f. Reactions of supporting lugs, rings, saddles or other type of supports
g. Temperature effects

9.2 Anchor Motion

Piping must have sufficient flexibility to accommodate the seismic motions of structures, equipment and headers to which it is attached. One of the most common causes of piping failure in strong motion earthquakes is Seismic Anchor Motion (SAM) resulting from:

a. Large displacement of unanchored tanks or equipment
b. Large displacement due to flexibility of the tank or equipment anchorage
c. Large differential motions of structures to which the piping is attached
d. Large motions of header piping induced into smaller branch piping

9.3 Suggested Remedies

a. Tanks and equipment, to which piping connects, must be properly anchored to prevent sliding, rocking or overturning.
b. Tanks and equipments to which piping connects, and the supports, should be relatively stiff to minimize SAM. When vibration isolators are present, they are a source of SAM. If there are no seismic stops built into the isolators, the equipment will require the addition of seismic restraints to limit motion.
c. Piping rigidly attached to two different buildings or substructures within a building must be sufficiently flexible to accommodate the differential motion. Attention may be paid to piping that has its axial motion restrained at supports points in two different structures.
d. Header motion imposed on small branches must be assessed, or the header must be restrained near the branch.

9.4 Equipment Nozzle Loads

Equipment and component nozzles, except for valves that are stronger than the pipe, should be protected, by appropriate restraints, from excessive seismic loads. Piping layout should be reviewed to evaluate that large seismic loads are not reacted at the equipment nozzle. One potential problem is a long axial run of pipe not restrained from axial movement except at the equipment nozzle.

Piping reaction loads at the nozzles of rotating equipment may affect their function. The seismic reaction loads imparted by the piping on the nozzle of the active (rotating) equipment shall be small (un-intensified bending stress less than 600 psi), or within the estimated capability of the equipment.

9.5 Interaction with Other Structures

A piping system subjected to seismic loads will displace or swing laterally, and may impact adjacent components. Lateral displacements or swing deflections of piping spans shall be estimated. Impact must be avoided if it affects the following components:

a. Active components
b. Instrumentation Tubing
c. Unstable or little weight structures
d. Electrical cabinets and panels

The load combinations to be adopted for ASME Class-1, 2 and 3 vessels, components and supports for design and service levels are given in Table 3. The combined stresses shall meet the relevant design code requirements of ASME B&PV Code Section III, Division-1, Subsection-NB-3000, NC-3000 or ND-3000 for class 1, 2 and 3 components respectively.

The combined stresses in supports shall meet the requirements of ASME Section III, division-1, Subsection-NF-3000. The combined stresses in the nozzle, as calculated using WRC-297/WRC-

107 or any other justifiable method shall meet the requirements of ASME Section-III, Division-1, Subsection NB, NC and ND.

The stresses in the base plate shall be checked. The stresses may be checked as per NUREG CR-6241, IS 800, IS 456 or any other applicable code. The foundation bolts shall be designed as per ACI 349.

The combined stresses in mounting bolts of pump casing to base frame between motor and base frame, coupling bolts, flanges, connecting bolts shall meet the relevant codal requirement viz., ASME section III, division-1, subsection NB-3000, NC-3000, ND-3000 and NF-3000.

Table 3 Load combinations for Design condition and service levels for equipment, piping and supports

Plant Classification	Design/Service level	Load combination
Design	Design Condition	Pressure, dead weight, sustained loads, temperature (1)
Normal	Service Level A	Normal operating transients (Pressure temperature, mechanical)
Upset	Service Level B	Pressure, dead weight, sustained loads upset condition transients (pressure, temperature (2) mechanical), including OBE (3)
Emergency	Service Level C	Pressure, dead weight, sustained loads temperature (1), emergency condition transients (Pressure, mechanical)
Faulted	Service level D	Pressure, dead weight, sustained loads, Temperature (1), faulted condition transients (pressure, mechanical), pipe rupture loads, SSE (3)

Note: 1) Temperature is used to determine allowable stress only
2) Thermal transients under Level C shall also be considered in Level B
3) OBE and SSE include both inertial and seismic anchor movement (SAM)

10 FATIGUE EVALUATION

Apart from satisfying these limits, a fatigue evaluation shall also be performed considering 50 cycles of peak-to-peak acceleration loading corresponding to S1 level earthquake (this effectively accounts for five S1 events with 10 cycles in each event). Due importance should be given to vibration loads generated due to internal events such as flow induced vibrations, internal missiles, rotating machinery vibrations, sudden electrical or mechanical systems failure and fatigue valuation need to be performed for the same.

Similar procedure for design of SSCs handling limited activity that is given in the IAEA TECDOC [2] may be adopted.

11 GENERAL DESIGN GUIDELINES

Following points [25] should be taken into account to improve the resistance to earthquake-induced vibration:

a. Care should be taken in the design of the supports to ensure that all the joints are designed to behave as assumed (e.g. hinge, fixed, guided etc.) in the analysis of the support and transmit the full range of loads determined in the connected members. If integral supports are used, they should be designed, manufactured and installed so as to minimize the danger that any unexpected failure or crack initiated in the supporting element propagates to the functional parts, such as pressurized shell and the primary piping.

b. Care should be taken in the design of devices for anchoring equipment, to ensure that all potential forces and moments are fully evaluated and that anchoring materials are suitable for their purpose. It is of particular importance to ensure that base plates are sufficiently stiff to avoid prying effects and that anchor bolts are adequately tightened to avoid rocking effects, lowered frequencies, increased response levels, higher-than-design loads and increased risk of loosening, pull-out or fatigue.

c. Equipment support legs should be cross-braced unless their dimensions warrant departure from this recommended practice. In most cases, stiffness can be increased to avoid resonance. If systems are made stiffer, the effect of thermal stresses, other dynamic loads and differential motions of supporting points should be addressed.

d. It is important to avoid, as far as practicable, resonance of equipment, piping, instrumentation and core internals at the frequency of the dominant modes of supporting structures. In some cases, where the response of equipment, although significant, cannot be reduced by other means, the damping of the system may be increased by suitable design modifications.

e. Particular attention should be paid to the possibility of collision between adjacent components, or between components and adjacent parts of a building, as a consequence of their dynamic displacement. It is also important to allow for flexibility of connections between such components, between components and building penetrations and underground connections to buildings, as well as between buildings.

12 QUALIFICATION OF ACTIVE EQUIPMENT- ANALYSIS AND TESTING

Active equipments, which need attention, are the electrical relays, switches, contacts; shut-off rod drives etc. whose functioning solely depends on the actuation of some parts within them. For these components, it is necessary to demonstrate their operability during an external event by testing. The procedures given in [26] for the assessment of integrity and operability may be adopted for testing. For heavy equipment it is acceptable to qualify by detailed finite element analysis wherein the bearing stiffness, gap between the casing and the impeller etc. must be considered in mathematical model.

13 ISSUES AND RESEARCH AND DEVELOPMENT EFFORTS

In the present design, clarity is needed related to design ground motion, soil-structure interaction, structure-equipment interaction, failure modes of piping systems and evaluation of capacity of RCC structures. These are considered conservatively in the design leading to high cost. However, Earthquake Engineering has gained the attention of many researchers throughout the world and extensive research work is performed. Linear behavior of structures, systems and components (SSCs) subjected to earthquake loading is clearly understood. Nevertheless, nonlinear behavior of SSCs subjected to earthquake loading attributed to the above issues need to be understood clearly and design methods need to be validated experimentally.

In view of this, three major areas in earthquake engineering are identified for research includes: design and development of passive devices to control the seismic response of SSCs, nonlinear behavior of piping systems subjected to earthquake loading and nonlinear behavior of RCC structures under seismic excitation. The work performed so far is helping for clearer understanding of nonlinear behavior of SSCs as well as in developing new schemes, methodologies and devices to control the earthquake response of SSCs. A brief overview of the works performed in the past and present is given below.

14 SEISMIC RESPONSE CONTROL IN STRUCTURE SYSTEMS AND COMPONENTS

Controlling of seismic responses in Structures, Systems and Components can be performed using active, semi-active, passive dampers and Seismic Base Isolators. In the first stage, work was taken up to understand and validate experimentally the performances of Elasto-plastic dampers (EPDs), Lead Extrusion Dampers (LEDs), Tuned Liquid Dampers (TLDs), Tuned Mass Friction Dampers (TMFDs) and Seismic Base Isolators such as Laminated Rubbers Bearings (LRBs), Lead Plug Bearings (LPBs). Shape Memory Alloy Dampers (SMADs) are being designed and tested. Some details of these devices and applications identified are described below

14.1 Elasto-Plastic Dampers (EPDs)

EPDs, based on plastic deforming steel plates, consist of X shaped plates (Fig 9(a)), which yield at small load and sustain many cycles of stable yielding deformation, resulting in high levels of energy dissipation or damping. The force displacement loop of the X- plate is shown in Fig 9(b).

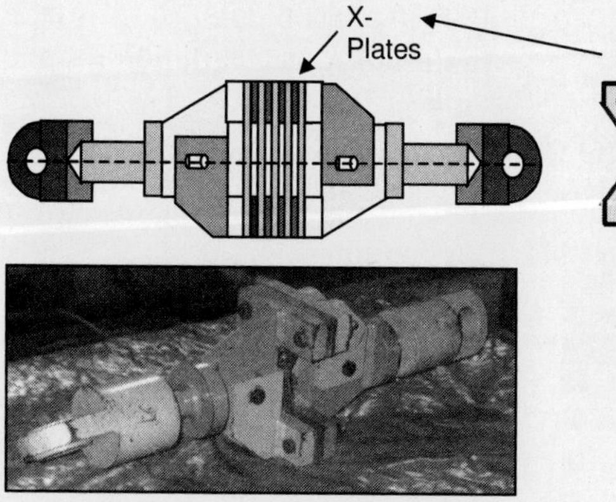

Fig. 9(a) Elasto-plastic Damper

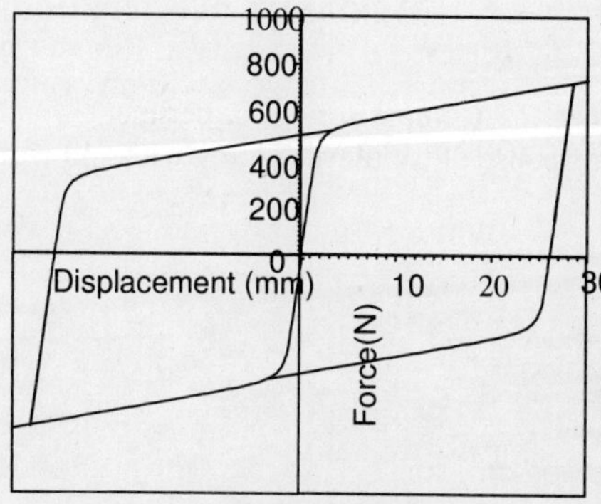

Fig. 9(b) Force Displacement Characteristics of EPD

The area of this force displacement relationship gives the energy dissipated by damper plate. Elasto-plastic dampers were fabricated in BARC and tested for its force-displacement characteristics. X-shaped plate in elasto-plastic energy absorber facilitates a constant strain over the height of the device, thus ensuring that yielding occurs simultaneously and uniformly over the full height of the damper. Cantilever (Fig. 10) and 3-D piping systems (Fig. 11) were tested with and without EPDs on shake table at SERC.

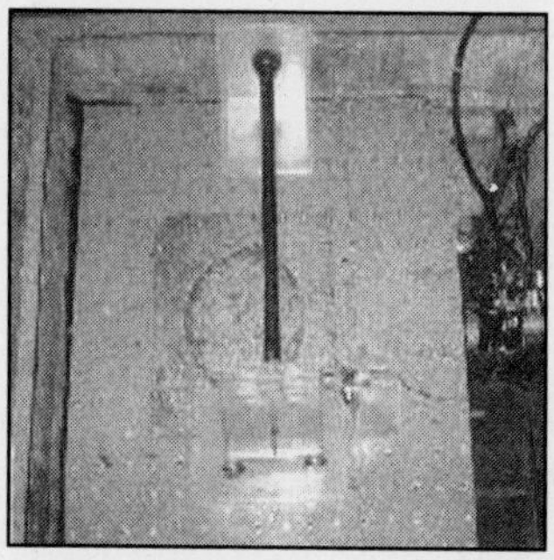

Fig. 10 Cantilever piping system

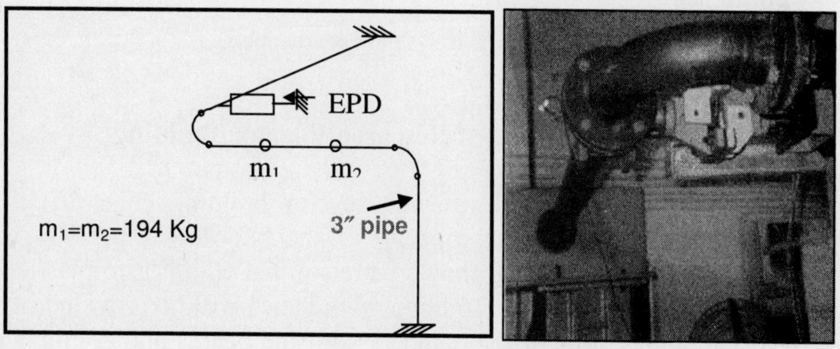

Fig. 11 3-D piping system

Using a finite element model of the piping systems, linear and nonlinear time history analysis is carried out using Newmark's time integration technique. The analytical maximum response displacement obtained at the Elasto-Plastic Damper support for the two piping systems is in good comparison with experimental values as shown in Figs. 12 and 13. Shake table tests were also conducted on complex piping system (Fig. 14) with two EPDs and significant reduction in response was observed (Fig. 15). Thus piping supported on multiple EPDs and subjected to earthquake will give significant response reduction.

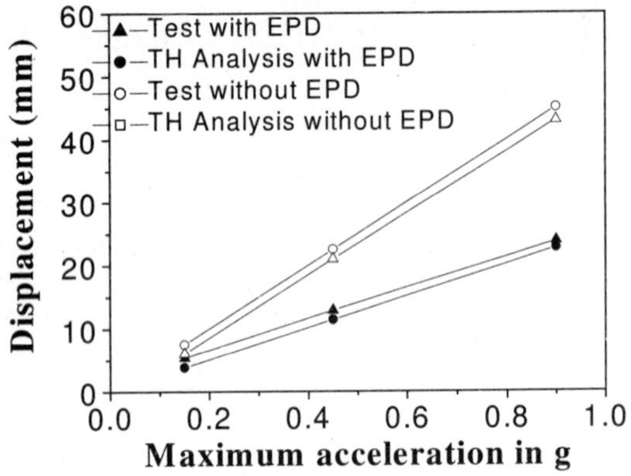

Fig. 12 Comparison of Experimental and analytical response of cantilever piping with and without EPD

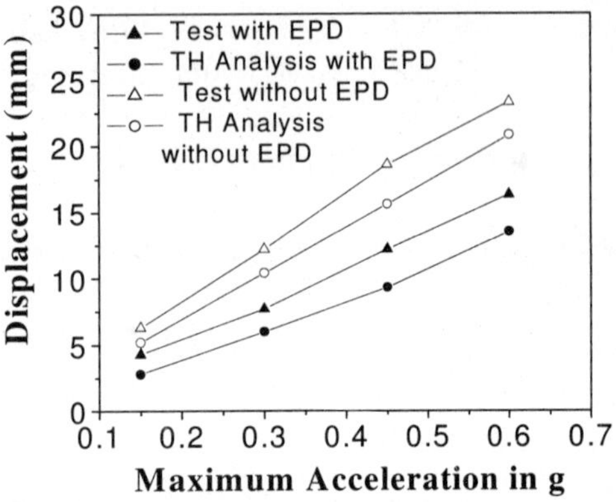

Fig.13 Comparison of Experimental and analytical response of 3-D piping with and without EPDs

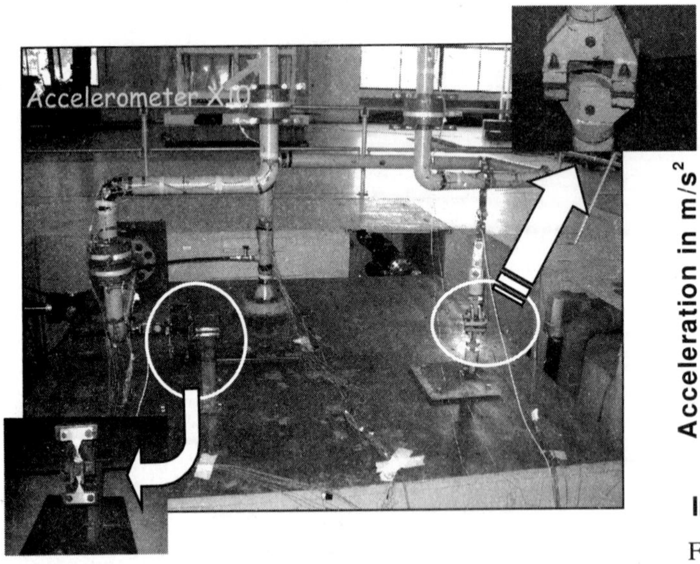

Fig. 14 Complex piping system with dampers

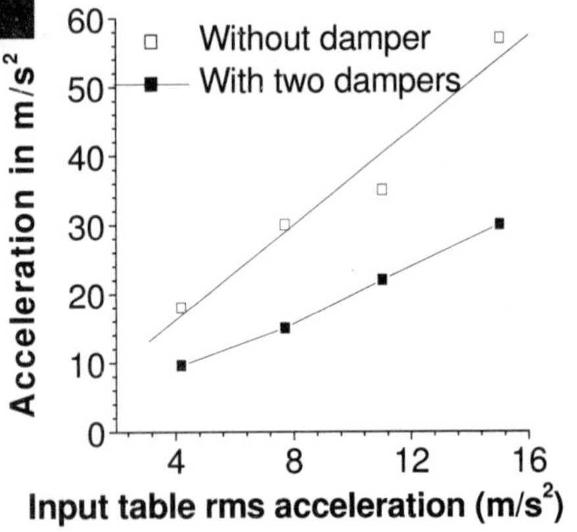

Fig.15 Response of complex piping system with dampers

14.2 Seismic Retrofitting of Research Reactor Building

Seismic re-evaluation of research reactor building (Fig. 16) is performed and it was found necessary to improve the capacity of the building, especially in footings. To meet this, it is proposed that the building should be retrofitted with Elasto-plastic Dampers. Elastoplastic dampers consisting of 15 nos. SS-316 6mm thick plates will be provided in the frames of the RCC building and the connections of the dampers with the beams and columns of the structure are to be made using ISMC 125 box sections. To prove the efficiency of dampers in reducing seismic response of RCC framed structure, shake table tests are performed on RCC building model (Fig 17 (a)) with

EPDs. Fig. 17 (b) shows the comparison of response without and with dampers when the model is subjected to sine sweep excitation.

Fig 16 Research Reactor Building

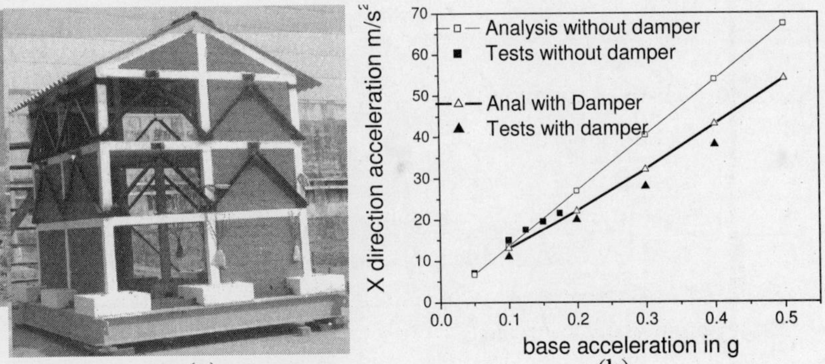

(a) (b)

Fig 17 Research Reactor Building Model

14.3 AHWR Down Comer Supported on EPD

Three possible Layouts of AHWR Down Comer as shown in Fig. 18 have been considered for codal qualification in order to ensure its operability and functionality during normal and occasional loads. Along with the codal qualification a good layout should have less number of bends and weld joints in order to reduce the in-service inspection. Less number of bends will reduce the pressure drop in natural circulation. Among the three layouts, Layout –2 is having lesser thermal stresses in comparison to Layout-1 but the number of elbows and weld joints remains similar as in Layout-1.

A simple layout-3 has been considered for analysis, which has almost half number of bends and weld joints as compared to earlier two layouts, but thermal stresses exceed the code allowable limit when conventional supports are employed to control the seismic stresses in layout-3. In order to qualify the layout-3 elasto-plastic damper (EPD) supports have been employed. EPD supports absorb large amount of earthquake energy without much affecting the thermal stresses in the piping. 400 Kg dampers are to be used at location-1 and at locations 2, 3 & 4 dampers of capacity 1350 kg are to be used at each support. The comparison of thermal and seismic stresses, required lengths and welds of three layouts are given in Table 1 from which it can be inferred that layout 3 will be most economical and meeting both thermal and seismic requirements.

Table 4 Comparison of maximum stress, number of welds and total length of pipe

Description	Thermal stresses (N/mm^2)	Seismic stresses (N/mm^2)	Total length of piping (m)	No. of welds
Layout-1	206.13	47.23	608.72	112
Layout-2	152.07	150.15	660.74	120
Layout-3	153.42	99.93	520	72

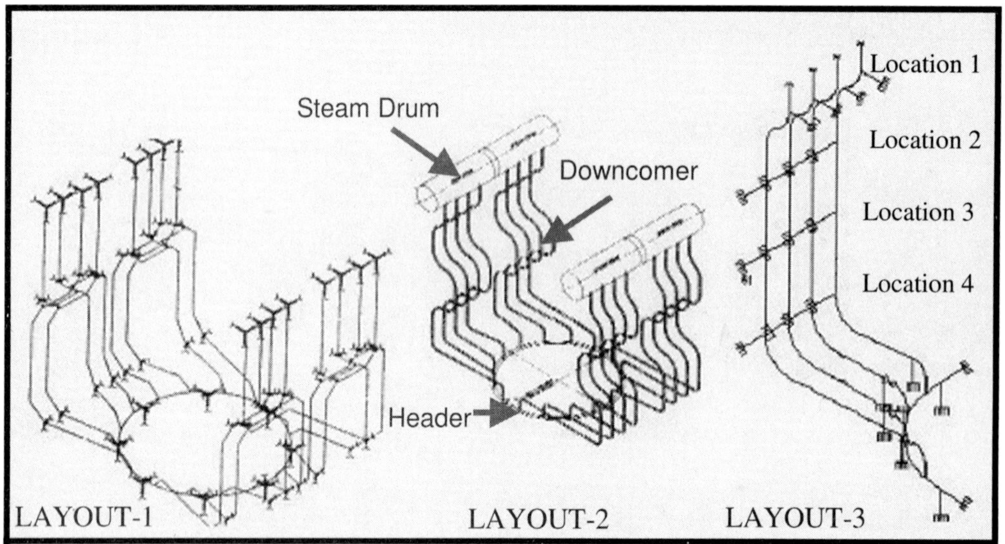

Fig. 18 Three proposed layouts for AHWR Down comer

14.4 Design of Melter cell structure of WIP, Kalpakkam using EPDs

Melter cell supporting structure of Waste Immobilization Plant (WIP), Kalpakkam is designed using EPDs. EPDs helped to reduce the loads on Embedded parts (EPs) and facilitated to use existing EPs. It also helped to meet the requirement of thermal loads.

14.5 Lead Extrusion Dampers (LEDS)

LEDs work on the principle of extrusion of lead. LED absorbs vibration energy by plastic deformation of lead and thereby mechanical energy is converted to heat. On being extruded lead re-crystallizes immediately and recovers its original mechanical properties before next extrusion. Lead extrusion Damper of 15 tones capacity was designed and fabricated (Fig.19). The area of force displacement relationship of the damper as shown in Fig. 20 gives the energy dissipated in the extrusion of lead.

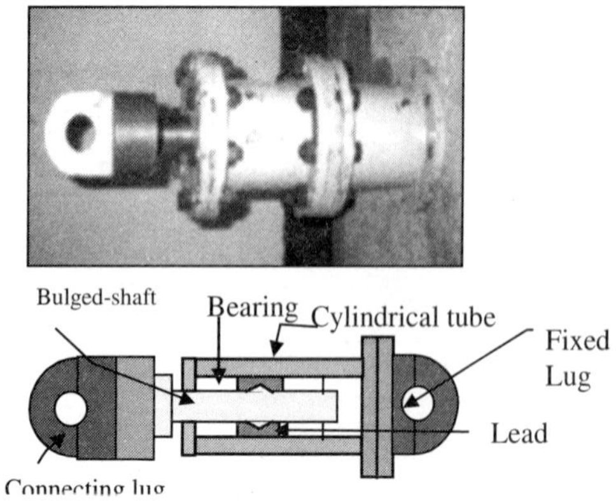

Fig 19 Lead Extrusion Damper (LED)

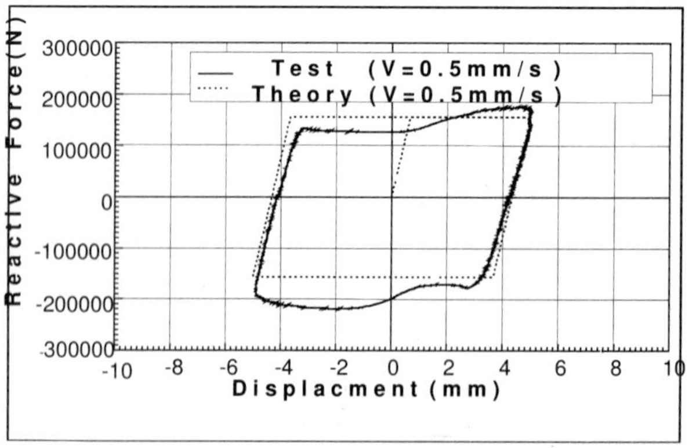

Fig 20 Hysteretic Load-Displacement Relationship for LED (Theoretical and Experimental)

Static and dynamic tests are carried out on the LEDs. The desired hysteretic characteristics were demonstrated by this test (Fig. 20). One possible application of the LED is in the seismic response reduction of coolant channel when it is coupled to the fuelling machine. The load transmitted to the coolant channel is significantly reduced when the LEDs are attached to the F/Ms. Shake table tests are planned on vessels restrained by Lead Extrusion Dampers.

14.6 Tuned Liquid Dampers (TLDs)

TLDs are rigid walled containers filled with liquid (mostly water) up to certain height, to match the sloshing frequency with structure frequency, and placed at rooftop of the structure. The device absorbs vibration energy through liquid sloshing principle. A series of tests have been conducted in BARC (Fig 21) on TLDs on a structural model excited by shake table. As the acceleration level increases, the roof displacement or the displacement at the base of tank also increases. Aspect ratio is defined as ratio of roof displacement to diameter of the tank. For aspect ratio less than 0.03, the sloshing is essentially linear whereas above that, wave breaking might occur. The experimental and analytical results show a significant reduction in the response of the structure with TLDs (Fig22).

Fig 21 Experimentation on TLDs

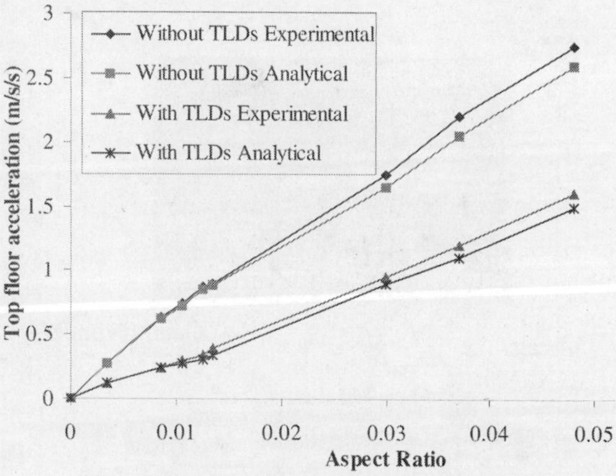

Fig. 22 Reduction in response with TLDs

14.7 Retrofitting of RCC building with Tuned Liquid Dampers

One RCC building did not meet the seismic requirements against the earthquake under a PGA level of 0.2g. As various facilities are already running in the building, it was not practically feasible to retrofit the structure using conventional methods because of space limitations The retrofitting was therefore suggested using Tuned Liquid Dampers (TLDs) to control the response of the structure against earthquake The analysis of the building (Fig. 23) was performed using TLDs and it showed that the responses reduce to a level within acceptable limits even if the tuning is lost by 15 % on either side of perfect tuning (Fig 24). The optimization of the design is being carried out.

14.8 Shape Memory Alloy Damper

Shape-memory alloy damper shown in Fig. 25 is made up of wires (e.g. Nickel Titanium wires) that dissipate energy when axially stressed. NiTi wires are very attractive for passive vibration control as they have pseudoelastic property (the property by which the alloy recovers its initial shape when external load is removed) and can sustain large amount of inelastic deformation. Cyclic tensile tests are carried out on wires of 1.2 mm dia for 1200 cycles and the force deformation relation is shown in Fig. 26. Shake table tests on piping with Shape memory alloy damper are also performed.

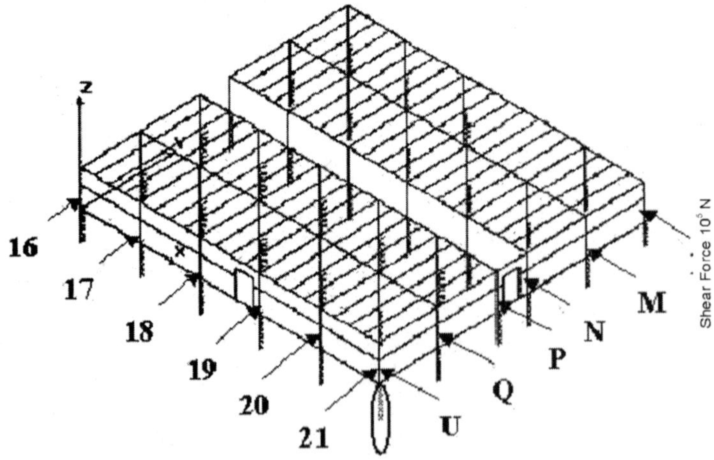

Fig. 23 FEM Model of Building

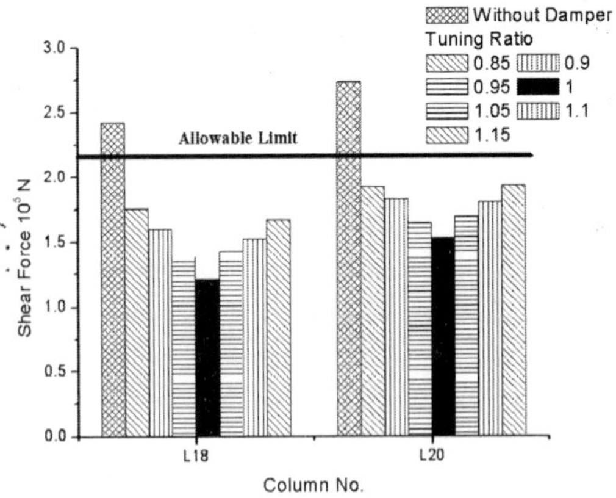

Fig 24 Response, without and with TLDs

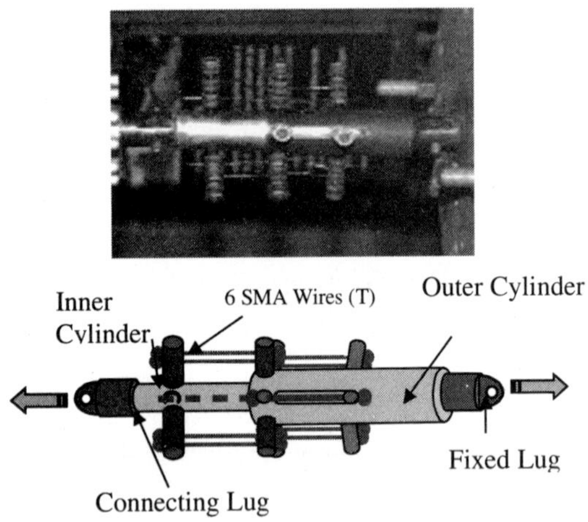

Fig. 25 SMA Damper

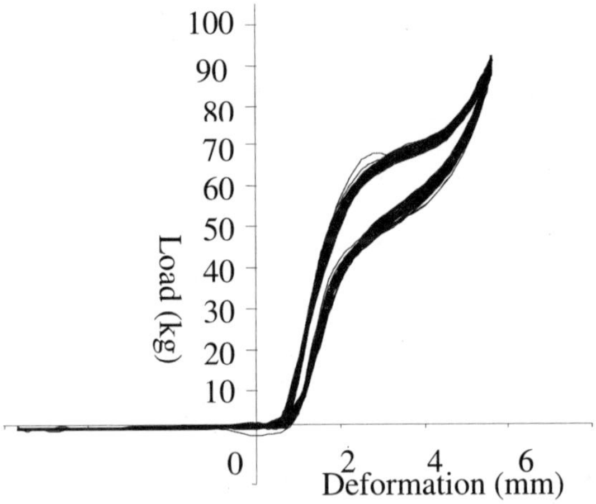

Fig. 26 Cyclic tensile tests of 1.2 mm dia wire for 1200 cycles

14.9 Base Isolators

The base isolation is aimed to attenuate the horizontal acceleration transmitted to the superstructure. The isolation system attempts to decouple the building or structure from the horizontal components of the ground motion by interposing structural elements with low horizontal stiffness between the structure and the foundation, which results the shifting of fundamental frequency ($\approx$ 0.5 Hz) of the structure out of the range of dominant earthquake energy and increasing the energy absorbing capacity.

14.10 Verification Test of a Prototype Base Isolated Three Storeyed - R.C.C. Framed Building Subjected to Actual Earthquake

A prototype of three storeyed R.C.C. framed building and a base isolated three storeyed- R.C.C. framed building has been built in IIT (Guwahati). These structures experienced actual earthquakes recently. The measured ground acceleration time history and the time histories recorded at the top of prototype building with fixed base and a prototype base isolated building are shown in Fig. 27. From the signal base isolation effects are clearly seen.

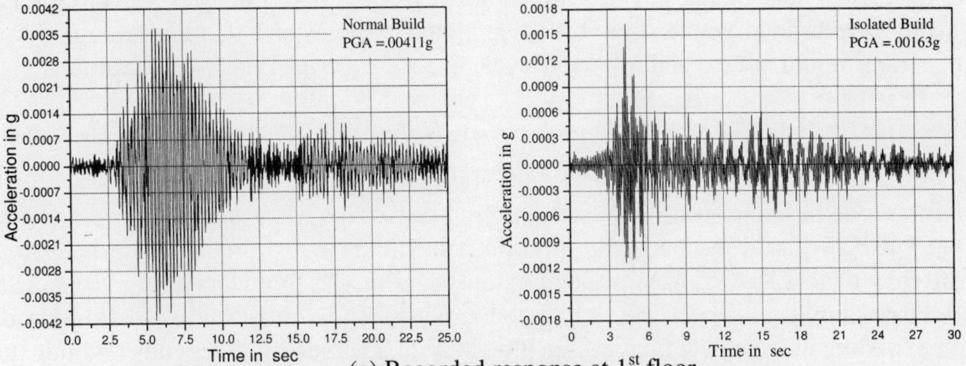

(a) Recorded response at 1st floor

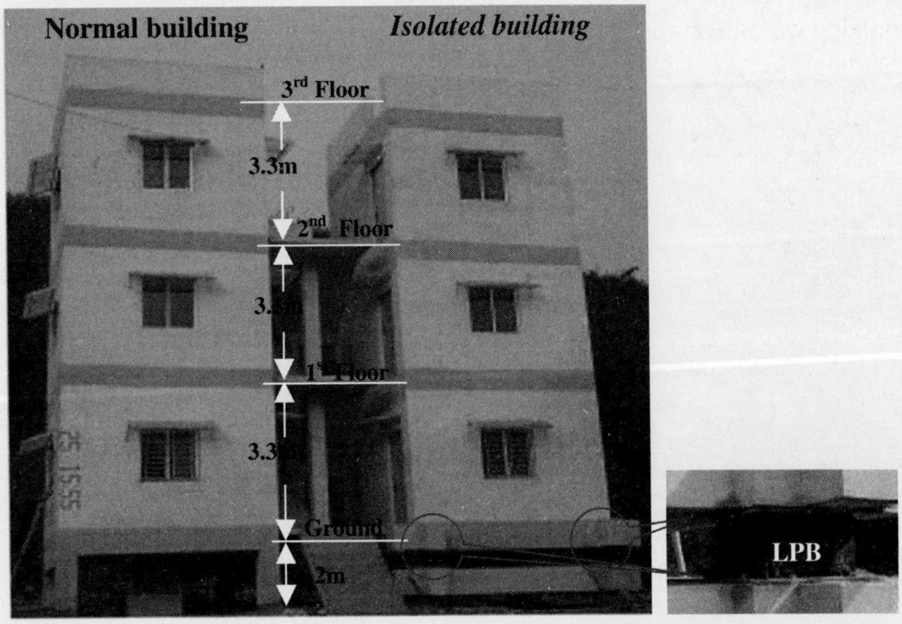

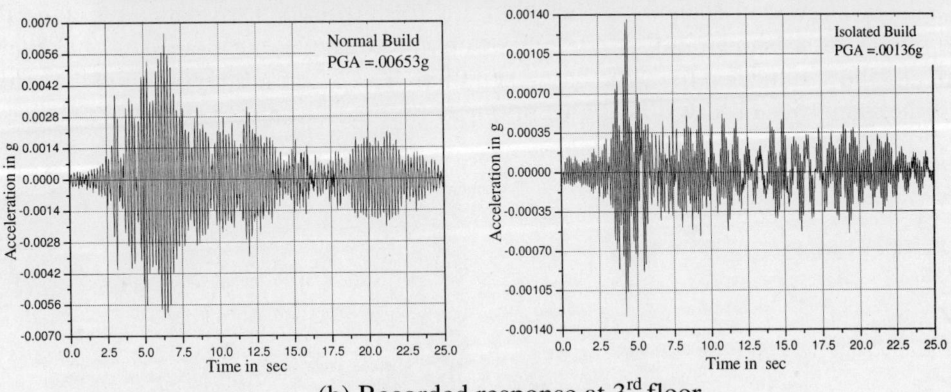

(b) Recorded response at 3rd floor

Fig 27 Buildings with fixed base foundation and base isolation

15 INVESTIGATIONS OF BEHAVIOR OF NUCLEAR PIPING SYSTEMS UNDER SEVERE EARTHQUAKE LOADING

The nuclear facility piping systems are designed for SSE and stresses are categorized under faulted load condition as per ASME code. ASME code rule for piping design with SSE loading was based on the principle of plastic collapse. Recently it is observed that during such events, fatigue ratcheting is most likely failure mode of piping. In view of this, ASME Boiler & Pressure Vessel code, Section-III (1995) has increased the allowable primary stress limit for Level-D service condition. It also provides an alternate strain based approach, which states that the average through wall ratcheting strain should not exceed 5% in 10 cycle event for Level-D service condition.

The ASME, Sec. III (2001) brought back the primary stress limit for Level-D loading. However, B_2 indices are modified as B_2' indices and strain based approach is not changed in this revision. The 2004 edition of the code has deleted strain based approach and stress based approach is not changed in this revision. The basis of these changes is still under debate by many regulatory agencies in US and Europe. In view of this, BARC has initiated extensive research to understand revised ASME design rule. A large number of experiments have been planned to validate the ASME code rule. Few ratcheting experiments conducted have shown that strain accumulation depends on both stress rate and stress ratio. In the biaxial test on straight pipe (Fig. 28), ovalization of pipe cross section was observed when the pipe was subjected to constant internal pressure and cyclic bending load. Local bulging was observed at higher loading as shown in Fig. 29.

Fig. 28 Experimental setup for straight pipe

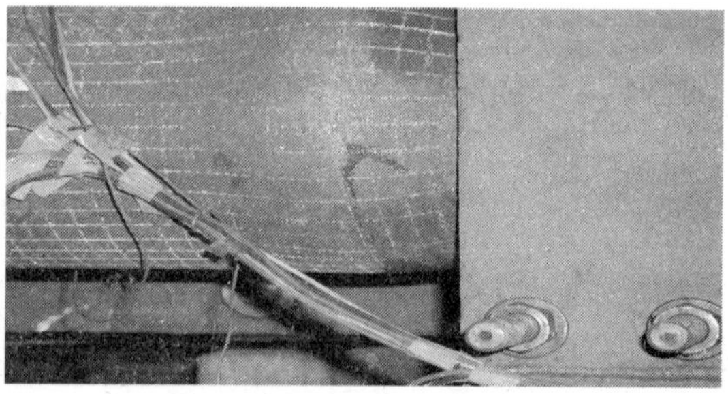

Fig. 29 Bulging of Pipe

The pipe did not show any shakedown behavior for the given cycles of loading and exhibited continuous ratcheting as shown in Fig. 30. Similar observations were made during shake table test on pressurized elbow subjected to cyclic base excitation. Test setup is shown in Fig. 31. Here also continuous ratcheting was observed as shown in Fig 32 before failure at the weld.

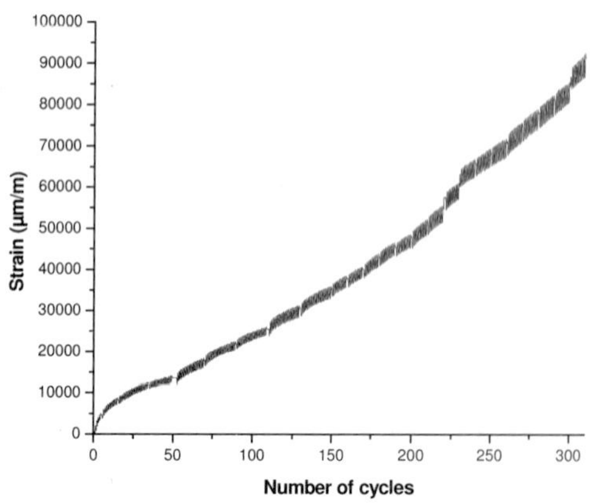

Fig 30 Hoop strain v/s number of cycles

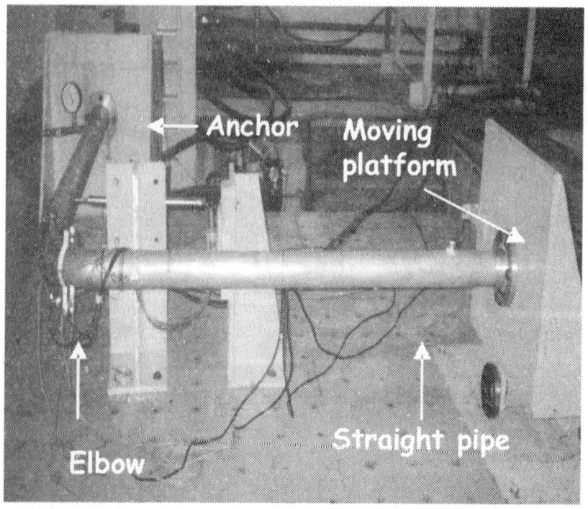

Fig 31 Experimental setup for elbow

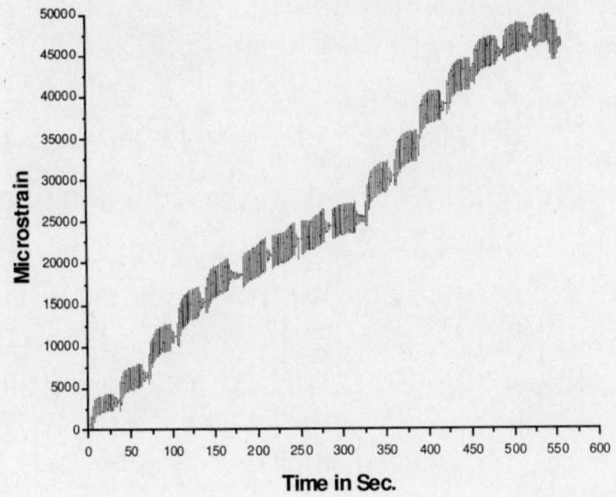

Fig 32 Hoop strain v/s time at crown

Fig 33 Experimental setup for 3D piping system

During shake table tests on three-dimensional piping systems (Fig 33), fatigue-ratcheting failure was observed in one test, (strain accumulation is shown in Fig. 34). In the next test, a weld failure was observed as shown in Fig. 35 after a large number of cycles. Further tests have been planned to find out strain limit to prevent ratcheting mode of failure in the piping system under SSE.

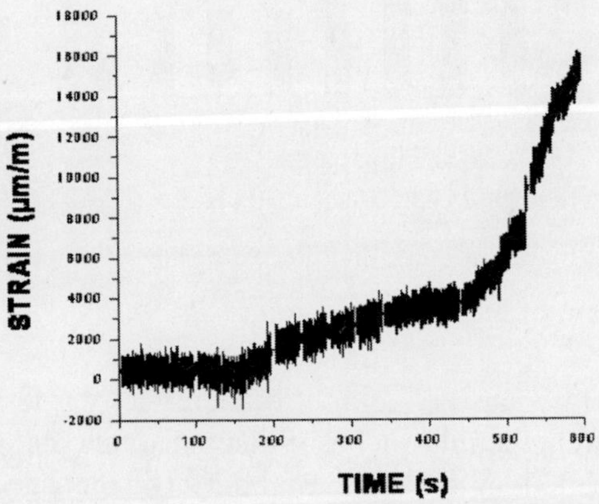

Fig. 34 Hoop strain time history at an elbow

Fig. 35 Failure of piping system

16 INVESTIGATIONS ON REINFORCED CONCRETE STRUCTURES

Behavior of concrete structures under seismic excitation still poses great challenges in front of civil engineers throughout the world. The vulnerability of the RC beam-column connections in a framed structure under seismic loads is also well known. Keeping these in view, a three-phase program has been designed to understand seismic behavior of RC structures under SSE loading.

The first phase of the program includes experimentation of RC beam-column connections under monotonic and quasi-cyclic loads, having varied cross section dimensions in a range of 100mm to 1000mm. Reinforcement detailing, shear span ratio, compressive strength of concrete and type of loading are among the parameters which are varied. In total thirty joints have been tested till date and further twelve joints under this program have been cast. Fig 36 shows the testing under progress and the damaged state of (a) Small Sized Joint, and (b) Large Sized Joint. Fig 37 shows results of experimental program on beam-column joints in form of (a) Typical hysteretic loops and (b) Variation of ductility with size of beam-column joints.

(a) Small-Sized Joint (b) Large-Sized Joint

Fig. 36 Testing under progress and damaged state for beam-column joints

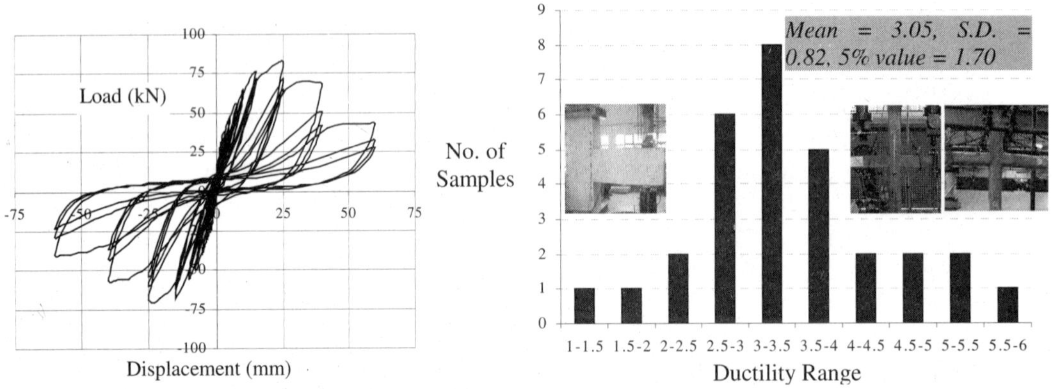

Load: 2100 kN Load:500 kN to 30 kN Load: 3.80 kN

(a) Typical Hysteretic Loops (b) Variation of ductility with size

Fig 37 Results of experimental program of beam-column joints

Retrofitting of beam-column joints of existing RCC structure may be required if the capacity is less than the demand. A retrofitting technology based on strengthening of beam-column joints using Carbon Fiber Reinforced Polymers (CFRP) was developed along with IIT, Bombay. The damaged joint (Fig 36 (a)) was retrofitted using CFRP and tested again (Fig 38(a)). The damaged state of the joint before repair is shown in Fig 38(b). The results (Fig 38(c)) show a significant increase both in strength and deformation capacity of the repaired joint over the control joint.

(a) Testing of Repaired Joint (b) Damaged State (c) Hysteretic Loops

Fig 38 Experimental Program on Testing of CFRP Repaired Joints

The second phase of the program includes experimentation of RC structures under monotonic pushover type loading. Pushover technique for RCC structures subjected to earthquake loading is a very useful technique to know its performance states such as initial cracking, yielding of bars, ultimate state and failure state. However, higher mode effects and damping effects are considered conservatively and hence requires experimental verification. Under this program, one scaled down model structure (Fig 39 (a)) has been tested and another full-scale prototype of existing RC structure (Fig 39 (b)) is under construction. Pre-test analysis of the prototype structure is performed (Fig 39 (b)). The experimentation on these structures will be carried out under gradually increasing monotonic pushover loading up to failure.

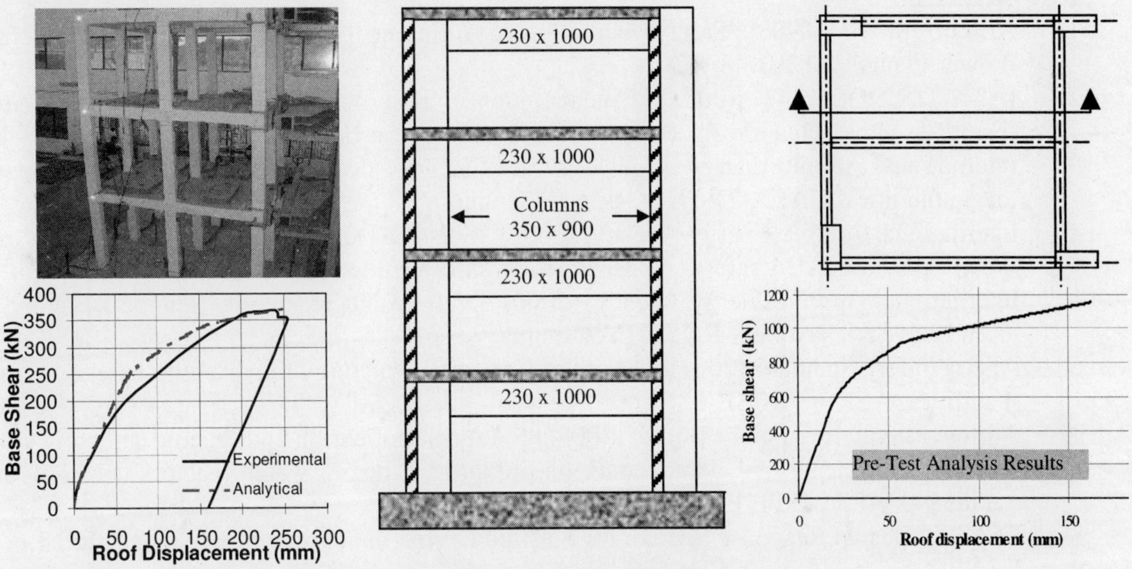

(a) Scaled Down Model Structure (b) Elevation and Plan of Prototype structure and pre-test analysis

Fig 39 Structures for Testing against Pushover Loading

The third phase of the program includes the experimentation of RC structures under dynamic loads using shake table. Under this program three scaled down models of RC structures will be tested on tri-axial shake table. One such structure tested on shake table is shown in Fig 40. Various failures observed in the structure are also shown in the figure. Analytical work is under progress.

Fig 40 Typical failures in the structure tested on shake table

17 CONCLUSIONS

Significant progress has made in the seismic analysis and design of nuclear power plant structures, equipment and piping. Personal computer and commercial computer codes have played a major role in simplifying the seismic analysis. However, large efforts are still required to verify and improve operability of safety related SSCs in nuclear facilities. There is also a need to develop fragility of the pumps, valves, relays and shutdown system by conducting shake table and multi support excitation tests. With the sufficient experimental data, the reliability of various SSCs under seismic motion can be ensured.

ACKNOWLEDGEMENT

Author acknowledges the efforts and support of various institutions like SERC (Chennai), CPRI (Bangalore), IIT (Guwahati), IIT (Madras) and IIT (Bombay) in the experimental work. Also the support provided by Shri P.N. Dubey and Shri Akanshu Sharma is highly appreciated.

REFERENCES

1. AERB/ SM-S-1, (1988) "Safety Manual for Civil Engineering and Building Works of Nuclear Power Plants", AERB.
2. IAEA-TECDOC-1347 , (2003) "Consideration of External Events in the Design of Nuclear Facilities Other Than Power Plants", With Emphasis on Earthquakes.
3. International Atomic Energy Agency, (2001) "Seismic design considerations of nuclear fuel cycle facilities", IAEA-TECDOC-1250, Vienna.
4. International Atomic Energy Agency, (2003) "Seismic Design and Qualification for Nuclear Power Plants", IAEA safety standards series, safety guide No. NS-G-1.6.
5. International Atomic Energy Agency, (2000) Safety of Nuclear Power Plants: Design, Safety Standards Series No. NS-R-1, IAEA, Vienna.
6. AERB Safety Standard (2001), Design of Concrete Structures Important to safety of Nuclear Facilities, AERB/SS/CSE-1.
7. Atomic Energy Regulatory Board, (2003) "Safety classification and seismic categorization for structures, systems and components of pressurized heavy water reactors", AERB Safety Guide No. AERB/NPP-PHWR/SG/D-1.
8. USNRC, Design Response Spectra for Seismic Design of Nuclear Power plants, RG 1.6.
9. USNRC, Seismic Design Classification Regulatory Guide 1.29.
10. JEAG 4601-1987, (1987) Technical Guidelines for Aseismic Design of Nuclear Power plants, Japan Electric Association publication.
11. International Nuclear Safety Advisory Group (1996), Defense in Depth in Nuclear Safety, INSAG Series No. 10, IAEA, Vienna.
12. International Nuclear Safety Advisory Group (1999), Basic Safety Principles for Nuclear Power Plants, INSAG-12, IAEA, Vienna.
13. International Atomic Energy Agency, (2005) "Safety of research reactors", IAEA safety standards, Safety requirements No. NS-R-4, Vienna.
14. IS 1893 (2002), "Criteria for earthquake resistant design of structures", BIS, New Delhi.
15. H.S.Kushwaha, (2006) "An overview of regulatory requirements for seismic qualification of nuclear facilities", Theme meeting on seismic qualification of nuclear facilities, BARC.
16. ASCE 4-98, (1998) Seismic Analysis of Safety Related Nuclear structures and Commentary.
17. Beredugo, Y.O, and Novak, M. (1972) Coupled Horizontal and Rocking Vibration of Embedded Footing, Canadian Geotechnical Journal, Vol.9, No.4, pp 477-497.
18. SRP 3.7.2, (1990) "Seismic system Analysis", U.S. Nuclear Regulatory Commission.
19. Reddy, G.R., Vaze, K.K., Kushwaha, H.S., Tandale, J.V. Subramanian, K.V. Kelkar, S.P. (1994) Procedure of Applying Decoupling Criteria for a Complex Primary and Secondary System. Proc. ASME Pressure Vessel and Piping conference. 272-1: 129-139.
20. Reddy, G.R., Kohei Suzuki, Kushwaha, H.S. Mahajan, S.C., (1997). Decoupling Criteria for Multi-Connected equipment. Journal of Pressure Vessel Technology, 120: 93-98.
21. ACI – 349 (1997), "Code requirements for nuclear safety related concrete structures".
22. IS 456 (2000), "Plain and Reinforced Concrete, Code of Practice", BIS.
23. IS 13920 (1993), "Ductile detailing of reinforced concrete structures subjected to seismic forces – code of practice", BIS, New Delhi.
24. IS 875 (1987), Code of Practice for design loads (other than earthquake load) for building and structures- wind loads. BIS, New Delhi.
25. DOE/EH-0545, (1997) Seismic Evaluation Procedure for Equipment in US.
26. IEEE, Standard 344-1987, 1987. (IEEE) Recommended Practice for Seismic Qualification of Class 1E Equipment for Nuclear Generating Stations.

SEISMOTECTONICS OF NORTHEASTERN INDIA WITH SPECIAL REFERENC E TO EARTHQUAKE HAZARD

D. R. Nandy

Retired Director, GSI and Emeritus Scientist, DST. nandy49@gmail.com

ABSTRACT

For proper assessment of earthquake hazard of an area/ region, geodynamics of the tectonic blocks, especially the interaction of their interface and nature of contemporary tectonic activities should be clearly understood. The whole of Northeastern India has been classed as zone V in seismic hazard zonation map of India (IS 1892:2002) though seismic hazard potential of the whole region may not fall in the same category. The region can be divided into four major tectonic blocks of distinct tectonic evolution and different styles of contemporary tectonic activities making them seismically active. They are (a) the Himalayan and the trans-Himalayan zone where thrusts and fold belts are cut across by transverse faults that are seismically active and may form the locales of moderately large earthquakes; (b) the Mishmi block, situated at the northeast corner of the area, is characterized by number of NW-SE strike-slip faults, the movements along one such fault (the Po Chu fault) had generated the 1950 great earthquake having magnitude >8.5, devastating a vast tract in upper Assam though some of the earthquake events occurring along the frontal thrusts of this block also yielded thrust solution; (c) active oblique subduction is going on along the Indo-Myanmar mobile belt with back arc, fore arc and outer arc accretion complex settings, attended with abundant seismicity, most of which have moderately deep seated origin and with many hinge faults in this domain producing strike-slip motion and (d) the Meghalaya Plateau and the Mikir Hill block, buttressed in between the Himalayan and the Indo-Myanmar mobile belts, is characterized by many active N-S, NW-SE, NE-SW and E-W (Dauki fault) faults and seismically very active. Of these the NW-SE Kopili fault, passing in between the Meghalaya Plateau and Mikir Hill cutting across the Himalayan domain, may produce large earthquakes in near future.

Keywords: Tectonic block/domain, Seismotectonic, Active/ seismogenic fault, Earthquake Hazard

INTRODUCTION

In India though lot of emphasis has been put by different organizations in instrumental monitoring of earthquakes during the last three decades no much effort has been given to identify the active seismogenic faults producing many earthquakes in our country, which is very important for earthquake hazard mitigation. Although some sporadic work are being carried out to identify such faults till now we are not unequivocally sure about the causative faults (exposed or hidden) generating many of the great, major and large Indian earthquakes. Mapping and characterization of such active seismogenic fault is of vital importance for assessment of earthquake hazard potential of a particular area. Even during the on going National Program on GPS measurement the average crustal velocity vectors are determined without giving due consideration to the differential movements between various adjoining tectonic blocks and nature of movements along their interface. Sometimes scanty data and cursory interpretation of existing data set has led to assigning some nonexistent fault (Oldham fault) to have produced great earthquake like 1897 great Assam earthquake (Bilham et al, 2001). The primary task for identification of such faults is to work out the geodynamics through geological past to arrive at the contemporary status of tectonics of individual tectonic blocks and their interface where the earthquakes occur and then closely map the identified active faults both geologically and geophysically to ascertain their present level of activity. Having this idea in mind the author and his coworkers have carried out research in seismotectonics of Northeastern India and the adjoining areas resulting in identification of many seismogenic faults in the region.

Northeastern India and it adjoining areas are located in the northern collision and eastern subduction domains of north-drifting Indian plate are characterized by active tectonic movements causing intensive seismicity in the region (fig. 1). The region consists of 4 distinct tectonic domains. They are i) the eastern Himalayan mobile belt and the trans-Himalayan Tethyan tectogenes; ii) the Mishmi block; iii) the Patkoi-Naga-Chin-Arakan Yoma (Indo-Myanmar) mobile belt and iv) the Meghalaya Plateau and the Mikir Hills block. Each of these domains has its own evolutionary history and distinct pattern of intra- and inter-domain dynamics of movements generating characteristic seismicity (fig. 2, Nandy, 2001).

ACTIVE TECTONICS AND SEISMICITY

Northeastern India and the adjoining region fall in the most intense seismic zone of the world. Out of the 5 great earthquakes (M>8) experienced in India 3 occurred in this region including the 1934 Bihar-Nepal event. Besides, there were incidences of 11 large earthquakes (M>7) during the period from 1869 to 1947. Of these 5 occurred in the Indo-Myanmar subduction zone, 4 in the Meghalaya Plateau and Mikir Hills, one in the eastern Himalaya and one in the northeastern corner of Bengal Basin in Bangladesh. The area between 88 and 98 degree east longitudes and 20 to 31 degree north latitudes has experienced 974 earthquake events having magnitude≥4 during the period from 1964 to 1993 (source: ISC, fig. 3). Though the region is highly seismic it has not experienced any major (M>7) earthquake during the last 60 years. It is interesting to note that first scientific study of Indian earthquake was under taken here for the 1869 Cachar earthquake (M>7) by T. Oldham (Memoir, GSI, Vol.19, pt. 1, 1882), whose shocks were felt severely at far off places like Kolkata and Patna.

Considering the past and present level of location accuracy of earthquake data it appears that the pattern of seismicity in the region is diffused except that occurs along the Indo-Myanmar subduction zone. Though in recent past and at present lots of studies have been and are being carried out to understand the nature of earthquake events and significant progress has been made in monitoring the seismicity of Northeastern India, many questions considering the location and exact nature of earthquake sources based on contemporary tectonics have remained largely unsolved. However, based on ground truth and analysis of seismicity pattern attempt has been made to broadly correlate the source zones of major earthquakes by the author and his coworker (Das Gupta and Nandy, 1982; Nandy, 86; Nandy and Das Gupta, 1991 and Nandy, 2001).

Eastern Himalayan Domain

The Eastern Himalayan Domain is characterized by three major thrusts viz. the Frontal Foothill Thrust (FFT), the Main Boundary Thrust (MBT) and the Main Central Thrust (MCT). It was contemplated that most of the Himalayan earthquakes occurred due to movements along these thrusts. But the 1988 Bihar-Nepal earthquake and 1980 Gangtok earthquake, both having magnitude greater than 6, had well constrained focal depth of 60km and 47km respectively yielding clear strike-slip solution (Das Gupta, et. al, 1987). These earthquakes neither could have been generated by thrust movements, nor that had been possible for any thrust tectonics in the Himalayan setting to produce such deep focus earthquake events. Seismotectonic analysis of the Eastern Himalayan zone (Das Gupta et al, 1987 and Nandy, et at, 1991) has clearly indicated that many of the transverse faults lying athwart to the Himalayan trend are at present active. The most important of them are, from west to east, East Patna, Kanchen Dzonga, Yadaon Gulu (graben), Tista, Jamuna, Dudhnoi, Kulsi, Gyau (graben), Kopili and Bomdila faults. Focal mechanism solutions for well constrained earthquake events occurring along these faults yielded strike slip solutions but those occurring in the higher Himalayan domain along the graben structures gave normal solutions (Nandy and Das Gupta, 1991). It may be mentioned here that the Dudhnoi and Kulsi faults cutting across the Meghalaya Plateau also traverse across the frontal Himalayan fold and thrust belt.

Mishmi Block

This tectonic block is characterized by NW-SE trending thrust belt of mainly diorite- granodiorite massif. Three important high angle thrust/reverse faults are, from foot hill to the higher Himalayan domain, Mishmi thrust, Tidding suture, Lohit thrust and Po Chu fault. The frontal Mishmi thrust overrides the NE trending fold and thrust belt of the eastern Himalaya as well as the ENE trending fold and thrust belt of the Indo- Myanmar mobile belt. The earthquake events occurring in the southeast extremity of the Mishmi thrust indicated right lateral strike-slip faulting along steeply dipping NE to ENE nodal plane. Other events occurring right inside the Mishmi block had also yielded strike slip solutions indicating thereby that present tectonic transport is mainly through strike-slip motion. It is interesting to note that the 1950 great earthquake (M>8.5) occurred in the Po Chu fault zone located in the NW extremity of the Mishmi massif. This event indicated both dextral slip along NW nodal plane and thrust movements by different workers (Nandy, 2001).

Indo-Myanmar Mobile Belt

This mobile belt and the Andaman-Mentawai arc had developed during the Cenozoic in response to the east/northeast directed sbuduction of the Indian lithosphere below the Myanmar- Andaman plate (fig.1). In this vast tectonogeologic set up the active regional and fundamental structures experiencing brittle failure are the subducting slab of Indian lithosphere and the overriding Myanmar plate, the geometry of which has been worked by many workers (see Nandy, 2001). Rupturing along this subduction interface in the Sumatran domain produced the 26th December, 2004 great earthquake (M 9.1) rupturing about 1200km of the interface that was accompanied by devastating tsunami. Other regional active structures are Shan-Sagaing fault delimiting the eastern boundary of Central Myanmar Molasse Basin, the Central Myanmar Volcanic Line passing southward through Narcondom and Barren Islands to the coastal zone of Sumatra, the Eastern Boundary Thrust representing the surface trace of the sbuduction interface, Basin Margin Fault of the Paleogene outer arc limiting the western margin of the Surma Basin, the frontal thrust belt of the Upper Assam Basin (Belt of Schuppen) and number of deep seated cross faults. Seismotectonic analysis of this mobile belt indicates ongoing active subduction though there is active dextral slip along the Shan-Sagaing fault. Most of the hypocentral data and focal mechanism solutions constrain the subducting slab (fig. 3). The relatively deep seated cross faults have developed due to stress partitioning along the curve-planer long subducting slab that indicated both normal and strike-slip motion (Nandy 2001). The Neogene outer arc represented by the Suma Basin south of the Meghalaya Plateau is delimited to the northwest by the seismically active NE Sylhet fault that passes along the northeastern corner of Bangladesh in Bengal Basin. This is a high angle reverse fault though a few of the earthquake events occurring in the Dauki fault zone north of the Sylhet fault yielded strike-slip motion.

Meghalaya Plateau and Mikir Hills

This tectonic block represents the northeastern most extension of the Indian shield element consisting of early Proterozoic granite gneiss and intrusive granites with late Proterozoic sediments of Shillong Series. The southern margin of the Plateau contains least disturbed sediment of Cretaceous to Quaternary age. Patches of Cretaceous ultramafic-carbonatite complex occurs along a NE-SW zone from the Plateau to the Mikir hill area.

The most important structural element is the 200 km long E-W Dauki fault defining the southern margin of the Meghalaya Plateau that had been active science Cretaceous to the Recent. At present this fault seems to be dormant (Nandy, 2001). The most active faults in this domain are N-S Dudhnoi and Kulsi faults, NW-SE Kopili fault demarcating the boundary between Meghalaya Plateau and Mikir Hills and the Bomdila fault cutting across the outer arc of the Indo-Myanmar mobile belt, Brahmaputra basin and the Himalayan mobile belt. The movements along the Dudhnoi fault produced the 1897 great Assam earthquake (M>8) and Kopili fault was responsible for the 1869 Cachar earthquake (M>7) as well as the 1943 event (M>7).

The earthquake events located in the Meghalaya Plateau, Mikir Hills and the adjoining areas also indicated strike-slip motion with N, NW and NE directed pressure axes (Nandy and Das Gupta, 1991; Chen and Molnar, 1990). 79 well constrained and precisely located earthquake events (coda magnitude varying from 2 to 4) by the Regional Research Laboratory, Jorhat during January to August, 1999 clearly showed that the wedge shaped Kopili-Bomdila fault zone is highly seismic having manifestation up to subcrustal depth. This work and seismotectonic analyses carried out earlier (Nandy and Das Gupta, 1991) indicate the probability of large earthquake in this zone in near future (Nandy, 2005).

Source of 1897 Great Assam Earthquake

Bilham and England (2001) has postulated an E-W fault, Oldham fault, along the northern margin of the Meghalaya Plateau which supposed to have generated the 1897 great Assam earthquake mainly based on the result of post earthquake geodetic survey though there is no physical existence of such fault. After detailed investigation of the great earthquake Oldham (1899) enunciated the theory that the earthquake was caused by a north dipping thrust plane measuring 160km in length and 80km in width at a depth of 8 to 14km, drawing inferences from the thrust tectonics of the Scottish Highland and Scandinavia, though he clearly stated that the epicentral tract ran in a N-S narrow zone from the Sylhet plain of Bangladesh to the high hills of Bhutan Himalaya (cutting across various geological milieu) lying just east of 91 degree meridian (page 169 0f Oldham, 1899). This field observation is contrary to his thrust hypothesis. Perhaps, guided by this thrust hypothesis of Oldham, subsequent workers like Seeber and Armbuster (1981), Khatri (1992), Molnar (1987) and others had visualized a northerly dipping midcrustal shallow active thrust below the Meghalaya Plateau which was supposed to join the E-W vertical Dauki fault. It is interesting to note here that Oldham (1926) himself had revised his idea of existence of a midcrustal causative thrust as he later realized that such interpretation was not satisfactory and not consistent with facts, nor were able to account for somc of the peculiarities of distribution of intensities of the earthquake shock. The cause of intense destruction and ground distortion along a long narrow N-S band was correlated by him with the movement along a deep seated N-S fault which is co-relatable with the Dudhnoi fault. Unfortunately, none of the later workers including Bilham and England (2001) have referred to the revised and up dated postulation of Oldham (1926). Moreover, the N-S section drawn by Bilham et al (2001, fig.3) is not factually correct as it does not corroborate with the topographical and geological ground truth (Nandy, 2005). The pattern in the post earthquake change in the level and coordinates of the different survey stations could be explained by strike-slip sinisteral movement along the Dudhnoi fault (Nandy, 2001).

CONCLUDING REMARKS

1. For earthquake hazard assessment and subsequent development planning including building design the active seismogenic faults/thrusts of the individual tectonic domains should be first identified and delineated and their potential for generating earthquake of possible highest magnitude should be ascertained through seismotectonic analysis and field observation.
2. It has been observed that in the eastern and northeastern Himalayan domains the zones of intersection of the transverse faults cutting the thrust-fold belt are experiencing intense seismicity having relatively deeper focus manifestations and may form locales for future earthquake.
3. In the Meghalaya Plateau and Mikir Hill domains the N-S and NW-SE faults traversing in to the Himalayan domain are very active at present. The NW Kopili fault passing in between the Plateau and Mikir Hills is highly seismic which had witnessed two large earthquakes in the past and may rupture in near future to generate large event.
4. Now there is a chance of rupturing along the Indian and the Myanmar Plates in the continental domain as huge amount of stress has been released in the far southern part of this interface having ruptured for length of 1200km from the tip of Banda arc to the Andaman Island giving rise to 9.1 magnitude earthquake on the 26th December, 204 followed by many moderate to large magnitude events.
5. Indo-Myanmar mobile belt in the continental domain may be closely studied for any eventual abnormal stress build up along any of the described dislocation zones.
6. One may be careful to design any structure based on the postulated Oldham fault whose existence is highly doubtful.

ACKNOWLEDGEMENT

The author is thankful to Shri Jibitesh Bhattacharya of GSI and Shri Jayanta Sen of Calcutta University for cartographic assistance.

REFERENCES

1. Bilham, R and England, P. (2001), "Plateau 'pop up' in the great 1897 Assam earthquake", Nature, **410** (6830), 806-809.

2. Chen, W.P. and Molnar, P. (1990), "Source parameters of earthquakes and intraplate deformation beneath Shillong Plateau and northern Indo-Burman ranges", Jour. Geophys. Res., **95** (B8), 12527-12552.

3. Das Gupta, S., Mukhopadhyay, M. and Nandy, D.R. (1987), "Active tectonic features in the central part of the Himalayas", Tectonophysics, **136**, 255-264.

4. Das Gupta, S. and Nandy, D. R. (1982), "Seismicity and tectonics of Meghalaya Plateau, Northeastern India", VII Sympo., Earthquake Engineering, University of Roorkee", **1**, 19-24.

5. Khattri, K.N. (1992), "Seismological investigation in Northeastern Region of India", Mem. Geol. Soc. India, **23**, 275-302.

6. Molnar, P. (1987), "The distribution of intensities associated with the great 1897 Assam earthquake and constraints on the extent of rupture". Jour. Geol. Soc. India, **30**, 13-27.

7. Nandy, D.R. (1986), "Geology and tectonics of Arakan Yoma- A reappraisal". Bull. Geol. Soc. Malaysia, **20**, 137-148.

8. Nandy, D. R. and Das Gupta, S. (1991), "Seismotectonic domains of Northeastern India and adjacent areas". Geology and Geodynamics of Himalayan collision zone, 2, Physics and Chemistry of Earth, **18**, 1-11, Pergamon Press, PLC, Oxford, 371-384.

9. Nandy, D. R. (2001), "Geodynamics of Northeastern India and the adjoining region". acb Publication, Lake town, Kolkata 700 089.

10. Nandy, D. R. (2005), "Geodynamics and seismicity of Northeastern India and the adjoining areas". KECS-2005, GSI Spl. Pub. **85**, 49-59.

11. Oldham, R. D. (1899), "Report of great earthquake of 12th June, 1897". Mem. Geol. Surv. India, **29**, 1-379.

12. Oldham, R. D. (1926), "The Cutch (Kachh) earthquake of 16th June, 1819". Mem. Geol. Surv. India, **46** (2), 118-141.

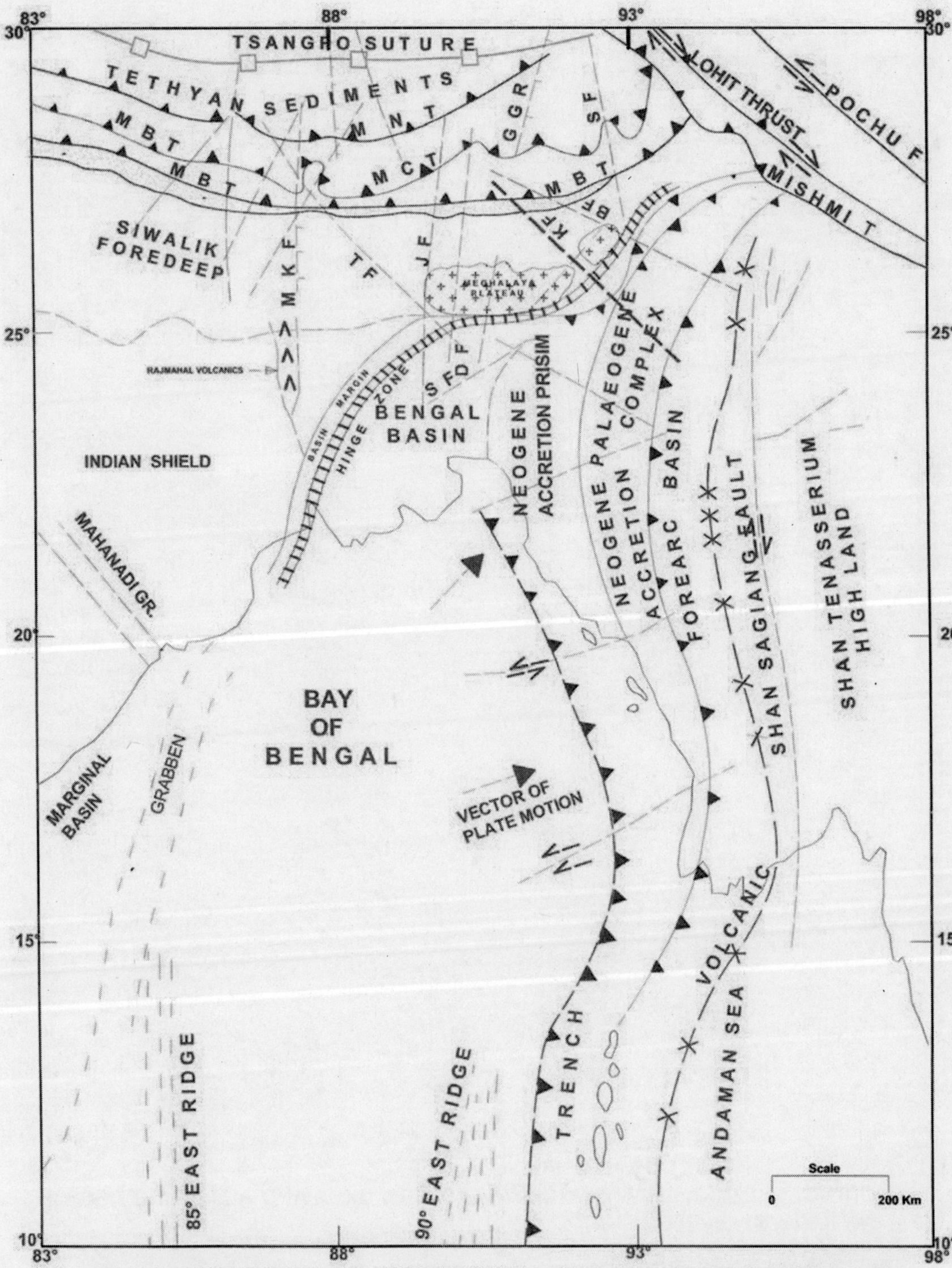

Figure 1 : Generalised Tectonic Map of India and the adjoining region. MCT-Main Central Thrust, MBTMainBoundary Thrust, MNT-Main Northem Thrust, YGR-Yadong Gulu Grabben,GGR-Gayo Grabben,SF- Siang Fracture,MKF-Malda Kisangunj Fault,TF-Tista Fault,JF-Jamuna Fault,,DF-Dudhnoi Fault and Dauki Fault,KF- Kolpili faultBF-Bomdila fault,SF-Sylhet fault,,EBT- Eastem Boundary Thrust,GR-Grabben,PL-Plateau,T-Thrust,Asterisks indicate Volcano

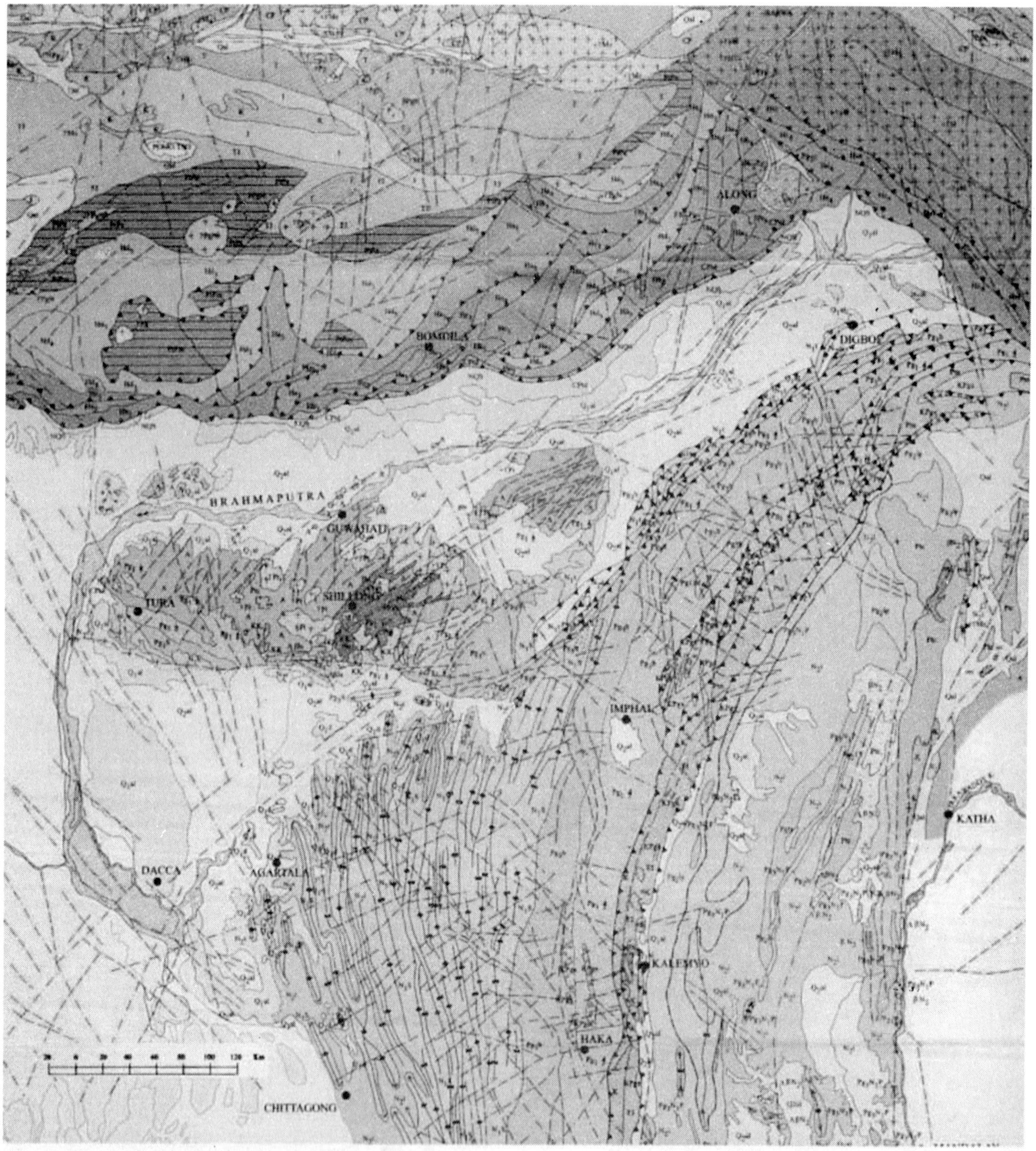

Fig.2. TECTONO-GEOLOGICAL MAP OF NORTH-EASTERN INDIA AND ADJOINING REGION (For detail, see Nandy,2001)

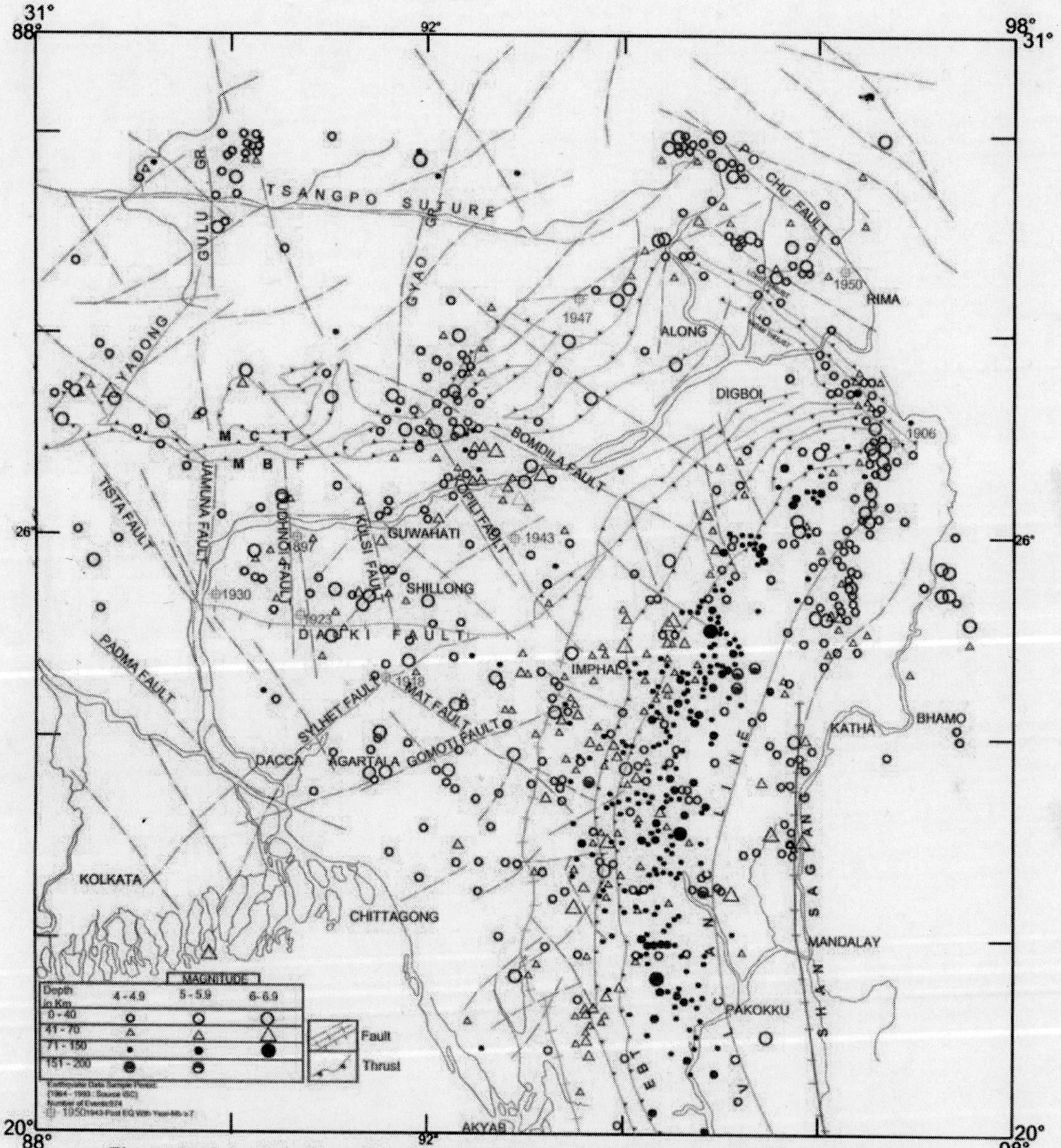

Figure 3 : Seismicity Map of Northeastern India and adjoining region. Earthquake Data from 1964-19993(Source ISC)\Plotted on the tectonic map of the area for detail see Nandy,2001.

Part-3

Engineering Seismology and Seismic Hazard

SITING OF LNG TERMINAL IN CHILE

Y. Zaczek[1] and R. Boroschek[2]
[1]*Tractebel Engineering, Belgium, yannick.zaczek@tractebel.com, yannick.zaczek@gmail.com*
[2]*Universidad de Chile., Civil Engineering Departments, e-mail: rborosch@ing.uchile.cl*

ABSTRACT

A new LNG terminal is to be built on the Pacific Coast of the Northern part of Chile. Due to subduction process, this region is one of the most active seismic areas of the world. In addition, the subsoil is made of more than 200m of diatomite covering volcanic bedrock. According to International Codes EN1473 and NFPA59A, a detailed siting process must be carried out to demonstrate the safety of the project. Seismic levels with response spectrum have to be determined. Absence of active faults has to be demonstrated. The paper describes the key factors of the siting methodology.

Keywords: Siting, Earthquakes, Faults, LNG Terminal

1 INTRODUCTION

Siting studies for a new LNG terminal to be built in North of Chile has been initiated in 2006 by the Tractebel Suez Group. Experts from Chilean Universities, engineering and geophysics consultant companies have been appointed to participate in the project. The site is located on the Pacific Coast close to Mejillones in the Antofagasta region. In order to analyze the feasibility and to demonstrate the safety of the project, siting criteria from International Codes EN1473 and NFPA59A are followed. They are very similar to the requirements from nuclear power plant siting codes US-NRC100 app A. After a description of the siting methodology, presentation of the preliminary steps is provided.

2 SITING METHODOLOGY

The siting methodology is developed according to the following 6 steps.

2.1 Data Collection

First, all available documents related to geology, hydrogeology, seismicity and geotechnics are collected for the region including the candidate area. Secondly, site visits and visual inspection of topography, geological and geotechnical features and buildings in the region are made. Meetings are organized with local companies to collect information about geotechnical data, foundation types and problems or damage due to previous earthquakes, soil conditions or other natural hazard. No field or laboratory tests are carried out at this step. Geology, tectonics, earthquake activity and local foundation information are summarized and potential site(s) is (are) selected. This step is of relatively low cost and short duration.

2.2 Rejection Criteria

The potential site(s) has (have) not to be bought yet, but must be accessible for some field investigation. It (they) should be located more precisely in order to focus the investigation within the potential site area(s) and therefore limit the investigation cost.

The first objective is to detect all geo-hazards that could make the site unsuitable. The geo-hazards could be large geological phenomenon like karst or important deposits of poor material like swellings clays or collapsible soils, but mainly tectonic accidents. Especially the active faults in the region are detected and potential sites are located outside of their influence.
Satellite image and aero-stereo photos are interpreted to detect any evidence of faults (lineations) in the site region, prior to realize complementary geophysical investigations. When deep enough waterways are found close to the site(s), cheaper and faster offshore seismic tests can be conducted.
The second objective in a high seismic risk region as the one considered here, is to evaluate preliminary values for the Operating Basis Earthquake (OBE) and Safe Shutdown Earthquake (SSE) levels on the site(s).The OBE is mainly an internal decision of the Owner while the SSE determination is ruled by international codes, such as NFPA59A and EN1473. Seismic risk analyses (probabilistic and deterministic) shall determine the SSE level, and help in OBE selection. The Tsunami risk is also evaluated to select a site not affected by wave's impact and flooding.

2.3 Global Site Characterization

The site(s) should be located at few hundreds of meters accuracy and be accessible for soil tests. Type of buildings, typical loads, settlement criteria and earthquake levels must be established. Detailed review of the earthquake level shall be carried out in order to issue accurate OBE and SSE levels to be applied for the site qualification. During this step, field tests (e.g.: borings, CPT, pressuremeter tests) are realized on site to characterize the typical soil layers affected by the construction. Liquefaction risk, bearing capacity, settlement and any geotechnical hazard detection are estimated. A general foundation methodology (design of shallow and deep foundation, soil improvement special structures considerations, etc.) shall be issued.

2.4 General Stability

The general stability of the buildings is studied, taking into account general layout, seismic and geotechnical data. The layout, sizes and loads of the buildings (including the tanks) are adapted in consequence, for both static and dynamic conditions; settlement criteria of BS7777 must be followed. Displacements of the buildings during earthquake must be controlled. Typical piles and soil improvement methods shall be established, and finally, general cost estimates are elaborated. No geotechnical tests are carried out in this phase.

2.5 Site Characterization

The site must have been bought, in case it is not previously own, as soon a clear indication of feasibility is found. Specific soil investigation (geotechnical, geophysical, cross-hole borings, etc) are realized on a regular grid mesh covering the future LNG infrastructure areas in order to elaborate a geotechnical report, to be used for the project predesign and to elaborate LNG initial plant budget. The plant layout could be adapted and technical specifications for foundations design and soil improvements shall be issued according to this report.

2.6 Detailed Soil Investigation

The plant layout must be fixed and approved. Detailed soil investigations are carried out either to analyze local soil hazards or to elaborate very specific foundation design recommendations. The better the site characterization is detailed in Step 2.3, the less uncertainty will be resolved in this step of the study.

3 MEJILLONES SITE

The site is located in the industrial area at 8km from the Mejillones village, in the Atacama Desert where the pluviometry is close to zero. Strong winds are blowing mostly according to 2 main directions. Nevertheless the velocity is not as high as to control the structural design of the components. Due to extensive presence of loose sands, wind shall be considered for the durability and protection of expose structures.

Fig. 1: Site General View

No rain or snow is expected so they are not considered for the design. The region has the typical geography of coastal desert without vegetation, Fig 1. Temperature could vary more than 25 ^{o}C degrees in twelve hours.

The subsoil is made of sediments (sands and diatomites) covering the volcanic bedrock. The top surface is made of 3 to 5m of sands and shells with heavy salt content. Diatomites interbedded with sandy layers are found up to a depth of 200m. Diatomites are siliceous microscopic fossils in a silty calcareous matrice.

Diatomites have a very low density (sometimes lower than 1) and a very high porosity, up to 300%. It has an elastic behavior mostly brought by the air content, up to a kind of overconsolidated limit where it collapses due to rupture of siliceous skeletons.

About 100m of weathered and fractured rocks are found below the diatomites, covering the sound volcanic bedrock.

4 SEISMOTECTONIC STUDY

Initial seismological studies were performed in the area. World and local networks were used to identify seismogenic sources, Fig 2. Historical events were studied establishing the maximum acceleration, intensity and possible rupture area.

The most important seismogenic sources are associated with the interaction of the South American and Nazca plates. The sources can be parameterized in two main mechanisms: Thrust area (or interplate) and Inslab rupture at intermediate depth. These two sources are used to develop deterministic and probabilistic demands. As expected subduction – trust event control the largest demands. Historically the area is susceptible of events larger than Mw= 8.0 and events over Mw=8.5 are possible.

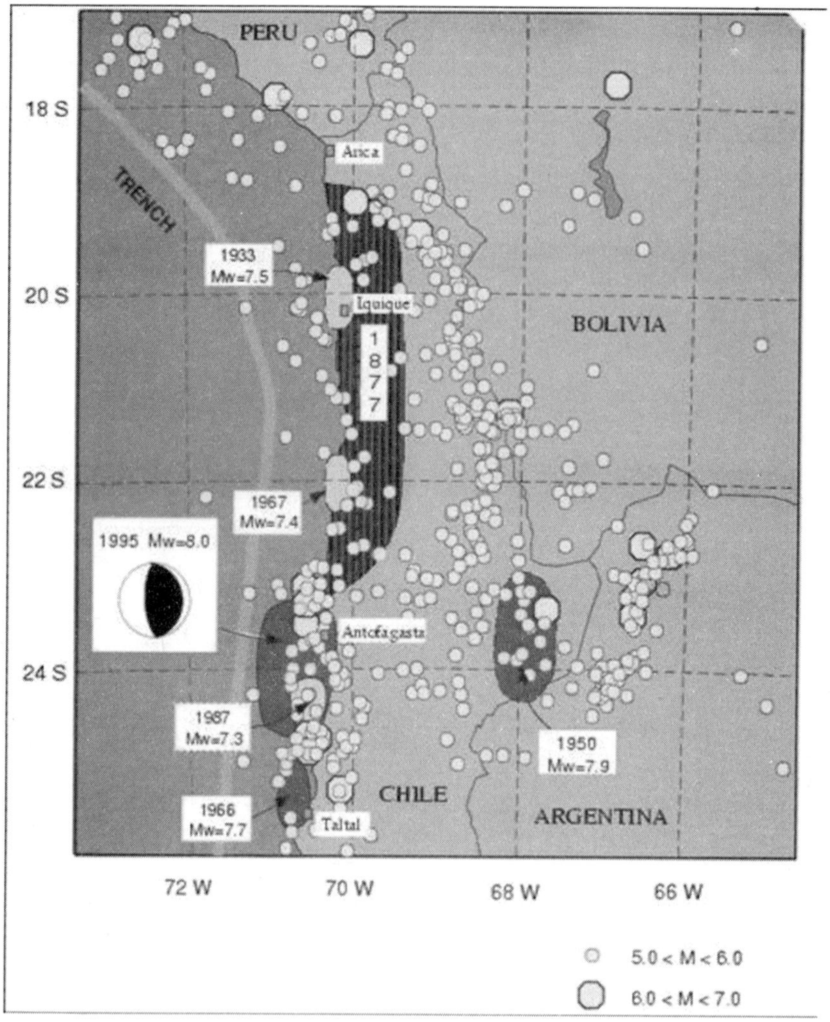

Fig. 2: Partial Area Seismicity (by Dr D. Conte)

Based on 475 and 2475 years return periods, the OBE and SSE Peak Ground Accelerations (PGA) have been determined with their corresponding spectrum.
Four 3D-accelerometers have been installed at different depths in deep borings in order to evaluate the site amplification effect induced by the particular diatomites layer.

5 FAULT INVESTIGATION

Exception made for the well known Antofagasta and Mejillones fault located at more than 8 km from the project, there are no other fault mentioned in the Mejillones plain. Detailed fault investigations have been therefore made around the site.

Aero stereo photo interpretation has been made on the site without discovering significant lineaments which could be correlated to fault activity. Most of the lineations parallel to the shore line could have been created by wind erosion and old sea level marks. There are traces of very old water flows perpendicular to the shore line.
Geophysical investigations have been realized: Transient Electromagnetic (TEM) investigation, Seismic Refraction and Seismic Reflection profiles.
In order to calibrate the TEM, a first profile was started at the outcropping foothills of the mountain at about 10km back of the site, Fig 3.

Rather homogeneous subsoil has been discovered with clear contrast between the rocks from the mountain and the sediments which seem to be saturated at the sea level.

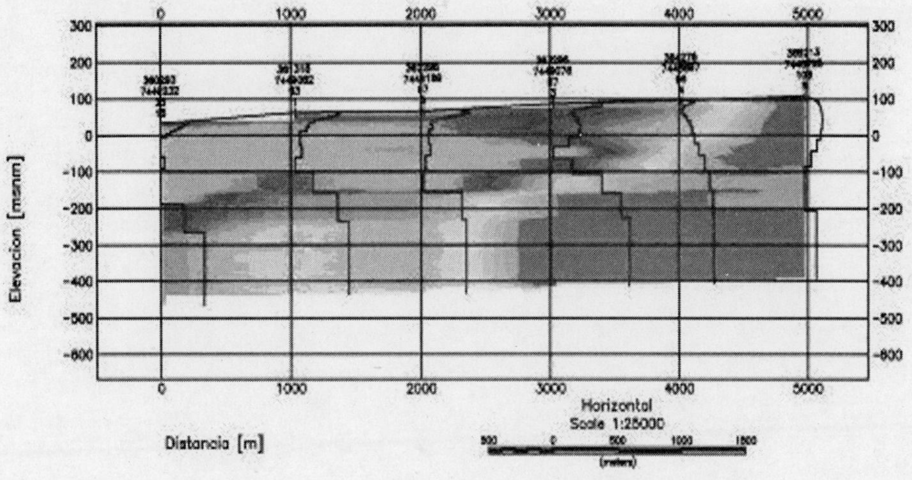

Fig.3: Typical TEM Profile

No clear picture of the bedrock has been obtained with only this method. Seismic refraction and reflection profiles were therefore requested to investigate deeper layers.
One 1.5km long seismic refraction profile was made to detect the different layers with their wave velocities, Fig. 4. Four successive layers have been discovered, Table 1.

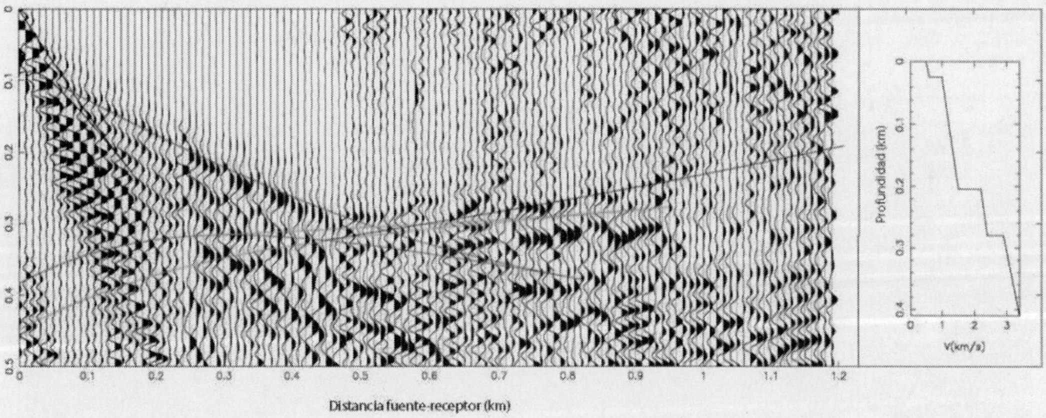

Fig. 4: Example of Refraction Profile

Table 1: Global Refraction Interpretation

Layers	Thickness	Vp
Sands & Dry Diatomites	25 m	550 m/s
Saturated Diatomites	180 m	1,250 m/s
Weathered Rock	75 m	2,300 m/s
Volcanic Bedrock	>120 m	3,150 m/s

A 13km of seismic reflection profiles has been realized on the site with 24 geophones line and explosives, Fig 5. Detail characterization of base rock has been found.

Fig. 5: Seismic Profiles

To improve the fault detection capability a local network of seismometers has been deployed. They are arranged in such way to determine the seismicity on the project especially on the top 20 km. This instrumentation will be monitoring small events for sufficient number of month in order to discard the presence of seismicity in the site area, Fig 6.

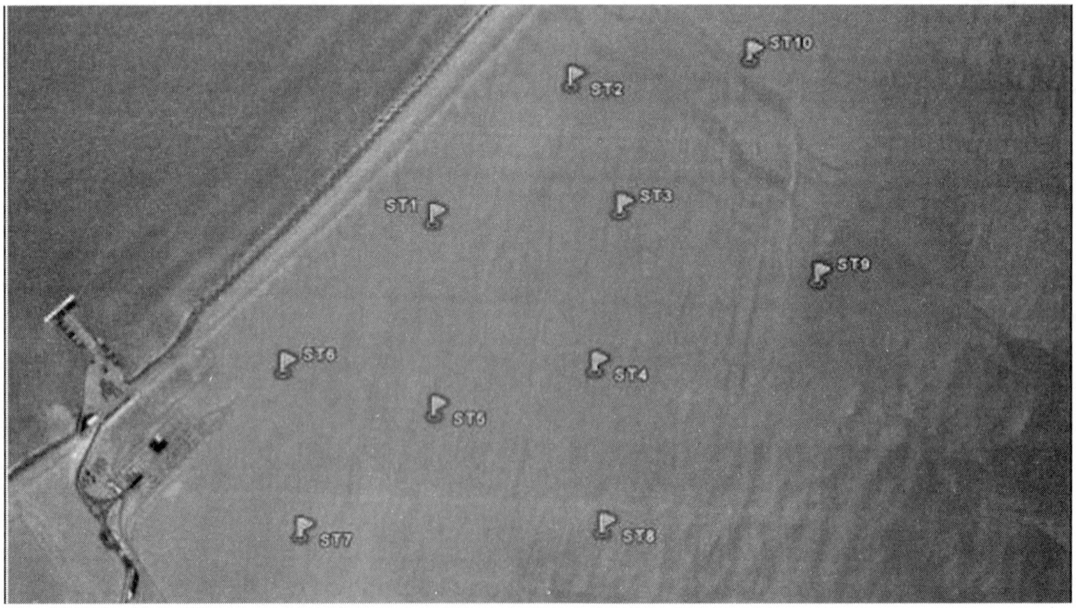

Fig. 6: Preliminary Seismic Network Arrangement

According historical seismicity of Chile, more than 10,000 events are expected for the region in the following 6 months which is quite enough to conclude for the site.

6 GLOBAL SITE CHARACTERIZATION

Base on preliminary site study the seismicity of the area has been refined and no active fault found. The soil characteristics consisting of a superficial layer of sand and a rather deep layer of diatomite with sand lenses and layers is found appropriate for plant predesign.Twenty one borings are drilled with realization of Standard Penetration Tests (SPT ASTM D1586) in the sandy layers and Pressuremeter Tests (ASTM D4719) in the diatomites layers.

Below a top loose sands covering, the SPT indicate a dense to very dense sand with N blows values regularly over 50. The pressuremeter values of "Em" modulus and limit pressure "pl" are quite high in the diatomites indicating a good bearing capacity and small settlements, as soon as the "over consolidation stress" is not superseded. There is no risk of liquefaction. Deep cross-hole(s) shall be realized at the tank location(s) in order to collect dynamic characteristics to issue the structural seismic design.

7 CONCLUSION

At this stage, the following conclusions can be drawn:

1. A potential site has been selected without hazards which could make the site rejected.
2. At the potential site, the subsoil seems to be composed of 200 meters of sediments (sands and diatomites), lying on volcanic rock basement.
3. Faults are known in the region but they are far enough from the tank site and would not prevent the construction of the terminal.
4. Preliminary Peak Ground Acceleration value and hazard spectrum for Operating Basis Earthquake (475-year return period) have been computed. The PGA of 0.65 g is very important but is in agreement with the very high seismicity in the region. Anyway, detailed probabilistic and deterministic analyses of the seismic hazard still have to be performed to confirm the OBE values and to determine the SSE values.
5. Final spectrum shall be defined after having analyzed site effect from measurements taken by four 3D-accelerometers installed in deep borings. Anticipated period of measurement shall be 6 months during which about 10,000 seismic events are expected in the region.
6. Considering the high horizontal and vertical accelerations, special foundation and stability design shall be applied for the safety and LNG-related structures.
7. Based on the previous items, the siting process (site characterization, general and detailed stability and general and detailed soil investigation) can be continued after administrative actions.

6 ACKNOWLEDGEMENT

The authors would like to thank Prof. Diana Comte for her support in seismological studies, and the Chilean investigation companies like Geodatos, Giantec, Idiem etc… for their professional works.

7 REFERENCES

1. NFPA 59A-2001: Standard for the Production, Storage, and Handling of Liquefied Natural Gas (LNG).
2. EN1473: European Standard – Installation and equipment for Liquefied Natural Gas – Design of onshore installations
3. US NRC 10CFR100: Nuclear Reactor Site Criteria

SEISMIC SCENARIO OF GUWAHATI CITY

Nath, S.K.[1#], K.K.S. Thingbaijam[1], A. Raj[1], K. Shukla[1], I. Pal[1], D.R. Nandy[2#], M.K. Yadav[5], B.K. Bansal[3], S. Dasgupta[2], K. Majumdar[2], J.R. Kayal[2], A.K. Shukla[4], S.K. Deb[6], J. Pathak[7], P.J. Hazarika[8], and D.K. Paul[9]

[1] *Indian Institute of Technology Kharagpur*
[2] *Geological Survey of India, # former*
[3] *Seismology Division, Department of Science and Technology*
[4] *India Metrological Department*
[5] *Assam Electronics Development Cooperation Ltd. (AMTRON)*
[6] *Indian Institute of Technology Guwahati*
[7] *Jorhat Engineering College*
[8] *Assam Engineering College*
[9] *Indian Institute of Technology Roorkee*
#Email: nath@gg.iitkgp.ernet.in

ABSTRACT

A seismic scenario of the Guwahati city is derived through integration of different geological and seismological thematic layers on a GIS platform. The geological hazard factor is drawn from four themes namely - Geological and Geomorphological, Basement or bed rock formations, Landslide, and Factor of Safety. The seismological aspects, on the other hand, include Shear Wave Velocity, Site Response, Predominant Frequency, and Peak Ground Acceleration. Five hazard zones are seen in the region namely very high, high, moderate, low and very low corresponding to average hazard index of 0.55, 0.44, 0.35, 0.26 and 0.16 respectively. Furthermore, the urban demographic distribution is considered to obtain a preliminary seismic population risk map having five major zones with average risk index of 0.19, 0.32, 0.47, 0.63, and 0.79 that corresponds to average PGA value of 0.15, 0.38, 0.52, 0.67, and 0.81 respectively.

Keywords: Guwahati, Peak Ground Acceleration, Seismic Hazard, Site Response

1 INTRODUCTION

A seismic scenario of a mega city necessitates detailed analytical as well as numerical treatment of Seismological, Geological, Geomorphological and Geotechnical concepts through a systematic data analysis (Pal et al., 2007; Nath, 2004). The Guwahati city, a natural gateway for the northeast Indian Territory is overwhelmed with burgeoning population and rapid urbanization. Furthermore, the city resides on the Brahmaputra basin with the Himalayan mobile belt to the north and the Shillong plateau to the southwest having complex tectonic and topographic features. In the present study, an amalgamation of different perspectives of the seismic scenario is presented with the development of thematic layers depicting spatial distributions of various contributing attributes, and integration thereof on a GIS platform using Analytical Hierarchy Process (AHP) - a multi criteria decision making method (Saaty, 1980).

2 DATA AND METHODOLOGY

In the present analysis, the following themes are considered: (i) Geological and Geomorphological (GG), (ii) Basement or bed rock formations (BM), (iii) Landuse (LU) (iv) Landslide (LS), and (v) Factor of Safety (FS) (vi) Shear wave velocity (V_S^{30}) (vi) Site Response (SR), (vii) Predominant Frequency (PF), (viii) Peak Ground Acceleration (PGA), and (ix) Demographic distribution (DM).

The GG map, as shown in Fig. 1, is synthesized from Survey of India (SOI) topographic maps with further inputs from IRS PAN and LISS III satellite images in digital format and extensive GPS based point survey to form a base map. The base map depicts the geological formations, river and water bodies. Major aggradational units identified include Active flood plain and Levee deposits, and the Digaru, Bordang, Sonapur, and Pediment surfaces respectively.

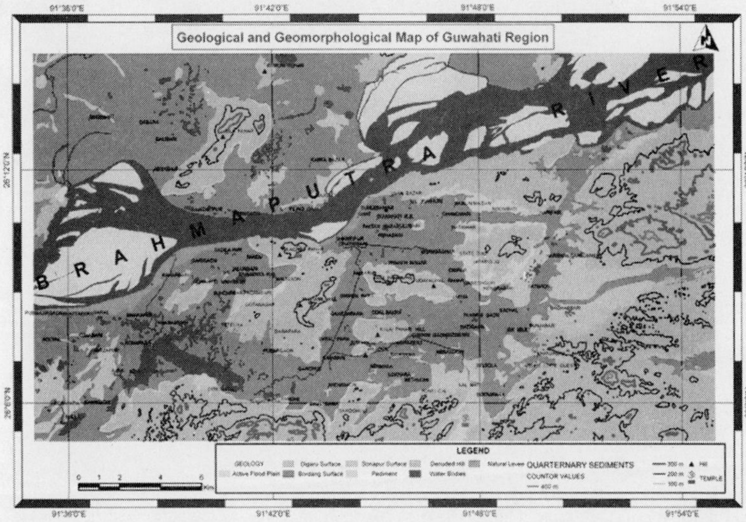

Figure 1. The geological and geomorphological map of Guwahati region.

The BM layer is prepared by examining the results of vertical electrical resistivity sounding surveys carried out by Geological Survey of India (GSI) during 1986-88 and 2002-03, and the data obtained from 30 boreholes. The zonation map thus obtained is presented in Fig. 2.

Figure 2. The basement zonation map of Guwahati region.

A Landuse (LU) map as shown in Fig. 3 is prepared from the land use pattern from the SOI toposheets of 1986-87 along with recent satellite imageries of the area. LS hazard zones as depicted in Fig. 4 are demarcated in the region using slope angle, lithological structures, relative relief, land-use cover, hydrological correlation, seismicity, rainfall, and landslide incidences. Fig.5 depicts 200 boreholes locations, 141 ambient noise survey and 5 strong motion recording stations.

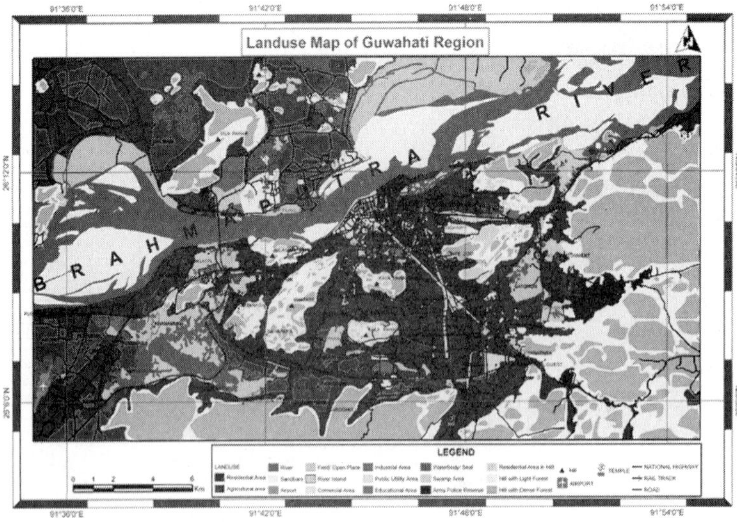

Figure 3. Landuse map of Guwahati region.

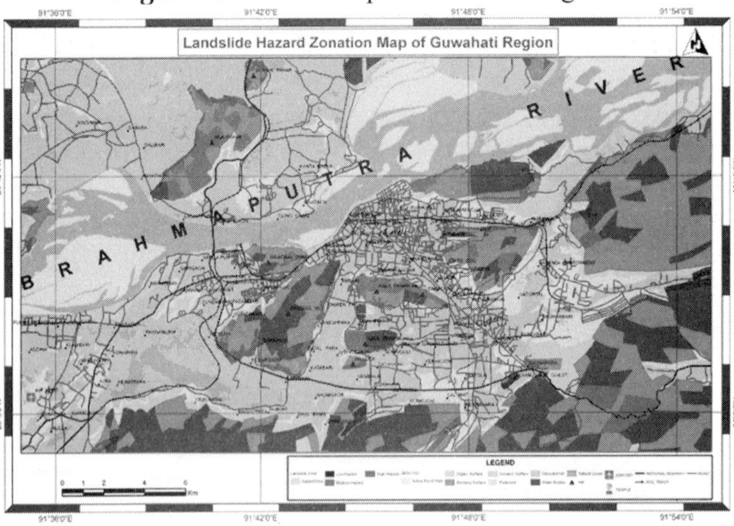

Figure 4. Landslide hazard zonation map of Guwahati region.

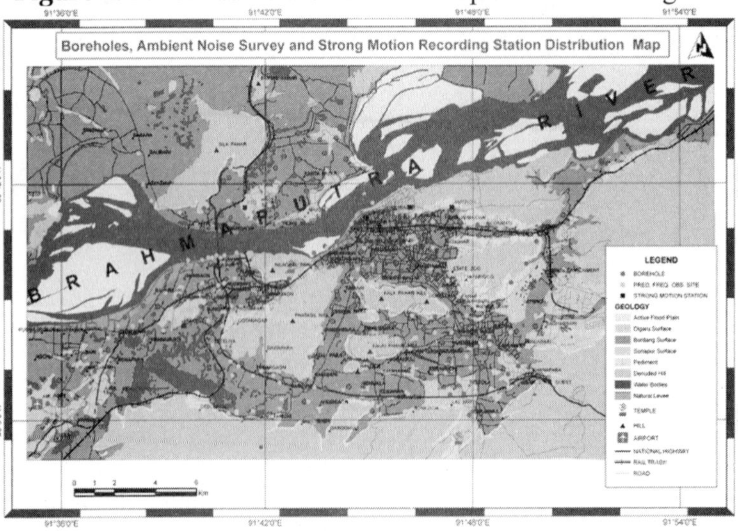

Figure 5. Boreholes locations along with ambient noise survey and strong motion recordings stations are presented.

The shear wave velocity V_S^{30} is estimated from the geotechnical data obtained at 200 borehole sites from the empirical relationships of Tonouchi et al. (1982), Fumal and Tinslay (1985), and Kayabali (1996) and is depicted in Fig. 6. The instrumental SR is studied with ambient noise survey (Nakamura, 1989) to determine Predominant Frequency (PF) covering almost the entire area at 141 sites. The distribution map is shown in Fig. 7. The site amplification factors are obtained using WESHAKE91 (Sykora and Wahl, 1992) from geotechnical data at the 200 borehole locations and corroborates with those from Horizontal-to-Vertical Spectral Ratio (HVSR) using the data of the 5 strong motion stations (Nakamura, 1989). The resulting SR contour map, as shown in Fig. 8, is found to corroborate the surface geology and the (V_S^{30}) distribution. The strong ground motion is synthesized to create a seismic hazard scenario for the 1897 Shillong Earthquake of M_W 8.7 using Frequency-Wave number (F-K) integration technique (Herrmann and Mandal, 1986). The Peak Ground Acceleration (PGA) as estimated with the Random Vibration Theory (Cartwright and Longuet-Higgins, 1956) defines a monotonic variation across the terrain with an average value between 0.1g -1.0g and the corresponding distribution is depicted in Fig. 9.

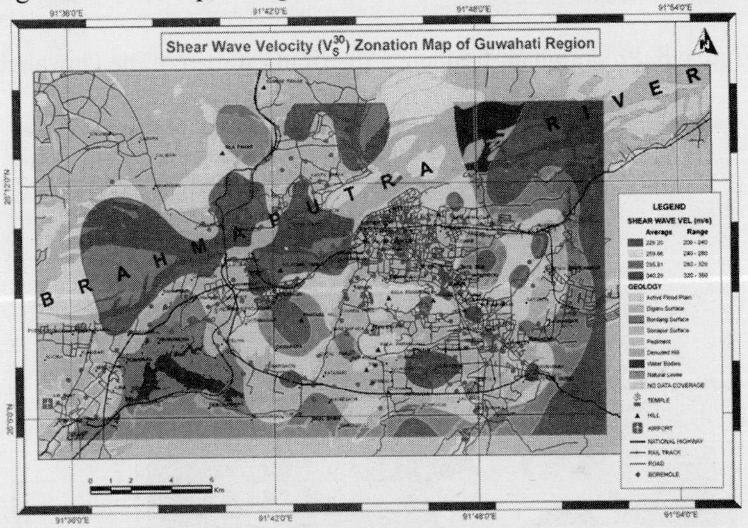

Figure 6. Shear wave velocity (V_S^{30}) distribution map of the study region.

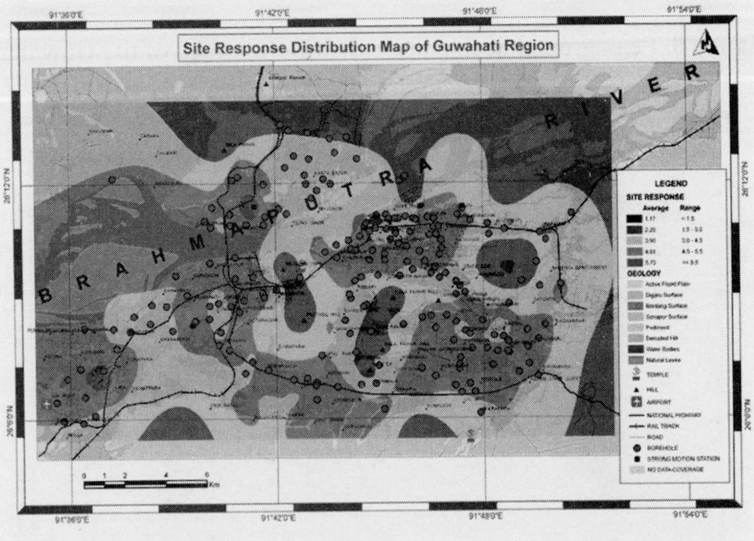

Figure 7. Site response distribution map of the study region from ambient noise data analysis.

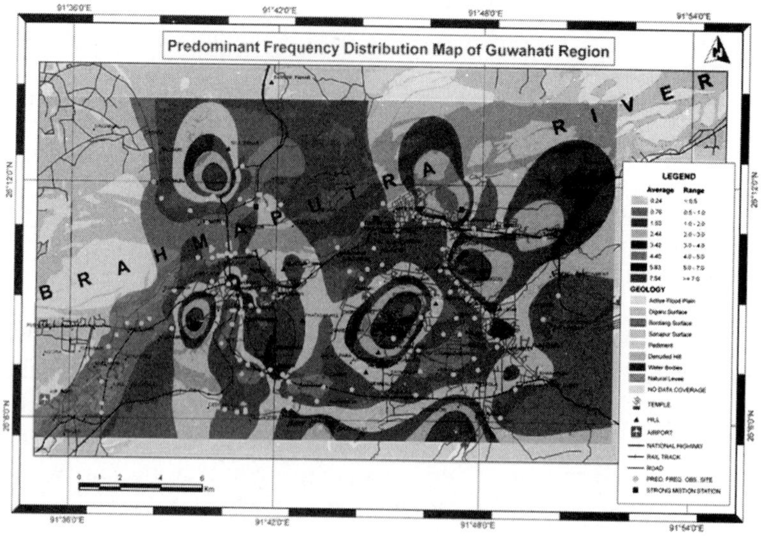

Figure 8. Predominant frequency distribution map of the Guwahati region

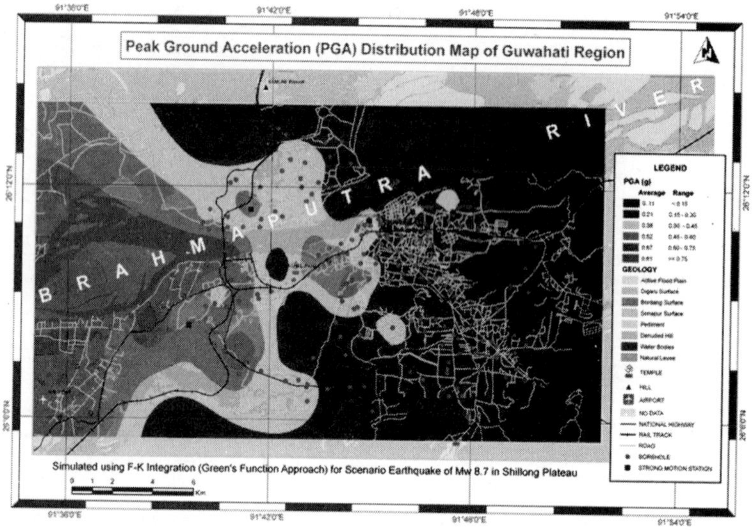

Figure 9. Peak ground acceleration distribution map of the Guwahati region

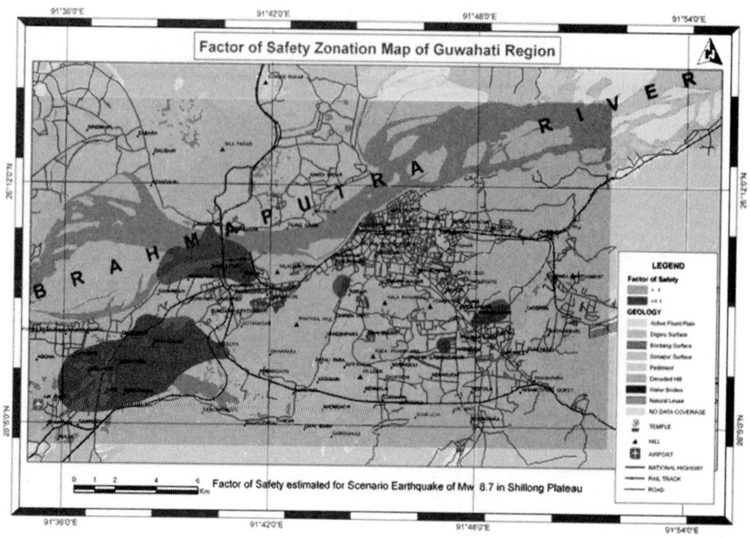

Figure 10. Factor of Safety Zonation Map of Guwahati Region

The FS is estimated from the following at each of the borehole locations giving safe and unsafe zones as shown in Fig. 10.

$$FS = \frac{\text{Maximum Load}}{\text{Actual Load}}, \tag{1}$$

Where the 'Maximum Load' is obtained from the PGA evaluated and the 'Actual Load' is obtained in situ. FS = 1 indicates that the site is on the point of failure, FS < 1 indicates a failed state, and FS > 1 indicates a safe state.

Following AHP, weights are assigned to each of the aforesaid themes according to their relative importance over another decided through a participatory mode in which decision makers decides the relative importance of factors on a consensus opinion. The various features within each theme are then assigned ranks or scores, normalized to ensure that no layer exerts an influence beyond its determined weight. Table 1 lists the themes along with normalized weights and ranks of the features thereof. Same approach can be seen in Pal et al. (2007), Mohanty et al. (2007), and Nath (2004).

Table 1: Normalized weights and ranks assigned to each feature of the thematic maps. The themes and features are presented in descending order according to the weights and ranks respectively.

Theme	Weight	Feature	Rank
Geology (GG)	0.2000	River ,Water Bodies & Swampy area	1.0000
		Active Flood Plain	0.8571
		Natural Levee	0.7143
		Pediment	0.5714
		Sonapur Surface	0.4286
		Digaru Surface	0.2857
		Bordang Surface	0.1429
		Denuded Hills	0.0000
Basement (BS) in metres	0.1778	> 600	1.0000
		500 - 600	0.8333
		400 - 500	0.6667
		300 - 200	0.5000
		200 - 300	0.3333
		100 - 200	0.1667
		< 100	0.0000
Landslide Hazard (LS)	0.1556	High Hazard Zone	1.0000
		Medium Hazard Zone	0.6667
		Low Hazard Zone	0.3333
		River, Sand Bar & Hazard Free Zone	0.0000
Landuse (LU)	0.1333	Residential Area	1.0000
		Educational, Army/Police Reserve, Commercial area	0.8333
		Sandbars ,River Island & Swampy area	0.6667
		Field/ open space & Agricultural Area	0.5000
		Residential Areas in Hill	0.3333
		Hill with dense & light forest	0.1667
		River, water bodies/Beel	0.0000
Shear Wave Velocity (V_s^{30}) in m/s^2	0.1111	200 - 240	1.0000
		240 - 280	0.6667
		280 - 320	0.3333

Theme	Weight	Feature	Rank
		320 - 360	0.0000
Peak Ground Acceleration (PGA) in g	0.0889	>= 0.75	1.0000
		0.60 - 0.75	0.8000
		0.45 - 0.60	0.6000
		0.30 - 0.45	0.4000
		0.15 - 0.30	0.2000
		< 0.15	0.0000
Site Response (SR)	0.0667	>= 5.5	1.0000
		4.5 - 5.5	0.7500
		3.0 - 4.5	0.5000
		1.5 - 3.0	0.2500
		< 1.5	0.0000
Predominant Frequency (PF) in Hz	0.0444	< 0.5	1.0000
		0.5 - 1.0	0.8571
		1.0 - 2.0	0.7143
		2.0 - 3.0	0.5714
		3.0 - 4.0	0.4286
		4.0 - 5.0	0.2857
		5.0 - 7.0	0.1429
		> 7.0	0.0000
Factor of Safety(FS)	0.0222	≤ 1	1
		> 1	0

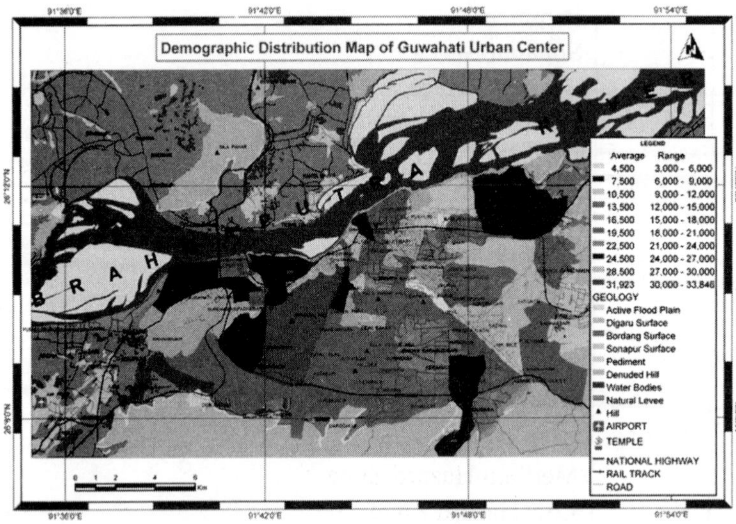

Figure 11. Demographic distribution of the Guwahati urban center.

A seismic hazard zonation map is obtained through the integration of all the aforesaid themes using the following relation

$$HAZ = GG_w GG_r + BM_w BM_r + LS_w LS_r + LU_w LU_r + V_s^{30}{}_w V_s^{30}{}_r + PGA_w PGA_r + SR_w SR_r + PF_w PF_r + FS_w FS_r / \Sigma_w \quad (2)$$

Additional subscripts have been added to the notation assigned to each theme to indicate normalized weight and rank with 'w' and 'r' respectively. Further integration of the map with urban demographic distribution given in Fig. 11 yielded a preliminary seismic population based risk map.

The demographic distribution is adapted from Census of India (2001). The integration is done on equal weightage. The corresponding normalized weights and ranks are presented in Table 2.

Table 2: The weights and normalized ratings for integrations of risk assessment

Theme	Weight	Feature	Rank
Microzonation (Hazard Index)	0.50	>= 0.50	1.0000
		0.40 - 0.50	0.7500
		0.30 - 0.40	0.5000
		0.20 - 0.30	0.2500
		< 0.20	0.0000
Demography (Number of Population)	0.50	30,000 - 33,846	1.0000
		27,000 - 30,000	0.8889
		24,000 - 27,000	0.7778
		21,000 - 24,000	0.6667
		18,000 - 21,000	0.5556
		15,000 - 18,000	0.4444
		12,000 - 15,000	0.3333
		9,000 - 12,000	0.2222
		6,000 - 9,000	0.1111
		3,000 - 6,000	0.0000

(a)

(b)

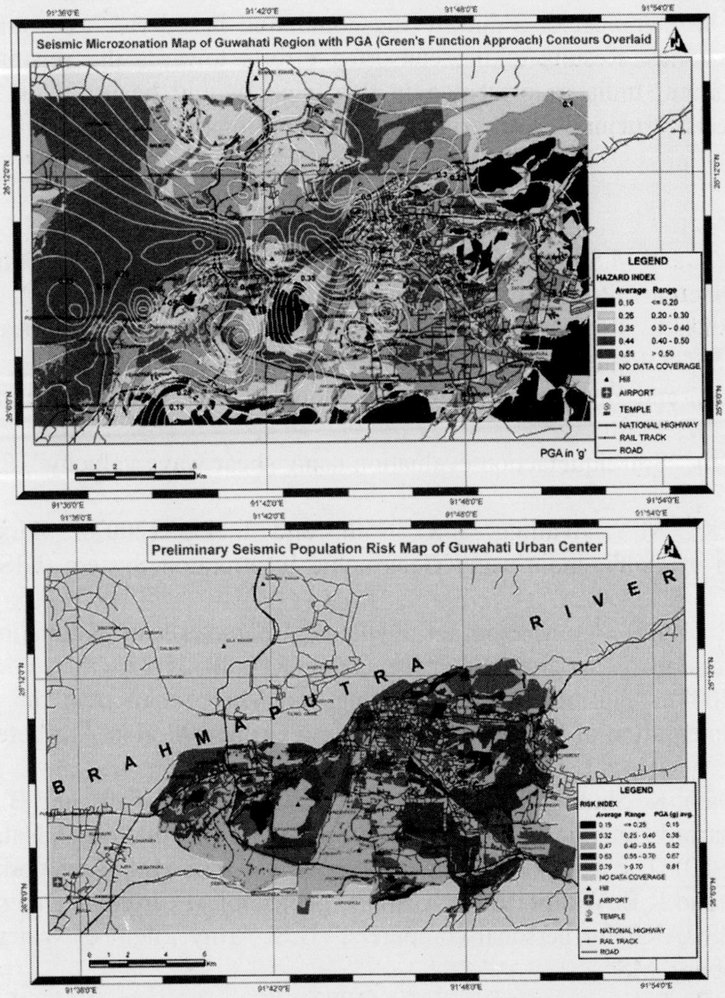

Figure 12. (a) Seismic hazard map overlaid with the peak ground acceleration contours. (b) A preliminary Seismic Population Risk Map of the Guwahati urban center.

3 RESULTS

The seismic hazard zonation map as shown is Fig. 12a exhibits five zones with average HAZ index corresponding to 0.55, 0.44, 0.35, 0.26 and 0.16. We termed these zones as very high, high, moderate, low and very low hazard regions. The preliminary population based risk map shows five major zones corresponding to average risk index of 0.19, 0.32, 0.47, 0.63, and 0.79 respectively with average PGA value of 0.15, 0.38, 0.52, 0.67, and 0.81 as depicted in Figure 12b.

The very high hazard zone covers the western part of the region wherein active flood plain with pediment, Sonapur and some patches of Bordang surfaces are the prevailing geology and the average PGA exhibits 0.81g. On the other hand, high hazard zone with average PGA of 0.52g is dominated by Sonapur surface, water bodies and pediments covering areas. Moderate hazard zone is predominantly Bordang surface with an average PGA of 0.38. Low and very low hazard zone exhibits relatively very low average PGA of 0.11g encompassing mostly hillock areas and some area in the eastern part of the region. The high risk zone covers the area namely Baharbari, Garigaon, Sadilapur, Dawapara, and Pubargaon. Some part of Phatasil hill also comes under high risk zone because of higher demographic distribution. However, most of the urban centre comes under moderate risk zone.

4 CONCLUSIONS

Significant variation in seismic hazard scenario is seen across the Guwahati region. Five zones of the hazard could be demarcated. Similarly the population based risk assessment could also be presented in five zones. Results of the consorted efforts towards the seismic scenario of the Guwahati city, Assam, India as presented in this paper would be useful in landuse planning, development projects, structural engineering, and hazard mitigation efforts.

REFERENCES

1. Cartwright, D.E. and M.S. Longuet-Higgins (1956). "The statistical distribution of the maxima of a random function", Proc. Royal Soc. Lon, Series A (Math. and Phy. Sc.), **237**, 212-232.
2. Fumal, T.E., and J.C. Tinsley (1985). "Mapping shear-wave velocities of near-surface geologic materials", U.S. Geological Survey Prof. Paper, **1360**, 127-149
3. Herrmann, R.B., and B. Mandal (1986). "A study of wave number integration techniques", Earthq. Notes, **57**, 33-40.
4. Kayabali (1996). "Soil liquefaction evaluation using shear wave velocity", Engg. Geology, **44**, 121-127.
5. Mohanty, W.K., M.Y. Walling, S.K. Nath, and I. Pal (2007). "First Order Seismic Microzonation of Delhi, India using Geographic Information System (GIS)", Nat. Haz., **40**, 245-260.
6. Nakamura, Y. (1989). "A Method for Dynamic Characteristics Estimations of Subsurface Using Micro tremors on the ground Surface, QR RTRI", **30**, 25-33.
7. Nath, S.K. (2004). "Seismic Hazard Mapping and Microzonation in the Sikkim Himalaya through GIS Integration of Site Effects and Strong Ground Motion Attributes", Nat. Haz., **31**, 319-342.
8. Pal, I., S.K. Nath, K. Shukla, D.K. Pal, A. Raj, K.K.S. Thingbaijam, and B.K. Bansal (2007). "Earthquake hazard zonation of Sikkim Himalaya using a GIS platform", Nat. Haz., (In Press).
9. Saaty, T.L. (1980). The Analytic Hierarchy Process, McGraw-Hill, New York, NY, USA.
10. Sykora, D.W. and R.E. Wahl (1992). "USACE geotechnical earthquake engineering software, Report 1, WESHAKE for personal computers," U.S. Army Engineer Waterways Experiment Station, Vicksburg, MS.
11. Tonouchi, K., T. Sakayama, and T. Imai (1982). "S wave velocity in the ground and the damping factor", Bull. Int. Assoc. Eng. Geologists, **26-27**, 327-333.
12. Census of India (2001). District Census Handbook, Kamrup District.

MAXIMUM CREDIBLE EARTHQUAKE PREDICTION- A NEURAL NETWORK APPROACH

Thingbaijam Kiran Kumar Singh[1] and Sankar Kumar Nath[2]
Department of Geology and Geophysics
Indian Institute of Technology
Kharagpur 721302
[1] *tkks@rediffmail.com*
[2] *nath@gg.iitkgp.ernet.in*

ABSTRACT

The Gutenberg-Richter law of magnitude distribution is extended to a generalized function correlating earthquake magnitude not only to its recurrence but also to the underlying zonal seismogenic attributes through an elementary neural network approach. Backpropagation feed forward neural network model is adopted. The approach is demonstrated with a case study conducted in the northeast Indian region.

Keywords: Maximum Credible Earthquake, Neural Network, Northeast India

1 INTRODUCTION

Maximum credible earthquake (MCE) is a key parameter in seismic hazard analysis often used for simulation of peak ground acceleration (Pal et al., 2007; Nath et al., 2005). This study presents an elementary neural network approach for the prediction of the parameter. A neural network is an interconnected system of processing nodes characterizing a computational and generalizable model which is based on the experience of a data set (Howard et al., 2006). In the recent times, neural network applications in various areas such as medical, engineering and business have been remarkable (Kemsley et al., 1992).

2 THEORY

The earthquake frequency magnitude distribution (FMD) is well accepted to follow the Gutenberg Richter (GR) law (Gutenberg and Richter, 1944),

$$\log_{10}(N(m)) = a\text{-}bm \qquad (1)$$

Where N (m) is cumulative number of events with magnitude $\geq$ m. The slope and the intercept are referred to as a- and b-value. The relation, however, has been observed to have significant deviations from linearity in several cases (Utsu, 1999). A magnitude of completeness or a threshold magnitude, m_t, is usually introduced for the lower magnitudes. However, the departures at higher magnitudes have been attributed to various physical constraints. The present approach implicates the deviations to the seismogenic attributes of the delineated seismic zone.

Eq. (1) can be further represented as,

$$m = F(N). \qquad (2)$$

Further, a general relation may be formulated for a seismicity, geophysical or geotechnical parameter, ϕ_i, as,

$$m=F(\phi_1, \phi_2, \phi_3, \dots), \quad i=1,2,3,..,n \tag{3}$$

A neural network framework can implicate Eq. (3) by connecting the seismogenic attributes as input parameters to an output corresponding to the earthquake magnitude. The basic principle is to classify the seismic zone and at same the time, associate each frequency of earthquake occurrences with m as given by Eq. (1). The prediction is, then, targeted for the earthquake recurrence with 10% probability of exceedance in 50 years, a standard for engineering purposes (Nath et al., 2005). The recurrence is assumed to follow Poisson's distribution given by

$$P=1-e^{-\lambda t}, \tag{4}$$

Where P is the probability of exceedance of a magnitude, m, with a frequency of occurrence of earthquakes, λ, in t years.

3 DATA

For similar purpose, various attributes have been suggested by Schenk et al. (1994). However, in the present analysis, the data considered owing to the feasibility include (i) b-value (ii) $\log_{10}(N(m))$, (iii) m, (iv) fractal correlation dimension, D_C, of the epicenters, (v) Box counting estimated fractal dimension, D, for the fault segments, (vi) Heat-flow rate, H_R, and (vii) Strain rate, S_R.

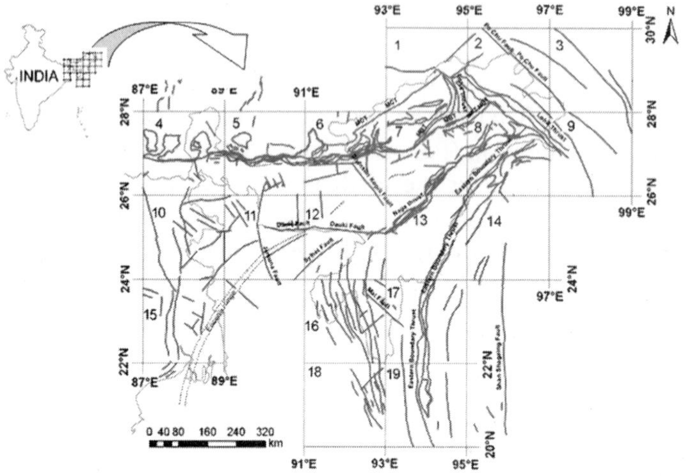

Figure 1. The northeast Indian region is divided into square cells or windows of 2°. The fault pattern is adapted from Dasgupta et al. (2000).

We take up the northeast Indian region for a case study. The region is accentuated by the Himalayan terrains where future great earthquake (M>8) is highly probable (Bilham et al., 2001). As such there have been incidents of two great earthquakes (M 8.7) - the Great Shillong Earthquake of 1897 and the Great Assam Earthquake of 1950 in the region. The study region is delineated into several square cells or source windows of 2° numbered up to 19 for the purpose (Fig.2). The parameters are, thereby, assessed for each of the source windows.

H_R is obtained from the global heat flow database provided by International Heat Flow Commission, the University of North Dakota (http://www.heatflow.und.edu, last assessed April, 2007), and that of Pollack et al. (1993) assessable at http://www.ngdc.noaa.gov/ (last assessed April, 2007). The data for second invariant of the model strain rate tensor field is obtained from the Global Strain Rate Map (GSRM) project (http://gsrm.unavco.org, last assessed April, 2007; Kreemer et al., 2003) wherein S_R is given in 1×10^{-9} $year^{-1}$. In both the cases, the mid-range values are employed to measure the dispersions thereof within a cell.

Table 1. The training data prior to preprocessing with MRSR and MRHR denotes the mid-range values of the strain rate in 1x 10^{-9} yr^{-1} and the heat flow in milli Watt per sq. metre respectively.

			Input			Target
b-value	D_C	D	MRSR	MRHRF	$\log_{10} N(m)$	m
1.383	1.216	1.501	52.694	110.000	1.875	4.9
1.383	1.216	1.501	52.694	110.000	1.724	5.0
1.383	1.216	1.501	52.694	110.000	1.505	5.1
1.383	1.216	1.501	52.694	110.000	1.505	5.2
1.383	1.216	1.501	52.694	110.000	1.146	5.3
1.383	1.216	1.501	52.694	110.000	1.146	5.4
1.383	1.216	1.501	52.694	110.000	1.041	5.5
1.383	1.216	1.501	52.694	110.000	1.000	5.6
1.383	1.216	1.501	52.694	110.000	0.845	5.7
1.383	1.216	1.501	52.694	110.000	0.477	5.8
1.383	1.216	1.501	52.694	110.000	0.301	5.9
1.383	1.216	1.501	52.694	110.000	0.301	6.0
1.383	1.216	1.501	52.694	110.000	0.000	6.3
0.919	1.697	1.253	93.656	70.000	1.724	4.6
0.919	1.697	1.253	93.656	70.000	1.623	4.7
0.919	1.697	1.253	93.656	70.000	1.591	4.8
0.919	1.697	1.253	93.656	70.000	1.544	4.9
0.919	1.697	1.253	93.656	70.000	1.415	5.0
0.919	1.697	1.253	93.656	70.000	1.322	5.1
0.919	1.697	1.253	93.656	70.000	1.279	5.2
0.919	1.697	1.253	93.656	70.000	0.954	5.3
0.919	1.697	1.253	93.656	70.000	0.954	5.4
0.919	1.697	1.253	93.656	70.000	0.699	5.5
0.919	1.697	1.253	93.656	70.000	0.699	5.6
0.919	1.697	1.253	93.656	70.000	0.699	5.7
0.919	1.697	1.253	93.656	70.000	0.301	5.8
0.919	1.697	1.253	93.656	70.000	0.301	5.9
1.185	1.568	1.394	67.244	120.000	2.064	4.8
1.185	1.568	1.394	67.244	120.000	1.968	4.9
1.185	1.568	1.394	67.244	120.000	1.857	5.0
1.185	1.568	1.394	67.244	120.000	1.724	5.1
1.185	1.568	1.394	67.244	120.000	1.568	5.2
1.185	1.568	1.394	67.244	120.000	1.447	5.3
1.185	1.568	1.394	67.244	120.000	1.301	5.4
1.185	1.568	1.394	67.244	120.000	1.176	5.5
1.185	1.568	1.394	67.244	120.000	1.176	5.6
1.185	1.568	1.394	67.244	120.000	1.079	5.7
1.185	1.568	1.394	67.244	120.000	0.903	5.8
1.185	1.568	1.394	67.244	120.000	0.845	5.9
1.185	1.568	1.394	67.244	120.000	0.477	6.0
1.185	1.568	1.394	67.244	120.000	0.301	6.1
1.185	1.568	1.394	67.244	120.000	0.000	6.3
0.845	1.670	1.216	44.581	150.000	2.299	4.5
0.845	1.670	1.216	44.581	150.000	2.255	4.6
0.845	1.670	1.216	44.581	150.000	2.170	4.7
0.845	1.670	1.216	44.581	150.000	2.121	4.8
0.845	1.670	1.216	44.581	150.000	2.045	4.9
0.845	1.670	1.216	44.581	150.000	1.934	5.0

			Input			Target
b-value	D_C	D	MRSR	MRHRF	$\log_{10} N(m)$	m
0.845	1.670	1.216	44.581	150.000	1.763	5.1
0.845	1.670	1.216	44.581	150.000	1.633	5.2
0.845	1.670	1.216	44.581	150.000	1.491	5.3
0.845	1.670	1.216	44.581	150.000	1.342	5.4
0.845	1.670	1.216	44.581	150.000	1.279	5.5
0.845	1.670	1.216	44.581	150.000	1.255	5.6
0.845	1.670	1.216	44.581	150.000	1.114	5.7
0.845	1.670	1.216	44.581	150.000	1.079	5.8
0.845	1.670	1.216	44.581	150.000	0.954	5.9
0.845	1.670	1.216	44.581	150.000	0.845	6.0
0.845	1.670	1.216	44.581	150.000	0.699	6.1
0.845	1.670	1.216	44.581	150.000	0.699	6.2
0.845	1.670	1.216	44.581	150.000	0.699	6.3
0.845	1.670	1.216	44.581	150.000	0.602	6.4
0.845	1.670	1.216	44.581	150.000	0.477	6.5
0.845	1.670	1.216	44.581	150.000	0.477	6.6
0.845	1.670	1.216	44.581	150.000	0.000	7.2
0.980	1.063	1.281	39.207	120.000	2.013	4.6
0.980	1.063	1.281	39.207	120.000	1.954	4.7
0.980	1.063	1.281	39.207	120.000	1.845	4.8
0.980	1.063	1.281	39.207	120.000	1.732	4.9
0.980	1.063	1.281	39.207	120.000	1.623	5.0
0.980	1.063	1.281	39.207	120.000	1.531	5.1
0.980	1.063	1.281	39.207	120.000	1.447	5.2
0.980	1.063	1.281	39.207	120.000	1.322	5.3
0.980	1.063	1.281	39.207	120.000	1.176	5.4
0.980	1.063	1.281	39.207	120.000	1.079	5.5
0.980	1.063	1.281	39.207	120.000	0.954	5.6
0.980	1.063	1.281	39.207	120.000	0.954	5.7
0.980	1.063	1.281	39.207	120.000	0.602	5.8
0.980	1.063	1.281	39.207	120.000	0.602	5.9
0.980	1.063	1.281	39.207	120.000	0.477	6.0
0.980	1.063	1.281	39.207	120.000	0.301	6.1
0.980	1.063	1.281	39.207	120.000	0.301	6.2
0.980	1.063	1.281	39.207	120.000	0.000	6.4
1.231	0.848	1.304	15.599	125.000	-2.676	8.7
0.919	1.697	1.253	93.656	70.000	-2.676	8.7

The moment magnitude based earthquake catalog (1964-2006) from Pal et al. (2007) is employed. The b-value is estimated from the maximum likelihood method with consideration of m_t at 90% confidence level (Aki, 1965; Bender, 1983; Utsu, 1999; Wiemer and Wyss, 2000). The D_C is computed from the correlation integral method of Grassberger and Procaccia (1983). The spatial gaps due to the lack of valid m_t are filled up by contouring. Since a- and b-values are positively correlated, only b-value is considered. As such spatial variations of b-value are associated to differential stress levels and earthquake mechanisms (Schorlemmer et al., 2005). The values of $\log_{10}(N)$ and the corresponding m are obtained only for source windows with valid m_t. D_C is a measure of earthquake clustering and can differentiate seismic zones (Kagan, 2007). The fault structures are quantified with D through box counting method (Hirata, 1989). While $\log_{10}(N)$ is used to implicate the GR relationship. The heat flow rate variation could indicate variations of pressure and temperature (Schenk et al., 1994). The strain rate, on the other hand, is related to crustal deformations and the plate motion tectonics.

Table 2. The input data for the prediction prior to the preprocessing with the corresponding source window is listed.

Source Window	b-value	D_C	D	MRSR	MRHFR	$Log_{10}N(m)$
1	1.252	1.040	1.365	13.854	140	-2.676
2	1.231	0.848	1.304	15.599	125	-2.676
3	0.965	1.420	1.207	11.954	125	-2.676
4	1.114	1.462	1.388	49.984	125	-2.676
5	1.031	1.371	1.465	49.367	125	-2.676
6	1.383	1.216	1.501	52.694	110	-2.676
7	0.803	1.322	1.463	51.217	70	-2.676
8	1.004	1.296	1.479	48.172	70	-2.676
9	0.842	1.357	1.204	36.280	190	-2.676
10	0.883	1.455	1.264	89.079	110	-2.676
11	0.904	1.603	1.254	87.769	70	-2.676
12	0.919	1.697	1.253	93.656	70	-2.676
13	1.185	1.568	1.394	67.244	120	-2.676
14	0.845	1.670	1.216	44.581	150	-2.676
15	0.676	1.463	1.242	00.000	105	-2.676
16	0.665	1.695	1.382	21.807	55	-2.676
17	1.207	1.058	1.331	35.657	125	-2.676
18	0.866	1.723	1.348	00.000	55	-2.676
19	0.980	1.063	1.281	39.207	120	-2.676

4 METHODOLOGY

The present problem falls in the purview of classification and prediction therewith. In many such cases, feed-forward backpropagation networks are generally employed (Howard et al., 2006). The backpropagation network represents a simple paradigm guided by the weight adjustments to minimize an error function involving the training input and desired output (Rumelhart, 1986). The MATLAB - neural network toolbox is used for the network simulation. The data are normalized within the range [-1, 1] in order to enable the input and output to become effectively a part of the network using a MATLAB function, mapminmax() on each individual dataset. The training data prior to the preprocessing is listed in Table 1. Two entries corresponding to the Great Earthquakes of 1897 and 1950 are made after initial exercises resulted in prediction of underestimated values in the respective zones. The input data for the prediction is listed in Table 2.

4.1. Network Model

Simpler the model the better it is to generalizing to the unseen data (Swingler, 1996). Therefore, the number of hidden units for a network must be sufficient for the correct representation, but sufficiently low. Hecht-Neilson (1990) derived an upper bound for the required number of hidden units as one greater than twice the number of inputs. In the present study, fully connected feed-forward backpropagation network model is adopted and 3 network architectures (Fig. 2) are tested. The first architecture is a single layered model with 13 hidden layers and is denoted as network 13:1. The second architecture has 2 hidden layers with 4 and 3 numbers of hidden nodes respectively and is denoted as network 4:3:1. The last network, on the other hand, has four hidden layers of 5, 4, 3, and 2 numbers of hidden nodes respectively. It is denoted as network 5:4:3:2:1. In all the cases, bias to the nodes is set to 0 for simplicity. The inputs to each node consist of data input or outputs from the preceding layer multiplied with the corresponding weights while the output is a summation of the inputs. A transfer function is required with each neural node to squash the output to an acceptable range. Tan-Sigmoid transfer function (Vogl et al., 1988) is employed in the hidden layers and linear transfer function is employed for the output layers.

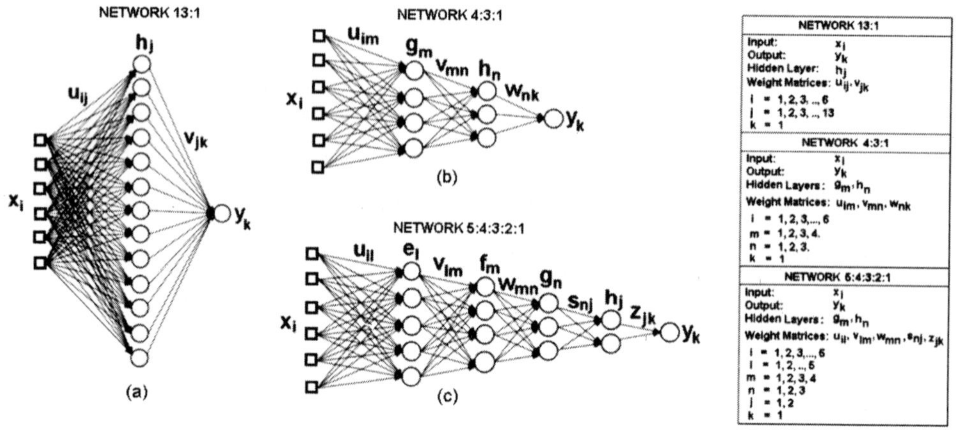

Figure 2. The three network models: (a) 6 inputs, 1 hidden layer with 13 nodes, and 1 output is depicted. (b) 6 inputs, 2 hidden layers with 4 and 3 nodes respectively, and 1 output, and (c) six inputs, 4 hidden layers with number of nodes in descending order from 5 to 2 and 1 output.

4.2. Training

The neural network needs to be trained for the purpose wherein the synaptic weights of the network are adjusted. In small to moderate networks, Levenberg-Marquardt algorithm is found to give good performance (Hagan and Menhaj, 1994) and, therefore, is adopted. To avoid overfitting i.e. small error with training set and large error with new data, Bayesian regularization (Mackay, 1992; Foresee and Hagan, 1997) is adopted. The Bayesian regularization minimizes as well as modifies a linear combination of squared errors and weights to achieve a good generalization. The training is performed using a MATLAB function, trainbr(). The adopted algorithm works only in batch training mode with the weights being updated only after all the inputs are presented. The network is trained until it reaches the convergence which is indicated when the sum squared error (SSE), the sum squared weights (SSW), and the effective number of parameters (# parameters) reaches constant values. Furthermore, the robustness of the network is checked by training with several initial conditions.

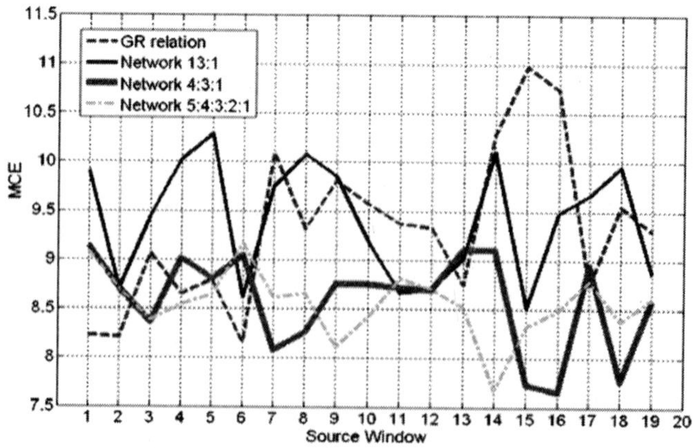

Figure 3. The predicted MCE is plotted against each corresponding source window.

5 RESULTS

The training parameters for each of the network architecture are listed in Table 3. The predictions from the GR relation are quite high as with the case of the network 13:1 (Fig. 3). Amongst the later models, the prediction of the network 4:3:1 is more consistent with the input data. Further taking into account the training parameters, the network is seen to be most effective. The corresponding weights obtained after the training is given in Table 4. The predicted values as depicted in Fig. 4 exhibits considerable leniency to the geophysical traits. The correlations between the input parameters with the predicted MCE exhibit tendencies towards positive correlation except in case of D and D_C which are seen with an overall negative correlation.

Table 3. The training parameters- number of epochs (#Epochs), Sum of Square Error (SSE), Sum of Square Weights (SSW), and effective number of parameters (# parameters) for the three network architectures is listed.

Architecture	#Epochs	SSE	SSW	# Parameters
13:1	325	0.0513486	15.1266	18.6014
4:3:1	200	0.0415548	22.7992	22.7334
5:4:3:2:1	1000	0.0388065	24.1901	25.8088

Table.4. The network 4:3:1 is described by (a) input weights, (b) weights of the hidden layer, and (c) weights of the output layer. The notations used are in accordance with those used in Fig. 5.

(a)

u_{im}	i=1	i=2	i=3	i=4	i=5	i=6
m=1	-0.5047	-0.4374	-0.4322	-0.2575	-0.8593	1.9170
m=2	0.1278	0.1090	0.0692	0.4596	-1.1825	-1.8508
m=3	-0.0531	0.0152	0.1861	0.2262	-0.1177	0.6282
m=4	0.1366	0.0836	0.1651	0.3315	0.0162	-0.4948

(b)

v_{mn}	n=1	n=2	n=3	n=4
m=1	0.7351	-0.6512	-1.0625	1.0185
m=2	1.3963	0.4548	0.3688	-0.4156
m=3	-0.2390	-1.4432	0.3326	-0.1566

(c)

w_{nk}	n=1	n=2	n=3
k=1	1.5663	-0.7967	-0.8688

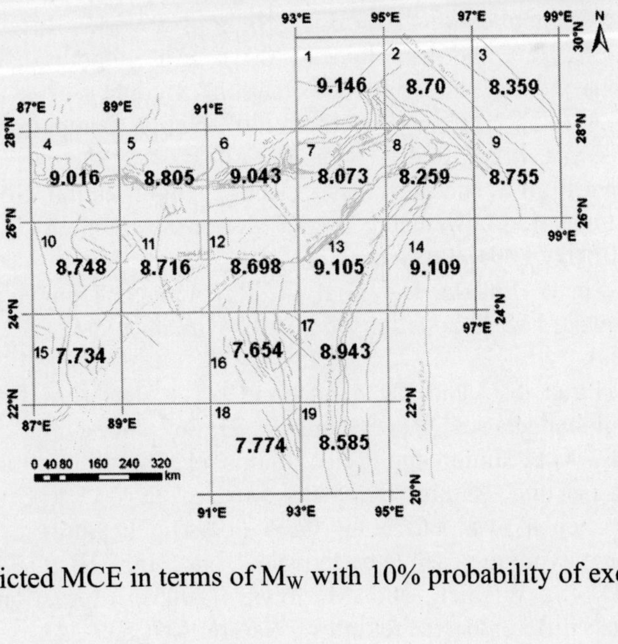

Figure 4. The predicted MCE in terms of M_W with 10% probability of exceedance in 50 years

6 CONCLUSIONS

An advantage of the neural network approach is easy incorporation of seismogenic attributes into an empirical based prediction technique. However, the results obtained in the present analysis are preliminary in nature in addition to its applicability only for the study region. The analysis investigates the feasibility of a neural network approach to establish a base for further rigorous exercises towards improvement of the technique.

REFERENCES

1. Aki, K. (1965). "Maximum likelihood estimate of b in the formula log N = a - bM and its confidence limits", Bull. Earthq. Res. Inst. Univ. Tokyo, **43**, 237-239.
2. Bender, B. (1983). "Maximum likelihood estimation of b-values for magnitude grouped data", Bull. Seism. Soc. Am, **73**, 831–851.
3. Bilham, R., V.K. Gaur and P. Molnar (2001). "Himalayan seismic hazard", Science, 293, 1442-1444.
4. Dasgupta, S., P. Pande, D. Ganguly, Z. Iqbal, K. Sanyal, et al. (2000). Seismotectonic atlas of India and its environs, Geol. Survey India.
5. Foresee, F.D. and M.T. Hagan (1997). "Gauss-Newton approximation to Bayesian regularization", Proc. 1997 Intern. Joint Conf. Neural Networks, 1930-1935.
6. Grassberger, P. and I. Procaccia (1983). "Measuring the strangeness of strange attractors", Physica D, **9**, 189-208.
7. Gutenberg, B. and C. F. Richter (1944). "Frequency of earthquakes in California", Bull. Seism. Soc. Am., **34**, 185–188.
8. Hagan, M.T. and M. Menhaj (1994). "Training feed-forward networks with the Marquardt algorithm", IEEE Trans. Neural Networks, **5**, 989-993.
9. Hecht-Neilson, R., (1990). Neurocomputing, Addison Wesley
10. Hirata, T. (1989). "Fractal dimension of fault systems in Japan: fractal structure in Rock fracture geometry at various scales", Pageoph, **131**, 157-170.
11. Howard, D., M. Beale, and M. Hagan (2006). Neural network toolbox, user's quide, version 5.0, The MathWorks, Inc.
12. Kagan, Y.Y. (2007). "Earthquake spatial distribution: the correlation dimension", Geophys. J. Int., **168**, 1175–1194.
13. Kemsley, D.H., T.R. Martinez, and D.M. Campbell (1992). "A survey of neural network research and fielded applications", Int. Jour. Neural Networks: Res. Appl., **2**, 123-133.
14. Kreemer, C., W.E. Holt, and A.J. Haines (2003). "An integrated global model of present-day plate motions and plate boundary deformation", Geophys. J. Int., **154**, 8-34.
15. MacKay, D.J.C. (1992). "Bayesian interpolation", Neural Computation, **4**, 415-447.
16. Nath, S.K., M. Vyas, I. Pal, and P. Sengupta (2005). "A Seismic Hazard Scenario in the Sikkim Himalaya from Seismotectonics, Spectral Amplification, Source Parameterization, and Spectral Attenuation Laws using Strong Motion Seismometry", J. Geophys. Res., 110, B01301, doi: 10.1029/2004jb003199.
17. Pal, I., S.K. Nath, K. Shukla, D.K. Pal, A. Raj, K.K.S. Thingbaijam, and B.K. Bansal (2007). "Earthquake hazard zonation of Sikkim Himalaya using a GIS platform", Natural Hazard (In Press).
18. Pollack, H.N., Hurter, S.J., and J.R. Johnson (1993). "Heat flow from the earth's interior: analysis of the global data set", Rev. Geophys., **31**, 267-280.
19. Rumelhart, D.E., G.E. Hinton and R.J. Williams (1986). "Learning representations by back-propagating errors", Nature, **323**, 533 – 536.
20. Schenk, V., Z. Schenkova, and V.G. Gitis (1994). "Features of geonomic pognostic functions for the maximum possible earthquake", Nat. Haz., **10**, 97-115.
21. Scholemmer, D., S. Wiemer, and M. Wyss (2005). "Variations in earthquake-size distribution across different stress regimes", Nature, **437**, 539-542.

22. Swingler, K. (1996). Applying Neural Networks: A Practical Guide, Morgan Kaufmann.
23. Utsu, T. (1999). "Representation and analysis of the earthquake size distribution: a historical review and some new approaches", Pageoph, **155**, 509-535.
24. Vogl, T.P., J.K. Mangis, A.K. Rigler, W.T. Zink, and D.L. Alkon (1988). "Accelerating the convergence of the backpropagation method", Biological Cybernetics, **59**, 257-263
25. Wiemer, S. and M. Wyss (2000). "Minimum magnitude of complete reporting in earthquake catalogs: examples from Alaska, the Western United States, and Japan", Bull. Seism. Soc. Am., **90**, 859–869.

SEISMIC HAZARD AND MICROZONATION METHODOLOGY FORFLAT TERRAIN

P. Anbazhagan.
Department of Civil Engineering,
Indian Institute of Science, Bangalore, Karnataka, India, 560012
E-mail: anbazhagan@civil.iisc.ernet.in, anbazhagan2005@gmail.com

ABSTRACT

In this paper, methodology as formulated for the seismic hazard and microzonation for flat terrain is presented. For the seismic hazard and microzonation, the procedure is subdivided into three major items: First step is evaluation of the expected input motion, second one to find local site effects and ground response analysis and finally, preparation of microzonation maps. The steps followed in seismic hazard and microzonation for flat terrain in the present investigation is illustrated in the form of a flow chart. In which, the first step illustrates the assessment of the expected ground motion using the deterministic and probabilistic seismic hazard analysis. The site characterization for the study area at local scale using geotechnical and shallow subsurface geophysical data are discussed in the next step. Third step is the study of local site effects using first and second part output data and producing the ground level hazard parameters. Forth step is the assessment of liquefaction potential in terms of factor of safety against liquefaction. The detailed methodology and flow chart is presented in this paper. As the procedure is formulated for the flat terrain, landslide possibility and tsunami prediction are considered as remote. This paper also presents about the level or/ grade of microzonation maps based on scale of the study area.

Keywords: Microzonation, site characterization, site response, liquefaction

1 INTRODUCTION

Rapidly growing cities with increasing population are most vulnerable to natural hazards due to agglomeration of the population at one place. Preparation of the geotechnical microzonation maps provides an effective solution to overcome to some extent from hazards. Seismic microzonation attempted to understand the effects of earthquake generated ground motions due to local soil or/and man-made structures. It is very important that the selected microzonation parameters should be meaningful for city planners as well as for public officials. Ansal (2004) recommend that the national seismic zoning maps are mostly at small scale level (1:1,000,000 or less) and are mostly based on seismic source zones defined at similar scales. The seismic microzonation for a town requires 1:5,000 or even 1:1,000 scale studies and needs to be based on seismic hazard studies at similar scales. The general trend in conventional microzonation studies in India was to simplify the applied methodology by adopting the macrozonation seismic hazard maps as the primary source to estimate the earthquake hazard. In addition, due to the lack of sufficient geological and geotechnical data, a second simplification is to define the site conditions with respect to local geological units. Even though these two simplifications may appear logical, to many engineers and scientists they are the main source of incorrectness in any microzonation study. Seismic microzonation requires an essential input from civil engineering, especially from the field of geotechnical engineering (DRM, 2003). Keeping these things in mind, in this study an attempt has been made to describe microzonation methodology for flat terrain.

2 SEISMIC MICROZONATION AND SCALE

The earthquake damage basically depends on three groups of factors: earthquake source and path characteristics, local geological and geotechnical site conditions, structural design and construction features. Seismic microzonation should address the assessment of the first two groups of factors. In general terms, seismic microzonation is the process of estimating the response of soil layers for earthquake excitations and thus the variation of earthquake characteristics is represented on the ground surface. Seismic microzonation is the initial phase of earthquake risk mitigation and requires multidisciplinary approach with major contributions from geology, seismology, geotechnical and structural engineering (Ansal, 2004).Seismic Microzonation falls into the category of "applied research". That is why it needs to be upgraded and revised based on the latest information. In Indian context, the microzonation defined as "*Microzonation is subdivision of a region into zones that have relatively similar exposure to various earthquake related effects*". This exercise is similar to the macro level hazard evaluation but requires more rigorous input about the site specific geological conditions, ground responses to earthquake motions and their effects on the safety of the constructions taking into consideration the design aspects of the buildings, ground conditions which would enhance the earthquake effects like the liquefaction of soil, the ground water conditions and the static and dynamic characteristics of foundations or of stability of slopes in the hilly terrain –DST Expert Group on Microzonation of Delhi Chaired by Dr. A.S. Arya in 1998 and the definition was endorsed by the DST subcommittee on Microzonation, chaired by Mr. P.K. Narula in 2001. For the present investigation the seismic microzonation has been subdivided into three major items:

1) Evaluation of the expected input motion
2) Local Site effects and ground Response analysis
3) Preparation of microzonation maps.

The microzonation is graded based on the scale of the investigation and details of the study carried out. The technical committee on earthquake geotechnical engineering (TC4) of the International society of soil mechanics and foundation engineering (TC4-ISSMGE 1999) states that the first grade (Level I) map can be prepared with scale of 1:1,000,000 – 1:50,000 and the ground motion was assessed based on the historical earthquakes and existing information of geological and geomorphological maps. If the scale of the mapping is 1:100,000-1:10,000 and ground motion is assessed based on the microtremor and simplified geotechnical studies then it is called second grade (Level II) map. In the third grade (Level III) map ground motion has been assessed based on the complete geotechnical investigations and ground response analysis with a scale of 1:25,000-1:5,000.

Expert groups chaired by Dr. N.C. Nigam and later on by Dr. A.S. Arya recommended seismic microzonation mapping on a scale of 1:25000 for India (Bhandari et al.2003). The present investigation was carried out with a scale of 1:20,000 and ground motion are arrived based on the detailed geotechnical/geophysical investigations and ground response analysis. The methodology followed for the seismic hazard and microzonation of Bangalore will be discussed in the next section.

3 STEPS FOR SEISMIC HAZARD ANALYSIS AND MICROZONATION

Even though the steps involved in seismic hazard analysis and microzonation grouped in to three major groups as discussed in previous section, it need to adopt step by step procedure to arrive at the final map of microzonation. The steps followed in seismic hazard and microzonation of Bangalore is discussed here. The first step for any microzonation is the assessment the expected ground motion using the deterministic and probabilistic seismic hazard analyses which are discussed in the section 4 and 5. The second step is to evaluate the sub soil properties by following site characterizations steps using available data or conducting field geotechnical experiments. The site characterized for the study area has to be carried out at local scale of 1:20,000 or less using geotechnical and shallow subsurface geophysical data. Third step is the study of local site effects using first and second part output data and producing the ground level hazard parameters; the methodology is discussed in the section 7. Forth step is the assessment of liquefaction potential in terms of factor of safety against liquefaction. Steps involved in this step discussed in section 7. As this procedure is formulated for the flat terrain, landslide and Tsunami possibility is considered as remote. Finally microzonation maps are prepared in terms of ground motion parameters and factor safety against liquefaction.

4 DETERMINISTIC SEISMIC HAZARD ANALYSIS

The methodology followed for DSHA of Bangalore is described in four steps as given below:

1. Source characterization which includes identification and characterization of all earthquake sources which may cause significant ground motion in the study area.
2. Selection of the shortest distance between the source and the site of interest.
3. Selection of controlling earthquake i.e. the earthquake that is expected to produce the strongest level of shaking. Controlling earthquake has been evaluvated based on past earthquake data and assummed subsurface fault ruputure length.
4. Defining the hazard at the site formally in terms of the ground motions produced at the site by the controlling earthquake.

5 PROBABILISTIC SEISMIC HAZARD ANALYSIS

The PSHA procedure can also be described in four steps:

1. Identification of earthquake sources such as active faults, which may affect the study area. Characterize the probability distribution of potential rupture locations within the source.
2. Characterization of the seismicity of each source zone using a recurrence relationship, which specifies the average rate at which an earthquake of some size will be exceeded (recurrence relationship).
3. Estimation of the ground motion produced at the site by earthquakes of any possible size occurring at any possible point in each source zone using predictive relationships.
4. Obtaining the probability that the ground motion parameter will be exceeded during a particular time period.

6 SITE CHARACTERIZATION

A complete site characterization is essential for the seismic site classification and site response studies, both of them can be used together for seismic microzonation. Site Characterization should include an evaluation of subsurface features, sub surface material types, subsurface material properties and buried/hollow structures to determine whether the site is safe against earthquake effects. Site characterization should provide data on the following:

- Site description and location
- Geotechnical data
- Soil conditions
- Geological data
- Hydrogeology/ ground water data
- Aquifer or permeable characteristics
- Presence and distribution of contaminants (if any)
- Climatic conditions (if needed)

Assessment of available data should include an analysis of the sufficiency and validity of the data in relation to the proposed application/ study.

As part of the site characterization, experimental data should be collected, interpolated and represented in the form of maps. The representation maps can be further used for the site classification and seismic studies. Here an attempt has been made to characterize the Bangalore site using Geotechnical and Geophysical experimental data. About 850 collected geotechnical borehole information with standard penetration "N" values data and 58 geophysical data of multichannel analysis of surface wave field test results are used for Bangalore site characterization. Figure 1 shows the steps followed in site characterization of Bangalore. The discussion About the site chracterization of bangalore is given in section 6.1. Sumarry of geophysical method of MASW, field testing, results, evaluation of dynamic properties of soil, mapping of shear wave velocity, site classification and correlation between SPT corrected "N" values and shear wave velocity is given in section 6.2.

6.1 Site Characterization Using SPT Data

The 3-D subsurface model with geotechnical data has been generated for site characterization of Bangalore. The base map of Bangalore city (220sq.km) with several layers of information (such as Outer and Administrative boundaries, Contours, Highways, Major roads, Minor roads, Streets, Rail roads, Water bodies, Drains, Landmarks and Borehole locations) has been generated. GIS database for collating and synthesizing geotechnical data available with different sources and 3-dimensional view of soil stratum presenting various geotechnical parameters with depth in appropriate format has been developed. In the context of prediction of reduced level of rock (called as "engineering rock depth" corresponding to about SPT "N" > 100) in the subsurface of Bangalore and their spatial variability evaluated using Artificial Neural Network (ANN). Observed SPT 'N' values are corrected by applying necessary corrections, which can be used for engineering studies such as site response and liquefaction analysis.

6.2 Site Characterization Using MASW Data

Site characterization has also been carried out using measured shear wave velocity with the help of shear wave velocity survey using MASW. MASW (Multichannel Analysis of Surface Wave) is a geophysical method, which generates a shear-wave velocity (Vs) profile (i.e., Vs versus depth) by analyzing Raleigh-type surface waves on a multichannel record. MASW system consisting of 24 channels Geode seismograph with 24 geophones of 4.5 Hz capacity were used in this investigation. The shear wave velocity of Bangalore subsurface soil has been measured and correlation has been developed for shear wave velocity (Vs) with the standard penetration tests (SPT) corrected 'N' values. About 58 one-dimensional (1-D) MASW surveys and 20 two-dimensional (2-D) MASW surveys has been carried out with in 220 sq.km Bangalore urban area. Dispersion curves and shear velocity 1-D and 2-D have been evaluated using SurfSeis software. Using 1-dimensional shear wave velocity, the average shear wave velocity of Bangalore soil has been evaluated for depths of 5m, 10m, 15m, 20m, 25m and 30m (Vs^{30}) depths. The sub soil classification has been carried out for local site effect evaluation based on average shear wave velocity of 30m depth (Vs^{30}) of sites using NEHRP (National Earthquake Hazard Research Programme) and IBC (International Building Code) classification. Bangalore falls into "site class D" type of soil. Mapping clearly indicates that the depth of soil obtained from MASW is closely matching with the soil layers in the bore logs. The measured shear wave velocity at 38 locations close to SPT boreholes, which are used to generate the correlation between the shear wave velocity and corrected 'N' values using a power fit. Also, developed relationship between shear wave velocity and corrected 'N' values corresponds well with the published relationships of Japan Road Association.

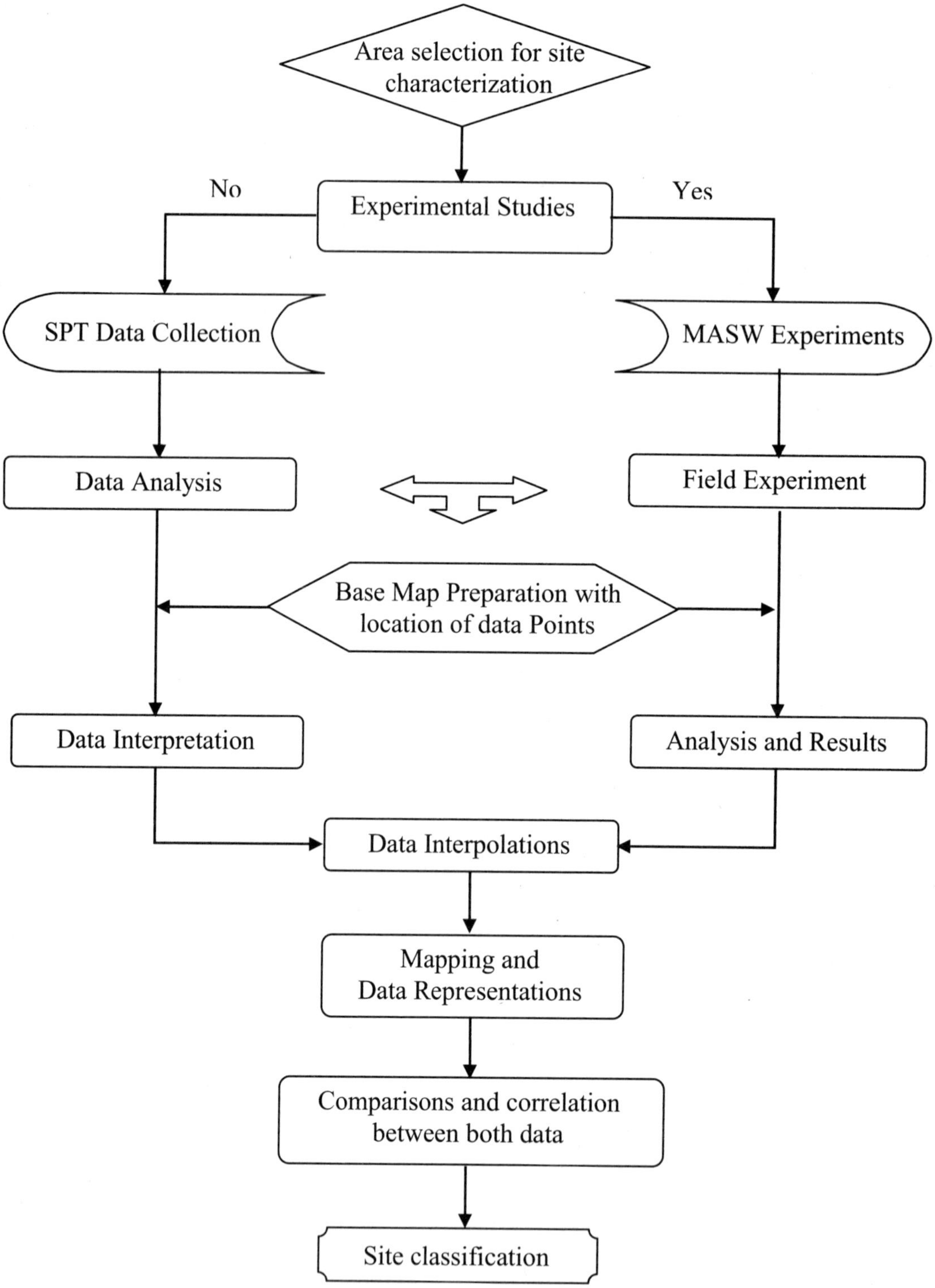

Figure 1: Steps involved for site characterization

7 LOCAL SITE EFFECTS AND SITE RESPONSE STUDY

The site response studies for microzation are carried out using one-dimensional ground response analysis and experimental method of microtremor. Site response analysis aims at determining the response of a soil deposit to the motion of the bedrock immediately beneath it. The overburden plays a very important role in determining the characteristics of the ground surface motion thus emphasizing the need for ground response analysis. A number of techniques have been developed for ground response analysis. These techniques can be grouped as one-, two-, and three-dimensional analyses according to the dimensionality of the problems they can address. A one-dimensional method can be used if the soil structure is essentially horizontal and is widely used in earthquake geotechnical engineering:

Assumptions in one-dimensional ground response analysis are:

- The soil layer boundaries are horizontal and extend infinitely in the horizontal direction.
- Response of the soil deposit is predominantly caused by SH-waves propagating vertically from the underlying bedrock.

A number of different techniques are available for one-dimensional ground response analysis. These methods can be broadly grouped into the following three categories:

- Linear analysis
- Equivalent linear analysis
- Nonlinear analysis

Of the above, the most popular method used in professional practice is the "equivalent linear" approach which is incorporated in the computer program SHAKE. One-dimensional ground response analysis software SHAKE2000 was used in the present study. It requires three input parameters such as bedrock motion, dynamic material properties and site specific soil properties. The peak surface acceleration, ground response spectrum and period of soil column are obtained as output from this analysis. These values are used to get map indicating zones of amplification potential, spectral acceleration maps at various frequencies and period of soil column map for Bangalore city.

Site response studies mainly deal with the determination of peak frequency of soft soil, amplification and the nature of response curve defining the transfer function at the site which forms an important input for evaluating and characterizing the ground motion for seismic hazard quantification. Peak frequency or Predominant frequency of Bangalore soil has been carried out using microtremor studies at 54 locations. Present study has been carried out by adopting the Nakamura method for obtaining the transfer function at various sites in Bangalore using Microtremor studies. The surface sources for the ambient noise generate Rayleigh waves which affect the vertical and horizontal motion equally in the surface layer. The spectral ratio of the horizontal component by the vertical component of the time series provides the transfer function at a given site. The dominant peak is well correlated with the fundamental resonant frequency.

8 LIQUEFACTION ANALYSIS

The first step in calculation of liquefaction potential is to determine if the soil has the potential to liquefy during the earthquake. Liquefaction potential analysis usually carried out by using simplified empirical procedure, originally developed by Seed and Idriss (1971). This simplified procedure represents the standard of practice in North America and in many other countries across the globe. Simplified procedures are reviewed periodically by groups of experts who make recommendations and changes according to data collected from new earthquakes and new developments in liquefaction hazard assessment. The potential for liquefaction is assessed with the aid of liquefaction charts or semi empirical equations, which are based on observations of whether liquefaction did or did not occur at specific sites during numerous past earthquakes (Chen and Scawthorn, 2003). This study simplified approach has been used with SPT bore log information collected. A simple spread sheet developed using Excel macros to calculate the factor of safety against liquefaction for 720 locations. Step adopted is shown in Figure 6 in the form of a flow chart. The earthquake loading is evaluated in terms of cyclic stress ratio using Seed and Idriss (1971) simplified approach. Cyclic resistance ratio (CRR) is arrived based on corrected "N" value using plots of CRR versus corrected "N" value from a large amount of laboratory and field data of earthquake using equation proposed by Idriss and Boulanger (2005).

9 CONCLUSIONS

Seismic microzonation studies need to include earthquake damage factors of earthquake source, path characteristics and local geological, geotechnical site conditions. Study of earthquake source and path characteristics should include the geology, geomorphology and seismological data to assess the seismic hazard of region considering deterministic as well as probabilistic approach. For the study of local geological and geotechnical site effects, site has to be characterized using geotechnical data, which is very much useful for assessing site amplification and liquefaction potential of the area. In this paper simple methodology followed for seismic microzonation of Bangalore is summarised. Seismic microzonation of Bangalore addresses the seismic hazard, site characterization, site response studies and liquefaction.

10 ACKNOWLEDGEMENTS

Authors thank the Department of Science and Technology (DST), Seismology Division, Govt. of India for funding the project "Seismic microzonation of Bangalore" (Ref no.DST/23(315)/SU/2002 Dated October 2003) and ISRO-IISc Space Technology Cell, Indian Institute of Science, Bangalore, India for funding the project titled "Use of remote sensing and GIS for Seismic Hazard Analyses of Bangalore Urban Area, (Ref. No. ISTC/CCE/TGS/176/2006).

REFERENCES

1. Ansal, A (ed) (2004), "Recent Advances in Earthquake Geotechnical Engineering and Microzonation," Kluwer Academic Publishers. Printed in Netherlands.
2. Bhandari, R.K., Rajarathnam, S., Janardhanan,G. and Ganapathy,G.P. (2003), "Seismic Microzonation" A Training progarmme guide, published by Centre for Disaster Mitigation and Management, Anna University, Chennai, India, P-140.
3. Chen. W.F and Scawthorn, C. (2003), "Earthquake Engineering Handbook" published by CRC Press LLC, 2000 N.W. Corporate Blvd., Boca Raton, Florida 33431.
4. DRM (2003); "Project Microzonation for Earthquake Risk Mitigation (MERM) - Microzonation of Pilot Areas: Adapazari, Gölcük, Ihsaniye and Degirmendere, Part 2C - CaseStudies,"http://geotechnics.ch/main.html/ publication=722
5. Kramer, S. L. (1996), "Geotechnical Earthquake Engineering ", Published by Pearson Education Pvt, Reprinted 2003, Delhi, India.
6. IBC, (2000), "International Building Code," International Code Council: Inc. 5th Edition, Falls Church, VA.
7. Idriss, I. M., and Boulanger, R. W. (2005), "Evaluation of Liquefaction Potential, Consequences and Mitigation" Invited Expert Lectures, Proc. Indian Geotechnical Conference-2005, 17-19 December 2005, Ahmedabad, INDIA, pp 3-25.
8. Seed, H. B., and I. M. Idriss, (1971), "Simplified procedure for evaluating soil liquefaction potential," Journal of the Soil Mechanics and Foundation Division, ASCE, **97**(9), pp 1249-1274.
9. TC4-ISSMGE (1999), "Manual for Zonation on Seismic Geotechnical Hazard," Revised edition, Technical Committee for Earthquake Geotechnical Engineering (TC4) of the International Society of Soil Mechanics and Geotechnical Engineering (ISSMGE), 209p.

SEISMIC HAZARD ASSESSMENT AND SOIL EFFECT STUDIES IN THE CITY OF TABRIZ

M. Hosseinpour[1], M. Hajialilue-Bonab[2] and M. Zare[3]
[1] Graduate student, University of Tabriz, Tabriz, Iran, m.hosseinpour@gmail.com
[2] Assistant Professor, University of Tabriz, Tabriz, Iran, hajialilue@tabrizu.ac.ir
[3] Associate Professor, IIEES, mzare@iiees.ac.ir

ABSTRACT

The investigation was performed in two categories. Seismic hazard analysis was done in a region centered by Tabriz city. Site response analysis has been performed along 8 km of the first line of Tabriz metro tunnel. Earthquake hazard analysis was performed by using deterministic and probabilistic approaches. All active faults around the city were recognized and their seismic and rupture parameters were determined. Classification of seismic zones for seismic hazard analysis was done with Arc GIS software. After determination of seismic parameters, probabilistic and deterministic hazard analysis were performed and resulted in seismic hazard curves and maps with different return periods and PGA values for TABRIZ city. In the next part, geological, geotechnical and dynamical characteristics of soil in TABRIZ site were determined and site response analysis has been performed on selected logs along first line of Tabriz metro tunnel. Different results of these analyses are presented.

Keywords: seismic hazard, site effect, Tabriz city, amplification

1 INTRODUCTION

Northwest region of Iran that study area is a part of it contains west and east Azerbaijan and Ardebil provinces and centered by Tabriz city is a tectonically active region and have different and complex structural trends within. Some of the most evident and seismotectonically active faults are North Tabriz, Tasuj and Bozghush fault that among them North Tabriz fault (NTF) with NW-SE trend that passes through the vicinity of Tabriz city is more dominant because of its obvious morphologic features and historical seismicity. Tabriz city was devastated completely during three destructive historical earthquakes related to North Tabriz fault. The region shows intense historical and moderate instrumental seismic activities and a significant 20th century seismic gap exist in 50km around the city. Tabriz city that is the second most populated city of Iran is located on the alluvial basin. All of stratigraphic units and formations surrounded the Tabriz urban area is limited to the after Miocene period. Main part of deposits in the city consists of silty sands which lies on bedrock made up of marl layers. The only reliable geotechnical data for soil effect analysis in the city can be obtained from metro boreholes. Seismic gap of major earthquakes in 20th century, and installation of the city on a alluvial basin that increase the risk of amplification of ground motion show the importance of performing seismic hazard assessments and soil effect studies in the city.

2 SEISMIC HAZARD ANALYSIS

2.1 Active Tectonic of Study Area

Study area was defined as a rectangular zone with area of 53000 km^2 centered by Tabriz city that lie in Azerbaijan seismotectonical sub province. Azerbaijan is a part of Central Iran seismotectonic province that stretches up to east of Turkey and Caucasia (Zare, 2005). There are different structural trends in this province. Faults with NW-SE and E-W trends, are the major trends of region and show reverse and trust mechanisms whereas N-S and NE-SW trend faults act as strike slip and normal faults in the region. NTF that is a major structural feature in the region is the part of Tabriz- Zanjan structural trend. This NW-SE system, act as right lateral strike slip fault with reverse component. The overall length of fault is 250km and average displacement along it is 3mm/year (Berberian&Arshadi, 1976).It is believed that NTF is an extension of North Anatolian Fault toward Iran, but as a discontinuous segments rather than a single fault (Jackson & McKenzie, 1984). The fault has a rich catalog of intense historical earthquakes. Other sesmotectonically important faults in the study region are Tasuj, Shabestar, Salmas, Bozghush and Ardebil faults that have different trends and mechanisms and historical and instrumental activities are recorded along them. In this study all of active faults in the region that length of them was more than 20km were recognized and depicted in active faults map. (figure1) Calculation of seismic potential of each fault has been done using Mohajer ashjaii(1984) and Zare(1995) empirical relationships. Estimation of rupture parameters of the faults like surface ruptures length. Rupture width and subsurface rupture area were performed using Wells & Coopersmith (1984) empirical relationships. Table 1 indicates the most hazardous faults and their parameters in the region. This information is employed later in seismic zonation of the region.

Table1: Seismic parameters of more dominant faults in the studied region (M-N: Mohajer ashjaii-Noruzi, Z:Zare, W-C:Wells & Coopersmith)

N	Source	F.L (km)	Rupture Length) km(			Maximum Magnitude			R.W (km)	R.A (km^2)
			W-C	M-N	Z	M-N	Z	Ave		
4	Tabriz (NW)	80	35	40	29	7.0	6.7	6.9	18	718
5	Tabriz (SE)	112	55	56	41	7.1	7.0	7.1	22	1228
6	Tasooj	58	23	29	22	6.9	6.5	6.7	14	440
8	S of Bozghush	72	30	36	26	7.0	6.6	6.8	16	607

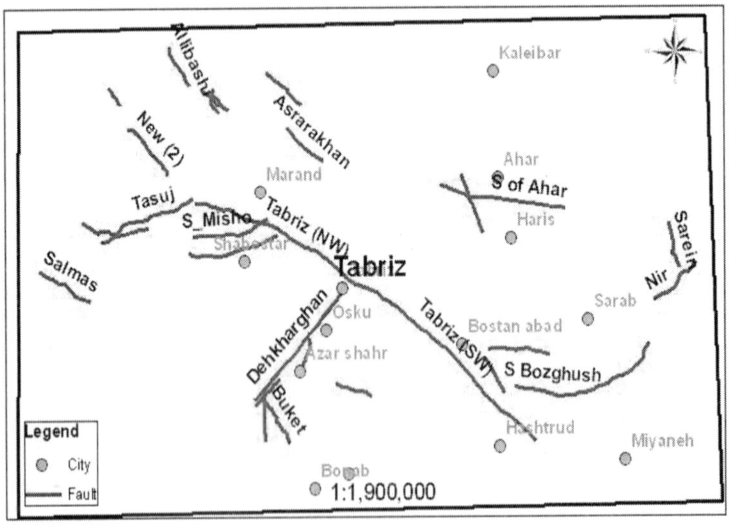

Figure1: active faults map in the study area

2.2 Seismicity of the Studied Region

Earthquake catalog of the region can be divided to historical (before1900) and instrumental (after1900) activities.

2.2.1 Historical Earthquakes

The most reliable sources for historical earthquakes of Iran and Azerbaijan are (Ambraseys & Melville, 1982 & Berberian, 1977). Most of the intense historical earthquakes were recorded along NTF. Base of different studies it can be concluded that earthquake with magnitude of 7.7 is the characteristics earthquake of North Tabriz Fault. Other faults like Tasuj and Dehkharghan faults show historical seimicity too (table2).

Table2: historical seismicity in the region (Ambraseys and Melville,1982)

Year (A.C)	) I_0(Intensity	) M_s(Magnitude	Human Lost	Other characteristics
858	VIII	6.0	-	
1042	VIII-X	7.6	40000	
1273	VII-IX	6.5	250	
1304	VII	6.7	-	
1550	-	5.9	-	Landslide
1641	VII	6.8	1200	Change in Water Table
1721	VIII-X	7.7	-	
1780	IX	7.7	-	Liquefactions, Landslide
1819	-	<5	-	
1883	-	<5	-	
1894	-	<5	-	

2.2.2 Instrumental Earthquakes

Instrumental catalog of the study was gathered form ISS (1900-1964), ISC (1964-2005), USGA, NEIC and seismic data of Tabriz Local seismic network (2000-2005). This network has more that 8 active station in the region and is able to record earthquakes with minimum magnitudes of 1. Plotting the coordinate of earthquake epicenters on fault map indicates a major seismic gap in 50km around Tabriz city from 1963 up to now. The maximum concentration of earthquake epicenters is in NW, SE and west limits of the area. The SW segment of NTF shows small earthquakes in 2000-2005 but the same seismic gap for intense activities continues around the city (Figure2).

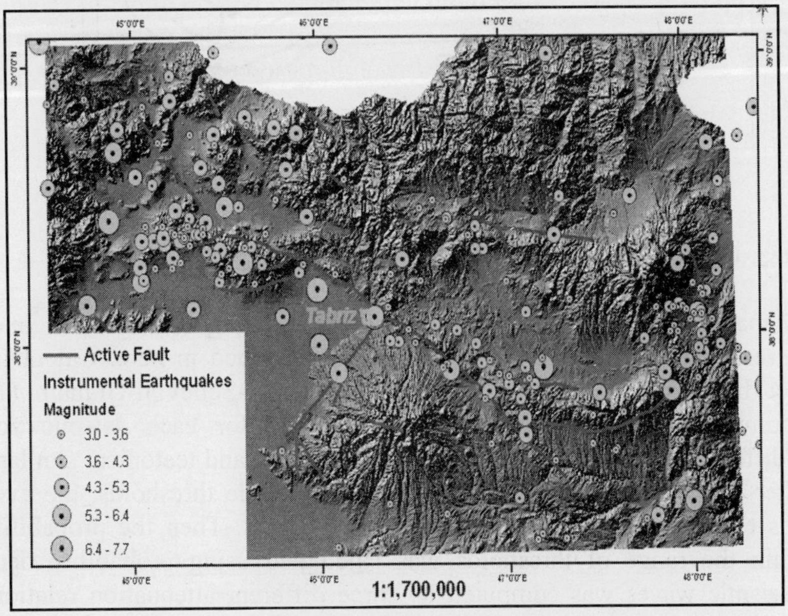

Figure2: Active faults and epicenter distribution

2.3 Seismic Zonation

Whereof most of the earthquakes in the region are related to faults so wherever was possible to directly associate the event to a particular fault, sources considered linear. In other conditions, regional sources were determined. Determination of seismic zones is done in Arc GIS environment. Characteristics of faults include rupture width and rupture area were prepared and weighted with respect to width and area of rupture and potential maximum magnitude of each fault. Earthquakes information contain spatial and time distribution of epicenters formed another data layer. All layers were inputted to software and were incorporated and classifies. After necessary corrections the seismic zonation map of the region was prepared. Figure3 to Figure5 show the procedure of zonation steps.

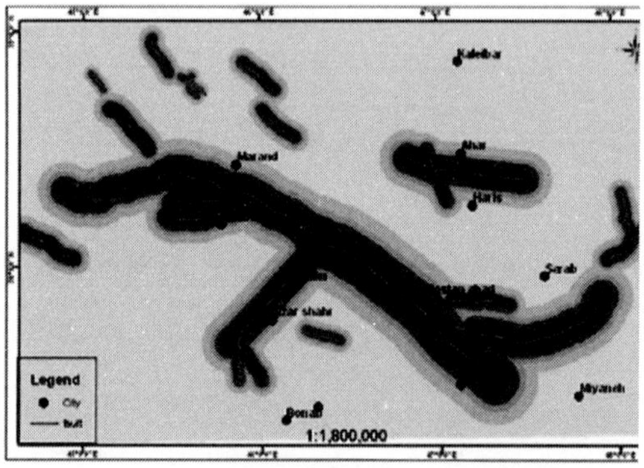

Figure3: Weighting of Faults on the basis of maximum magnitude and rupture width

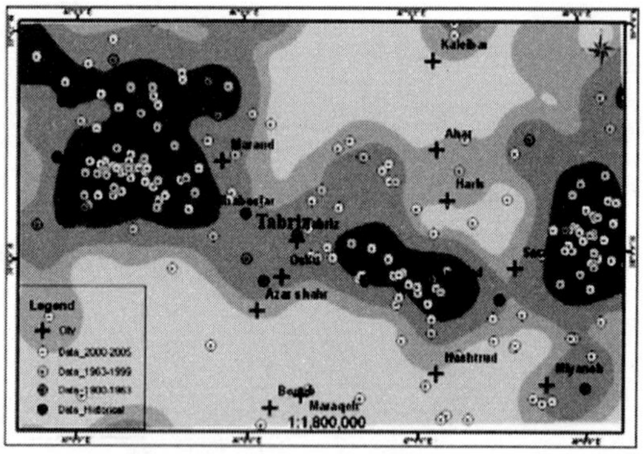

Figure4: Density of spatial distribution of earthquake epicenters

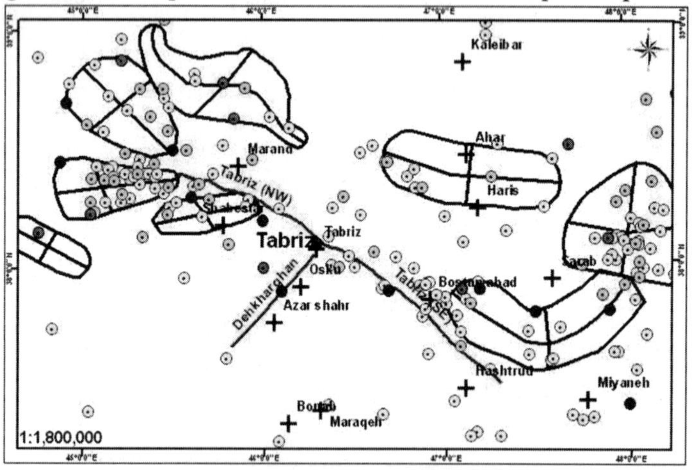

Figure5: Final zonation of seismic sources

2.4 Seismic Hazard Approaches

Seismic hazard analysis was performed in the region applying deterministic and probabilistic approaches and results of probabilistic approach were described in here. Calculation of seismic parameters (a & b) was done for each zone using Kijko-Sellevoll-Graham (1981) method. Afterwards maximum possible magnitude was estimated for each seismic zone using the combination of different methods such as historical seismicity and tectonical similarities (Wang & Law, 1994). After determining the upper and lower magnitude thresholds, the average range of occurrence (λ) is calculated for each magnitude in the zones. Then the probability of different magnitudes within the range of thresholds was calculated using probability density function. Weakening of seismic waves was computed by three different attenuation relationships. Joiner, Boore & Fumal (1993), Zare (1999) and Sinaian(2006) relationships were applied. The probability of exceedance of PGA from acc for all magnitude and distance intervals was calculated for all 11 zones and finally all the probabilities for linear and area zones were joined to show the overall hazard for Tabriz city. Seismic hazard curves were depicted using Zare(1999) and Sinaian(2006) relationships and seismic hazard maps were prepared for 50, 75 and 475 return periods. The values of PGA for different return periods in the city are very high and put the city in the region with very high risk (Figure6).

Table3: PGA values for Tabriz

Site	Attenuation law	Return period(year)	HPGA (g)
Tabriz	Joiner-Boore-Fumal (1993)	50	(0.28-0.3)g
		75	(0.29-
		475	(0.53-0.6)g
	Zare (1999)	50	(0.25-0.3)g
		75	(0.28-
		475	(0.5-0.64)g
	Sinaian (2006)	50	(0.23-
		75	(0.25-
		475	(0.48-0.6)g

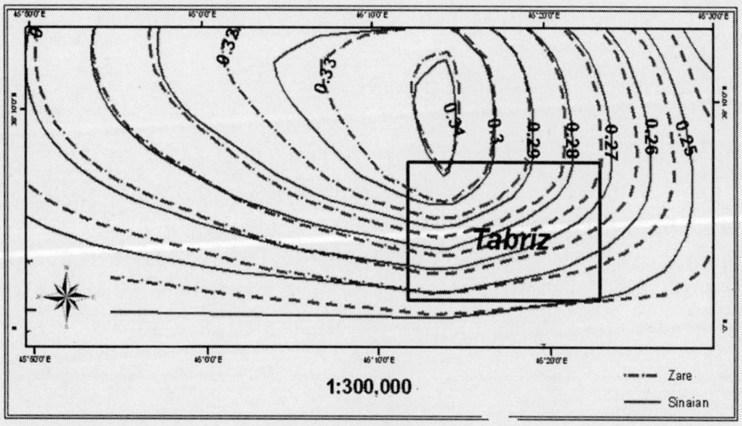

Figure6: Seismic hazard map for recurrence period of 75 years

3 SITE EFFECT ANALYSIS

3.1 Ground Characteristics of Tabriz

Tabriz city is surrounded with mountains consist of Upper Red Formation of Miocene in North and Northeast sections. Southern and southeastern part of the city are bordered by a series of low latitude hilly terrains consisting predominantly of Plio- Pleistocene conglomerates with some outcrops of green- gray marl and tuff in the east. Central and western parts of the city were developed on quaternary deposits (Haghshenas, 2006). NTF forms a well-marked boundary between Miocene URF and quaternary deposits, upthrusting the Miocene rocks against the alluvial deposits especially in NE of Tabriz airport (Berberian&Arshadi, 1976). Information about geotechnical characteristics of the city restricted to boreholes and engineering geology sheet of the Tabriz prepared by Geological Survey of Iran (GSI) (1994) that didn't give any information about the thickness of alluvial deposits. The best information source that were used in this study, are the boreholes excavated for Tabriz metro geotechnical investigations. Overall 101 boreholes were drilled with depths change from 25 to 60 meters. The site effect analyses were performed along the firs line of metro tunnel. The length of this line is about 8km and its path is E-W up to city center and then change to become NE-SW (Figure7).

Tunnel path consist of alluvial deposits with irregular layering and lens form shapes that lie on marl layers. Major coarse grained soil types through tunnel path in unified classification are: SM, SP-SM, SW-SM, SW, GP-GM, SP, GP, GM-GW, and SW. Fine grained types are classified as: CL, CH, ML, and MH that form surface soils and upper parts of marl layers. Marl layers that belong to

Miocene and Pliocene formations and lie in the depth varying between 20 to 70m, form the geological bedrock of the city. Seismic bedrock of the city according with studies of (Hajialilu 1995) is a layer with V_s>1000m/s that lies in a depth corresponds with marl layers of Pliocene formations. This layer was considered as seismic bedrock of the city. In this investigation 20 borehole were selected among 61 bores drilled for first line of deep metro tunnel. In all of these boreholes unified soil classification is applied and SPT tests were performed every 1 or 2meter.

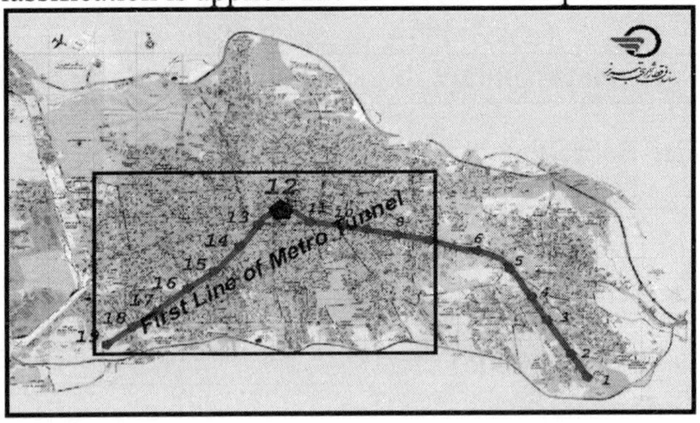

Figure7: Path of Tabriz metro, 12: Bazaar station

3.2 Ground Motion

Local seismographs of strong motions were not available in the city so it was decided to use the graphs belong to same tectonic regimen and their PGA and duration were scaled to correspond with characteristics of probable earthquake happen in the city. Two ground motions were chosen for site effect analysis: component L1 of Ardebil earthquake (1999) recorded in Nir station and L1 component of Kojur- Firoozabad earthquake (2004) recorded in Moalem Kelayeh station. Time histories and Fourier spectrums of these motions are illustrated in (Figure8).

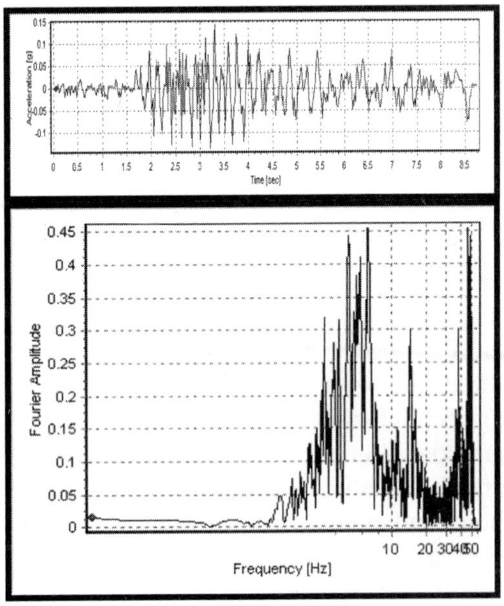

Ardebil (1999)

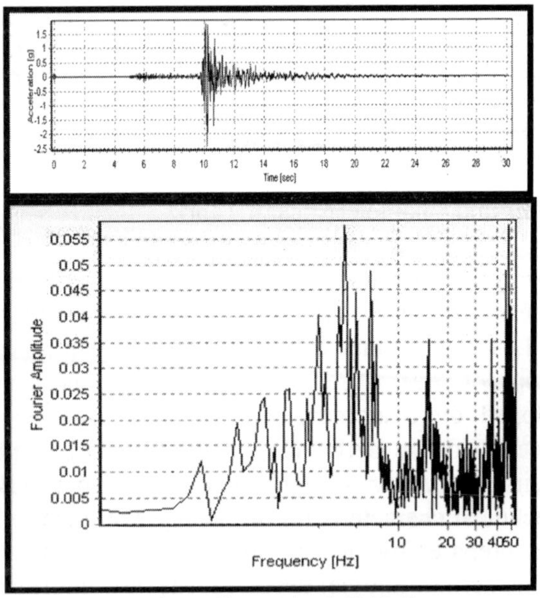

Kojur- Firoozabad (2004)

Figure8: Time histories and Fourier spectrums of ground motions

3.3 Methods of Analysis

Site effect analysis was performed using EERA software. EERA is a modern implementation of the well- known concepts of equivalent- linear earthquake site response analysis and was developed from the same basic concept as SHAKE. To start working with the program soil profiles of each borehole was prepared. Each profile consist of soil material type, thickness of the layer(m), depth at middle of layer(m), shear wave velocity(m/s) that was computed from SPT values and maximum shear modulus G_{max} (MPa) that was calculate by the program itself. Other requirements that were inputted to program were earthquake input motions and material stress- strain damping- strain curves that here we used (Seed & Sun 1989) for clay, (Seed & Idriss 1970) for sand and (Seed et al) for gravel and rock. Analyses were performed applying the PGA value calculated for return period of 75 years (0.3g) in hazard analysis section. From the huge amount of outputs some parameters and spectrums were selected for presentation and investigation that include: natural period and average shear wave velocity of each profile, maximum acceleration at the surface, acceleration profile and amplification spectrum

3.4 Results of Representative Profiles

Investigation of soil profiles and outcomes of the program show that the depth of the alluvial increases from west to east and central part of the city has the thickest deposits. Density of alluvial decrease in the same manner and in the central and southeastern parts in the depth of 15-25m SPT values slump intensely that is a sign of loose soil in these sections that affect the seismic response of the soil seriously .Metro investigations prospect the probability of liquefaction in these parts. Shear wave velocity of profiles reduces from east to west that is in accordance with decreasing of SPT values (Figure9). Natural period of selected profiles changes in the range of 0.49-0.8s (Figure10) Changes in dominant periods of profiles differ from 0.56 to 5.0s that as expected. Acceleration profiles except for some deamplification cases show the increasing of acceleration toward surface but there are some differences in the soil response applying each ground motion. The profiles that situated in the zone with high potential of liquefaction reflect the characteristics of the loose layer of sand in the form of sudden decrease in acceleration in this layer. Results indicate that the average value of amplification factor is 2 and this amount increases to the west and southwestern parts. This part corresponds with increases in alluvial depth toward western parts. The maximum value of amplification factor is 2.6 (Figure11). Low values of this factor are a proof of existence of shallow alluvial basin in the study area. Deamplification is observed in some profiles too.

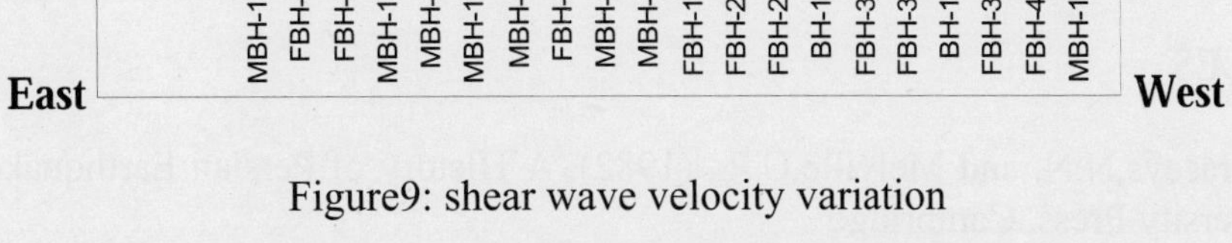

Figure9: shear wave velocity variation

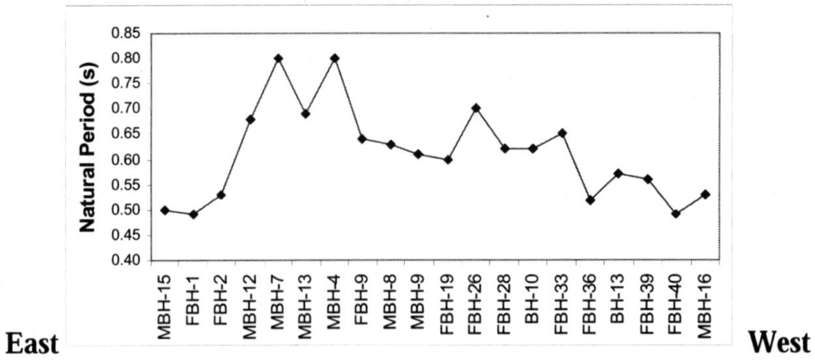

Figure10: Natural period variation

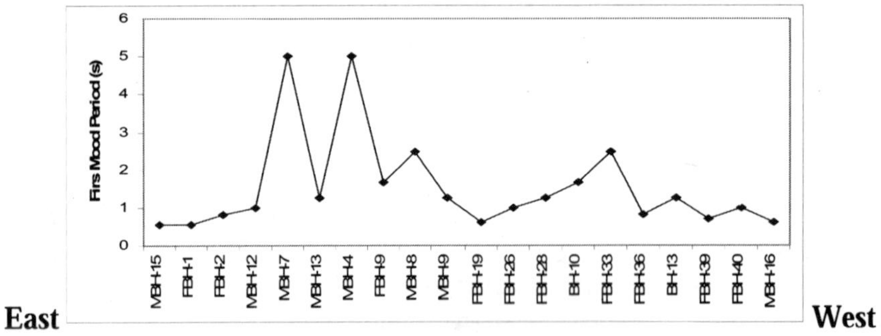

Figure9: Amplification Factor variation

4 CONCLUSION

The city of Tabriz is situated on the shallow alluvial basin in an active tectonical region and surrounded with several active and hazardous faults that most of them especially NTF has many catastrophic historical earthquakes. Seismotectonical investigations show a major seismic gap around Tabriz city in instrumental period. Hazard analyses were performed with probabilistic approach and by applying two different attenuation relationships. Arc GIS software was used for zonatioin step. Results indicate the high amount of PGA values for different return periods in the city. Geotechnical assessments and site effect analysis were performed along Tabriz metro line and results were presented in different tables. Investigations show that the thickness of alluvial increases from east to vest that cause the increase in shear wave velocity and amplification factor. The average amount of amplification Factor is 2 in the city that corresponds with shallow quaternary deposit. Estimations reveal the risk of liquefaction for western part of the city.
The papers presented at the International Workshop on Earthquake Hazards and Mitigations will be printed in form of the workshop proceedings in hard copy along with a CD and be distributed during the workshop. The following is the description, and is formatted in, of the desired format for the papers. Use 11 TNR size for complete text of the paper.

REFERENCES

1. Ambraseys,N.N. and Melville,C.P., (1982). A History of Persian Earthquakes. Cambridge University Press, Cambridge
2. Berberian. M, and Arshadi, S., (1976). "On the evidence of the youngest activity of the Nrth Tabriz Fault and seismicity of Tabriz city" Geological Survey of Iran, Contribution to the seismotectonics of Iran (Part II), Report No. 39
3. Geological Survey of Iran (1994). Report on engineering geological map of Tabriz area.

4. Haghshenas. E, Jafari. M., (2006). "Preliminary Results of Site Effect Assessment in the City of Tabriz (Iran) Using Earthquake Recordings, Third Internation Symposium on the Effects of Surface Geology on Seismic Motion, Grenoble, France, P:106-115
5. Hajialilue-Bonab M., (1995). "The effects of site condition on seismic motion characteristics and design spectra in the city of Tabriz," Ms.C. Thesis; Sharif University of Technology, Tehran, Iran (In Persian)
6. Jackson, J. and McKenzie, D., (1984) " Active tectonics of the Alpine- Himalayan Belt between western Turkey and Pakistan," Geophys.J.R.astr.Soc77,185-246
7. Kijko-Sellevoll-Graham Method. Program written by A.Kijko. Council for Geosciences, Geological Survey of South Africa, 2001
8. Wang J. and Law, K., (1994) Sitting In Earthquake Zones.A.A. Balkema, Rotterdam.
9. Wells, D.L. and Coppersmith, K.J., (1994). "New Empirical Relationship among Magnitude, Rapture Length, Rapture width, Rapture Area and Surface Displacement," Bulletin of Seismological Society of America. Vol.84, No.4,pp.974-1002
10. Zare. M., (2005). "Intrudaction to applied seismology," First edition. IIEES press.

EFFECT OF SHILLONG TOPOGRAPHY ON GROUND MOTION

Kumar Pallav[1], STG Raghukanth[2] and Konjengbam Darunkumar Singh[3]
[1]Scientific officer, Indian Institute of Technology Guwahati , kpallav@iitg.ernet.in
[2]Assistant Professor,Indian Institute of Technology Guwahati, rk@iitg.ernet.in
[3]Assistant Professor,Indian Institute of Technology Guwahati, darun@iitg.ernet.in

This paper investigates the effect of Shillong plateau topography on the ground motion using a two dimensional spectral element method. During the 1897 earthquake which had its epicenter in the Shillong plateau, it was observed that the damage at Shiilong was more than that at Guwahati although the epicenter was nearly equidistant (~ 60 km) from both the places Oldham [1]. In the first instance, this could be attributed to the ground motion amplification due to the topographical terrain of the Shillong plataeu which is at an elevation of 1.5 km above that of Guwahati city. This observation is also consistent with those reported in the literature for several past earthquakes especially 1971 San Fernando earthquake, wherein the amplification of peak ground acceleration at Pacoima dam site was about 50 % Boore [2]. Surface topography plays a siginificant role in amplifying the ground motion when the wavelength of the incoming seismic waves are smaller than the topographical irregularities (e.g., basewidth and slopes of hillsGali *et al.*,[3]. To date, no systematic study has been reported to analyse the effect of Shillong topographical terrain on the earthquake damage. In this paper, an attempt is made to study the topographical effect using an anti-plane strain idealisation.

The governing equation for shear-horizonatal (SH) wave propogation in a 2-D medium can be written as

$$\left(\frac{\partial^2 u}{\partial x^2}+\frac{\partial^2 u}{\partial z^2}\right)=\frac{1}{V_s}\left(\frac{\partial^2 u}{\partial t^2}\right) \tag{1}$$

where u and V_s denote the SH wave displacement and shear wave velocity respectively. In the spectral element method the Shillong region under consideration is meshed using quadrilateral elements. After discretization into spectral elements, the resulting system of ordinary differential equations can be written in matrix form as,

$$\mathrm{M}\ddot{u}+\mathrm{K}u=\mathrm{F} \tag{2}$$

where M and K are mass and stiffness matrices respectively. Expressions for elemental mass and stiffness matrix are available in Schuberth [5].

The topography of NE India is available from the Shuttle Radar Topography Mission (SRTM) website maintained by NASA (http://srtm.csi.cgiar.org/). A section passing through Guwahati and Shillong is shown in Figure 1. The locations of Guwahati and Shillong are shown in Figure 2. The domain is descretised with 160 quadrilateral spectral elements of size 6.25 km in horizontal and 2.5 km in vertical directions. Absorbing boundary conditions are applied for the boundaries other than free surface to nullify wave reflections, simluating near real situation. A Ricker wavelet forcing function shown in Figure 3 is applied at a depth of 5 km below the ground surface consistent with the focal depth of the 1897 earthquake (Figure 2). The shear wave velocity model of the region is taken from the work of Mitra et al [4]. The velocity time histories obtained by solving Equation 2 at Guwahati and Shillong are shown in Figures 4 and 5. To analyze the effect of topography, ground motion time histories are also simulated at both the two stations without considering the topography as shown in Figures 4 and 5. It can be observed that as the source is placed approximately equidistant from both receivers (Figure 2), the source activates both receivers at the same time (~ 9 sec (without topography), see Figure 4 and 5). The Phase difference between the simulated velocity time histories at Guwahati is smaller compared to that at Shillong i.e. time lag is high. This can be attributed to the fact that the seismic waves has to travel 1.5 km more in the case of topographical domain at Shillong. It is seen that amplification factor, which is defined as the ratio of peak velocities with and without topography for Shillong is 1.1. At Guwahati, the amplification is negligable. This amplification at Shillong is perhaps due to multiple reflections caused by the topographical irregularities. It can be observed from Figures 4 and 5 that in case of flat surface the velocities are decreasing gradually with time after the receivers get activated for the first time. However, at Shillong, in case of topography, peak velocity gets further amplified after some time due to multiple reflections from surface undulations.

These results indicate that the presence of massive topographical structures like hills increases the time duration of ground motion and also amplify the peak velocity, and thereby increasing the effects of seismic hazard, which is consistent with the observations of the 1897 Shillong earthquake. This study will be extended considering 3D terrain, including inhomogenity in material properties.

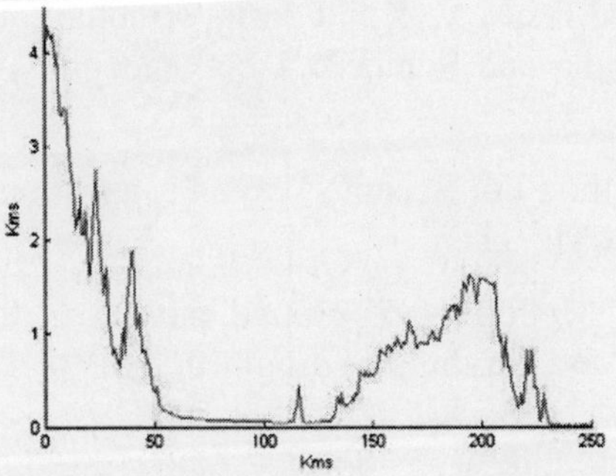

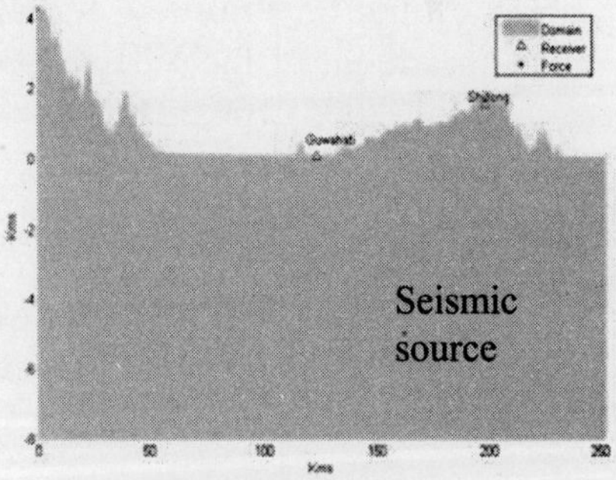

Figure 1 Topography of NE region | Figure 2 Location of Receiver and Input Excitation

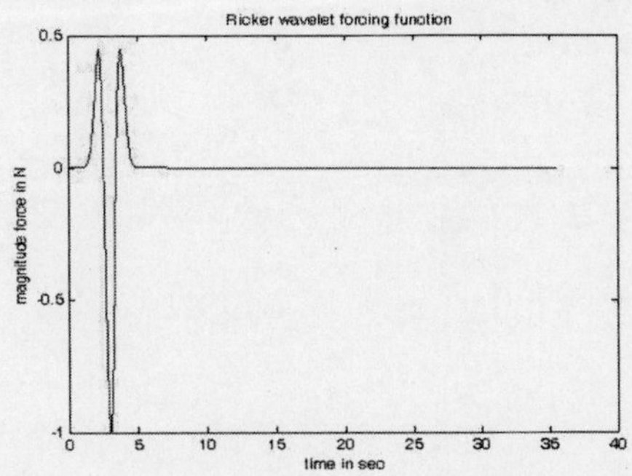

Figure 3 Ricker Wavelet Forcing Function

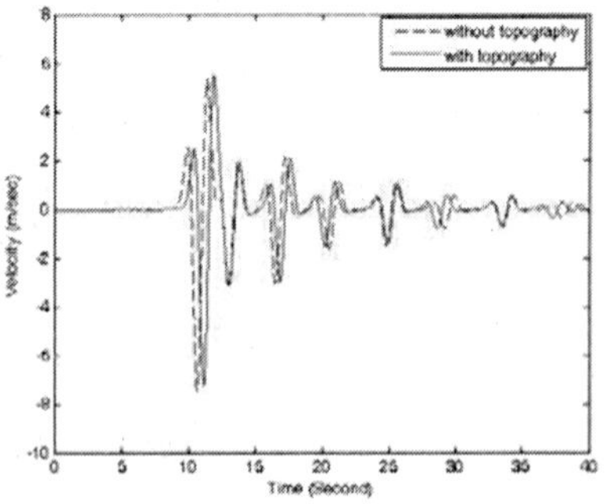

Figure 4 Velocity time histories at Guwahati

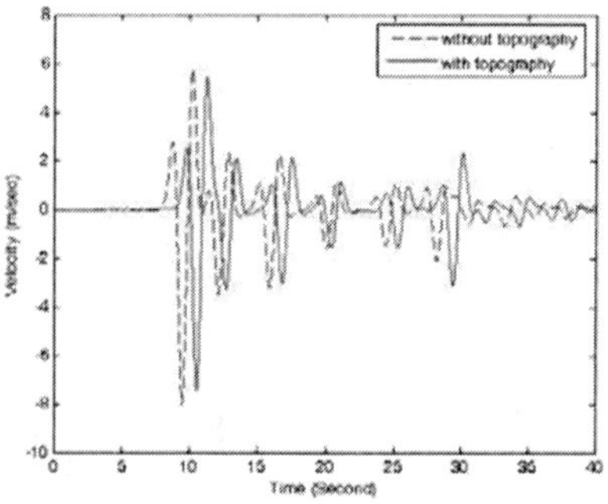

Figure 5 Velocity time histories at Shillong

REFERENCE

1. Oldham, R. D. (1899), "Report on the great earthquake of 12 June 1897, Mem. Geol. Soc. of India, 29.
2. David m. Boore (1972), "A note on the Effect of simple Topography on Seismic SH waves Bulletin of the Seismological Society of America. Vol. 62, No. 1, pp. 275-284.
3. Geli ,L., Bard, P. and Jullien, B., (1988) "The Effect Of Topography On Earthquake Ground Motion: A Review And New Results Bulletin Of The Seismological Society Of America, Vol. 78, no. 1, pp. 42-63.
4. Mitra. S., Priestley, K., Bhattacharyya, A. K. and Gaur, V. K., "Crustal structure and earthquake focal depths beneath Northeastern India and Southern Tibet" Geophys. J. Int.(2005) 160,227-248.
5. Schuberth, B. (2003), "The Spectral Element Method For Seismic Wave Propagation", Diploma Thesis, Ludwig-Maximilians Universitat Munchen.
6. Priolo, E., Grimaz ,S., Chiaruttini, C., (1996) "Modelling of ground motion in the vicinity of massive structures", Soil Dynamics and Earthquake Engineering Vol-15, 75-82

Part-4

Geotechnical Earthquake Engineering

EFFECTS OF EMBANKMENT RIGIDITY ON BEHAVIOR OF NATURALLY DEPOSITED SOILS DURING/AFTER EARTHQUAKES

Noda, T.[1], Asaoka, A.[2], Nakano, M.[3], Nakai, K.[4] and Takeuchi, H.[5]
[1] Department of Civil Engineering Nagoya University, noda@civil.nagoya-u.ac.jp
[2] Department of Civil Engineering Nagoya University, asaoka@civil.nagoya-u.ac.jp
[3] Department of Civil Engineering Nagoya University, nakano@civil.nagoya-u.ac.jp
[4] Department of Civil Engineering Nagoya University, nakai@civil.nagoya-u.ac.jp
[5] Fudo Tetra Corporation, hidekatsu.takeuchi@fudotetra.co.jp

ABSTRACT

The effects of the rigidity of earth structures such as embankments on the deformation behavior of sand-clay laminated soil (top layer of sand, under-layer of clay) and embankment coupled systems during and after earthquakes were investigated. The investigation was carried out through static and/or dynamic soil-water coupled finite deformation analysis using the SYS Cam-clay model as the constitutive equation of soils. The results showed that the large-scale deformation behavior of the top sand layer during and after seismic activity is different from that of the clay under-layer depending on the rigidity of the embankment. If the embankment is made up of soft soil, the occurrence of liquefaction of the top sand layer during an earthquake is accompanied by large-scale deformation of the embankment itself, but the clay under-layer remains almost unaffected even after the earthquake. In contrast, if the embankment is rigid, while the sand layer is again subjected to liquefaction during the earthquake, the embankment load is transferred to the lower clay layer, leading to large-scale ground (clay layer) settlement after the earthquake has passed.

Keywords: Soil-water coupled analysis, Embankment, SYS Cam-clay model, Liquefaction

1 INTRODUCTION

It would not be an overstatement to say that in the area of investigations dealing with earthquake resistance of soils only or of coupled systems consisting of soil and earth structures such as embankments, few or no analytical studies have been extended to include post-earthquake behavior. In dynamic analysis such as analysis of liquefaction behavior, attention has been focused on the unstable behavior (liquefaction) that occurs only during the earthquake. This situation seems to have arisen from the existing line of thought that it would be sufficient to investigate only sandy soils and embankments whose seismic resistance is considered to be problematic in order to carry out the necessary soil improvements. The SYS Cam-clay model (Asaoka et al., 1998, 2002) allows for description of soils ranging from clay or intermediate soils to sand within a single theoretical framework (Nakano et al., 2003, 2005; Yamada and Nakano, 2007). By mounting this model onto the constitutive equation of soils, the authors carried out soil-water coupled finite deformation analysis (Asaoka and Noda, 2006) in order to investigate both the dynamic and static deformation behavior of sand-clay laminated soil (top layer of sand, under-layer of clay) and embankment coupled systems during and after earthquakes. This paper deals expressly with the effect of rigidity of embankments on the behavior of such soil systems.

2 CALCULATION CONDITIONS

Fig.1 shows the finite element mesh of the loaded soil structure (embankment) used in the calculations, while Table 1 shows the material constants (elasto-plastic parameters, evolutional parameters) and initial values of the soil layers.

The layered soil structure targeted for the calculations is a typical sand-clay laminated structure in the city of Yokkaichi (located in Mie Prefecture, Japan). It is a horizontal sedimentary soil structure consisting of a top layer of silty sand (As layer: 5m), a sand layer (As' layer: 5m) and a clay under-layer (Ac layer: 10m). Beneath the clay layer is an engineering bedrock having N-values higher than 30. It was assumed that the embankment was saturated and that the pore water pressure in the ground as well as that on the upper surface of the embankment was zero. The lateral surfaces of the ground were treated as undrained, and the lower surface was assumed to be the drainage boundary. In addition, the components of all finite element nodal points located a given height along the lateral surfaces of the ground were allotted equal displacement conditions (periodic boundaries). The material constants were determined from the results of isotropic consolidation tests and undrained triaxial compression tests carried out using samples taken from the site (see the grain size distribution in Fig.2). Sufficient consideration was given to the (ideal) sampling process (Noda et al., 2005) in order to ensure good simulation of all types of mechanical behavior. The elasto-plastic parameters were determined from tests carried out on remolded samples, whereas undisturbed samples were used in the tests for determining the evolutional parameters. Figs. 3, 4, and 5 show the results of undrained triaxial compression tests of the undisturbed samples (dotted lines) together with the results of simulation using the SYS Cam-clay model (solid lines).
The material constants of each type of soil are discussed below. The elasto-plastic parameters indicate that the compression index $\tilde{\lambda}$ and swelling index $\tilde{\kappa}$ were larger for the clayey soil (having a higher content of fine particles) and smaller for the sandy soils (having a lower content of fine particles). In other words, clay has higher compressibility than sand. Of the two sand layers, although As was classified as sand, it can be considered to be a highly compressible soil because it had a very high compression index $\tilde{\lambda}$. Furthermore, the

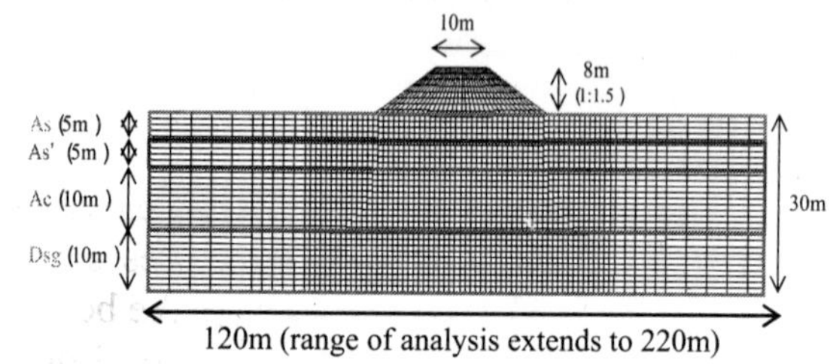

Fig.1 Finite element mesh

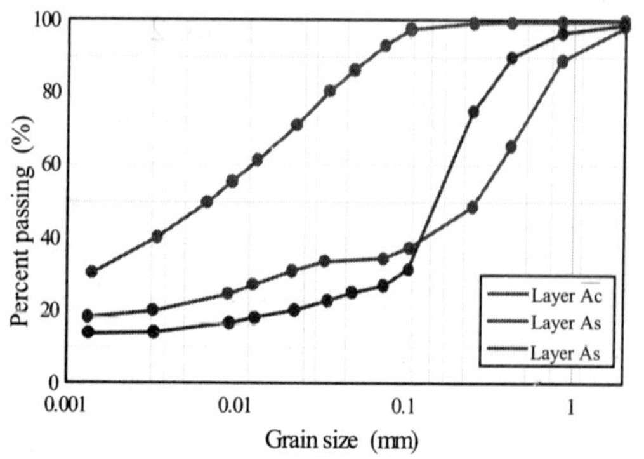

Fig.2 Accumulated particle size distribution

Table 1 Material constants and initial values of soils

< Elasto-plastic parameters >	As	As'	Ac	Dsg
Compression index $\tilde{\lambda}$	0.120	0.040	0.150	0.050
Swelling index $\tilde{\kappa}$	0.012	0.004	0.03	0.012
Critical state constant M	1.45	1.45	1.50	1.99
Intercept of the normal consolidation line N	1.65	1.71	2.08	1.00
Poisson's ratio ν	0.40	0.40	0.10	0.30
< Evolutional parameters >	**As**	**As'**	**Ac**	**Dsg**
Degradation index of structure a (b=c=1)	20.0	10.0	0.45	2.63
Ratio of $-D_v^p$ to D_s^p cs	1.0	1.0	0.3	1.0
Degradation index of OC m	0.03	0.01	2.50	0.08
Rotational hardening index b_r	0.05	0.10	0.05	0.514
Rotational hardening limit constant m_b	0.65	0.65	1.0	0.50
< Initial values >	**As**	**As'**	**Ac**	**Dsg**
Specific volume v	1.668	1.586	2.416	1.790
Stress ratio η_0	0.55	0.55	0.55	0.55
Degree of structure $1/R^*_0$	1.18	1.30	25.7	1.26
Overconsolidation ratio $1/R$	5.6–20	70–83	1.3–130	100-175
Degree of anisotropy ζ_0	0.4	0.55	0.55	0.55
Permeability k (cm/sec)	5×10^{-6}	1×10^{-4}	5×10^{-7}	4×10^{-2}

evolutional parameters showed that layer

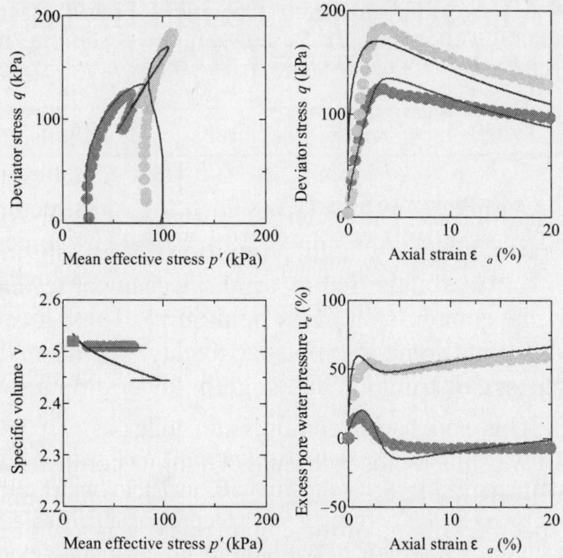

Fig.3 Results of simulation of the undrained triaxial compression test using Ac

Ac was a typical clayey soil with slow decay of structure, fast loss of overconsolidation and slow growth/extinction of anisotropy. In contrast, the As and As' layers were sands with fast decay of structure, slow loss of overconsolidation and fast growth/extinction of anisotropy. The decay of structure of layer As, which had a more gradual particle size distribution, was faster than that of As'. The measured material constants also allowed for estimation of the initial conditions of the soil skeleton structure (structure, overconsolidation and anisotropy) of each type of soil at the sampling site. Since the specific volume is nearly homogeneous within a given soil layer, the initial distributions of structure and anisotropy were assumed to be nearly the same for each soil layer. Based on this assumption, the overconsolidation ratios were distributed in the vertical direction in proportion to the earth cover pressure.

In the calculations, the embankment was loaded step-wise up to a height of 8 m using the finite element addition method (Takeuchi et al., 2006). After soil consolidation ended, the earthquake vibration patterns were input in the horizontal direction to all nodal points at the bottom surface of the ground, and the calculations were continued until settlement subsided. The input values for acceleration were 1/2 of those shown in Fig. 6. At the nodal points of the lower edges of the ground corresponding to the engineering bedrock, the horizontal direction along the

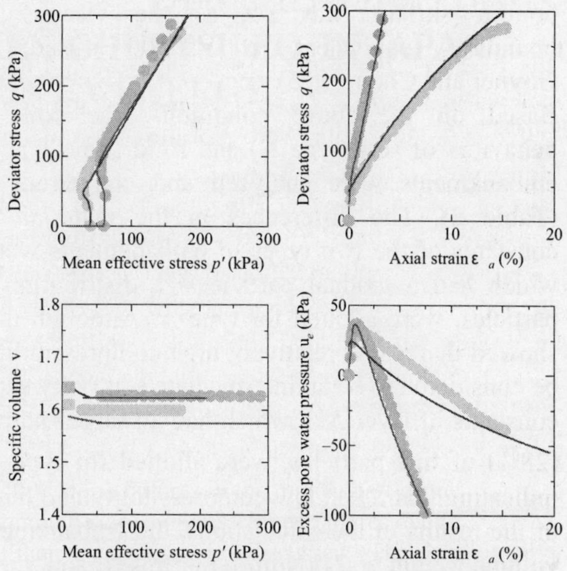

Fig.4 Results of simulation of the undrained triaxial compression test using As'

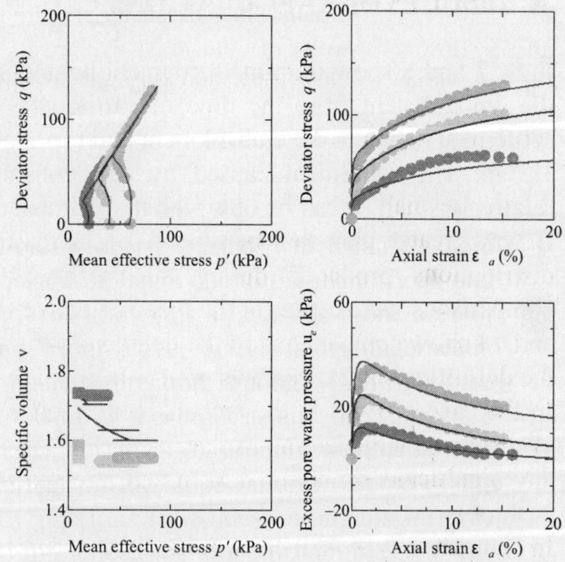

Fig.5 Results of simulation of the undrained triaxial compression test using As

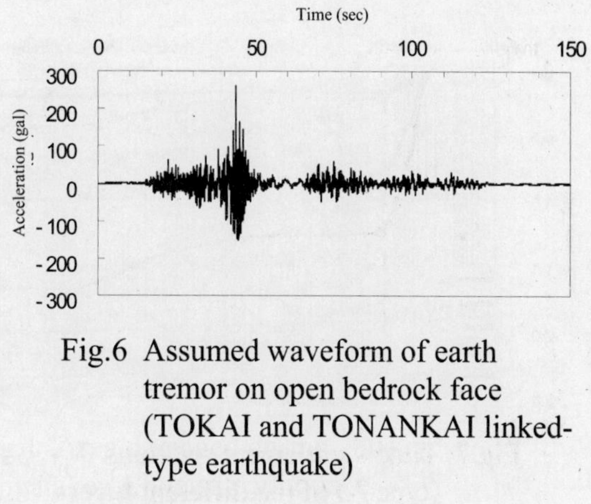

Fig.6 Assumed waveform of earth tremor on open bedrock face (TOKAI and TONANKAI linked-type earthquake)

bottom surface was set as the viscous boundary (equivalent to Vs=300 m/sec) (Joyner and Chen, 1975) .

Based on the above conditions, the soil behaviors of soft (type A) and rigid (type B) embankments were analyzed and compared (Table 2). The differences in the material constants of the two types of embankments were as follows. The material constants of layer As, which had a gradual particle size distribution and a relatively high proportion (35%) of fine particles, were allotted for type A. Although this soil was classified as sand, mechanical testing showed that it had relatively high compressibility, the compression index being 0.12. Thus, it can be considered to be an intermediate soil (silty sand) having some resemblance to clay. The material constants of layer As', which had a sharper particle size distribution and slightly lower proportion (28%) of fine particles, were allotted for type B. This soil had a compression index $\widetilde{\lambda}$ of 0.04, indicating that it had low compressibility and that it was much closer to sand. To allow comparison of the results of the calculations, the embankment loads for types A and B need to be equal (unit volume weight γ_t=21kN/m3). For this reason, the specific volume of the embankment was assumed to be 1.5 for both cases, type A being the softer embankment compared with type B.

Table 2 Analytical cases

	Embankment material	Specific volume	Embankment rigidity
TypeA	As	1.50	Soft
TypeB	As'	1.50	Rigid

3 RESULTS OF CALCULATIONS

Figs. 7 and 8 show the time-settlement behavior of the respective soil layers in the central part of the embankment from the time of earthquake occurrence. As soon as the earthquake occurred, settlement began in the embankment and top sand layer. The overall amount of settlement exceeded 1.5 m. The settlement caused by soil consolidation that followed this initial settlement was relatively small. It can be observed that compaction of the clay under-layer (Ac) in the case of type B was greater than that of type A by at least 30 cm. Figs. 9 and 10 compare the shear strain distributions produced during the earthquake and after the deformation subsided. Similar comparisons were made for the mean effective stress distribution in Figs. 11 and 12, while Figs. 13 and 14 show comparisons of the degree of structure (1/*R**). Please refer to Asaoka et al. (2002) for the definitions of the symbols used.

In the case of type A, the embankment itself exhibited large deformation, and the top sand layer also showed large lateral displacement. For this reason, the clayey under-layer was not subjected to any significant strains or transmitted stresses (the liquefied layer seems to have functioned as a cushion).

In contrast to type A, there was hardly any deformation of the embankment itself in the case of type B. However, large stress concentrations occurred at the boundary between the top sand layer and the clayey under-layer that induced the long-term settlement observed later in the clayey layer. However, no significant differences were seen in the decay of structure of the clay under-layers.

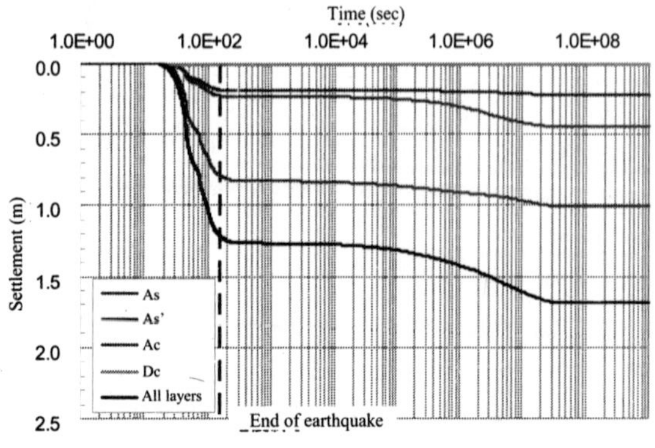

Fig.7 Time - settlement relations (type A) of the different layers

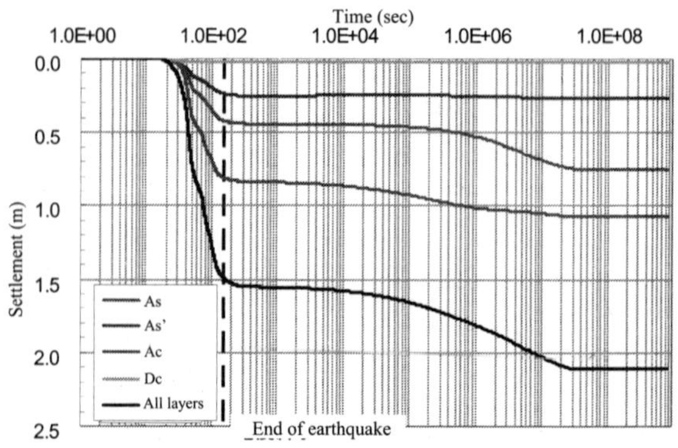

Fig.8 Time - settlement relations (type B) of the different layers

(a) Before the earthquake

(b)150 seconds after commencement of earthquake

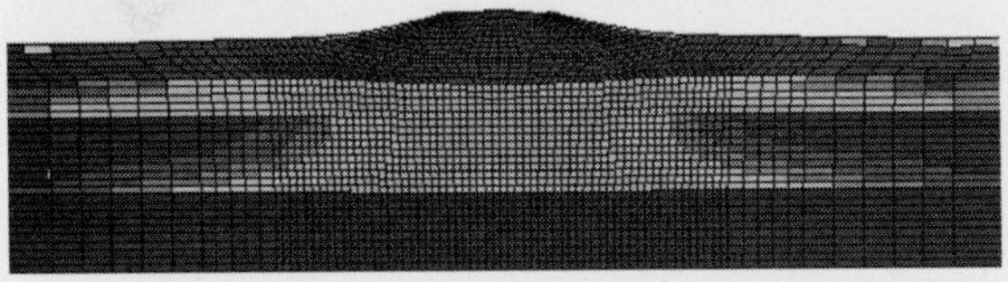

(c) After consolidation has ended

(a) Before the earthquake

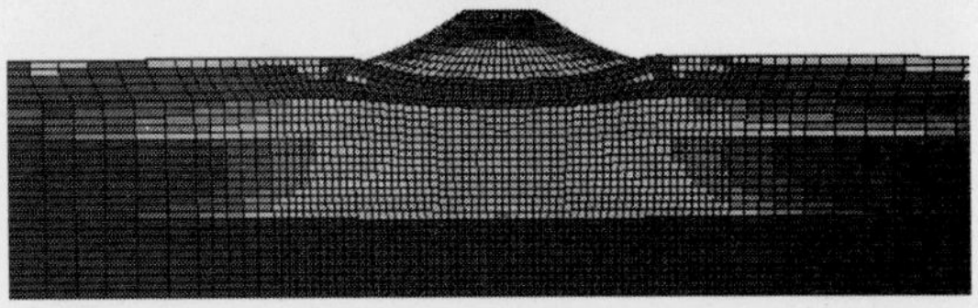

(b)150 seconds after commencement of earthquake

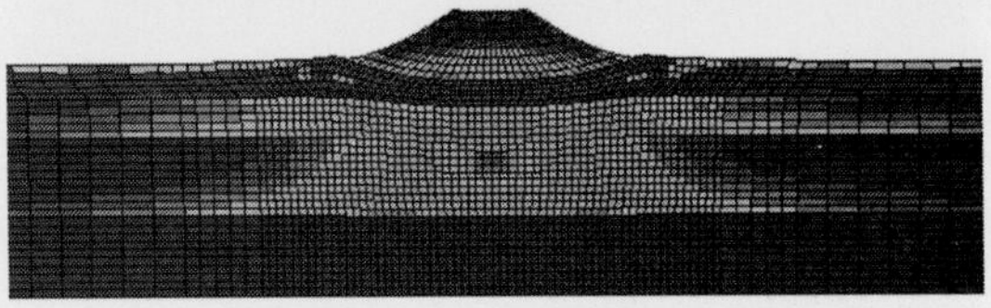

(c) After consolidation has ended

0%  15% (All values higher than 15% are counted as 15%)

Fig.9 Shear strain distribution (type A)

Fig.10 Shear strain distribution (type B)

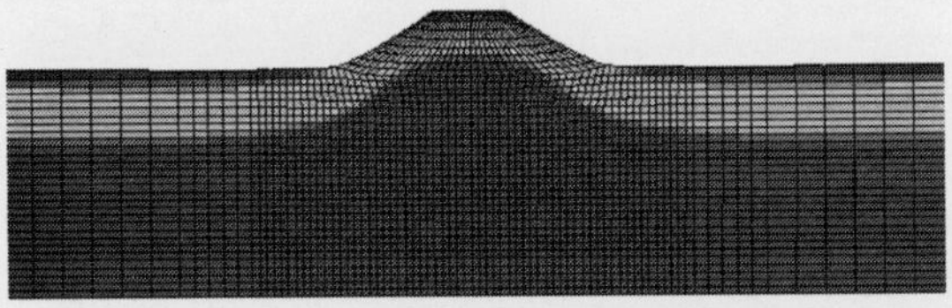

(a) Before the earthquake

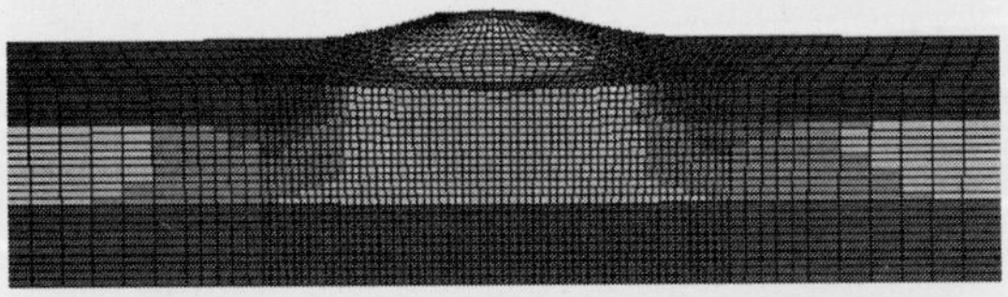

(b)150 seconds after commencement of earthquake

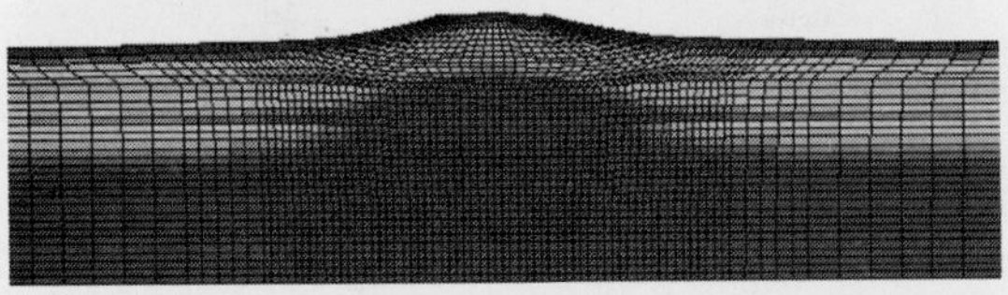

(c) After consolidation has ended

(a) Before the earthquake

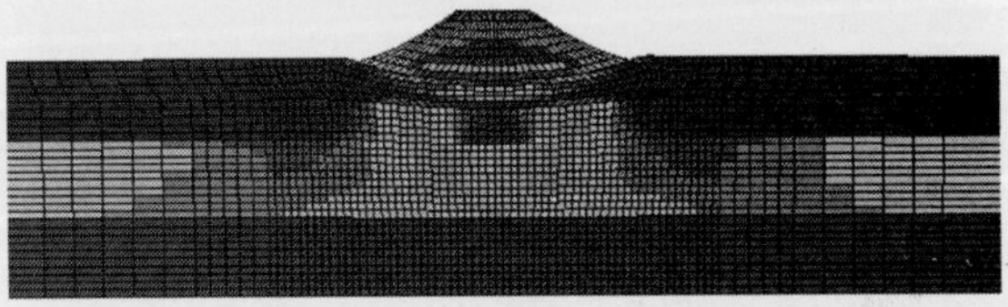

(b)150 seconds after commencement of earthquake

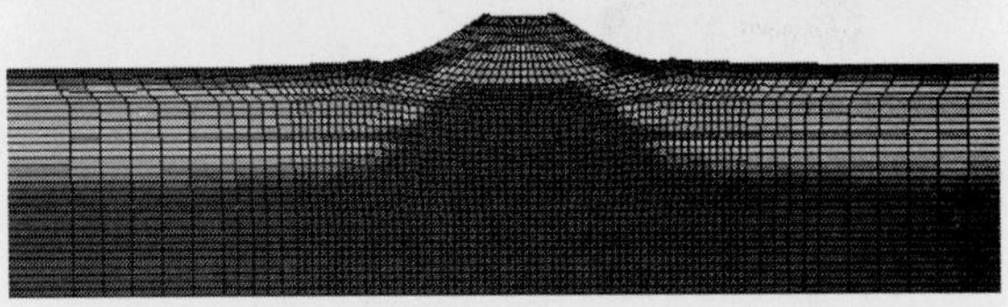

(c) After consolidation has ended

0kPa  80kPa (All values higher than 80kPa are counted as 80kPa)

Fig.11 Mean effective stress distribution (type A)

Fig.12 Mean effective stress distribution (type B)

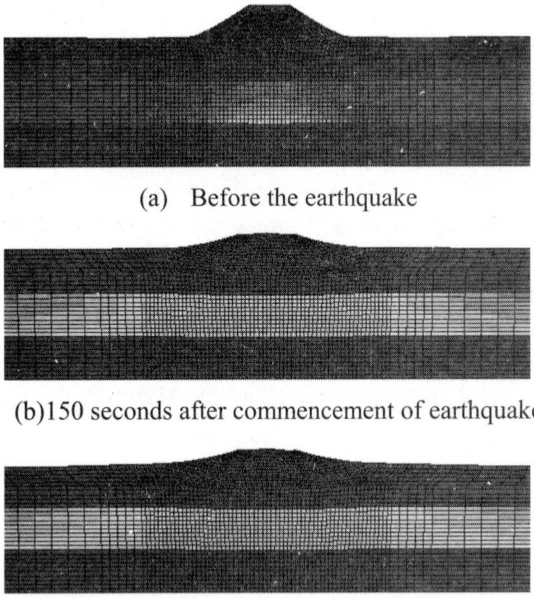

(c) After consolidation has ended

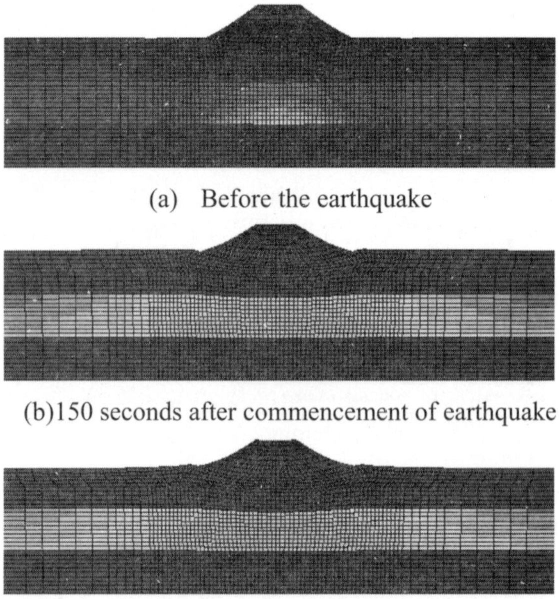

(c) After consolidation has ended

1.0 ▬▬▬ 25.7

Fig.13 Degree of structure ($1/R^*$) distribution (type A)

Fig.14 Degree of structure ($1/R^*$) distribution (type B)

In other words, if the embankment is made up of soft soil, although it would move together with the liquefied top layer during an earthquake and exhibit large deformation, the clay under-layer would remain unaffected even after the earthquake. In contrast, although liquefaction of the sand layer occurs during an earthquake, even if the embankment is rigid, the vertical stresses that are transmitted to the lower parts are not reduced. Because of this, the ground (clay layer) exhibits large-scale settlement after the earthquake. The above results made it clear that the deformation behavior of the top sand layer and the clay under-layer during and after an earthquake differ depending on the differences in the rigidity or deformation characteristics of the embankment.
The behaviors of the elements of the sand layers (As, bottom of embankment slope, GL-2m) are

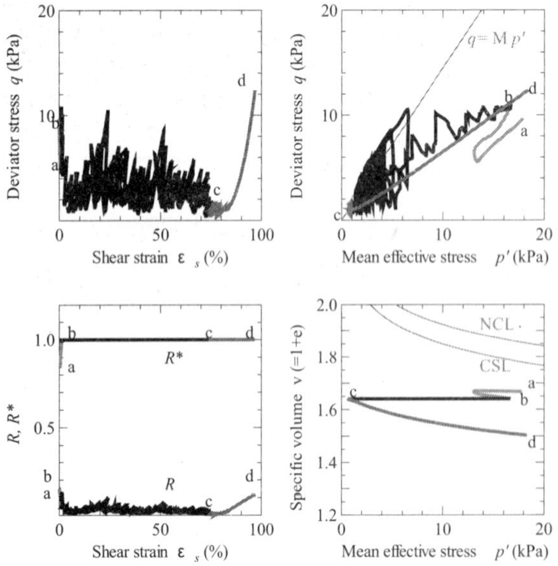

Fig.15 Behavior of the elements of the sand layer (type A)

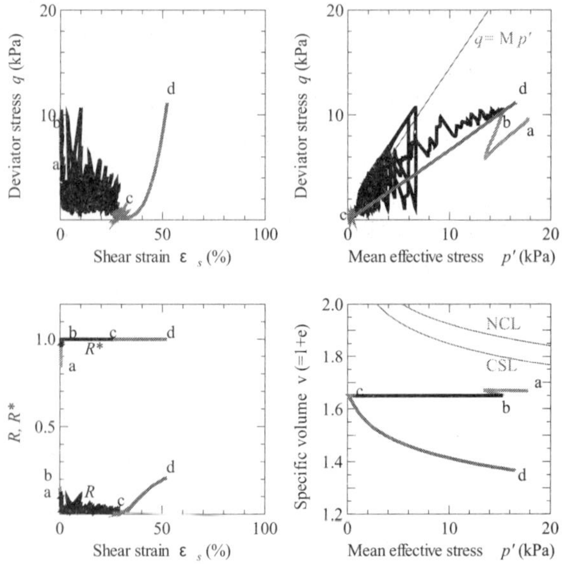

Fig.16 Behavior of the elements of the sand layer (type B)

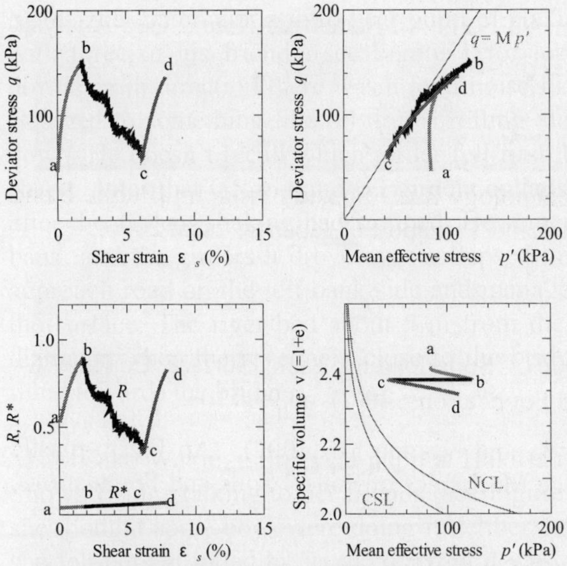

Fig.17 Behavior of the elements of the clay layer (type A)

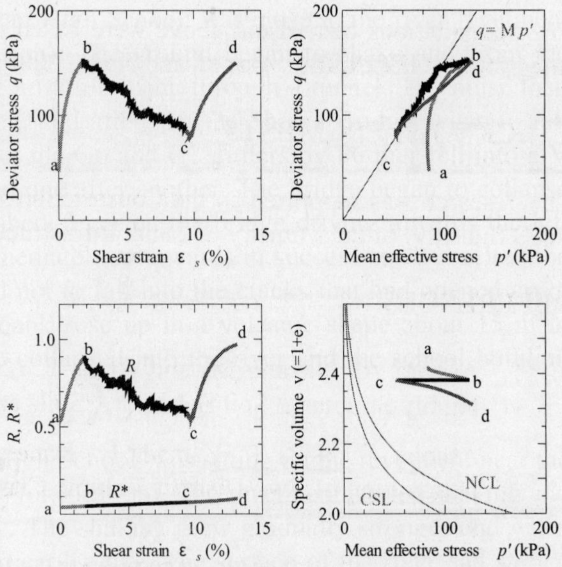

Fig.18 Behavior of the elements of the clay layer (type B)

illustrated in Figs. 15 and 16. Those of the clay layers (Ac, directly under the embankment, GL-12m) are shown in Figs. 17 and 18. The blue lines(a-b) within the figures refer to the behavior at the time of embankment construction, and the black lines(b-c) indicate that during the earthquake. The red lines(c-d) denote the behavior after the earthquake, up to the time at which deformation subsided. It can be observed that there was some loss of structure of the sand at the slope bottom at the time of embankment construction (R^*=1). Naturally, those elements that were not subjected to the embankment load retained their structure. Looking at the effective stress paths during the earthquake, it can be seen that the shear stress q and mean effective stress p' approach the origin (p'=0), indicating the occurrence of liquefaction. Furthermore, liquefaction was accompanied by a large amount of shear strain. The effective stress paths of type A (soft embankment) and type B (rigid embankment) were very similar, but the shear strain produced during the earthquake in particular was large in the case of type A. As has already been mentioned, this indicates that for a relatively soft embankment, large lateral flow is produced by liquefaction of the sand layer. The specific volume ratio v is almost constant during this period, and the soil exhibits undrained behavior. Once the excess pore water pressure drops after the earthquake passes, the mean effective stress increases again, and the specific volume decreases (consolidation of the sand occurs).

In contrast to the above, the shear strains produced in the clay layer directly underneath the embankment were larger in the case of type B. This can be explained by the fact that the minimum value of the shear stress was lower (60kPa) for type A compared with type B (80kPa). Because of the higher shear stress, type B experienced a larger repeated shearing load.

4 CONCLUSIONS

The effects of an earthquake are not local but extend over a wide area, attacking the weaker locations of an earth structure. Thus, if an embankment is less rigid, the embankment is affected. If the embankment is more rigid, the ground is affected. The results of this study show that if the embankment is made rigid with the intent of improving its stability during earthquakes, instances of unexpected settlement and creation of large strains in the clay layer below the embankment may occur. In the engineering field of steel structures, it is well known that if a particular member of the structure is strengthened in order to improve structural seismic resistance, unexpected damage may occur to other components during earthquakes. Therefore, when planning to strengthen and/or improve earth structures that contain soil layers prone to liquefaction, it is important to give sufficient consideration to not only the liquefaction-prone layer, but also the condition of the entire structure, including the other soil layers that comprise the structure.

The computations carried out above were based on All Soils All States All Round Geo-Analysis Integration (***GEOASIA***; Asaoka and Noda, 2006)

5 ACKNOWLEDGMENTS

This research was supported by the Construction Technology R&D Subsidy Program (Public Basic R&D Subsidy) of the Ministry of Land, Infrastructure and Transport of Japan in fiscal 2006.

REFERENCES

1. Asaoka A., Nakano M. and Noda T. (2000). "Superloading Yield Surface Concept for Highly Structured Soil Behavior," Soils and Foundations, **40**(2), pp.99-110.
2. Asaoka A., Noda T., Yamada E., Kaneda K. and Nakano M. (2002). "An Elasto-plastic Description of Two Distinct Volume Change Mechanics of Soils," Soils and Foundations, **42**(5), pp.45-57.
3. Asaoka A. and Noda T. (2007). "All Soils All States All Round Geo-analysis Integration," International Workshop on Constitutive Modelling - Development, Implementation, Evaluation, and Application, Hong Kong, China, pp.11-27.
4. Nakano M., Noda T., Asaoka A. and Nakai K. (2003). "Compaction Behavior of Sand with Degradation of Structure and Overconsolidation," Proc. of Sino-Japanese Symposium on Geotechnical Engineering, Beijing, China, pp.455-462.
5. Nakano M., Nakai K., Noda T. and Asaoka A. (2005). "Simulation of Shear and One-dimensional Compression Behavior of Naturally Deposited Clays by Super/subloading Yield Surface Cam-clay model," Soils and Foundations, **45**(1), pp.141-151.
6. Noda T., Yamada S. and Asaoka, A. (2005). "Elasto-plastic Behavior of Naturally Deposited Clay during/after Sampling," Soils and Foundations, **45**(1), pp.51-64.
7. Takeuchi H , Takaine T. and Noda T. (2006). "Effect of Nonlinearity in Geometry on The Consolidation Deformation of Saturated Clay Soil," Journal of Applied Mechanics, JSCE, vol.9, pp.39-550. (in Japanese).
8. Yamada E. and Nakano M. (2007). "Improvement of Manmade Island Filled with Intermediate Soils by SCP method," Proc. 13th Asian Regional Conference on Soil Mechanics and Geotechnical Engineering, to appear.
9. Joyner, W. B. and Chen, A.T.F. (1975). "Calculation of Nonlinear Ground Response in Earthquakes," Bulletin of Seismological Society of America, **65**(5), pp.1315-1336.

UNDRAINED CYCLIC TORSIONAL SHEAR TESTS ON SAND UP TO EXTREMELY LARGE STRAIN LEVELS

Koseki, J.[1], Kiyota, T.[2], Sato, T.[3] and Mohammad, A.M.[4]
[1]*Institute of Industrial Science, the University of Tokyo, Japan, koseki@iis.u-tokyo.ac.jp*
[2]*ditto, kiyota@iis.u-tokyo.ac.jp*
[3]*ditto, tsato@iis.u-tokyo.ac.jp*
[4]*Tehran University, Iran*

ABSTRACT

In order to study the behavior of liquefied sands at extremely large strain levels, a series of undrained cyclic torsional shear tests was performed on saturated Toyoura sand up to double amplitude shear strain of about 100 %, while correcting for the effect of membrane force on the measured shear stress. After exceeding a certain level of overall strain, localization of specimen deformation was observed, which was associated with the changes in the cyclic amplitude of deviator stress and the increment of shear strain? The limiting value of overall shear strain to initiate strain localization increased with a decrease in the relative density of the specimen.

Keywords: Liquefaction, Torsional shear test, large strain, Membrane force

1 INTRODUCTION

Extremely large deformation of liquefied soils has been observed in several case histories, such as the 1964 Niigata, 1983 Nihonkai-Chubu and 1995 Hyogoken-Nanbu earthquakes in Japan (Hamada et al., 1993; Ishihara et al., 1996). However, properties of liquefied soils under large deformation are not understood well.

In the present study, in order to study the behavior of sands during the process of liquefaction up to extremely large strain levels, a series of undrained cyclic torsional shear tests was performed on saturated Toyoura sand up to double amplitude shear strain of about 100 %, while correcting for the effects of membrane force on the measured shear stress during the test.

2 TEST APPARATUS

The test apparatus employed in the present study is schematically shown in Fig. 1. In order to enlarge the effective range of torsional displacement, an improvement was made on the one employed by HongNam and Koseki (1995) by modifying the belt-driven torsional loading system connected to an AC servo motor through electro-magnetic clutches and reduction gears (Fig. 2a). The details of the torque-transmission part after modification are shown in Fig. 2b. The linear-motion guide has a stroke of 60 cm, which can rotate the loading shaft by 1.2 rounds through a pulley having a diameter of 15 cm.

In addition, as compared to the specimen dimensions employed by HongNam and Koseki (1995), the outer and inner diameters of the specimen were reduced from 20 and 12 cm, respectively, to 15 cm and 9 cm, while keeping the specimen height at 30 cm. This modification was made in order to conduct tests on in-situ frozen specimens that were available with a size of 15 cm in diameter. The test results on in-situ frozen specimens are reported elsewhere (Kiyota 2007).

As a result of the above modifications, it was made possible to achieve double amplitude torsional shear strain levels exceeding 100 %. In order to evaluate such large torsional deformation, a potentiometer with wire and pulley was employed as shown in Fig. 2b. The capacities of torque and vertical load are 0.15 kN*m and 8 kN, respectively. They were measured with a two-component load cell that is installed inside the pressure cell (Fig. 1).

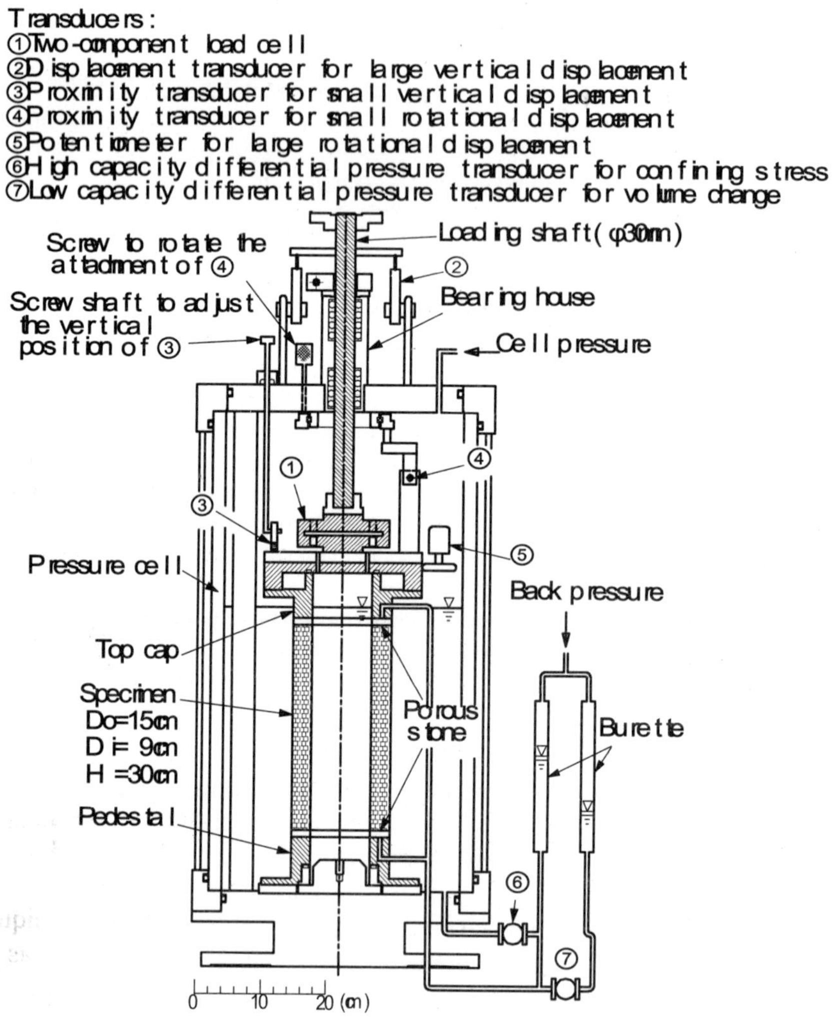

Fig. 1 Torsional shear test apparatus on hollow cylindrical specimen

3 TEST PROCEDURES

Test material was Toyoura sand, a uniform sand with subangular particles (specific gravity G_s=2.635, maximum void ratio e_{max}=0.975, minimum void ratio e_{min}=0.561, mean diameter D_{50}=0.20 mm with no fines content under 75 μm). Several specimens were prepared by pluviating air-dried sand particles through air, while controlling the values of initial relative density in the range of 5 to 90 % by changing the height of pluviation. After saturating the specimens with pouring carbon dioxide through the void among the sand particles and then pouring de-aired water, they were consolidated to an isotropic effective stress state of 100 kPa with a back pressure of 200 kPa.

Using the modified loading device, the torsional torque was applied at a constant shear strain rate of 5.5 %/min. The loading direction was reversed when the shear stress amplitude that was corrected for the effects of membrane force reached the specified value. In order to improve the accuracy of cyclic torsional loading, the shear strain rate was reduced by half at its early stage, since the shear strain amplitude at each loading reversal was extremely small. Throughout the cyclic torsional loading, the vertical strain of the specimen was maintained to be zero, by using a mechanical locking device for the vertical displacement of the top cap.

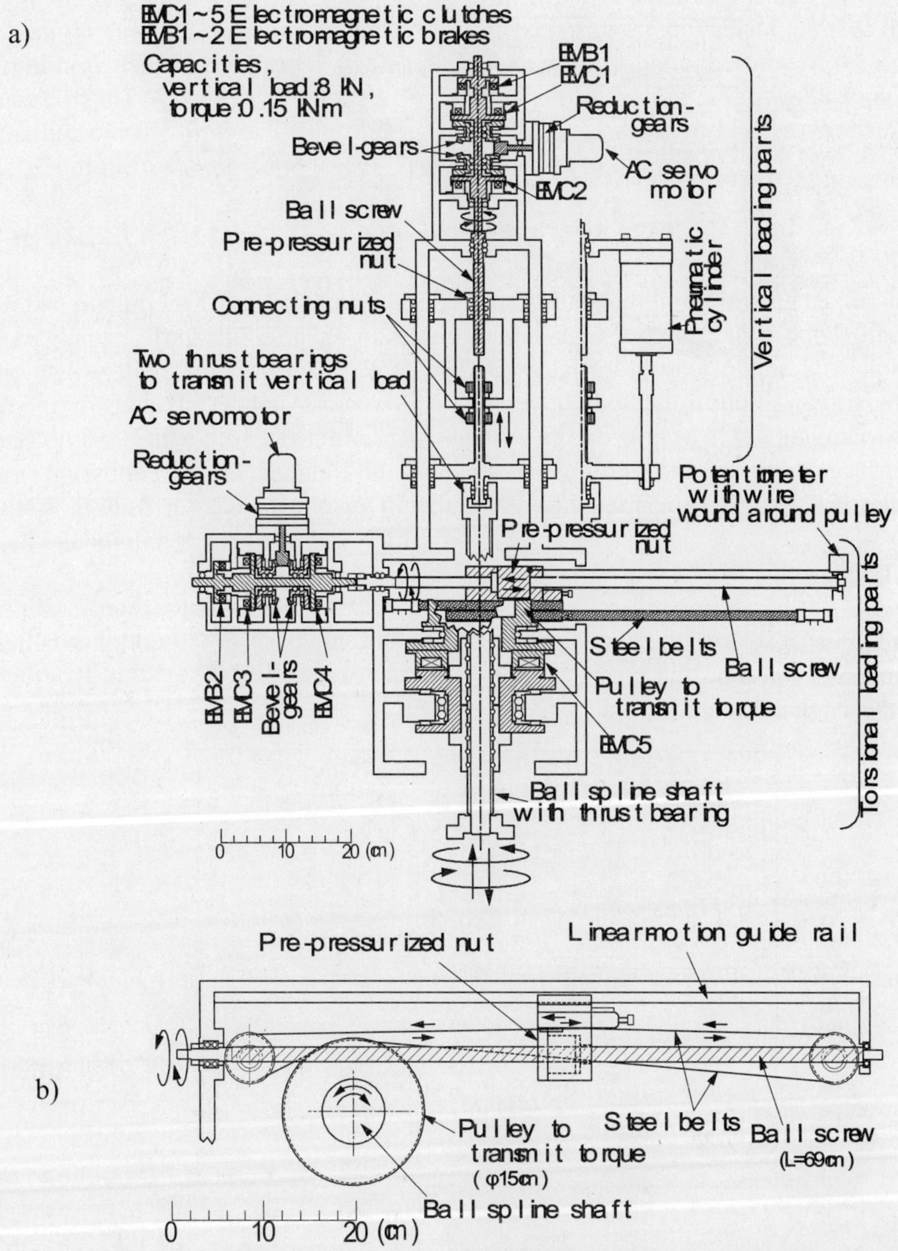

Fig. 2 loading device; a) general cross-section and b) plan view of torque-transmission part

In order to correction for the effects of membrane force on the measured shear stress, the theoretical values of the apparent shear stress τ_m that is mobilized by the forces in outer and inner membranes due to torsional deformations were evaluated as:

$$\tau_m = t_m E_m (r_o^3 + r_i^3)\theta / \{(r_o^3 - r_i^3)h\} \tag{1}$$

where θ is rotational angle of the top cap measured with a potentiometer shown in Fig. 2a; h is the height of the specimen; r_o and r_i are outer and inner diameters of the specimen; and t_m and E_m are thickness (=0.3 mm) and Young's modulus (=1492 kPa), respectively, of the membrane. In Fig. 3, these values are compared with those measured by filling water between the outer and inner membranes and shearing it under undrained condition. Due possibly to the difference in the deformation modes of the membranes between the assumption in the theory and the reality, some discrepancy was observed under large strain levels. Therefore, the measured relationship between τ_m and θ is employed for the correction of membrane force hereafter.

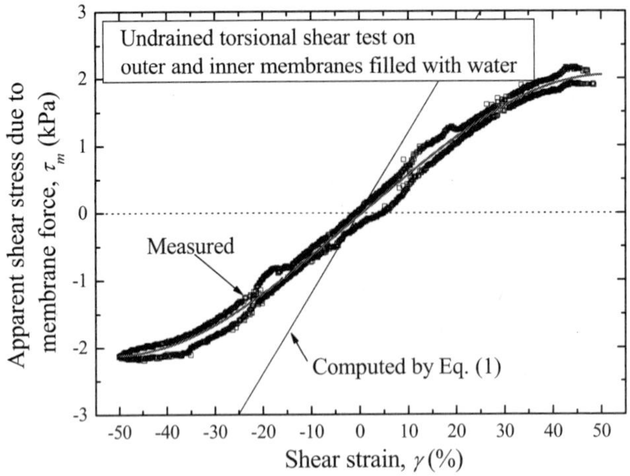

Fig. 3 Relationships between apparent shear stress due to membrane force and shear strain

4 TEST RESULTS

Typical test result on a loose specimen is shown in Fig. 4. As shown in Fig. 4a, by correcting for the effect of membrane force on the measured shear stress, reasonable stress strain relationships could be obtained even under large strain levels.

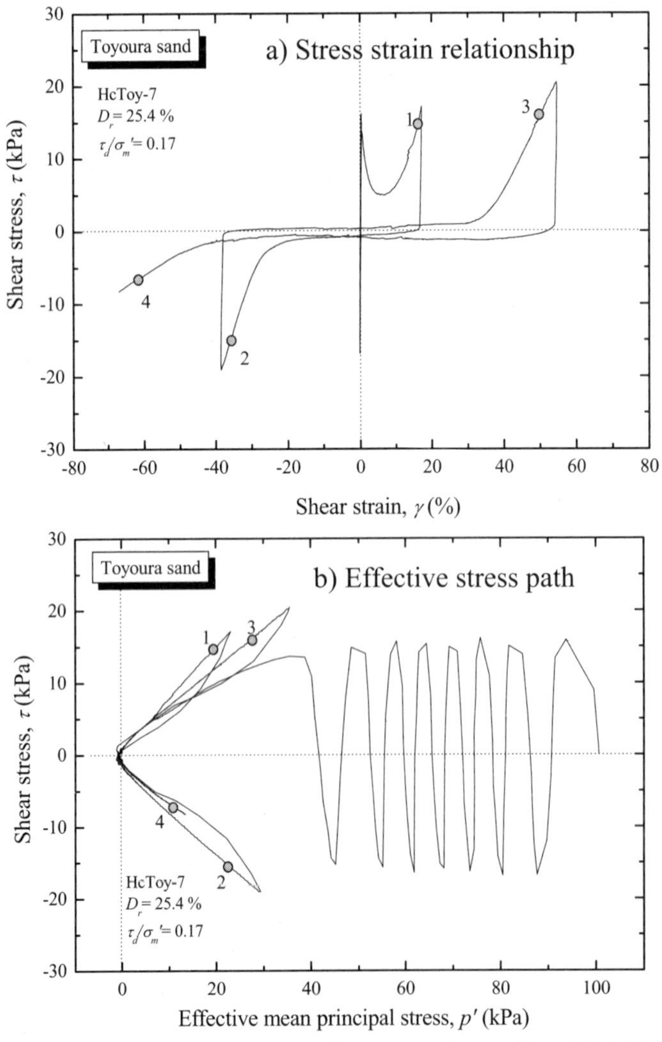

Fig. 4 Typical test result on loose specimen (Dr=25.4 %)

The specimen deformation at several states as numbered 1 through 4 in Fig. 4 is shown in Photo 1. At state 1 with an overall shear strain γ of about 15 %, the deformation was rather uniform except for the regions near the top cap and pedestal that are affected by the end restraint. At state 2 with the γ value of about -35 %, the outer membrane wrinkled at several locations, in particular in the upper part of the specimen. At state 3 with the γ value of about 60 %, the specimen deformation was localized in the upper part of the specimen. At state 4 with the γ value of about -65 %, the specimen was twisted extensively.

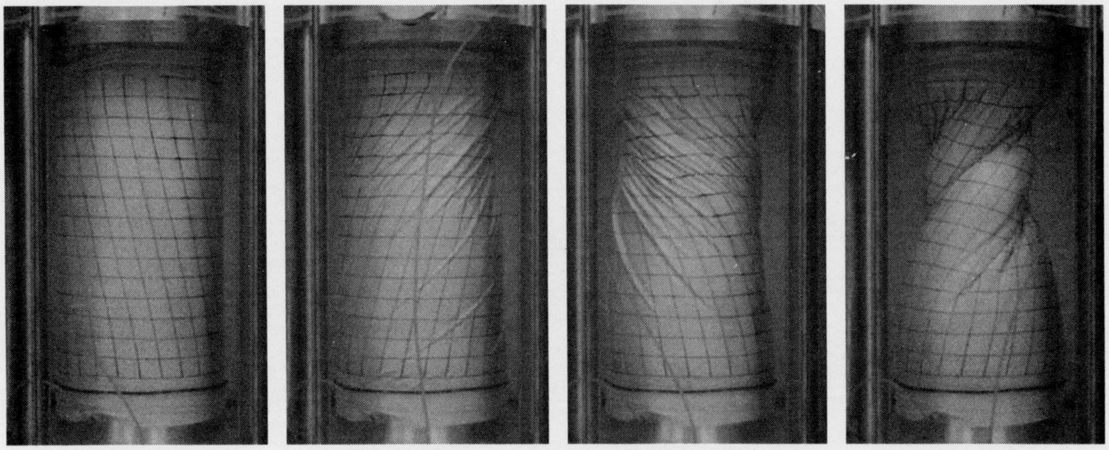

a) State 1 (γ=15%) b) State 2 (γ=-35%) c) State 3 (γ=50%) d) State 4 (γ=-65%)

Photo 1 Specimen deformation at states 1 through 4 shown in Fig. 4

Typical test result on a moderately dense specimen is shown in Fig. 5. Since the initiation of strain localization could not be clearly defined based on the visual observation of specimen deformation, it was defined based on the change in response of the deviator stress. For example, after passing state A shown in Fig. 5c, the amplitude of the cyclic change in the deviator stress decreased suddenly. It was accompanied with an increase in the increment of single amplitude of shear strain Δγ, as shown in Fig. 5d. It should be noted that, before reaching state A, the value of Δγ decreased with an increase in the number of cycles. Therefore, such state A was considered in the present study as the limiting state to initiate strain localization.

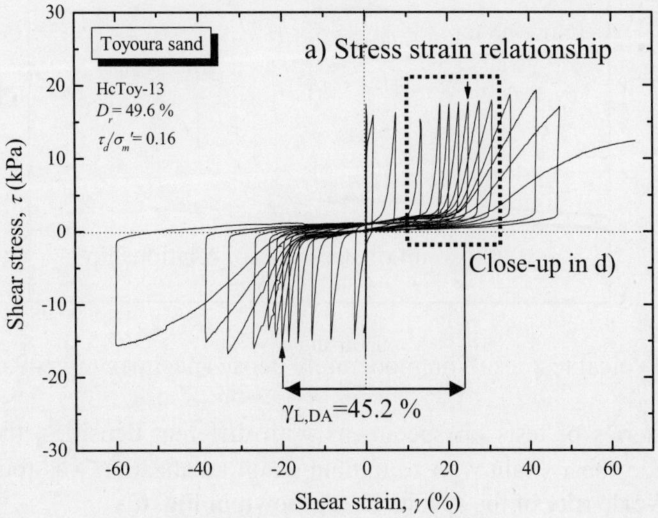

Fig. 5 Typical test result on moderately dense specimen (Dr=49.6 %) (to be continued)

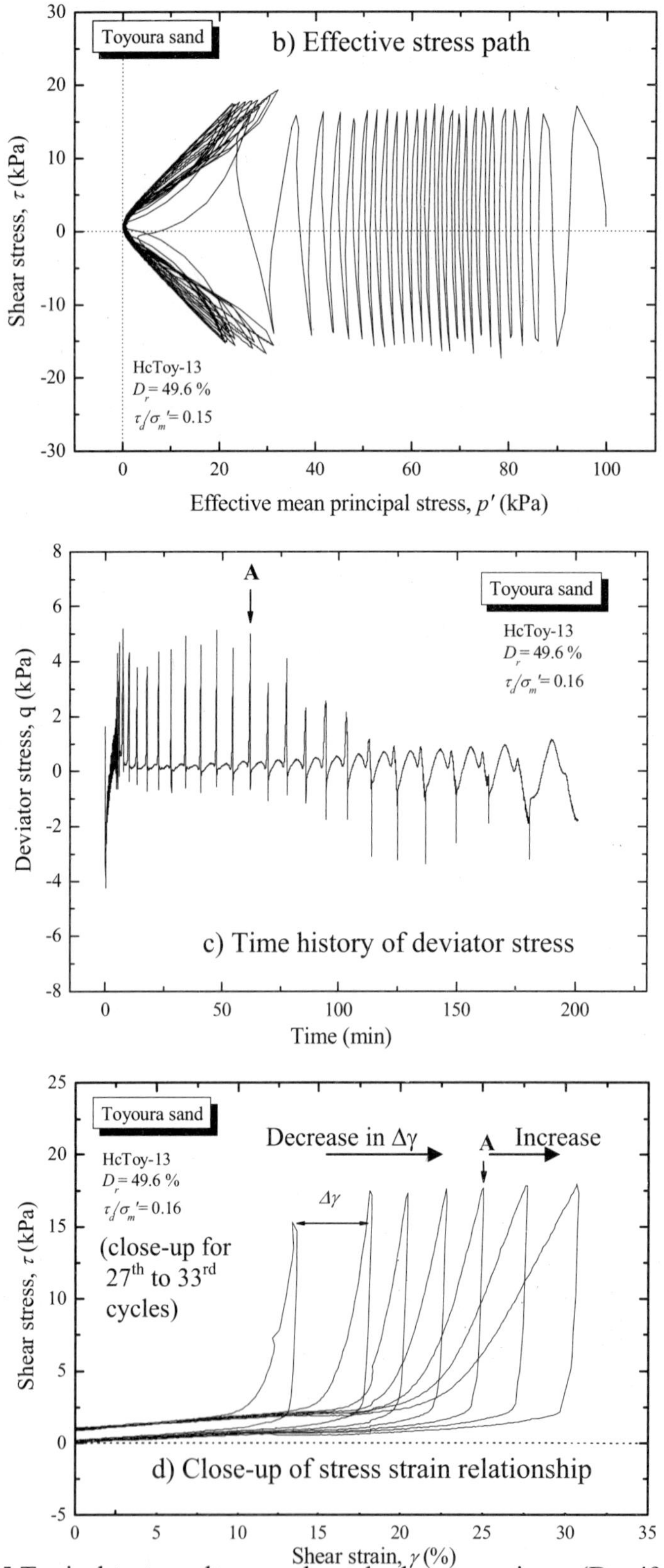

Fig. 5 Typical test result on moderately dense specimen (Dr=49.6 %)

As a result from the series of tests on specimens with different densities, the limiting value of overall double amplitude shear strain $\gamma_{L,DA}$ to initiate strain localization was found to increase with a decrease in the relative density of the specimen, as shown in Fig. 6.

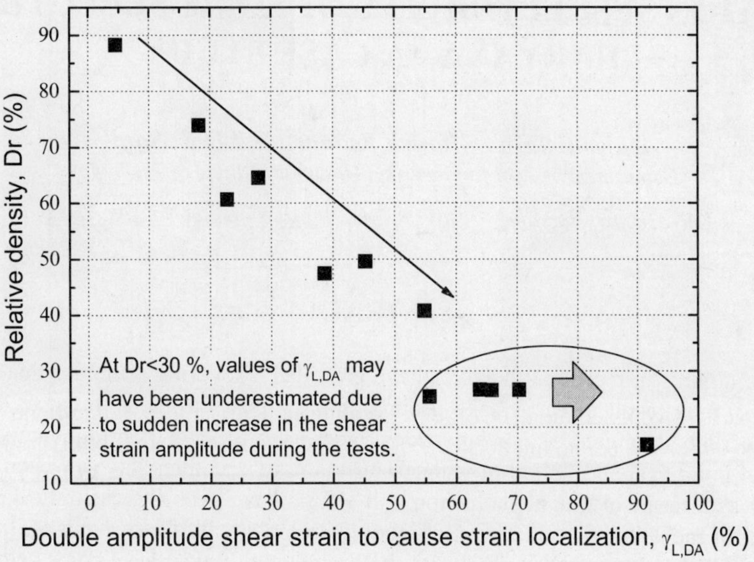

Fig. 6 Relationship between limiting value of overall shear strain and relative density of specimen

5 CONCLUSIONS

The following conclusions were drawn from the present study:

1. Using the torque loading device that was modified for enlarging the effective range of torsional displacement, undrained cyclic torsional shear tests could be conducted on saturated Toyoura sand having different densities up to double amplitude shear strain of about 100 %.
2. By correcting for the effect of membrane force on the measured shear stress, reasonable stress strain relationships could be obtained even under large strain levels.
3. After exceeding a certain level of overall strain, localization of specimen deformation was observed. Initiation of such strain localization was associated with the changes in the cyclic amplitude of deviator stress and the increment of shear strain
4. The limiting value of overall shear strain to initiate strain localization was found to increase with a decrease in the relative density of the specimen.

REFERENCES

1. Hamada, M., O'Rourke, T.D. and Yoshida, N. (1993). "Liquefaction-induced large ground displacement, Performance of Ground and Soil Structures during Earthquakes," 13th ICSMFE, JGS, 93-108.
2. HongNam, N. and Koseki, J. (2005). "Quasi-elastic deformation properties of Toyoura sand in cyclic triaxial and torsional loadings," Soils and Foundations, **45**(5), 19-38.
3. Ishihara, K., Yasuda, S. and Nagase, H. (1996). "Soil characteristics and ground damage," Soils and Foundations, Special Issue on Geotechnical Aspects of the January 17 1995 Hyogoken-Nambe Earthquake, 109-118.
4. Kiyota, T. (2007). Liquefaction strength and small strain properties of in-situ frozen and reconstituted sandy soils, PhD thesis, the University of Tokyo (in Japanese).

LIQUEFACTION SUSCEPTIBILITY OF SOME SELECTED SITES IN HARYANA - A CASE STUDY

Ashwani Jain[1], Manish Kumar[2] and D.K. Soni[3]

[1 2 3]Department of Civil Engineering, National Institute of Technology,

Kurukshetra (Haryana) – 136119, E-mail: ashwani.jain66@yahoo.com

ABSTRACT

Catastrophic failures in recent earthquakes have provided a sobering reminder that liquefaction of sandy soils as a result of earthquake ground shaking poses a major threat to the safety of Civil Engineering Structures. Major landslides, lateral movements of bridge supports, settling and tilting of buildings, and failure of waterfront retaining structures have all been observed in recent years as a result of this phenomenon and efforts have been increasingly directed to the development of methods of evaluating the liquefaction potential of soil deposits. In the present study, an attempt has been made to evaluate the liquefaction potential of some selected sites in Haryana using standard penetration resistance value 'N'. These sites have been selected keeping in mind their increased susceptibility to liquefaction due to favourable soil & water table conditions. According to IS: 1893 (Part I): 2002, Haryana lies in the Seismic Zone IV, which is likely to be subjected to an earthquake of maximum magnitude of 7.5 on Richter's Scale and the same magnitude value has been adopted to evaluate the liquefaction susceptibility of the sites. The standard penetration test value and the other relevant data have been collected from the field studies conducted at these sites for consultancy purpose by Geotechnical Engineering Division of N.I.T. Kurukshetra. Seed et al. (1985) approach has been used for evaluation of liquefaction potential of the sites.

Keywords: earthquake, liquefaction, standard penetration resistance, cyclic shear stress.

1 INTRODUCTION

Geotechnical Earthquake Engineering has advanced significantly, but earthquakes continue to cause destruction of life and property year after year. Although, the total duration of earthquakes during the 20th century was less than an hour but the damages were extensive; more than 2 lakh people lost their lives during these earthquakes. One of the major causes of earthquake-induced damages to structures has been liquefaction of saturated loose cohesionless soil deposits. It is manifested either in the form of sand boils and mud spouts at the ground surface formed by seepage of water, or in some cases by the development of quicksand like condition where the liquefaction occurs. Building may sink substantially into the ground or tilt excessively; lightweight structures and foundations may get displaced laterally causing structural failures. Recognition of liquefaction as a significant problem in Geotechnical Engineering was led primarily by major failures observed during Niigata and Alaska earthquakes of 1964, but undoubtedly, liquefaction is best illustrated in Niigata earthquake (Seed et al. 1967).

In the present study, an attempt has been made to evaluate the liquefaction potential of some selected sites in Haryana using standard penetration resistance value 'N'. The standard penetration test value and the other relevant data have been collected from the field studies conducted at these sites for consultancy purpose by Geotechnical Engineering Division of NIT Kurukshetra. Seed et al. (1985) approach has been used for evaluation of liquefaction potential of sites.

2 SOIL LIQUEFACTION

When a suitably intense earthquake shakes a deposit of loose, saturated, coarse-grained soil, the grain structure is triggered towards more compact packing. It is unable to do so because pore water does not have sufficient time during the earthquake to drain off. The pore water pressure then shoots up as the incompressible water takes up the applied stress and the effective stress approaches zero - the deposit liquefies with zero effective stress and therefore with no shear strength, disastrous consequences follow liquefaction.

The soil deposit that is susceptible to liquefaction must be loose, saturated and coarse grained. A loose state is essential since soil in that state suffers decrease in volume during drained shear and during undrained loading this contracting nature increases the pore water pressure. The soil should be fully saturated, i.e. the soil should be below ground water table. If the soil deposit is above the water table, then the vibrations from the earthquake will succeed in increasing the soil density because there is no time lag for air to escape, and finally the soil must be coarse grained, and this can be interpreted as implying that clays do not liquefy because vibrations are not effective in bringing clay particles together into closer packing and the shear strength of clays does not become zero when the effective stress becomes zero. Liquefaction only occurs in saturated soils; its effect is most commonly observed in low-lying area near bodies of water such as rivers, lakes, bays and oceans. Sand most susceptible to liquefaction has coefficient of permeability in the range of 1×10^{-5} to 1×10^{-3} m/s. Sand deposited with rounded and sub-rounded grains are more susceptible to liquefaction than those with angular grains. Liquefaction is more likely for clean sand having a relative density lesser than about 40%, the susceptibility with in the range of 40-60 % depends on the confining pressure and sands with relative density more than 60 % are less likely to liquefy.

3 STANDARD PENETRATION TEST

One of the oldest and most common in-situ tests is the standard penetration test or SPT. It was developed in the late 1920s and has been used extensively in North and South America, the United Kingdom, Japan, India and elsewhere. Because of this long record of experience, the SPT is well established in engineering practice. It is performed inside exploratory boring using inexpensive and readily available equipment, and thus adds little cost to a site characterization programme.

Although the SPT also is plagued by many problems that affect its accuracy and reproducibility, it probably will continue to be used for the foreseeable future, primarily because of its low cost. However, it is partially being replaced by other test methods, especially on larger and more critical projects.

Seed et al. (1985) recommended the following additional criteria for conducting standard penetration test:

- Use the rotary wash method to create a boring that has a diameter between 200 and 250 mm. The drill bit should provide an upward deflection of the drilling mud.
- If the sampler is made to accommodate liners, then these liners should be used so that the inside diameter is 35 mm.
- Use A or AW size drill rods for depths less than 15 m, and N or NW size for greater depths.
- Use a hammer that has an efficiency of 60 percent.
- Apply the hammer blows at a rate of 30 to 40 per minute.

Inspite of certain disadvantages, the SPT does have at least three important advantages over other in-situ test methods. First, it obtains a sample of the soil being tested. This permits direct soil classification. Most of the other methods do not include sample recovery, so soil classification must be based on conventional sampling from nearby borings and on correlations between the test results and soil type. Second, it is very fast and inexpensive because it is performed in borings that would have been drilled anyway. Finally, nearly all drill rigs used for soil exploration are equipped to Perform this test, whereas other in-situ tests require specialized equipment that may not be readily available.

4 EVALUATION OF LIQUEFACTION POTENTIAL (Seed et al. Method, 1985)

According to Seed et.al (1971, 1983), the average equivalent uniform shear stress τ_{av} is about 65 percent of the maximum shear stress τ_{max}. Therefore

$$\tau_{av} = 0.65\ .\gamma h.\ r_{d.}\ a_{max}/g \quad \ldots\ldots 4.1$$

Where,

$a_{max}/g = 0.24$ for zone IV.

$r_d =$ Stress reduction factor = (1-0.015h).

h = Depth in meters below the ground surface.

γ = Unit weight of soil.

The procedure of locating liquefaction zone can be summarized in following steps:

- Establish the design earthquake, and obtain peak ground acceleration a_{max}. Also obtain number of significant cycles N_s corresponding to earthquake magnitude.
- Using Equation 4.1, determine τ_{av} at depth 'h' below ground surface.
- Determine the value of standard penetration resistance value $(N_1)_{60}$ at depth 'h' below ground surface. Using the equation 4.2, obtain corrected $(N_1)_{60}$ value after applying overburden correction to N (Liao et al. 1986, Skempton 1986).

$$(N_1)_{60} = C_{N.}ER_m.N/60 \quad \ldots\ldots 4.2$$

Where,

C_N= Coefficient for overburden pressure (from Fig. 4.1)

ER_m= Actual energy efficiency delivered to the drill rod.

= 45, for donut hammer with cathead release mechanism (Kovacs et al. 1982).

N = Observed SPT value.

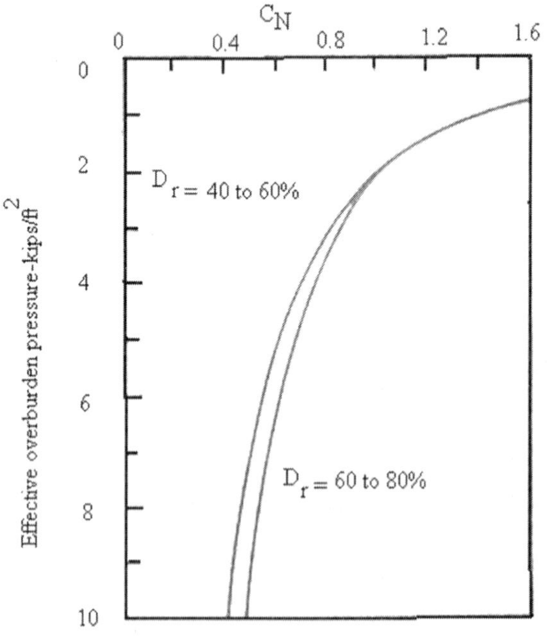

FIG. 4.1 CHART FOR VALUES OF C_N

- Using the Figure 4.2 (a) & (b), determine stress ratio $\frac{\tau_h}{\sigma_v}$ for different value of $(N_1)_{60}$.

Multiplying $\frac{\tau_h}{\sigma_v}$ with effective stress at depth 'h' below ground surface, one can obtain the value of shear stress τ_h required for causing liquefaction.

- At depth h, liquefaction will occur if, $\tau_{av} > \tau_h$.
- Repeat steps for other values of 'h' to locate the zone of liquefaction.

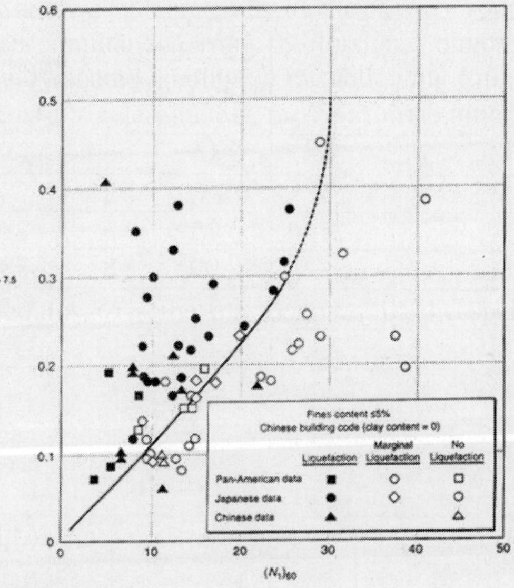

FIG. 4.2(a) RELATIONSHIP BETWEEN CYCLIC STRESS RATIO CAUSING LIQUEFACTION AND $(N_1)_{60}$ VALUES FOR CLEAN SAND IN M=7.5 EARTHQUAKES (after Seed et al., 1985)

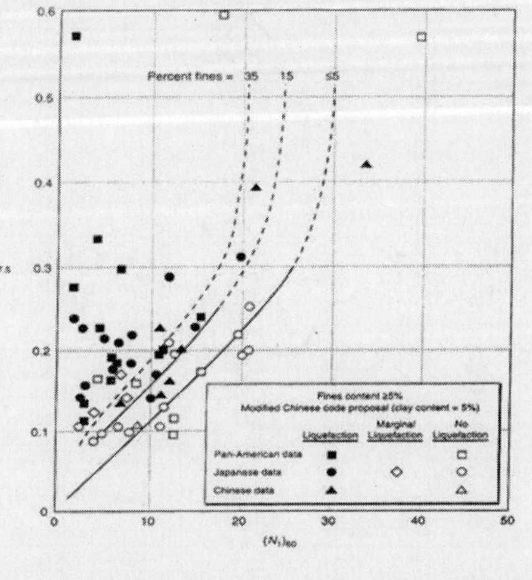

FIG. 4.2(b) RELATIONSHIP BETWEEN CYCLIC STRESS RATIO CAUSING LIQUEFACTION AND $(N_1)_{60}$ VALUES IN M = 7.5 EARTHQUAKES (after Seed et al., 1985)

5 INTERPRETATION OF DATA

An attempt has been made to evaluate the liquefaction potential of some selected sites in Haryana using standard penetration resistance value. The sample calculations for Site 1 have been given and for rest of the sites only bore hole data and soil properties have been reported with interpreted results as per Seed et al. approach (1985) (reported elsewhere, Kumar 2007).

SITE 1. RD 171499 OF DELHI CARRIER CHANNEL (PANIPAT DIVISION)

Bore Hole Data & Soil Properties

S.No.	Depth (m)	IS Classification of soil	ϕ in degree	D_{50} (mm)	N-Value	D_R (%)	Unit weight (γ) (kN/m^3)	Remarks
1	0.75	ML	25	0.16	8	28	20	
2	1.5	ML	25	0.155	8	28	20	W.T at 1.5 m
3	3	ML	25	0.15	8	28	20	
4	4.5	ML	31	0.09	8	28	20.2	
5	6	CL-ML	31	0.15	13	39.5	20.2	
6	7.5	ML	33	0.14	13	39.5	20.1	

Computation of τ_{av}

S.No.	Depth (m)	σ_v (kN/m^2)	$r_d = (1-0.015h)$	τ_{av} (kN/m^2)
1	0.75	15	0.98875	2.3136
2	1.5	30	0.9775	4.5747
3	3.0	60	0.955	8.9388
4	4.5	90.3	0.9325	13.1359
5	6	120.6	0.910	17.1203
6	7.5	150.75	0.8875	20.8713

Computation of τ_h

S.No	Depth (m)	N-Value	D_R (%)	C_N	$\overline{\sigma}_v$ (kN/m^2)	$(N_1)_{60} = C_N.ER_m.N/60$	Stress ratio $(\tau_h/\overline{\sigma}_v)$	τ_h (kN/m^2)
1	0.75	8	28	1.6	15	9	0.165	2.475
2	1.5	8	28	1.5	30	9	0.165	4.95
3	3.0	8	28	1.3	45	7	0.145	6.525
4	4.5	8	28	1.25	60.3	7	0.145	8.7465
5	6.0	13	39.5	1.1	75.6	10	0.175	13.23
6	7.5	13	39.5	1.05	95.75	10	0.175	15.8812

Result: Liquefaction will occur below 3.0 m.

SITE 2. RD190818 AT BISHANGARH VILLAGE (AMBALA DIVISION)

Bore Hole Data & Soil Properties

Sl. No.	Depth (m)	IS Classification of soil	φ in degree	D_{50} (mm)	N-Value	D_R (%)	Unit weight(γ) (kN/m^3)	Remarks
1	0.75	CL-ML	29	0.18	3	11	19.0	
2	1.5	CL-ML	29	0.19	3	11	19.0	
3	3.0	ML	29	0.165	3	11	19.0	W.T.at 3m.
4	4.5	ML	29	0.135	12	38	19.0	
5	6.0	ML	31	0.155	12	38	19.5	
6	7.5	ML	31	0.185	9	31	19.5	
7	9.0	CL-ML	31	0.19	9	31	19.5	

Result: Liquefaction will occur below 3.0 m.

SITE 3. JALMANA KURLAN ROAD (ASSANDH DIVISION)

Bore Hole Data & Soil Properties

S.No.	Depth (m)	IS Classification of soil	φ in degree	D_{50} (mm)	N-Value	D_R (%)	Unit weight (γ) (kN/m^3)	Remarks
1	0.75	ML	27	0.16	3	11	20.2	
2	1.5	ML	27	0.16	3	11	20.2	W.T at 1.5 m
3	2.25	ML	27	0.145	5	18	20.2	
4	3.0	ML	30	0.175	5	18	20.5	
5	4.5	ML	30	0.185	10	35	20.5	
6	6.0	ML	30	0.175	11	36.5	20.5	
7	7.5	ML	30	0.185	13	39.5	20.5	

Result: Liquefaction will occur below 0.75 m.

SITE 4. RD 3477 OF LINK DRAIN CROSSING UNDER SIRSA BRANCH (KAITHAL DIVISION)

Bore Hole Data & Soil Properties

S.No.	Depth (m)	IS Classification of soil	φ in degree	D_{50} (mm)	N-Value	D_R (%)	Unit weight (γ) (kN/m^3)	Remarks
1	1.5	ML	28	0.155	5	18	18.8	
2	3	ML	28	0.165	5	18	18.8	
3	4.5	ML	30	0.175	10	35	18.8	W.T at 4.5 m
4	6	ML	30	0.19	10	35	19.1	
5	7.5	ML	30	0.175	11	36.5	19.1	
6	9.0	ML	30	0.16	11	36.5	19.0	
7	10.5	ML	31	0.17	10	35	19.0	

Result: Liquefaction will occur below 1.5 m.

SITE 5. ABUBSHEHAR (SIRSA DIVISION)

Bore Hole Data & Soil Properties

S.No.	Depth (m)	IS Classification of soil	ϕ in degree	D_{50} (mm)	N-Value	D_R (%)	Unit weight (γ) (kN/m^3)	Remarks
1	0.75	ML	27	0.155	3	11	18.7	
2	1.5	ML	27	0.135	3	11	18.7	
3	2.25	ML	27	0.165	5	18	18.7	W.T at 2.25 m
4	3.0	ML	30	0.165	5	18	18.7	
5	3.75	ML	30	0.20	8	28	18.7	
6	4.5	ML	30	0.195	10	35	18.7	
7	6.0	ML	30	0.170	10	35	18.7	

Result: Liquefaction will occur below 2.25 m.

6 CONCLUSION

In the present study, evaluation the liquefaction potential of some selected sites in Haryana has been done using standard penetration resistance value 'N'. These sites have been selected keeping in my mind their increased susceptibility to liquefaction due to favourable soil and water table conditions. Seed et al. (1985) approach has been used for evaluation of liquefaction potential at a site. It should be noted that although the results have been presented for 7.5-magnitude earthquake, these could be easily extended to locations of other earthquake magnitudes. It is hoped that the results presented will serve to clarify questions and concerns, which have been raised in regard to the different practices used to measure SPT N-values for use in liquefaction correlations. Clearly the results could readily be converted to any given energy standard, but the $(N_1)_{60}$ value seems to be a reasonable standard for use in many countries at the present time. However, it should be noted that the standard penetration test cannot be performed at all depths (say greater than 30.5 m or through large depths of water) and in soils containing a significant proportion of gravel particles. Thus, it is desirable that it be supplemented by other in-situ test methods, which can also be correlated with soil liquefaction potential. In many cases, the static cone test, which can be performed more rapidly and more continuously, may provide a good means for evaluating liquefaction potential, especially if it is correlated on a site dependent basis with SPT results. However, this procedure is limited also to sands and silty sands. In dealing with soils containing large particles, or in difficult environments, other in-situ characteristics, such as the shear wave velocity or the electrical characteristics of the soil may provide a more suitable means for assessment of liquefaction potential.

REFERENCES

1. IS: 1893 (Part I) (2002), "Criteria for Earthquake Resistant Design of Structures – Part 1: General Provisions and Buildings", Bureau of Indian Standards, New Delhi.
2. Kovacs, W. D. and Salamane, L. A. (1982), "SPT Hammer Energy Measurement", Journal Geotechnical Engineering Division. ASCE, **108** (GT4), 599-620.
3. Kumar, Manish (2007), "Evaluation of Liquefaction Potential using Standard Penetration Resistance – A Case Study", M.Tech. Dissertation, N I T Kurukshetra.
4. Liao S. S., Whitman, R.V. (1986), "Overburden Correction Factors for SPT in Sand", Journal of Geotechnical Engineering, ASCE, **112** (3), 373-377.
5. Poulos, S.J, Castro, G., and France, J.W. (1985), "Liquefaction Evaluation Procedure", Journal of Geotechnical Engineering, ASCE, **111** (6), 772- 792.

6. Seed, H.B., and Idriss I.M, (1967), "Analysis of Soil Liquefaction: Niigata Earthquake", Journal Soil Mechanics and Foundation Division, ASCE, **93** (SM3), 83-108.
7. Seed, H.B.and Idriss I.M., (1971), "Simplified Procedure for Evaluating Soil Liquefaction Potential", Journal Soil Mechanics and Foundation Division. ASCE, **97** (SM9), 1249-1273.
8. Seed, H.B., and Peacock, W.H., (1971), "Test Procedures for Measuring Soil Liquefaction Characteristics", Journal of the Soil Mechanics and Foundation Division, ASCE, **97** (SM8), 1099-1119.
9. Seed, H.B, Idriss, I.M., and Arango, I. (1983), "Evaluation of Liquefaction Potential using Field Performance Data", Journal of the Geotechnical Engineering Division, ASCE, **109** (3), 458-482.
10. Seed, H.B, Tokimatsu. K, Harder, L. F. and Chung R.M. (1985), "Influence of SPT Procedures in Soil Liquefaction Resistance Evaluation", Journal of the Geotechnical Engineering Division, ASCE, **111**(12), 1425-1445.
11. Skempton, A.W. (1986), "Standard Penetration Test Procedures and the Effects in Sands of Overburden Pressure, Relative Density, Particle Size, Ageing and Overconsolidation", Geotechnique, **36** (3), 425-447.

STRESS DISTRIBUTIONS IN EARTH AND ROCKFILL DAMS

B.K. Maheshwari[1] and P. Anuradha[2]
[1]Assistant Professor, Dept. of Earthquake Engineering, IIT Roorkee, e-mail: bkmahfeq@iitr.ernet.in
[2]M.Tech. Student, Dept. of Earthquake Engineering, IIT Roorkee, e-mail: anurppeq@iitr.ernet.in

ABSTRACT

In a conventional method of slope stability analysis of dams a sliding surface is assumed and a quantitative estimate of the factor of safety is obtained by examining the equilibrium conditions at incipient failure comprising the strength of the fill material on the entire surface. These procedures indicate the factor of safety of the dam with respect to instability but provide no information regarding the deformations of the dams. In recent years, there has been a growing realization of the need to determine the stress distribution and deformations inside the dam section.

In this paper, elastic stress distributions, inside the dam section have been obtained considering plane stress condition in a finite element model. The computation technique has been validated comparing the results with those available in literature. An earth and rock fill dam without and with foundation soil are considered for further analysis. Effects of hydrostatic and hydrodynamic forces (along with self weight of the dam) are investigated on the horizontal and vertical stresses along the centre line and the base of the dam. Finite element software ANSYS 10.0 has been used for all the computation.

Keywords: Stress Distribution, Earth and Rockfill Dam, Finite Element Method

1 INTRODUCTION

Dams retaining huge quantity of water play an important role in the modern development of society. Their safe performance is vital paricularly during adverse natural conditions such as earthquakes. The failure of a dam may lead to disastrous consequences (Newmark 1965, Kramer 1996). Earth and rockfill dams are the most ancient type of dams, as they can be built with the natural construction materials with a minimum cost. The main structural element of rockfill dam is quarry run rock dumped or placed in layers. This material is highly pervious and water tightness is obtained either by a membrane of manufactured materials or by an earth core. Rockfill dams have high inherent stability as there is very few recorded case of sliding failure of rockfill dam. They have high degree of resistance due to their damping and flexibility (Singh and Sharma 1976). But the rockfill dam founded on solid rock when subjected to earthquake (due to the extremely high stiffness of foundation) behaves like a rigid structure with motion at the base equal to the free field motion. On the other hand, the same structures would respond differently if supported on as soft soil deposit. Also the behavior of dam under hydrodynamic pressure (due to earthquake) is different (Sarkar 2006).The inclusion of effect of hydrodynamic pressure in the study of seismic response of dams is very important and requires a through understanding.

In the present study, considering plane stress conditions a two-dimensional finite element model of a dam has been created and elastic stress distribution inside the dam section has been obtained. ANSYS software has been used for computation. The computation technique has been validated comparing the stress distributions due to self weight of the dam with the results available in literature. An earth and rock fill dam without and with foundation soil are considered for further analysis. It is assumed to respond as a two dimensional system under plane stress conditions. The two dimensional assumption is also extended to model the compressible reservoir water, a body impounded upstream of the dam. The dam is analyzed for three loading cases i.e., Gravity (G); Gravity and Hydrostatic (G + H_s); Gravity, Hydrostatic and Hydrodynamic loading (G +H_s + H_d). Hydrodynamic loading considered is due to seismic forces. It was observed that there is a significant effect of hydrostatic and hydrodynamic forces on elastic stress distribution.

2 VALIDATION OF COMPUTATION TECHNIQUE

As a rigorous method (finite element) is used its validation is important. Acar et al. (1988) analyzed an embankment (Fig. 1a) using Structural Analysis Program (SAP), in this present study the same model (Fig. 1b) is analyzed using ANSYS-10.0. Nine-nodded plane finite element is considered in the former while eight-nodded finite element is used in the latter. In present study, the boundaries are selected at 12 times the loading width (R). The nodes along the centre line of the embankment and the rightmost boundary are restrained to horizontal displacements, while the nodes at the bottom are restrained to vertical displacements, simulating a smooth base. Foundation is assumed weightless in both the analyses.

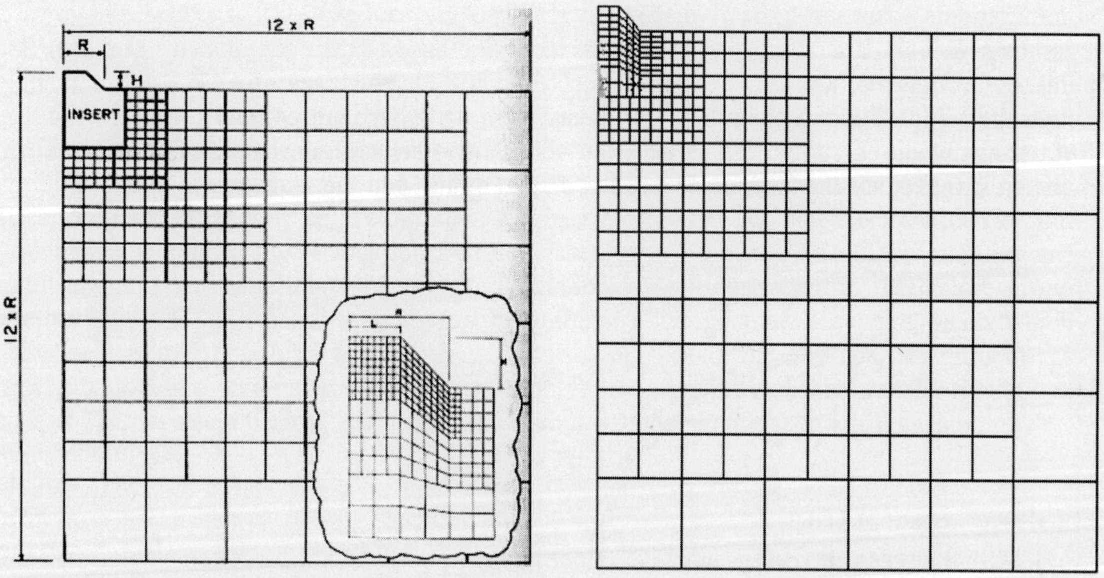

(a) Acar et al. (1988) (b) Present Study

Fig. 1: Finite Element Discretization of Embankment

In following sections, stresses along symmetry axis of embankment and base of embankment obtained from the present study are compared with those presented by Acar et al. (1988) and found to be in good agreement, verifying the computation technique.

2.1 Stresses along Symmetry Axis of Embankment

Fig. 2 shows the comparison of vertical stresses along the symmetry axis of the embankment obtained from ANSYS and those presented by Acar et al. (1988) using structural analysis program (SAP-8.0). The results are presented in dimensionless form; stresses are normalized with respect to unit weight (γ) of soil and depth Y is normalized with depth (H) of the embankment. In Fig. 2a results are shown by Acar et al. (1988) i.e., lines a, b, c as well as by Perloff et al. (1967). The result obtained from the present study is compared with line c. It can be observed from Fig. 2 that vertical stresses from the present study are in very good agreement with those obtained by Acar et al. (1988). Similar verification has been performed for horizontal stresses also (Anuradha 2007).

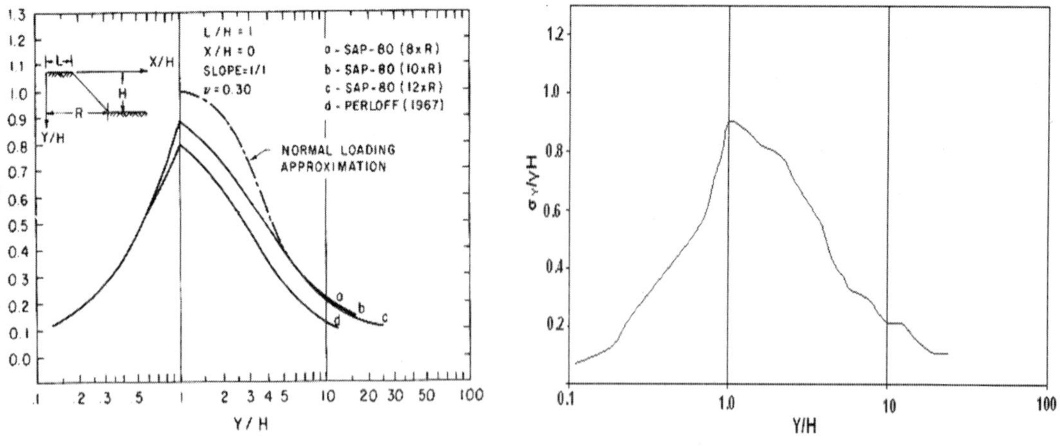

(a) Acar et al. (1988) (b) Present Study

Fig. 2: Comparison of Vertical Stresses along Symmetry Axis of Embankment

2.2 Stresses along Symmetry Base of Embankment

Fig. 3 shows the comparison of vertical stresses along the base of the embankment. Fig. 3(a) also contains the results of Verruijt (1969), Perloff (1967) and FEM analysis. The present study compares the vertical stresses along base of embankment with those presented by Acar et al. (1988) - FEM. It can be observed that the variation of vertical stresses with horizontal distance obtained from the presented study show the same trend as shown by FEM in Fig. 3a.

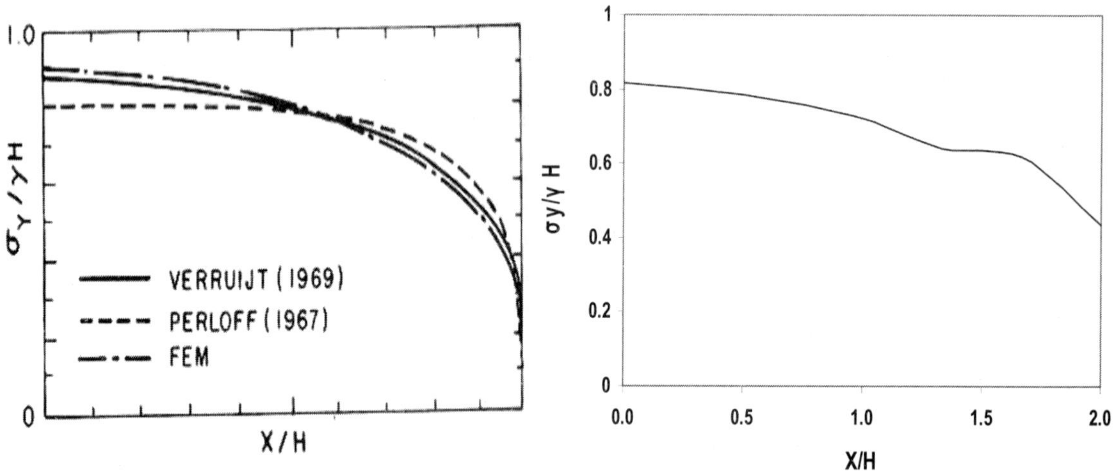

(a) Acar et al. (1988) (b) Present Study

Fig. 3: Comparison of Vertical Stresses along Base of Embankment

3 MODEL AND INPUT DATA CONSIDERED FOR ANALYSIS

A 2D finite element model of earth and rock fill dam is considered. The cross-sectional details of the dam is as as follows (Fig. 4):

Width of dam	= 650 m
Height of dam	= 150 m
Maximum water level	= 147 m
Depth of soil considered below ground level	= 650 m

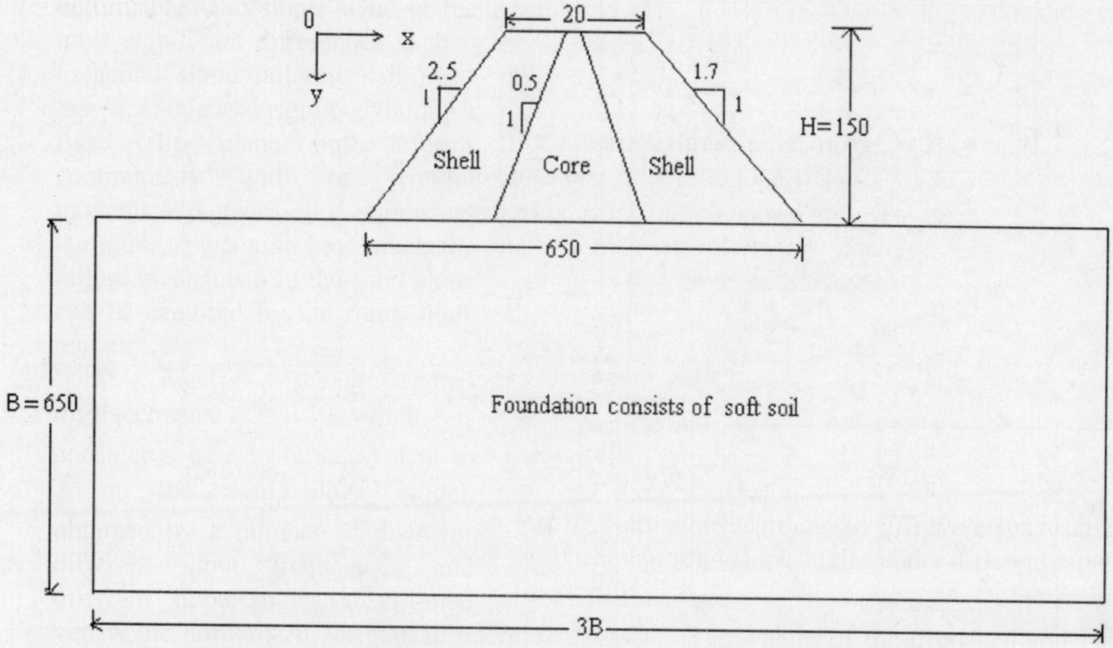

Fig. 4: Design cross section of Dam with Foundation (all dimensions are in meters - not to scale)

A four nodded quadrilateral element having two degrees of freedom at each node i.e., U_x, U_y is used for meshing the embankment. The base of the foundation soil is assumed to be restrained to vertical displacements and sides of the foundation soil is restrained to horizontal displacements (Acar et al. 1988).

Table 1: Material Properties of Dam and Foundation Soil

Material Properties	Unit Weight (kN /m^3)	Poisson's Ratio	Young's Modulus (MPa)
Pervious Shell Sandy Soil	17	0.3	25
Impervious Clay Core	19	0.4	55
Foundation consists of soft soil	Weightless	0.45	400

Behaviour of a structure is different depending on whether it is founded on solid rock or on soft soil. Therefore, the dam is analyzed considering without and with foundation soil for the following three cases of loading:

Case 1: Gravity (G) (Empty reservoir condition)
Case 2: Gravity (G) + Hydrostatic Pressure (H_s)
Case 3: Gravity (G) + Hydrostatic Pressure (H_s) + Hydrodynamic Pressure (H_d)

Fig. 5 shows how the hydrostatic and hydrodynamic forces are acting on the upstream face of the embankment. Earthquake forces shall not be included in stability analysis for the construction stage or for the reservoir empty condition as per IS 1893-1984. Due to horizontal acceleration of the foundation, there is an instantaneous hydrodynamic pressure (or suction) exerted against the dam in addition to hydrostatic forces. The direction of hydrodynamic force is opposite to the direction of earthquake acceleration. Based on the assumption that water is incompressible, the hydrodynamic pressure at a depth below the reservoir surface is determined based on IS: 1893-1984 as follows:

$$H_d = C_S \, \alpha_h \, \gamma_w \, h_w \qquad (1)$$

Where H_d = Hydro dynamic pressure in kN/m^2
C_S = Coefficient varies with shape and depth
α_h = Design horizontal seismic coefficient
γ_w = Unit weight of water kN/m^3
h_w = Depth of reservoir (m)

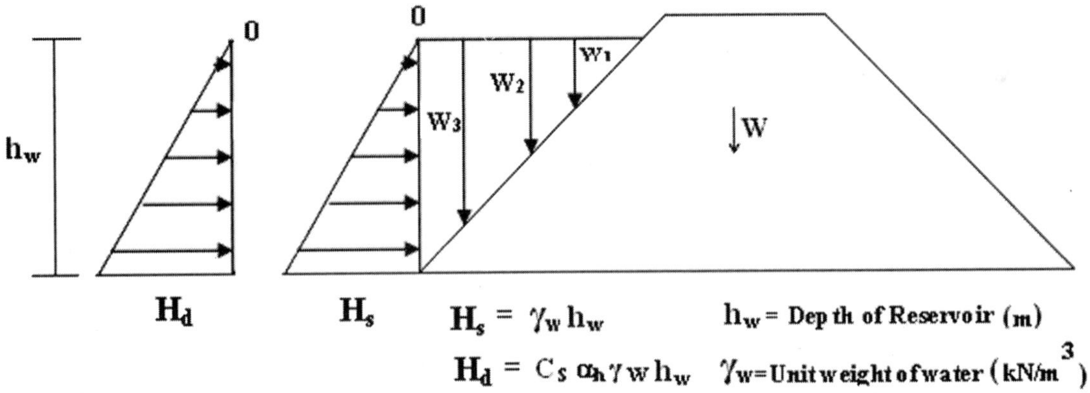

Fig. 5: Loading on the Dam (Self weight +Hydrostatic Pressure+ Hydrodynamic pressure)

4 RESULTS AND DISCUSSION

In the analysis a rockfill dam (Fig. 4) is considered without foundation soil and with foundation soil. Variation in stresses and displacements along centre line and along base of the dam are computed for all the three cases of loading. Few of the results are presented here, further detail can be found in Anuradha (2007).

4.1 Rock Fill Dam without Foundation Soil

Here it is assumed that dam is founded on solid rock hence foundation shown in Fig. 4 is not considered in the analysis. Thus depth considered in the analysis will be only the height of the dam. Variation in stresses along centerline and base of the dam is computed and presented here.

4.1.1 Stresses along Centreline of Dam

Fig. 6 shows variation in vertical stress along the centre line of rock fill dam resting on rigid base for all three loading cases. Results are presented in dimensionless form. Depth (Y) is normalized with the height of embankment (H) and the stress (σ_y) is normalized with respect to unit weight of soil (γ). It is observed that with the increasing depth, vertical stresses are also increases for all the three loading cases. It can be observed that the vertical stresses are higher for second and third case compared to gravity loading only. Thus hydrostatic loading increases stresses all along the depth which is further increased due to addition of hydrodynamic forces. Similar trend was observed in variation of horizontal stresses with respect to depth.

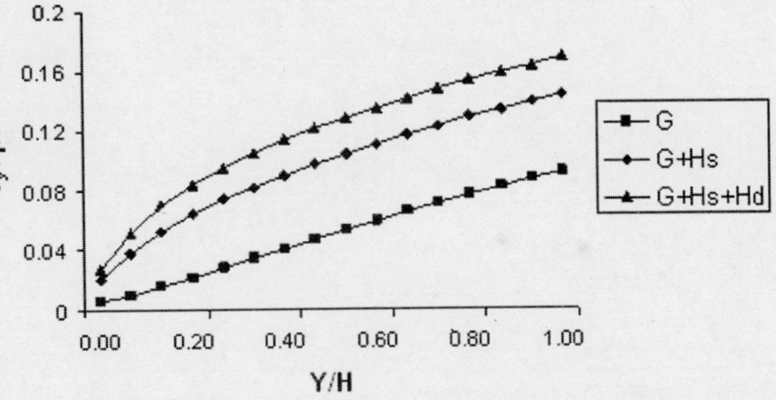

Fig. 6: Vertical Stresses along Centreline of Dam (without foundation)

4.1.2 Stresses along Base of Dam

Fig. 7 shows the variation in vertical stress along the base of dam for all three loading conditions. It can be observed that the vertical stress is maximum at the centre of embankment for all the three cases. In fact, stresses increases gradually from zero at both extreme ends and become maximum at the centre. Stresses are showing a symmetrical pattern. Effect of hydrostatic and hydrodynamic forces on the vertical stresses are insignificant. It appears that at the same depth influence of gravity loading on the vertical stresses is dominating.

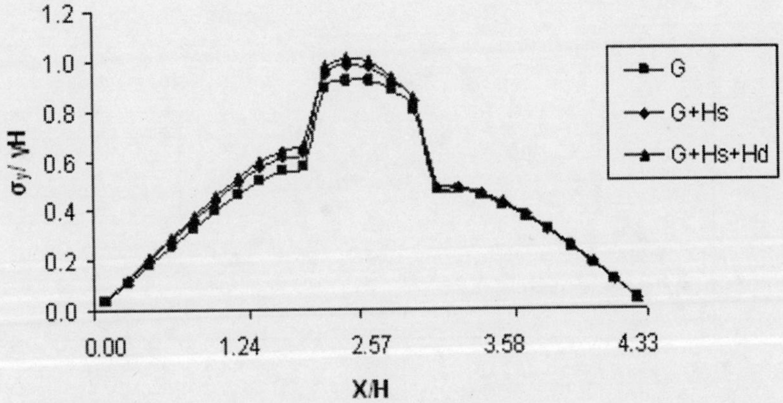

Fig. 7: Vertical Stresses along Base of Dam (without foundation)

4.2 Rock Fill Dam with Foundation Soil

Effect of presence of foundation soil on the stresses is investigated in this section. Results are presented considering foundation soil (Fig. 4).

4.2.1 Stresses along Centreline of Dam

Fig. 8 show the variation in vertical stress along the centre line of rock fill dam resting on soft soil for all three loading cases. The stresses initially increase with depth which may be attributed to the fact that confining pressure increases with depth. This increase of stresses takes place up to the base of dam and then these values are decreasing because of weightless foundation. These results are in good agreement with those presented by Acar et al. (1988). Similar trend was observed for horizontal stresses.

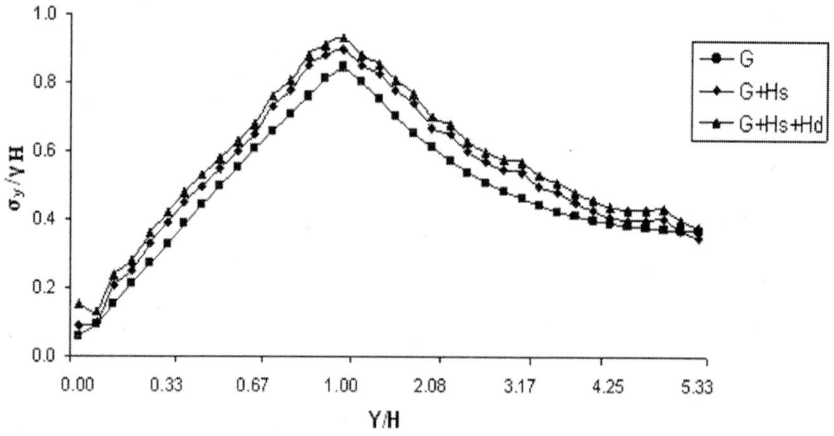

Fig. 8: Vertical Stresses along Centreline of the Dam (with foundation)

4.2.2 Stresses along Base of Dam

Fig. 9 show the variation in vertical stress along the base of dam, for different loading conditions considering foundation soil. It can be observed that the vertical stresses are maximum at the centre line of embankment for all the three cases. Stresses increases gradually from zero at both extremes and become maximum at the centre. Stresses are showing a symmetrical pattern. Effect of hydrostatic and hydrodynamic on the gravity loading is insignificant.

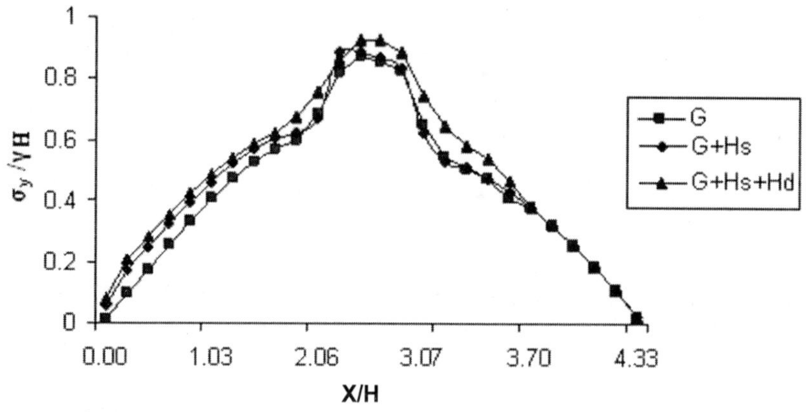

Fig. 9: Vertical Stresses along Base of the Dam (with foundation)

5 CONCLUSIONS

A simple but useful two-dimensional analysis using ANSYS software has been presented. The computation technique has been verified by comparing few of the results of the present study with those available in literature. Effects of foundation soil on the stresses are investigated. The following conclusions can be derived from the results presented in this paper:

1. When embankment is founded on a solid rock, there is significant effect of hydrostatic and hydrodynamic forces on the vertical stresses. Due to increase in loading, stresses increases all along the depth. However, stresses along the base of the embankment is not significantly changed due to increase loading.

2. When embankment is supported on a soft soil foundation, the vertical stresses along centre line of embankment increases with increase in depth up to the bottom of the embankment and then show decrement below the embankment. However, effect of increase loading (due to hydrostatic and hydrodynamic forces) is not as significant as observed without foundation.

It can be concluded that the effects of hydrostatic and hydrodynamic forces (in addition to gravity forces) are smaller on the embankment with foundation soil, as compared to the embankment without foundation soil. It appears that effect of soft soil dominates the stress distribution. However, this conclusion is based on the results presented here and more cases are need to be studied before it can be generalized. Such generalized results will be very helpful in seismic slope stability analyses of embankment.

REFERENCES

1. Acar, Y. B., H. Hilal and M.T. Tumay, (1988). "FEM Analysis of Elastic Stress Distributions in Embankments," J. of Geotechnical Engineering, ASCE, **114**(6), 711-718.

2. ANSYS manual: A general finite-element program, version 10.0.

3. Anuradha P. (2007). Seismic Slope Stability Analysis of Embankments, M.Tech. Dissertation, Dept. of Earthquake Engineering, Indian Institute of Technology, Roorkee.

4. IS: 1893-1984. "Indian Standard Criteria for Earthquake Resistant Design of Structures," Bureau of Indian Standards, New Delhi.

5. Kramer, S.L., (1996). Geotechnical Earthquake Engineering, Prentice-Hall, Inc., NJ.

6. Newmark, N., (1965). "Effects of Earthquakes on Dams and Embankments," Geotechnique, **15**(2), 139-160.

7. Perloff, W.H., G.K. Baladi and M.E. Harr, (1967). "Stress Distributions within and under Long Elastic Embenkment," Highway Research Record, **181**, 12-40.

8. Sarkar Rajib, (2006). Seismic Response of Concrete Gravity Dam Including Dam-Reservoir-Foundation Interaction, M.Tech. Dissertation, Dept. of Earthquake Engineering, Indian Institute of Technology, Roorkee.

9. Singh B. and H.D. Sharma, (1976). Earth and Rockfill dams, Sarita Prakashan, Meerut, India.

10. Verruijt, A. (1969). Stresses due to gravity in an elastic half-plane with notches or mounds, Ph.D. thesis, Delft Univ., Holland.

TESTIMONY FROM EYEWITNESSES OF SHOWA BRIDGE COLLAPSE IN THE 1964 NIIGATA EARTHQUAKE

K. Wakamatsu[1], ***T. Tazoh***[2], ***N. Yoshida***[3], ***H. Nakazawa***[4], ***H. Kiku***[5], ***S. Yasuda***[6], ***and I. Towhata***[7]

1 National Research Institute for Earth Science and Disaster Preventions, wakamatsu@bosai.go.jp
2 Shimizu Corp., tazoh@shimz.co.jp
3 Tohoku Gakuin University, yoshidan@tjcc.tohoku-gakuin.ac.jp
4 Port and Airport Research Institute, nakazawa@pari.go.jp
5 Kanto Gakuin University, kiku@kanto-gakuin.ac.jp
6 Tokyo Denki University, yasuda@g.dendai.ac.jp
7 The University of Tokyo, towhata@geot.t.u-tokyo.ac.jp

ABSTRACT

Testimony from eyewitnesses of the Showa Bridge collapse was collected with the objective of pinning down the cause of the collapse. The 16 June 1964 Niigata earthquake is well known as the earthquake during which liquefaction causes serious damage to structures and is thus important in geotechnical engineering. The collapse of the Showa Bridge was one of the damage caused by the earthquake that attracted considerable attention, but even now, more than 40 years later, no common understanding is obtained in identifying the cause of the collapse, although many studies took place since the earthquake. For this study, therefore, we collected testimony from eyewitnesses of the Showa Bridge collapse. On the basis of this testimony we established a chronology of events, including the timing of the strong shocks of the earthquake, the bridge collapse and the occurrence of liquefaction-induced flow. As a result the following conclusions have been reached: the time of the onset of the collapse of the Showa Bridge can be estimated as about 70 seconds after the first shock of the earthquake; significant liquefaction-induced flow occurred after the Showa Bridge collapsed in the vicinity of the Showa Bridge.

Keywords: Liquefaction, Liquefaction-induced flow, Bridge, Niigata earthquake

1 INTRODUCTION

The 16 June 1964 Niigata earthquake is well known as the earthquake during which liquefaction causes serious damage to structures and is thus important in geotechnical engineering. The collapse of the Showa Bridge was one of the damage caused by the earthquake that attracted considerable attention (Photo 1), but even now, more than 40 years later, no common understanding is obtained in identifying the cause of the collapse. Many studies took place immediately after the earthquake to investigate the cause of collapse, because the Showa Bridge was designed based on the most advanced earthquake-resistant engineering at that time and completed just before the earthquake.

The causes set out past researches for the earthquake damage to the Showa Bridge can be classified as follows from the point of view of external load:
Inertia force: significant inertia force worked at the time of the earthquake.
Liquefaction: liquefaction reduced lateral subgrade reaction, or pile displacement increased when ground displacement occurred as apparent length above the ground surface became large.
Liquefaction-induced flow: liquefaction-induced flow affected the pile foundations causing considerable deformation to the piles, or the horizontal movement of the left bank pushed bridge girders towards the center of the river.

Photo 1 Collapsed Showa Bridge
(Modified from National Information Service for Earthquake Engineering)

Considering the relation between these causes and the time of the bridge collapse, it can be predicted that if inertia force had caused the collapse, the collapse would have started from the time of largest acceleration, if liquefaction were the cause, the collapse would have started from the time ground displacement increased, and if liquefaction-induced flow were the cause it would have started after flow had occurred. However, it is unclear at what time the bridge collapsed and liquefaction-induced flow occurred

2 OUTLINE OF DAMAGE TO SHOWA BRIDGE

The Showa Bridge was a first-class bridge crossing the Shinano River, 303.9 m in length and 24.0 m in width, with 12 spans. It was a simple steel girder bridge, and the spans were 27.64 m, with 13.75 m spans G1 and G12 at the ends of the bridge (JSCE, 1966). The bridge piers and abutments were of multiple column structure, with one row of 9 steel columns of 609 mm outer diameter.

The bridge was hit by the earthquake just 15 days after completion, and 5 of the 12 spans fell down (G3~G7) as shown in Photo 1. Eyewitness testimonies of the collapse were recorded in several reports including the JSCE earthquake damage survey report (JSCE, 1966; Horii and Miyahara, 1966; Miyamura, 1964). According to the majority of eyewitnesses first the G6 girder fell down taking down pier P6, and caused great splashing. Next the G5 and G7 girders fell down in tandem, followed by G4 and G3. A survey of the raised girders confirmed that G7 had fallen on top of G6, backing up the validity of the eyewitness reports (JSCE, 1966). G6 and G7 were supported by P6 pier, and clearly the fact that both girders were roller supported contributed to the collapse of G6 (Kubo, 1981).

The embankment of the approach road on the left bank side suffered subsidence, many fissures appeared on the road surface, and the left bank abutment slid and tilted. Hardly any damage was caused to the abutment and road on the right bank side; compared with the right bank, the left bank side suffered much greater damage. Damage indicating the pounding between girders was identified from G6 to the final girder on the left bank side. But the girders on the right bank side suffered no major damage except deformation of handrail above pier P9 (JSCE, 1966).

3 EYEWITNESS TESTIMONY

3.1 Eyewitnesses

For this study, 27 people who witnessed the collapse of the bridge and the ground deformation in the vicinity of the bridge provided testimony about the earthquake. Table 1 shows the location of the main eyewitnesses at the time of the earthquake and what they were doing, and their major experiences after the earthquake. The distance from each of these witnesses to the Showa Bridge was less than 200 m, with the exception of one at Niigata Meikun High School (850 m). All the testimony given below, with the exception of that of Mr Yoshida, was heard by the authors directly from the eyewitnesses, however some parts overlap with the content of JSCE (1966) and Horii and Miyahara (1966).

Table 1 Location of eyewitnesses at the time of the earthquake

No.	Location of eyewitnesses	Eyewitnesses	Circumstances at time of earthquake
1	On the Showa Bridge	a Fukuda Corporation employee and a Fuji taxi driver	Traveling in vehicles
2	Under the Showa Bridge near left bank abutment	5 people employed by or associated with Fukuda Corporation	About to restart water pipe installation work after lunch break
3	Hakusan High School concrete building (left bank)	6 teachers of Niigata City Hakusan High School	On lunch break in staff room
4	Hakusan High School wooden building (left bank)	2 pupils of Niigata City Hakusan High School	On lunch break in classroom
5	River bank in front of Hakusan High School (left bank)	7 pupils of Hakusan High School	Relaxing with friends
6	Niigata University residence (right bank)	a lecturer of Niigata University	Preparing lecture in study
7	Meikun High School (left bank)	1 pupil of Niigata Meikun High School	Relaxing on 4th floor veranda

3.1.1 Testimony of Eyewitnesses at the Showa Bridge

At the time of the earthquake water pipes were being installed under the Showa Bridge. For this purpose scaffolding had been set up under the bridge, and pipe with 5 m long and 300 mm in diameter had been placed on top of the scaffolding. On the day of the earthquake, these were being welded together, the procedure being that at each welding location materials were lowered from a small truck on the top of the bridge. Mr Teiji Shuto was driving the truck at low speed from left to right bank towards the welding location, and Mr Shigeru Yoshida was waiting on the bridge at the welding location.

As Mr Shuto drove across the G3 girder closer to the left bank he felt the truck tilt and heard a loud noise behind him. He got out and checked the back of the truck but couldn't identify the cause of the noise. He then saw up and down motion at the expansion joint and realized there was an earthquake. He drove the truck at a speed of 20~30 km/s and when he reached the G7 girder he saw the stationary Fuji taxi and he too stopped. But he felt another powerful shake and so drove off in panic for the right bank. When he reached the approach road on the right bank he stopped the truck, and by the time he had walked back to the approach road of the bridge, the bridge had already fallen down.

Fuji taxi driver Mr Saburo Yamada experienced the earthquake on the Showa Bridge, on his return trip after taking a passenger to the golf course on the left bank side. On the G7 girder he felt a big jolt as though all four tires had punctured and he stopped the vehicle. The vehicle was stopped directly over an expansion joint. He got out and peered under the vehicle to examine his tires. He saw that the expansion joint was on the point of separating, realized the danger and quickly jumped back in the vehicle. But the vehicle was shaking violently and he couldn't proceed. At that moment the light truck of Fukuda Corporation went past. He drove in low gear over the groaning bridge towards the right bank. In low gear his speed was likely to have been about 10 km/h. Driving was very difficult because the bridge was moving violently from side to side and up and down, but he managed to reach the bank. After crossing the bridge he looked in his rear view mirror and the bridge had already fell down. After that he heard the Showa oil storage tank explosion and saw the column of thick smoke.

Mr Shuzo Hidano (supervisor), Mr Mito Yokoyama, Mr Masaru Abe, Mr Koji Wada, Mr Seisaku Kita and several women employees of Fukuda Corporation were directly under the Showa Bridge in the vicinity of the P1 pier (Figure 1, point F), involved in the pipe laying work. With the resumption of work at 1 pm several of them were about to enter the scaffolding area under the bridge. A road runs under the G1 girder, as shown in Photo 2, and from P1 pier under the bridge towards P2 pier there was scaffolding for this work.

From the abutment side of the river revetment Mr Hidano was watching Mr Wada climb a ladder set up by P1 pier. Mr Yokoyama, nearby, was also watching. As Mr Wada climbed and reached out to the scaffolding he thought someone was shaking the ladder for fun and shouted, 'Don't shake the ladder!' But no-one was near it, and Mr Yokoyama answered, 'There's no-one there!' At that moment one of the women yelled, 'It's an earthquake!' Mr Kita, who was following Mr Wada up the ladder, told exactly the same story.

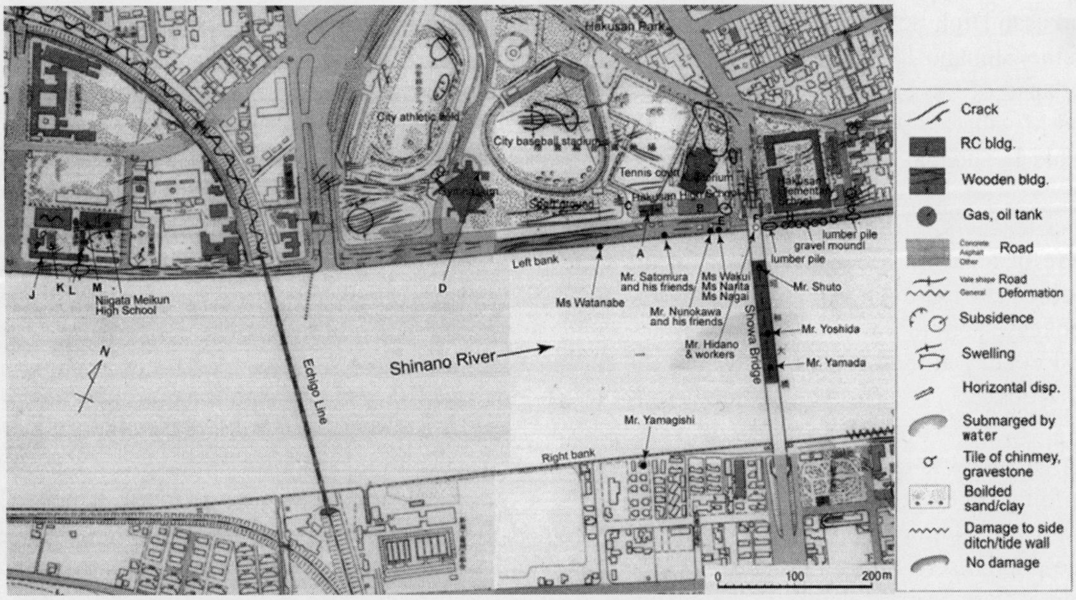

Figure 1 Location of eyewitnesses and facilities at the time of the earthquake (modified from Niigata Univ. and Fukada Geological Institute, 1964)

Photo 2 Ground failure of left bank immediately upper portion of the Showa Bridge (Modified from The Japanese Geotechnical Society, 2004)

Mr Wada heard the bridge abutment groaning so he jumped down from the ladder and moved out from under the bridge. At that moment there was a loud noise and he witnessed the bridge collapse, one section falling after another starting from around the center and sending up a lot of spray. When Mr Wada jumped down from the ladder, Mr Yokoyama witnessed concrete of the AL abutment of the Showa Bridge breaking and falling off. Afterwards, fissures opened up in the road and water came spouting up. At that point Mr Wada took shelter on the bank at the side of Hakusan Elementary School by a pile of lumber for making rafts (point G, Figure 1). As Mr Wada moved away from the bridge on the downstream (Hakusan Elementary School) side, the revetment collapsed towards the river quickly.

3.2 Eyewitnesses in the vicinity of Hakusan High School on the left bank of the Shinano River

Hakusan High School was at a distance of only about 30 m from the Showa Bridge on the left bank of the Shinano River. On the day the earthquake struck many pupils and teachers witnessed the collapse of the bridge from the school buildings or nearby (Miyamura, 1964; JSCE, 1966; Horii and Miyahara, 1966). The authors obtained testimony from a total of 15 eyewitnesses, as shown in Table 1. The main points from that testimony are as follows:

Four teachers, Mr Hitoshi Kobayashi, Ms Masako Otsuka, Mr Masao Aida, and Mr Kimiro Sano, were in a second-floor staff room in the central section of Hakusan High School's RC building (point A, Figure 1). At the time of the earthquake, Mr Kobayashi was standing in front of a bookcase about 2 m tall positioned in the center of the room. The three others were sitting at their desks. They all described the onset of the earthquake in the same way, as everything lurching, and Mr Kobayashi and Mr Aida remembered that the bookcase looked as though it would fall over.

Ms Otsuka was eating lunch at her desk some distance from the windows on the river side of the building. The building suddenly shook strongly two or three times. She rushed to the window and saw 15 or 20 pupils on the river bank who seemed to be jumping up and down on the revetment. The pupils, screaming, ran back towards the school building, but cracks were opening up in the road at their feet. At that time there were choppy waves on the surface of the river, as if someone were shaking a basin. She looked up when she heard someone shout, 'The bridge has collapsed!' and she saw the spray of water thrown up by the collapse. It seemed as if 'there was shaking, shaking and more shaking, and an aftershock brought the bridge down'. One of the pupils referred to above who was on the revetment at the time of the earthquake stated, 'For a moment I thought my friends had pushed me from behind. We got to safety by jumping across the top of the revetment, careful not to fall in the river. Afterwards I realized it had been caused by an earthquake.'

Mr Kazuei Satomura, a third year pupil of Hakusan High School, was close to the river revetment with three of his friends (see Figure 1 for location). The ground began to shake and from the downstream direction there was a loud noise like a train passing through a tunnel. The noise then changed to something like oil drums rolling along, and after that the Showa Bridge began to fall down. He doesn't recall which girder fell first, but the G6 and G7 girders by P6 pier fell into a V shape. After that G5, G4 and G3 girders collapsed one after another. The bridge began to collapse after the violent shaking had stopped. He remembers a car on the bridge driving towards the left bank, and the girders it drove over collapsed immediately afterwards in succession. It reached the approach road on the left bank side and managed not to fall into the cracks that had opened up in the surface. The river bed about 5 m from the bank rose up in a volcanic shape about 15 m in diameter. Then the revetment close to the bridge collapsed into the river and the school building tilted towards the bridge.

Ms Misako Wakui, a first year pupil at Hakusan High School, was sitting on the revetment near the Showa Bridge talking to her friends (see Figure 1 for location). The revetment started shaking and she thought some boys were doing it deliberately. The shaking grew gradually stronger and when she felt herself jolted by it, she realized it was an earthquake. The surface of the river had started getting choppy, and the Showa Bridge was swaying considerably to left and right. She saw a car driving across the bridge, and after it had crossed, the girders from around the center began to collapse one after another. She set off at a run towards the passage that crossed between the concrete and wooden buildings of the school. At her feet the revetment subsided, falling towards the river, and she was afraid that the river water would come surging up. When she started to run the passage seemed to undulate, and the passage and wooden building tilted over and sank.

Ms Sumiko Watanabe, a first year pupil at Hakusan High School, was on a gravel barge moored in front of the school, talking with her friends. The barge began to rock and she thought someone was doing it deliberately as a prank. The rocking grew stronger and stronger and the barge moved about 1 m away from the bank. When the barge moved back to the bank she quickly jumped onto the bank. Immediately afterwards the Showa Bridge collapsed with a crashing noise. She stood watching the bridge collapse to the end, staring in shock, and then quickly ran back towards the school building. Earth fissures about 20 cm wide were opening up at her feet and she was afraid she might fall into one as she ran. She didn't notice at what point the revetment collapsed, but she believes it was after she turned her back to the river and started to run to safety, because she hardly jumped onto the bank if it had collapsed.

Ms Emiko Narita, a first year pupil at Hakusan High School, was sitting on the revetment in front of the wooden school building relaxing with her classmates Ms Nagai and Ms Wakui (see Figure 1 for location). A small boat was tied up at the river bank in front of her. She felt herself being shaken but because she was watching the boat rocking, she thought her eyes were playing tricks. But then the roof of the wooden school building started a rolling movement and she realized it was not a hallucination but an earthquake. She noticed a small truck driving on the Showa Bridge from right to left bank. She thinks it was immediately afterwards that the girder at the center started to collapse. She thinks the girder to the left of center collapsed first and then the girder to the right, and she saw 5 girders fall in total. She couldn't be sure how much time had passed to that point but it felt like a minute. As she didn't see the revetment collapse, it must have happened after she ran off towards the sports ground.

3.3 Testimony of eyewitness on the right bank of the Shinano River

At the time of the earthquake Mr Hiromasa Yamagishi, a lecturer of Niigata University, was at his university residence, a wooden, single-story house behind the revetment on the right bank of the river (see Figure 1 for location). It was about 170 m away from the Showa Bridge. He was sitting at his desk in his study at the back of his house (see Figure 2 for location) preparing a lecture. He felt a sudden strong shock, an up-down shake, as if struck from below. He realized the earthquake's epicenter must be close and thought it dangerous to stay inside, so he headed for the front door. But the key was the old-fashioned screw-in type, and in his panic he couldn't get the door unlocked. He gave up going out that way and as he returned to his study the plaster was falling from the walls. He thought he would escape through the study window and climbed onto the window frame. To get outside he had to get over the window frame, which was about waist-high, and outside, broken glass had fallen to the ground, so he hesitated momentarily to get out that way. As a result it took a little time to get outside.

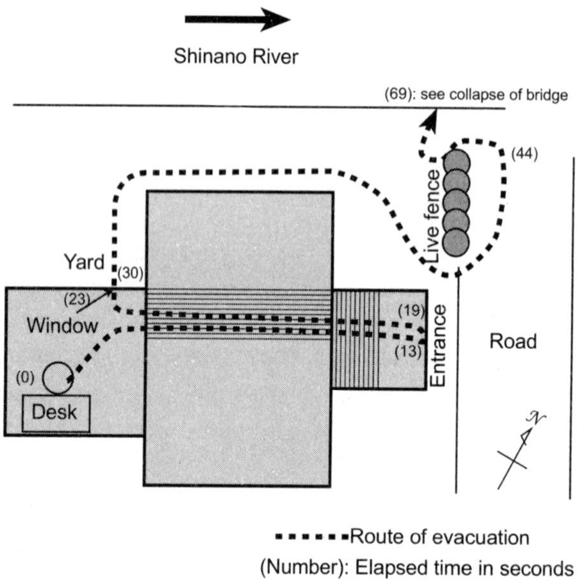

Figure 2 Floor plan of the Mr. Yamagishi's house (see Figure 1 for location)

He went as far as the bank of the river, but the ground was shaking violently and cracks were visible in the ground so he thought it dangerous to be on the river bank and moved back to a hedge growing in front of the house (Figure 2). Grasping hold of the hedge he looked towards the river and saw cracks open in the revetments and Hakusan High School and the river surface shaking violently.Towards the estuary he saw black smoke rising. He hadn't heard any explosions so he didn't think this was the result of an explosion and fire. At this point the Showa Bridge had not collapsed. It was as the shaking died down and he began to feel reassured that he heard a great crashing noise. Wondering what it was, he went towards the river bank and looked in the direction of the bridge. The bridge began to collapse really slowly, as though he was watching a film in slow motion. The sections shook as they fell, like falling leaves. He thought it took several seconds to the end.

The authors visited with Mr Yamagishi to the location, now an empty lot, and got him to re-enact his movements. This was recorded on video, and the amount of time that each action took was measured. According to those measurements, the Showa Bridge began to collapse about 69 seconds after the first jolt of the earthquake, and finished collapsing about 91 seconds after the first jolt.

3.4 Photographs taken from Niigata Meikun High School

Mr Yutaka Takeuchi, a third year student at Niigata Meikun High School, was in the corridor of his 4th floor classroom (building M, Figure 1) when the earthquake struck. He had brought his camera to school that day and when the shaking seemed to be subsiding he went back in the classroom to get his camera and took a series of shots of muddy water spurting up in the school yard from the 4th floor corridor. Then he moved to the school building L and went on taking photos of earthquake damage. All the photographs have recently been made public (JGS, 2004).

The authors felt sure that establishing the times that Mr Takeuchi's photos were taken would offer important clues for determining the times when sand boiling, liquefaction-induced flow, etc., occurred. Mr Takeuchi twice re-enacted his behavior at Meikun High School at that time and we established the times when all the photographs were taken (Wakamatsu et al., 2004). According to the time measurement, the ground water started spurting up from the schoolyard after 3 minutes and 15 seconds from he felt the shaking. The depth of the water in the yard reached approximately 50cm after 5 minutes and 40 second. After 5 minutes and 49 second, the buildings L and M in Figure 1 had separated 2 meters, which were connected before the earthquake.

4 CONCLUSIONS

Based on a study of eyewitness testimony, we established a chronology of events, including the timing of the strong shocks of the earthquake, the bridge collapse and the occurrence of liquefaction-induced flow. As a result the following conclusions have been reached: the time of the onset of the collapse of the Showa Bridge can be estimated as about 70 seconds after the first shock of the earthquake, i.e., long after the strong shock; significant liquefaction-induced flow occurred after the Showa Bridge collapsed in the vicinity of the Showa Bridge.

REFERENCES

1. Horii, K. and Miyahara, H (1966): "Bridges Spanning over the Lower Course of the River Shinano," Bulletin of Science and Engineering Laboratory, Waseda University, 34, Special Issue on Niigata 2 2 Earthquake, Science and Engineering Laboratory, Waseda University, 151-170.
2. Japan Society of Civil Engineers (JSCE) Investigation Committee for Niigata Earthquake, (Ed.) (1966). Report of the damage investigation of the 1964 Niigata Earthquake, 376-395, (in Japanese).
3. The Japanese Geotechnical Society (Ed) (2004). Photographs and Motion Picture of the Niigata City Immediately After the Earthquake in 1964, Tokyo, Maruzen Publishing Division.
4. Kubo, K. (1981). Earthquakes and Civil Engineering Structures, Tokyo, Kajima Publishing Co., 66-67 (in Japanese).
5. Miyamura, K. (1964). Experience Notes of the Niigata Earthquake, Niigata Hakusan High School (in Japanese).
6. National Information Service for Earthquake Engineering. EQIIS Image Database Niigata, Japan, http://nisee.berkeley.edu:8080/images/servlet/EqiisBrowse?group=Niigata1964-01.
7. Niigata Univ. and Fukada Geological Institute (1964). Maps Showing Ground Failure Caused by Niigata Earthquake (in Japanese).
8. Wakamatsu, K., Tazoh, T., Yasuda, S., Towhata, I., Yoshida N., Kanatani, M., Kiku, H. and Nakazawa, H. (2004): Special Session for the 40th Anniversary of the 1964 Niigata Earthquake, Witnesses of the Soil Liquefaction, Tsuchi-to-Kiso, 52(12) 15-20, (in Japanese).

CAUSES OF SHOWA BRIDGE COLLAPSE IN THE 1964 NIIGATA EARTHQUAKE

N. Yoshida[1], T. Tazoh[2], K. Wakamatsu[3], S. Yasuda[4], I. Towhata[5], H. Kiku[6], and H. Nakazawa[7]

[1] Tohoku Gakuin University, yoshidan@tjcc.tohoku-gakuin.ac.jp
[2] Shimizu Corp., tazoh@shimz.co.jp
[3] National Research Institute for Earth Science and Disaster Preventions, wakamatsu@bosai.go.jp
[4] Tokyo Denki University, yasuda@g.dendai.ac.jp
[5] The University of Tokyo, towhata@geot.t.u-tokyo.ac.jp
[6] Kanto Gakuin University, kiku@kanto-gakuin.ac.jp
[7] Port and Airport Research Institute, nakazawa@pari.go.jp

ABSTRACT

Causes of collapse of the Showa Bridge that fell down during the 1964 Niigata earthquake is investigated. Firstly, time when the bridge collapsed is evaluated to be about 70 seconds from the beginning of the ground shaking by analyzing the behavior of the people who were on or near the bridge. Then time when horizontal movement of the quay wall, which is identical to the time when liquefaction-induced flow occurred, is evaluated from the eyewitness of the people who were on the river dike or near the river. It is concluded that liquefaction-induced flow occurred after the collapse of the bridge. Therefore, the collapse of the bridge is caused by the kinematic interaction between the ground and structure with high possibility, and loads such as inertia force, liquefaction, and liquefaction-induced flow are not direct cause although they are one of the factors to trigger the bridge collapse.

Keywords: Liquefaction, Liquefaction-induced flow, Bridge, Kinematic interaction

1 INTRODUCTION

Time sequence is one of the important factors in the design of a structure if soil liquefaction is to be considered. In the design of a pile foundation, for example, inertia force of a superstructure acts at the pile head, and ground displacement, which can be significantly large to control the design of the pile in the liquefied ground, acts in the whole pile. However, as they may not act simultaneously because liquefaction-induced ground displacement usually occurs after the main shock. Therefore, considering both maximum loads in the design yields over or costly design.

On the other hand, considering each maximum loads individually may yields dangerous design, because damage caused by inertia force may reduce strength of the structure and it result in severer damage under subsequent load (e.g., Nanjo et al, 2000) than the case that the load is applied singly.

It is, therefore, important to consider both loads in relevant combination, not both maximums, and not individual maximum. It can be done by making the time sequence of the phenomena clearly, but there is little research on this topic, probably because of the difficulty to obtain sufficient eyewitness.

The Showa Bridge was a modern bridge that came to completion about two weeks before the 1964 Niigata earthquake, and collapsed during the earthquake. As schematically shown in Figure 1, several girders fell down and pile deformed as shown in Figure 2. Liquefaction and liquefaction-induced flow were observed as shown in Photo 1 in the vicinity of the bridge.

The authors compiled eyewitness of people who saw the collapse of the Showa Bridge and who were near the bridge (Wakamatsu et al, 2007). It seems an effective data to make the time sequence occurred near the Showa Bridge. In this paper, we will evaluate the time of the event and identify the cause of the Showa Bridge based on the evaluation. Figure 3 shows location of eyewitnesses and facilities at the time of the earthquake. See the accompanying paper (Wakamatsu et al, 2007) for details of the testimony and it will be referred without notice.

2 PAST RESEACH ON CAUSE OF BRIDGE COLLAPSE

The Japan Society of Civil Engineers (JSCE) study showed the following as possible causes of the bridge collapse, from the damage observed to the superstructure and foundation structure of the bridge and to the ground (JSCE, 1966):1) Horizontal movement of the left bank side ground of about 3 m towards the center of the river, 2) liquefaction of the ground, 3) horizontal external force worked on the bridge girder resulting from horizontal movement of the abutment on left bank side, 4) large horizontal displacement of the foundation resulting from ground displacement, 5) reduction in lateral ground resistance due to liquefaction, and 6) loss of bearing capacity due to poorly welded pile joints.

After that many research have been made, and the causes for damage to the Showa Bridge can be classified as follows from the point of view of external load:

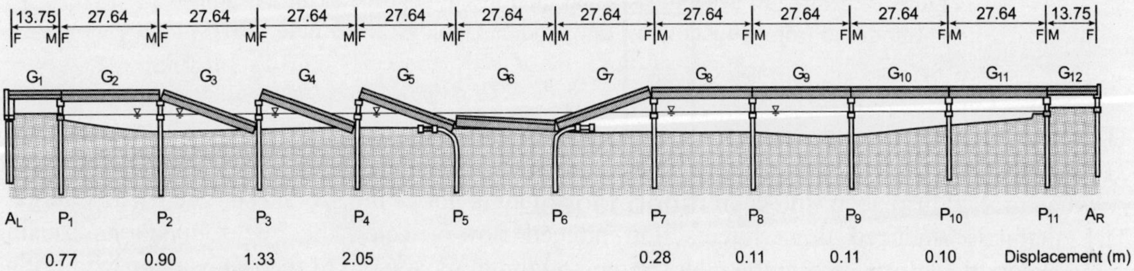

Figure 1 Elevation view of the collapsed Showa Bridge (modified from JSCE,1966) (unit : m)

(M: moveable bearing, F: fixed bearing, numbers at the bottom of the figure represent horizontal displacement of the ground at each pier

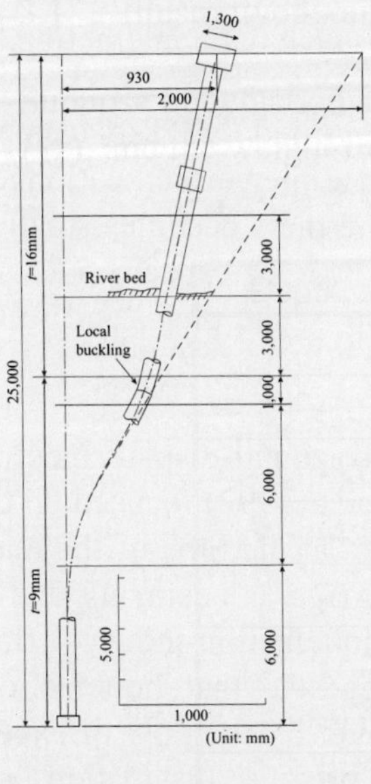

Figure 2 Restored deformed shape of the pulled out pile (PWRI, 1965)

Photo 1 Ground failure of left bank (modified from Japanese Geotechnical Society , 1999)

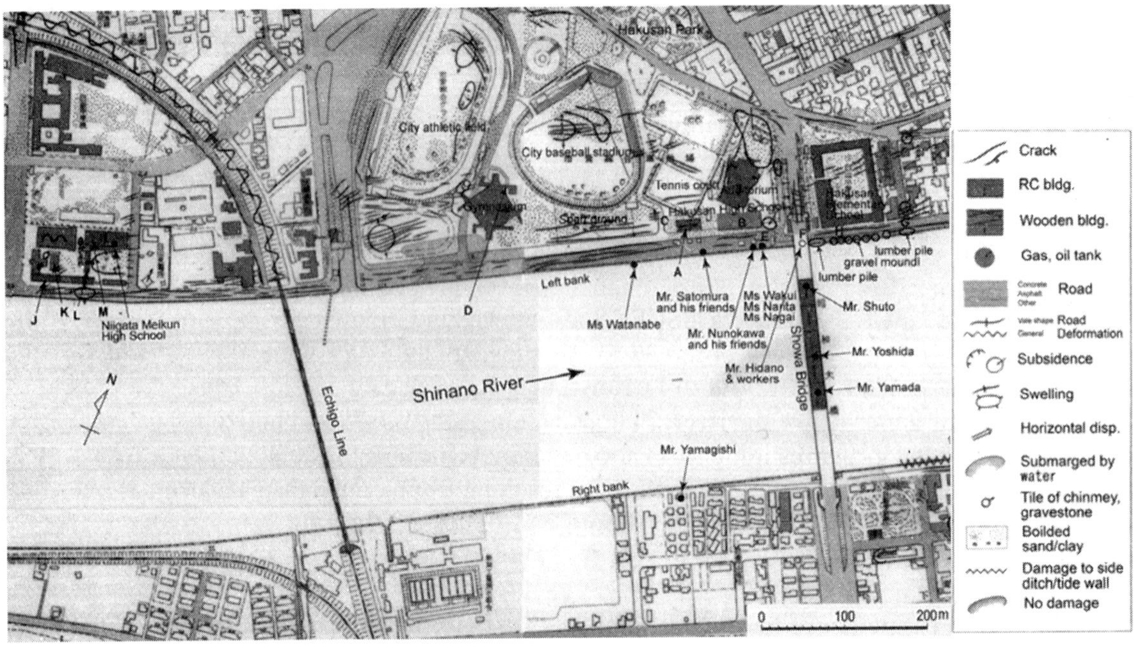

Figure 3 Location of eyewitnesses and facilities at the time of the earthquake (modified from Niigata Univ. and Fukada Geological Institute, 1964)

1) Inertia force: significant inertia force worked at the time of the earthquake.
2) Liquefaction: liquefaction reduced lateral subgrade reaction, or pile displacement increased when ground displacement occurred as apparent length above the ground surface became large.
3) Liquefaction-induced flow: liquefaction-induced flow affected the pile foundations causing considerable deformation to the piles, or the horizontal movement of the left bank pushed bridge girders towards the center of the river.

Among the researches that inertia force is responsible, Kuribayashi (1987) suggests yielding of the pile due to inertia force. The Public Work Research Institute (PWRI) (1965) and Arakawa et al (1982) pointed out difference of predominant period between P6 pier and P5, P7 piers because of difference to support girder. Tazoh et al (1985) suggested the possibility of kinematic interaction.

As researches attributing the cause to liquefaction, Okubo et al (1966) suggested that displacement of the pier becomes larger as the predominant period of the ground, hence pier, becomes longer. Okamoto (1971) has argued that the length of the embedded section of the piers becomes smaller. Kubo (1981) has pointed out several competing factors: a reduction in bearing capacity of sand layer due to liquefaction, insufficient width to carry the girder at the top of the piers, and roller support of G6 and G7 girders on P6 pier.

Hamada et al (1986) showed that the earthquake caused considerable horizontal displacement of the revetments along the Shinano River (liquefaction-induced flow). Hamada (1992) focused on the fact that, from the testimony of witnesses, the Showa Bridge collapsed not at the time of the main shock but after the main shock, which makes it easier to explain the damage as occurring due to liquefaction-induced flow. But he does not mention exactly how liquefaction-induced flow influenced on the collapse of the bridge. Sento et al (2001) have suggested that liquefaction-induced flow occurred because of the failure at the top of the liquefied layer by the flow of excess pore water pressure from downward in the process of excess porewater pressure dissipation, and that this mechanism also caused the collapse of the Showa Bridge. But their research assumes a gradient in the liquefied layer in the direction perpendicular to flow, while ignoring the gradient in the direction of flow.

The existing research mentioned above only suggests possible causes of the damage without substantiation. Furthermore, where analytical studies have been undertaken, they state only how the collapse might be explained through hypothetical mechanisms, and have not considered how this relates to the actual timing of the collapse. Thus, although a number of causes have been offered on the collapse of the Showa Bridge, none has reached the stage of gaining universal acceptance.

3 TIMING TO FEEL QUAKE

In the estimations of timing based on testimony, the time the eyewitnesses felt the earthquake is the baseline. However, feeling an earthquake depends on circumstances. For example, according to the Japan Meteorological Agency (JMA)'s seismic intensity scale (JMA, 1949; 1978), at an intensity 2 'most people indoors feel the shake', which would apply to Mr Yamagishi. At an intensity 4 'people walking feel the shake. Some people driving may also notice it', and so this would apply to the pupils of Hakusan High School. It can be assumed that Mr Shuto also noticed the earthquake at this stage. According to the JMA an intensity 2 is equivalent to an acceleration of 2.5~8 cm/s^2 and an intensity 4 to 25~80 cm/s^2. Figure 4 shows the acceleration records obtained at the Kawagishicho prefectural government apartment located 1.3 km west of the Showa Bridge. Referring to this, we can assume that Mr Yamagishi felt the earthquake immediately after it began, while the pupils of Hakusan High School and the Fukuda Corporation workers felt it as the shaking grew stronger 5 seconds or so later.

In order to confirm people's awareness of seismic motion, a 3-dimensional shaking table was undertaken. The seismic records shown in Figure 4 were input without warning and 9 people (6 in 20s, 1 in 30s, 1 in 40s, and 1 in 50s) were asked to raise their hands when they felt the earthquake first time and when they felt strong shakes. They were positioned so that they couldn't see each other and 6 tests were undertaken with them standing and sitting. In all cases they felt the earthquake first time immediately after the acceleration began, and the strong shakes 3.5~7.7 seconds (averaging 6.1 seconds) later. From this we can assume that most people felt the earthquake around 6 seconds after it started, and we hypothesize below that the people on the bridge felt the shakes 6 seconds after the earthquake started.

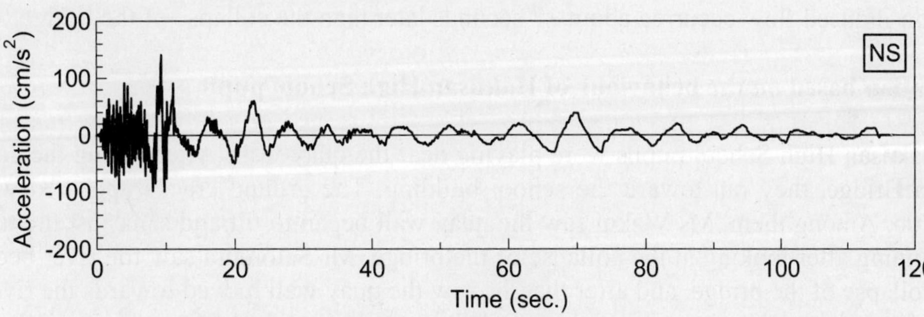

Figure 4 Acceleration time history observed at Kawagishi-cho apartment house

4 ESTIMATION OF THE TIMINGS OF THE SHOWA BRIDGE COLLAPSE

(1) Estimation based on Mr Shuto's testimony

Mr Shuto was driving the Showa Bridge at the time of the earthquake. He felt the car shook and stopped to check the car, which took 15 to 20 seconds. After feeling subsequent large shaking, he returned to the car and rushed to the right bank with an average speed of 20 to 30 km/h (5.6 to 8.3 m/s). He took 31 to 41 seconds to run the distance of 257 meters to the right bank. The time required for him to stop at the Mr Yamada's place and to start the car again was about 10 seconds. After crossing the bridge, he stopped the car, returned back to the right bank and found that the bridge already fell down, which took about 30 seconds. We need to add 6 seconds from the beginning of the earthquake to the time when he felt jolt as described in the preceding. In total, it took 92 to 112 seconds from the beginning of the earthquake to the time when he saw bridge collapse. This indicates that the bridge fell down at 92 seconds from the beginning of the earthquake at latest.

(2) Estimation based on Mr Yamada's testimony

Mr Yamada, a taxi driver, stopped the car near the intersection between G6 and G7 girder when he felt strong jolt. He felt that the joint of the decks would separate and he drove the car to the right bank. He started after Mr Muto left and reached the right bank with safe. The speed is about 10 km/h (2.78 m/s) according to his testimony because the car was in low gear. The girders that collapsed were from G3 to G7, which means that G7 girder collapsed after Mr Yamada had passed. It takes about 10 seconds until he passed G7 girder. In combination with the above estimated time, the bridge did not collapse until 56 seconds from the beginning of the earthquake.

(3) Estimation based on Mr Yamagishi's testimony

According to the time measurement of Mr Yamagishi's major actions, it took 69 seconds from the first shake of the earthquake until the bridge began to collapse and 91 seconds until it finished collapsing.

5 ESTIMATION OF THE TIMING OF THE LIQUEFACTION-INDUCED FLOW

(1) Estimation based on testimony of workers under the Showa Bridge on the left bank

Mr Wada felt the earthquake when he was climbing the pier P1 for construction of the water supply pipeline. On the way to escape from there to the Hakusan Elementary school, he saw the Showa Bridge fell down with a sheet of spray. Then when he came to the lumber dump in the downstream side, he saw the quay wall begins to collapse towards the river. This collapse is associated with liquefaction-induced flow. He ran about 50 m from looking at the collapse of the bridge to arrive at the lumber dump with a speed of about 3m/s. It indicates that collapse of the quay wall or liquefaction-induced flow occurred about 17 seconds later than the collapse of the bridge.

(2) Estimation based on the behaviour of Hakusan High School pupil

Several Hakusan High School pupils were playing near the quay wall. After seeing the collapse of the Showa Bridge, they ran toward the school building. The ground cracks appeared when they began to run. Among them, Ms Wakui saw the quay wall began to tilt and sank just the time when he start running after looking at the collapse of the bridge. Mr Satomura saw the river bed uplifted after the collapse of the bridge, and after that, he saw the quay wall moved towards the river. These ground deformation and movement of quay wall is associated with the liquefaction-induced flow. They were observed after the collapse of the bridge.

(3) Estimation based on photographs taken by Mr Takeuchi

Mr Takeuchi, a pupil of the Meikun High School, did not see the bridge collapse, but he took many photos that capture the damage to the ground and buildings after the earthquake. As the Meikun High School is located about 850 meters from the Showa Bridge, similar behavior is supposed to occur.

Photo 2 is an aerial photograph of the area around Meikun High School taken on 19th June, three days after the earthquake. It shows the buildings after they had shifted horizontally towards the

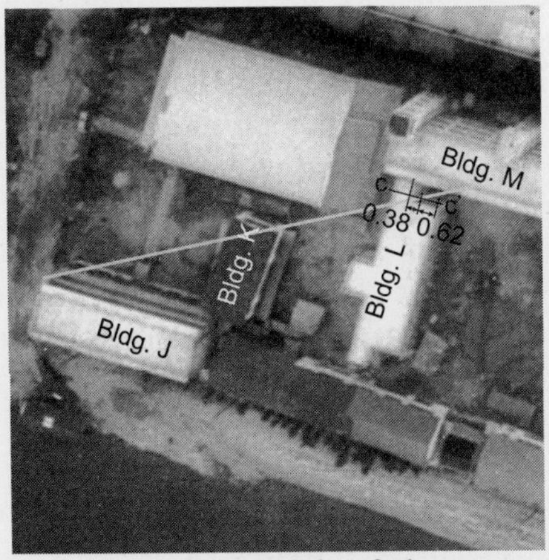

Photo 2 Aerial photograph of the area around Meikun High School (modified from photo taken by Kokusai Kogyo Co. Ltd.)

Photo 3 Photo taken by Mr Takeuchi at about 1.08 pm on the day of the earthquake

Photo 4 Photo taken by Mr Takeuchi at about 1.05 pm on the day of the earthquake

Photo 5 Photo of wave front on the Shinano River taken by Mr Takeuchi immediately before Photo 2, at about 1.05 pm (contrast enhancement)

river due to liquefaction-induced flow. Looking at buildings J and K, a line linking the corners of these buildings intersects a line c-c' along the top of the parapet on the roof of building L in the ratio 0.62: 0.38. Photo 3 is a photo taken by Mr Takeuchi at about 1.08 pm on the day of the earthquake. The buildings L and M in this photo are the same as in Photo 2. The space that can be seen between buildings L and M in Photo 3 did not exist before the earthquake and occurred as a result of liquefaction-induced flow. This means that 6 minutes after the earthquake struck liquefaction-induced flow had already occurred in the school grounds.

Of the photos Mr Takeuchi took before Photo 3, Photo 4 (taken at about 1.05 pm) shows buildings J and K, seen in Photo 2. A line drawn through the corners of buildings J and K, as in Photo 2, intersects the line c-c' in the ratio 0.19: 0.81. If building L did not move, and only building K moved towards the river, the c' side value should be relatively higher, and at the time Photo 4 was photographed we can infer that either building L had not shifted towards the river due to liquefaction-induced flow, or that the degree of shift was still small. Thus it can be deduced that at the time Photo 2 was photographed (about 1.05 pm), which was about 3 minutes after the earthquake occurred, no liquefaction-induced flow that had a significant effect on the structures had yet occurred.

The argument above assumes either that buildings J and K did not move due to liquefaction-induced flow, or that they shifted horizontally without any relative displacement between them. As a result the above conclusion is not watertight. But considering that the ratio changed significantly between Photos 4 and 6, we can conclude that after Photo 4 was photographed at about 1.05 pm some considerable liquefaction-induced flow took place that had a major effect on the structures. We cannot, however, deduce when the liquefaction-induced flow began from these photos.

Photo 5 was photographed immediately before Photo 4, at about 1.05 pm. It shows the wave front on the Shinano River to be parallel to the revetment (evaluation of wave front by Japanese Identification Center of Forensic Science Co., Ltd.). This type of wave front parallel to flow is not normally seen. Thus, assuming that this wave front resulted from the movement of the revetment due to liquefaction-induced flow, it means that liquefaction-induced flow had occurred within 3 minutes after the earthquake. This wave formation occurred in an area about 50 m out from the revetment, and the time taken for the wave front due to the revetment shift to arrive at a distance 50 m from the revetment can be calculated as follows:

According to shallow water theory, flow speed is expressed as the square root of the product of depth and gravitational acceleration. Taking the water depth as 3 m, the speed of transmission of the wave movement is thus 5.4 m/s. It would therefore take about 10 seconds to move about 50 m out from the revetment, and so we may assume that this wave movement was generated about 10 seconds before the time the photo was taken (which was about 3 minutes after the earthquake).

6 DISCUSSION AND CONCLUSION

The collapse of the Showa Bridge starts between 50 to 92 seconds from the testimony of Mr Shuto and Mr Yamada. This time agrees with the testimony from Mr Yamagishi, 69 seconds. Among the testimony, that by Mr Yamagishi is the most accurate because same time was reported in JSCE (1966) although he did not know it before we told him. Looking at Figure 4, acceleration time history at the Kawagishicho prefectural government apartment, the maximum acceleration is about 160 cm/s^2 becomes around 10 seconds and liquefaction is supposed to occur about 12 seconds (Yoshida and Kudo, 2000). Maximal acceleration around 70 seconds is about 70 cm/s^2. This indicates that inertia force cannot be a predominant cause of the collapse.

Many testimonies suggest that liquefaction-induced flow occurred after the collapse of the Showa Bridge. Therefore, liquefaction-induced flow does not seem to be the cause of the collapse of the bridge. The deformed pile shown in Figure 2 is used to suggest the cause of bridge collapse by liquefaction-induced flow (e.g., Hamada, 1992), but this residual deformation may occur as the fallen girder pushed the pile or by ground deformation; deformed pile cannot be an evidence of liquefaction-induced flow for the cause of bridge collapse.

Considering them, it can be said that there is a high possibility that the cause of the Showa Bridge collapse was due to increased displacement of the ground in circumstances where pile deformation occurred more easily due to liquefaction. In other words, kinematic interaction between the softened ground and pile-structure system is the cause of the collapse of Showa Bridge.

REFERENCES

1. Arakawa, N., Kimata, T., Kondo, M., Tokida, K. and Takamatsu, S. (1982): "Earthquake response analysis of the Showa Bridge considering soil liquefaction," Proc. 17th Japan National Conference on Soil Mechanics and Foundation Engineering, pp. 1941-1944 (in Japanese).

2. Hamada, M., Yasuda, S., Isoyama, R., and Emoto, K. (1986): "Observation of permanent ground displacement induced by soil liquefaction," Journal of Japan Society of Civil Engineers , No.376 , III-6, pp. 211-220, 1986 (in Japanese with English Abstract).
3. Hamada, M. (1992): "Large Ground Deformations and Their Effects on Lifelines: 1964 Niigata Earthquake," Case Studies of Liquefaction and Lifeline Performance During Past Earthquakes, Volume 1 Japanese Case Studies, edited by M. Hamada and T. D. O'Rourke, February 17, pp. 3-28.
4. Japanese Geotechnical Society (1999): Air photographs of the Niigata City immediately after the earthquake in 1964, 24pp.
5. Japan Society of Civil Engineers (JSCE) Investigation Committee for Niigata Earthquake, (Ed.) (1966): Report of the damage investigation of the 1964 Niigata Earthquake, pp. 376-395, (in Japanese).
6. Japan Meteorological Agency (1949, 1978): Seismic Intensity Scale.
7. Kubo, K. (1981) : Earthquakes and Civil Engineering Structures, Tokyo, Kajima Publishing Co., pp. 66-67.
8. Kudo, K., Uetake, T., and Kanno T. (2000): "Re-evaluation of Nonlinear Site Response during the 1964 Niigata Earthquake using the Strong Motion Records at Kawagishi-cho, Niigata City, " Proceedings 12WCEE, Auckland, New Zealand, Paper No. 0969.
9. Kuribayashi, E. (1987): "Lessons learned from 1964 Niigata earthquake (bridge) -disaster of Showa Bridge-," Kisoko, Vol. 15, No. 5, pp. 24-30 (in Japanese).
10. Nanjo, A., Yasuda, F., Fujii, T., Tazoh, T., Ohtsuki, A., Fuchimoto, M., Nakahira, A. and Kuroda, C. (2000): Analysis of the damage to the pile foundation of a highway bridge due to the 1995 Great Hanshin earthquake and study of effective countermeasures, Jour. of Structural Mechanics and Earthquake Engineering, Proc. of JSCE, No. 661/III-53, pp. 195-210 (in Japanese)
11. Niigata Univ. and Fukada Geological Institute (1964). Maps Showing Ground Failure Caused by Niigata Earthquake.
12. Okamoto, S. (1971): Seismic Engineering, Tokyo, Ohmsha, pp. 272-273.
13. Okubo, T., Kuribayashi, E. and Iwasaki, T. (1966) : "On the damage and response analysis of the Showa-Ohashi Bridge," Report of the National Research Institute for Earth Science and Disaster Prevention, No. 12, pp. 11-20 (in Japanese).
14. PWRI, Report on Niigata earthquake, Report of the PWRI, No. 125, 1965, 330pp.
15. Sento, N., Kazama, M., Akahori, K. and Ohmura, H. (2001): "Mechanism, and prediction of flow deformation due to seepage after earthquake," Proc., Lessons learned from recent strong earthquakes, Earthquake Geotechnical Engineering Satellite Conference, XVth ICSMGE, Istanbul, Turley, pp. 369-374.
16. Tazoh, T., Nakahi, S., Shimizu, K., Kimata, H., and Kogawa, T. (1985): "Considerations on Seismic Damage to Foundation Piles," Proc. 20th Japan National Conference Soil Mechanics and Foundation Engineering, pp. 773-776, (in Japanese).
17. The Japanese Society of Soil Mechanics and Foundation Engineering (1966): Brief explanation on pictures taken at the moment of Niigata earthquake, Soil and Foundation, Vol.6, No.1, pp. i-vi.
18. The Public Works Research Institute, Ministry of Construction (1965): "Report on Niigata Earthquake," Report of The Public Works Research Institute, Ministry of Construction, No. 125, pp. 226-239 (in Japanese).
19. Wakamatsu, K., Tazoh, T., Yoshida, N., Nakazawa, H., Kiku, H., Yasuda, S. and Towhata, I. (2007): Testimony from eyewitnesses of Showa Bridge collapse in the 1964 Niigata earthquake
20. Yoshida, N. and Kudo, K. (2000): Geotechnical re-evaluation of liquefaction at Kawagishi-cho site during 1964 Niigata earthquake, Summaries of the Technical Papers of Annual Meeting of AIJ, Tohoku, Vol. Structure II, pp. 293-294 (in Japanese)

LIQUEFACTION RESISTANCE AND DYNAMIC SHEAR MODULI OF IN-SITU FROZEN AND RECONSTITUTED SANDY SOILS

KIYOTA. T[1], KOSEKI. J[2] and SATO. T[3]
[1]Institute of Industrial Science, University of Tokyo, Japan, kiyota@iis.u-tokyo.ac.jp
[2]ditto, koseki@iis.u-tokyo.ac.jp
[3]ditto, tsato@iis.u-tokyo.ac.jp

ABSTRACT

In order to investigate the difference in liquefaction behaviours between in-situ frozen samples taken from Holocene and Pleistocene deposits and their reconstituted samples, a series of undrained cyclic triaxial tests was performed. During the test, the dynamic shear moduli which would reflect the soil structure were measured by using a trigger and accelerometer system, and attempts were made to evaluate the correlation between the liquefaction resistance and dynamic shear moduli which were measured immediately before the undrained cyclic triaxial tests. In addition, although the dynamic shear moduli were decreased as decrease in effective stress during undrained cyclic teriaxial tests, the different trends of the reduction of dynamic shear moduli between Holocene and Pleistocene samples were observed. This feature may be linked with the difference in the effects of natural aging effect which frozen samples originally had.

Keywords: Liquefaction, Dynamic measurement, Shear moduli, Aging effect

1 INTRODUCTION

Liquefaction resistances of sand deposits have been evaluated by laboratory tests on in-situ undisturbed samples, and that the in-situ liquefaction resistance correlates well with the N-value of the standard penetration test (SPT). Meanwhile, several investigations about relationship between liquefaction resistance and small strain characteristics have been also carried out (e.g., Tokimatsu et al., 1986; Tokimatsu & Hosaka, 1986; Yoshimi et al., 1989). The small strain characteristic is one of the important parameters which describe the soil behaviour under working loads including small to moderate earthquake motions. And the measurements of small strain characteristics have some positive aspects in the following points; (1) they are obtained in a non-destructive test, while the SPT-N values are measured in a destructive test. (2) Although they indicate different values between static and dynamic measurement (e.g., AnhDan et al., 2002), they would reflect the current soil structure. (3) The quality of an undisturbed sample would be examined through comparisons of them obtained from laboratory tests and the wave velocity measured in the field (e.g., Tokimatsu & Hosaka, 1986). (4) They are calculated with physical meaning (small stress-strain relation or wave velocity), whereas the SPT-N values don't correspond to such properties directly. In addition, it is well known that both the liquefaction resistance and the small strain characteristics are influenced by aging effect of the sample (Koseki et al, 1999 and 2001). Hatanaka et al., (1985) reported that the liquefaction resistances of in-situ frozen samples were much larger than those of reconstituted samples even if specimen densities were similar to each other. Moreover, the change of liquefaction resistance due to sample disturbance has been also investigated (e.g., Goto, 1993).

In the present study, the dynamic shear moduli, G_d, of two kinds of in-situ frozen samples which were taken from Holocene and Pleistocene deposits, and those of reconstituted samples before and during undrained cyclic triaxial tests were measured. The liquefaction resistances of in-situ frozen sample and reconstituted sample were also shown, and attempts were made to evaluate the correlation between liquefaction resistances and G_{d0} which were measured immediately before the undrained cyclic triaxial tests. Finally, differences of liquefaction behaviour between Holocene and Pleistocene deposits were discussed based on the change in the G_d during liquefaction.

2 TEST APPARATUS

An automated triaxial apparatus illustrated in Fig. 1 was used for this study. The cylindrical specimen was approximately 50 mm in diameter and 100 mm in height. The deviator load was measured with a load cell fixed between the loading piston and the specimen top cap in the cell. The axial strain of the specimen was measured with a normal external transducer. In addition, a set of accelerometers was used to measure the arrival of S wave at two different heights on the side surface of specimen (Fig.2). S wave was generated by a pair of wave sources (triggers) attached on the top cap, which were excited simultaneously in the torsional direction. The shear moduli by dynamic measurement, G_d, were evaluated employing the following Eq. (1). To evaluate the travel time of the S wave between the two accelerometers, the rise to rise time lag of the measured wave was taken as shown in Fig.3.

$$G_d = \rho \cdot V_s^2 \tag{1}$$

where G_d is dynamic shear modulus, ρ is the mass density of the specimen, and V_s is S wave velocity.

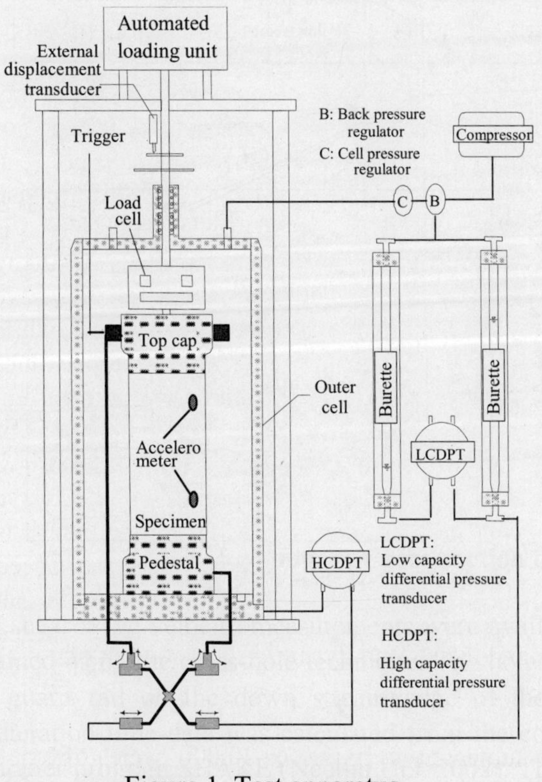

Figure 1. Test apparatus

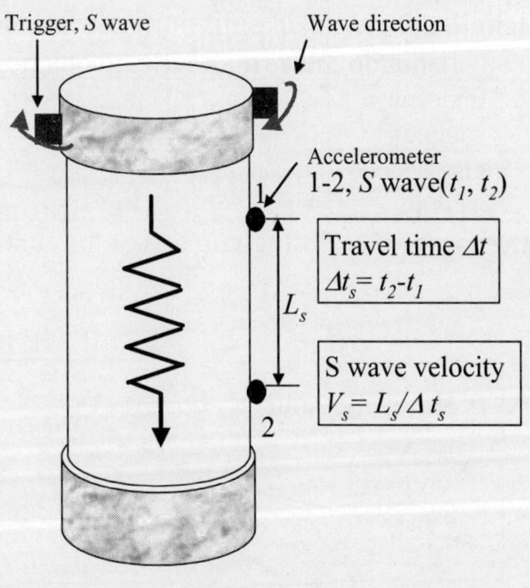

Figure 2. Diagram of S wave triggers and accelerometers on a specimen

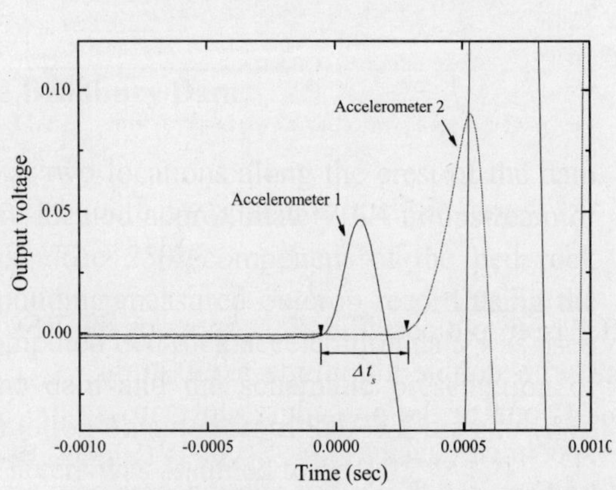

Figure 3. Definition of wave travel time

3 TEST PROCEDURES

Two kinds of in-situ frozen samples (FSs) were taken from Holocene and Pleistocene deposits. Both of FSs are sandy soils and the fine contents are less than three percents as shown in table 1. To avoid disturbance of FSs before the tests, sample preparations were carried out carefully in this study. The cylindrical specimens with a diameter of 5 cm were cored out of the frozen core samples having a diameter of 15 cm with pouring a calcium chloride solution of -12 degrees on the specimens. The specimens were transported to the freezer immediately after the coring. FSs were set into the triaxial cell and applied confining pressures of 98 kPa which is equivalent to the in-situ overburden stress at the depth of sampling, in the freezer at temperature of -20 degrees. After the preparation, the triaxial cell was transported to the testing room. FSs were thawed under atmospheric temperature of about 25 degrees, and saturated at the keeping confining pressure of 98 kPa. After confirming that the B value of the specimens is not smaller than 0.96, the confining pressure was increased by $\sigma_v' = \sigma_h' = 100$ kPa. The reconstituted samples (RSs) were prepared at almost same dry densities as the each FS and saturated at a confining pressure of 30 kPa. After confirming the B value of the specimens, RSs were subjected to isotropic consolidation to $\sigma_v' = \sigma_h' = 100$ kPa. The G_d were measured on the both FSs and RSs at $\sigma_v' = \sigma_h' = 100$ kPa, and then undrained cyclic triaxial tests were performed.

Table 1. Testing samples

Sample	Depth (GL- m)	D_{50} (mm)	F_c (%)	U_c	e_{max}	e_{min}	V_s* (m/s)	*Geological age*
Holocene sample	11.8~ 12.1	0.188	1.2	2.0	1.066	0.675	240	8•10^3 yr
Pleistocene sample	10.3~11.0	0.561	3.0	4.3	1.043	0.710	270	130•10^3 yr

*: V_s measured by PS logging test at the field

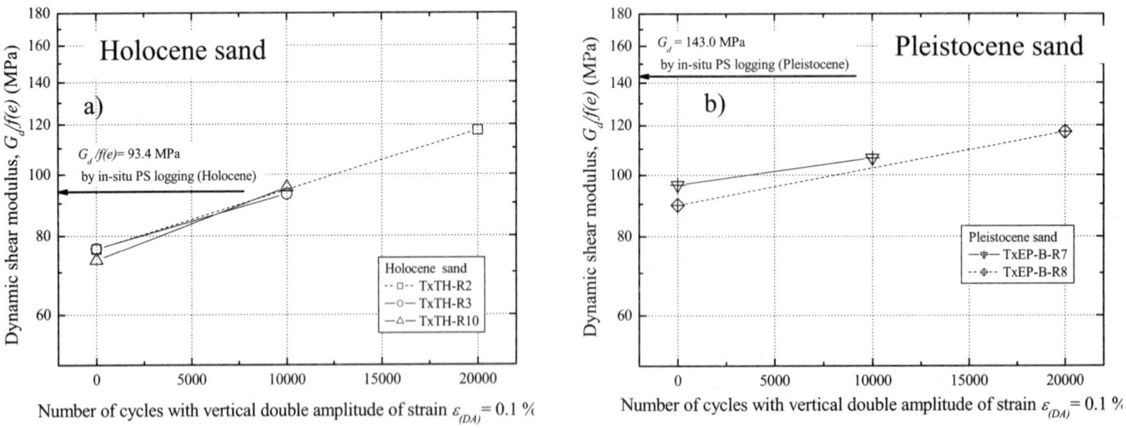

Figure 4. Relationships between number of drained cycles and dynamic shear moduli of a) Holocene, b) Pleistocene sands

After isotropic consolidation, some of the RSs were subjected to 10,000 or 20,000 cycles of axial load with double amplitude axial strain, $\varepsilon_{v(DA)}$, of approximately 0.1 % under drained condition. This is one of the procedures to increase the liquefaction strength without significantly changing the specimen density (e.g., Seed, 1979), and the RS which has stress history is called RSCL in this study. Figure 4 shows the relationships between number of drained cycles and the G_d of RSCLs of

Holocene and Pleistocene samples. In order to correct for the effects of change in void ratios, the following Eq. (2) proposed by Hardin & Richart (1963) is applied to normalize the G_d.

$$f(e) = (2.17 - e)^2/(1+e) \qquad (2)$$

Increase in the values of G_d was observed with increase in the number of drained cycles. The G_d values of the RSCL became similar to or larger than the result of in-situ PS logging test in case of Holocene sand. Although the G_d values of RSs of Pleistocene sand could not reach to result of PS logging test even though the number of cycles exceeded 20,000 cycles, the G_d values were approximately 30 % increased from the values which were measured before drained cyclic loadings.

4 TEST RESULTS

4.1 Liquefaction resistance of FSs and RSs

Considering the process of natural deposition of sandy soils, the strength of the ground would be increased with time. Specifically, the inter-locking between soil particles is structured due to change in the environments; creep with time, earthquake, change in the water table, change in the landform and so on. In addition, cementation is taken place due to chemical reaction and change in the particle shape as time passes by. These hardening processes are usually called aging effects, and difference in the aging effects were reflected on the test results significantly.

Undrained cyclic triaxial tests with constant stress amplitude were performed on FSs and RSs of Holocene and Pleistocene deposits. Figure 5 compares the relationships between shear stress ratio, $\sigma_d/2\sigma_c'$, and number of cycles required to cause double amplitude of vertical strain, $\varepsilon_{v(DA)}$= 3 % for both FSs and RSs. The liquefaction resistances defined as the shear stress ratio at 15 cycles, R_{L15}, of the RSs were approximately 40 to 50 % of those of FSs even though both of the specimens had similar relative densities. Similar results to this study were obtained by previous research (Hatanaka, et al., 1985). These results may be explained that FSs would have a natural aging effect which causes soil structure to be strengthened with inter-loking or cementation between soil particles, while RSs wouldn't have any such effects.

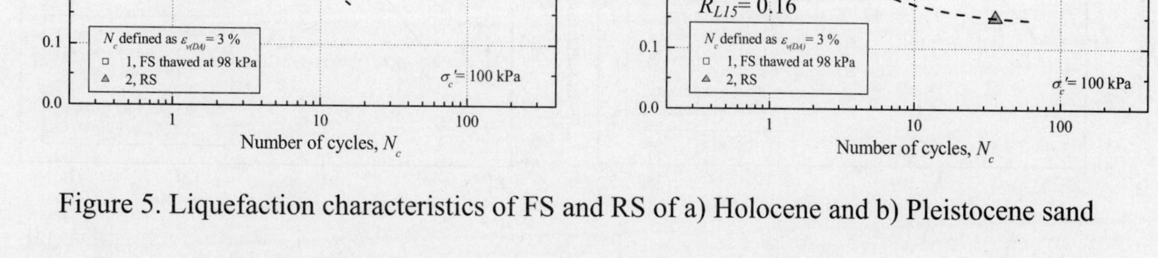

Figure 5. Liquefaction characteristics of FS and RS of a) Holocene and b) Pleistocene sand

4.2 Correlation of Liquefaction resistance and dynamic shear moduli

It is well-known that the stiffness and liquefaction resistance of the specimen are increased with stress history before liquefaction tests. In this study, to obtain similar G_d and R_{L15} with RS to those with FS, drained cyclic loading with $_{v(DA)}$= 0.1 % were applied. Although the increase in G_d due to drained cyclic loading was already shown in Fig. 4, the volumetric strains of RSCLs of Holocene and Pleistocene sands during the drained cyclic loading were relatively small (less than 0.6 % and 1.3 % with 20,000, respectively). Figure 6 shows the comparisons of liquefaction curves between FSs, RSs, and RSCLs of Holocene and Pleistocene sands. As shown Fig.6, the R_{L15} values of RSCLs were increased significantly and approached the values of the FSs after applying drained cyclic loadings with RSs. In addition, the R_{L15} values seem to be linked with the number of drained cyclic loading before undrained cyclic triaxial tests.

Figure 7 shows the relationships between R_{L15} and G_{d0} values which were measured immediately before the undrained cyclic triaxial tests. The tests results from FSs which were thawed at 30 kPa were also shown in the figure. Although the correlations between R_{L15} and G_{d0} values were not exactly matched each sample, the larger R_{L15} was observed as larger G_{d0} values consistently. Especially, In the case of RSCL with 10,000 cycles of Holocene sand, almost same G_{d0} and R_{L15} values were obtained as those of FSs which were thawed at 98 kPa. It may be considered that the increment in G_{d0} and R_{L15} values of RSCL was caused by formation of inter-locking of the soil particle due to number of drained cyclic loading, and these results suggest that G_{d0} which measured before liquefaction may be useful in estimating the liquefaction resistance, which is affected by the aging effect.

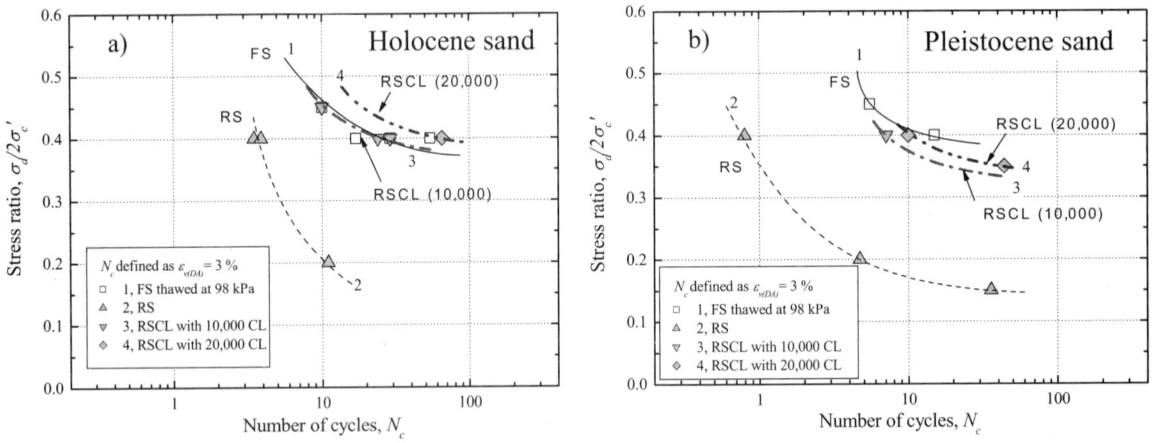

Figure 6. Liquefaction characteristics of FS, RS and RSCL of a) Holocene and b) Pleistocene sand

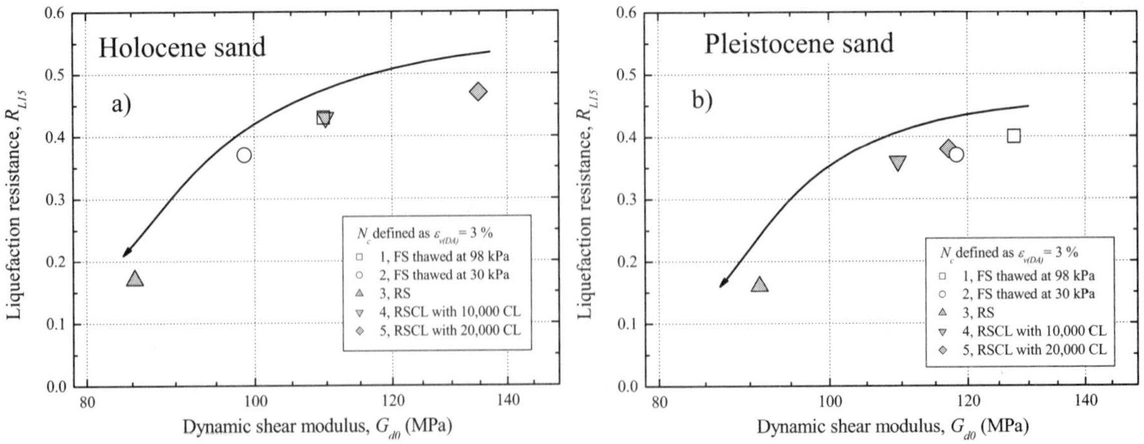

Figure 7. Relationships between liquefaction resistance and G_{d0} of a) Holocene and b) Pleistocene sand

4.3 Dynamic shear moduli during liquefaction

Note again that good correlations of liquefaction resistance, R_{L15}, with dynamic shear moduli, G_{d0}, were obtained, and RSCL could have similar R_{L15} to that of FS when similar G_{d0} values were obtained each other. Hence, in order to investigate liquefaction behaviours of FSs and RSCLs of Holocene and Pleistocene samples, G_d values during undrained cyclic triaxial tests were compared. As indicated on the stress-path and stress-strain relation of typical liquefaction test on FS of Holocene sand in Fig.8, the G_d values were measured at several stress states. Figure 9 shows the G_d values of FSs and RSCLs of Holocene and Pleistocene sands, which were measured under the isotropic state plotted versus an effective stress parameter, $(\sigma_v' \bullet \sigma_h')^{0.5}$. Note that the G_d was normalized by *f(e)* proposed by Hardin & Richart (1963). The decrease in G_d during liquefaction was observed with the reduction of the effective stress level on all specimens. In the case of Holocene sand, the change in G_d values of RSCL seems to correspond to those of FSs. On the other hand, in the case of Pleistocene sand, although a sudden reduction of G_d was observed at an initial part of the liquefaction process on RSCL, a slower reduction of G_d was observed during liquefaction with FSs. This contrastive tendency between Holocene and Pleistocene sands may indicate that two samples have different types of aging effect. Take account of the geological age of each sample, cementation effect of Pleistocene sand could be higher than that of Holocene sand. Consequently, it seems to be difficult to simulate the behaviour of FS which have a cementation effect; Pleistocene deposit, by employing RSCL even if the G_{d0} values were adjusted each other by drained cyclic loading. Meanwhile, it may be possible to adopt RSCL in substitution for FS in the case of un-cemented sample; Holocene deposit.

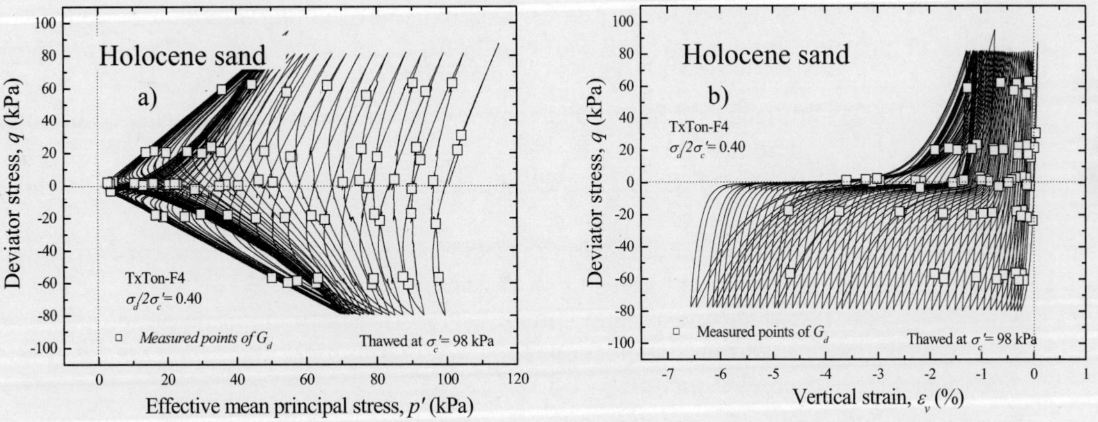

Figure 8. a) Stress-path and b) stress-strain relation of Holocene frozen sample

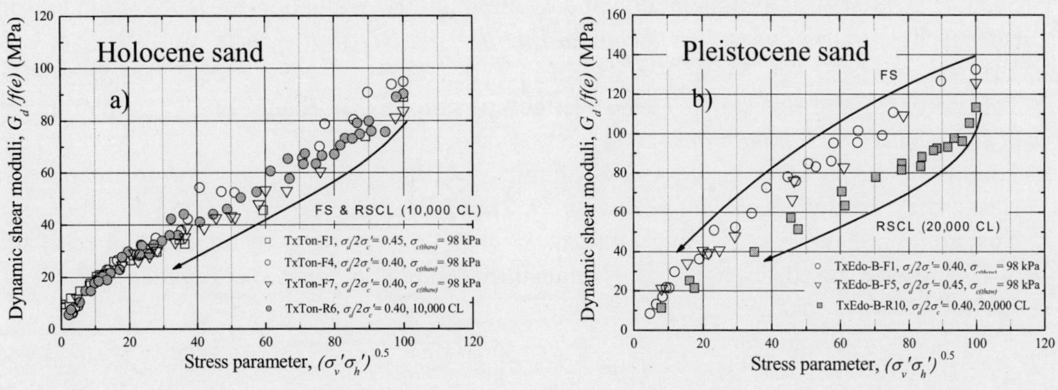

Figure 9. Comparison of G_d between FS and RSCL on a) Holocene and b) Pleistocene sands

5 CONCLUSIONS

The results from a series of undrained cyclic triaxial tests on in-situ frozen samples (FSs) which were taken from Holocene and Pleistocene deposits, and their reconstituted samples (RSs) could be summarized as follows:

1. The liquefaction resistance of the FS was larger than that of RS even though both of the specimens had similar relative densities.
2. The liquefaction resistance and shear moduli evaluated by S-wave measurement, G_d, of RS were increased with number of drained cyclic loading history (RSCL), and approached the values of the in-situ frozen sample.
3. The G_d which measured before liquefaction (G_{d0}) may be useful in estimating the liquefaction resistance, which is affected by the aging effect.
4. In the case of Holocene sand, the change in G_d values of RSCL corresponds to those of FS. On the other hand, in the case of Pleistocene sand, although a sudden reduction of G_d was observed at an initial part of the liquefaction process on RSCL, a slower reduction of G_d was observed during liquefaction with FS. Therefore, it seems to be difficult to simulate the behaviour of FS which have a cementation effect by employing RSCL even if the G_{d0} values were adjusted each other. Meanwhile, it may be possible to adopt RSCL in substitution for FS in the case of un-cemented sample.

REFERENCES

1. AnhDan. L, Koseki, J. and Sato, T. (2002): Comparison of young's moduli of dense sand and gravel measured by dynamic and static methods, *Geotechnical Testing Journal*, **25** (4), 349-368.
2. Goto, S. (1993): Influence of freeze and thaw cycle on liquefaction resistance of sandy soils, *Soils and Foundations*, **33** (4), 148-158.
3. Hardin, B. O. and Richart, F.E. (1963): Elastic wave velocities granular soils, *Journal of ASCE*, **89** (1), 33-65.
4. Hatanaka, M., Sugimoto, M. and Suzuki, Y. (1985): Liquefaction resistance of two alluvial volcanic soils sampled by in situ freezing, *Soils and Foundations*, **25** (3), 49-63.
5. Koseki, J., Sato, T., Maeshiro, N. and Urano, I. (1999): Elastic deformation properties of sands containing fines during liquefaction, *Proc. of International Workshop on the Physics and Mechanics of Soil Liquefaction*, 121-132.
6. Koseki, J. and Ohta, A. (2001): Effects of different consolidation conditions on liquefaction resistance and small strain quasi-elastic deformation properties of sands containing fines, *Soils and Foundations*, **41** (6), 53-62.
7. Seed, H.B. (1979): Soil liquefaction and cyclic mobility evaluation for level ground during earthquakes, *Journal of the Geotechnical Engineering Division*, ASCE, Vol. 105, No. GT2, 201-255.
8. Tokimatsu, K. and Hosaka, Y. (1986): Effects of sample disturbance on dynamic properties of sand, *Soils and Foundations*, **26** (1), 53-64.
9. Tokimatsu, K., Yamazaki, T. and Yoshimi, Y. (1986): Soil liquefaction evaluations by elastic shear moduli, *Soils and Foundations*, **26** (1), 25-35.
10. Yoshimi, Y., Tokimatsu, K. and Hosaka, Y. (1989): Evaluation of liquefaction resistance of clean sands based on high-quality undisturbed samples, *Soils and Foundations*, **29** (1), 93-104.

SHAKING TABLE MODEL TESTS ON RESIDUAL DISPLACEMENTS OF EARTH DAMS FOR THEIR PERFORMANCE-BASED DESIGN

***S.Sendir*[1], *J.Sato*[2] *and I.Towhata*[3]**

[1]*PhD Student, University of Tokyo, sendir@geot.t.u-tokyo.ac.jp*

[2]*MSc., University of Tokyo, junichi@geot.t.u-tokyo.ac.jp*

[3]*Professor, University of Tokyo, towhata@geot.t.u-tokyo.ac.jp*

ABSTRACT

Shaking table tests of earth dam models were performed to monitor the residual displacements caused by seismic effects. Two different types of earth dams, such as central impervious core dam and asphalt or concrete lining dam, were modeled. The asphalt or concrete lining dam showed a better performance than central impervious core dam. In central core dam model, greater deformations were observed with the decrease in relative density. Importance of the degree of compaction during construction of earth dams was also proved. In performance based design defining the performance level of the structure is an important issue. By using the observed residual displacements obtained from the shaking table test, the ratio of the settlement and height at the crest were determined. The damage levels that can be used in performance based design were also defined by classifying this ratio obtained from shaking table tests. There is a good consistency of damage level between predictions of model tests.

Keywords: Performance-based design, earth dam, shaking table test, residual displacements

1 INTRODUCTION

In recent decades, importance of researches about seismic performance of geotechnical structures such as retaining walls, embankments, earth fill dams and abutments, are increasing. The main reason for this is to obtain a better design solution which meets appropriate performance objectives that were determined before starting the design. This design method is called performance based design. A safe but more economical design can be proposed by using this method.

In this method, the allowable damage level of a geotechnical structure is decided by considering the needs of the users. The residual displacements of the geotechnical structure are evaluated by the effect of a design earthquake motion. After the end of the calculation, the calculated residual displacement is compared with the allowable damage level. If the calculated value is greater than the allowable limit, the original design has to be modified (Towhata, 2008). Therefore, in this design methodology, good prediction of residual displacement of the structures and choosing the allowable damage level that will be used in the design are the key parameters.

Previously, researchers (Shi et al., 1989, Kuwano et al., 1988) studied the prediction of residual displacements of an earth dam by the effect of an earthquake by using FEM method and laboratory experiments (as triaxial and hollow cylindrical torsional shear tests). Shi et al. were able to predict the deformations of Liyutan dam before its construction and the dam survived the 1999 ChiChi earthquake (Towhata, 2008). Also some researchers (Seed et al., 1963, Masukawa et al., 2004, Yang et al., 2004) studied residual deformation caused by earthquakes in earth dams with a clay core by modeling them in shaking table and centrifuge.

The present research aims to model two different types of earth dams, one with a central clay core and the other with asphalt or concrete lining by including the effect of reservoir. Also an attempt was made to define the damage levels that can be used in performance-based design of earth dams.

2 SHAKING TABLE TESTING PROGRAM

2.1 Dam Model Geometry, Material and Instrument Configuration

Two different dam models were investigated in this study. Fig. 1 shows the central impervious core dam and Fig.2 shows the dam model whose upstream face covered by a membrane. Impervious core was constructed by Kibushi clay with a water content of 55%. Toyoura sand was used for dam body with relative densities of 70%, 50% or 20%. The gradients of the upstream and downstream slopes were 1:2.5 and 1:2 respectively. Dam models were planned to have a 10cm width of crest, 190cm width of bottom and 40 cm height. Moreover at the foundation 10cm height of Toyoura sand was placed with a relative density of 70%.

Accelerometers, pore water pressure transducers, laser displacement transducers and inclinometers were placed as shown in Figs. 1 and 2. Inclinometers were prepared by mounting accelerometers on 7cm height of acrylic plates that were connected with each other by hinges making a column. Colored sand layers were also placed to observe the displacements in the dam body.

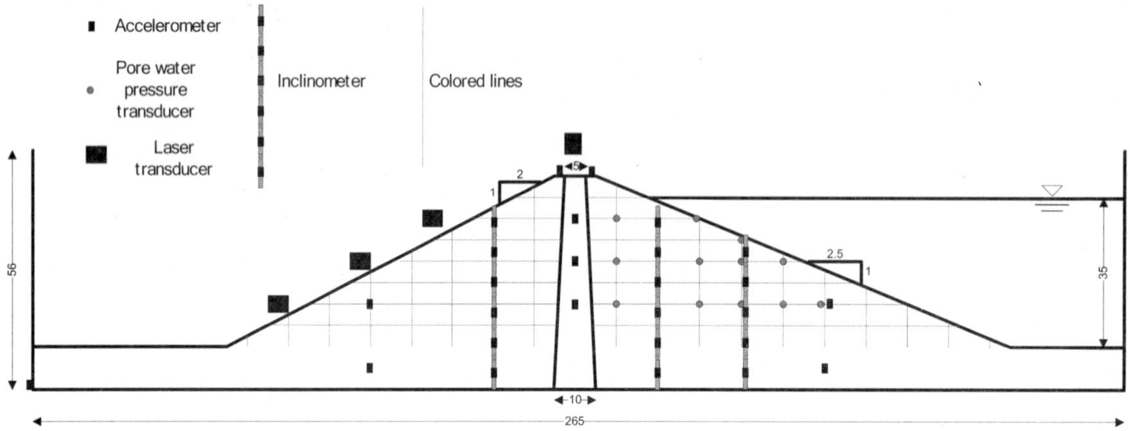

Fig. 1 Dam model with an impervious clay core

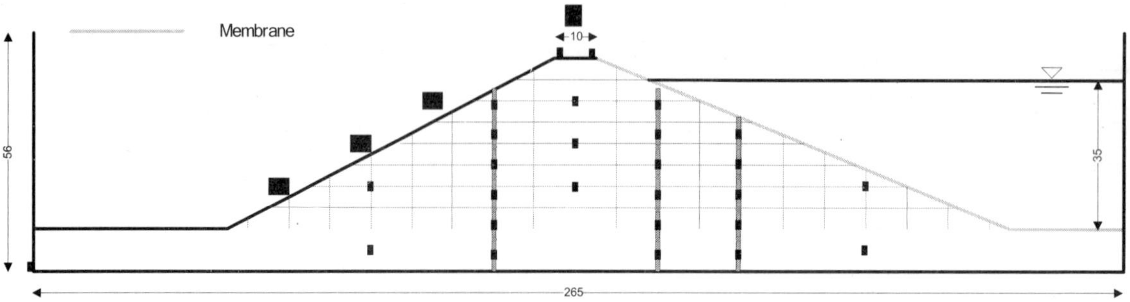

Fig. 2 Dam model that upstream face covered with membrane

2.2 Model Preparation and Input Motion

An impervious clay core model (Fig. 3) was prepared starting from the core material. Kibushi clay was mixed with water at water content of 55%, placed in mould and compacted by hand (Fig. 4). Later on mould was removed and the face was trimmed to prepare a core with 5cm width of crest, 10cm width of bottom and 50cm of height. Toyoura sand was placed by wet tamping method.

Firstly Toyoura sand was mixed with 5% of water and placed by 5cm each by compacting with 10kg of weights (Fig. 5). Sand was compacted to the desired density such as 70%, 50% or 20%. After placing of dam body was finished, the faces of dam were trimmed to have the designed slope ratios. Upstream slope face was also covered with 5mm level of gravels (Fig. 6).

Fig. 3 Impervious clay core model

Fig. 4 Placement of the clay core

Fig. 5 Placement of dam body

Fig. 6 Upstream face covered with gravel

Fig. 7 Upstream face covered with membrane

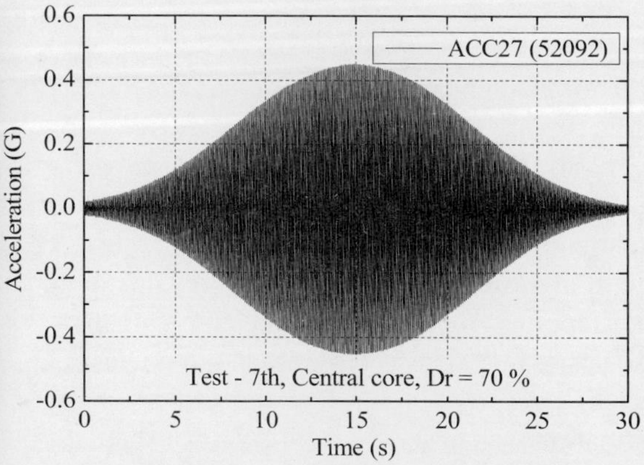

Fig. 8 Time history of shaking

In the case of the surface pavement dam, the dam body was prepared in the same way and later on the face of the upstream was covered with a membrane (Fig. 7) having a thickness of 0.3mm.

After the placement of material was finished, CO_2 was circulated through the model for 6 hours. Then reservoir was filled with water and 24 hours were spent for fully saturation. The physical models were all subjected to the same input motion, as shown in Fig. 8. The input motion was

10Hz sinusoidal wave. The amplitude of the wave was increased from 0 to 0.5G (500 gal) gradually.

3 RESULTS OF EXPERIMENTS

Figs. 9 to 12 show the shape of earth dams after the experiment. Visual inspection was done by the observation of the colored sand layers. Furthermore displacements at different depths were calculated by using the data from the inclinometers.

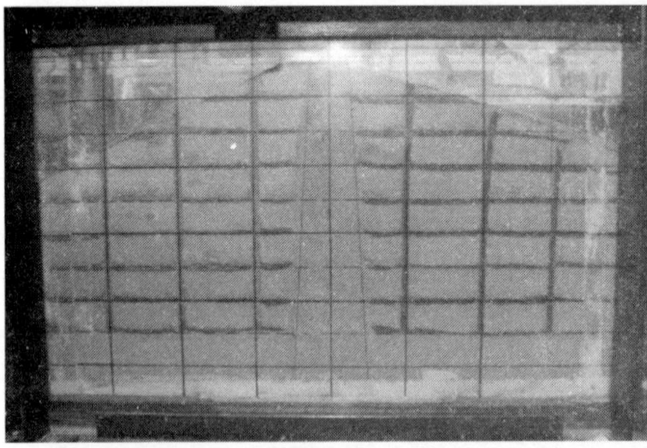

Fig. 9 Dam body Dr 70%, with central core

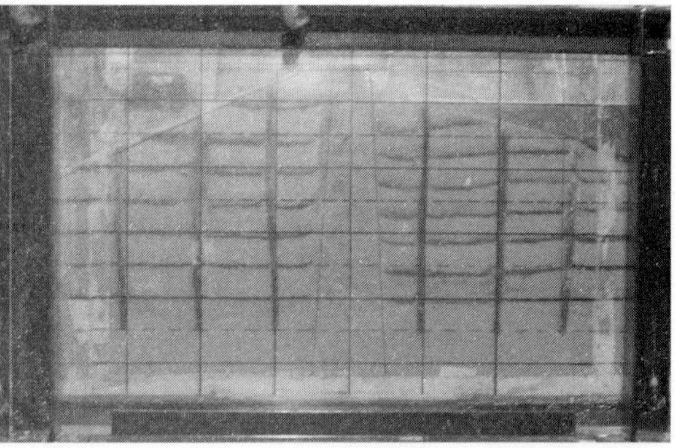

Fig. 10 Dam body Dr 50%, with central core

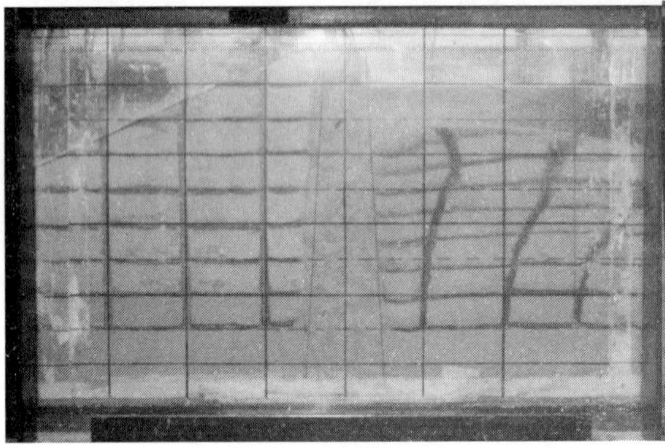

Fig. 11 Dam body Dr 20%, with central core

Fig. 12 Dam model with membrane covered face

Fig. 9 to 11 demonstrates an increase in the deformation of upstream side with the decrease of relative density of the dam body. There is not any deformation observed in the clay core part for all three cases. The deformation in upstream part was grater than the core part, while the core was supported by dam body on both sides. Moreover, the parts near the surface of the upstream and downstream easily deformed due to the low confining pressure that caused low rigidity and strength in these parts. Settlement in the upstream part was larger than in the downstream part even if the slope was milder in the upstream part than downstream part. Note that the effective stress of in the upstream part was lower than the downstream part due to the effect of water reservoir. By the effect of pore water pressure, the effective stress in the upstream part was lowered and more deformation was observed in this part of the dam body. In the case of the model with the upstream face covered by membrane (Fig. 12), no deformation was observed. In this case, water in the reservoir did not cause the effective stress in the upstream to decrease. Because of the membrane on the face, water in the reservoir did not seep and water pressure was converted to effective stress. Therefore, stability was improved.

Similar results were exhibited by inclinometers (Figs. 13 to 15). However in the downstream slope with 50% relative density, greater deformation than the downstream slope with 70% relative density was observed. This is because during the water saturation, water seeped between the side wall of the box and plastic plate that the lines were drawn. That caused the leakage through the downstream part and the decrease of the effective stress. Consequently, greater displacement was induced unintentionally.

In Fig. 16, crest settlement and dam height ratios that were evaluated from the shaking table tests, were plotted against the change in the relative density and also the data for the dam with membrane lining was included. Deformations were greater in the case of lower relative density and model with the upstream face covered by membrane showed the best performance with the effect of earthquake loading.

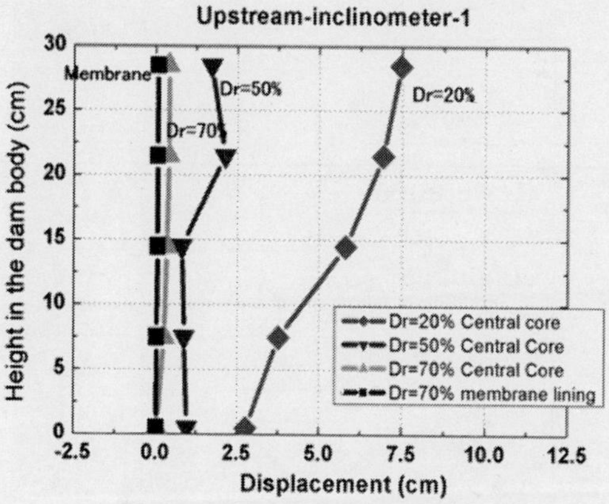

Fig. 13 Displacements in the upstream (inc1)

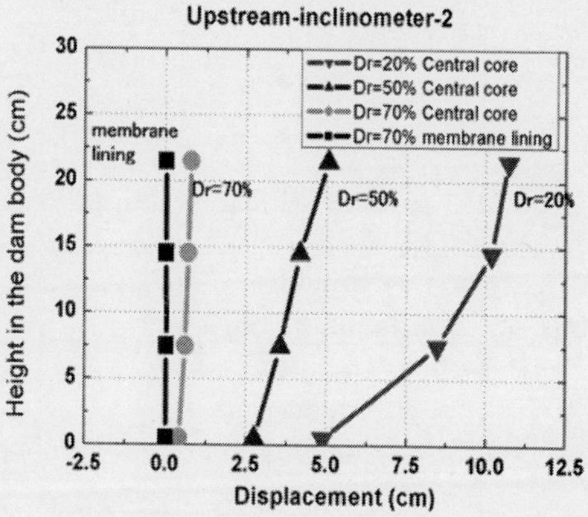

Fig. 14 Displacements in the upstream (inc2)

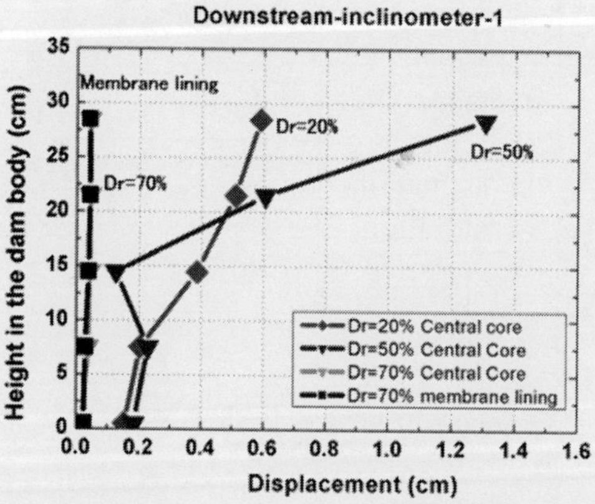

Fig. 15 Displacements in the downstream

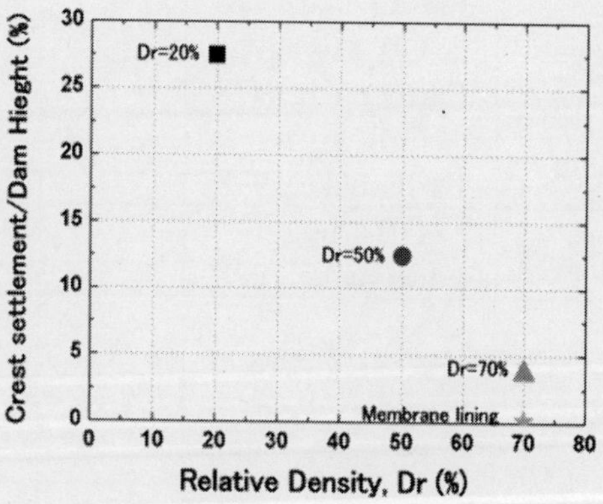

Fig. 16 Crest settlement/Dam height with respect to Dr

4 DISCUSSION

In Table.1, previous damages that were observed in earth dams by the effect of earthquakes were listed. As it can be seen from the table, greatest damage occurred in Unoki Dam whose dam body was flowed by the reservoir water after earthquake. There was no loss of human life or animals while the water was stopped by other dams in the downstream side. In the case of Kanyuu dam, there was a small amount of displacement and restoration took place quickly. The information that is gained from the past experience is important to be able to define the damage levels. It is also worth to mention that the effects of earthquakes in Table.1 change in a logarithmic manner. Restoration of the dam can be possible right after the damage occurred, 1-2 days later or will take longer while the damage was devastating.

X values in Table.2 are the ratio of crest settlement and dam height. This value is used as performance index to define the damage levels. Damage levels were described by categorizing the previous damages in earth dams by the effect of earthquakes, shaking table test results and also the information obtained from the authorities related with earth dam construction (Sato et al., 2007). X values are calculated for shaking table experiments (Table. 3). For all experiments damage levels are also defined by using Table. 2.

Table1. Previous Damages caused by Earthquakes on Earth Dams (Sato et al., 2007)

Dam	Earthquake	Dam Height (m)	Settlement (cm)	Situation after earthquake	X value(Crest settlement /dam height)	Damage level
Yugawara	Niigata Earthquake, 2004	35	70	Restoration was conducted after the water drained out of reservoir in 1 day.	1.9 %	3
Shin Yamamoto	Niigata Earthquake, 2004	27.22	> 50		1.8 %	3
Yamamoto	Niigata Earthquake, 2004	44.5	50		1.1 %	3
Unoki	Nihonkai Chubu Earthquake, 1983	4.9	340	Overflow of reservoir	69.4 %	4
Koromogawa No.1	Off-Miyagi Earthquake,1978	35.5	A little	There was a small damage. After the investigation, restoration was started.	0-0.1 %	1
Koromogawa No.5	Off-Miyagi Earthquake,1978	19.1	A little		0-0.1 %	1
Kanyuu	Off-Miyagi Earthquake,1978	27.6	A little		0-0.1 %	1
Yadonosawa	Off-Miyagi Earthquake,1978	26	A little		0-0.1 %	1
Makio	Nagano-ken Seibu (West) Earthquake, 1984	105	40	The crest settled. Quick restoration was performed	0.38 %	2

Table2. Classification of damage levels (Sato et al., 2007)

Damage level	X value(Crest settlement /dam height)	Description
4	10%<X	Collapse, High damage
3	1%<X≤ 10%	Unable to serve, need to be repaired as soon as possible
2	0.1%<X≤ 1%	Can serve with minimum performance, need to be repaired
1	0%<X≤ 0.1%	Minor damage, need to be repaired
0	0%	No damage

Table3. Crest settlement/Dam height (Sato et al., 2007)

Type of earth dam	X value (%)	Damage level
Central core, dam body Dr=20%	27.5	4
Central core, dam body Dr=50%	12.5	4
Central core, dam body Dr=70%	4.8	3
Membrane covered face	0.0	0

5 CONCLUSIONS

As a conclusion, it has been shown that asphalt or concrete lining dam had a better performance than central impervious core dam. The main reasons were the presence of pore water in the upstream of central impervious core dam and the water pressure in the reservoir was converted to effective stress in the upstream part of the asphalt or concrete lining dam.

In the central core dam model, greater displacements were observed with the decrease in relative density. Importance of the degree of compaction during construction of earth dams has also been proved. Also the observed displacements were larger in the upstream side, because of decrease in effective stress by the effect of pore water pressure.

Damage levels were described by examining previous damages on earth dams by the effect of earthquakes, the deformations that were observed in shaking table tests and the information that was gained by the authorities related with earth dam construction.

6 ACKNOWLEDGMENTS

The authors express their sincere gratitude to the Ministry of Education, Science, Sports and Culture, Government of Japan (MONBUSHO) and New Energy and Industrial Technology Development Organization (NEDO) for their financial support.

REFERENCES

1. Kuwano, J. and Ishihara, K., (1988). “Analysis of Permanent Deformations of Earth Dams due to Earthquakes”, Soils and Foundations、28(1), 41-55
2. Masukawa, S., Yasunaka M., and Kohgo, Y., (2004). “Dynamic Failure and Deformations of Dam Models on Shaking Table Tests”, 13th World Conference on Earthquake Engineering, Vancouver, Canada, Paper No. 2359
3. Sato, J., Sendir, S., Honda, T. and Towhata, I., (2007). “Shaking Table Tests and Numerical Analysis for Performance-Based Design on Earth Dams”, 29th Earthquake Engineering Conference, Fukuoka Japan.
4. Seed, H.B. and Cough, F., (1963). “Earthquake Resistance of Sloping Core Dams”, Journal of Soil Mechanics and Foundation Division, ASCE, SM1, 209-242
5. Shi, L.P., Towhata, I. and Wieland, W., (1989). “Prediction of Seismically Induced Deformation of Liyutan Dam, Taiwan, by means of Cyclic Triaxial Test and Finite Element Analysis”, Computers and Geotechnics, 7(3), 202-222
6. Towhata, I., (2008). Geotechnical Earthquake Engineering, Section 13-5, 25-1, Springer-Verlag
7. Yang, Z., Elgamal, A., Adalier, K. and Sharp, M.K., (2004). “Earth Dam on Liquefiable Foundation and Remediation: Numerical Simulation of Centrifuge Experiments”, Journal of Engineering Mechanics, 130(10), 1168-1176

POTENTIAL FLAWS IN COMPUTING LIQUEFACTION POTENTIAL INDEX, P_L

Chihping Kuo[1] and Muhsiung Chang[2]
[1]Department of Construction Engineering, National Taiwan University of Science and Technology, D9005202@mail.ntust.edu.tw
[2] Department of Construction Engineering, National Yunlin University of Science and Technology, changmh@yuntech.edu.tw

ABSTRACT

A widely used method for estimating the possibility of soil liquefaction, called Liquefaction Potential Index (P_L), was proposed by Iwasaki in 1982. The index is computed in form of integration by multiplying the danger function (1-safety factor) and the depth weight function. Theoretically, the computation should adequately reflect the liquefaction potential for the depth interval calculated. If the danger function is calculated based on CPT logs, the integration would be more desirable due to the continuity in the data. However, SPT logs are much more common and generally used in the engineering practices today. The danger function calculated by SPT logs would become erroneous (sometimes overestimated) in cases where the integration is performed by summing up the discontinuous SPT values with irregularly-spaced depth intervals. If the SPTs are uniformly spaced, the calculation should be correct. A misjudgment would be made and sometimes lead to an overestimated P_L threshold value if the SPT log is non-uniformly distributed. Accordingly, the study herein proposes a simple method in order to correct the above mentioned issue. The result shows that the correction can reduce the computation error and have the calculated P_L values fall into the right ranges.

Keywords: Liquefaction Potential Index, Soil Liquefaction, SPT

1 INTRODUCTION

The method for liquefaction damage assessment and liquefaction potential prediction currently adopted in the Taiwanese Building Codes (Chap.10) is based on Iwasaki's (1982) liquefaction potential index, P_L, which was also suggested by ISSMFE TC (1993). This method was first proposed by Iwasaki (1978) and used for liquefaction damage assessment during earthquakes. The assessment was based on a collected database from several post-earthquake liquefaction incidents in Japan, which was constantly augmented and revised (Tatsuoka et al. 1980, Iwasaki et al. 1982).
The Chichi Earthquake of 1999 had caused serious soil liquefaction phenomena in the mid-west plains of Taiwan. Most of the liquefaction locations could be corrected estimated based on Iwasaki's procedure after the earthquake. However, some liquefaction sites were inconsistently computed with liquefaction potentials too low. It appears that the Iwasaki procedure or its database would probably need to be slightly adjusted for use in Taiwan. Previous studies (Kuo et al. 2003) had shown the liquefaction evaluation would be influenced by the amount and completeness of borehole data.
Based on Iwasaki (1982), the value of P_L is computed through the integration of the soil liquefaction danger function (1 minuses liquefaction factor of safety) with the depth weight function. If the computed soil liquefaction factor of safety is continuous with depth, such as for CPT logs, the P_L should confirm with its definition. For the SPT logs, however, the integration is approximated by the summation with respect to several depth intervals. Since depth intervals may vary in response to the change in material strata, the P_L computation would therefore be misleading, and sometimes overestimating the liquefaction potential index, as originally proposed by Iwasaki (1982). The paper herein will discuss the potential flaws in this regard.

2 REVIEW SUMMARY OF P_L DERIVATIONS

2.1 P_L Definition

Liquefaction potential index, P_L, was developed and defined by Iwasaki et al. (1978, 1982) and Tatsuoka et al. (1980), and the reviews of the documents are summarized as follows.

- Based on liquefaction case histories, distributions of the computed liquefaction safety factor (F_L) were found generally increasing with depth. When the computed liquefaction safety factor $F_L <$ 1.0, the sites were generally evidenced by liquefaction damages.
- P_L was adopted for quantifying the soil liquefaction potential for a single borehole. The depth was considered from the ground surface to 20m below. P_L was defined with the consideration of liquefaction safety factor at given depths as well as the influence of the depths. P_L ranges from 0 to 100; in which 0 represents no liquefaction damage and 100 mean a complete liquefaction damage.
- The weight function, $W(z)$, was used to consider the influence of depth where liquefaction factor of safety is computed. The function was assumed to be a reverse triangle distribution within the 20m below the ground.
- The liquefaction potential index, P_L, was defined as follow, with reference to Fig.1:

$$P_L = \int_0^{20} F(z) \cdot W(z) \cdot dz \tag{1}$$

Where $F(z) = 1 - F_L$ (2)

$W(z) = 0.5 \cdot (20 - z)$ (3)

F_L = factor of safety against liquefaction (considered only $\leqq$ 1.0)

z = depth (m)

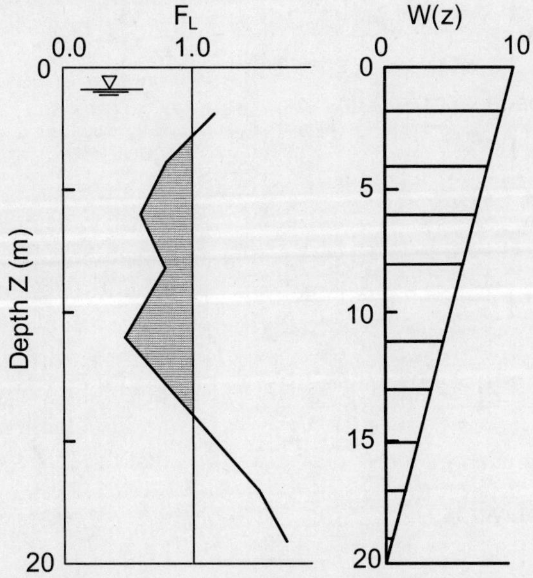

Fig.1. Schematic illustration of F_L and $W(z)$ distributions (after Iwasaki et al. 1982)

- For non-liquefaction sites, most of the computed P_L's were less than 15; and about 70% of them were less than 5. For liquefaction sites, only 20% cases were computed with P_L less than 5; and 50% of them higher than 15.
- The liquefaction damage thresholds were proposed as: (i) $P_L = 0$ stands for very low liquefaction risk; (ii) $P_L \leq 5$ stands for low liquefaction risk; (iii) $5 < P_L \leq 15$ stands for high liquefaction risk; (iv) $P_L > 15$ stands for very high liquefaction risk.

- A comparison of two liquefaction safety factor estimation procedures (Seed & Idriss 1971, Tatsuoka et al. 1080) indicated that different P_L would be resulted. The authors suggested that P_L should be computed using Iwasaki's procedure for liquefaction safety factor.
- The method seemed to be more reliable due to completeness of database. However, the database was established by the cases in Japan.

2.2 Expansion of P_L's Integration

In computing, the P_L's integration needs to expand due to discontinuous distribution of F_L with depth. However, adoption of depth range for each F_L depends on the distribution of SPTs with depth and to some degrees on personal judgment. The boundary between depth interval is generally considered to be the bisector of two corresponding locations of F_Ls. Accordingly, Eq.(1) can be derived and shown in the following, with reference to Fig.2:

$$P_L = \int_0^{20} F(z)\cdot W(z)\cdot dz = \sum_0^{20m} P_{L(i)} \tag{4}$$

Where $$F(z) = 1 - F_{L(i)} \tag{5}$$

For the first layer (i=1), P_L can be expressed as:

$$\begin{aligned} P_{L(1)} &= \int_0^{\frac{(z_1+z_2)}{2}} F(z)\cdot W(z)\cdot dz \\ &= \int_0^{\frac{(z_1+z_2)}{2}} \left(1-F_{L(1)}\right)\cdot(10-0.5\cdot z)\cdot dz \\ &= \left(1-F_{L(1)}\right)\cdot\left(10\cdot\left(\frac{z_1+z_2}{2}\right)-\frac{1}{4}\cdot\left(\frac{z_1+z_2}{2}\right)^2\right) \end{aligned} \tag{6}$$

Similarly, the P_L of the second layer (i=2) is:

$$\begin{aligned} P_{L(2)} &= \int_{\frac{(z_1+z_2)}{2}}^{\frac{(z_2+z_3)}{2}} F(z)\cdot W(z)\cdot dz \\ &= \left(1-F_{L(2)}\right)\cdot\left(10\cdot\left(\frac{z_2+z_3}{2}\right)-\frac{1}{4}\cdot\left(\frac{z_2+z_3}{2}\right)^2\right) \\ &\quad -\left(1-F_{L(2)}\right)\cdot\left(10\cdot\left(\frac{z_2+z_1}{2}\right)-\frac{1}{4}\cdot\left(\frac{z_2+z_1}{2}\right)^2\right) \end{aligned} \tag{7}$$

In general, the P_L of the i-th layer is:

$$\begin{aligned} P_{L(i)} &= \int_{\frac{(z_{i-1}+z_i)}{2}}^{\frac{(z_i+z_{i+1})}{2}} F(z)\cdot W(z)\cdot dz \\ &= \left(1-F_{L(i)}\right)\cdot\left(10\cdot\left(\frac{z_i+z_{i+1}}{2}\right)-\frac{1}{4}\cdot\left(\frac{z_i+z_{i+1}}{2}\right)^2\right) \\ &\quad -\left(1-F_{L(i)}\right)\cdot\left(10\cdot\left(\frac{z_i+z_{i-1}}{2}\right)-\frac{1}{4}\cdot\left(\frac{z_i+z_{i-1}}{2}\right)^2\right) \end{aligned} \tag{8}$$

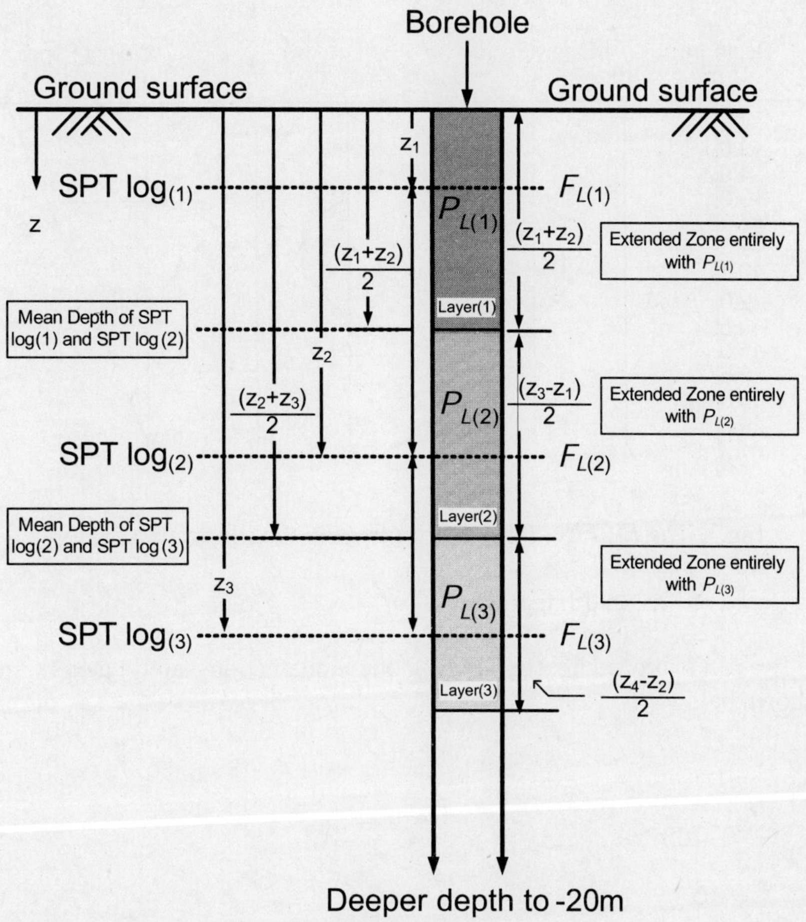

Fig.2 SPT logging and depth intervals for P_L computation (after Kuo et al. 2004)

3 POSSIBLE FLAWS IN P_L COMPUTATION

3.1 Illustrative Example

Suppose liquefaction safety factors of a site are listed in Table 1. The liquefaction danger function, $F(z)$, and the weight function, $W(z)$, are shown in Fig.3.

Table 1. Computed safety factors at different depths

Depth, z (m)	Safety factor, F_L	Depth, z (m)	Safety factor, F_L
0	-	-12.0	0.75
-1.5	1.5	-13.5	-
-3.0	1.5	-15.0	-
-4.5	1.5	-16.5	1.5
-6.0	1.0	-18.0	1.75
-7.5	0.5	-19.5	2.0
-9.0	-	-20.0	-
-10.5	0.5	-	-

Note: "-" indicates no SPT data available.

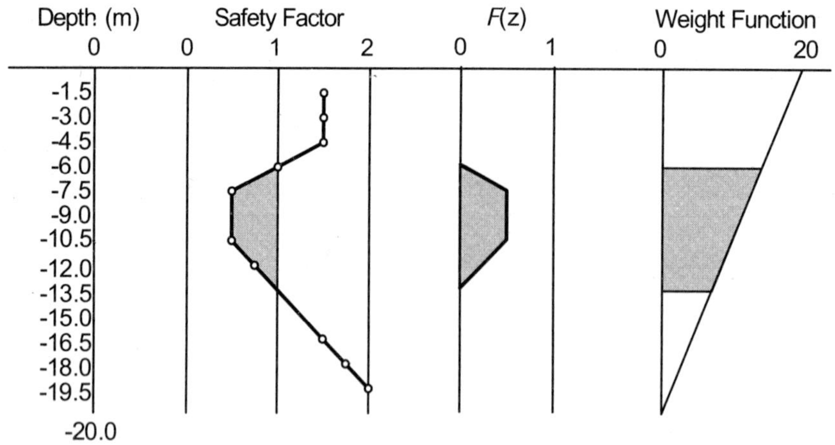

Fig.3. The F_L, $F(z)$, $W(z)$ distributions in the illustrative example.

3.2 Results Based on Original Integration

By following Iwasaki's procedure (i.e. Fig.1), the liquefaction safety factors in various depth intervals are shown below:

$$F_L(z)\Big|_{z=-4.5}^{z=-1.5} = 1.5 \quad (9)$$

$$F_L(z)\Big|_{z=-7.5}^{z=-4.5} = -0.33 \cdot z + 3 \quad (10)$$

$$F_L(z)\Big|_{z=-7.5}^{z=-6.0} = 0.5 \quad (11)$$

$$F_L(z)\Big|_{z=-19.5}^{z=-10.5} = 0.167 \cdot z - 1.25 \quad (12)$$

The above equations are then converted to the danger functions as:

$$F(z)\Big|_{z=-7.5}^{z=-6.0} = 0.33 \cdot z - 2 \quad (13)$$

$$F(z)\Big|_{z=-10.5}^{z=-7.5} = 0.5 \quad (14)$$

$$F(z)\Big|_{z=-13.5}^{z=-10.5} = -0.167 \cdot z + 2.25 \quad (15)$$

Finally, the P_L of the borehole was computed as:

$$P_L = \int_{6}^{7.5} F(z)\Big|_{z=-7.5}^{z=-6.0} \cdot W(z) \cdot dz + \int_{7.5}^{10.5} F(z)\Big|_{z=-10.5}^{z=-7.5} \cdot W(z) \cdot dz + \int_{13.5}^{10.5} F(z)\Big|_{z=-10.5}^{z=-13.5} \cdot W(z) \cdot dz = 13.88 \quad (16)$$

3.3 Results Based on Expanded Integration

In order to compute P_L by the expanded integration procedure (Fig.2), the information in Table 1 needs to be revised. As shown in Table 2 and Fig.4, the computed P_L is **15.09**.

3.4 Remarks on Discrepancy

The P_L computed by the expended integration (15.09) is higher than the original integration (13.88) by about 8.7%. Although the error appears to be small, the result would bring the liquefaction risk level from "high" to "very high", as proposed by Iwasaki et al. (1982), and would increase extra costs for liquefaction mitigation works. Causes of the discrepancy in the computed P_L could be explained through Fig.4. The problem appears to occur in the depth range of -12.0m to -16.5m. Since SPT was missing at depths of -13.5m and -15.0m, no liquefaction safety factors were computed at these depths; and therefore, discrepancies resulted.

Table 2. Computing procedure and computed result of expanded integration

Depth, z (m)	Boundry of depth range (limit of segmental integration)	Safety factor, F_L	Dangerous factor, F F=1-F_L	$P_{L(i)}$
0	-	-	-	-
	-0.75			0
-1.5		1.5	0	
	-2.25			0
-3.0		1.5	0	
	-3.75			0
-4.5		1.5	0	
	-5.25			0
-6.0		1.0	0	
	-6.75			
-7.5		0.5	0.5	**6.82**
	-8.25			
-9.0		-	-	
	-9.75			
-10.5		0.5	0.5	**5.55**
	-11.25			
-12.0		0.75	0.25	
	-12.75			
-13.5		-	-	**2.72**
	-14.25			
-15.0		-	-	
	-15.75			0
-16.5		1.5	0	
	-17.25			0
-18.0		1.75	0	
	-18.75			0
-19.5	-	2.0	0	-
				P_L = **15.09**

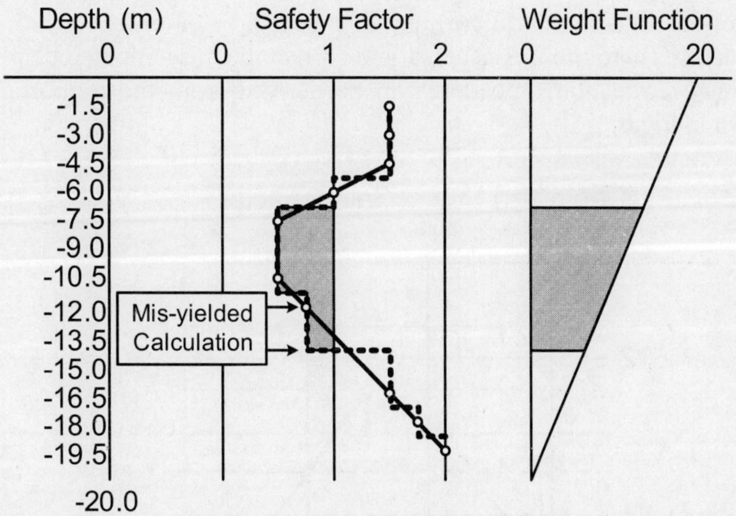

Fig.4. Comparisons of safety factor distributions used in original and expanded integrations.

4 SUGGESTED CORRECTION PROCEDURE

Due to the missing data in SPT logs, four situations of the adjacent depth data are possible for a given depth with known F_L (<1). As shown in Fig.5, the four situations are illustrated and discussed in the following:

① Circumstance describes while the adjacent safety factors stretch from safe side to unsafe side, i.e. F_L<1 and $(1\text{-}F_{L(i)})=(F_{L(i\text{-}1)}\text{-}1)$, the miscomputed result will not be induced and vice versa.

② Circumstance describes while the safety factors are all in unsafe side, no matter adjacent or not (the interval depth of SPT is not uniform), the miscomputed result will not be induced.

③ Circumstance describes while the safety factors are all in unsafe side, no matter adjacent or not (the interval depth of SPT is not uniform), and one of them the value is 1 (F_L=1), the miscomputed result will not be induced.

④ Circumstance describes while the safety factors, no matter adjacent or not, that strength from unsafe side to safe side, and $(1-F_{L(i)}) \neq (F_{L(i-1)}-1)$, the miscomputed result will be induced and vice versa.

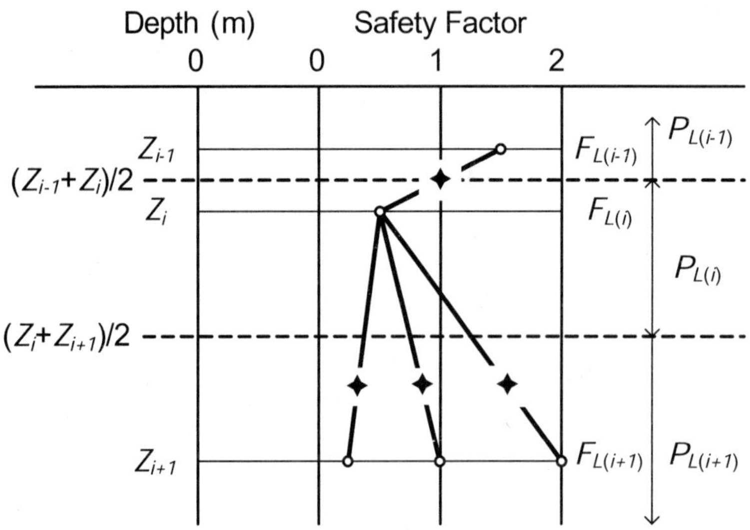

Fig.5. Situations of safety factor distributions at a depth with adjacent SPT data missing.

To conclude above four situations, while $F_{L(z)}$=1, but $z \neq (z_{i-1}+z_i)/2$ or $(z_i+z_{i+1})/2$, the miscomputed result must be induced. Therefore it is necessary to correct the miscomputed depth into proper one. The above mentioned correction procedure can be derived from the proportion relationship in geometry, as shown in Fig.6.

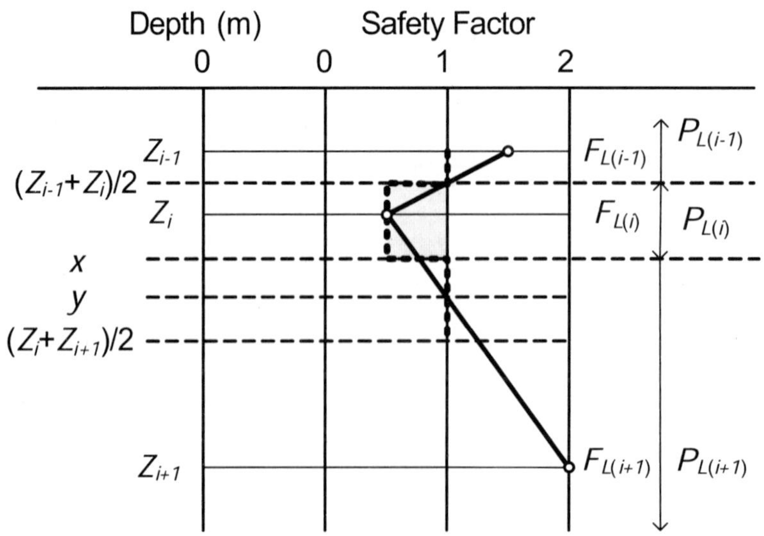

Fig.6. Correction of missing data for liquefaction potential index calculation.

In this case, the necessary correction of depth range is from $(z_{i-1}+z_i)/2$ to $(z_i+z_{i+1})/2$. By assuming $F_{L(z)}=1$ (corresponding to $z=y$ and $x=(y-z_i)/2$), the correct calculation range of $F_{L(i)}$ should be $(z_{i-1}+z_i)/2$ and x. Accordingly, the correction equations are shown below:

$$\frac{y-z_i}{z_{i+1}-y}=\frac{1-F_{L(i)}}{F_{L(i+1)}-1} \tag{17}$$

$$\Rightarrow y-z_i=\frac{1-F_{L(i)}}{F_{L(i+1)}-1}\cdot\left(z_{i+1}-y\right) \tag{18}$$

$$\Rightarrow y=\frac{z_{i+1}\cdot\left(1-F_{L(i)}\right)+z_i\cdot\left(F_{L(i+1)}-1\right)}{\left(F_{L(i+1)}-F_{L(i)}\right)} \tag{19}$$

$$\Rightarrow x=\frac{y-z_i}{2}=\frac{z_{i+1}\cdot\left(1-F_{L(i)}\right)+z_i\cdot\left(F_{L(i+1)}-1\right)-z_i\cdot\left(F_{L(i+1)}-F_{L(i)}\right)}{2\cdot\left(F_{L(i+1)}-F_{L(i)}\right)} \tag{20}$$

In the original case, the depth range for the safety factor at depth -12m is from -11.25m (the upper limit) to -14.25m (the lower limit), in which the lower limit was miscomputed. According to Fig.6 and Eq.(20), the lower limit of the depth interval should be corrected to -13.125m, as shown Fig.7. By replacing the corrected depth interval into Table 2, the corresponding $P_{L(i)}$ at this depth was revised to be 1.33; and the overall P_L became 13.70. Compared with the original result (13.88), the error was reduced to -1.3%.

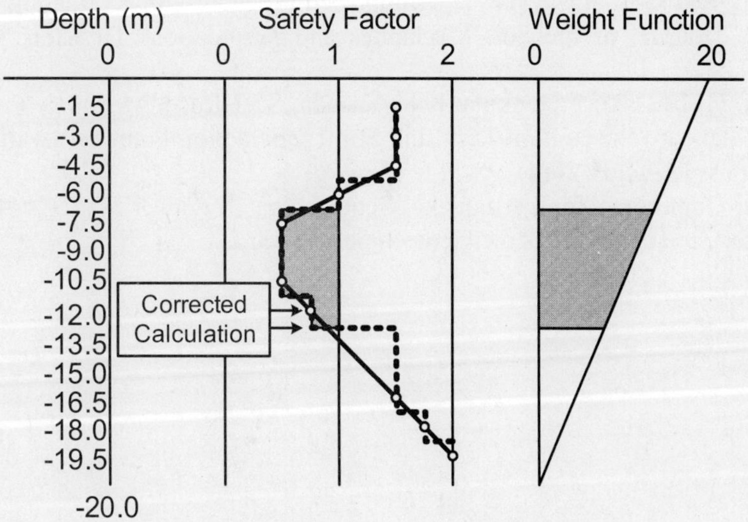

Fig.7. Corrected calculation for the illustrative example.

5 CONCLUSIONS AND SUGGESTIONS

1. Iwasaki's liquefaction potential index, P_L, has been considered as a convenient and easy measure for liquefaction damage assessment. However, the illustrative example appeared to clarify the computed error that may be induced as a result of the discontinuity and non-uniformity in the SPT logging.
2. Possible miscomputed situation may occur due to segmental integration ranges that can not properly reflect the F_L distribution and lead to an overestimation of the F_L value.
3. The proposed correction procedure in this paper could properly adjust the segmental ranges of integration and the results were reasonably close the original integration values.

4. In practice, the calculation method by expanded integration is generally adopted. However, the method may cause miscomputation in the liquefaction potential index and then impact the engineering design and construction cost. It is therefore suggested that the results by the original and the expanded integration methods be compared with care.
5. Iwasaki indicated uncertainties remained in the proposed ranges of liquefaction risk. It appears that more attentions are needed for the computed liquefaction potential indices fall close to the threshold values, as the one shown in the illustrative example of this paper.

REFERENCES

1. Iwasaki, T., Tatsuoka, F., Tokida, K., and Yasuda, S., (1978). "A Practical Method for Assessing Soil Lliquefaction Potential," Proceedings of 2nd. International Conference on Micro-Zonation for Safer Construction Research and Application, 885-896
2. Iwasaki, T., Arakawa T., and Tokida, K., (1982). "Simplified Procedures for Assessing Soil Liquefaction during Earth-quakes," Proceedings of Soil Dynamics & Earthquake Engineering Conference, Southampton, 925-939.
3. Kuo, C.P., Chang, M.H., Hsu, R.Y., and Shau, S.H., (2003). "A Study on Localization of Iwasaki's Liquefaction Potential Index (P_L) in Taiwan," Proceedings of the 10th. Conference on Current Researches in Geotechnical Engineering in Taiwan, Taoyuan, 717-720.
4. Kuo, C.P., Chang, M.H., Shau, S.H., and Hsu, R.Y., (2004). "Suitability of Using Iwasaki's PL in the Characterization of Liquefaction Damages During the 1999 Chi-Chi Earthquake, Taiwan," Proceedings of ISC'2–2004 International Site Characterization, Porto, Portugal, 713-719.
5. Seed, H.B. and Idriss, I.M., (1971). "Simplified Procedure for Evaluating Soil Liquefaction Potential," Journal of the Soil Mechanics and Foundations Division, ASCE, **97**(SM9), 1249-1273.
6. Tatsuoka, F., Iwasaki, T., Tokida, K.I., Yasuda, S., Hirose, M., Imai, T. and Kon-no, M., (1980). "Standard Penetration Tests and Soil Liquefaction Potential Evaluation," Soils and Foundations, JSSMFE, **20**(4), 95-111.
7. Technical Committee for Earthquake Geotechnical Engineering, TC4, ISSMFE, (1993). Manual for Zonation on Seismic Geotechnical Hazards.

MODEL TESTS ON SEISMIC PERFORMANCE OF REINFORCED SOIL RETAINING WALLS BY USING DIFFERENT GEO-GRIDS

Nakajima, S.[1], Hong, K.[2], Mulmui, S.[3], Koseki, J.[4], Watanabe, K.[5] and Tateyama, M.[6]

[1]*University of Tokyo, snakaji@iis.u-tokyo.ac.jp*
[2]*University of Tokyo, hkimhor@iis.u-tokyo.ac.jp*
[3]*University of Tokyo, smulmi@iis.u-tokyo.ac.jp*
[4]*Institute of Industrial Science, University of Tokyo, koseki@iis.u-tokyo.ac.jp*
[5]*West Japan Railway Company, kenji-watanabe2@westjr.co.jp*
[6]*Railway Technical Research Institute, tate@rtri.or.jp*

ABSTRACT

Results from a series of shaking table model tests on geo-grid reinforced soil retaining walls are discussed so as to investigate into the effects of material properties of geo-grid (i.e. pull out resistances, rapture strength and tensile rigidity) on seismic performance of the reinforced soil retaining walls. Residual displacements of wall facing and settlement of backfill were almost equal to each other even though the material properties obtained from pullout tests and direct tension tests of geo-gird models were largely different.

Keywords: Reinforced soil retaining wall, geo-grid model, residual displacement, material property

1 BACKGROUND

Reinforced soil retaining walls with full height rigid facing showed good seismic performance during recent earthquakes in Japan (Tatsuoka et. al.[1]). Because of their good performances, it is requested to shift our seismic design procedure to performance based design from conventional limit equilibrium method. In the performance based design procedure, seismic performances of the retaining walls are typically evaluated by verifying whether the residual displacement would exceed the allowable one or not.

A series of shaking table model tests under gravitational field were conducted by the authors (Watanabe et. al.[2]) so as to investigate the seismic behavior of the reinforced soil retaining walls. Based on the analysis of the results from these model tests, the authors found that sliding displacement of the wall facing was induced by shear deformation of subsoil just beneath reinforced backfill while tilting of the wall facing was induced by shear deformation of the reinforced backfill. The authors also developed a procedure to predict residual displacements of the reinforced soil retaining walls (Koseki et. al.[3]) based on these observations. In these model tests, geo-grid models made of phosphor bronze strips were used without consideration of the similitude because of several reasons. Firstly, use of the phosphor bronze as a reinforcement models made it possible to measure the tensile forces mobilized during the shaking by pasting strain gages on the surface of the model. Secondly, internal stabilities of the reinforced soil retaining walls have been focused in our studies.

Therefore, effects of the material properties of the geo-grid (pullout resistances, rapture strength and tensile rigidity) on the seismic performance of the reinforced soil retaining walls weren't yet studied by the authors. Based on the above background, shaking table model tests on reinforced soil retaining walls by using different geo-grid models were carried out. Their results were analyzed in this study with the consideration of the material properties of the geo-grid models by conducting pullout tests and direct tension tests of the geo-grid models.

2 MATERIAL PROPERTIES OF GEOGRID MODELS

2.1 Geo-grid models

Two types of geo-grid models were used in this study. For the one type of the model, which will be called as PB hereafter, phosphor bronze strips having a width of 3mm and a thickness of 0.1mm were prepared in a lattice shape, and their cross points were fixed by soldering as shown in Fig.1. Sand particles were pasted on the surface of the reinforcement PB so as to mobilize the frictional resistances effectively. For the other type of model as also shown in Fig.1, a polyester mesh sheet having a thickness of 0.6mm and a mesh size of 3mm was used. The latter model will be called as PE hereafter.

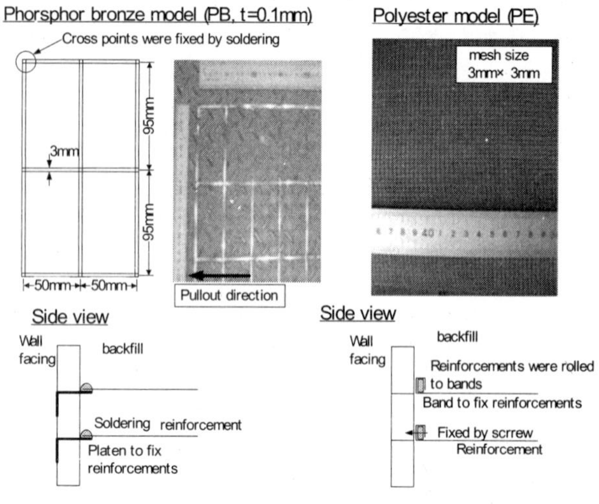

Fig.1 Geo-grid models

2.2 Pullout tests

Pullout tests were carried out so as to evaluate the pullout resistances of the geo-grid models. Test apparatuses are shown in Fig.2. A specimen of the reinforcement model having a width of about 27cm and a length of 50cm (the total length including the portions in air was about 90cm) was placed on a model ground consisting of air dried dense Toyoura sand having a relative density of about 90%. Density of the model ground in pullout tests was set equal to the backfill layers in the shaking table model tests which will be discussed later.

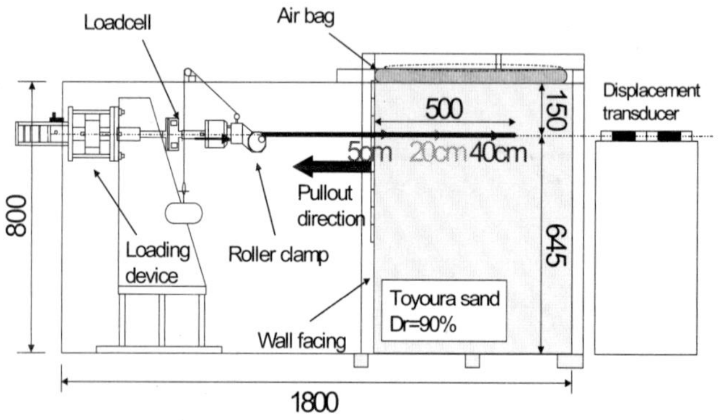

Fig.2 Apparatuses for pullout tests (unit in mm)

The edge of the reinforcement model was held by a roller clamp. After applying an overburden pressure by air bag, the reinforcement was pulled out from the ground at a constant speed of about 1mm/min. So as to simulate the stress state of the backfill layers in shaking table model tests, overburden pressures of the pullout tests were set equal to 5kPa(middle height of the backfill) and 10kPa(bottom of the backfill). Pullout displacements were measured at a horizontal distances of 5cm, 20cm and 40cm by using displacement transducers and wires which were connected to the reinforcements. Pullout resistances are plotted versus the pullout displacement measured at 5cm from the wall facing in Fig.3. With increase in the overburden pressure, the peak resistance T_{peak}

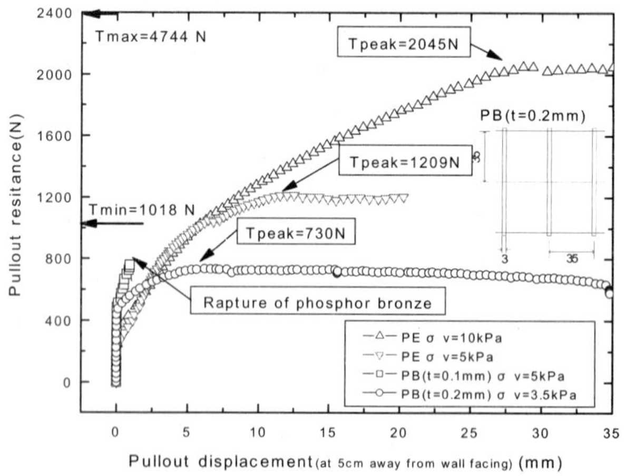

Fig.3 Relation ship between pullout resistance and pullout displacements

was increased in case of the model PE. Rapture of the phosphor bronze strips occurred at the pullout resistance of about 800N in case of the PB (t=0.1mm) as shown in Fig.4. Test result on the model PB (t=0.2mm)
having a different shape and thickness (Nojiri et. al. [4]) is also plotted in Fig.3. The peak pullout resistance of the model PB(t=0.1mm) was thought to be lower than that of the model PE by referring to the test results on the model PB(t=0.1mm) although the shapes and the test conditions were slightly different.

Kuwano et. al. [5] summarized results from the pullout and direct tension tests on various types of commercial geo-grids. Based on the results, the pullout resistances of the commercial geo-grids at the overburden pressure of 5kPa could be evaluated by using Eq.1.

$$T = 2 \times A\ (\sigma_v \tan\varphi + c) \cdots \text{Eq.1}$$

where A is the area of the reinforcement located in the ground, σ_v is the overburden pressure, φ is frictional angle between the reinforcement and the ground and c is the cohesion.

The maximum and the minimum values (T_{max} and T_{min}) reported by Kuwano et. al. are also indicated by horizontal arrows in Fig.4. It was found from this comparison that the pullout resistances of the geo-grid models used in this study were relatively lower than those of commercial geo-grids.

The strain distributions, when the pullout displacements reached to 0.5, 1.0, 5.0 and 10mm at the 5cm from the wall facing, are shown in Fig.5. The strains were obtained by normalizing relative displacements between two measured locations by their distances. In case of the model PE, tensile strain of the geo-grid model was increased when the pullout displacement exceeded about 1mm. In contrast to the model PE, strain of the geo-grid models weren't accumulated during the whole test in case of the model PB. These behaviors imply that the model PE was pulled out from the model ground with its tensile strain mobilized, while the model PB was pulled out from the ground with mobilizing much smaller tensile strains.

2.3 Direct tension tests

To evaluate the rapture strength and tensile rigidity of the geo-grid models, direct tension tests of the geo-grid models were conducted

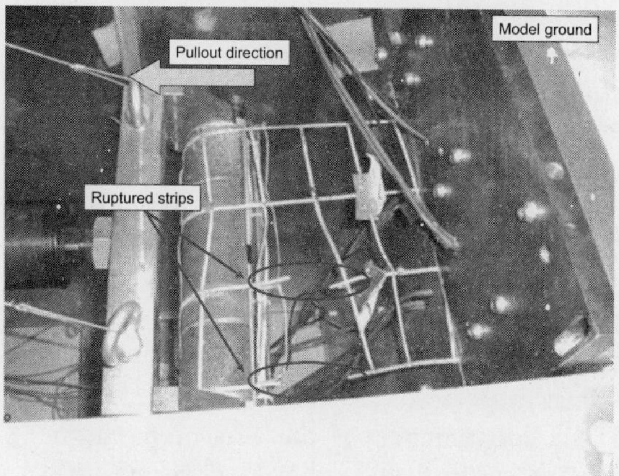

Fig.4 Rupture of reinforcements of the PB

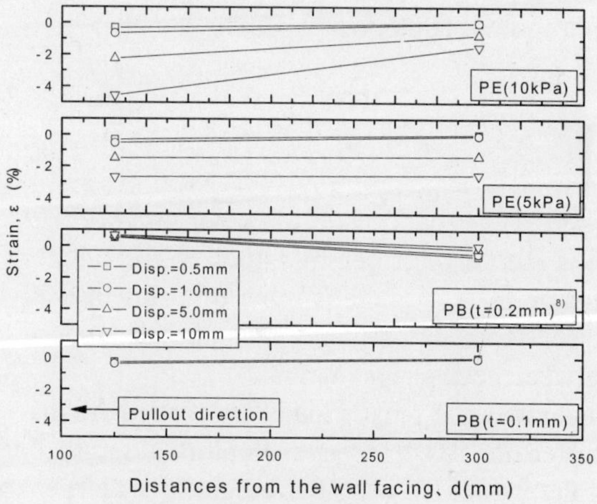

Fig.5 Strain distribution in the reinforcements

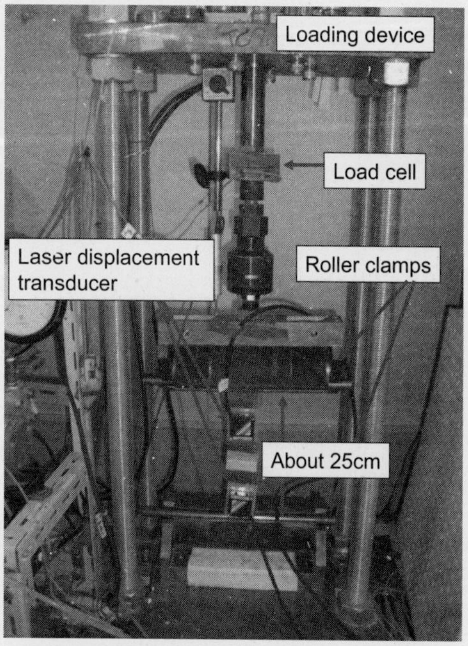

Fig.6 Apparatuses for direct tension tests

as shown in Fig.6. The both edges of the geo-grid models were held by a pair of the roller clamp, while the distance between two clamps was about 25cm (total length of the geo-grid model was about 90cm). A pair of laser displacement transducers was used to evaluate tensile strain of the geo-grid models. The width of the geo-grid model in the direct tension tests was set equal to 5cm in case of the model PE, while one strip having a width of 3mm was tested in case of the model PB for the convenience of the test preparation. Constant strain rate loading at 1%/min was applied in both cases Tensile loads obtained from the tests were converted to the values per unit width(=1000mm) by Eq.2.

$$Te = \frac{Ti}{n} \times (\frac{1000}{d} + 1) \cdot \cdot \cdot \cdot \quad \text{Eq.2}$$

where Te is the Tensile force per unit width, Ti is the tensile force measured in the direct tension tests, n is the number of strips in the direct tension tests and d is the distances between each strips in mm

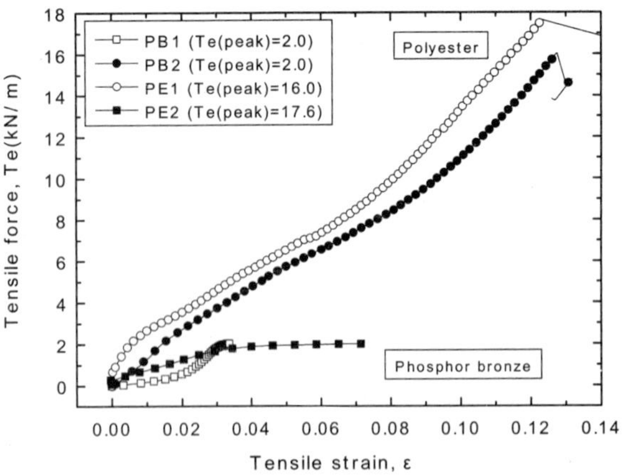

Fig.7 Relationship between tensile force and tensile strain

Table.1 Comparison of tensile rigidities

	T_i(peak)/2 (kN/m/ε)	T_i(peak) (kN/m/ε)	T_i(peak)/2 (kN/strip/ε)	T_i(peak) (kN/strip/ε)
PE1	120.5	143.1	0.36	0.43
PE2	105.1	130.0	0.31	0.39
PB1	65.7	60.5	65.7	5.25
PB2	65.7	29.0	65.7	2.52

The converted tensile loads are plotted versus the tensile strain in Fig.7. Rapture strength $Te_{(peak)}$ are also indicated in Fig.7. The rapture strength of the model PE was about eight times larger than the ones of the model PB because the number of the strips parallel to the tensile direction of the model PE was much larger than that of the PB. The first term of the right side in Eq.2 means the rapture strength of the single strip. This value of the model PB was about 200N, while in the pullout tests of the model PB having four strips, the rapture of the strips occurred when the pullout resistances exceeded about 800 N. It was found from these behaviors that the maximum mobilized resistance by the model PB would be determined by the rapture strength under the present test conditions. Equivalent tensile rigidities per unit width and single strip of the two models, when the Te(peak) values and half of them were mobilized, are summarized in Table 1. The stiffness of single strip of the model PB was stiffer than that of the model PE made of polyester, while the stiffness per unit width of the model PE was greater than that of the model PB due to larger number of the strips of the model PE.

3 SHAKING TABLE MODEL TESTS

3.1 Test procedure

As shown in Fig.8, two models of the geo-grid reinforced soil retaining walls PE and PB were tasted. They had a height of 500mm and they were placed on a horizontal subsoil layer having a thickness of 200mm. During a preparation of the backfill layers, the two geo-grid models as

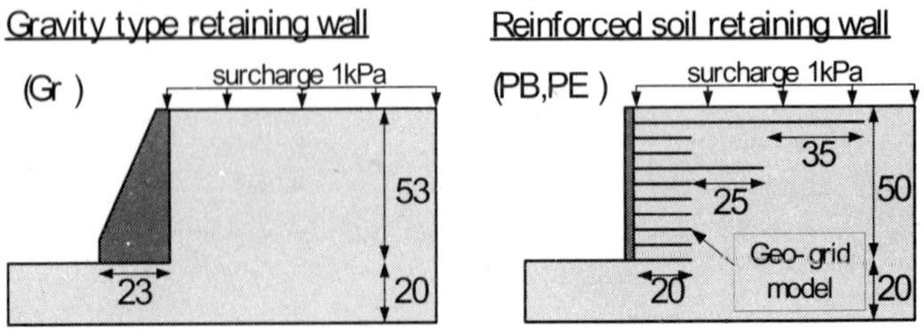

Fig.8 Cross sections of the model retaining walls (unit in cm)

discussed in Chapter 2 were fixed with the wall facing at the vertical spacing of 50mm as shown in Fig.1. The cross section of the model test on gravity type retaining wall (Gr, Watanabe et. al.[6]), which will be compared with the seismic performance of the reinforced soil retaining wall, is also shown in Fig.8.

Both the subsoil and backfill layers consisting of air dried Toyoura sand having a relative density of about 90% were prepared by air pluviation using a mechanical hopper.

Seismic loads were applied by shaking the soil container horizontally with irregular waves as shown in Fig.9, while the maximum base acceleration was gradually increased from 0.9G to 1.5G at an increment of about 0.3G. The final shaking step having the maximum acceleration of about 1.5G was applied twice in both tests.

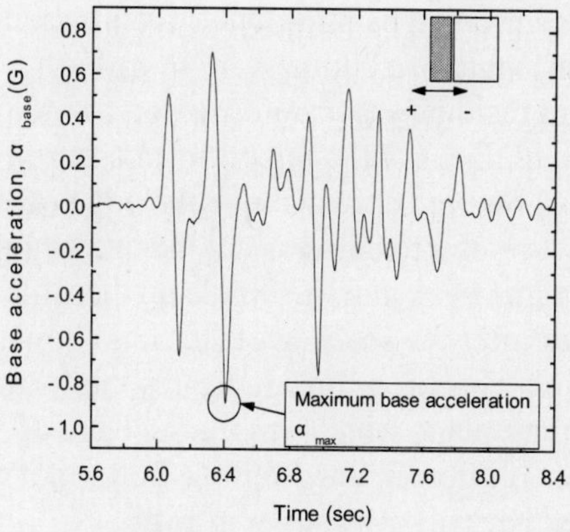

Fig.9 Typical time history of irregular excitation

3.2 Residual displacements

Residual displacements after each shaking step are plotted versus the maximum acceleration in Fig.10. Test results on the gravity type retaining wall, which was conducted under the same model ground conditions while it was subjected to shaking steps at the maximum acceleration of about 1G twice, are also plotted in Fig.10. Sliding displacement, tilting angle of the wall facing, and the settlement of the backfill layers at the horizontal distances of 100mm from the wall facing are concerned as the representative residual displacements in this study.

As clearly shown in Fig.10, residual displacements of the gravity type retaining wall were about two times larger than those of the two reinforced soil retaining walls. It was also observed that the increments of the residual displacement after the first shaking of the reinforced type retaining walls were much smaller than those of the gravity type due to their ductile performances against earthquake.

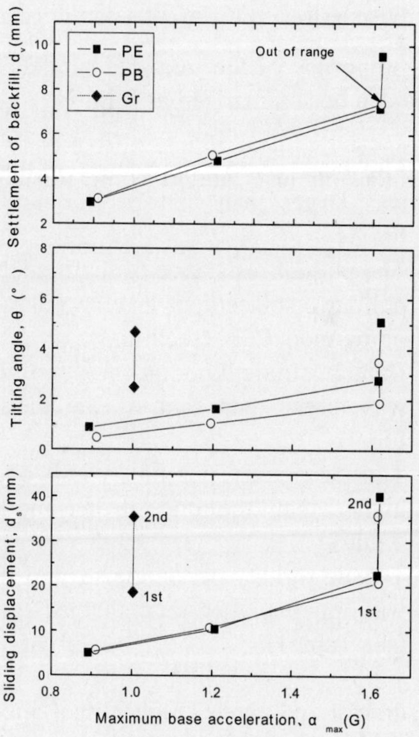

Fig.10 Comparison of residual displacements

Seismic performance of the two reinforced soil retaining walls was almost the same even though the material properties like pullout resistances, rapture strength and tensile rigidity were largely different from each other. Tilting angle of the wall facing of the model PB was slightly smaller than that of the model PE while larger mobilized resistances could be expected by the model PE according to the test results from the pullout tests and direct tension tests as shown in Chapter 2. The possible reason for this will be discussed in Section 3.4.

3.3 Failure processes of the model and deformations of backfill

Photos taken after the shaking at the 1.2G and 1.5G in the model PE are shown in Figs. 11(a) and (b). The length of the reinforcements (i.e. 200, 450 and 800mm) is also indicated by the horizontal arrows. Reinforced backfill, where reinforcements with their length of 200mm were installed, underwent residual shear deformation as indicated in Fig.11. Settlement of the uppermost coloured sand layer was observed at just above the edge of the uppermost reinforcements as emphasized by the circle in Fig.11 (a), and this tendency became more significant after the shaking of 1.5G as shown in Figs. 11(b). This behaviour indicates the formation of the failure plane which passed the edge of the uppermost reinforcement from the bottom of the reinforced backfill. As shown in Figs 12(a) and (b), similar tendencies were observed in case of the model PB while the localized settlement of the uppermost coloured sand layer wasn't so clear as compared with the case of the model PB even after the shaking of 1.5G.

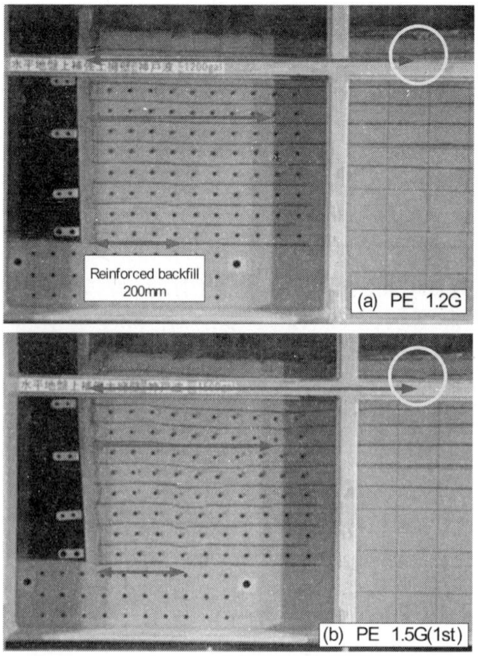

Fig.11 Failure process of model PE

3.4 Discussion on different seismic performances

The properties of the geo-grid models used in this study can be summarized as follows.

1. The peak pullout resistance of the model PE was larger than that of the model PB. (see Fig.3)
2. The pullout resistance of the model PB at small pullout displacement level was larger than that of the model PE. (see Fig.3)
3. Rapture strength per unit width of the model PE was larger than that of the model PB. (see Fig.7)
4. Tensile rigidity per unit width of the PE was higher than the one of the model PB. (see Table1)
5. Tensile rigidity per single strip of the model PB was much higher than that of the model PE. (see Table1)

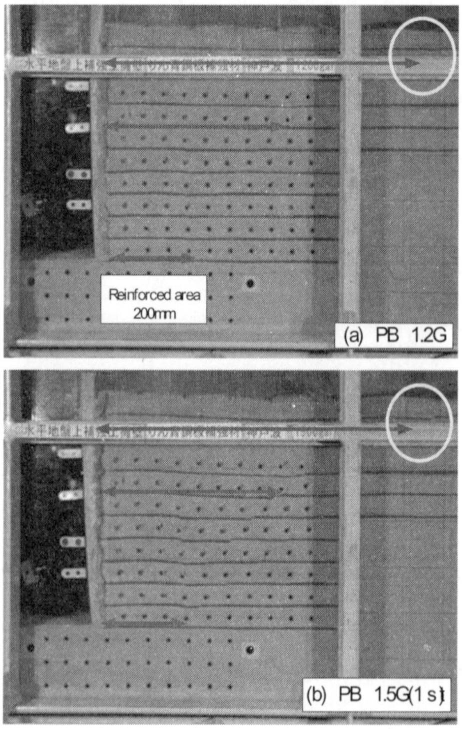

Fig.12 Failure process of model PB

As summarized above, higher resistance could be expected by the model PE than by the model PB, while the seismic performances of the wall weren't so largely different from each other based on the shaking table model tests. Rather, tilting angle of the model PB was smaller than that of the model PE. The rapture strength didn't affect the seismic performance in this study because the rapture of the reinforcements weren't observed in both tests. It was observed in above model tests that the sliding displacement of the reinforced soil retaining wall was induced by the shear deformation of the subsoil beneath the reinforced backfill and tilting of the wall was induced by the shear deformation of the reinforced backfill. Sliding displacements of the two model tests were almost similar because the subsoil conditions were the same in this study. On the other hand, tilting angle of the model PB was smaller than that of the model PE. Two factors might have affected this behaviour. Firstly, the shear deformation of the reinforced backfill itself of the model PB was larger than that of the model PE because the rigidity of the reinforced backfill of the model PB was lower than the model PE as summarized in Table 1. Secondly, tilting of the wall facing due to the shear deformation of the reinforced backfill was restricted by the pullout resistances of the reinforcement having the length of 45cm and 80cm. Pullout resistance of the model PB before its rapture was higher than those of the model PE as shown in Fig.3. Because the latter factor overcame the former factor in this study, tilting angle of the model PB was thought to be lower than that of the PE.

4 SUMMARY

Results from this study can be summarized as follows.

1. Even though the material properties of the geo-grid models were largely different, only a little differences of the seismic performance of the reinforced retaining walls were observed.
2. The larger pullout resistance before the rapture of the model PB might be the reason for the smaller displacement of the model PB than the model PE.

REFERENCES

1. Tatsuoka et. al. (1996), Performance of soil retaining walls for railway embankments, Soils and Foundations, Special issue, pp.311-324
2. Watanabe, K., Munuf, Y., Koseki, J., Tateyama, M. and Kojima, K. (2003), Behaviors of several types of model retaining walls subjected to irregular excitation, Soils and Foundations, Vol.43, No.5, pp.13-27
3. Koseki, J., Tatusoka, F., Munuf, Y., Tateyama, M. and Kojima, K.(2004), Evaluation of the seismic displacement of reinforced walls, 3rd Asian regional conference on Geosynthetics, Seoul, pp.217-224
4. Nojiri et. al.(2007), Effects of pullout resistance on performance of geosynthethics reinforced-soil structure(in Japanese), The 42nd Japan National Conference on Geotechnical Engineering, pp.1573-1574
5. Kuwano, J., Takahashi, A. and Kimura, H.(1999), Material properties and pullout characteristics of geogrids used in Japan (in Japanese), Geosynthetics Engineering Journal, Vol.14, pp.195-204
6. Watanabe et. al.(2005), Application of high-speed digital cameras to observe static and dynamic deformation characteristics of sand, Geotechnical Testing Journal, ASTM, pp.423-435

EXPERIMENTAL STUDY ON BORE-TYPE TSUNAMI WAVE ACTING ON A COASTAL DIKE

***Kazuo Tominaga*[1], *Susumu Nakano*[2], *Takeshi Okabe*[3] *and Seiji Amou*[4]**

[1]*Department of Civil and Environmental Engineering, Tokushima University,* tommy@ce.tokushima-u.ac.jp
[2]*Department of Civil and Environmental Engineering, Tokushima University,* nakano@ce.tokushima-u.ac.jp
[3]*Department of Civil and Environmental Engineering, Tokushima University,* okabetak@ce.tokushima-u.ac.jp
[4]*Department of Civil and Environmental Engineering, Tokushima University,* amou-s@yonkon.co.jp

ABSTRACT

Characteristics of tsunami wave pressures acting on a coastal dike were studied in a tsunami experimental tank. The results obtained in this experimental study were as follows; 1) Greatest impulsive pressure occurs when a crest of bore-type tsunami just overflows at an upright coastal dike. 2) Greatest continuous pressure of tsunami generally exceeds the value estimated by Tanimoto's formula (1984). 3) Impulsive pressure of tsunami is not so much affected by coastal bed slope. 4) In addition, installation of a breakwater in the front of a coastal dike is very effective for decreasing wave pressure.

Keyword: tsunami bore, coastal dike

1. INTORODUCTION

In Japan, coastal structures have been mainly constructed as countermeasures against storm surge and shore erosion. Of course, as countermeasures against tsunami, after the Chili earthquake tsunami that occurred in 1960, construction of tsunami breakwaters and seawalls has been carried out along the Sanriku coast facing to the Pacific Ocean. Unfortunately, design of seawalls built so far did not adequately take into account factors such as trench-type earthquake, quake-resistance and tsunami wave force. In the cases of the Niigata Earthquake Tsunami (1964) and Indian Ocean Tsunami (2004), it was observed that the front of tsunami waves took the form of a bore. There are few coastal structures designed taking into account the behavior of bore-type tsunami with impulsive wave force. For countermeasure against earthquakes, such as the Nankai Earthquake which is predicted to occur in the near future, correct evaluation of tsunami wave pressure is an important issue. Including Fukui (1962), a number of experiments on bore-type tsunami have been carried out by Tanimoto (1984), Matsutomi (1994), Mizutani (2000, 2002), Murakami (2005), Arikawa (2005, 2006). Recently, three-dimensional numerical simulation methods have been developed and are utilized in the evaluation of tsunami power acting on coastal structures. However, characteristics of bore-type tsunami wave and its hydrodynamic force have not been thoroughly investigated to date. In this study, our attention is focused on bore-type tsunami propagating in coastal areas with steep slopes as there is little data regarding such areas. In addition, the dynamic pressure which is caused on coastal dike by that type of tsunami wave after flowing up a steep slope is experimentally examined. On the other hand, form a post-event survey on the Indian Ocean tsunami disaster that occurred in 2004, it was found that the characteristics of damage were different between the background of harbors and other land areas and it was guessed that coastal structures such as breakwaters played an important role to reduce tsunami wave force. However, up to the present, quantitative estimation has not been done with regard to tsunami damage reduction effect of breakwaters. In this study, a series of model experiments were also carried out in order to clarify the characteristics of dynamic wave pressure produced by bore-type tsunami waves that have been affected by breakwaters.

2. OUTLINE OF EXPERIMENT

2.1 Experimental Set-Up

The experiment was carried out in a wave tank measuring 20m long, 0.8m wide and 1m deep (Fig.1). Tsunami waves were generated by quickly opening an electronically controlled sluice gate. The distance between the gate and the foot of the slope part, length of slope modeling a coastal bed and the platform modeling a background area were 4m, 5m, and 6m, respectively. The gradient of the slope was fixed at 1/10. Models of different-size coastal dikes were placed on the platform 15cm apart from its outer edge.

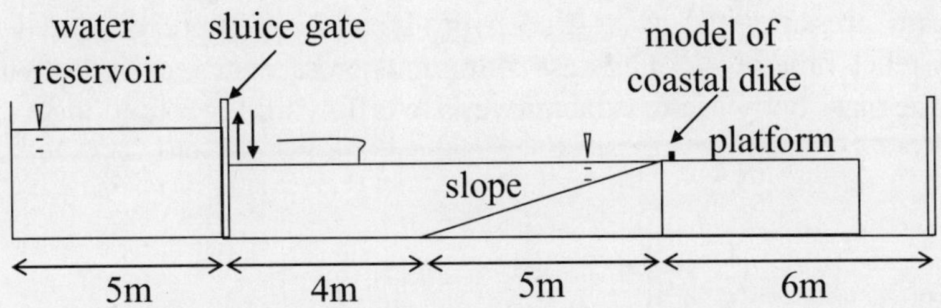

Fig.1. Experimental tank for bore-type tsunami

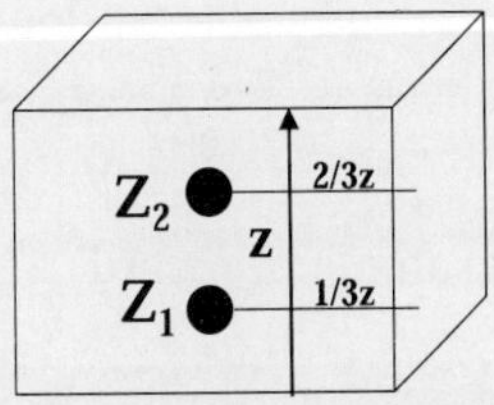

Fig.2. Location of wave pressure meters on the front surface of coastal dike

Table 1 Experimental condition

case name	A1	A2	B1	B2
existence of coastal dike	not exist	not exist	not exist	exist
existence of break water	not exist	exist	not exist	exist
slope inclination	1/10			
wave height(cm)	2,4,6,8,10			
water depth(m)	0.5			

2.2 Experimental Method

Experimental conditions are summarized in Table 1. All experimental runs are divided into 4 groups, namely, case A1, A2, B1 and B2 depending on whether coastal dikes and breakwater exist or not. In each experimental run, a single bore-type tsunami wave was generated by quick gate opening. Subsequently, height and celerity of the wave in the offshore region were measured at a point 4m downstream from the gate. Measurements were also taken with regard to both profile of the wave overflowing a model dike and dynamic wave pressure acting on it. Dynamic wave pressure was measured at two points on the front surface of the dike as can be seen in fig.2, the two being located, 1/3 and 2/3 of total dike height above the surface of platform. The depth of water at the offshore region was kept at 0.5m and the offshore wave height was changed in 5 values ranging from 0.02-0.1m.

3. RESULTS OF EXPERIMENT

3.1 Definition of Representative Wave Pressure

An example of wave pressure record is shown in Fig.3 with respect to a model dike with height and width of 10cm and 4cm, respectively. Four representative quantities of wave pressure were defined in this study as indicated in Fig.3. Impulsive pressure, P_i, is the first peak-pressure observed that is assumed to be caused by a collision of a water-mass which forms the wave front and may be regarded as a solid body in a kinematical consideration. Greatest impulsive pressure P_{dm}, is the second peak-pressure that seems to be caused by impulsive flow acceleration following sudden and local rise of the water surface that takes place successively after the collision of wave front. Bore-type tsunami pressure, P_d, is relative hydrodynamic pressure observed in a short period around the occurrence time of P_{dm}. Greatest continuous pressure, P_{ds}, is hydrodynamic pressure that is generated by the main body of the tsunami wave overflowing for some longer time.

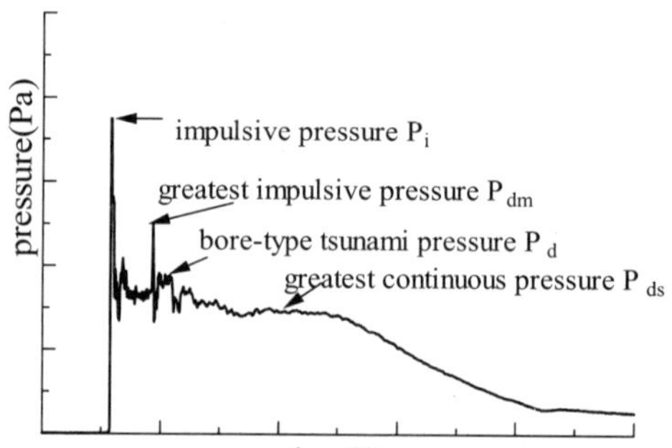

Fig.3. An example of record of wave pressure

Photo.1. Wave profile at the moment of occurrence of great impulsive pressure

3.2 Wave Celerity

A comparison of wave celerity measured in this experiment with those calculated by Fukui (1962) theoretical formula was done in order to examine the physical adequacy of bore-type waves generated in these experiments. Fukui derived Eq.1 for c theoretically based on continuity equation and momentum conservative equations (Fig.4).

$$c = \sqrt{\frac{gH(H+h)}{2(H-\xi\eta)}} \qquad (1)$$

In Eq. 1, c = wave celerity, g = acceleration of gravity, H = water depth of bore-type tsunami,

h= depth of still water in offshore region, ξ = resistance coefficient, η = wave height of bore-type wave. Fig.5 presents a comparison of theoretical values calculated by Eq.1 with those obtained in the present experiment. The calculated and measured wave celerity do not show very close agreement. However, as the difference is insignificant, the wave generated in this experiment may be regarded as bore-type wave.

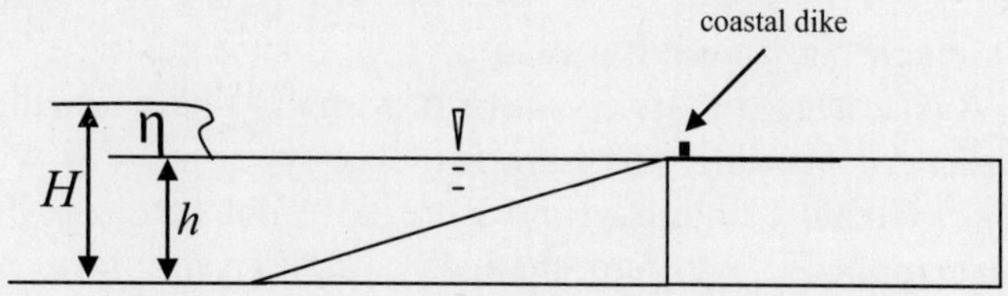

Fig.4. Definition sketch of bore-type wave

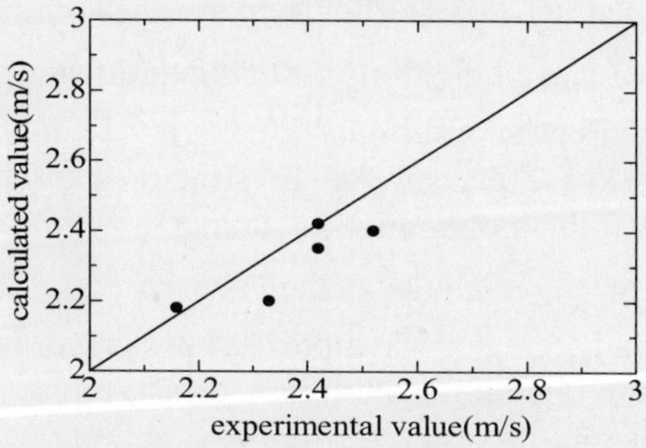

Fig.5. Comparison between experimental value and calculated value of wave celerity

3.3 Impulsive Pressure P_i

Fukui (1962) presumed that impulsive pressure would be uniquely related to wave celerity, and based on results of his experiment found impulsive pressure to be proportional to forth power of wave celerity, as follow.

$$P_i = k\frac{w_o c^4}{g^2 \eta} \tag{2}$$

where P_i= impulsive pressure, k= wave pressure coefficient and Fukui gave 0.51 for vertical wall, w_o= specific weight of water. Fig.6 shows comparison between Eq.2 and experimental result of this study and Fukui's. It is seen that P_i does not only on c, but tend to vary with wave height and other hydraulic parameters.

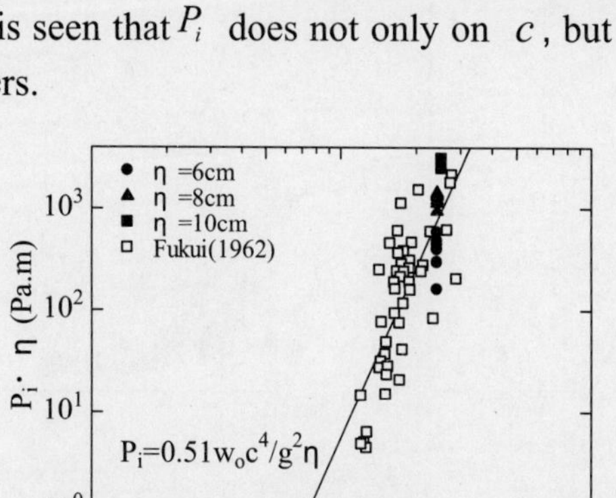

Fig.6. Comparison of relationship between measured P_i and c that of Eq.2

3.4 Vertical Distribution of P_{dm}

Non dimensional vertical distribution of greatest impulsive pressure can been seen in fig.7. Also, z/η in a vertical axis is non dimensional vertical height, $P_{dm}/\rho g\eta$in a cross axis is non dimensional. Figs.7 and 8 show vertical distribution of P_{dm} observed in this study, where measured height, z, is non dimensionalized with η andη_{max} (wave height in front of a coastal dike), respectively. Greatest impulsive pressure has much bigger wave pressure value than other type of wave pressure except of impulsive pressure. A characteristic of this wave pressure is that it has almost the same value as that of impulsive pressure depending on shape of bore-type tsunami when bore-type tsunami overflows a coastal dike. Greatest impulsive pressure dose not take significant values under any conditions, also greatest impulsive pressure changes complexly for form of the coastal dike and wave forms when overflowing a coastal dike. A comparison of greatest impulsive pressure measured in this experiment with Tanimoto(1984) theoretical formula was done, and found that our results greatly exceeded those of Tanimoto's as is shown in fig.7. Our results show greatest impulsive pressure distribution at the coastal dike front as is seen in fig.8. With regard to evaluation of the biggest wave pressure of wave height η_{max} of structure front, Asakura(2000) suggests wave pressure expression acting on structure after overflowing revetment. According to Asakura(2000), it is assumed that the greatest wave pressure at the structures front occurs less than 3 times for hydrostatic pressure. However, this finding does not coincide with in greatest impulsive pressure in this experiment at the maximum. As a whole, as the value of z/η_{max} decreases, the value of $P_{dm}/\rho g\eta_{max}$ tends to increase. In this study, greater wave pressure was recorded at the lower parts of model dikes rather than at upper parts. Greatest continuous pressure base on wave height of bore-type tsunami with offshore wave height or structure front was analyzed under experimental conditions of 1/30, 1/50 incline by Arikawa(2005). Arikawa's(2005) experiment, focused on offshore wave height, Arikawa's greatest continuous pressure agreed with Tanimito's proposed expression. Also, with regard to wave height in front of a coastal dike.Such pressures proportionate hydrostatic pressure. Although, our results do not agree with those of Tanimoto(1984) which have non dimensional vertical distribution for offshore wave height and coastal dike front wave height. In addition, our results show a value bigger than Tanimoto's in the coastal dike lower part. Greatest continuous pressure for wave height at the coastal dike front showed a value bigger than that of hydrostatic pressure.

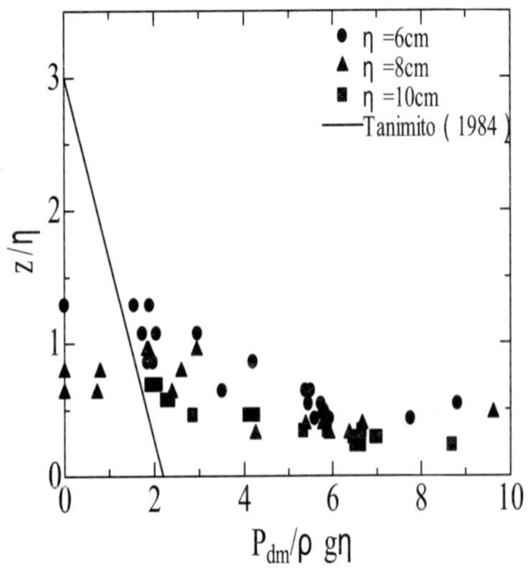

Fig.7.Non dimensional vertical distribution of greatest impulsive pressure

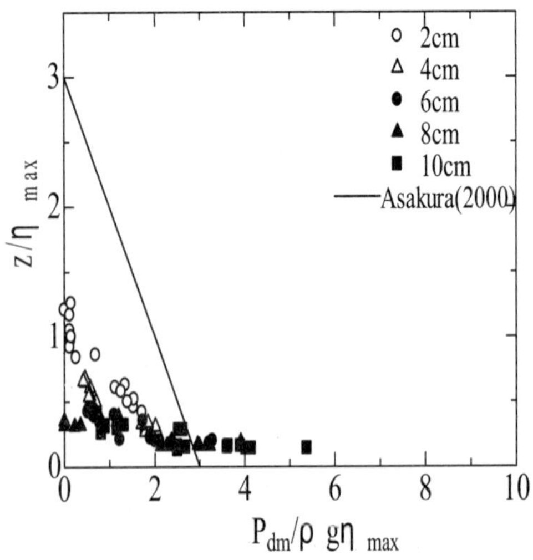

Fig.8.Non dimensional vertical distribution of greatest impulsive pressure

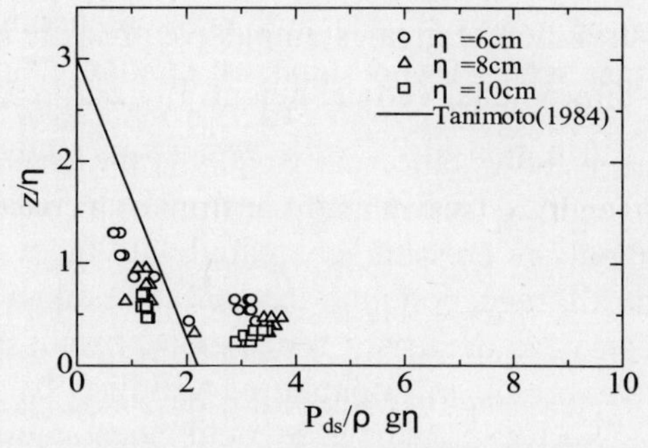

Fig.9. Non dimensional vertical distribution of greatest continuous pressure

3.5 Relations between Greatest Impulsive Pressure and Greatest Continuous Wave Pressure

Relations of greatest impulsive pressure and greatest continuous pressure are shown in fig.10, with the greatest bore wave pressure (Arikawa2005). The maximums of all waves pressure from his experiment results, P_{max}, and the maximum continuous wave pressure, P_{ds}, that with the experiment condition with a slope incline of 1/30, 1/50 are shown. The relations were found to be P_{max}=1.17P_{ds} and P_{max}=1.27P_{ds} from experimental results with an incline of 1/30, 1/50. In addition, the relations of greatest impulsive pressure and greatest continuous pressure showed a value of P_{dm} =1.97 P_{ds} in this study. The value in the dotted line in parenthesis of fig.10 showed a value bigger than average value of greatest impulsive pressure and greatest continuous wave pressure. As for relations of greatest impulsive pressure and greatest continuous pressure, the very big wave pressure value such as shown by the dotted line in parenthesis are shown. Also, this wave pressure is confirmed in 1/30 slope experiments of Arikawa (2005). Wave pressure shown in fig.10.'s dotted line in parenthesis was not confirmed in all experiment results.

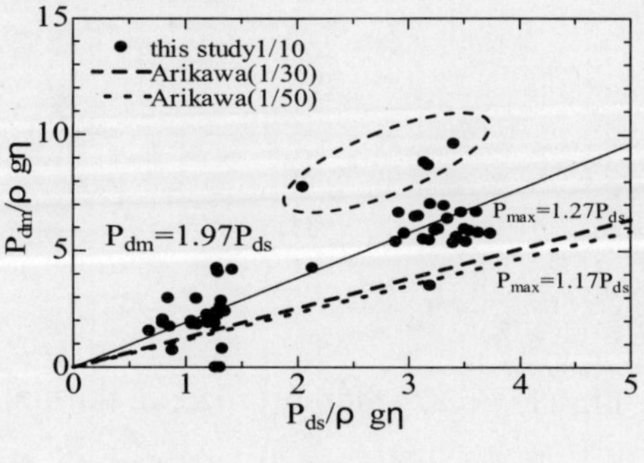

Fig.10. Relationship between P_{dm} and P_{ds}

4. INFLUENCE OF BREAKWATER ON TSUNAMI

Many breakwaters have been installed along the coast of Japan. However, in regard to tsunami which overflow a breakwater, it is not clarified what kind of damage have. Therefore, the experiment cases was carried out with the same condition of outlined in section 2 in order to measure wave height, wave celerity around shoreline or reduction ratio of wave pressure acting on coastal dike after having overflowed breakwater (hcb=3cm). Compared with the actual breakwater, the bank body width of breakwater is a value of 1/10 less. However, we think no problem related to investigating reduction effect of wave pressure. The set up of the breakwater is shown in fig.11.

4.1 The Reduction Ratio of the Bore-Type Tsunami due to Setting of Breakwater

Fig.12 shows reduction ratio of bore-type tsunami pressure acting on the coastal dike under conditions of having breakwater setting or not. Greatest continuous pressure, its reduction ratio is reduce to a half of in the case of being η/h_{cb} from 2 to 3, around 20% fall when they surpass a value of 3. Greatest impulsive pressure shows ratio nearly equal to the greatest continuous pressure in the case where η/h_{cb} is 2.67. However, reduction ratios seems to decrease when offshore wave height increase. In the case of the impulsive pressure, compared with other representative pressure, when wave height in the offing is small, reduction ratio increases remarkably. Also, when wave height in the offing increases, reduction ratio decreases remarkably. From studying of the Indian Ocean tsunami, breakwaters which were built in Republic of Maldives by financial support from Japan reduced the tsunami damage. This was confirmed by field investigation in the Maldives. According to the study of Arikawa (2005), it is assumed that impulsive pressure is related to damage or destruction of building in the area along the shore. As a result of this study, it is assumed that existence of breakwater decreases impulsive pressure in the Maldives.

4.2 Influence of Wave Height and Wave Celerity Due to Breakwater Setting

With regard to an inundation disaster, the influence of inundation depth is significant. There are relationships between size of tsunami and tsunami damage. The runup height at the structure front has a very big influence demanding on design of coastal structure and influences relations with overtopping rate of coastal structure's rear area very much. Figs.13 and 14 show the reduction ratio of wave height and wave celerity at shoreline after overtopping breakwater. Under the condition of having setting or not of coastal dike, wave height in shore line where there was no coastal dike, it was found that the reduction ratio of wave height was smaller than in the case of setting coastal dike. Also, with regard to wave celerity, the reduction ratio of wave celerity was from 0.8 to 1. And the reduction ratio of wave celerity with installing breakwater was not significantly large.

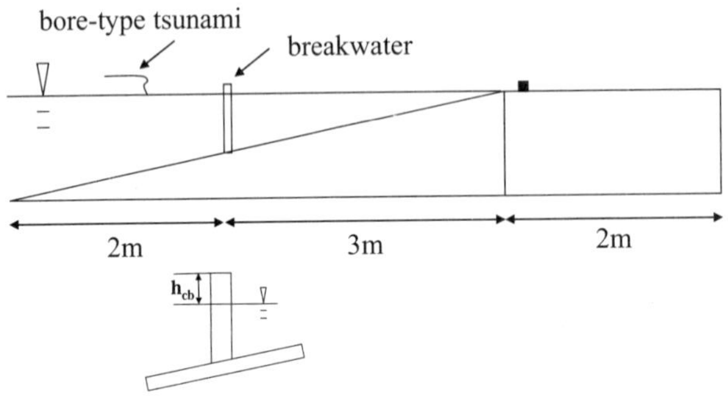

Fig.11. Breakwater installed on the slope

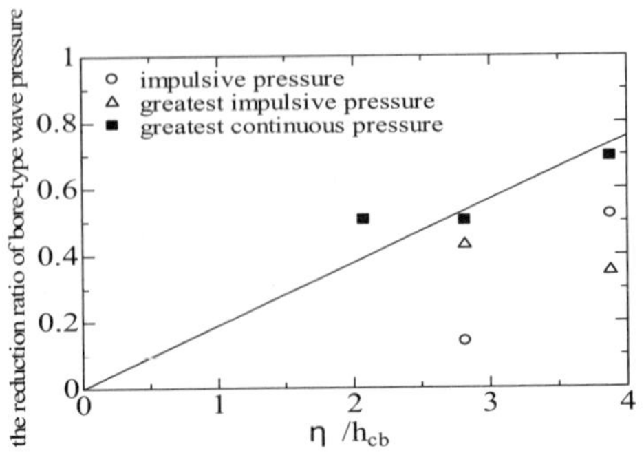

Fig.12. Reduction ratio of bore-type tsunami pressure due to breakwater

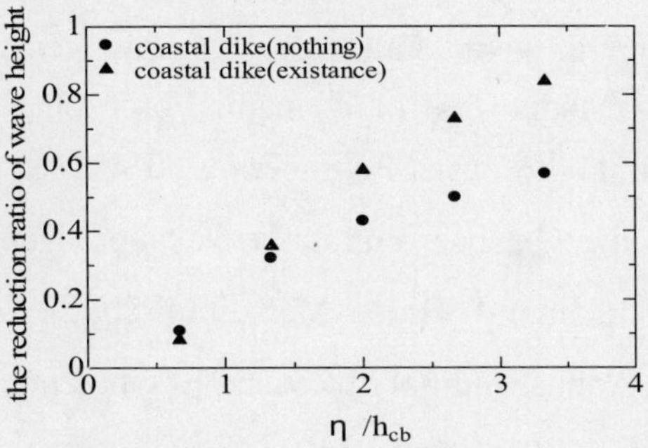

Fig.13. Reduction ratio of wave height due to breakwater

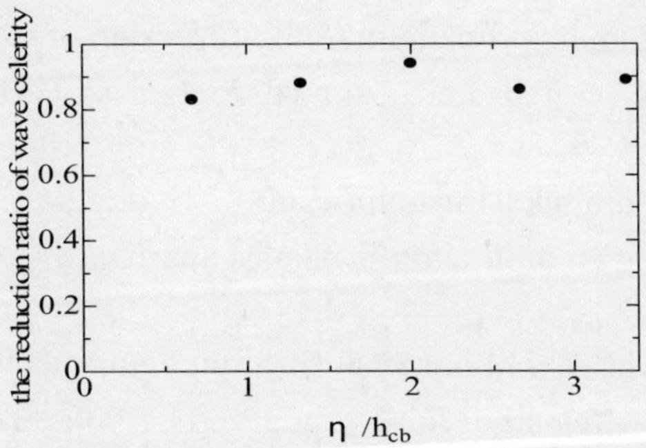

Fig.14. Reduction ratio of wave celerity due to breakwater

5. CONCLUSION

In this study, we measured wave pressure of bore-type tsunami which acted on a coastal dike running up a steep slope. In addition, a breakwater was placed on the slope at the front of the coastal dike and reduction of wave pressure was examined due to the breakwater, As a result, the following was concluded.

1. Impulsive wave pressure occurs when a crest of bore-type tsunami just overflows at an upright coastal dike. Great pressure acting on coastal dike is confirmed.

2. Greatest impulsive pressure and greatest continuous pressure generally were found to be underestimated when it was compared with Tanimoto's formula (1984).

3. When the value of impulsive pressure is based on tsunami wave height in the offshore, our results show an approximately equal value even if the slope at the front of the coastal dike is a horizontal or a slanting incline.

4. Greatest continuous pressure, the wave height of the background decreases when a breakwater is placed, but wave celerity does not decrease significantly.

REFERENCE

1. Tarou Arikawa· Masamitu Ikebe· Fumimori Yamada· Kenichirou Shimosako· Fumihiko Imamura(2005)."Large Model Test of Tsunami Force on a Revetment and on a Land Structure", Annual Journal of Coastal Engineering , JSCE, **52** , pp.746-750.
2. Tarou Arikawa· Daisuke Otsubo· Fumitake Nakano· Kenichirou Shimosako· Shigeo Takahashi· Fumihiko Imamura· Hideo Matsutomi(2006). "Large Model Test on Surge Front Tsunami Force", Annual Journal of Coastal Engineering, JSCE , **53**, 796-800.
3. Toshirou Fukui· Hidehiko Shiraishui· Mitsuru Nakamura· Yasuo SASAKI(1962) : The study of Tsunami , Annual Journal of Coastal Engineering , JSCE, **9** , pp.44-54.
4. Masashi Mizutani· Fumihiko Imamura (2002), "Design of Coastal Structure Including the Impact and Overflow on Tsunami", Annual Journal of Coastal Engineering, JSCE, **49**, 731-735.
5. Masashi Mizutani· Fumihiko Imamura (2000),"Hydraulic Experimental Study on Wave Force of a Bore Acting on a Structure", Annual Journal of Coastal Engineering, JSCE, **47**, 946-950.
6. Keisuke Murakami· Masaaki Murakami· Toshirou Yamaguchi(2005),"Control of Inundated Water Depth and Flow Velocity With Using a Vertical Tide Barrier against Tsunami Attack "Annual Journal of Civil Engineering in the Ocean, JSCE ,**22**, 151-156 .

SEISMIC PERFORMANCE OF EARTH EMBANKMENTS

Prasad, S. K[1]., Chandradhara, G. P[2]. and Nanjundaswamy, P[3].
[1]Assistant Professor, S. J. College of Engineering, Mysore, India, prasad_s_k@hotmail.com
[2]Assistant Professor, S. J. College of Engineering, Mysore, India, chandu_gpc@yahoo.com
[3]Assistant Professor, S. J. College of Engineering, Mysore, India, pnswamy@yahoo.com

ABSTRACT

Seismic performance of earth embankments has been a hot topic of discussion since several decades. Many failures of earth embankments are reported even during the recent earthquakes. It has still not been possible to ensure whether the available analytical approaches for the analysis of embankments under earthquake loading are accurate enough for usage. The main focus of this paper is to use Finite Element technique treating the earth embankment as a continuous two dimensional plane strain idealization from equivalent linear approach. The results are compared with the field results of performance of Bradbery dam during the Santa Barbara earthquake of 1978. A comparison also is made with the solutions from modified SHAKE and FLUSH. It was heartening to observe that equivalent linear approach from Finite Element technique produced comparable results to the data obtained during the actual earthquake.

Keywords: Equivalent linear approach, Seismic performance of earth dam, Finite Element Method, Earthquake motion

1 INTRODUCTION

Earth embankments are the most important geotechnical infrastructural facilities that form components of bridges, dams, canals, highway and railway. Their proper performance during unexpected loadings and natural calamities is extremely important as they form a part of the life line. Past experience reveals that there have been many instances of failure to earth embankments during earthquake leading from minor to major levels of damage. Even in India, many failures of embankments during earthquake have been reported (Towhata, 2001, Prasad, 2002, Chandradhara, 2005). It is therefore very essential to understand the performance of earth embankments during earthquakes. Further, from the point of view of design, it is necessary to ensure that the available methods of analysis are acceptable. In order to gain confidence about our understanding of nature and to ensure proper analysis is carried out, it is important to compare the existing data during actual earthquakes with analysis results. In this connection, the present paper focuses on comparing the results of Finite Element Analysis with the measured data during the actual earthquake (Santa Barbara earthquake of 1978) from Bradbury earth dam in U.S.A.

2 SITE RESPONSE ANALYSIS

There are several methods and related computer procedures available for studying the dynamic response of embankments to seismic loading. Site response analyses began making their way into geotechnical engineering practice in the early 1970s with the development of SHAKE (Schnabel et al., 1972). Within a few years, additional site response programs such as QUAD-4 and FLUSH (Lysmer et al., 1975) were available. A significant number of computer programs are now available for site response analyses. These programs can be divided in different ways, but it is useful to distinguish them on the basis of general soil model (equivalent linear vs. nonlinear), dimensions (1-D, 2-D, or 3-D), and interface (DOS- or Windows based).

Table 1 Geotechnical computer codes available for site response analysis.

Idealization	Operating system	Equivalent Linear analysis	Nonlinear analysis
1-D	DOS	Dyneq, Shake91	AMPLE, DESRA, DMOD, FLIP, SUMDES, TESS
	Windows	Shake Edit, ProShake, Shake2000, EERA	Cyber Quake, Deep Soil, NERA, FLAC, Shear Beam
2-D / 3-D	DOS	FLUSH, QUAD4/QUAD4M, TLUSH	DYNAFLOW, TARA-3, FLIP, VERSAT, DYSAC2, LIQCA
	Windows	QUAKE/W, SASSI2000	FLAC, PLAXIS

Table 1 shows that geotechnical engineers have a variety of equivalent linear and nonlinear codes to choose from for one- and multi-dimensional site response analyses. Some of the codes are very widely used, for example SHAKE and its numerous DOS- and windows-based derivatives. Some of the windows-based equivalent linear codes operate essentially as pre-processors and post-processors for SHAKE 91, while others were written completely from scratch. The windows-based codes generally offer convenient plotting of input data (soil profile and ground motions) and resulting output. FLUSH and QUAD4 are well established and widely used two-dimensional, DOS-based, equivalent linear codes; QUAKE/W (GEOSLOPE, 1993) is similar to QUAD4, but has a windows GUI. Nonlinear codes are somewhat less well established with DESRA remaining popular for one-dimensional analysis and a variety of DOS-based programs for nonlinear 2-D and 3-D analyses are available. General purpose geotechnical analysis codes such as FLAC and PLAXIS now have dynamic analysis capabilities and are becoming increasingly popular for such applications. Depending on the type of soil, accuracy of solution desired, the strain level experienced by ground, generation of excess pore water pressure and available input parameters, a suitable soil model is chosen.

Equivalent linear models use an iterative approach to approximate the nonlinear, inelastic behavior of soils. An average shear modulus, G_{sec}, is used over an entire cycle of loading to approximate the hysteresis loop. While equivalent linear material parameters that are iteratively adjusted to be consistent with an effective level of shear strain can be found, the analysis is nevertheless linear. The method is computationally efficient and provides reasonable results for many cases, especially for those where small strains (< 1-2%) and modest accelerations (<0.3-0.4 g) develop. The linear approach allows computation of the bedrock motion from a given free surface motion, or deconvolution. The reliability of deconvolved motions should be carefully evaluated on a case by case basis.

The actual nonlinear behavior of the shear modulus and damping ratio under dynamic loading conditions can be simulated approximately by an equivalent linear analysis. In an equivalent linear analysis, constant G and damping ratio are used during a dynamic loading analysis. New modulus G and damping ratio are computed from the cyclic or an equivalent cyclic shear strain obtained in the dynamic analysis. Then a new dynamic loading analysis starts with the new modulus and damping ratio. This iterative process will repeat until changes in displacements within two successive iterations are small.

There are a few limitations of equivalent linear model. Because the model is linear, it cannot be used to calculate permanent displacements. The shear strain returns to zero after loading is complete. The inherent linearity of the soil can also lead to spurious resonances that would not occur in the field. Moreover, the equivalent linear model is not capable of modeling pore pressures because a total stress approach is used in the analysis. Proper selection of an effective shear strain is required to prevent over or under softening of the response.

3 PERFORMANCE OF BRADBURY DAM DURING SANTA BARBARA EARTHQUAKE

The Bradbury dam is a zoned earth fill structure with maximum height of 62.8 m above stream bed. The cutoff trench beneath the dam has a maximum depth of about 19.8 m. The embankment consists of three main zones, i.e., the impervious core (zone 1), sand and gravel shells (zone 2) and a special zone 3 within the downstream shell. The upstream face of the dam is protected by riprap and downstream face is protected by a cobble zone. Fig 1 shows the cross-sectional details of Bradbury embankment dam and its foundation at the maximum section. Fig 2 shows the locations and orientation of the strong motion accelerometers. These instruments were installed after the completion of the dam in 1950.

The instruments recorded the ground motion of the Santa Barbara earthquake on 13th August 1978 which had a magnitude of 7.25. A number of dynamic analyses were performed after the earthquake to check the accuracy of the computational procedures. There is a measured record of time versus ground acceleration at the crest of the Bradbury dam and at the outcropping base rock layer during the Santa Barbara earthquake. Two horizontal components of ground acceleration in orthogonal directions (340° components along the axis and 250° components normal to the axis of the dam) were recorded at both instrument sites. Out of four acceleration time histories, the digital data of only two components (250° component at the out crop and crest of dam) were obtained.

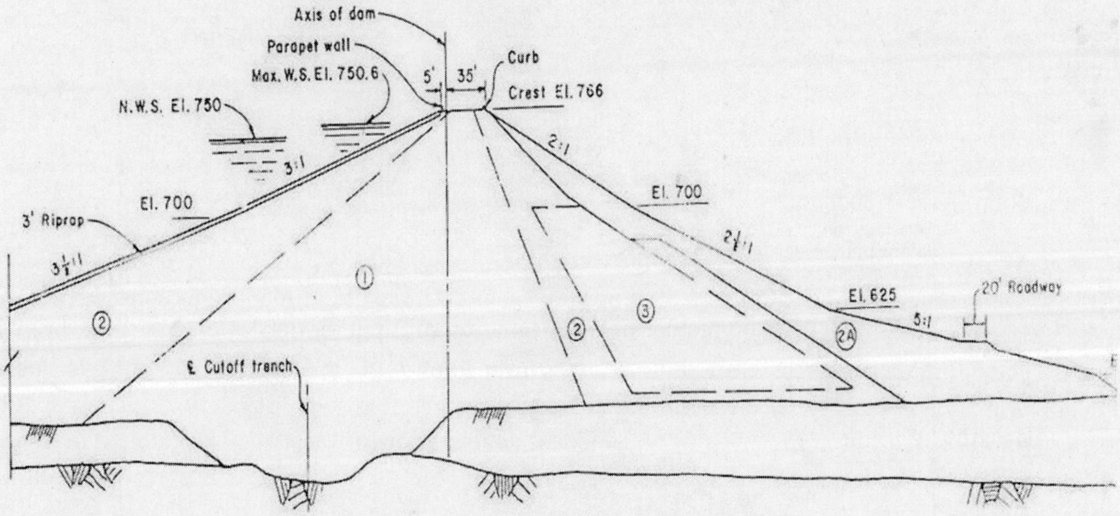

Fig. 1: Cross-section of the Bradbury Dam

The shear wave velocity measurements were available at two locations along the crest of the dam obtained using the cross-hole technique. The holes were located approximately 1.4 m upstream of the guard rail on the down stream edge of the crest. The 250° component of the bed rock acceleration time data was calculated from the corresponding measured outcrop record using the computer program SHAKE (Neethu Urs, 2002). The computed bed rock acceleration data was used as an input to obtain the response at the crest of the dam and the schematic presentation of determination of input motion and computed bed rock motion obtained from SHAKE are shown in Fig. 3. The initial critical damping ratio for each of two layers was assumed to be 0.05.

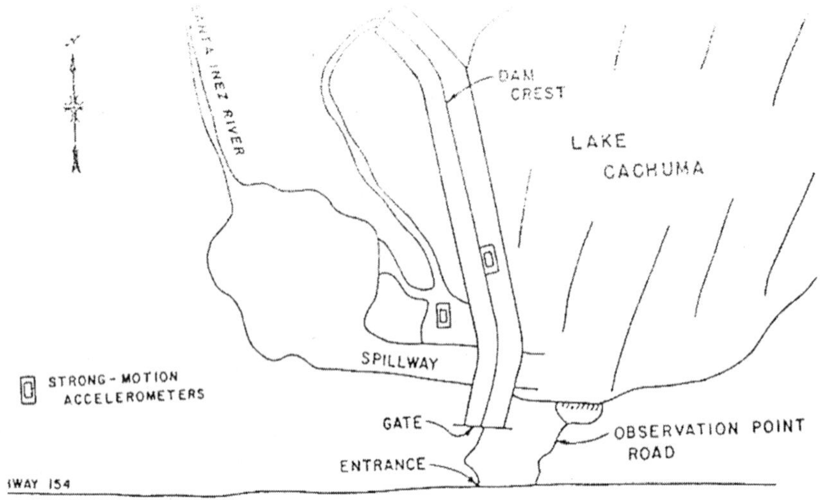

Fig. 2: Locations of strong motion accelerometers and a general plan view of Bradbury dam

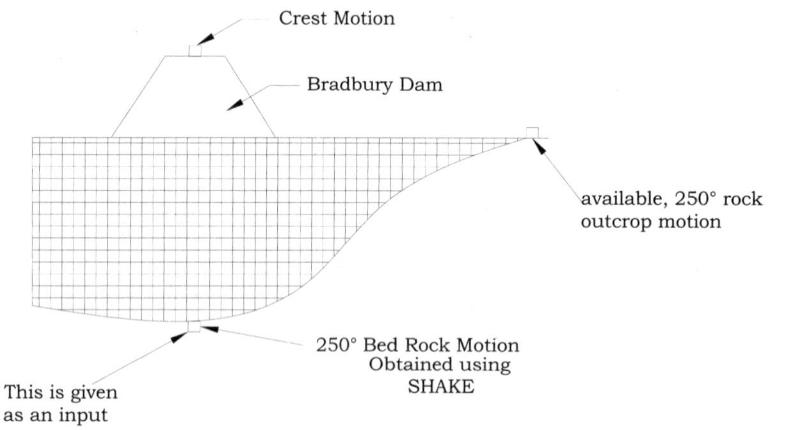

a. Schematic picture of determination of input motion

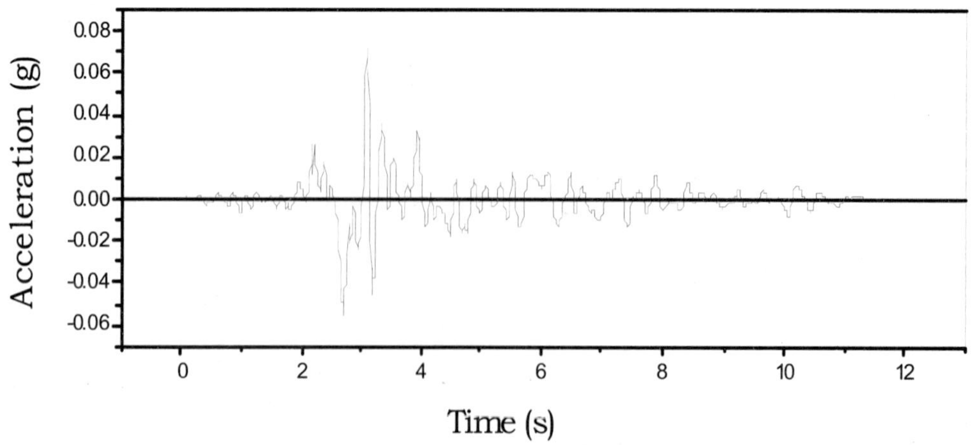

b. 250° component of bed rock motion from SHAKE

Fig. 3: Details of input motion at the base of dam for the analysis

4 FINITE ELEMENT ANALYSIS OF BRADBURY DAM

The analysis of Bradbury Dam using GEO-SLOPE (1993) was performed using the 250° component of the bed rock motion record as the input motion. The layer thickness, material density and the Shear Modulus for each layer were based on the available data (Chugh, 1985). The variations of shear modulus and damping ratio with effective shear strain were considered for the analysis. For each layer, maximum shear stress and maximum acceleration were calculated. The acceleration, velocity and displacement spectra were plotted for the crest motion using 5 % damping ratio. Analysis carried out using equivalent linear models were compared with those from Chugh (1985) using SHAKEM and FLUSH and the measured field results.

The crest motion of Bradbury dam was obtained using equivalent linear analysis from GEOSLOPE in the present study. These results are compared with the actual measured data and the analysis using FLUSH and SHAKEM (Chugh, 1985) as shown in Fig. 4. The crest acceleration from different approaches are comparable with the measured data. The spectral acceleration of Bradbury dam obtained using equivalent linear method is shown in Fig. 5. The results from GEOSLOPE using Equivalent linear analysis are in good agreement with the measured data and those from FLUSH and SHAKEM. The spectral displacement of Bradbury dam obtained using equivalent linear analysis are presented in in Fig. 6.

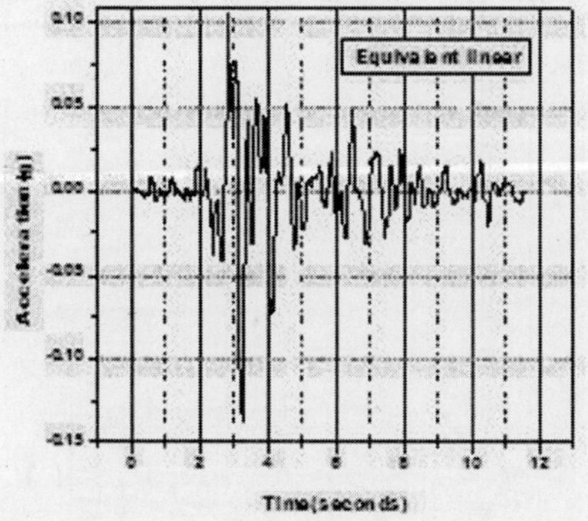

a. Results from GEOSLOPE

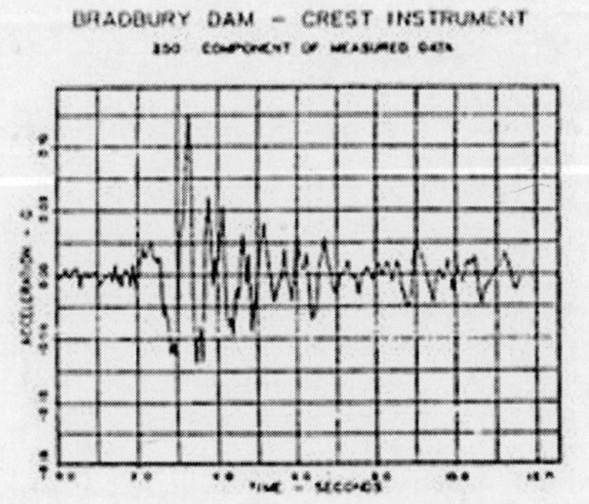

b. Measured data

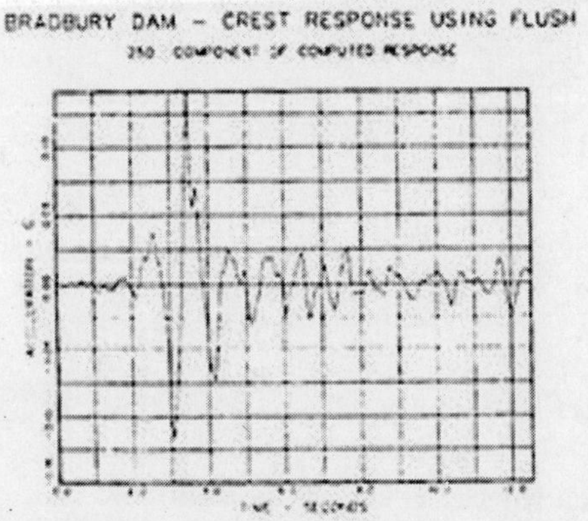

c. Results from Flush (Chugh, 1985)

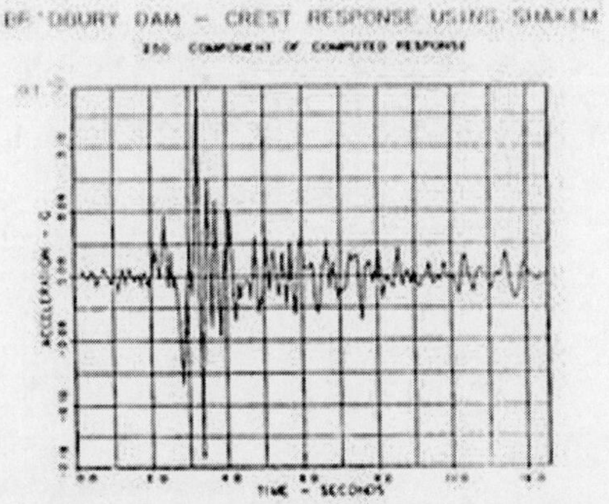

d. Results from SHAKEM (Chugh, 1985)

Fig 4: Time-acceleration response at the crest of the dam

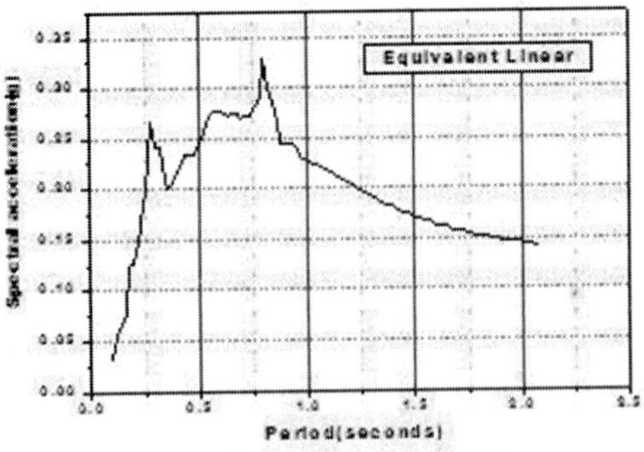

a. Results from GEOSLOPE

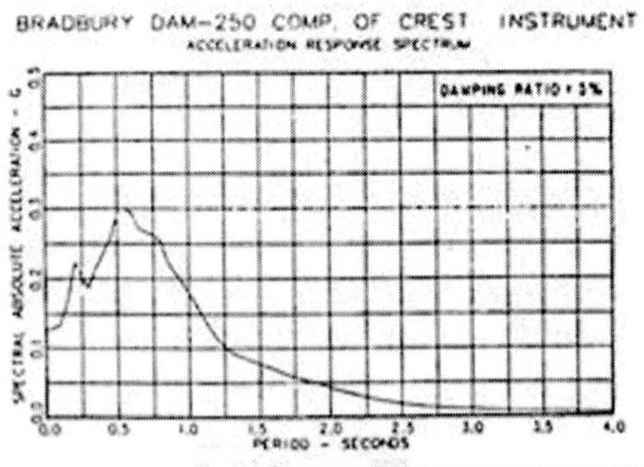

b. Measured data

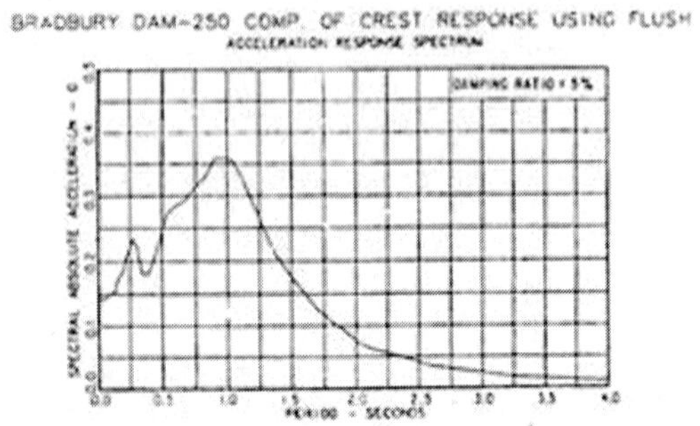

c. Results from FLUSH (Chugh, 1985)

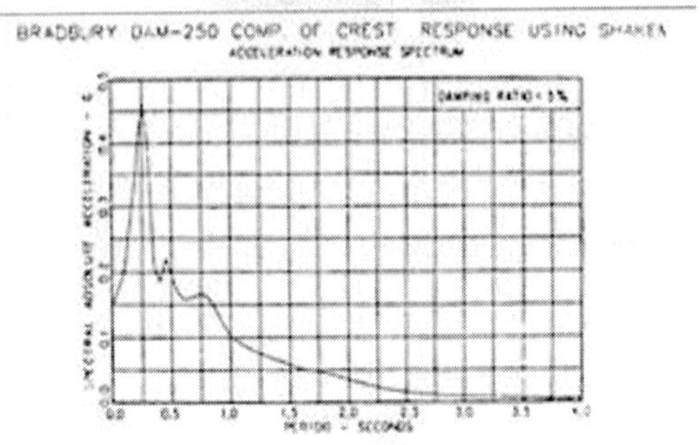

d. Results from SHAKEM (Chugh, 1985)

Fig. 5: Acceleration Response spectra for the Crest of the dam

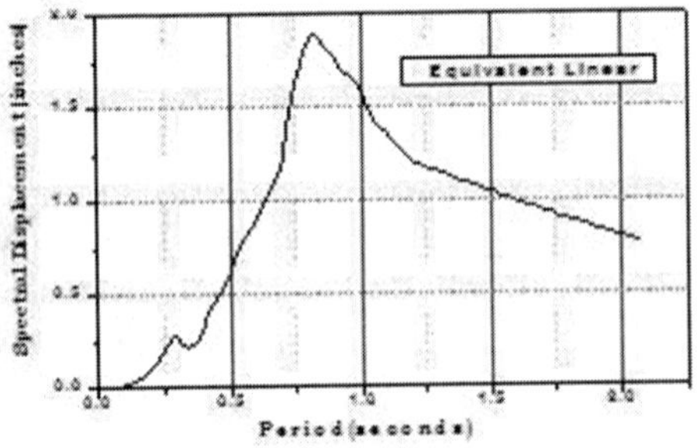

a. Results from GEOSLOPE

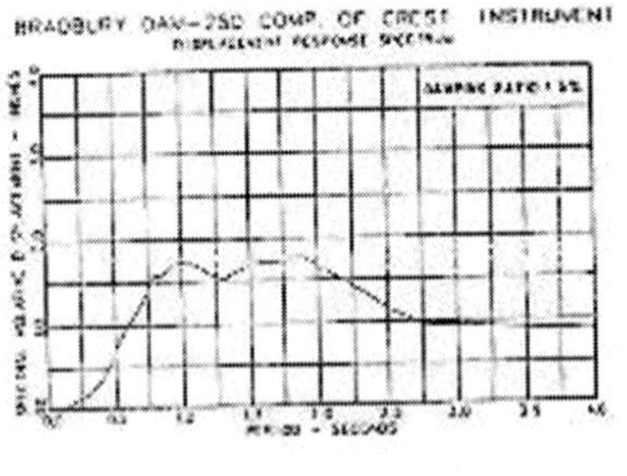

b. Measured data

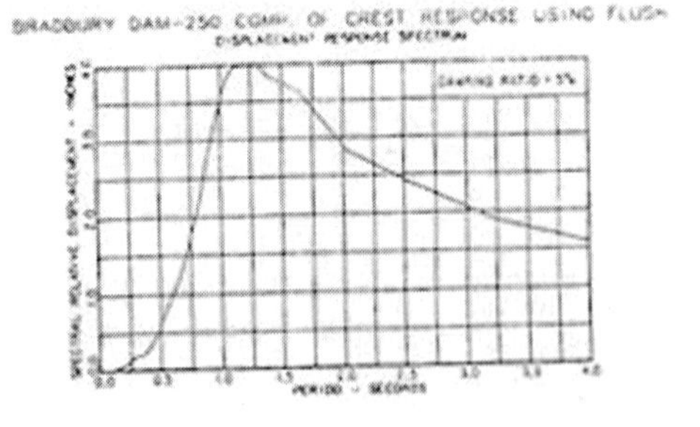

c. Results from Flush (Chugh, 1985)

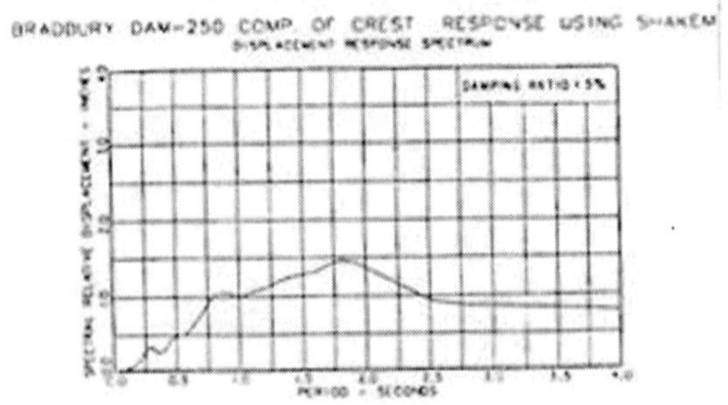

d. Results from SHAKEM (Chugh, 1985)

Fig. 6 Displacement response spectra for the crest of the dam

There is a good agreement with the measured data and with the results from FLUSH and SHAKEM. The maximum acceleration and maximum shear stress of Bradbury dam were obtained along the centre line of the dam in vertical direction using equivalent linear methods in GEOSLOPE. The results are shown in Fig. 7.

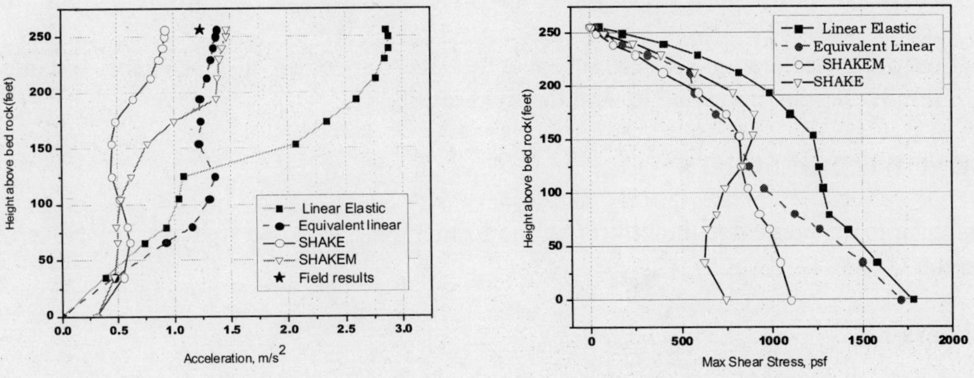

a. Maximum acceleration along the depth b. Maximum shear stress along the depth

Fig. 7 : Peak response along the depth of dam

The results of maximum shear stresses obtained in GEOSLOPE are higher than the results from SHAKEM at greater depth. The maximum acceleration obtained from SHAKEM slightly matches with the equivalent linear analysis. The present results are comparable with the field results and the analysis from SHAKEM and FLUSH and are tabulated in Table 2.

Table 2 Comparison of results from present study with measured data from the field and results from SHAKEM and FLUSH

Peak at the crest of dam	GEOSLOPE Equivalent linear	Field results	FLUSH (Chugh, 1985)	SHAKEM
Crest acceleration response (g)	0.1372	0.1261	0.1418	0.1468
Spectral acceleration (g)	0.3280	0.3010	0.3600	0.4750
Spectral velocity (m/s)	0.3560	0.3304	0.5663	0.2484
Spectral displacement (m)	0.0481	0.0432	0.0725	0.0328

From Table 2, it is clear that the equivalent linear analysis results obtained from GEOSLOPE are in good agreement with the measured field data.

5 CONCLUDING REMARKS

In the present work, the dynamic analysis of Bradbury dam was performed using equivalent linear approach available in GEOSLOPE. The instrumentation at the site recorded ground motion data during Santa Barbara earthquake in the year 1978. The results of the analysis were compared with the available field measurements and those from popular geotechnical codes (FLUSH and SHAKEM). The following were the conclusions drawn from the present study.

- Equivalent linear method using Finite Element Analysis provided results that match well with the recorded ground motion during earthquake at moderate strain levels.

- SHAKEM being one dimensional analysis, gave results slightly far from the actual measured data.
- Spectral acceleration, Spectral velocity and Spectral displacement obtained from GEOSLOPE are comparable with the field results.

6 ACKNOWLEDGEMENTS

Authors wish to acknowledge the efforts by the former post graduate students Mr. Goverdhan and Ms. Neethu Urs.

REFERENCES

1. Chandradhara, G. P., Prasad, S. K. and Nanjundaswamy, P. "Seismic Response Of Model Embankments From Shaking Table Tests", 12SEE-2002, December, 2002.
2. Chugh, A. K., (1985), "Dynamic Response analysis of Embankment Dams", International Journal for Numerical and Analytical methods in Geomechanics, vol. 9, 101-124
3. Elton, D.J., Snie, C.F. and Hads-Hamou, J., (1991) " one and two dimensional analysis of earth dams", second international conference on recent advances in geotechnical earthquake engineering and soil dynamics, St Louis, Missouri., pp 7-16.
4. GEOSLOPE (1992), "User's Manual", GEO-SLOPE OFFICE, GEOSLOPE Intl, Calgary, Alberta, Canada.
5. Goverdhan, (2007), "Construction of design spectrum for earth embankments" , M. Tech. Thesis, Dept. of Civil Engg., S.J.C.E., Mysore.
6. Kramer. S. L., (1996), "Geotechnical Earthquake Engineering". Prentice Hall, Inc., USA.
7. Lysmer, J., Udaka, T.,Hwang, R., and Seed, H,B.(1974). "LUSH:a computer program for complex response analysis of soil structure systems report EERC 72-12, Earthquake Engineering Research Centre, University of California, Berkely
8. Lysmer, J., Udaka, T., Tsai, C.F., and Seed, H,B.(1975). "FLUSH:a computer program for approcimate 3-D analysis of soil structure systems report EERC 72-12, Earthquake Engineering Research Centre, University of California, Berkely.
9. Neethu Urs, B.A., (2002), "Seismic behavior of half space and Embankment Using Wave Propagation Theory" , M. Tech. Thesis, Dept. of Civil Engg., S.J.C.E., Mysore.
10. Prasad, S. K. , Towhata, I. and Chandradhara, G. P., "Field Investigation On Damage Suffered By Earth Dams During The Gujarat Earthquake", 12SEE - 2002, December, 2002
11. Schnabel, P.B., Lysmer, J., and Seed, H.B.(1972). "SHAKE: a computer program for earthquake response analysis of horizontally layered sites," report EERC 72-12, Earthquake Engineering Research Centre, University of California, Berkely.
12. Towhata, I., Prasad, S. K., Tsuyoshi Honda and Chandradhara, G. P., "Geotechnical Reconnaisance study on damage caused by 2001 Gujarat earthquake, India", Soils and Foundations, Vol. 42, No. 4, August 2002.

STUDY OF THE EFFECT OF EXCESS PORE PRESSURE DISSIPATION ON LIQUEFACTION USING SHAKING TABLE TEST

Kazuhiro Kaneda[1], Hiroyuki Yamazaki[2], Kenji Nagano[3] and Hemanta Hazarika[4]
[1]*Soil Dynamics Division, Geotechnical & Structural Engineering, Port and Airport Research Institute, kaneda@pari.go.jp*
[2]*Ditto, yamazaki@pari.go.jp*
[3]*Ditto, nagano-k89ag@pari.go.jp*
[4] *Graduate School of System Science and Technology Akita Prefectural University, hazarika@akita-pu.ac.jp*

ABSTRACT

Shaking table tests of both horizontal ground and a caisson type quay wall were performed to investigate the excess pore water pressure dissipation after liquefaction. In the case of loose sand, the excess pore pressure stays high for a while after liquefaction. On the other hand, in the case of dense sand, the excess pore pressure dissipates promptly after liquefaction. The caisson type quay wall with the sandy backfill moves during vibration, but in the case of loose sand and short continuation time, after vibration it also moves because of the high excess pore pressure. This paper shows the experimental results on the dissipation of excess pore pressure.

Keywords: Liquefaction, Shaking table test, Caisson type quay wall

1 INTRODUCTION

The 1995 Hyogo-ken Nanbu earthquake that damaged Kobe City, Japan, was an inland earthquake characterized by high acceleration and comparatively short duration. Such earthquakes can devastate harbor facilities. Under liquefaction, the excess pore pressure increased, the ratio of excess pore pressure exceeded 0.95, and the soils liquefied. The pile and marine structure were damaged, and lateral flow was observed. In this study, the shaking table test was used to determine the fundamental characteristics of liquefaction. The focus of the study was the rise of excess pore pressure. First, the shaking table test of horizontal ground was performed and the effect of the sand density was investigated. In the experiment performed on the caisson type quay wall (called "caisson" hereinafter), lateral flow occurred after vibration in the case of loose sand, but this result was not observed in the case of dense sand. The results are reported in this paper.

2 MODEL DESCRIPTION AND PROCEDURE

Figure 1 shows the soil box models with the locations of various devices (pore-water pressure cells, accelerometers, and displacement gauges) used in the testing program. Figure 1(a) shows the cross-section and front view of the horizontal ground and Figure 1(b) shows the cross-section and front view of the caisson. The soil box was an acrylic container 850 mm long, 360 mm wide and 550 mm deep. All of the monitoring devices were installed at the center of the soil box to eliminate the effect of sidewall friction on the measurements. The sand was Sohma sand (No. 6). Table 1 lists the physical properties of the sand. (Yamazaki, Kaneda, Nagano and Hazarika (2006), Yamazaki, Kaneda and Nagano (2007)).

A 150-mm-thick foundation of crushed stone was placed in the bottom of each soil box to simulate bedrock. A 20-cm-thick foundation of dense, saturated sand was prepared in four layers on top of the crushed stone using a vibratory compaction technique. The relative density of this sand foundation was approximately 100%. Next, in the case of Figure 1(a), the horizontal ground, and in the case of Figure 1(b), the caisson, backfill was prepared in four layers to a depth of 200 mm using a free-falling technique. Compaction was also achieved by shaking the soil box. Three relative densities (Dr; 40%, 50% and 80%) of the sand were prepared. Table 2 shows the law of similitude (Kagawa (1978)). From the similarity ratio, the height of the caisson is 200 mm high in the model; therefore, one caisson is 10 m high in the prototype. Figure 2 shows the input accelerations. Sinusoidal loading was applied to the soil-structure system for 1.0 and 0.25 seconds with a frequency of 20 Hz and the maximum amplitude 100 and 200 Gal. For a given acceleration, we measured the response acceleration, displacement of the caisson, and the pore water pressure in various locations of the sand or backfill. The experimental cases are listed in Table 3.

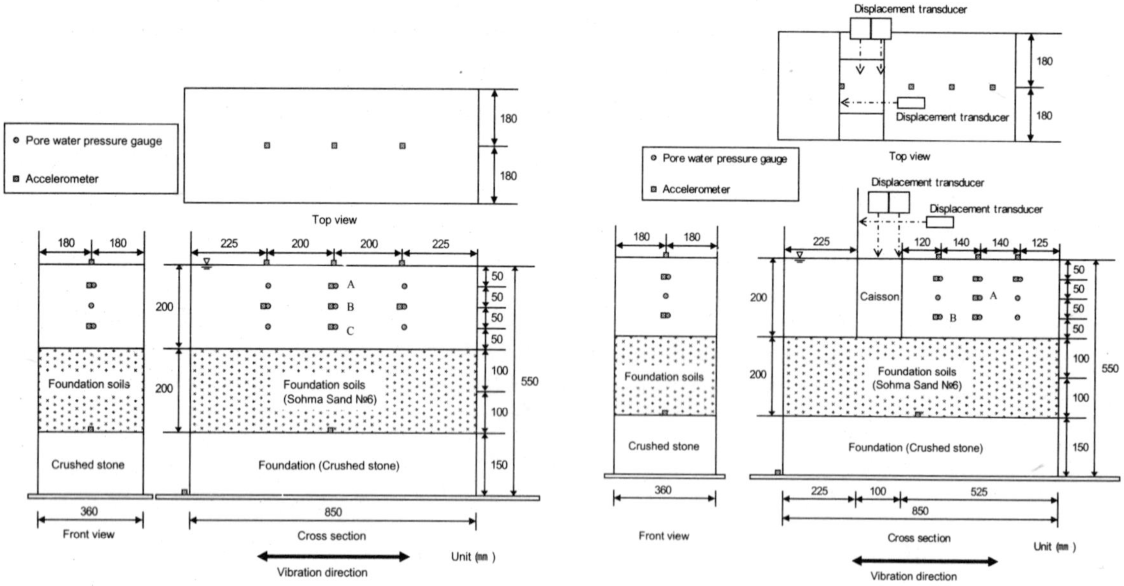

(a) Horizontal ground (b) Caisson

Figure 1. Soil boxes: (a) Cross-section and front view of horizontal ground, (b) cross-section and front view of the caisson

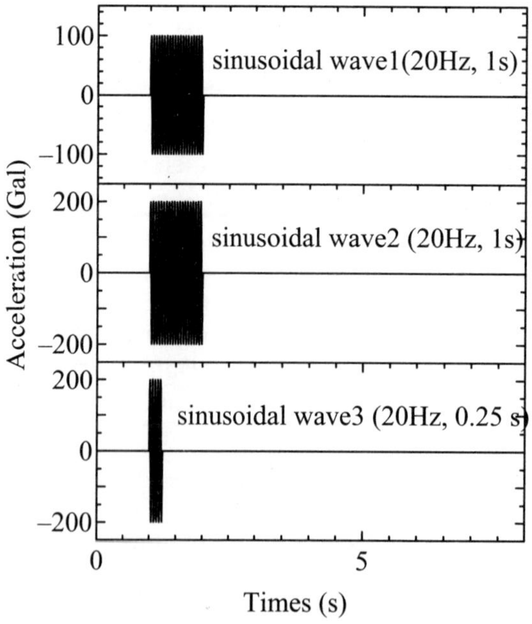

Figure 2. Input accelerations

Table 1. Physical properties of Soma sand

Soil particle density ρ_s (g/cm^3)	2.66
Maximum dry density ρ_{dmax} (g/cm^3)	1.72
Minimum dry density ρ_{dmin} (g/cm^3)	1.52
Permeability k (cm/s)	1.0×10^{-2}

Table 2. Law of similitude

	Prototype/model
Distance	50
Continuation time	$50^{3/4}$
Acceleration	1
Density	1

Table 3. Experimental cases

Case No.		Input wave	Relative density (%)	Continuation time (s)	Maximum Acceleration (Gal)
1	Horizontal ground	Sinusoidal wave 1	50	1.00	100
2		Sinusoidal wave 2	80	1.00	200
3	Caisson quay wall	Sinusoidal wave 3	40	0.25	200
4		Sinusoidal wave 3	80	0.25	200

3 TEST RESULTS

Horizontal ground case

Figure 3 shows the results of the ratio of excess pore pressure and acceleration with Dr 50% and 80% at points A (shallow), B (middle) and C (deep), which are shown in Figure 1(a), for sinusoidal waves 1 and 2. The amplitude of the sinusoidal wave of Dr 80% (200 Gal) is larger than that of Dr 50% (100 Gal). The liquefaction is defined as occurring when the ratio of excess pore pressure exceeds 0.95. In the case of Dr 50%, liquefaction of all points occurred during vibration and the pore pressure subsequently remained high for some time. The dissipation of excess pore pressure began at point C, because pore water moves from a deep area to a shallow area. In the case of Dr 80%, liquefaction at both points B and C occurred, after which the pore pressure dissipated quickly. As in Dr 50%, the dissipation of excess pore pressure begins from the deepest area.

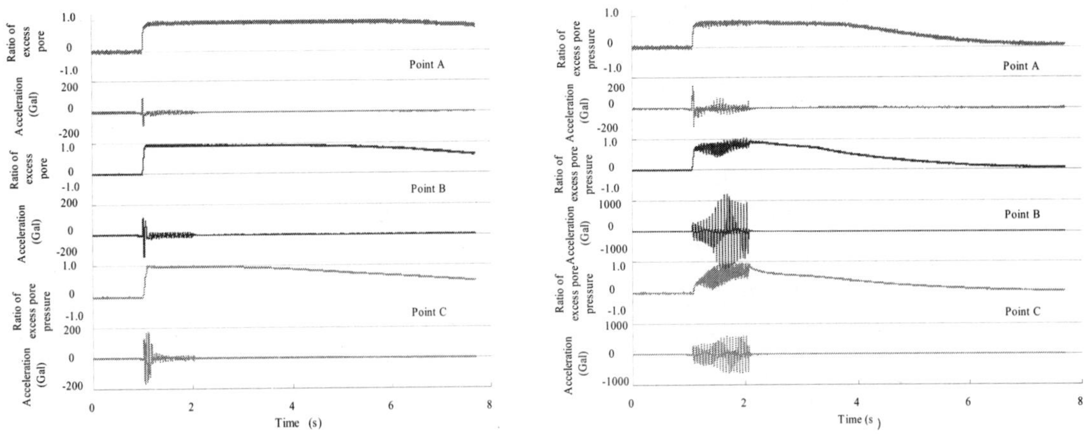

(a) Dr 50%

(b) Dr 80%

Figure 3. Time histories of the ratio of excess pore pressure and response accelerations of the horizontal ground

<u>Caisson quay wall case</u>

Next, the behavior of the caisson is shown after vibration using the shaking table. Figure 4 shows the relationship between the resultant residual displacement and density for sinusoidal wave 3, shown as Figure 2. The equation of the resultant residual displacement is as follows:

$$\delta_{vh} = \sqrt{\delta_h^2 + \delta_v^2} \tag{1}$$

Where δ_{vh} is the resultant residual displacement and δ_v,δ_h are the vertical and horizontal displacements, respectively, recorded at the top of the caisson. Figure 5 shows the time history of the horizontal displacement of the caisson during and after vibration. At this point, the vertical displacement is minor compared with the horizontal displacement, and so the results of the vertical displacement are abbreviated. The time histories of the ratio of excess pore pressure and acceleration at point A, which is shown in Figure 1(b), are shown in Figure 6.

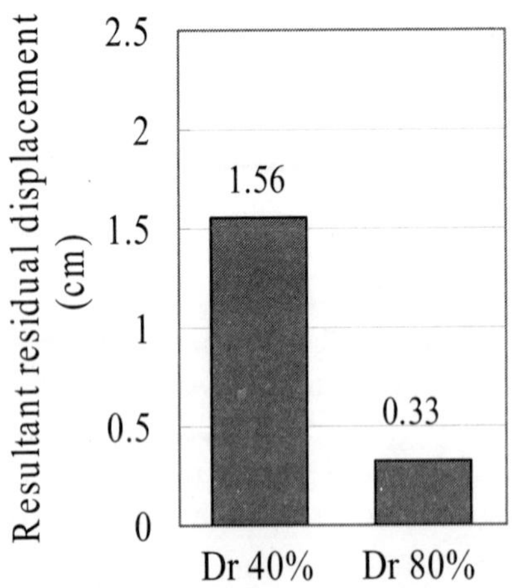

Figure 4. Relationship between resultant residual displacement and relative density

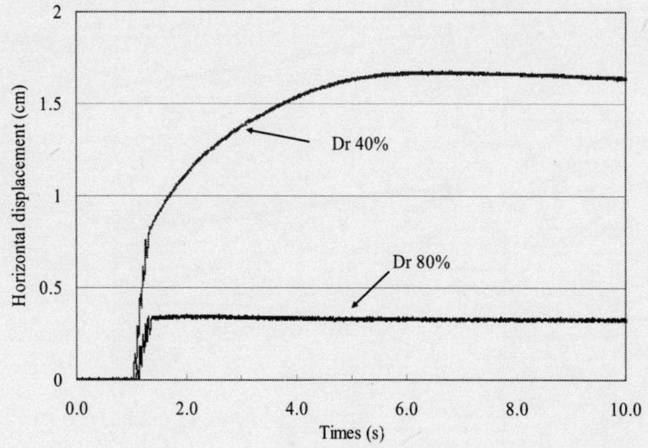

Figure 5. Time history of horizontal displacement of the caisson

In the case of loose sand (Dr 40%), after vibration (continuation time, 0.25 second), horizontal displacement (such as the lateral flow) occurred gradually. From the Figure 6, the time history of the ratio of excess pore pressure for this case shows that after vibration, high values of the ratio of excess pore pressure were maintained for approximately 5 seconds after liquefaction. The high value of this ratio is considered to be the cause of the generation of lateral flow. This supposition is described later. On the other hand, in the case of dense sand (Dr 80%), the ratio of excess pore pressure fell rapidly after vibration, and therefore horizontal displacement did not occur after vibration.

Figure 7 shows the excess pore pressure, ratio of excess pore pressure, response acceleration at point B (see Figure 1(b)) and horizontal displacement of the caisson in order to further investigate the relation between dissipation of the pore pressure and displacement of the caisson in the case of Dr 40%. The horizontal displacement occurred for approximately 6 seconds after vibration with dissipation of the excess pore pressure. When the backfill sand liquefied under the vibration and the high excess pore pressure continued, the isotropic excess pore pressure acted on the wall of the caisson after vibration. When the dissipation of excess pore pressure began, the effective stress recovered and the total pressure acting on the wall began to decrease. It is considered that the movement of the caisson with lateral flow occurs according to the balance between the friction of the soils and the caisson, total horizontal pressure and vibration continuation time, and so on.

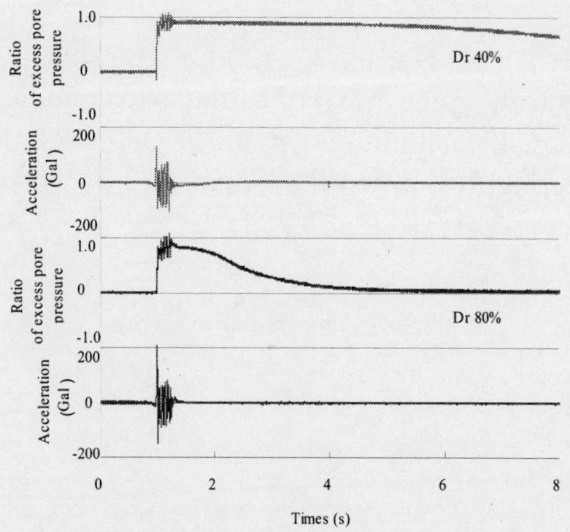

Figure 6. Time histories of ratio of excess pore pressure and response acceleration in the caisson wall case

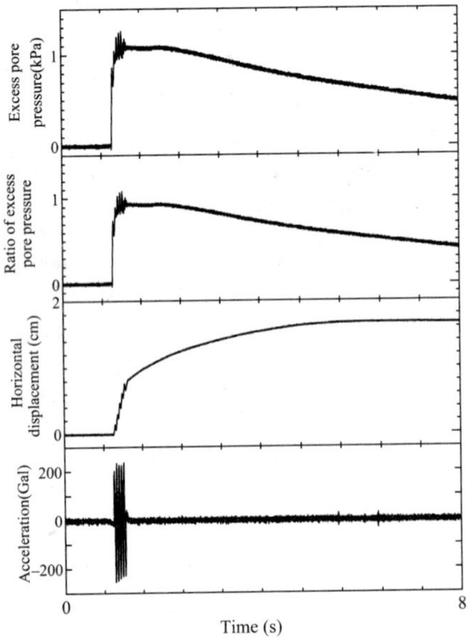

Figure 7. Relationship between dissipation of excess pore pressure and horizontal displacement in the caisson wall case

4 CONCLUSIONS

Shaking table tests were performed with a focus on the dissipation of excess pore pressure after liquefaction. For loose sand, the excess pore pressure is maintained at high levels for a while after liquefaction, and horizontal displacement of the caisson occurs after vibration. For dense sand, the excess pore pressure dissipates rapidly following shaking, and therefore the horizontal displacement of caisson stops after vibration. In the case of loose sand, the lateral flow occurs under a certain condition, which needs to be further investigated with respect to earthquake continuation time.

REFERENCES

1. Yamazaki H., Kaneda K., Nagano K. and Hazarika H., (2006). Shaking table test of caisson on the effect of earthquake continuation time, In Proc. of International Conference on New Developments in Geoenvironmental and Geotechnical Engineering, Incheon, Korea, pp.283-290.
2. Yamazaki H., Kaneda K. and Nagano K., (2007). The shaking table test on liquefaction of the caisson type quay wall on the effect of earthquake continuation time, JGSE.
3. Kagawa T.,(1978). On the similitude in model vibration tests of eath-structure, J. of Geotechnical Engineering, No. 275, Vol. 7, pp. 69-77 (in Japanese).

ON THE CHOICE OF INPUT MOTION FOR SITE - SPECIFIC EARTHQUAKE RESPONSE ANALYSIS

L.Govindaraju[1] and Subhamoy Bhattacharya[2]

[1] Department of Civil Engineering, R.V.College of Engineering, Bangalore
lgr_civil@yahoo.com
[2] Department of Engineering Science, University of Oxford (U.K)
Subhamoy.bhattacharya@eng.ox.ac.uk

ABSTRACT

Earthquake Geotechnical Engineers often need to develop a site-specific design ground motions for designing structures such dams, retaining walls, pile-supported structure or a raft or to analyse complex soil-structure interaction problems. This is required in the form of an acceleration time history series. The only data available to the engineers through their seismologist colleagues comprises of the earthquake source mechanisms in the vicinity of the region (location of the fault and the likely rupture characteristics) and in some cases the recorded data of past earthquakes in the region. For the cases, where some recorded data is available – manipulation of these records can generate the required site–specific motion. However, where these records are unavailable, artificial ground motions has to be generated. In this case, the concerned engineers are left with the country-specific code of practice which will provide with the response spectra and expected peak bed rock acceleration at the location. This is indeed the problem with most parts of India and other developing seismic hazard countries where recorded motion is not available. Several methods are used in practice for the generation of these artificial ground motions at these places. These approaches include a) Direct use of strong motion recorded elsewhere having similar expected earthquake magnitude at the site under consideration; b) Scaling of strong motion records to the expected P.B.R.A (Peak Bed Rock Acceleration) expected at the location; c) Code-specified Spectrum–compatible ground motion. This paper presents a comparative study on the performance of these three approaches for a site in Kolkata (India). The site is layered having fine sand underlying silty clay. Acceleration time-series as input motions have been generated by each method and the ground motion parameters at the surface of layered soil deposit are evaluated and compared.

Keywords: Amplification ratio, Artificial ground motion, Site response, Spectrum compatible

1 INTRODUCTION

Appropriate rock motions (either natural or synthetically generated) are often required to represent the design rock motion for a site. The rock motion should be associated with the specific seismotectonic structures, source areas or provinces that would cause most severe ground motion at the site under currently known tectonic framework. If recorded natural time histories are used, it is preferable to use a set of few natural time histories (typically 3) that have ground motion characteristics similar to those predicted for the design rock motions for the site under consideration. In seismically active regions it may be possible to reduce earthquake damage by conducting detailed specific prediction of seismic ground motion. With the knowledge of earthquake source mechanisms and path effects, detailed ground motion at any specific site of interest may be determined. This approach is of great importance, especially in the regions where ground motion records of engineering interest are totally absent. In the absence of recorded motions, artificial motions can be generated. Several methods are in practice for the generation of artificial ground motions. However, the response of a structure due to input ground motion depends on the choice of the method of generating synthetic ground motions.

This paper presents the results of a study on the seismic response of layered ground in a region subjected to input ground motions that are generated artificially by three methods. The first method involves the selection of strong motion records that are close to expected earthquakes in the study area. In the second method, the selected strong motion records are scaled to obtain appropriate ground motion parameters of the expected earthquakes. The third method involves the development of ground motions that are compatible with the target response spectrum. The selected study area is considered to be in moderate seismic zone (zone III) with a zone factor of 0.16 according to seismic hazard map of India, Bureau of Indian Standards (I.S:1893 part I - 2002). The design ground motion parameter for the study area has been obtained from Global Seismic Hazard Assessment Programme (GSHAP) map which is based on 10% probability of exceedance in 50 years. Hence the corresponding maximum bed rcok acceleration (PGA) of 1.6 m/s^2 (0.163 g) has been selected for the region corresponding to an expected earthquake of magnitude (M) in the range of 6.0 to 6.5. Figures 1 and 2 show the geotechnical bore hole data and average Standard Penetration (SPT) 'N' values at the selected region. The bulk density of soil layers vary from 16 -19 kN/m^3 and platicity index from 0 - 48%.

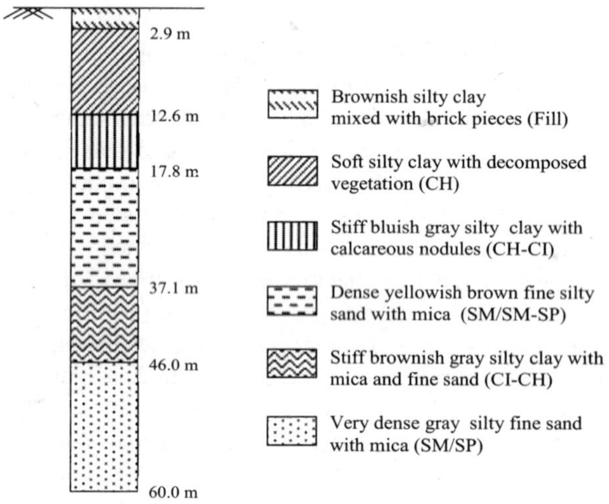

Figure 1 : Selected soil profile

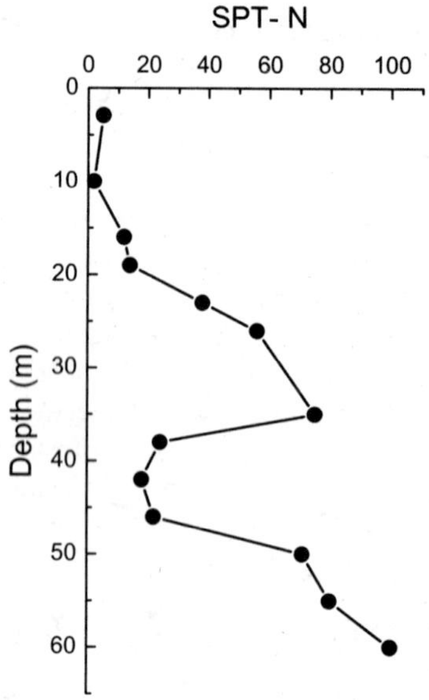

Figure 2: Typical Standard Penetration Test (SPT) values

2 SEISMIC RESPONSE ANALYSIS

In the present study the parameters of interest in the seismic response analysis involves the fundamental frequency of the soil layers, amplification factor of ground parameters, time history of ground motion parameters at the ground surface and response spectra. These parameters are evaluated using the computer program SHAKE2000 based on equivalent linear approach. The input parameter for the model such as shear wave velocity of each soil layer has been obtained from the empirical relation proposed by Japanese Road Association (Lee, 1992). To account for soil behaviour under irregular cyclic loading, the dynamic properties of soils such as modulus reduction and damping vs. shear strain curves proposed by Vucetic and Dobry (1991) were used based on plasticity characteristics of respective soil layers.

3 SELECTION OF INPUT EARTHQUAKE MOTION

While selecting a suitable strong motions several important factors should be considered. These include similar magnitude, peak acceleration close to the target value, similar fault distance and similar site conditions (Kramer, 1996). Based on this, earthquakes of magnitude (M) in the range of 6.0 to 6.9 recorded at rock sites (site class 'A' as per USGS classification for which shear wave velocity, V_s > 750 m/s) were selected from the database compiled by the Pacific Earthquake Engineering Research Center (PEER), University of California at Berkeley.

3.1 Input Motion based on Method - 1 (Direct use of strong motion records)

Table 1 illustrates five earthquake records and their characteristic selected from the database compiled by the Pacific Earthquake Engineering Research Center (PEER), University of California at Berkeley. The selected earthquakes represent nearly the similar magnitude of 6 to 6.5 and PGA for the study area. Figure 3 shows the time history of selected strong motions.

Table 1: Characteristics of Selected Earthquake Records
(Site Conditions : USGS – (A), V_s > 750 m/s)

Earthquake	**Northridge**	**SanFernando**	**Whittier Narrows**	**LomaPrieta**	**Northridge**
Date	17/01/1994	09/02/1971	01/10/1987	18/10/1989	17/01/1994
Recording Station (ST)	127 Lake Hughes # 9	127 Lake Hughes # 9	24399 Mt Wilson-CIT Seista	47379 Gilroy Array #1	90017 LA - Wonderland Ave
Magnitude	M = 6.7 M_L= 6.6 M_s= 6.7	M=6.6 M_s=7.6	M= 6.0 M_L= 5.9 M_s= 5.7	M =6.9 M_s = 7.1	M = 6.7 M_L= 6.6 M_s= 6.7
Peak Acceleration (a_{max})	0.165 g	0.157 g	0.186 g	0.209 g	0.172 g
Closet to fault rupture (km)	26.8	23.5	21.2	11.2	22.7
Closet to surface projection rupture (km)	28.9	20.2	-	10.5	-

M = Moment magnitude, M_L = Local magnitude, M_s = Surface wave magnitude

The results of the site response analysis using five strong motion records as input motions are shown in Figs. 4 and 5. Due to the wide range of input peak accelerations (0.157 to 0.21g) , the maximum acceleration at the ground surface varies over a wide range of 0.18 to 0.28 g. Fig. 4 shows the variation of amplification ratio with frequency. The maximum amplification ratio (A) varies over a range of 5.1 to 9.0 for the corresponding frequency (f) range of 0.88 to 1.13 Hz. Fig. 5 shows the variation of spectral acceleration with period corresponding to 5% damping for various locations in Kolkata based on five earthquakes. The maximum spectral acceleration varies from 0.75 to 1.0 g. However, based on the average of all five earthquakes, the maximum spectral acceleration of 0.62g may be obtained.

3.2 Input Motion based on Method - 2 (Scaling of strong motion records)

Table 2 shows two earthquake records and their characteristics selected from the database compiled by the pacific Earthquake Engineering Research Center (PEER), University of California at Berkeley (site class ‘A’ as per USGS classification). The selected earthquakes represent nearly the similar magnitude (M) of 6 to 6.5 but different peak accelerations (0.411g and 0.09 g). Figs. 6 and 7 show the selected strong ground motions from Loma Prieta and Northridge earthquakes respectively. These strong motion records are scaled to the required peak acceleration of 0. 163 g using appropriate scaling factor with out altering the duration of strong motions. Figs. 8 and 9 show the modified input motions with scaling.

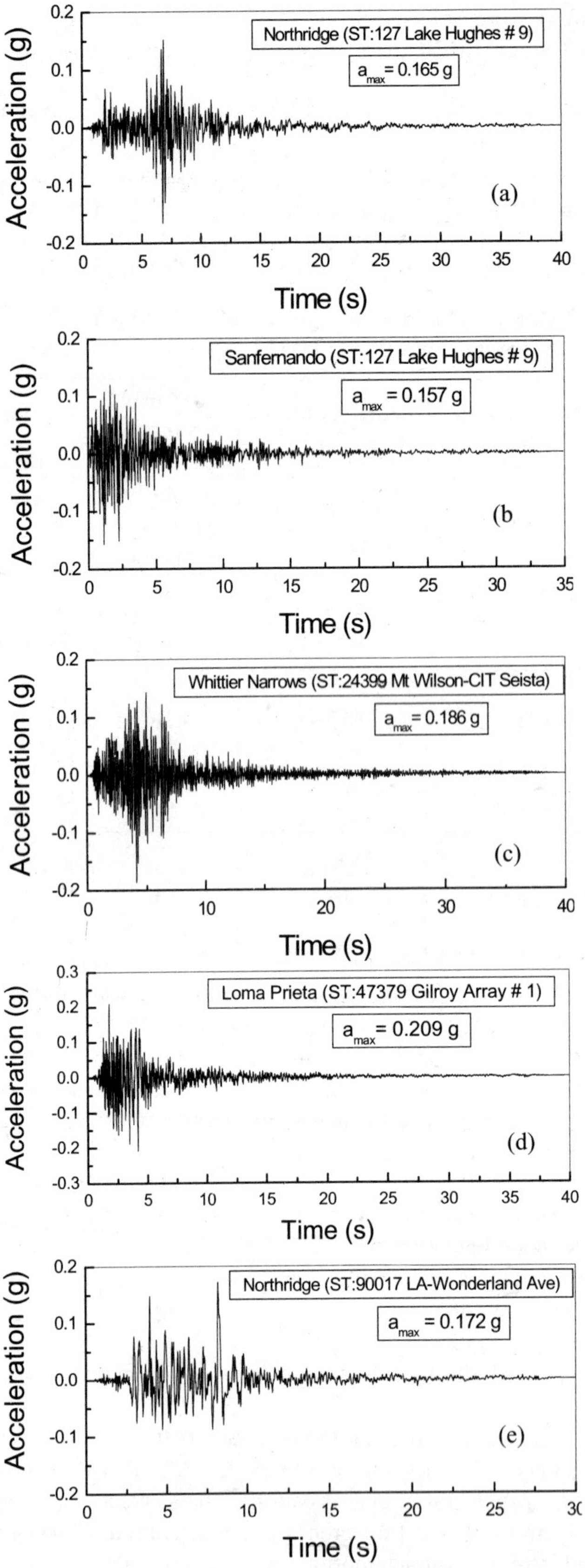

Figure 3: Time history of selected strong motion records

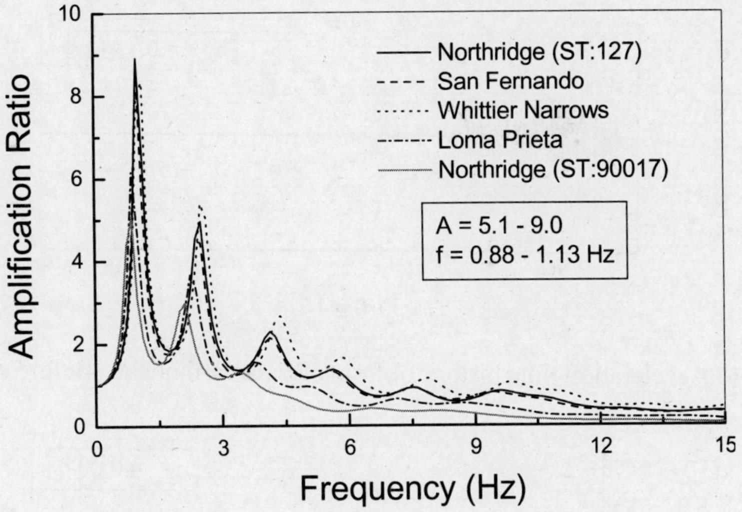

Figure 4 : Amplification ratio (damping 5%)

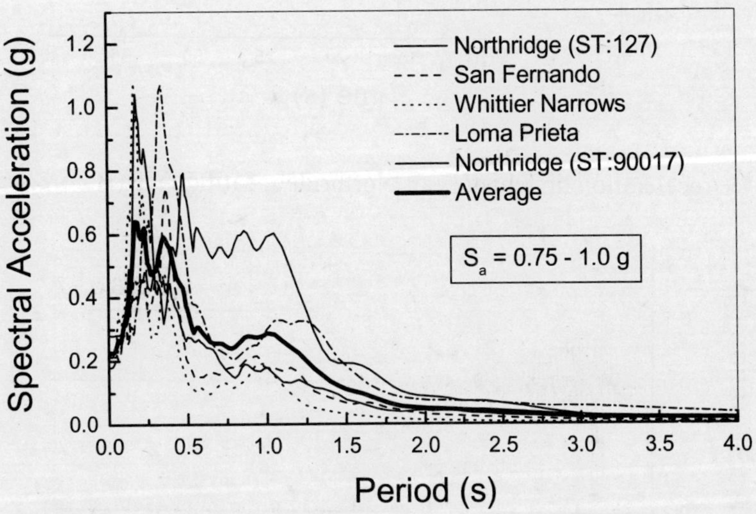

Figure 5 : Spectral accelerations at ground surface (5% damping)

Table 2: Characteristics of Selected Earthquake Records (Site Conditions : USGS – (A), V_s > 750 m/s)

Earthquake	**LomaPrieta**	**Northridge**
Date	18/10/1989	17/01/1994
Recording Station (ST)	47379 Gilroy Array #1	24644 Sandberg Bald Mtn
Magnitude	M =6.9, M_s = 7.1	M = 6.7, M_L= 6.6, M_s= 6.7
Peak Acceleration (a_{max})	0.411g	0.09 g
Closet to fault rupture (km)	11.2	43.3
Closet to surface projection rupture (km)	10.5	-

M = Moment magnitude, M_L = Local magnitude, M_s = Surface wave magnitude

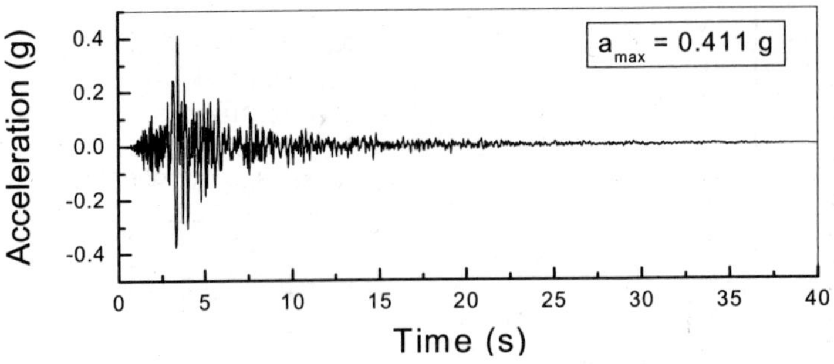

Figure 6: Acceleration time history of Loma Prieta earthquake (Before scaling)

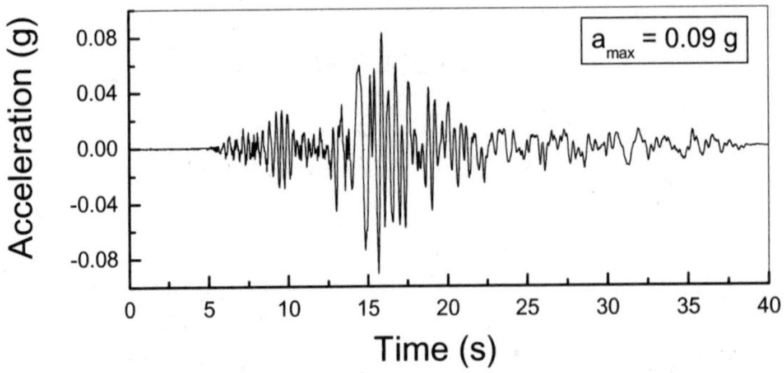

Figure 7: Acceleration time history of Northridge earthquake (Before scaling)

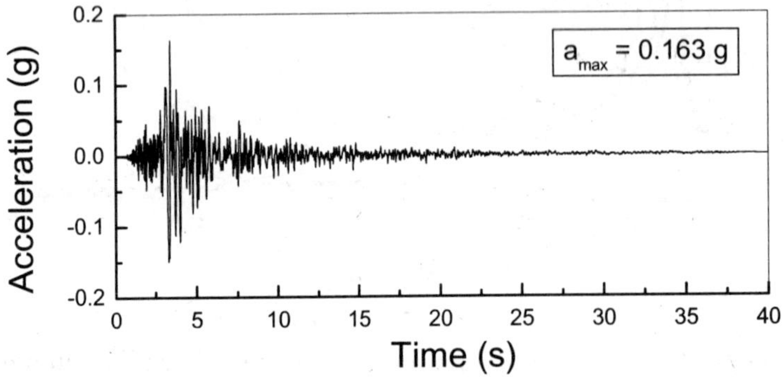

Figure 8: Acceleration time history of Loma Prieta earthquake (After scaling)

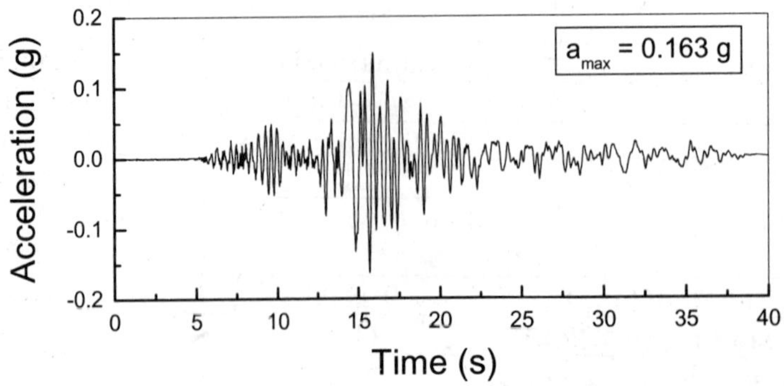

Figure 9: Acceleration time history of Northridge earthquake (After scaling)

The results of site response analyses based on the scaled strong motion records are shown in Figs. 10 and 11. As observed from Fig. 10, the amplification ratios differ enormously at different fundamental frequencies even though two earthqukes are scaled to yield a maximum input acceleration of 0.163 g. Similar variations in the spectral acceleration values can be noticed from Fig. 11. Higher spectral accelerations can be seen at lower amplification ratios as in the case of scaled Northridge earthquake. Further the peak accelerations of 0.33g and 0.24 g can be obtained at the ground surface while using Loma Prieta and Northridge scaled motions respectively. This clearly demonstrates that the peak ground acceleration alone do not predict completely response of a site during an event.

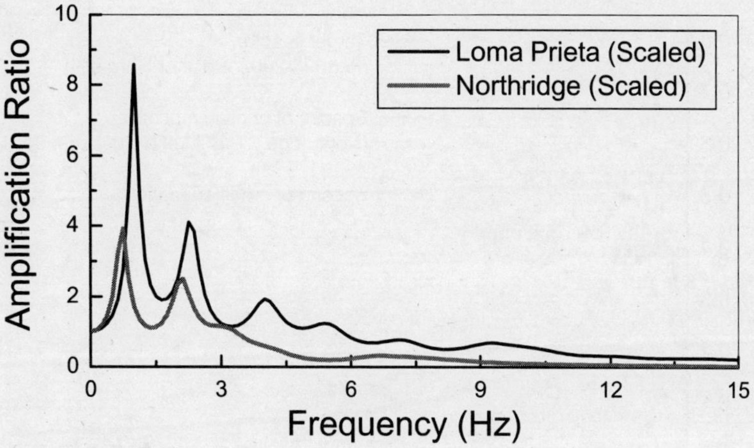

Figure 10 : Amplification ratio (damping 5%)

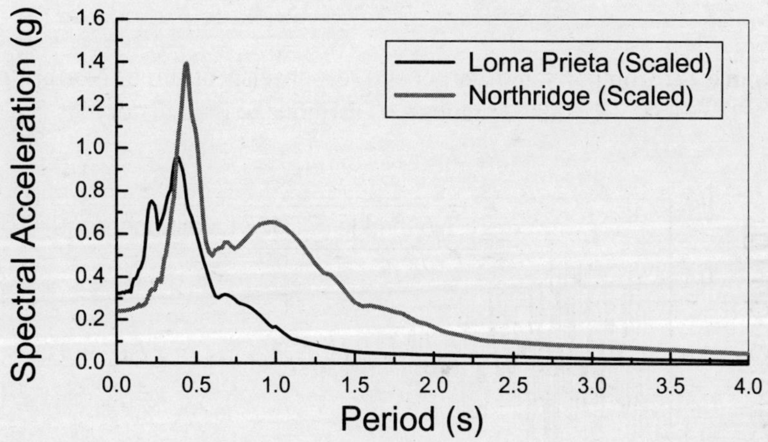

Figure 11 : Spectral accelerations at ground surface (5% damping)

3.3 Input Motion based on Method - 3 (Wavelet based generation of spectrum-compatible time-histories)

In the present study, a wavelet-based procedure has been used for the generation of spectrum-compatible time-histories. WAVGEN is one such computer program developed by Mukherjee and Gupta (2002). This wavelet-based procedure uses the decomposition of recorded accelerogram in to desired number of time-histories with non-overlapping frequency contents, and then each of the time-history has been suitably scaled for matching of the response spectrum of the revised accelerogram with the specified design spectrum. Based on the desired PGA (0.163g) for the study area as per GSHAP and target response spectra at 5% damping for rock or hard soil according to

I.S. 1893 (part I): 2002, the program WAVGEN was used to generate spectrum-compatible time-histories. The strong ground motion records as mentioned in Table 1 are used for response spectrum matching. As an example, the response spectrum of Northridge (ST:127) earthquake before and after matching the target spectra is shown in Fig. 12. Figure 13 shows the spectral matched time history of acceleration with modified maximum acceleration of 0.154 g. Similarly, the modified accelerations of 0.172 g, 0.167 g, 0.158 g and 0.21 g can be obtained for San Fernando, Whittier Narrows, Loma Prieta and Northridge (ST: 90017) earthquakes respectively. The spectral matched time histories of acceleration are then used as input motions to conduct site response analyses.

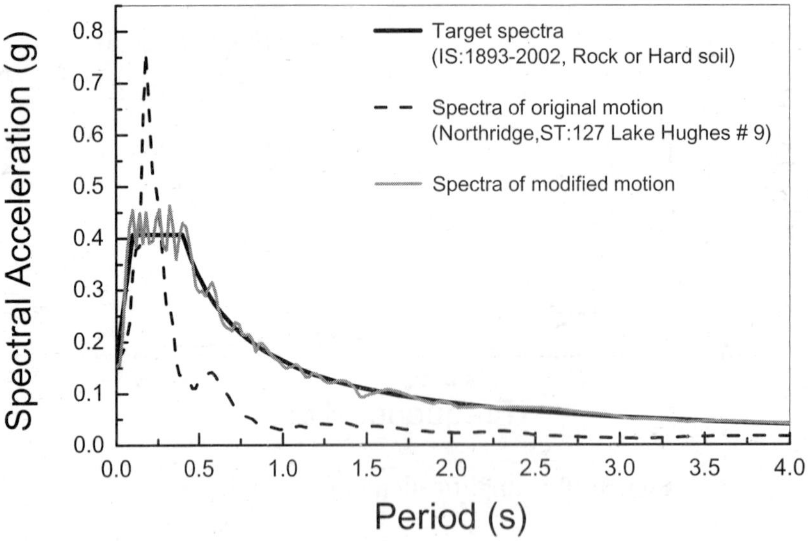

Figure 12: Comparison of target and response spectrum of Northridge (ST: 127 Lake Hughes # 9) earthquake time history

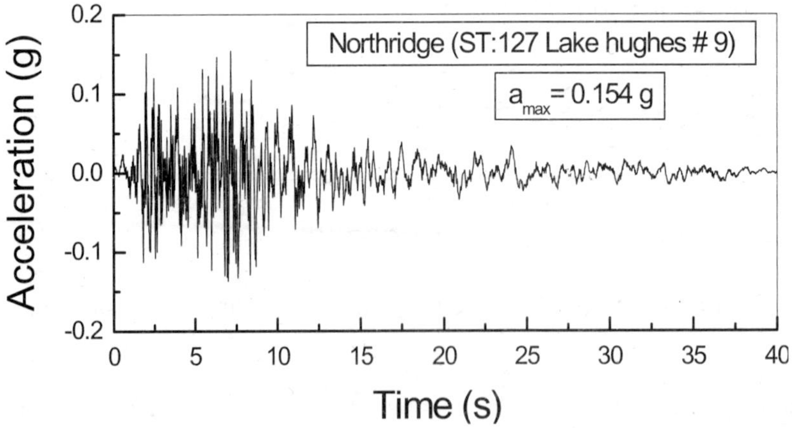

Figure 13: Target spectra matched time history of acceleration

The results of the site response analysis using spectrum compatible time histories developed from five strong motion records are presented in Figs. 14 and 15. Despite the wide range of peak accelerations (0.154 to 0.21g) of the spectrum compatible time histories of input motions, the maximum acceleration at the ground surface varies over a small range from 0.220 to 0.278 g. The maximum amplification of ground motion parameters considering five spectrum compatible input motions is shown in figure 14. As observed from this figure, the maximum amplification ratio (A) varies over a small range of 4.1 to 5.0 for the corresponding frequency (f) range of 0.75 to 0.87 Hz.

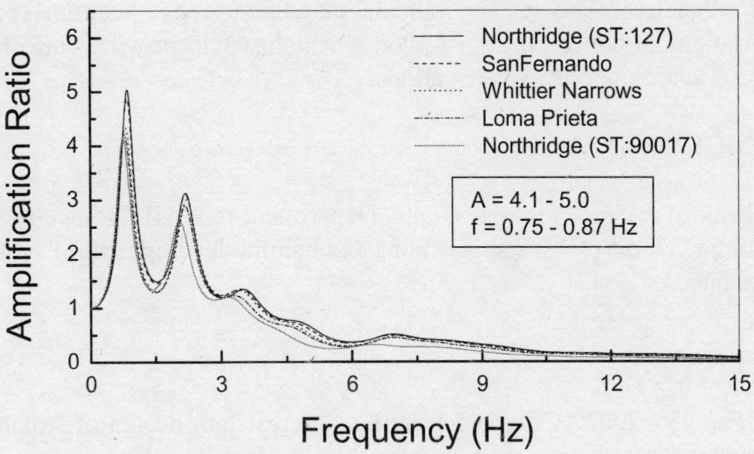

Figure 14: Amplification ratio (damping 5%)

Fig. 15 shows the variation of spectral accelerations with period corresponding to 5% damping based on spectral matched five earthquakes. The maximum spectral acceleration due to five events vary over a small range from 0.80 to1.10 g. However, based on the average of all five events, the maximum spectral acceleration of 0.947 g may be obtained.

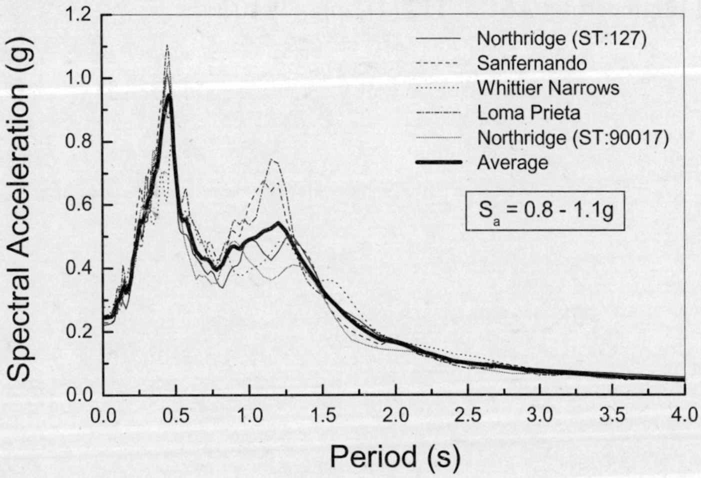

Figure 15: Spectral accelerations at ground surface (5% damping)

4 CONCLUDING REMARKS

This paper presents a comparative study on a site-specific ground response analysis for which an input ground motion is required. The motion was artificially generated using three approaches: a) Direct use of appropriate strong motion records; b) Scaling of strong motion records to the appropriate level; c) Spectrum–compatible synthetic accelerograms. Equivalent linear site response analysis was performed using SHAKE2000. Based on the results of the study, it has been shown that the ground motion parameters at the surface varied widely with the choice of input motion. Noticeable variations in the response parameters exist when strong motion records having similar magnitudes and PGA are used directly. Further, large variation in the response parameters is also possible when scaled strong motion records are used. This clearly indicates that the response analyses are subjective. However, use of Spectrum–compatible wavelet-based synthetic accelerograms for obtaining response parameters provides least variations in the response parameters compared to other approaches. This clearly demonstrates that the method is straight forward and not subjective. The results indicate that the use of spectrum–compatible ground provides least variations in the response parameters which would provide uniformity amongst the designers and would avoid unnecessary litigations.

5 ACKNOWLEDGEMENT

The authors are thankful to Prof. Vinay K.Gupta, Department of Civil Engineering, Indian Institute of Technology, Kanpur (India) for his suggestions and help in development of synthetic earthquake motions for our studies.

REFERENCES

1. IS 1893 (Part 1) : 2002 "Criteria for earthquake resistant design of structures", Bureau of Indian Standards
2. Manual on SHAKE2000 (2004) - "A computer program for the 1-D analysis of geotechnical earthquake engineering problems", University of California, Berkeley.
3. Sushovan Mukherjee and Vinay K. Gupta (2002) "Wavelet-based generation of spectrum-compatible time-histories", Soil Dynamics and Earthquake Engineering, **22**, 799-804.
4. Shannon Hsien-Heng Lee (1992) "Analysis of the multicollinearity of regression equations of shear wave velocities", Soils and Foundations, 32, 205–214.
5. Vucetic, M. and Dobry, R. (1991). "Effect of soil plasticity on cyclic response," Journal of Geotechnical Engineering, ASCE, **117 (1),** pp. 89-107

EFFECT OF LATERAL LOAD ON PILES EMBEDDED IN SANDY SLOPE

N. Almas Begum[1] *and Dr.K.Muthukkumaran*[2]
[1]*Research Scholar, National Institute of Technology, Trichirapalli,n_almas@rediffmail.com*
[2] *Lecturer, National Institute of Technology, Trichirapalli,*

ABSTRACT

Pile foundation is one of the important areas in geotechnical practices and is generally adopted for high-rise structures. When piles are constructed in sloping ground, the behaviour of piles under lateral load is differ from piles in horizontal ground. Very few research works were carried under lateral load and some analytical approaches have been developed in recent years for the response analysis of laterally loaded piles. The main objective and scope of the present investigation is to study the behaviour of piles on sloping ground under lateral loads. A series of model test had been carried out in the instrumented pile to study the behaviour of pile on horizontal ground, 1V: 1.5H and 1V: 2H slopes of relative density of sand as 30% and varying the embedment length to diameter ratio as 25,30,35 under lateral load.. Based on the sub grade reaction approach results have been produced for the effects of slope with different steepness, embedment length to diameter ratio and bending moment, deflection are discussed The increase in embedment length of pile have been found to increase the lateral load carrying capacity of pile significantly. This study will help us to learn about soil structure interaction of laterally loaded piles on sloping ground in sandy soil.

1. INTRODUCTION

Pile foundations are subjected to lateral loads and moments, which come up due to wind loads and seismic forces in buildings, earth pressure in retaining walls and abutments and water pressure in water front structure. When used under tall chimneys, high rise structures, piles are also subjected to significant amounts of lateral loads and overturning moments beside axial loads. Many transmission towers, high rise buildings and bridges constructed on steep slope are supported by piles. These structures may be subjected to large lateral loads such as violent winds and earthquakes. When piles are constructed in sloping ground, the behaviour of piles under lateral load is differ from piles in horizontal ground. The lateral resistance of piles on slopes was very small due to the reduction of soil mass when comparing to piles located on horizontal ground. And also the initial horizontal confining pressure acting on the piles on the slope is smaller than in horizontal ground.

A number of different approaches are available to determine the behaviour of laterally loaded piles on horizontal ground, but in sloping ground very few research were carried out particularly under lateral soil movement is not studied in details. The governing criterion in designing pile foundation to resist lateral loads in most cases is the maximum deflection of the foundation rather than its ultimate capacity. The maximum deflection at the pile head and the distribution of bending moment along the pile are important information for the successful design of piles that supported lateral loads. Knowing the maximum deflection at the pile head is important to satisfy the serviceability requirements of the super structure while the bending moment is required to structural sizing of pile.

2. LITERATURE REVIEW

Matlock and Reese (1960) gave a generalized solution for the laterally loaded pile. Basic equations and method of computations were given for both elastic and rigid behaviour. Soil modulus variation with depth in different forms was considered. Broms (1964a, 1964b) developed solutions for the ultimate lateral resistance of a pile assuming the distribution of lateral soil pressure and considering static of the problem.

Many analytical approaches have been developed in recent years for the response analysis of laterally loaded piles. Poulos and Davies (1980) modified the elastic solution to account for nonlinearity using yield factors. The modulus of subgrade reaction approach was extended to account for the soil nonlinearity. This was done by introducing p-y curves (Matlock 1970, Reese et al. 1974) Chae et.al (2004), described the results of several numerical studies performed with a three – dimensional finite element method.

Experimental studies were performed to examine the performance of piles under lateral loads. Prakash and Kumar (1996) developed a method to predict the load – deflection relationship for single piles embedded in sand and subjected to lateral load. Tak Kim et. al (2004), described the results of a model testing of single pile embedded in Nak – Dong River sand located in South Korea, under monotonic lateral loadings. P – Y curves for laterally loaded piles in Nak-Dong river sand were developed and evaluated the effect of relative density, pile head restraint conditions & pile installation method on p – y curves. Load – displacement response for predicted and measured curves was studied by Kumar and Lalvani (2004), using P-Y curve approach and K_{hmax} approach. Comparison between both approaches was presented.

However, all these studies have been directed towards the response of single pile or group of piles subjected to lateral load in horizontal ground with little attention being paid to pile subjected to lateral load in sloping ground.

3. EXPERIMENTAL TESTING PROGRAM

The model pile load tests were conducted in a test tank of dimension 2mx1mx1m deep filled with dry river sand shown in fig.1. The size of tank was designed large enough to minimize the influence of box boundaries. The bored pile is modeled as the preinstalled model pile.

The model pile tip was placed with a minimum clearance of at least 4 times the pile diameter from the bottom of the tank. Lateral loads were applied to the pile by using mechanical deadweights at an eccentricity of 75mm above the ground surface. The horizontal displacement of the pile head was measured using dial gauges.

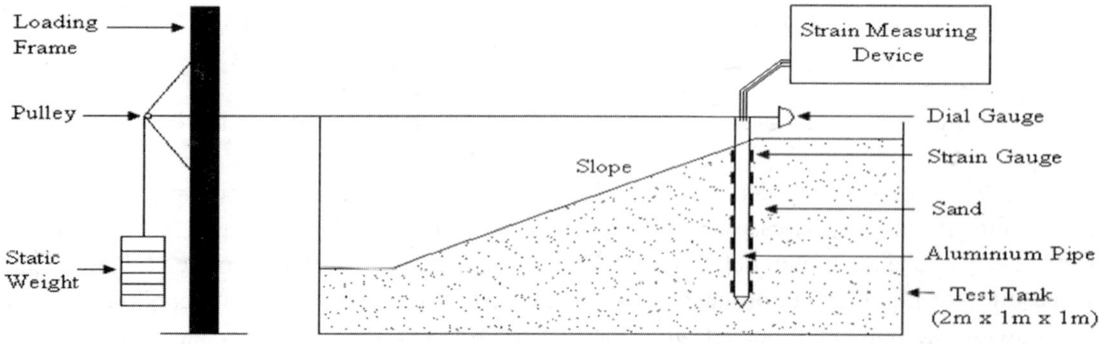

Fig 1 Experimental Setup.

3.1 Model piles and instrumentation

The dimensions of the model pile testing were determined by a dimensionless analysis (Buckingham π theorem). This theorizes the relation between the number of basic dimensions in a phenomenon, number of dimensionless parameters defining the phenomenon and the number of variables. There are six variables, which are displacement (y), pile diameter (D), pile length (L), load (W), Young's Modulus of Elasticity (E) and moment of inertia (I). According to π theorem two of the variables W and L may be common to all the π terms. In order to be complete similitude between the model testing and the prototype, the following relations between the prototype pile and the model pile was obtained and was given in the eqn.1 to 4.

$$D_s = k\,D_m \quad (1)$$
$$W_s = k^2 * (E_s/E_m) * W_m \quad (2)$$
$$I_s = k^4 * I_m \quad (3)$$
$$y_s = k * y_m \quad (4)$$

Where,

k – Scale Factor = L_s / L_m
W – Load applied on the pile (N)
E – Modulus of elasticity of the pile (N/mm^2)
y – Lateral deflection of the pile (mm)
I – Moment of Inertia of the pile material (mm^4)

Suffixes, 's' represents the prototype pile and 'm' represents the model pile.

An aluminium pipe having an outer diameter of 25.4mm with 1mm wall thickness was used as a model pile. Pile was instrumented for measuring bending stress by pasting electrical resistance type strain gauges having gauge length of 3 mm, gauge factor (K) of 2 and resistance (R) of 120 ohms. The strain gauges were pasted with its axis parallel to the pile axis. They formed quarter bridges with the dummy gauges placed in a separate strain gauge box. The were placed at a minimum spacing of 50mm apart with top gauge 25mm below the pile cap. Small holes were made on the side of the pile slightly above each gauge location to pass the wires inside the pile. Strain gauges were coated with about silicone gel for protection.

The bottom of the pile was closed by a 60° conical shoe. The model pile is modeled as an equivalent prototype pile (1:40) scale with an outer diameter of 1018mm.

Strain gauges were calibrated for bending moment along the pile length by conducting a bending test on a pile and it is given in eqn.5. The strain response was found to be linear with the bending moment applied as shown in Fig. 2. The calibration constant was obtained from the slope of the straight line and found to be 30.67 N.mm per micro strain. (1×10^{-6} mm/mm).

$$\textbf{Bending Moment 'M'} = C * \varepsilon \quad (5)$$

Where,

M – Bending Moment at Various depths (Nmm)
C - Bending Constant (1 micro strain = 30.67 Nmm per microstrain)
ε – Strain (microstrain)

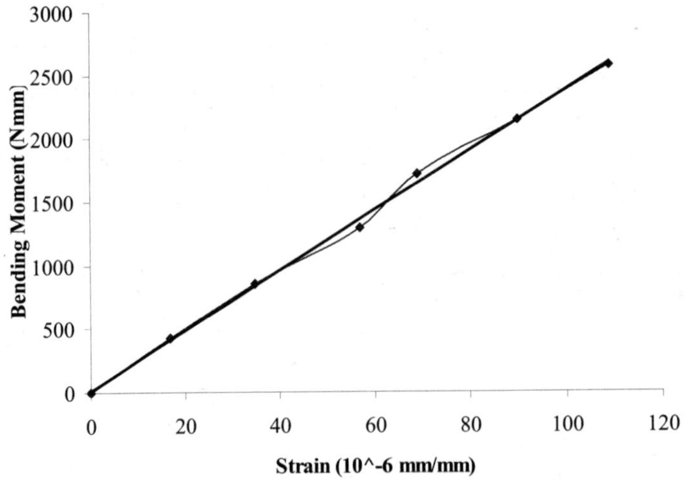

Fig. 2 Calibration of sand measuring device.

3.2 Soil Description and Model Preparation

The dry river sand was used for testing. Testing of sand was carried out as per IS specifications. It is classified as poorly graded sand (SP) of which the uniformity coefficient obtained as 2.7 and the co efficient of curvature was 1.2. The other Typical properties of the sand were D_{10} = 0.23mm, D_{50} = 0.50mm, G_s= 2.68, γ_{max} = 18.74 kN/m^3, γ_{min} = 14.72 kN/m^3. A model ground with three types of slope namely zero slope (horizontal ground), 1V; 1.5H, 1V:2H slope were made with soil of Density Index 30%. Sand Raining Device was used to achieve uniform density in the tank.

3.3 Measurements

The horizontal displacement of the pile head and the strain at various levels in the piles were measured for each tests. Mechanical dial gauges were used to measure the horizontal displacement. A manual compensating strain meter with quarter bridge circuit was used to measure bending strain along pile length at various elevations. Lateral deflection and strain readings were recorded for each increment of load. Pile was loaded up to allowable limit. It was observed that there was no increase in lateral movement of pile head for sustained load in sand. The lateral load was applied till the lateral deflection reaches 20% of pile diameter (5mm) the corresponding load was taken as lateral load capacity of the pile.

3.4 Tests Carried Out

A series of experimental tests were carried out by varying the parameters such as slope, L/D ratio and relative density of the soil as shown in the Table 1.

4. RESULTS AND DISCUSSIONS

The lateral load behaviour of the pile was studied by using lateral load – deflection curves. The curves were drawn between the lateral load applied and the lateral deflection at the soil surface. Lateral load was applied at the distance of 75mm above the ground surface and lateral deflections were measured at the soil surface. Tests were done on the pile of embedment length to the diameter ratios of 25, 30 and 35 on sandy soil of relative density 30% on both the horizontal ground and the sloping ground of 1V:1.5H and 1V:2H.

Table 1 List of experiments

SL.NO.	SOIL SURFACE	L/D RATIO	ID (%)
1	HORIZONTAL	25	30
		30	
		35	
2	1V:1.5H	25	30
		30	
		35	
3	1V:2H	25	30
		30	
		35	

Fig. 3 to 5 shows the comparison of lateral load capacity of pile with deflection for L/D ratio of 25, 30, 35 and relative density of 30. It can be seen that load-deflection curve for each of the three embedment length to diameter ratio when ground surface changing from horizontal to 1V:1.5H slope or 1V:2H, the lateral load capacity was significantly reduced. This reduction was due to the reduction in passive resistance of the soil in front of the pile. The increase in slope from horizontal to 1V:2H reduces the lateral load capacity by 12% for relative density of 30% for L/D ratio of 25. When the ground surface changes from horizontal to 1V:1.5H, the lateral load capacity reduces by 26% for relative density of 30% for L/D ratio of 25.

For L/D ratio of 30 the increase in slope from horizontal to 1V:2H reduces the lateral load capacity by 17% and when the ground surface changes from horizontal to 1V:1.5H, the lateral load capacity reduces by 28% for the relative density of 30%.
For L/D ratio 35, the increase in slope from horizontal to 1V:2H reduces the lateral load capacity by 19% and when the ground surface changes from horizontal to 1V:1.5H, the lateral load capacity reduces by 31% for the same density.

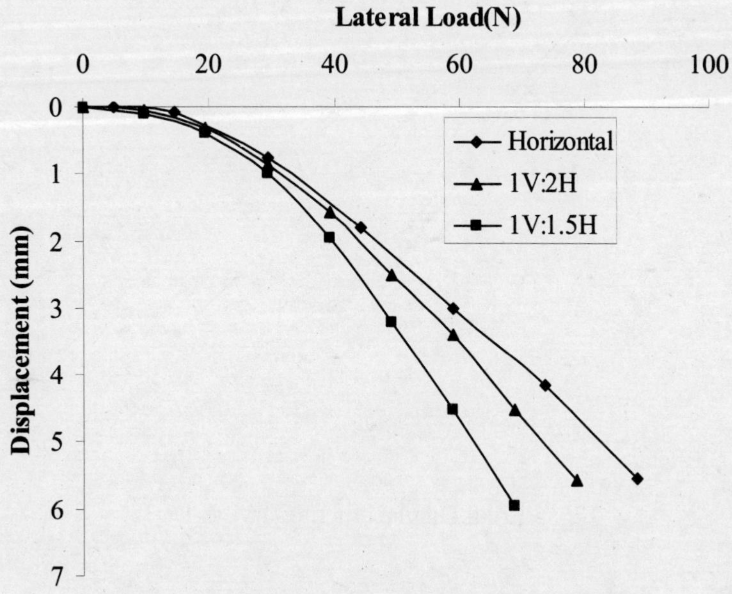

Fig. 3 Load Displacement Curve L/D=25.

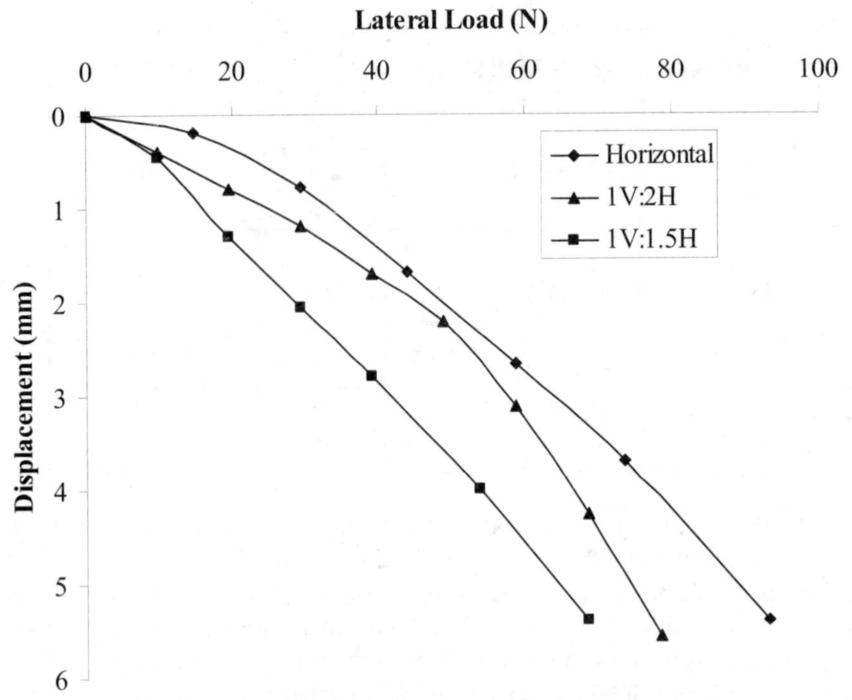

Fig. 4 Load Displacement Curve L/D=30.

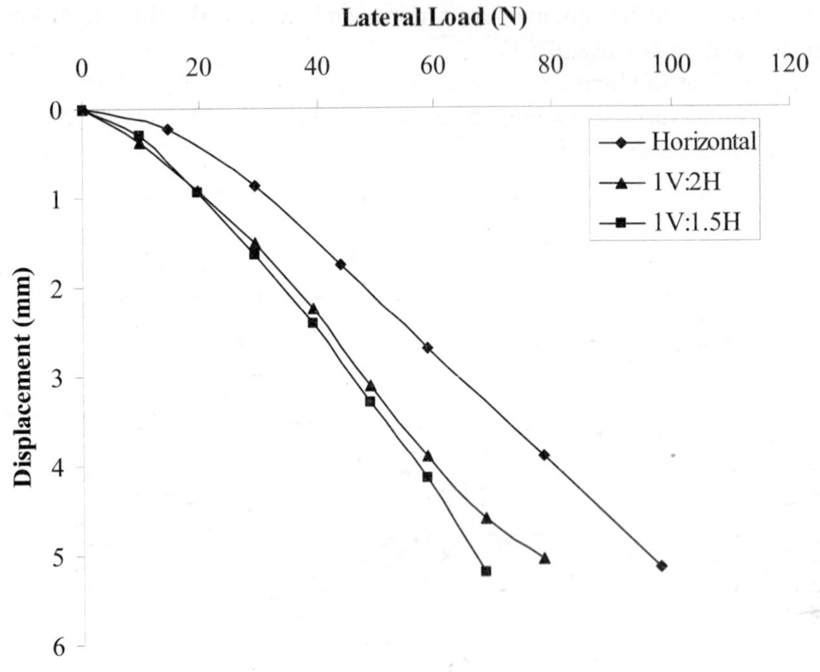

Fig. 5 Load Displacement Curve L/D=35.

4.1. Effect Of Slope

Fig. 6 shows the comparison of lateral load capacity of pile with slope for different L/D ratio and relative density of 30%. It can be seen that for L/D ratio 25 the lateral resistance offered by the pile in sloping ground of 1V : 2H reduces the lateral load carrying capacity of pile by nearly by 12% as the maximum load changes from 83.3 N to 73N. Further when the ground surface changes from 1V:2H to 1V:1.5H, the lateral load capacity reduces by 15% as the maximum load varies from 73N to 62N.

And for L/D ratio 30 the lateral resistance offered by the pile in sloping ground of 1V: 2H reduces the lateral load carrying capacity of pile from 89 N to 74.2N. Further when the ground surface changes from 1V:2H to 1V:1.5H, the lateral load capacity reduced from 74.2N to 64N.

And for L/D ratio 35 the lateral resistance offered by the pile in sloping ground of 1V: 2H reduces the lateral load carrying capacity of pile from 96 N to 78N. Further when the ground surface changes from 1V:2H to 1V:1.5H, the lateral load capacity reduced from 78N to 66.5N.

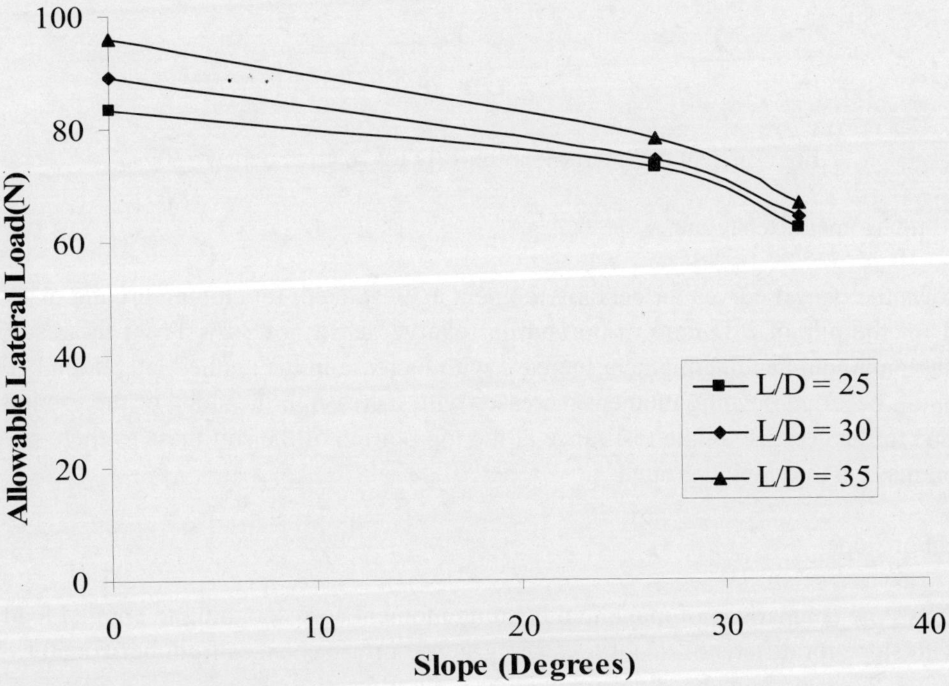

Fig. 6 Effect of Lateral load on slope for different L/D ratio.

4.2. Effect Of L/D Ratio

Fig. 7 shows the comparison of lateral load capacity of pile with L/D ratio in horizontal ground, sloping ground of 1V:1.5H and 1V:2H for relative density of 30%. The increase in L/D ratio increases the lateral load capacity.

This was due to increase in the resistance of the pile and the soil due to increase in the embedded length of the pile. when the embedment length of the pile to diameter ratio(L/D) increases from 25 to 30, the lateral load capacity increases by 6%,2%,3% for horizontal ground, sloping ground of 1V:2H and 1V:1.5H respectively for the relative density of 30%. And lateral load capacity increases by 7%, 5%, 4% for horizontal ground, sloping ground of 1V:2H and 1V:1.5H respectively when L/D increases from 30 to 35.

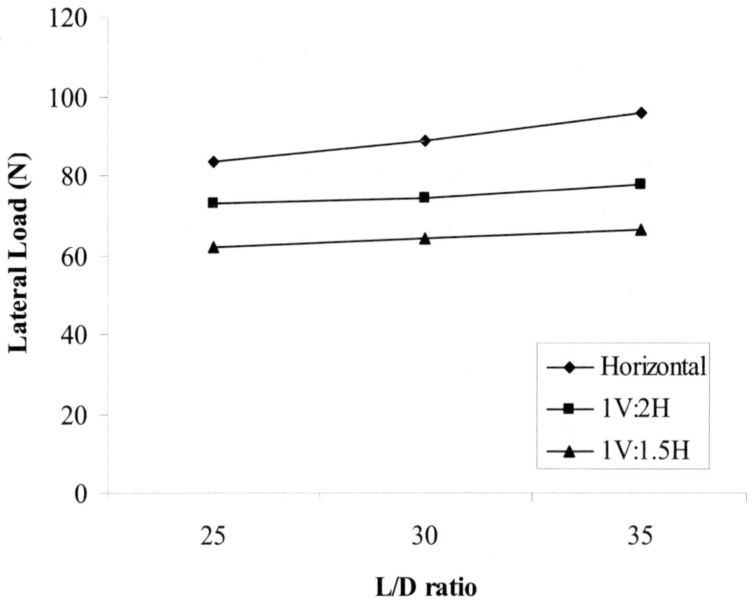

Fig. 7 Effect of Lateral load on L/D for relative density 30%.

4.3. Bending Moment Behaviour

Fig.8 shows the typical curves for bending moment against depth for sloping ground of 1V:2H and 1V:1.5H for the pile of L/D ratio of 30 and on relative density of 30%. From the results, it was found that maximum bending moment increase with increase in the applied lateral load. The depth of maximum occurred bending moment increases with increase in the slope of the ground surface. This is due to the decrease in the resistance at the top portion of the soil mass as there is reduction in the soil mass in the sloping ground.

4.4. Effect of slope

Fig. 9 shows the comparison of maximum bending moment corresponding to applied load of 60 N to pile with slope for different L/D ratio. When ground surface changes from horizontal to 1V:1.5H slope or 1V:2H, maximum bending moment was significantly increased. The increase in slope from horizontal to 1V:2H increases the maximum bending moment by 15% for relative density of 30% and for L/D ratio of 25. When the ground surface changes from 1V:2H to 1V:1.5H, the maximum bending moment increases by 10% relative densities of 30% for L/D ratio of 25.

4.5. Effect of L/D ratio

Fig.10 shows the comparison of maximum bending moment corresponding to the load of 6kg on pile with L/D ratio in horizontal ground, sloping ground of 1V:1.5H and 1V:2H for relative density 30%. The increase in L/D ratio decreases the maximum bending moment. This was due to decrease in the depth of fixidity and increase in the relative stiffness of the pile and the soil due to increase in the embedded length of the pile. When the embedment length of the pile to diameter ratio (L/D) increases from 25 to 30, the maximum bending moment decreases by 9%, 7%, 2% for horizontal ground, sloping ground of 1V:2H and 1V:1.5H respectively for the relative density of 30%. And maximum bending moment decreases by 6%, 9% and 5% for horizontal ground, sloping ground of 1V:2H and 1V:1.5H respectively when L/D increases from 30 to 35 for the relative density of 30%.

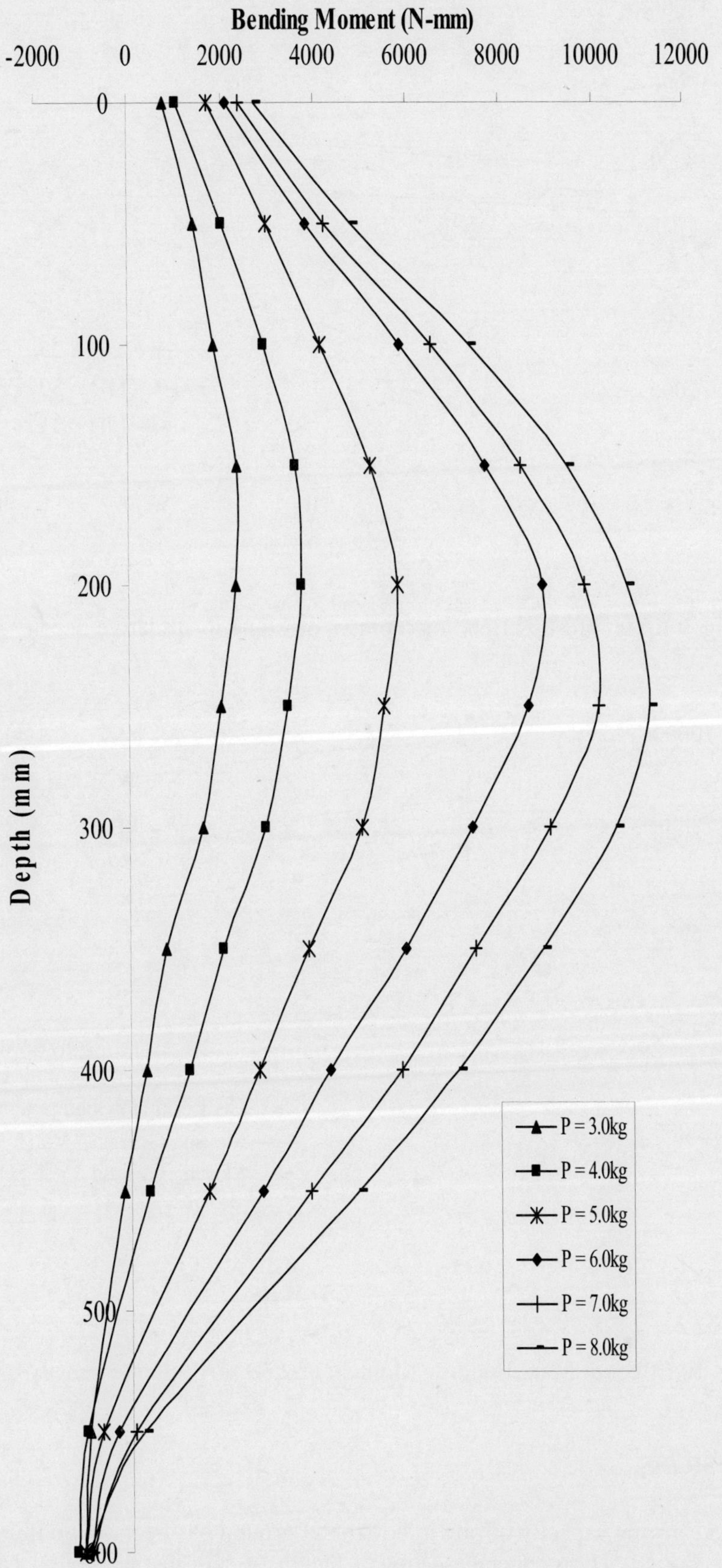

Fig. 8 Bending moment variation against Depth for various loading condition (1V:2H, L/D = 30, I_D = 30%).

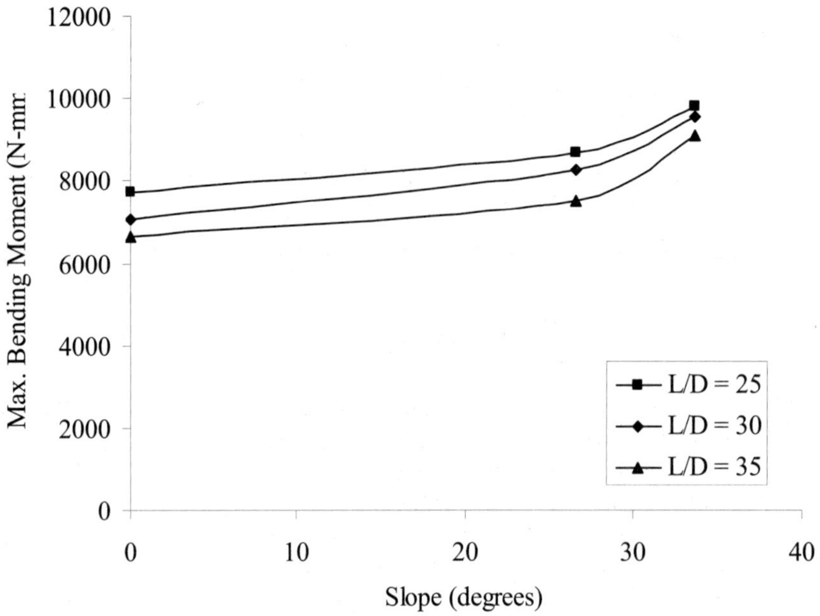

Fig. 9 Effect of Max. Bending moment on slope for Relative density 30%.

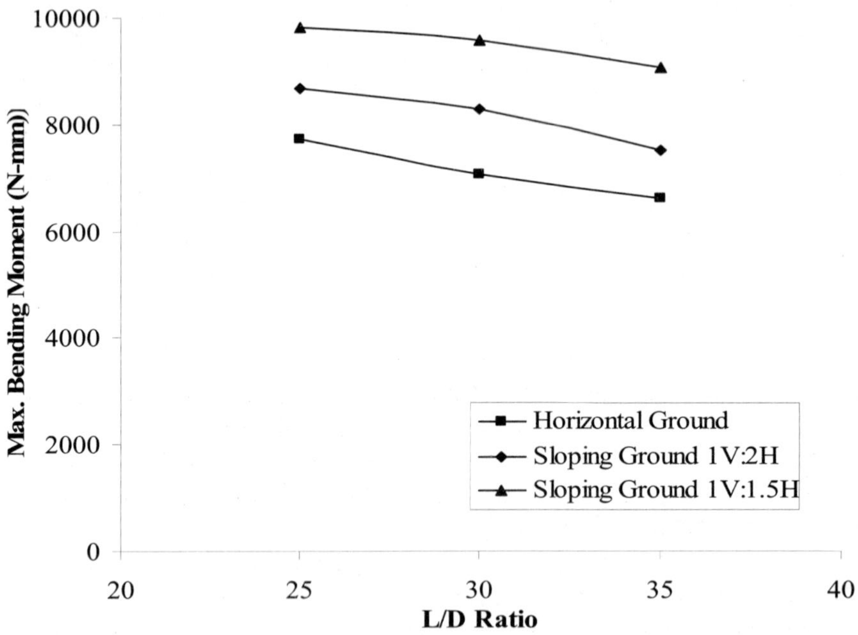

Fig. 10 Effect of Max. Bending Moment on L/D for Relative density of 30%.

5. CONCLUSIONS

The lateral load carrying capacity of pile in horizontal ground has been found more than that of the pile in sloping ground. The greater embedment length of pile in the ground has been found to increase the lateral load carrying capacity of pile significantly. The lateral load resistance of the pile can be increased by increasing the embedment length of the pile.

REFERENCES

1. Broms (1964a). "The lateral resistance of piles in cohesive soils". Journal of the Soil Mechanics and Foundation Div., 90(2), 27 - 63.
2. Broms (1964b). "The lateral resistance of piles in cohesionless soils". Journal of the Soil Mechanics and Foundation Div., 90(3), 123 – 156
3. Byung Tak Kim, Nak – Kyung Kim, Woo Jin Lee, Young Su Kim, (2004) "Experimental Load – Transfer Curves of Laterally Loaded Piles in Nak – Dong River Sand" - Journal of Geotechnical and Geoenvironmental Engineering, Vol.130, No.4, pp.416-425.
4. Chae, K.S., Ugai, K., Wakai, A., (2004) "Lateral Resistance of Short Single Piles and Pile Groups Located Near slopes" - International Journal of Geomechanics, Vol.4, No.2, pp.93-103.
5. IS: 1498 (1970) "Classification and Identification of Soils for General Engineering Purposes", Bureau of Indian Standards.
6. IS: 2720 - Part 13 (1986) "Methods of Test for Soils - Direct Shear Test", Bureau of Indian Standards.
7. IS: 2720 - Part 14 (1983) "Methods of Test for Soils -Determination of Density Index (Relative Density) of cohesionless soils", Bureau of Indian Standards.
8. IS: 2720 - Part III / Sec 1 (1980) "Methods of Test for Soils - Determination of Specific Gravity ", Bureau of Indian Standards.
9. IS: 2720 - Part XXVIII (1974) "Methods of Test for Soils - Determination of Dry Density of Soils in-place, by Sand Replacement Method", Bureau of Indian Standards.
10. Kumar, S., Lalvani, L., (2004) "Lateral Load – Deflection Response of Drilled Shafts in Sand" - IE (I) Journal-CV, Vol.84, pp. 282-286.
11. Matlock, H. (1970). "Correlations for design of laterally liaded piles in soft clay". Proc., 2nd Annual Offshore Technol. Conf., Houston, Tex., Paper OTC 1204
12. Matlock, H., and Reese, L.C (1960). "Generalized solution for the laterally loaded piles" Journal of the Soil Mechanics and Foundation Div., ASCE, 86(5), 63 - 91.
13. Poulos, H.G., and Davies, E.H (1980) Pile foundation and design. John Wiley and Sons, New York.
14. Prakash, S., and Kumar.S (1996) "Nonlinear lateral pile prediction in sand". Journal of Geotechnical Engineering 122 : 130 - 138
15. Reese, L.C., Cox W.R., and Koop, F.D (1974). "Analysis of laterally loaded piles in sand." Proc., 6th Annu. Offshore Technol. Conf., Richardson, Tex., 473 – 483.

EMPIRICAL CORRELATION OF LIQUEFACTION INDUCED SETTLEMENT OF COROMANDEL COASTAL SAND DEPOSITS

S. Senthamilkumar[1] Dr. C. Natarajan[2] Dr. K. Muthukumaran[3]

[1] Research Scholar, Department of Civil Engineering, Periyar Maniammai University, Vallam- 613 403, Thanjavur, Email: senthamilkumar@yahoo.co.in

[2] Prof. & Head, Department of Civil Engineering, National Institute of Technology, Trichy –620 015 Email: nataraj@nitt.edu

[3] Lecturer, Department of Civil Engineering, National Institute of Technology, Trichy –620 015 Email: kmk@nitt.edu

ABSTRACT

A study was carried out by conducting a detailed investigation on the dynamic geotechnical properties of the soil deposits with respect to the liquefaction potential in the Coromandel coastal line of Nagapattinam town, Tamilnadu, where major constructions related to the Tsunami relief works and Sethu canal project are undertaken over loose sandy deposits. It is mandatory to take into account the effect of liquefaction-induced settlements in the structural design and planning. Since, the Indian standard grouped this area under seismic zone III. With the intention of study the dynamic geotechnical properties with respect to liquefaction-induced settlements, a series of field tests were conducted using Standard Penetration Test (SPT) and Cone Penetration Test (CPT) at 51 locations. From the results, a microzonation map for the liquefaction potential and related settlement was developed. Empirical correlations using multilinear regression analysis between settlement, relative density, mean particle size, factor of safety, coefficient of uniformity, coefficient of curvature and CPT-SPT ratio for Nagapattinam town have been proposed to help the planning and construction of the structures.

Key words: Liquefaction, Standard Penetration Test, Static Cone Penetration Test, Settlement

1 INTRODUCTION

Liquefaction and the related settlement or ground displacement along the coastal line were observed in many parts of the world in the past. The structures like harbour, dock, lighthouse, other onshore structures etc. were severely damaged by liquefaction-induced settlement in the few known earthquakes such as the 1983 Nihonkai Chubu, the 1907 North Dik, the 1978 Pacific coast (Miyagi, Japan), the 1979 Enmedio Island, the 1993 Hokkaido Nansei-oki, the 1994 Northridge, the 1995 Hyogoken-Nambu, 1999 Turkey. Liquefaction is a condition and it occurs only in a saturated loose cohesionless soil deposits with poor drainage condition. The liquefaction occurs when the excess pore water pressure builds up during the earthquake and the shear strength decreases possibly to zero.

The development of roadways (NH-65A & NH-67), the Sethu Canal Project and the calamity caused by Tsunami paved the way to the development of new industries, townships and other major reconstruction activities along the Coromandel Coastal line of Tamilnadu, India. Due to the Sethu Canal Project, the development of many harbours and docks in the important cities along the coastal line can be foreseen. The Coromandel coastal line consists of poorly graded sand, silty sand and loamy soils, which are most vulnerable to liquefaction. The purpose of this paper is to present the Standard Penetration Test (SPT) and Cone Penetration Test (CPT) based estimation of liquefaction and the related ground settlements for the Nagapattinam Coastal line and the adjoining harbour areas. The effect of fines content, maximum surface accelerations etc. were also considered for the estimations. The correlations between mean grain size (D_{50}), relative density (D_r), fines content (F_c), coefficient of uniformity (C_u), coefficient of curvature (C_c), settlement (S) and the ratio of CPT and SPT were developed for this area, in order to use SPT values for estimating many engineering properties of the soil.

2 METHODOLOGY

From the SPT and CPT values, the factor of safety against liquefaction was calculated by the methods proposed by Seed and Idriss (1971), Robertson and Campanella (1985) and Idriss and Boulanger (2006). The factor of safety (FS) was calculated as the ratio of cyclic resistant ratio (CRR). (capacity of the soil to resist liquefaction) and cyclic stress ratio (CSR), (seismic demand to induce liquefaction). The following equations were used to calculate the cyclic stress ratio (CSR) on a soil layer at a depth of z below the ground level.

$$CSR = \frac{\tau_{cyc}}{\sigma'_{v0}} = 0.65 \frac{\sigma_{v0}}{\sigma'_{v0}} \frac{a_{max}}{g} r_d \qquad \text{(Seed and Idriss, 1971)} \qquad (1)$$

Where $Ln(r_d) = \alpha(z) + \beta(z)M$ for $z \leq 34$m (Idriss and Boulanger, 2006) (2)

$$\alpha(z) = -1.012 - 1.126 \sin\left[\frac{z}{11.73} + 5.133\right], \; \beta(z) = 0.106 + 0.118 \sin\left[\frac{z}{11.28} + 5.142\right] \qquad (3)$$

z = depth in m and M = earthquake magnitude, for z > 34 m the reduction factor, r_d was calculated by the expression, $r_d = 0.12 \exp^{(0.22M)}$. But for the purpose of using standard curves, the CSR values calculated from Eq. (1) were modified by using magnitude scaling factor (MSF) to the magnitude of M = 7.5 by using the relation proposed by Idriss and Boulanger (2006).

The cyclic resistance ratio (CRR) was estimated from SPT- N and CPT- q_c values. The observed values of SPT and CPT are corrected for various factors, such as fines content, field procedures, overburden pressure and sloping ground orientations (Idriss and Boulanger, 2006). The study area belongs to seismic zone III as per IS-1893-Part-1 (2002) and a_{max} is taken as 0.16g. From charts recommended by Youd and Idriss (2001), the CRR was calculated for the anticipated magnitude of 7.5 earthquakes.

Ishihara and Yoshimine (1992) proposed a volumetric strain based approach for liquefaction-induced settlement from q_{c1} or $(N_1)_{60}$. Ishihara and Yoshimine (1992) proposed the charts only for clean sand. However, Zhang et. al. (2002) proposed corrections for the account of the effect of grain sizes. It was made either by estimating the normalized equivalent clean sand CPT values $(q_{c1N})_{cs}$ suggested by Robertson and Wride (1998) or by estimating the relative density of the silty sand using the CPT value. The volumetric strain was estimated based on the relative density and factor of safety by using the curves proposed by Ishihara and Yoshimine (1992). In this study, Zhang et. al. (2002) approach was followed and volumetric strains were determined from the $(q_{c1N})_{cs}$ for the various depths and cumulative settlement has been estimated. The SPT-N values were converted to CPT-q_c values using the developed correlations for the study area.

The correlation between SPT and CPT for the fine or silty loose to medium dense sand was proposed by Meyerhof (1965). The ratio of observed cone resistance (q_c) and observed SPT values (N) was proposed by Schmertmann (1978). Various researches like Youd and Bennett (1983), Robertson and Campanella (1985), Seed and De Alba (1986), Kayen et al. (1992) etc. have shown that the existing SPT-CPT relation is linear and is a function of mean grain size. Andrus et. al. (2004) and Chin et. al. (1988) proposed correlations.

3 TEST LOCATIONS

In this paper, the coastal line of Nagapattinam, Tamilnadu, India (Fig. 1) has been considered for geotechnical investigation. The study on the active faults in the tectonic plates of South India reveals that the fault systems are oriented along NE-SW in N51-60°E and N71-80°E and their intersection with an angle of 10-30° is more vulnerable to seismicity in South India (Ramasamy et. al., 2006). Very active faults of North India like N-S, NE-SW and NW-SE faults have connectivity with these South Indian faults. Many earthquakes in North India were experienced in few parts of South India also.

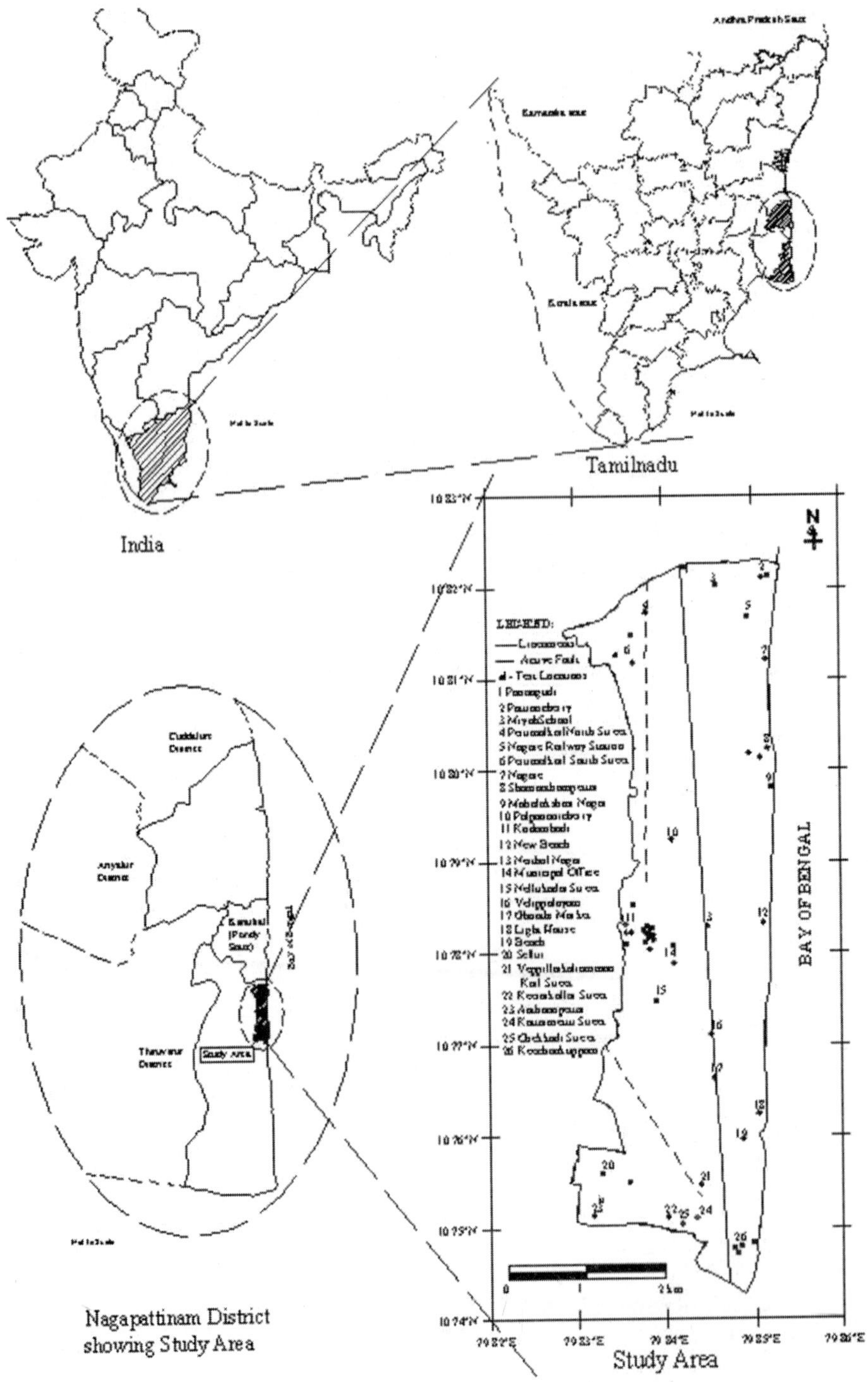

Fig. 1: Location of Study Area and Test Locations

3.1 Field and Laboratory Investigations

SPTs were conducted as per ASTM (D1586). The SPT test is liable to fourteen errors, which can occur either in nature or during testing (Seed et. al., 1985, Skempton, 1986).. The deviations in expected results were partially compensated by converting the SPT-N values recorded in the field to $(N_1)_{60}$ and $(N_1)_{60cs}$ based on the procedure proposed by Youd and Idriss (2001) and Robertson and Wride (1998). The SPTs were conducted in a 150mm dia borehole at 1m intervals. The hand dropped donut type hammer (63.5 kg drive weight) and standard split-spoon sampler with brass liner were used to measure the penetration resistance of soil. The observed penetration resistance values were corrected for hammer energy by taking hammer energy ratio for hand dropped donut hammer was 0.6, C_B for 150mm dia borehole was 1.05 and C_S was1 for sampler with liner. For the static cone penetration test, an engine driven 100 kN capacity hydraulic pump type static cone penetrometer was used with mechanical type (Begemann) cone and tests were performed in conformance with ASTM (D6066). For the liquefaction analysis, the CPT tip resistance values were corrected based on the recommendations by Youd and Idriss (2001) and Robertson and Wride (1998). The various engineering and index parameters were calculated from the laboratory tests conducted as per standard procedures. The grain size analysis for the determination of effective mean grain size (D_{50}) and fines content (F_c) were conducted using the air-dried soil samples. Field permeability tests were conducted as per ASTM (D 2573) at 1m depths. The obtained field permeability results were compared with laboratory permeability test. The field and laboratory permeability results were comparable.

4 RESULTS AND DISCUSSIONS

The CPT was conducted very close to the SPT locations (max 3m vicinity) and observed resistant values were compared with SPT-N values. It was found that the variation of both SPT and CPT values are in same pattern. Based on the CPT results tip resistance (q_c), friction ratio (f_s), factor of safety and probability of liquefaction (P_L) are estimated. The encountered soil profile was identified based on the CPT-q_c values vs friction ratio charts proposed by Robertson (1990). The soil profile was also identified based on the laboratory tests conducted on collected soil sample from SPT. The field tests (SPTs and CPTs) conducted in the study area shows that the soil deposit in the top layer is very loose poorly graded fine sand followed by dense sand deposit. In most of the area, the layer between 2 m and 10 m has loose silty sand with a net settlement more than 500 mm. The probabilistic liquefaction analysis showed that the probability of liquefaction is varying from 30 to 100% in this layer. The coastal area of Nagapattinam town was classified with respect to liquefaction potential (Fig. 2 (a)) and liquefaction induced settlements (Fig. 2 (b)). Since, the FS was calculated for different depths from the ground level, the probability based evaluation of liquefaction potential (P_L) was adopted for liquefaction mapping as proposed by Juang et. al. (2002) and Juang et. al. (2006). If $P_L \geq 0.85$, it was classified as liquefiable, if $0.35 \leq P_L < 0.85$, it was considered as moderately liquefiable and if $P_L < 0.35$ it was treated as non-liquefiable. For the study of the behaviour of the soil deposit along the Coromandel Coast, different correlations are developed from the data collected from 51 test locations and compared with various properties. An empirical Eq. (4) for calculating settlement (S) was developed based on the multilinear regression analysis by considering settlement as a dependent variable and relative density and mean grain size as independent variables. Fig. 3 (a) shows the scatter plot (R^2=0.9183) for the settlements estimated from analysis and regression model between the S, D_r and D_{50}. It is clear that an increase in D_r and D_{50} cause a decrease in settlement (S).

$$S = -1.10732 D_r - 12.8075 D_{50} + 94.0082 \tag{4}$$

The relative density is an important factor, which governs the liquefaction probability of a sandy soil deposit. The relative density is a function of voids ratio. However, the voids ratio is a depending factor of grain size distributions. By treating the relative density as dependent variable and mean grain size, coefficient of uniformity and coefficient of curvature as independent variables, the following Eq. (5) was developed from multilinear regression analysis.

$$D_r = -42.477 D_{50} + 5.5574 C_U - 2.5776 C_C + 57.7616 \tag{5}$$

The scatter plot (Fig. 3.b) between the D_r Values determined from both regression model and actual has been drawn and found good agreement ($R^2 = 0.8998$).

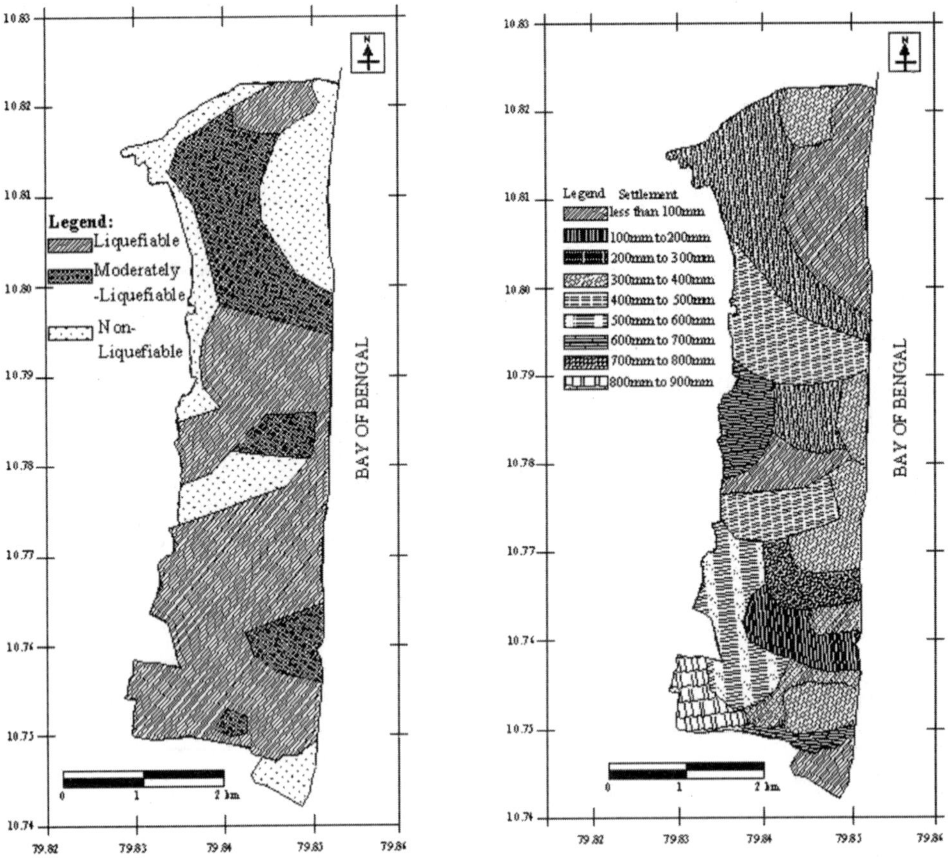

(a) Liquefaction Potential Microzonation Map (b) Liquefaction Induced Settlement Map

Fig. 2: Maps Showing the Final Results from the Analysis

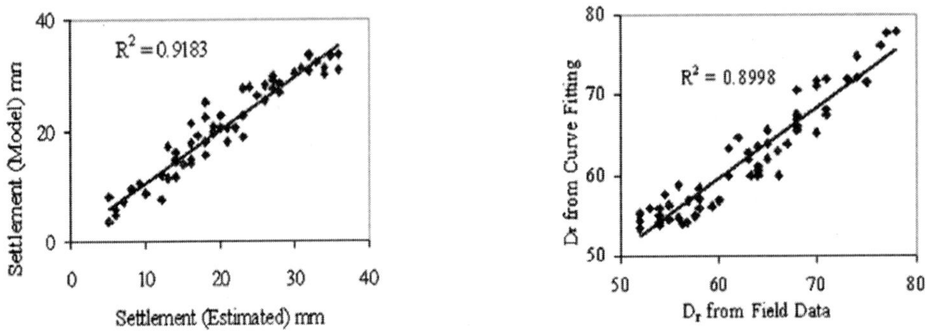

(a) Scatter Plot for the Equation (4) (b) Scatter Plot for the Equation (5)

Fig. 3: Proposed Relations between Various Parameters

From the SPT-CPT correlation analysis, it was found that the relationship between mean grain size (D_{50}) and q_c/N has behaviour of increase in the size of the particles increase in the q_c/N ratio as Eq. (6),

$$q_c / N = 8.0847 D_{50}^{0.3397} \tag{6}$$

The curve proposed by Robertson and Campanella (1985) follows Eq. (7)

$$q_c / N = 7.8335 D_{50}^{0.2885} \tag{7}$$

In addition to the mean particle size, the CPT and SPT values should be influenced by relative density (D_r) and fines content (F_c). Since these parameters are interrelated in determining the behaviour of any soil deposits. It is well known fact that the q_c/N goes on decreasing with increase in the amount of fines content, while increase with increase of relative density. Multilinear regression analyses were carried out to incorporate the effect of fines content and relative density with mean particle size. By grouping available data, based on the type of soil such as sand and clay, the Eqs. (8 and 9) were developed. From the scatter plot (Fig. 4), it was found that the equations are having 15% variation from Schmertmann's (1978) recommendations. Eq. (8) is valid for $D_{50} > 0.2$mm and Eq. (9) is valid for $D_{50} < 0.2$mm.

a). For sandy soil,

$$\frac{q_c}{N} = 2.19 + 2.4205 D_{50} + 0.0261 D_r - 0.0182 F_c \tag{8}$$

b). For clayey soil,

$$\frac{q_c}{N} = 1.535 + 2.5072 D_{50} + 0.0109 D_r - 0.0043 F_c \tag{9}$$

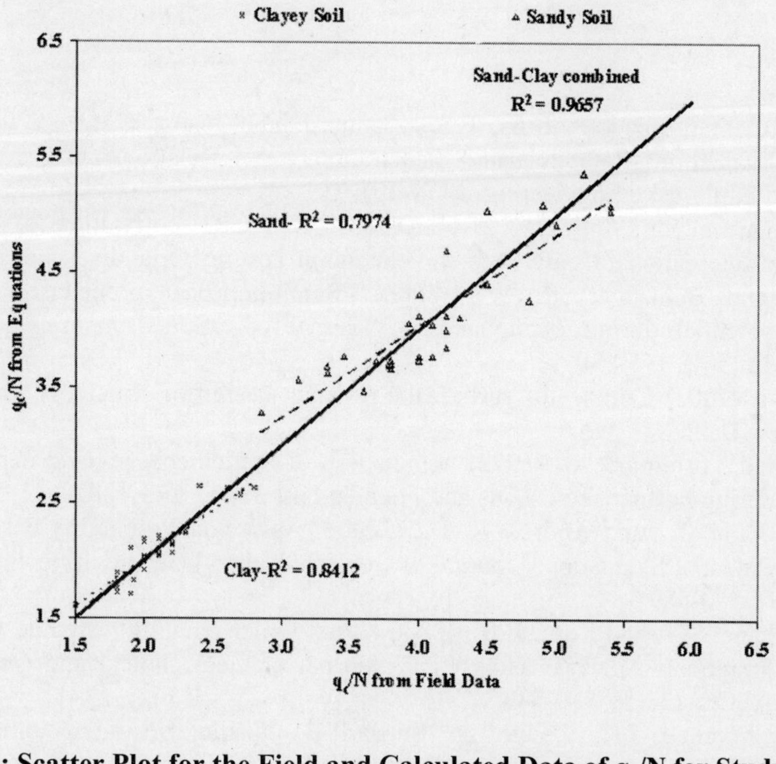

Fig. 4: Scatter Plot for the Field and Calculated Data of q_c/N for Study Area

The study on the effect of fines content, relative density and mean grain size on the SPT and CPT relation gives significant improvement for the creation of CPT database from the available SPT database in the study area and this will be more useful for any design problem. The equations are developed only for specific soil types and this concept may be a guidance for further investigation.

5 CONCLUSIONS

The settlement properties of soil deposits along the coastal region of study area tend to increase towards the seashore and moderate in areas away from the shore. The microzonation map shows that areas vulnerable for liquefaction lie more or less on river belts, canal belts and major drainage course belts. It may be due to the scouring action of water during floods and redepositing the fine clean sand at the estuary after the floods subside. During the design of quay wall, wharf, etc., in the harbour or the dock, special considerations have to be given to avoid failures due to liquefaction induced lateral displacement and settlement of ground during the earthquake. Pile foundations may be provided with proper pile heads to avoid failures of harbour structures. It was found that the non-liquefiable area is only 24%, the moderately liquefiable area 23% and the liquefiable area being 53%. Therefore, it is very essential to consider the liquefaction and its related settlement for the design of any major structure. The existing harbour area is prone to liquefaction and the techniques to avoid effect of liquefaction should be incorporated in the planning and designing of extension works. Relationships between mean grain size (D_{50}), relative density (D_r), fines content (F_c), coefficient of curvature (C_c), coefficient of uniformity (C_u) and q_c/N have been developed and it is found that q_c/N is directly proportional to the mean grain size and the relative density and the fines content is inversely proportional.

6 ACKNOWLEDGEMENTS

The authors acknowledge with a sense of indebtedness, the help rendered by the All India Council for Technical Education, New Delhi, for funding this research work.

REFERENCES

1. Andrus. R. D, Piratheepan, P, Ellis, B.S., Zhang, J. and Juang. C.H, (2004) "Comparing liquefaction Evaluation Methods using Penetration – Vs Relationships." Journal of Soil Dynamics and Earthquake Engineering, **24** (9-10), 713-721
2. Chin, C.T., Duann S. And Kao, T.C. (1988) "SPT-CPT correlation of granular soil", Proceedings of International Conference on Penetration Testing, Orlando, 1, pp. 335-339.
3. Idriss, I.M. and Boulanger, R. W., (2006) "Semi-empirical procedures for evaluating liquefaction potential during earthquakes." Journal of Soil Dynamics and Earthquake Engineering, **26** (2-4), 115-130
4. IS-1893-Part-1, (2002) Criteria for earthquake resistant design of structures, Bureau of Indian Standards, New Delhi
5. Ishihara, K, and Yoshimine, M, (1992) "Evaluation of Settlements in sand deposits following Liquefaction during Earthquakes" Soils and Foundations, **32** (1), 173-188
6. Juang, H.C., Jiang, T. and Andrus, R. D., (2002) "Assessing Probability-based methods for Liquefaction Potential Evaluation." Journal of Geotechnical and Geoenvironmental Engineering, ASCE, **128** (7), 580-589
7. Juang, H.C., Fang, S.Y. and Khor, E.H., (2006) "First-Order reliability Method for Probabilistic liquefaction Triggering Analysis using CPT" Journal of Geotechnical and Geoenvironmental Engineering, ASCE, **132**(3), 580-589
8. Kayen, R.E., Mitchell, J.K., Lodge, A., Seed, R.B., Nishio, S. and Countinho, R. (1992) "Evaluation of SPT-CPT and Shear Wave Based Methods for Liquefaction Potential Assessment Using Loma Prieta Data", Proc. of the 4th Japan U.S. Workshop on Earthquake Resistant Design

of Lifeline Facilities and Countermeasures for Soil Liquefaction, Technical Report NCEER-94-0019, 1, M. Hamada and T.D. O-Rourke (eds), 177-204

9. Meyerhof, G.G., "Shallow Foundations", Journal of Soil Mechanics and Foundation Division, ASCE, **91**(SM2), 1965, 21-31
10. Ramasamy, S. M. (2006), Remote Sensing and Active Tectonics of South India, International Journal of Remote Sensing, **27** (20), 4397 – 4431
11. Robertson P.K. (1990) "Soil Classification using the Cone Penetration Test," Canadian Geotechnical Journal, 27, 151-158
12. Robertson. P.K and Campanella. R.G. (1985) "Liquefaction Potential of Sands using the CPT" Journal of Geotechnical Engineering, ASCE. **111**(3). 384-403
13. Robertson, P.K. and Wride, C.E. (1998) "Evaluating Cyclic Liquefaction Potential using Cone Penetration Test", Canadian Geotechnical Journal, **35**(3), 442-459
14. Schmertmann, J.H. (1978) "Guidelines for Cone Penetration Test: Performance and Design", U.S. Department of Transportation
15. Seed, H.B. and De Alba, P. (1986) "Use of SPT and CPT tests for Evaluating the Liquefaction Resistance of Sands", Use of In-situ Tests in Geotechnical Engineering, Geotechnical special Publication, **6**, 281-302
16. Seed, H.B. and Idriss, I.M. (1971) "Simplified procedure for evaluating soil liquefaction potential", J. SMFD, ASCE, **97** (9), 1249-1273
17. Seed. H.B., Tokimatsu. K, Harder. L. F and Chung. R. M. (1985) "Influence of SPT Procedures in Soil Liquefaction Resistance Evaluations", Journal of Geotechnical Engineering, **111**(12), 1425-1445
18. Skempton, A.W. (1986) "Standard Penetration Test Procedures and its Effect in Sands of Overburden Pressure, Relative Density, Particle Size, Aging and Overconsolidation," Geotechnique, **36**(3), 425-447
19. Youd, L.T. and Bennett, M.J. (1983) "Liquefaction Sites, Imperial Valley, California", Journal of Geotechnical Engineering, ASCE, **109** (3), 440-458
20. Youd. T. L. and Idriss. I. M. (Editors) (2001) "Liquefaction Resistance of Soils: Summary Report from the 1996 NCEER and 1998 NCEER/NSF Workshops on Evaluation of liquefaction Resistance of Soils." Journal of Geotechnical and Geoenvironmental Engineering, ASCE, **127**(10), 817-833
21. Zhang, G., Robertson, P.K. and Brachman, R.W.I. (2002) "Estimating liquefaction- induced ground settlements from CPT for level ground", Canadian Geotechnical Journal, **39**(5), 1168-1180

ASSESSMENT OF SEISMIC SITE CHARACTERISTICS OF NAGAPATTINAM COROMANDEL COAST SAND DEPOSITS

***S. Senthamilkumar*[1] *Dr. C. Natarajan*[2] *Dr. K. Muthukumaran*[3]**

[1] *Research Scholar, Department of Civil Engineering, Periyar Maniammai University, Vallam- 613 403, Thanjavur,*
Email: senthamilkumar@yahoo.co.in

[2] *Prof. & Head and Lecturer, Department of Civil Engineering, National Institute of Technology, Trichy –620 015.*
Email: nataraj@nitt.edu

[3] *Prof. & Head and Lecturer, Department of Civil Engineering, National Institute of Technology, Trichy –620 015.*
Email: kmk@nitt.edu

ABSTRACT

Aim of this paper is to investigate, in detail, the seismic site characteristics of the deposits in the Coromandel coastal line of Nagapattinam town, Tamilnadu, where major constructions related to the Tsunami relief works and Sethu Canal Project are undertaken over the loose sand deposits. It is mandatory to take into account the local soil conditions and effects on ground motion, liquefaction and related deformations for the planning and construction. It also covers possible earthquake damages such as tectonic rupture, fault displacement, liquefaction, settlements, etc. With the intention of studying the dynamic geotechnical properties, a series of field tests were conducted using Standard Penetration Test (SPT) and Cone Penetration Test (CPT). By carrying out laboratory investigation relevant dynamic parameters were estimated. The shear wave velocity and shear modulus were calculated from the field penetration tests. The amplification and deamplification of waves passing through a soil deposit is mainly based on the properties of that soil deposit. Hence, in this paper correlations between the normalized values of peak ground accelerations, velocities and displacements, which were obtained from the equivalent linear ground motion analysis using SHAKE software, with fines content, relative density, effective mean particle size, factor of safety against liquefaction and corrected penetration resistance values. From the results it was found that at some of the locations as the ground becomes softer, amplification of the wave becomes larger. However, the deamplification occurs in both soft and hard soil deposits.

Key words: Peak ground acceleration, Peak ground displacement, Peak ground velocity, Standard Penetration Test, Static Cone Penetration Test

1 INTRODUCTION

The seismic response of a soil deposit is responsible for change in the characteristics, liquefaction, in soil stiffness, damages in underground structures, ground deformations (Youd and Keefer 1994, Pease and O'Rourke 1997). As a result of above changes structural collapse could occur. Strong ground motion records, which were obtained from different earthquakes, were analyzed by various researchers (Finn et al. 1993, Zeghal and Elgamal 1994, Elgamal et al. 1996). The effects on the soil properties and structures were well established. The evaluations of transient shear strain and strong motion characteristics at earthquake-affected areas might be used to create a model for a reduction in the soil stiffness due to liquefaction (Shibata et al. 1996, Pease and O'Rourke 1997, Youd et al. 2001).

The development of roadways (NH-65A & NH-67), the Sethu canal project and calamity caused by Tsunami paved the way for the development of new industries, townships and other major reconstruction activities along the Coromandal Coastal line of Tamilnadu, India. As a part of the Sethu canal project, the possibilities of the development of many harbours and docks in the important cities along the coastal line can be foreseen. In this paper, the seismic response of Coromandal coastal soil deposits is discussed from the field test results, laboratory test results and analytical results obtained from the equivalent linear model created in SHAKE software. Standard penetration tests (SPT), Cone penetration test (CPT) and plate load test were conducted in the field for the exploration of soil deposit and properties. Collected disturbed sample were tested in the laboratory. The obtained results were verified with earthquake geotechnical background. Some correlations were developed between the properties such as normalized acceleration (a_{output}/a_{input}), velocity (v_{output}/v_{input}),, displacement (d_{output}/d_{input}), response spectra, thickness of soil layer (T) , relative density (D_r), shear wave velocity (V_s), fines content (F_c) and penetration resistant values (N & q_c).

2 EFFECTS OF LOCAL SOIL CONDITIONS AND GEOTECHNICAL FACTORS

The acceleration of the soft ground, such as soft clay, loose sand deposits, could be amplified. The soft ground effect contributed to damage of structures in many of the earthquakes. The softer site could amplify the low frequency bedrock motions more than the stiff site (Kramer 1996). The thickness of the soft layers also significantly increased the peak ground acceleration (PGA) by a factor of 3 to 5 times. Soft bayshore deposits were predominantly contributed for the amplification of peak ground acceleration and damages of structures in many of the past earthquakes. The site situated over the rock outcrop was not amplified the peak ground acceleration. The soft ground can also increase the period of ground surface shaking, leading to resonance of structures (Seed et al. 1990). The response spectra, which were produced from frequency content of surface motions, were also influenced by local site conditions. The spectral amplification at the period of 0.5 second was much higher for soil than for rock. However, the spectral amplification increases with decreasing surface profile stiffness at longer periods (Seed et al. 1976). Idriss (1990) developed an empirical correlation between rock outcrop and soft soil peak accelerations for the San Francisco bay area. It was concluded that the amplification at soil site decreases with the increase in amplitude of the input motion. In addition, it was proved that larger amplification of acceleration in soil with higher relative density and lower pore pressure generation. However, the structural response is greatly influenced by the layering and a localized hidden soft patch in a dense soil (Ghosh and Madabhushi 2003).

3 GEOLOGICAL SETTINGS OF STUDY AREA

The area, which was considered for geotechnical investigation along Coromandal coastal line at Nagapattinam, Tamilnadu, India (Fig.1) lies between the latitudes of 10°44'50" and 10°49'30" North and longitudes of 79°49'40" and 79°51'10" East. The study on the active faults on the tectonic plates of South India reveals that the fault systems are oriented along the north east-south west in N51-60°E and N71-80°E and that their intersection with an angle of 10-30° is more vulnerable to seismicity in South India (Ramasamy 2006). Very active faults of North India like north-south, north east-south west and north west-south east faults have connectivity with these South Indian faults. Many earthquakes occurred in North India were experienced in few parts of South India also. This study area is lying over an active fault running east west and north south (Fig. 2). The existing gentle sloping ground surface is having 6.1 m elevation from the mean sea level (MSL) for 2.7 km horizontal distance in the east-west direction. In the north-south direction, it is varying from 0 to 5 m. However, some of the areas are below the MSL and there are many steep slopes in the ground surface.

The topographical study clearly shows the topography of the study area. More than 30% of study area has been covered with structures having various intensity of loading. The soil profile consists of alluvial and marine sediments in the east, along the coast and fine-grained sandy deposits with clay towards the west. The alluvium was deposited by floodwaters of the rivers, which are running from west. During the test, water table was found at 0.5m to 2 m below the ground level in the south end, which is having more undulations than north. At north side of Nagapattinam, the water table was above the ground level. Two rivers are running at south and north ends of this study area. The backwater in these rivers is always keeping the area in saturated condition.

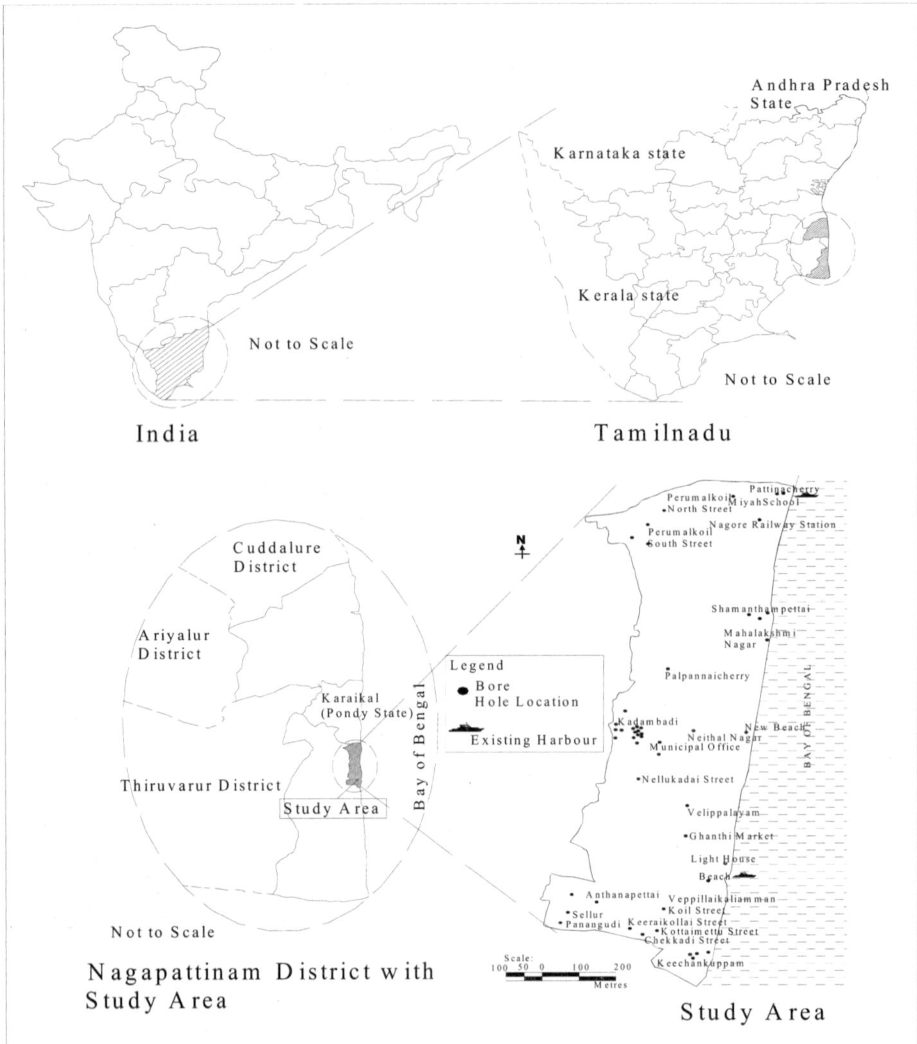

Fig. 1: Showing the Location of Test Area and Test Locations in the Study Area

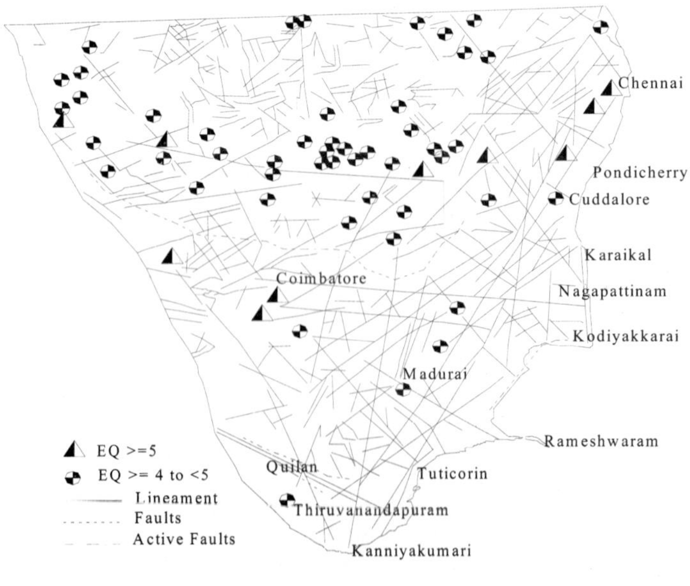

Fig. 2: Seismicity and Faults System of South India

4 FIELD TESTING

SPTs were conducted as per ASTM (D1586). The SPT test is liable to fourteen errors, which can occur either in nature or during testing (Seed et. al., 1985, Skempton, 1986). The deviations in expected results were partially compensated by converting the SPT-N values recorded in the field to $(N_1)_{60}$ and $(N_1)_{60cs}$ based on the procedure proposed by Youd and Idriss (2001). The SPTs were conducted in a 150mm dia borehole at 1m intervals. The hand dropped donut type hammer (63.5 kg drive weight) and standard split-spoon sampler with brass liner were used to measure the penetration resistance of soil. The observed penetration resistance values were corrected for hammer energy by taking hammer energy ratio for hand dropped donut hammer was 0.6, C_B for 150mm dia bore hole was 1.05 and C_S was1 for sampler with liner. For the static cone penetration test, an engine driven 100 kN capacity hydraulic pump type static cone penetrometer was used with mechanical type (Begemann) cone and tests were performed in conformance with ASTM (D6066). For the liquefaction analysis, the CPT tip resistance values were corrected based on the recommendations by Youd and Idriss (2001) and Robertson and Wride (1998). Plate load test was conducted at 1m from the ground level to verify the shear modulus (G_{max}) calculated from the relations proposed for the field penetration tests The various engineering and index parameters were calculated from the laboratory tests conducted as per standard procedures. The grain size analysis for the determination of effective mean grain size (D_{50}) and fines content (F_c) were conducted using the air-dried soil samples.

5 RESULTS AND DISCUSSION

The observed SPT and CPT results and laboratory tests were analyzed for the estimation of relative density (D_r), fine contents (F_c), mean grain size (D_{50}), liquefaction potential, factor of safety (FS_L) and maximum shear modulus (G_{max}). G_{max} was estimated from SPT and CPT values using relations as under,

$$G_{max} = 957.606\left(N_1\right)_{60}^{0.333}\left(\sigma'_m\right)^{0.5} \quad kN/m^2 \text{ (Seed et al. 1986)} \tag{1}$$

where $(N_1)_{60}$ is corrected SPT N value for field procedure and overburden pressure. (Youd et al. 2001), σ'_m is Mean Principle Effective Stress in kN/m^2

$$G_{max} = 1634\left(q_c\right)^{0.250}\left(\sigma'_v\right)^{0.375} \quad kN/m^2 \text{ (Rix and Stokoe 1991)} \tag{2}$$

where q_c is observed Tip resistance from CPT and σ'_v is effective vertical pressure in kN/m^2.

The estimated results such as density, shear modulus etc. were used as input parameters in the SHAKE analysis. SHAKE software is capable to create one-dimensional equivalent linear seismic model of horizontally layered soil deposit (Schnabel et. al. 1972) and accepted by many researchers like Idriss (1991), Zeghal and Elgamal, 1994, Pease and O'Rourke (1997), even though, the discrepancies exist on the stimulations. The time histories of acceleration, velocity and displacement for various layers were estimated from the analysis.

The ground motion records were obtained from the analysis using SHAKE software. The observed test results or estimated properties like density, thickness of layer, shear modulus, shear wave velocity water table were used. The available strong motions records of the past earthquakes such as El Centro ground motion record (1940 Imperial valley earthquake, M=7.0), Treasure island and Yerba Buena island ground motion record (1989 Loma Prieta earthquake, M=7.0), Topanga ground motion record (1994 North Ridge earthquake, M=6.7), Petrolia ground motion record (1992 Cape Mendocino earthquake, M=6.7) and Taft Kern country ground motion record (1952) were used as input ground motion.

Figs. 3 and 4 show the typical analytical ground motions at the surface and at the various layers of liquefiable (loose deposit) and non-liquefiable (dense deposit) layers at Nagapattinam town. These graphs were obtained for El Centro ground motion. The same type of behaviour was observed for all the input motions.

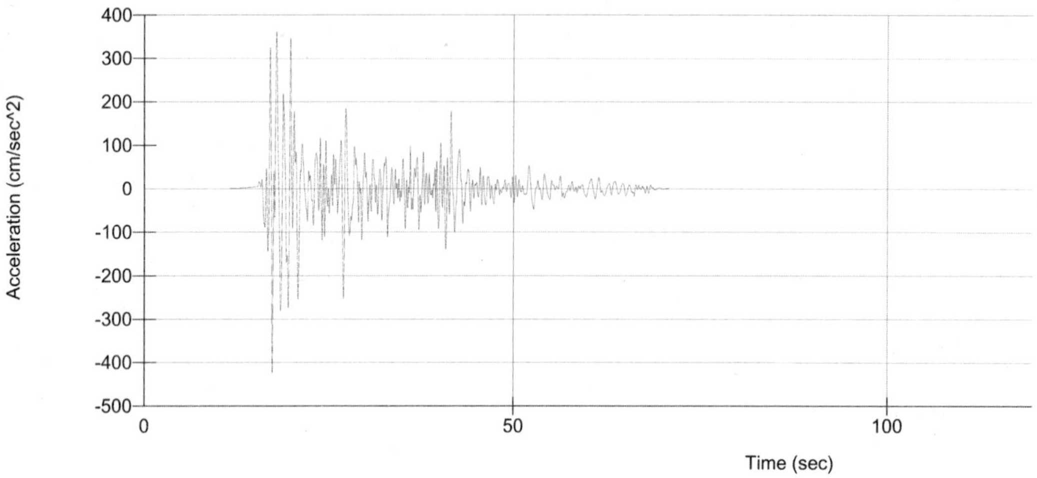

Fig. 3: Typical Acceleration Time History of Loose Sand Deposits at Study Area

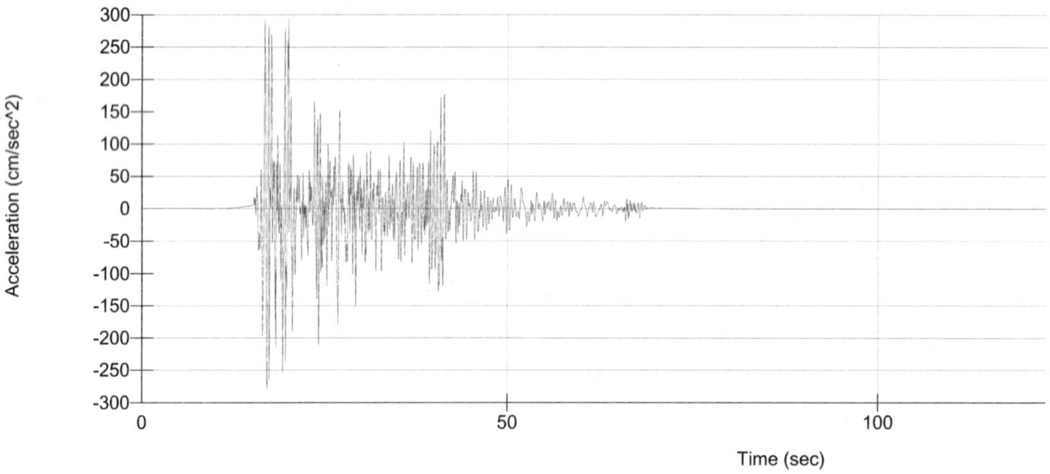

Fig. 4: Typical Acceleration Time History of Dense Sand Deposits

From the analysis, it clearly shows that the loose sand and soft clay deposits amplify the acceleration by a factor 1.2 to 2.2. Similarly, dense sand and stiff clay deamplify the acceleration by a factor 0.6 to 0.96. The thicknesses of the layers were also identified as major affecting factor of the amplification. The amplification of any soil deposits would be influenced by density of soil deposits, relative density of sandy deposits and fines content. The penetration resistance values might be indirectly indicates the compactness or stiffness of soil deposits, which affect the amplification. From the results observed in ground motion analysis, field tests and laboratory tests, the Eq. (3) was developed using multilinear regression analysis for sandy soil deposits in the Coromandal coastal line.

$$\frac{a_{output}}{a_{input}} = 0.0396V_s + 0.0115\gamma + 0.0095D_r - 0.0178F_c - 0.331N - 2.6084 \quad (3)$$

where, a_{output} is output acceleration from SHAKE analysis in cm/sec/sec, a_{input} is input acceleration from past earthquake in cm/sec/sec, V_s is shear wave velocity in cm/sec, γ is density of soil in kN/m^3, D_r is relative density of sand in percentage, F_c is fines content in percentage and N is observed SPT values. The developed regression equation was checked with the help of scatter plot. It was found that there was a good agreement between the actual values and calculated from regression model. However, the Eq. (3) is valid only for the sandy deposits. The amplification of acceleration was mainly defends on the input ground motion characters and type of soil layers. The increase in input acceleration leaded to increase in output acceleration. The increase in thickness of layer, increase either in amplification or deamplification. It is essential to note that any lenses or layer of different soil between either sandy layers or clayey layer significantly affected the amplification behaviour of entire soil profile. Since, for any soil profile with this condition, it should be properly analyzed. The ground motion behaviour of this study area significantly followed the same pattern of changing as shown in fig.5 which shows the comparison of the output acceleration of El Centro ground motion based analysis with other outputs.

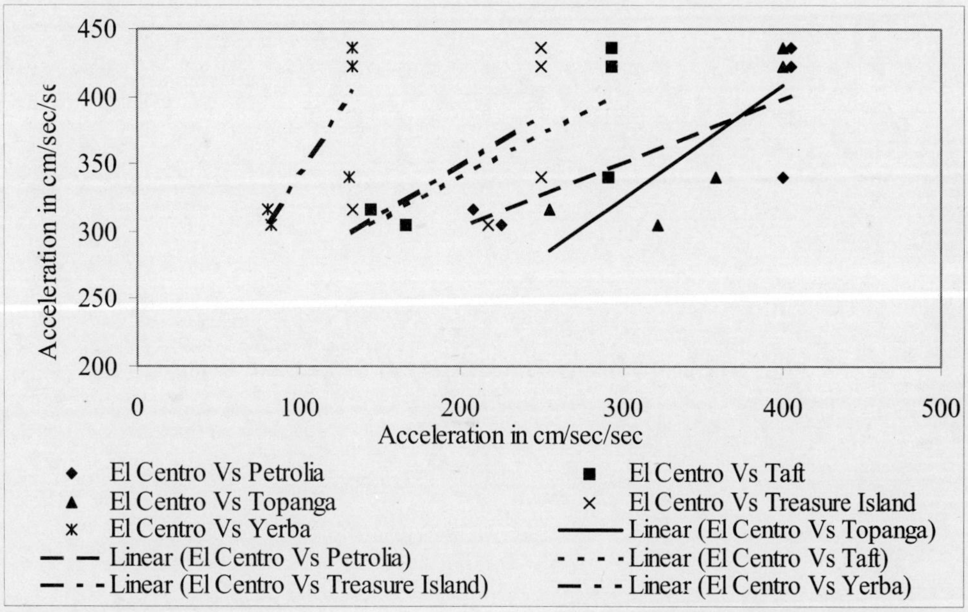

Fig.5: Comparison of Amplification Acceleration for various Input Ground Motions

Similarly, the velocity and displacement of the various soil layers were studied and concluded that the amplification or deamplification of these properties are also influenced by the properties that are mentioned earlier. The figures, from (6) to (9), show the typical variation of velocity and displacement of soil layers in the study area. From the multilinear regression analysis, the following equations were developed for the sandy soil deposits.

$$\frac{v_{output}}{v_{input}} = 0.0148V_s - 0.0205\gamma + 0.0071D_r - 0.0275F_c - 0.133N + 0.6771 \quad (4)$$

$$\frac{d_{output}}{d_{input}} = 0.0239V_s + 0.01\gamma + 0.0047D_r - 0.0094F_c - 0.1962N - 0.9822 \quad (5)$$

The Eqs. (4 and 5) are also valid only for sandy deposit. The response spectral acceleration was studied for various layers by adopting damping ratio of 5%. When the natural period was from 0.6 to 0.65 sec, the soil layers were highly accelerated by a factor of 3 to 5 compare to corresponding peak ground acceleration. It is the case for the loose soil only and for the dense sand the factor was very low. The spectral velocity and displacement also followed the same type of amplification, however, the natural period was shifted to 2 to 2.5 sec.

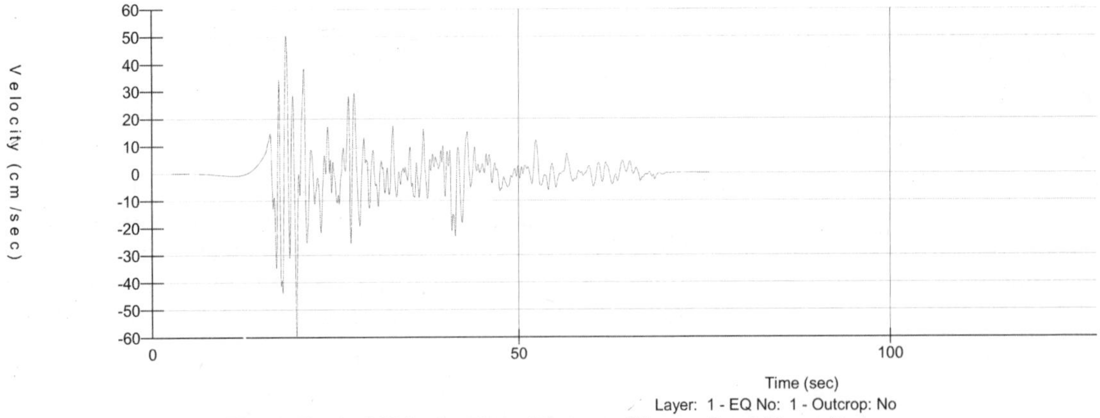

Fig. 6: Typical Velocity Time History of Loose Sand Deposits

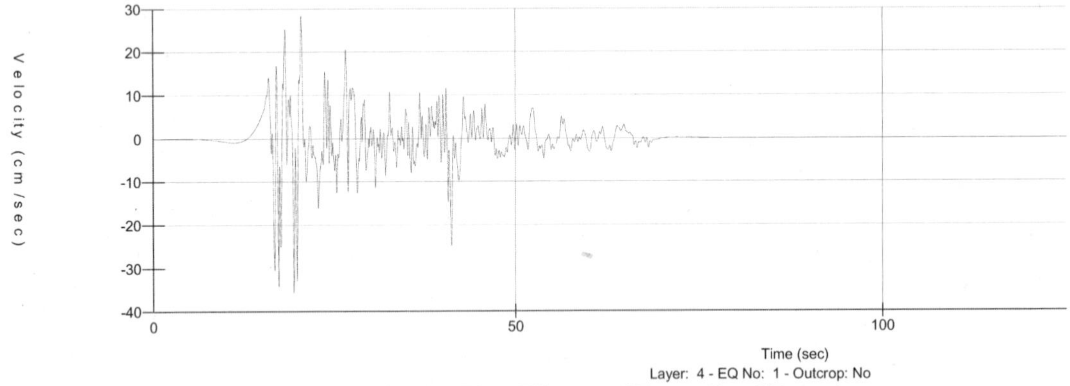

Fig. 7: Typical Velocity Time History of Dense Sand Deposits

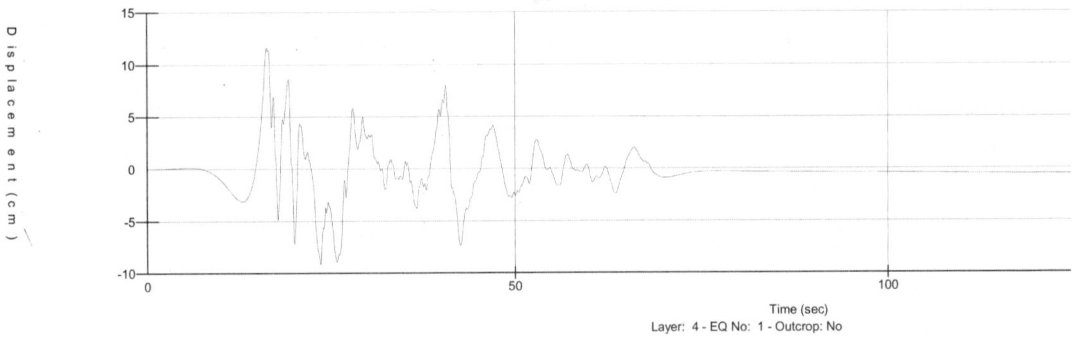

Fig. 8: Typical Displacement Time History of Loose Sand Deposits

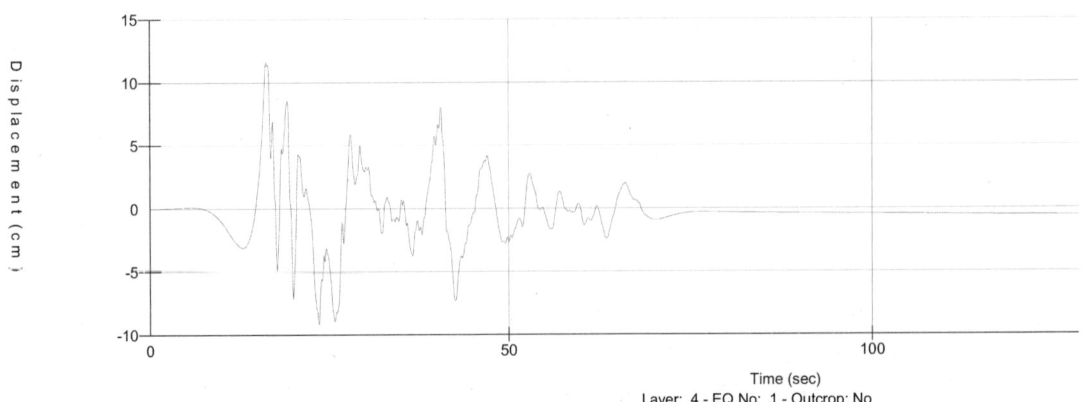

Fig. 9: Typical Displacement Time History of Dense Sand Deposits

In future, the outcome of this study could be incorporated in the analysis and design of any structures along the Coromandal Coast. The town planners, engineers, architects and administrators should be instructed to follow the conclusions during the planning of construction activities in the future.

6 CONCLUSIONS

From the ground motion analysis for the sand deposits at Coromandal coastal line of Nagapattinam the following conclusions were summarized.

- The loose sand or soft clay deposits amplified the peak ground acceleration by a factor of 1.2 to 2.2 and dense sand and stiff clay deamplified the acceleration by a factor 0.6 to 0.96.
- For the natural period from 0.6 to 0.65 sec, the soil layers were highly accelerated by a factor of 3 to 5 compare to corresponding peak ground acceleration.
- By using multilinear regression analysis the correlations between ground motion parameters and engineering properties were developed from the test results for sandy soil deposits in the Coromandal coastal line.
- The response spectral acceleration with damping ratio 5% for sandy deposit layers were highly accelerated by a factor of 3 to 5 compare to corresponding peak ground acceleration for the natural period varies from 0.6 to 0.65 sec, the same for loose sand and dense sand the amplification factor is very low.
- The response spectral velocity and displacement with damping ratio 5% for sandy deposit layers was accelerated by a very low factor with the natural period from 2 to 2.5 sec.

ACKNOWLEDGEMENT

The authors acknowledge with a sense of indebtedness, the help rendered by the All India Council for Technical Education, New Delhi, for funding this research work.

REFERENCES

1. ASTM (2007). *Annual Book of ASTM Standards, 04.02: Soil and Rock (I).* Standard No. C 136-06, "Standard Test Method for Sieve Analysis of Fine and Course Aggregates." West Conshohocken, PA.
2. ASTM (2007). *Annual Book of ASTM Standards, 04.08: Soil and Rock (I).* Standard No. D 1586-99, "Standard Test Method for Penetration Test and Spilt-Barrel Sampling of Soils." West Conshohocken, PA.
3. ASTM (2007). *Annual Book of ASTM Standards, 04.08: Soil and Rock (I).* Standard No. D 2487-06, "Standard Practice for Classification of Soils for Engineering Purposes (unified soil classification system)." West Conshohocken, PA.
4. ASTM (2007). *Annual Book of ASTM Standards, 04.08: Soil and Rock (I).* Standard No. D 3441-05, "Standard Test Method for Mechanical Cone Penetration Tests of Soils." West Conshohocken, PA.
5. Finn, W. D. L., Carlos, E. V. & Guoxi, W. (1993) "*Analysis of ground motions at Treasure Island site during the 1989 Loma Prieta earthquake*", Soil Dynamics and earthquake Engineering, 12: 383- 390.
6. Ghosh, B. and Madabhushi, S.P.G. (2003), "Effects of Localized Soil Inhomogeneity in Modifying Seismic Soil Structure Interaction", 16th Engineering Mechanics Conference in Seattle, ASCE, www.ce.washington.edu/em03/ proceedings/papers/ 633. pdf (31.07.2007)
7. Idriss, I.M., (1990) "Response of Soft Soil Sites During Earthquakes", *Proc.of H.B.Seed memorial symposium, Bi-Tech Publishers, Vancouver, British Columbia (Editor J.M. Duncan),* **2**, 273-289

8. Idriss, I.M., 1991. "Earthquake Ground Motions at soft soil Sites", *Proc. Second International Conference on Recent Advances in Geotechnical Earthquake Engineering and Soil Dynamics*, **.3**, 2265-2271

9. Kramer S. L. (1996). Geotechnical earthquake engineering, Pearson Edition, Second Indian Reprint (2004), New Delhi.

10. Pease.J.W and O'Rourke.T.D, (1997) *"Mapping Liquefiable Layer Thickness For Seismic Hazard Assessement."* Journal of Geotechnical and Geoenvironmental Engineering, ASCE, **123** (1), 46- 56.

11. Pease.J.W and O'Rourke.T.D, (1997) *"Seismic Response of liquefaction Sites."* Journal of Geotechnical and Geoenvironmental Engineering, ASCE, **123**(1), 37- 45.

12. Ramasamy, SM., and Balaji, S., (1995) "Remote Sensing and Pleistocene Tectonics of Southern Indian Peninsula", *Int. journal of Remote Sensing, tailors and Francis London,* **16**, (13), 2375-2391.

13. Robertson, P.K. and Wride, C.E., (1998)*"Evaluating Cyclic Liquefaction Potential using Cone Penetration Test",* Canadian Geotechnical Journal, **35**(3), 442-459.

14. Rix, G.J. and Stokeo, K.H., (1991) *"Correlation of Initial Tangent Modulus and Cone Penetration Resistance",* International Symposium on Calibration Chamber Testing, Elsevier Publishing, New York (Editor A.B.Huang), 351-362.

15. Schnabel, P. B., Lysmer, J. & Seed, H. B. (1972) *"SHAKE A Computer program for earthquake response analysis of horizontally layered sites",* Report No. EERC72-12, University of California, Berkeley.

16. Shibata, T., Oka, F. and Ozawa, Y., (1996) "Characteristics of Ground Deformation Due to liquefaction". Soils and Foundations, Japanese Geotechnical Society, Special Issue. 65-79.

17. Skempton, A.W., (1986) "Standard Penetration Test Procedures and its Effect in Sands of Overburden Pressure, Relative Density, Particle Size, Aging and Overconsolidation," Geotechnique, **36** (3), 425-447.

18. Seed, H.B. and De Alba, P., (1986) "Use of SPT and CPT tests for Evaluating the Liquefaction Resistance of Sands", *Use of In-situ Tests in Geotechnical Engineering, Geotechnical special Publication,* **6**, 281-302.

19. Seed, R.B., Dickenson, S.E., Riemer, M F., Bray, J.D., Sitar, N., Mitchell, J.K., Idriss, I. M., Kayen, R. E., Kropp, A.L., Harder, L.F. and Power, M.S. (1990) "Preliminary report on the principal geotechnical aspects of the October 17, 1989 Loma Prieta earthquake" Report No. EERC 90-05, University of California, Berkeley.

20. Seed, H.B., Murarka, R., Lysmer, J. and Idriss, I.M., (1976) "Relationships of maximum Acceleration, maximum velocity, distance from source and local site conditions for moderately strong earthquakes". Bulletin of the Seismological Society of America, **66**(4), 1323-1342.

21. Seed. H.B., Tokimatsu. K, Harder. L. F and Chung. R. M, (1985)*"Influence of SPT Procedures in Soil Liquefaction Resistance Evaluations",* Journal of Geotechnical Engineering, **111**(12), 1425-1445.

22. Youd, T. L., and Idriss, I. M. (Editors) (2001). "Liquefaction Resistance of Soils: Summary Report from the 1996 NCEER and 1998 NCEER/NSF Workshops on Evaluation of Liquefaction Resistance of Soils." *Journal of Geotechnical and Geoenvironmental Engineering,* ASCE, **127**(10), 817-833.

23. Youd. T. L. and Keefer, D. K. (1994) *"Liquefaction during the 1977 San Juan Province, Argentina earthquake (Ms=7.4)"* Engrg. Geol., ASCE, **37** (3), pp. 211-233.

24. Zeghal, M. and Elgamal, A.W., (1994) *"Analysis of site liquefaction using earthquake records"*, Jour. GT, ASCE, **120** (6), 996-1017.

ELALUATION OF LIQUEFACTION POTENTIAL OF GUWAHATI CITY

R. Ayothiraman, S. T.G. Raghu Kanth and S. Sreelatha
Department of Civil Engineering, Indian Institute of Technology Guwahati, India
raman@iitg.ernet.in, rk@iitg.ernet.in

ABSTRACT

The liquefaction potential of the Guwahati city is estimated using hundred boreholes data located at different places of city the with a design peak ground acceleration of 0.36g. The results are presented in form of factor of safety contours at several depths below the ground surface. These contour maps indicate that all the sites in Guwahati city area are susceptible for liquefaction and hence this aspect has to be considered in earthquake resistant design of foundations/structures in Guwahati city.

Keywords: Liquefaction potential, Standard Penetration Test (SPT), Guwahati.

1 INTRODUCTION

Soil liquefaction is a major design problem for large earth structures such as mine tailings, impoundments and earth dams. Liquefaction induced ground failures include loss of bearing strength, lateral spreading, and flow failures, which may cause many engineering problems such as foundation failures, damage to utilities, slope failures, land slides, and large displacements of earth dams. Structures can settle or tilt in the liquefied soil or be ripped apart as the ground spreads laterally or flows when subsurface sand layers loose strength because of liquefaction, lateral spreading can occur in overlaying sediments allowing them to move down even in the very mild slopes. In the case of flow failures, it can move large soil mass at greater velocities. Some of the spectacular examples of earthquake damage have occurred when soil deposits have lost their strength and appear to flow as fluids. Mogami and Kubo [1953] used the term "liquefaction" to describe similar phenomenon observed during earthquakes. The 1964 Good Friday earthquake (M =9.2) in Alaska followed by the Niigata earthquake (M =7.5) in Japan attracted the attention of many researchers on the phenomenon of soil liquefaction. It is found that liquefaction was the major cause of damage to the port facilities and settlement of pavements in Kobe port during 1995 Kobe earthquake. 2001 Bhuj earthquake caused significant damage due to liquefaction and reconnaissance study just after the earthquake showed that liquefaction occurred over more than 15,000 sq.m area. The extent of damage caused by liquefaction can be substantial, which is observed from the recent past earthquakes. Hence it is essential to evaluate the liquefaction potential of a city or a region, which is seismically active.

North East region of India comes under zone V as per IS: 1893(2002) and seismic activities in this region are very high. Many great earthquakes had occurred in this region. For example, 1897 Great Assam earthquake and 1950 Assam earthquake are few of them. During 1897 Great Assam earthquake liquefaction was observed at many places in Assam .Although it was great earthquake and large area of liquefaction was observed, it was found that the extent of liquefaction induced damage was very less. It might be due to the fact that the infrastructural facilities and population density was very less at that time. But in the present scenario significant development has been taking place in this region, in particular, Assam. Assam has many oil refineries, industries and advanced infrastructural facilities. Hence, the seismic risk of Assam in the present day is very high and degree of damage would be very high, if such great earthquake occurs again. Guwahati is a major city in the North eastern territory, at which all the developmental activities are concentrated now. This city has population of 0.81 million, very important structures such as multistoried buildings, flyovers, retaining walls, sports centers etc. So it is very important to assess the liquefaction potential of this city, which will guide the design engineer to take up suitable measures, if required. Hence the liquefaction analysis is carried out to find out the factor of safety against liquefaction (F_{SL}) using 100 boreholes data and the results are presented in this paper.

2 METHODS OF EVALUATION

Global Research on causes and effects of liquefaction was accelerated after the 1964 Niigata and Alaska earthquakes. The analysis of evaluation of liquefaction potential of a soil deposit involves the characterization of the intensity of seismic loading that the soil will be subjected to and the characterization of the liquefaction resistance of the soil. By characterizing both loading and resistance in common terms, the two can be compared to determine the liquefaction potential of the soil. A number of approaches to evaluate potential for initiation of liquefaction have been developed over the past three decades. There are two broad groups of analysis. The first group (Seed and Idriss, 1971) involves estimating the shear stress level likely to develop in the field under a certain design earthquake. By comparing the induced shear level and the liquefaction resistance, liquefiable soil zones are identified. The second group of analysis is based on field observations of performance of sites subjected to earthquakes in the past. Data on earthquake characteristics and soil resistance measured with standard penetration or cone penetration test are compiled to establish an empirical relationship for new sites.

2.1 Simplified Procedure

In the original Seed and Idriss (1971) simplified procedure, a soil column is considered as a rigid body. As the seismic loading is excited at the base of the soil column, the shear wave propagates to the ground surface. The shear stress generated in the soil column can be calculated by the following equation:

$$\left(\tau_{\max}\right)_r = \sigma_v \frac{a_{\max}}{g} \qquad - (1)$$

Where $\left(\tau_{\max}\right)_r$ the maximum shear stress for rigid body is, σ_v is the total overburden pressure, $a_{\max}$ is the peak horizontal acceleration on the ground surface, and g is the acceleration due to gravity. In reality, soil behaves as a deformable body instead of as a rigid one. Hence, the rigid body shear stress is reduced with a correction factor to give the deformable body shear stress $\left(\tau_{\max}\right)_d$. The correction factor is called the stress reduction factor, r_d and measures the attenuation of peak shear stress with depth due to the non-elastic behavior of soil, i.e.

$$r_d = \frac{\left(\tau_{\max}\right)_d}{\left(\tau_{\max}\right)_r} \qquad - (2)$$

It was found that the parameter r_d reduces from a value of unity at the surface to typically between 0.3 and 0.7 at a soil depth of about 30 m, as shown in Fig. 1. Thus, in the simplified procedure, the maximum shear stress is computed as,

$$\tau_{\max} = r_d \sigma_v \frac{a_{\max}}{g} \qquad - (3)$$

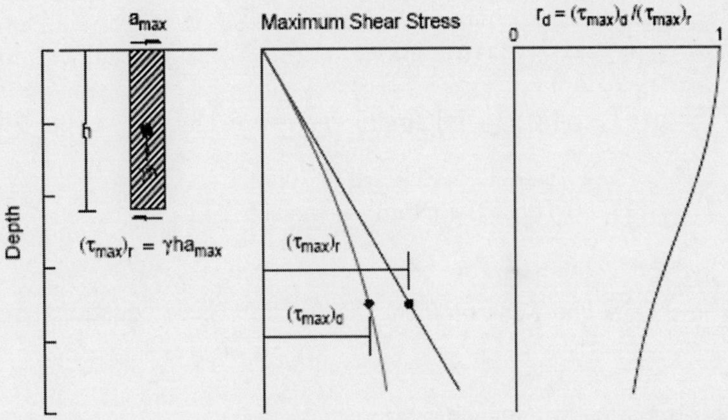

Fig. 1. Schematic for determining maximum shear stress, τ_{max}, depth reduction factor, r_d.

A factor of 0.65 is then typically employed to reduce the (single, one time) peak cyclic stress to the 'equivalent uniform cyclic shear stress' as,

$$\tau_{avg} = 0.65\tau_{max} \qquad \text{- (4)}$$

$$\tau_{avg} = 0.65\frac{a_{max}}{g}\sigma_v r_d \qquad \text{- (5)}$$

When this equivalent uniform cyclic shear stress is normalized by the initial effective overburden stress, the result is an estimate of the 'equivalent uniform cyclic stress ratio (CSR)'as,

$$CSR = 0.65\frac{a_{max}}{g}\frac{\gamma h}{\sigma'_v}r_d \qquad \text{- (6)}$$

Adjustment for the equivalent number of stress cycles in different magnitude earthquakes

The values of CSR calculated using equation (6) pertain to the equivalent uniform shear stress induced by the earthquake ground motions generated by an earthquake having a moment magnitude M. It has been customary to adjust the values of CSR calculated by equation (6) so that the adjusted values of CSR would pertain to the equivalent uniform shear stress induced by the earthquake ground motions generated by an earthquake having a moment magnitude $M = 7.5$, i.e.$(CSR)_{M=7.5}$. Accordingly, the values of $(CSR)_{M=7.5}$ are given by:

$$(CSR)_{M=7.5} = \frac{CSR}{MSF} = 0.65\left(\frac{\sigma_v a_{max}}{\sigma'_v g}\right)\frac{r_d}{MSF} \qquad \text{- (7)}$$

Where r_d = Stress reduction coefficient

MSF = Magnetic scaling factor

The shear stress induced at any point in a level soil deposit during an earthquake is primarily due to the vertical propagation of shear waves in the deposit. These stresses are particularly dependent on the earthquake ground motion characteristics, the shear wave velocity profile of site, and the dynamic soil properties. Idriss (1999), in extending the work of Golesorkhi (1989), performed several hundred parametric site response analyses and concluded that, the parameter r_d could be adequately expressed as a function of depth and earthquake magnitude (M). The following relation was derived using those results:

$$Ln(r_d) = \alpha(z) + \beta(z)M \quad \text{- (8a)}$$

$$\alpha(z) = -1.012 - 1.126\sin\left(\frac{z}{11.73} + 5.133\right) \quad \text{- (8b)}$$

$$\beta(z) = 0.106 + 0.118\sin\left(\frac{z}{11.28} + 5.142\right) \quad \text{-(8c)}$$

In which z is depth in meters and M is moment magnitude. These equations are applicable to a depth of $z \leq 34$ m;

Where as the following equation is applicable, for $z > 34$

$$r_d = 0.12\exp(0.22M) \quad \text{- (8d)}$$

The uncertainty in r_d increases with increasing depth such that the equation (8) should be applied for depths less than about $20 \pm$ m. Liquefaction evaluations at greater depths often involve special conditions for which more detailed analyses can be justified. For these reasons, it is recommended that CSR (or equivalent r_d values) at depths greater than about 20m should be based on site response studies.

The magnitude scaling factor, MSF, accounts for earthquake magnitudes different from M_w =7.5. The magnitude scaling factor, MSF, has been used to adjust the induced CSR during earthquake magnitude (M =8.1) to an equivalent CSR for an earthquake magnitude, M=7.5. The MSF is thus defined as,

$$MSF = \frac{CSR_M}{CSR_{M=7.5}} \quad \text{- (9)}$$

Thus, MSF provides an approximate representation of the effects of shaking duration or equivalent number of stress cycles. Values of magnitude scaling factors were derived by combining: correlations of the number of equivalent uniform cycles versus earthquake magnitude, and laboratory-based relations between the cyclic stress ratios required to cause liquefaction and the number of uniform stress cycles. Seed and Idriss (1982) introduced correction factors termed "Magnitude Scaling Factors (MSFs)". From that time the magnitude scaling factor given by different investigators is shown in Fig. 2.

Idriss proposed a new set of MSFs that are compatible with, and are only to be used with, the magnitude –dependent r_d that he also proposed. These new MSFs have lower values than the revised MSFs given in Fig. 2, but slightly higher values than the original Seed and Idriss (1982) MSFs.

Equation given by Idriss for calculating MSF is

$$MSF = 6.9\exp\left(\frac{-M}{4}\right) - 0.058 \leq 1.8 \quad \text{- (10)}$$

The r_d and MSF relation described in equations (10) and (8) are recommended for use in practice because they incorporate the primary features of behavior identified by analytical and experimental studies, without becoming too complex or implying undue accuracy.

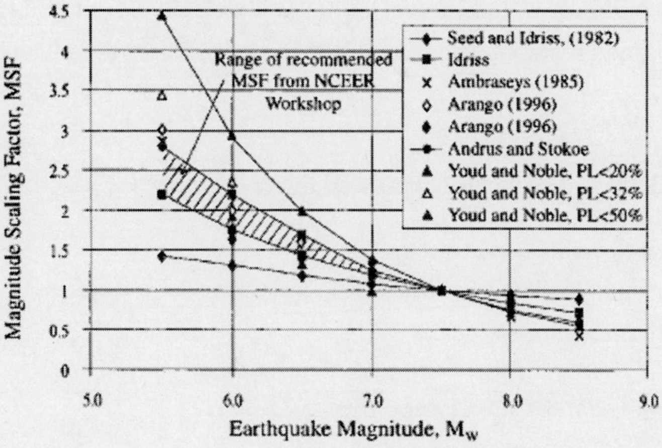

Fig. 2. Magnitude Scaling Factor derived by various investigators.

Overburden Correction Factor, K_σ

Cyclically loaded laboratory test data indicate that liquefaction resistance increases with increasing confining stress. The rate of increase however is nonlinear. To account for the nonlinearity between CRR and effective overburden pressure, Seed et al. (1983) introduced the correction factor K_σ to extrapolate the simplified procedure to soil layers with overburden pressures >100kPa. Cyclically loaded, isotropically consolidated triaxial compression tests on sand specimens were used to measure CRR for high-stress conditions and develop K_σ values. By taking the ratio of CRR for various confining pressures to the CRR determined for approximately 100 kPa (1 atm.). Seed (1983) developed the original K_σ correction curve. Other investigators have added data and suggested modifications to better define K_σ for engineering practice.

The K_σ factor is normally applied to the capacity side of the analysis during design, but it must also be used to convert the site CSR to a common σ'_v value for the empirical derivation of a CRR-$(N_1)_{60}$ curve. This is accomplished as:

$$(CSR)_{M=7.5,\sigma=1} = 0.65\left(\frac{\sigma_v a_{\max}}{\sigma'_v g}\right)\frac{r_d}{MSF}\frac{1}{K_\sigma} \qquad -(11)$$

Such that the values of CSR correspond to an equivalent σ_v of 1 atm, and thus the liquefaction correlation also corresponds to an equivalent σ'_v of 1 atm. The equation for K_σ given by Idriss & Boulanger (2006) is,

$$K_\sigma = 1 - C_\sigma \ln\left(\frac{\sigma'_v}{p_a}\right) \le 1.0 \qquad -(12a)$$

$$C_\sigma = \frac{1}{18.9 - 17.3 D_R} \le 0.3 \qquad -(12b)$$

Where C_σ is the coefficient

p_a is normally taken as 100 kPa

D_R is the relative density index

Idriss and Boulanger (2006) re-evaluated correlations between $(N_1)_{60}$ and D_R for the purpose of liquefaction evaluations, and recommended the following expressions for clean sands:

$$D_R = \sqrt{\frac{(N_1)_{60}}{46}} \quad - (13)$$

Idriss and Boulanger (2006) subsequently expressed the coefficient C_σ in terms of $(N_1)_{60}$ as

$$C_\sigma = \frac{1}{18.9 - 2.55\sqrt{(N_1)_{60}}} \le 0.3 \quad - (14)$$

With $(N_1)_{60}$ limited to a maximum values of 37.

Criteria for determining cyclic resistance ratio from SPT data

Semi-empirical procedures for liquefaction evaluations originally were developed using the Standard Penetration Test (SPT), beginning with efforts in Japan to differentiate between liquefiable and non-liquefiable conditions in 1964 Niigata earthquake. Subsequent developments have included contributions from many researchers, especially in the investigations of individual case histories where surface evidence of liquefaction was observed. The procedures recommended by Seed et al. (1971) to obtain and adjust the SPT blow count and to obtain the values of CRR are particularly note worthy as they have set the standard for almost three decades of subsequent engineering practice.

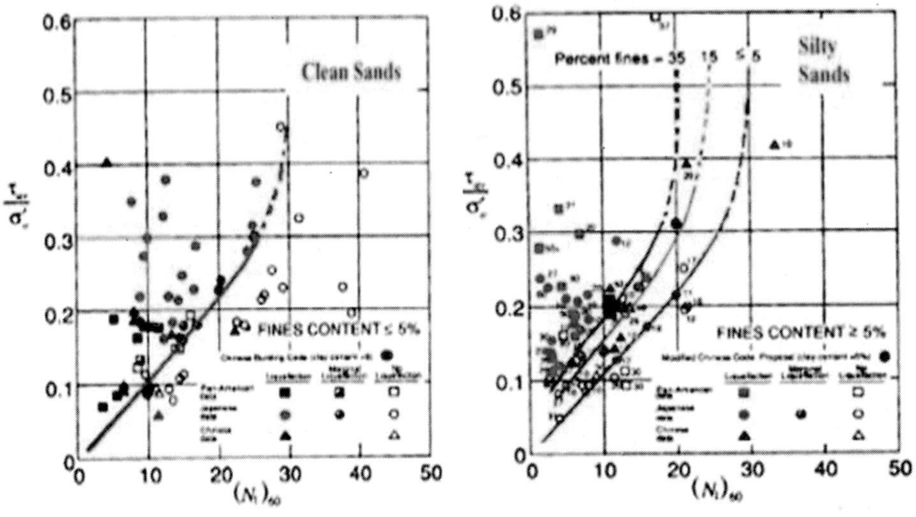

Fig. 3. Relation between cyclic stress ratio causing liquefaction and $(N_1)_{60}$ for M=7.5, Seed & Idriss (1971).

A revised boundary curve was proposed by Idriss & Boulanger (2006) and this boundary curve can be conveniently expressed in analytical equation for determining the CRR based on SPT results as follows:

$$CRR = \exp\left\{\frac{(N_1)_{60cs}}{14.1} + \left(\frac{(N_1)_{60cs}}{126}\right)^2 - \left(\frac{(N_1)_{60cs}}{23.6}\right)^3 + \left(\frac{(N_1)_{60cs}}{25.4}\right)^4 - 2.8\right\} \quad - (15)$$

where $(N_1)_{60cs}$ is the clean sand corrected N-value.

The use of these equations provides a convenient means for calculating the cyclic stress ratio required to cause liquefaction for a cohesionless soil with any fine content. In order to obtain equivalent, clean-sand corrected N-values or $(N_1)_{60cs}$, Idriss & Boulanger (2006) recommend the following equation:

$$(N_1)_{60cs} = (N_1)_{60} + \Delta(N_1)_{60} \qquad \text{- (16a)}$$

$$\Delta(N_1)_{60} = \exp\left(1.63 + \frac{9.7}{FC+0.1} - \left(\frac{15.7}{FC+0.1}\right)^2\right) \qquad \text{- (16b)}$$

where $(N_1)_{60}$ is the SPT N value corrected for field procedures and overburden stress.

$\Delta(N_1)_{60}$ is the variation of $(N_1)_{60}$ with FC (percent)

FC is the fine content present in the soil in percentage

Other corrections

Several factors in addition to fines content and grain characteristics influence SPT results. When determining the standard penetration test N-value to be used as outlined in the NCEER 1997 Liquefaction procedure, a series of correction factors are considered. The equation used for determining the corrected N-value, $(N_1)_{60}$ is as follows:

$$(N_1)_{60} = N.C_N.C_E.C_R.C_S.C_B \qquad \text{- (17a)}$$

$$N_{60} = C_E.C_R.C_S.C_B.N \qquad \text{- (17b)}$$

$$(N_1)_{60} = C_N N_{60} \qquad \text{- (17c)}$$

where N= measured standard penetration resistance

N_{60} = N value corrected for the field procedures to an average energy ratio of 60%

C_N =factor to normalize N to a common reference effective over burden stress

C_E =correction for hammer energy ratio (ER)

C_R =correction factor for rod length

C_S =correction for samplers with or without liners

C_B =correction for bore hole diameter

SPT penetration resistances are routinely normalized to an equivalent σ'_v=1 atmosphere to obtain quantities that more uniquely relate to the relative density, D_R of sand (i.e. no longer depend on σ'_v). One of the most commonly used expressions for the overburden correction factor proposed by Liao and Whitman (1986):

$$C_N = \left(\frac{P_a}{\sigma'_v}\right)^{0.5} \qquad \text{- (18)}$$

Boulanger and Idriss [2006] subsequently used the relation in equation (4.13) to obtain the following expression for determining C_N:

$$C_N = \left(\frac{p_a}{\sigma'_v}\right)^{\alpha} \leq 1.7; \quad \alpha = 0.784 - 0.0768\sqrt{(N_1)_{60}} \qquad \text{- (19)}$$

Factor of safety against liquefaction

Factor of safety against liquefaction potential is given by,

$$F_{SL} = \frac{CRR}{(CSR)_{M=7.5,\sigma=1}} \qquad -(20)$$

If F_{SL} is greater than 1.0, the soil is safe against liquefaction potential. But, if F_{SL} is less than 1.0, then the soil is susceptible for liquefaction.

3 LIQUEFACTION POTENTIAL OF GUWAHATI CITY

Evaluation of liquefaction potential is done for Guwahati city using hundred borehole data collected from Guwahati Metropolitan Development Authority(GMDA). These locations scattered in an area of about 9 km by 7 km respectively along with river Brahmaputra as shown in Fig. 4. The 'N' values at every location have been determined at an equal interval of 1.5m up to a maximum depth of 15m. At some locations, SPT has been done up to a maximum depth of 14m and also up to maximum depth of 20m respectively. To illustrate variation of standard penetration resistance of soils with depth, two samples of *N*-value profile at locations 1 and 3 are shown in Fig. 5. The average standard penetration resistance $(\bar{N})$ at all the boreholes can be computed by the expression (21).

$$\bar{N} = \frac{\sum_{i=1}^{n} d_i}{\sum_{i=1}^{n} (d_i / N_i)} \qquad -(21)$$

where d_i is the thickness of each layer and n is the total number of layers. Fig.6 shows the spatial distribution of average *N*-value calculated for Guwahati city. High $\bar{N}$ can be observed near Chandmari area. Many building codes uses $\bar{N}$ for classifying a site for purposes of incorporating local site conditions in estimation of design ground motion. The site classification based on average $\bar{N}$ value given by IBC (2000) is presented is Table 1.

Table 1. Site Classification for Seismic Purposes (IBC, 2000)

Range of $\bar{N}$	Site class
$\bar{N}<15$	E
$15<\bar{N}<50$	D
$\bar{N}>50$	C

As per IBC(2000) , E-type sites with low $\bar{N}$ are susceptible for liquefaction. Out of 100 boreholes in Guwahati city, 92 boreholes are classified as E-type, 7 belong to D-type and the remaining one with $\bar{N}$ greater than 50 can be classified as a C-type site. It is observed that most of sites in Guwahati city belong to E-type and hence vulnerable for liquefaction failure. Since most of the sites in the Guwahati city are susceptible for liquefaction, further analysis on evaluation of factor of safety against liquefaction potential is must and accordingly carried out.

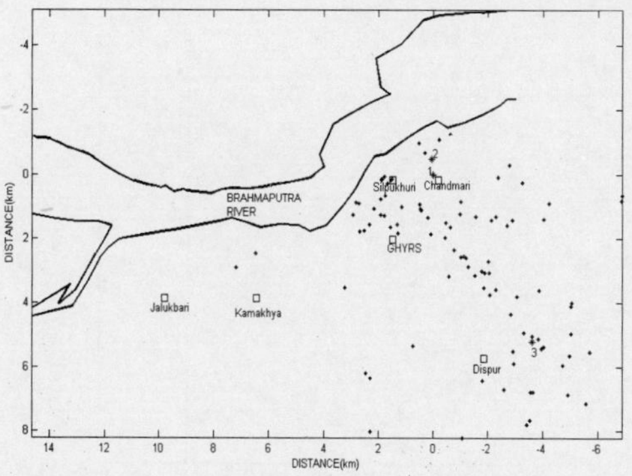

Fig. 4. Locations of Bore holes in Guwahati city.

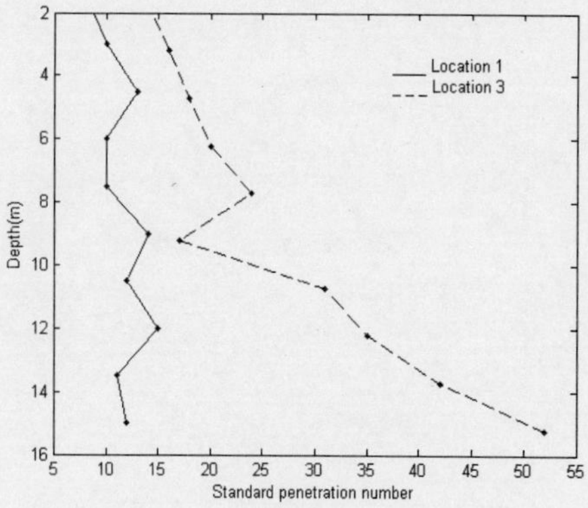

Fig. 5. Sample 'N' profile.

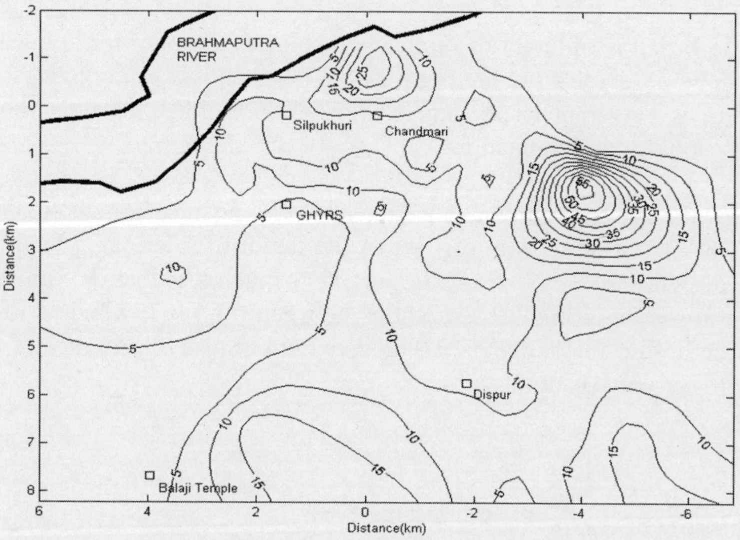

Fig. 6. Spatial distribution of Average 'N' value.

The peak ground acceleration recorded is 1897 Assam earthquake for Guwahati city is 0.36 g and has been considered in the present analysis. This PGA is consistent for an 8.1 magnitude earthquake occurring on a fault at an epicentral distance of 50 Km from Guwahati city. The Factor of safety against liquefaction is determined for all the hundred boreholes using Seed et al. (1975) and Idriss and Boulanger (2006). Although, the factor of safety against liquefaction is evaluated for all boreholes using both methods Fig. 7 shows the variation of factor of safety estimated by both methods for one borehole only, just for the sake of comparison. It is observed from the figure that the approach by Seed et al. (1975) overestimate the factor of safety against liquefaction by 10-20% at shallow depth, by about 40% at deeper depths, thus underestimating the liquefaction potential. This observation is consistent with other boreholes also, but only the degree of under estimation varies. This might be due to the errors associated with stress reduction coefficient, magnitude scaling factor and other important correction factors which were not rigorously considered by the Seed et al (1975). Hence the factor of safety of all boreholes estimated by the recent approach (Idriss and Boulanger, 2006) is presented herein.

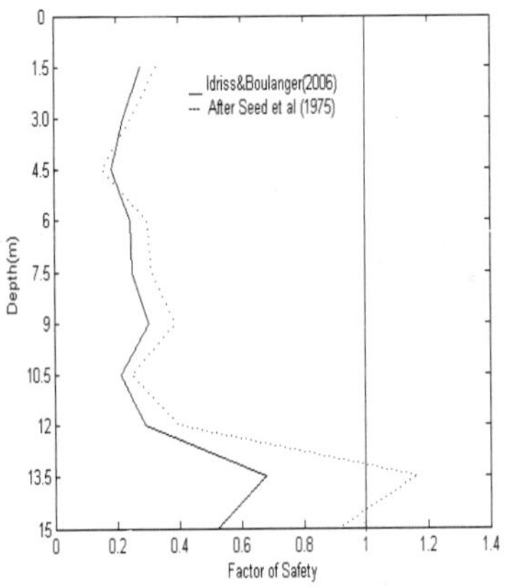

Fig. 7. Comparison of Factor of Safety estimated by Seed et al. (1975) and Idriss & Boulanger (2006).

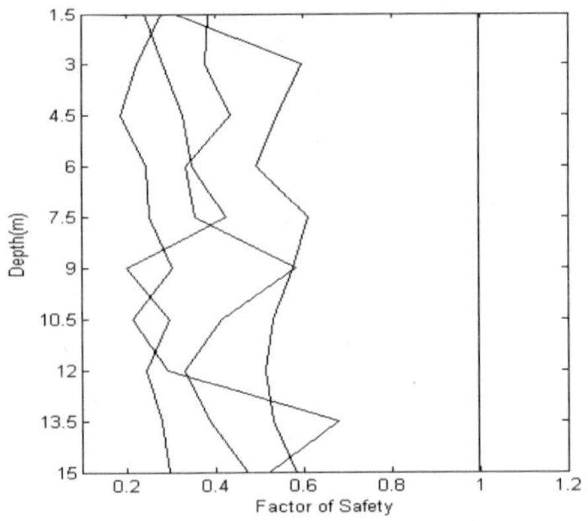

Fig. 8. FS along with depth for bore holes at GS road.

Fig. 8 presents the variation of factor of safety against liquefaction of four typical boreholes located in GS road. It can be observed from the Fig.8 is that the factor of safety against liquefaction for all the four boreholes is very much less than 1 upto 15 m depth, thus indicating that all the four boreholes are susceptible for liquefaction.

Fig. 9-11 presents the typical contour of Factor of safety against liquefaction at 3.0 m, 7.5 m, & 12.0 m depths. One can observe from Fig. 9 that the factor of safety at a depth of 3 m varies from 0.1 - 0.5 for the design PGA and these values are very much less than 1. Similarly factor of safety at a depth of 7.5 m shown in Fig. 10 also varies between 0.1 - 0.7, which is less than 1.0. Similar observation is seen in Fig. 11 for depth 12.0 m also, but because of presence of less data after 12 m depth contours are not very accurate.

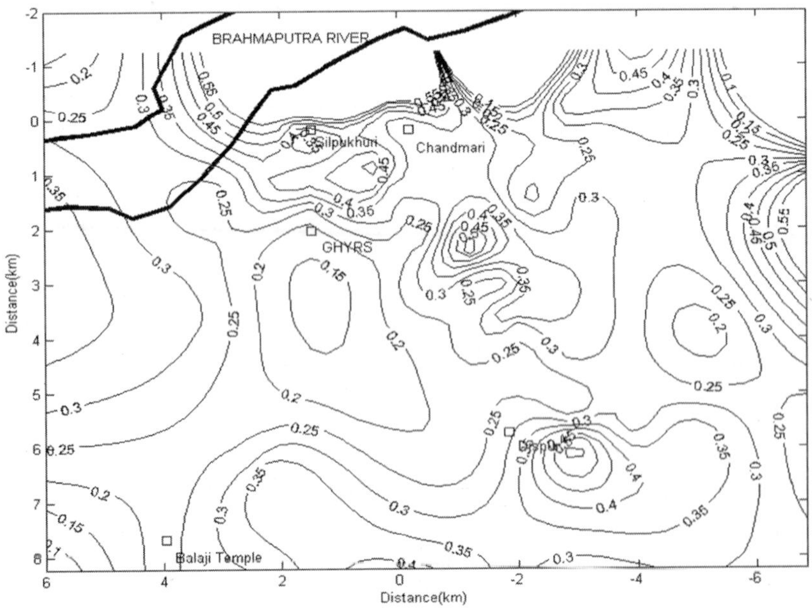

Fig. 9. F_{SL} contour at a depth of 3 m.

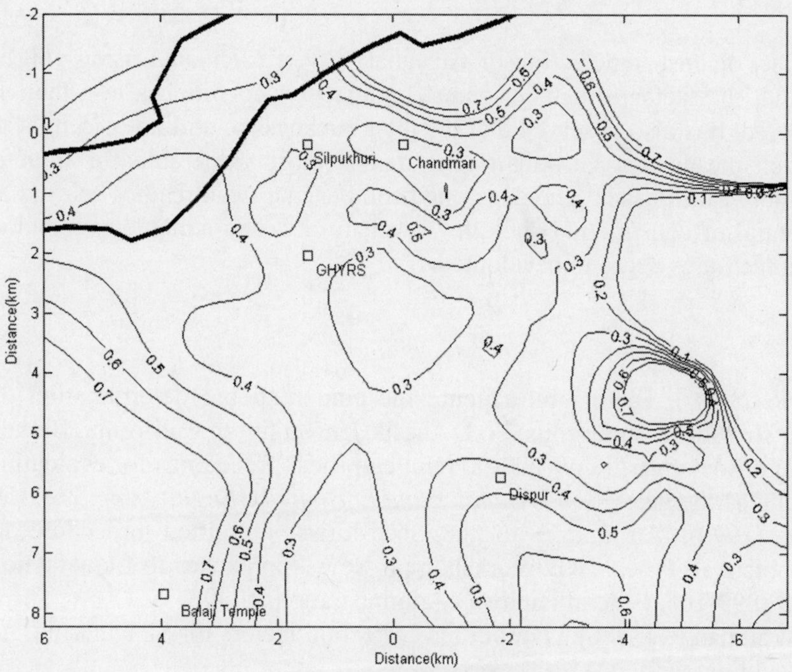

Fig. 10. F_{SL} contour at a depth of 7.5 m.

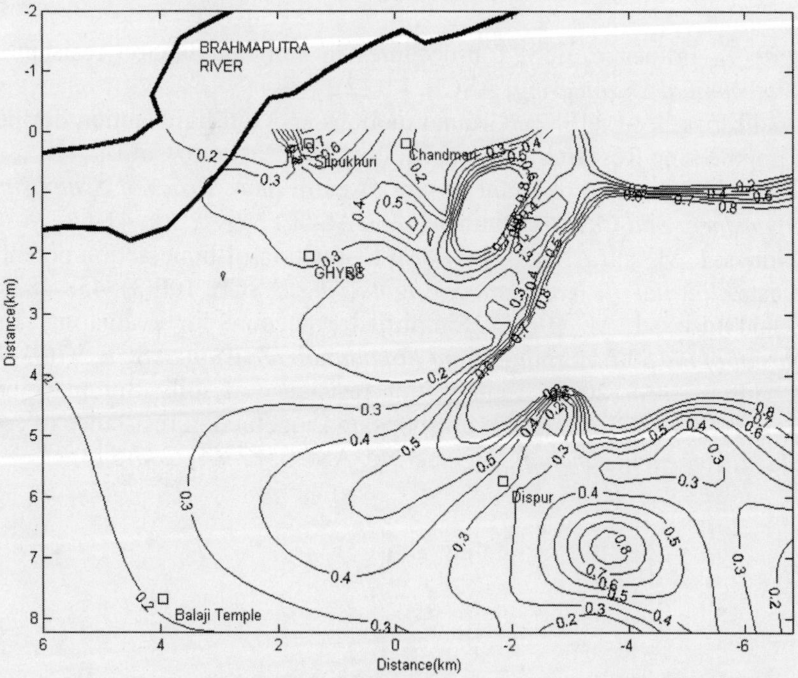

Fig. 11. F_{SL} contour at a depth of 12 m.

From the contours of F_{SL} shown in Figs. 911, it can be observed that the factor of safety at all depths is less than one, which means that the Guwahati city area is susceptible for liquefaction for an event with PGA of 0.36g. It is found from the soil exploration that the soil in Guwahati city is predominantly silty clay having fine content (60-100%) out of which the silt is more than 70%. Although the fine content is high, the city is susceptible for liquefaction, because of higher silt content soil. Hence a great care is required in selecting the suitable type of foundation or remedial soil improvement measures for various structures to be constructed in Guwahati city.

4. SUMMARY

The factor of safety against liquefaction of Guwahati city is calculated using 100 boreholes data and it is found that the factor of safety against liquefaction potential is less than 1.0 for all the boreholes considered. It is also observed that the total thickness of soil upto 15 m is susceptible for liquefaction, which means that Guwahati city area is most vulnerable for liquefaction–related hazards during future earthquakes. The developed contour maps can readily be used to evaluate the liquefaction potential of construction sites in Guwahati city, and suitable foundation or remedial soil improvement measures needs to be adopted.

REFERENCES

1. Golesorkhi, R. (1989). Factors influencing the computational determination of earthquake-induced shear stresses in sandy soils, Ph.D. thesis, University of California, Berkely, 395 pp.
2. Idriss, I. M. and Boulanger, R.W (2006). Semi empirical procedures for evaluating liquefaction potential during earthquakes, *Soil Dynamics and Earthquake Engineering*, 26,115-130.
3. Idriss, I. M. (1999). An update to the Seed-Idriss simplified procedure for evaluating liquefaction potential, Proc., TRB Workshop on New Approaches to Liquefaction, Publication No. FHWA-RD-99-165, Federal Highway Administration, 1999.
4. Liao S. and Whitman R. (1986). Overburden correction factors for SPT in sand, ASCE Journal of Geotechnical Engineering, 112(3), 373-7.
5. Mogami, T. and Kubu, K. (1953). The behavior of soil during vibration, *Proceedings, 3rd International Conference on Soil Mechanics and Foundation Engineering*, Zurich, Vol. 1, 152-155.
6. Seed et al. (1975). Influence of SPT procedures in soil liquefaction resistance evaluations, *Journal of Geotechnical Engineering*, ASCE, 11(12), -----.
7. Seed. H. B. and Idriss, I. M. (1982). Ground motions and soil liquefaction during earthquakes, Earthquakes Engineering Research Institute Monograph, Oakland, Calif.
8. Seed. H. B.(1983) Earthquake resistant design of earth dam. *Proc. of Symposium on Seismic design of Embankment and Caverns*, Philadelphia, ASCE, New York, 41-64.
9. Seed. H. B., Idriss, I. M. and Arnago. I (1983). Evaluation of liquefaction potential using field performance data. *Journal of Geotechnical Engineering,* ASCE, 109(3), 458-483
10. Seed, H. B. and Idriss, I. M. (1971). Simplified procedure for evaluating soil liquefaction potential. *Journal of the Soil Mechanics and Foundation Division*, ASCE, 97(9), 1249-1273.
11. Yound. T. L., Idriss, I. M (2001). Liquefaction resistance of soils: Summary report from the 1996 NCEER and 1998 NCEER/NSF workshops on Liquefaction resistance of soils, *Journal of Geotechnical and Geoenvironmental Engineering,* ASCE, 127(4), 297-313.

EFFECT OF PILE DIAMETER ON DEFLECTION BEHAVIOR OF PILES UNDER SEISMIC LOADS

R. Ayothiraman and G. Chandra Prakash
Department of Civil Engineering, IIT Guwahati, Assam, India.
E-mail: raman@iitg.ernet.in

ABSTRACT

The behavior of the pile foundation under seismic loads is determined by complex interaction effects that prevail in soil-pile-structure system, and an understanding of these effects is essential for a reliable design of pile foundations. The influence of these interactions is demonstrated and discussed in this paper, by performing the three-dimensional Finite Element analysis. The study investigates to bring out the influence of various parameters of soil and pile on the seismic behavior of single pile, in particular the deflection behaviour of pile. The proposed numerical model has been validated through the experimental/ numerical results reported in the literature and then a comprehensive parametric study was carried out by varying soil and pile parameters. The results of the numerical study are interpreted to estimate deflection of piles under seismic loads, i.e. the magnitude of maximum deflection and effective pile length under seismic loads. The effect of soil and pile characteristics on these quantities are discussed in this paper and it is found that the soil/pile characteristic has pronounced effect on seismic response of pile foundations in clay. An equation is proposed to estimate the effective pile length of single piles subjected to seismic loads.

Keywords: Pile, Earthquake, Clay, Effective Length, Deflection

1 INTRODUCTION

Investigation into the analysis of damages to civil engineering structure experienced during past and recent earthquakes reveals that the extent of damage to the structure is predominantly dependent on the type of foundation and soil condition, in particular to the behavior of liquefaction and/or strain softening of cohesionless and cohesive soil respectively. Design of pile foundation to resist lateral load is primarily based on the limiting deflection criteria considering the safe operation of structure. This deflection may be amplified or de-amplified under dynamic loads, which depends upon the dynamic characteristic of soil-pile system. Thus it is essential to investigate the deflection behavior of pile under seismic loads.

Based on the critical assessment of the literature, it is observed that though a few seismic/dynamic experiments on full-scale and model-scale single pile and pile group embedded in sand are documented, there is less no of literature of systematic studies reported on seismic response of full-scale piles/pile groups that have been fully instrumented to study its seismic bending and deflection behavior in clay. It is well known that the effective pile length is an important parameter in the design of piles subjected to lateral loads, but more studies are required to investigate the effects of intensity of seismic shaking and various soil-pile properties on the effective pile length. Therefore, a 3-D finite element model is developed which is validated with the two experimental data reported in literature. Subsequently a comprehensive parametric analysis was performed by varying soil-pile properties such as pile diameter, pile length and soil modulus etc and their effects on seismic response of single pile were investigated and discussed in this paper.

2 MODELING OF SEISMIC SOIL-PILE INTERACTION

Finite element method is well suited for analyzing problems, such as a pile/pile groups in layered soil, which is not easily handled with analytical or semi-analytical formulations. Also, hysteretic damping, Rayleigh damping and wave absorbing lateral boundaries conditions can be introduced to account for damping characteristics and appropriate boundary conditions. In the present study, 3D finite element formulation has been considered to model the seismic soil-pile-superstructure interaction using ANSYS. The pile and soil have been discretized as eight-noded brick elements. Each of the 8 nodes has three translational degrees of freedom in the nodal x, y, and z-directions. Full three-dimensional geometric models were used to represent the soil-pile system. The pile is completely embedded in the soil. The system is idealized as an assemblage of eight-node brick elements. These elements were selected for simplicity, ease in modeling, and computation. Eight-node elements may not be suitable for modeling the response of a system dominated by bending deformations. However, it is assumed that the responses of soil and pile are dominated by shear deformations rather than bending stresses. Bentley and El Naggar (2000) and Maheshwari et al. (2005) had used these elements successfully for modeling the dynamic response of piles; hence the similar model elements are used in the present study.

To simulate an infinite soil medium, springs and dashpots are attached on the side walls of the foundation which provide the proper boundary conditions. It is noted that these are used in all three directions along the boundary. The coefficients of the springs and dashpots are derived separately for the horizontal and vertical directions based on the predominant frequency of loading. The constants of the springs and dashpots in the two horizontal directions were calculated using the solution developed by Novak and Mitwally (1988):

$$k_r^* = \frac{G}{r_0}\left[S_1(a_r,\upsilon,D) + iS_2(a_r,\upsilon,D)\right] \tag{1}$$

where k_r^* = complex stiffness; G = shear modulus of soil; S_1 and S_2 = dimensionless parameters from closed-form solutions; D = material damping ratio; υ = Poisson's ratio; and i = imaginary unit = Î−1. Also, r_0 = distance in plan from the center of the pile to the node where these elements are attached. Further, a_r = dimensionless frequency = $r_0\omega/V_s$; where ω is the angular frequency of excitation and V_s is the shear wave velocity of the soil. The real and imaginary parts of Eq. (1) represent the stiffness and damping, respectively, i.e.

$$k_r = \frac{GS_1}{r_0} \; and \; c_r = \frac{GS_2}{\omega r_0} \tag{2}$$

For the static case, the damping term vanishes and the element reduces to a spring only. Similarly, the constants for the vertical direction are given by (plane strain case) Novak et al. (1978):

$$k_w^* = \frac{G}{r_0}\left[S_{w1}(a_r,D) + iS_{w2}(a_r,D)\right] \tag{3}$$

where the subscript w is used to represent the vertical direction and the other parameters are the same as in Eq. (1). Stiffness and damping for the vertical direction are determined in a similar fashion as mentioned in Eq. (2), i.e.

$$k_w = \frac{GS_{w1}}{r_0} \; and \; c_w = \frac{GS_{w2}}{\omega r_0} \tag{4}$$

To determine the stiffness and damping of the elements, the constants given by Eqn. (4) are multiplied by the area of the element face (normal to the direction of loading). It should be noted that for the vertical direction, the dimensionless parameters S_{w1} and S_{w2} are independent of Poisson's ratio, and for the static case both the spring and dashpot terms vanish. Thus, for the low-frequency range, the spring and dashpot constants are adjusted to match more rigorous solutions by choosing a minimum cutoff frequency $a_r = 0.3$ below which the stiffness is taken as constant and the damping is taken as linear. To adequately represent damping in the system, both radiation and material damping are considered in the analysis. Thus, the damping matrix C consists of two parts, radiation damping C_r and material damping C_m, i.e.

$$C = C_r + C_m \tag{5}$$

Radiation damping C_r = diagonal matrix and has nonzero terms only at the nodes on the boundary. At a particular node where dashpots are attached, damping coefficients for the horizontal and vertical directions are found using Eqs. (2) and (4) respectively. For conceptual and computation reasons, material damping C_m is taken as proportional to stiffness and is given by,

$$C_m = \alpha K \quad where\ \alpha = 2D/\omega_0 \tag{6}$$

where D = material damping ratio; and ω_0=predominant circular frequency of loading. Additional hysteretic damping may develop due to the nonlinearity but dissipation of seismic energy through inelastic deformation tends to overshadow the dissipation of the energy through hysteretic damping and is therefore neglected. The developed 3-D FE model is shown in Fig.1.

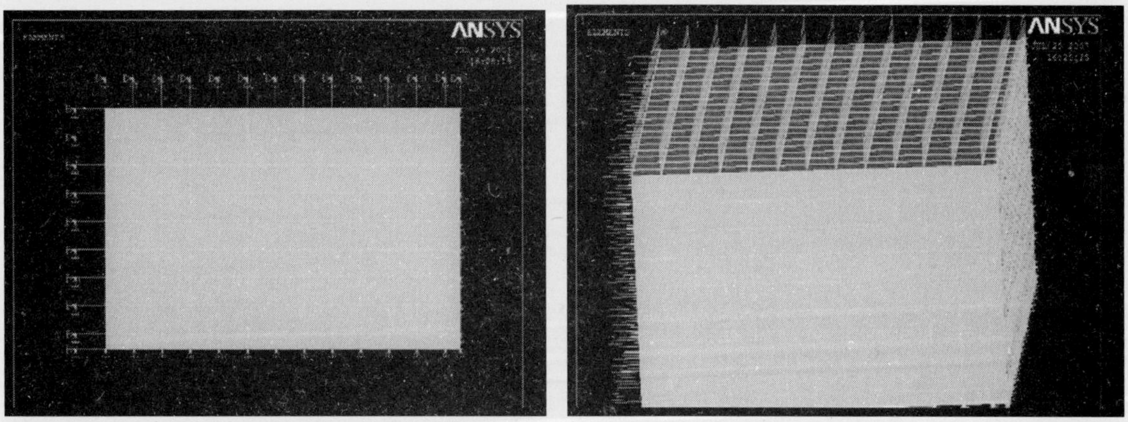

(a) (b)

Fig. 1. 3D –Finite element model (a) Plan (b) Elevation.

3 VALIDATION OF MODEL

The efficacy of the proposed model is demonstrated by comparing the predicted response of pile with the experimentally measured single pile response. Three well instrumented full-scale experiments are selected where the pile response was recorded beyond the elastic limit. Kramer et al. (1990) conducted lateral monotonic and cyclic tests on a cylindrical steel pile installed in the Mercer Slough peat near the eastern shore of Lake Washington in Bellevue, Washington. Jennings et al (1984) and Brown et al. (1988) conducted dynamic experiments on solid and hollow piles embedded in medium dense and saturated silty sand deposits respectively. The soil properties and pile characteristics are summarized in Table 1, and these properties are used in the analysis to predict the pile response.

Table 1: Properties of Soil and Pile Considered

Test/Year	Soil properties	Pile characteristics
Kramer et al. (1990)	νs= 0.49; Vs (tip) = 30 m/s ρs =1120kg/m3; c_u=14.4kN/m2	L = 14.9 m ;d = 0.20 m $E_pI_p = 3.81 \times 10^3$ kN-m^2
Jennings et al. (1984)	ν_s = 0.49; V_s (tip) = 125 m/s ρ_s = 1600 kg/m^3; $\phi'=38.5^0$	L = 6.75m ;d = 0.45 m $E_pI_p = 0.8 \times 10^5$ kN-m^2
Brown et al. (1988)	ν_s = 0.48; Vs (tip) = 160 m/s ρ_s = 1600 kg/m^3	L = 13.1 m ;d = 0.273 m t = 0.0093 m; $E_pI_p = 7.3 \times 10^3$ kN-m^2

The pile response, such as load-deflection behavior at pile head, pile deflection and bending moment profiles at120 kN load and p-y curves at different depths are obtained from the analysis. The results are compared with the experimental data of the respective literature and the results of numerical analysis reported by Badoni and Makris (1996). Fig. 2 and 3 shows the comparison of predicted and measured pile response and it is found from the figures that the agreement of the prediction with the measured values is very well.

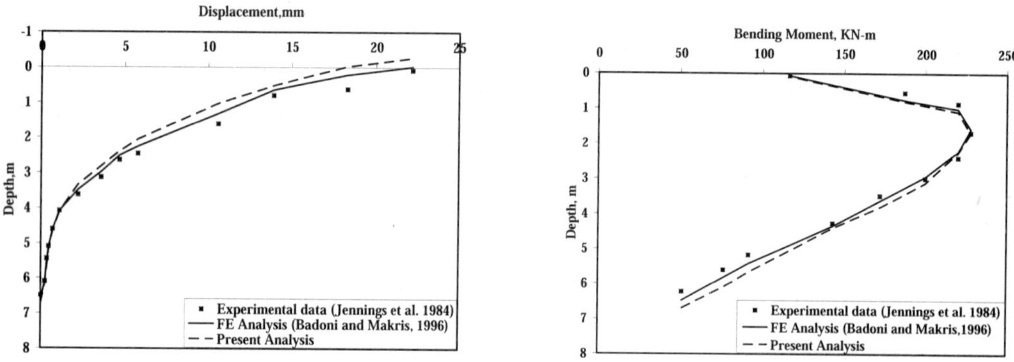

Fig. 2. Comparison of predicted and measured pile deflection and bending moment for lateral force of 120KN

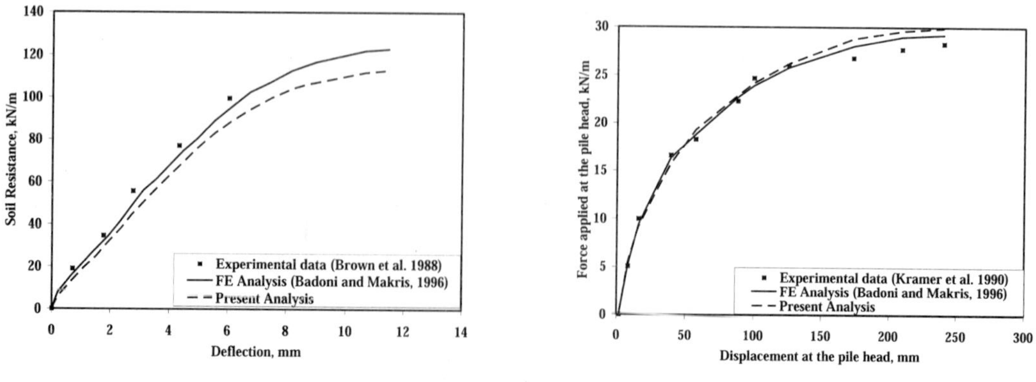

Fig. 3. Comparison of predicted and measured p-y curve at a depth, 1.22 m and force-displacement response at pile head.

4 PARAMETRIC ANALYSES

Parametric analysis for single pile under seismic loads has been performed by varying the different parameters like pile diameter, pile slenderness ratio and soil modulus etc using the proposed FE model. The various parameters which are used in the parametric analysis are summarized in Table 2. A comprehensive parametric analysis has been carried out by varying the parameters given the Table 2 and the results are presented, analyzed and discussed.

Table: 2. Parameters Considered in Analysis

Cases	Young's modulus of soil	Pile Diameter			
		d = 0.3 m	d = 0.5 m	d = 0.7 m	d = 1.0 m
Case-1	20 MPa	L/d = 10, 20, 30 and 40	L/d = 10, 20, 30 and 40	L/d = 10, 20, 30 and 40	L/d = 10, 20, 30 and 40
Case-2	40 MPa	L/d = 10, 20, 30 and 40	L/d = 10, 20, 30 and 40	L/d = 10, 20, 30 and 40	L/d = 10, 20, 30 and 40
Case-3	60 MPa	L/d = 10, 20, 30 and 40	L/d = 10, 20, 30 and 40	L/d = 10, 20, 30 and 40	L/d = 10, 20, 30 and 40
Case-4	80 MPa	L/d = 10, 20, 30 and 40	L/d = 10, 20, 30 and 40	L/d = 10, 20, 30 and 40	L/d = 10, 20, 30 and 40

The seismic loading is applied at the base as transient motion. For the transient motion, an acceleration time history for the El Centro Earthquake 1940 (N-S Component), with peak ground acceleration equal to 0.32g has been used.

5. RESULTS AND DISCUSSION

5.1 Pile Deflection

Pile deflection behavior under the given seismic excitation has been predicted for different pile diameters: 0.3 m,0.5 m, 0.7 m and 1.0 m, for different length to diameter ratios of piles: L/d = 10 and 40 and different soil modulus values: E_s = 20, 40, 60, and 80 MPa. The predicted pile deflection profiles are plotted against the normalized depth of pile (z/d), i.e. the ratio of desired depth along the pile (z) to pile diameter (d). The typical pile deflection profiles for L/d = 10 & 40 for E_s = 20 and 40MPa are shown in Fig. 4 and 5 respectively.

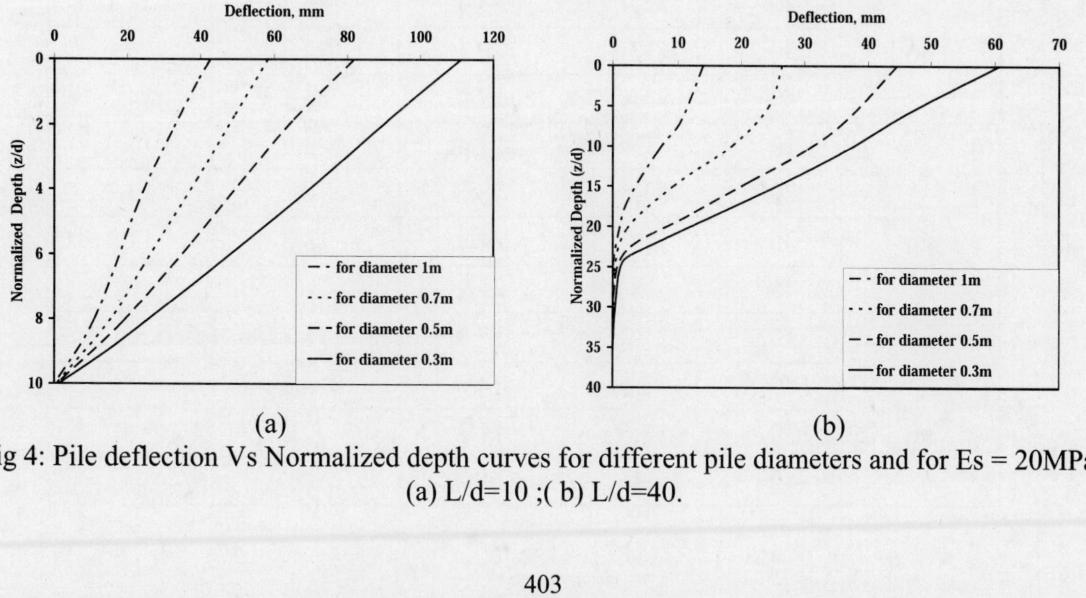

(a) (b)

Fig 4: Pile deflection Vs Normalized depth curves for different pile diameters and for Es = 20MPa: (a) L/d=10 ;(b) L/d=40.

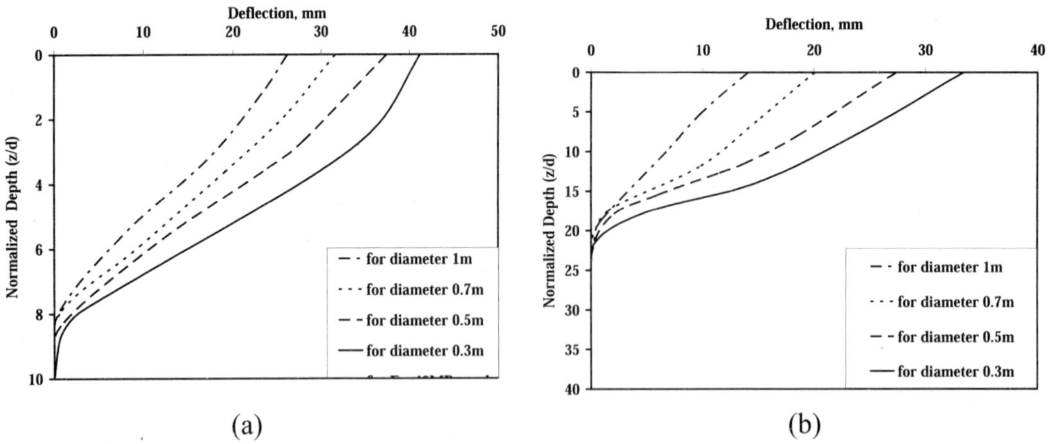

Fig. 5. Pile deflection Vs Normalized depth curves for different pile diameters and for Es = 40MPa: (a) L/d=10; (b) L/d=40.

It is observed from above Figs. 4 and 5 that the pile deflection is reducing with an increase of pile diameter and L/d ratio for all soil modulus. This is due to fact that the passive resistance zone increases with an increase of pile diameter and pile length. It is also noticed from these figures that increase of soil modulus causes reduction of pile deflection, because the increase of soil modulus cause increase in the stiffness of the soil which ultimately resist the pile against deflection.

The maximum deflection of the pile is measured from the pile deflection profiles of different cases and is presented in Table 3.

Table: 3. Maximum Deflection Measured from Deflection Profiles.

Young's Modulus E_s (MPa)	L/D Ratio	Maximum Deflection (mm)			
		d = 0.3m	0.5m	0.7m	1m
20	10	111.1	81.8	58.3	42.7
	20	82.3	62.9	42.2	26.8
	30	71.5	55	34.9	18.8
	40	60.6	44.6	26.4	14
40	10	41.2	37.4	31.6	28
	20	38.4	33.4	26.7	21.1
	30	37	29.1	22.3	16.2
	40	33.4	27.4	20.1	14.1
60	10	35	27.4	22.7	17.9
	20	33.5	28.5	19	13.3
	30	29	24.8	18.3	11.2
	40	25	23.2	16	10
80	10	21.2	17.8	11.3	8.3
	20	17.5	14.6	10.1	8
	30	13.6	11.3	8.1	6
	40	11	9	6.5	5

It is seen from the table that the maximum deflection decreases with increase of pile diameter, pile length and soil modulus. The variation of maximum deflection with L/d ratio of pile for different diameters at Es = 20 and 80 MPa is shown in Fig. 6. It is inferred from the figure that the maximum deflection decreases substantially with an increase of pile diameter and length at both soil modulus conditions. This may be due to increase of zone of passive resistance under lateral seismic shaking. This observation is consistent with pile embedded in other soil modulus conditions (soft to medium stiff clay). However it is noted from Fig. 6 that at higher soil modulus values, the rate of reduction of maximum deflection is relatively less when compared to the one in very low soil modulus (Es = 20 MPa). This is probably because of higher hysteretic damping in addition to radiation damping that prevails in very soft clay (Es = 20 MPa), thus leading to high reate of reduction of maximum deflection.

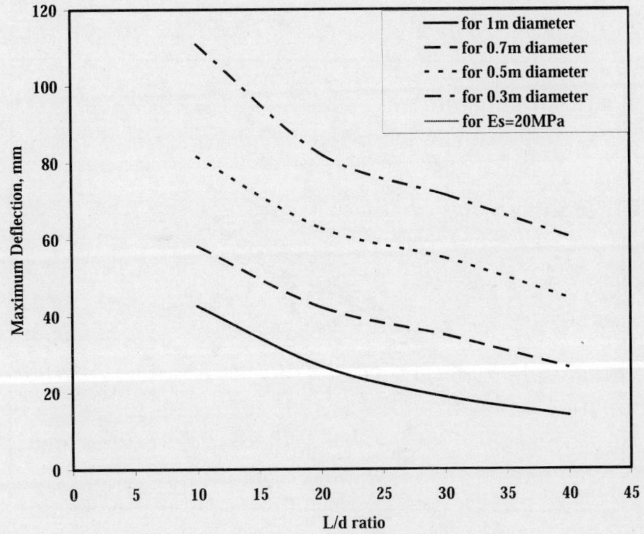

(a)

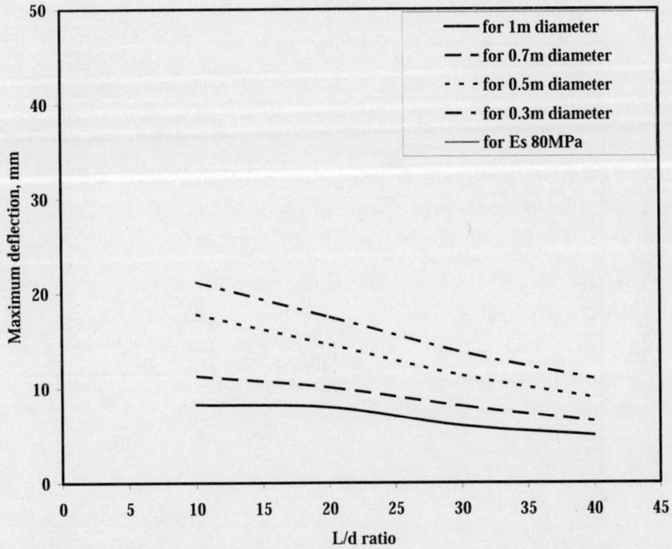

(b)

Fig. 6. Maximum deflection Vs L/d ratio for different diameters for soil modulus:
(a) Es = 20MPa and (b) Es = 80MPa

5.2 Effective Pile Length:

The effective pile length is defined as a pile length from the ground surface at which the pile deflection is zero. The effective length values, i.e. the normalized depth of zero deflection measured from the pile deflection profiles, which is normalized effective pile length. The variation of normalized effective pile length with L/d ratio of pile for different diameters at Es = 20 MPa and 80 MPa is shown in Fig. 7. It is inferred from the figure that the effective pile length increases with pile length for all soil modulus conditions, but the rate of increase is high for low soil modulus. This is because that the almost full length of pile undergoes vibration during seismic shaking for piles embedded in very soft clay having low soil modulus. It is also inferred from Fig. 7 that the pile diameter does not have significant effect on the effective pile length of short piles (L/d < 20), but has a significant effect for long piles (L/d > 30) embedded in all soil modulus conditions considered in the analysis.

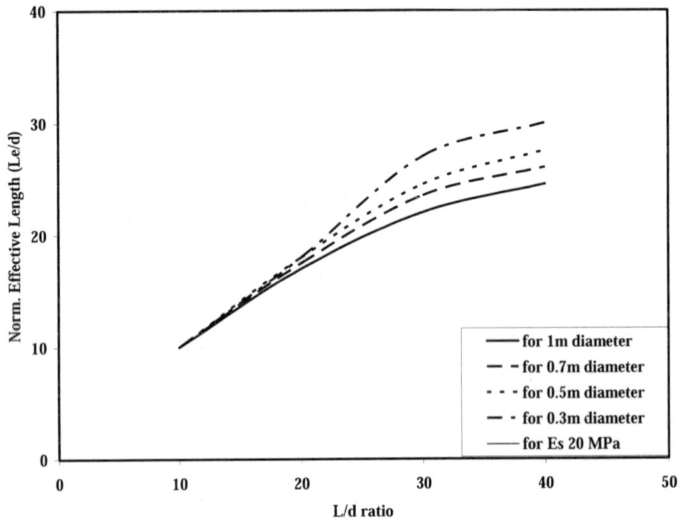

(a)

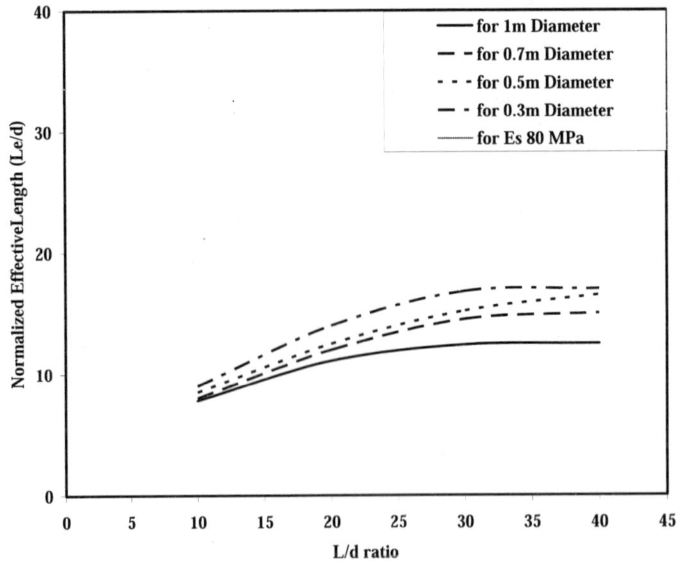

(b)

Fig. 7. Normalized effective pile length Vs L/d ratio for different diameters for soil modulus: (a) Es = 20MPa and (b) Es = 80MPa

The variation of normalized effective pile length with modulus ratio (Ep/Es), for different L/d ratio of pile and constant diameter of pile of 0.5 m is given Fig. 8. It is inferred from the figure that the effective length significantly reduces with decrease of modulus ratio, i.e. with an increase of soil modulus for all L/d ratios. The effective length estimated using existing analytical and semi-analytical expressions suggested by various authors (Krishnan et al. 1983; Gazaetas, 1984; Gazaetas, 1991) are also presented in Fig.8.

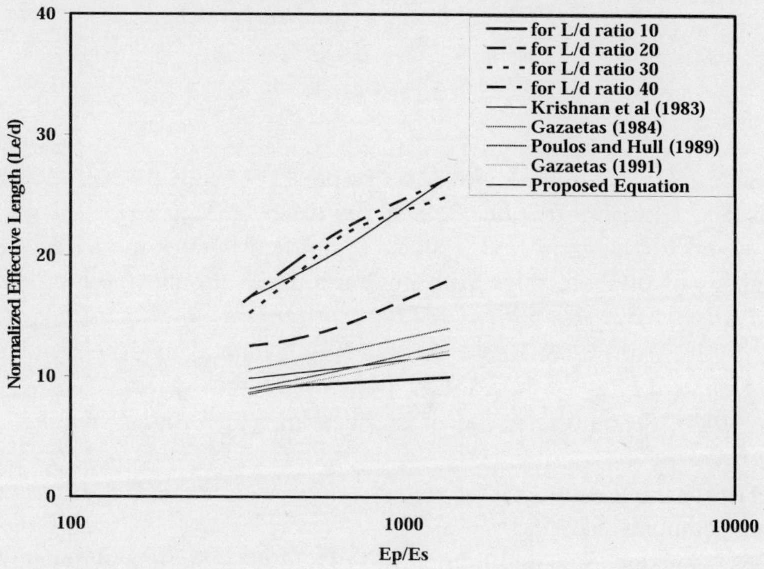

Fig. 8. Normalized effective pile length Vs Modulus ratio for different L/d ratios and constant diameter of 0.5 m.

It is found that these equations underestimating the effecting pile length of long flexible piles. It is also seen from the Fig. 8 that the effective length does not increase proportionally with L/d ratio for piles with L/d beyond 30 in all soil modulus values considered. Hence an equation is developed by performing multiple regression analysis to estimate the effective pile length of long piles under seismic loads as given below:

$$\frac{L_e}{d} = 2.20\left(\frac{E_p}{E_s}\right)^{0.345} \quad (7)$$

In which, L_e = Effective pile length, d is pile diameter, E_p is Young's modulus of pile and E_s is Young's modulus of soil. The regression analysis gives a correlation coefficient of 0.9465. The proposed equation can be used to find out the effective length of single piles subjected to seismic loads.

6 CONCLUSIONS

Based on the comprehensive parametric analysis carried out on the seismic response of piles embedded in clay, the following major conclusions are drawn:

- Pile diameter, L/d ratio and soil modulus significantly influences the deflection behavior of piels under seismic loads.
- Maximum pile deflection is consistently reducing with increase in the pile length, pile diameter and soil modulus. However, it is found that the rate of reduction of maximum deflection with increase of pile length at higher soil modulus is not significant.

- The effective pile length is significantly influenced by the soil modulus and pile length. However, beyond, L/d =30, the pile length does not have significant influence. Pile diameter has only marginal effect on effective pile length for long piles, whereas it does not have practically any effect for short piles.
- An equation is proposed to estimate the effective pile length under seismic loads for single piles. It is to be noted that the effect of seismic excitation on the proposed equation is not investigated in the present study and hence the equation may need to be modified, if necessary based on further investigations.

REFERENCES

1. Badoni, D., and Makris, N. (1996). Nonlinear response of single piles under lateral inertial and seismic loads. *Soil Dynamics and Earthquake Engineering*, 15, 29–43.
2. Bentley, K. J. and El Naggar, M. H. (2000). Simplified BNWF model for nonlinear seismic response analysis of offshore piles with nonlinear input ground motion analysis. *Canadian Geotechnical Journal*, 32, 183–196.
3. Brown, D. A., Morrison, C. & Reese, L. C. (1988). Lateral load behavior of pile group in sand. *Jl. Geotechnical Engineering*, ASCE, 144, 30-45.
4. Gazetas, G. (1984). Seismic response of end-bearing piles. *Soil Dynamics and Earthquake Engineering*, 3, 82 – 93.
5. Gazetas G. (1991). Foundation vibrations. *Foundation Engineering Handbook*, 2nd Edition, Van Nostrand Reinholds, 553-593.
6. Jennings D. N., Thurston, S. J. and Edmonds, F. D. (1984). Static and dynamic lateral loading of two piles. *Proc 8th WCEE, San Francisco, CA.* (3), 561-68.
7. Kramer, S. L., Satari, R. and Kilian, A. P. (1992). Evaluation of in situ strength of a peat deposit from laterally loaded pile test results. *Transportation Research Record No. 1278*, Transp. Res. Board, Washington, D.C., 103-109.
8. Krishnan, R., G. Gazetas, and A. Velez. (1983). Static and dynamic lateral deflection of piles in non-homogeneous soil stratum, *Geotechnique*, 33, 307 – 325.
9. Maheshwari, B. K., K. Z. Truman, P. L. Gould, and M. H. El Naggar. (2005). Three-dimensional nonlinear seismic analysis of single piles using finite element model: Effects of plasticity of soil. *Journal of Geomechanics*, ASCE, 5, 35-44.
10. Novak, M., and Mitwally, H. 1988. Transmitting boundary for axisymmetrical dilation problems. Journal of Engineering Mechanics, ASCE, 114(1): 181–187.
11. Novak, M., Nogami, T. and Aboul-Ella, F. Dynamic soil reaction for plain strain case. *Jl. Engineering Mech. Division*, ASCE, 1978, 104, 1024-41.
12. Poulos, H. and Hull, T. (1989). The role of analytical geomechanics in foundation engineering. In *Foundation Engg: Current Principles and Practices*, ASCE, 2, 1578-1606.

BEHAVIOR OF KAOLINITE CLAYS UNDER CYCLIC LOADING

Rakesh Pillai J.[1], _Robinson R.G._[2] _and Boominathan A._[3]
[1] _Research Scholar, Indian Institute of Technology Madras, rakeshpilla@iitm.ac.in_
[2] _Assistant Professor, Indian Institute of Technology Madras, robinson@iitm.ac.in_
[3] _Professor, Indian Institute of Technology Madras, boomi@iitm.ac.in_

ABSTRACT

A series of undrained cyclic triaxial tests and followed by post cyclic monotonic tests are performed on kaolinite specimens to study the post cyclic behavior of clays. Tests are carried out for different cyclic stress levels at wide range of confining pressures. Test results are compared with those of monotonic tests without prior cyclic loading. It is found that there is accumulation of pore pressure and strain with number of cycles for high cyclic stress levels while non failure equilibrium exists in the case of low cyclic stress levels for all confining pressures. Post cyclic test results show dilative behavior which confirms the apparent over consolidation due to cyclic loading. Stress paths of monotonic loading of samples without prior cyclic loading is similar to general normally consolidated clay behavior while the stress paths of post cyclic tests resemble over consolidated clay behavior.

Keywords: Cyclic Triaxial Tests, Kaolinite, Post cyclic behaviour of clays

1 INTRODUCTION

Saturated clay deposits are subjected to cyclic loading due to earthquakes, wave action, wind, pile driving etc. Clays are generally believed to be more resistant to cyclic loading when compared to sands, but many cases of excessive deformation due to cyclic loading on clayey deposits are reported in the literature (Yilmaz et al 2004, Boulanger and Idriss 2007). Accumulation of pore pressure and deformation occurs as a result of cyclic loading on clays and they may exceed the permissible limits if the magnitude of cyclic load is high. Many researchers have studied the behavior of clays under cyclic loading by carrying out cyclic triaxial tests on undisturbed and remoulded samples (Thiers and Seed 1968, Sangrey et al 1969, Andersen et al 1980, Matsui et al. 1980, Yasuhara et al. 1982, 1992). If the cyclic shear stress ratio, i.e. ratio of the cyclic shear stress to the undrained static shear strength, is more than the critical level of repeated stress the sample will fail due to excessive deformations while it will reach a state of non failure equilibrium if the cyclic stress ratio is less.

Effective stress considerations are necessary to study the cyclic stress-strain behavior of clayey soils which necessitates accurate measurements of pore pressure (Andersen et al 1980, Matsui et al. 1980, Yasuhara et al. 1982). For accurate pore pressure measurements the test has to be conducted sufficiently slowly for equalization of pore pressure to occur (Wood 1982, Zergoun and Vaid 1994). Andersen (2004) has reported that the loading period will not affect the fundamental behavior of the soil and only the numerical values of the different parameters may change with the load period.

Accumulation of pore pressure as a result of cyclic loading will cause a reduction in the mean effective stress which will leave the clay sample apparently over consolidated (Togrol and Guler 1984, Hyde and Ward 1985). These samples are found to follow a stress path similar to that of over consolidated samples when subjected to post cyclic monotonic loading. There is a controversy regarding the behavior of clays subjected to post cyclic monotonic loading. Castro and Christian (1976) rules out any chance of strength reduction in post cyclic static loading due to pore pressure accumulation during cyclic loading, while many researchers like Andersen et al (1980), Koutsoftas et al (1978) reports a small reduction in strength of not more than 10 – 20%. According to many authors the post cyclic strength reduction is largely a strain dependent phenomenon. Therefore, in the present study, normally consolidated kaolinite samples were subjected to undrained monotonic strain controlled loading and stress controlled one way cyclic loading. The samples which had not failed under cyclic loading were subjected to undrained monotonic loading to study the post cyclic behavior.

2 EXPERIMENTAL PROGRAMMES

2.1 Test Equipment for Cyclic Loading

Consolidated undrained stress controlled cyclic triaxial tests were carried out using InfraTest make cyclic triaxial testing apparatus. Cyclic load was applied by a pneumatic piston cylinder actuated through a timer controlled single acting spring return pneumatic solenoid valve. A specially designed timer, with a frequency range of 0.05 Hz to 1Hz controls the frequency of loading. Pressure input to the solenoid valve can be adjusted with the help of a pressure regulator and a pressure gauge, to apply the required deviatoric stress. Loads were applied as trapezoidal pulses. Trapezoidal pulses are a good approximation to the sinusoidal shape and the shape of pulse has little effect on the behavior of soil (Andersen et al 1980, Andersen 1988). The pore water pressure developed and axial deformation of the sample were measured using a base fitted pore pressure transducer and LVDT respectively. A dedicated data acquisition system was used to collect the pore pressure and displacement responses during testing.

2.2 Sample Preparation and Test Procedure

In this study commercially available kaolinite was used. It has a specific gravity of 2.6. The particle size distribution shows that 38% of the particles are in the silt size and the clay content is 62%. It has a liquid limit of 52% and a plastic limit of 25%.

Samples for triaxial tests were prepared by slurry consolidation method. Clay slurry was prepared at a water content of 104% (two times the liquid limit) and poured in to a large cylindrical mould of 15 cm diameter and 23 cm height. The slurry was then gradually loaded to a consolidation pressure of 100 kPa. During this stage the drainage was allowed from both top and bottom. Once the consolidation was over, samples of 38 mm diameter and 76 mm height were extruded from the consolidated clay block.

The samples were then transferred to the triaxial cell. Filter paper strips were provided on the sides of the sample for radial drainage in addition to the drainage through the two ends as per IS specifications (IS: 2720, part-12). Full saturation of the samples was ensured by applying back pressure for a B value greater than 0.95. Then the samples were consolidated isotropically at the required confining pressure. Static and cyclic tests were carried out for the confining pressures of 100 kPa, 150 kPa and 200 kPa. After consolidation, the samples were subjected to monotonic loading or cyclic loading. For undrained monotonic loading the rate of shearing was obtained from the time taken for 100% consolidation to ensure the equalization of pore pressure in the sample. Samples were sheared at a rate of 0.07 mm/minute. The undrained monotonic strength, stress strain and pore pressure response were obtained for comparing with the behavior of the soil after cyclic loading.

One way cyclic loading was applied at a frequency of 0.1 Hz with different cyclic stress amplitudes. The deviatoric cyclic stress is fixed from the undrained strength obtained from the monotonic tests. Tests were carried out with cyclic stress ratios (ratio of cyclic deviatoric stress to the undrained strength) of 0.25, 0.5 and 0.75. Each sample was subjected to 5000 cycles of compression load under undrained condition. The samples subjected to cyclic loading were then monotonically loaded under undrained conditions to study the post cyclic behavior. No drainage was allowed between cyclic and post cyclic monotonic loading.

3 RESULTS AND DISCUSSIONS

3.1 Behavior under Undrained Monotonic Loading

Undrained strain controlled monotonic tests were carried out on the samples to study the stress strain behavior and to compare with the post cyclic behavior. The stress strain and pore pressure response for samples subjected to undrained monotonic loading under different confining pressures are shown in Fig.1. The figure shows ductile stress-strain behaviour, which is a typical characteristic of normally consolidated soils. Positive pore pressure developed during shearing that shows the contractive nature of the samples.

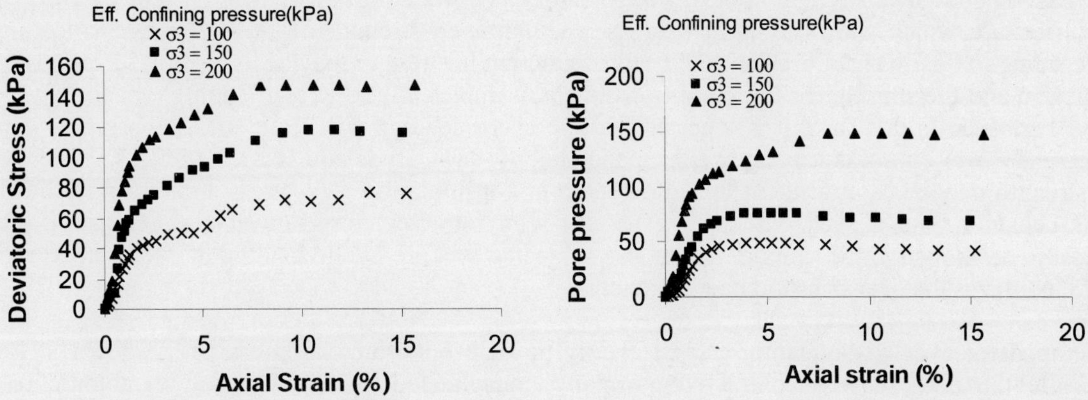

Fig. 1 Stress strain and pore pressure response of kaolinite samples during undrained static tests

3.2 Behavior under Cyclic Loading

The clay samples were subjected to undrained stress controlled one way cyclic loading with different confining pressures and cyclic stress ratios. One way cyclic loading will lead to an increase in the average shear strain, and may cause failure. Typical graphs showing the variation of pore pressure and axial strain with number of cycles for a confining pressure of 150 kPa is shown in Fig. 2. It can be seen that for cyclic stress ratio of 0.5 axial strain and pore pressure become constant after certain number of cycles and reach a non failure equilibrium state. For a cyclic stress ratio of 0.75 strain and pore pressure seem to increase even after 5000 cycles and may lead to failure at large number of cycles. For a higher cyclic stress ratio the samples may fail after a few cycles which will be a concern in earthquake point of view.

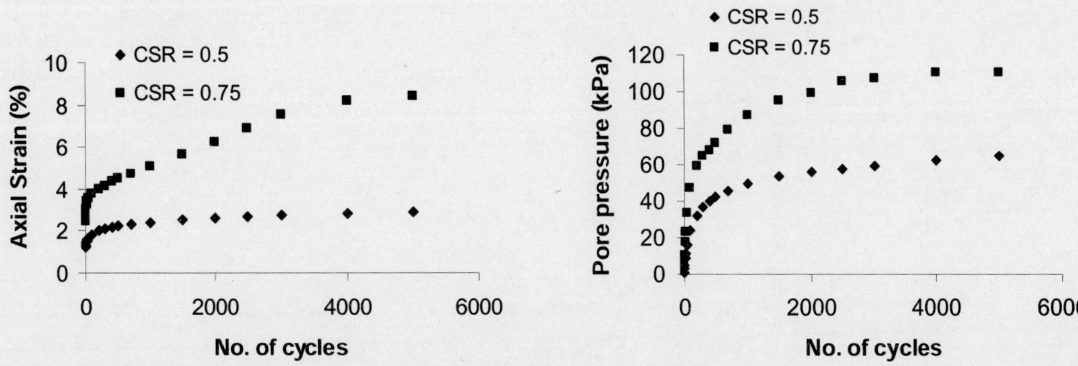

Fig. 2 Typical Strain and Pore Pressure Response Curves under Cyclic Loading under an effective confining pressure of 150 kPa

3.3 Behavior under Post Cyclic Monotonic Tests

After subjecting to 5000 cycles of compression load, the samples were kept for six hours for equalization of pore pressure. No significant change in pore pressure was observed for cyclic stress ratios up to 0.75. Then the samples were subjected to undrained monotonic test to study the post cyclic behavior. Typical stress and pore pressure variation with strain for tests carried out under an effective confining pressure of 150 kPa is shown in Fig.3. The curves for monotonic test without pre cyclic loading are also shown in Fig.3 for comparison. It can be found from the figure that the undrained strength for samples with pre cyclic loading is more for samples subjected to only monotonic loading. Similar results were obtained by Moses (2003) for marine clay at higher confining pressures. Austin (1979), as reported by Hyde and Ward (1985), also obtained an increase in undrained strength after repeated loading. This is not in agreement with many works in the literature which shows a slight decrease in undrained strength of clay after cyclic loading. According to Castro and Christian (1976) this reduction in strength may be as a result of collapse of structure and the subsequent remolding of the clay samples due to cyclic loading, which may not have happened in this case. It is reported that the cyclic loading will leave the clay samples in an apparently over consolidated state (Togrol and Guler (1984), Hyde and Ward (1985)). The increase in strength may be as a result of this apparent over consolidation. The stress strain curves show a stiff behavior with a peak, which is generally seen for over consolidated clay. Negative pore pressure developed in the sample, which was subjected to a pre-cyclic loading of cyclic stress ratio 0.75, confirms the over consolidated behavior.

The apparent over consolidation can be clearly brought out from the stress paths shown in Fig. 4.While the clay showed contractive normally consolidated behavior during monotonic tests without pre cyclic loading, it showed over consolidated behavior during post cyclic tests. It seems from the stress paths that the degree of over consolidation increases with the increase in cyclic stress ratio.

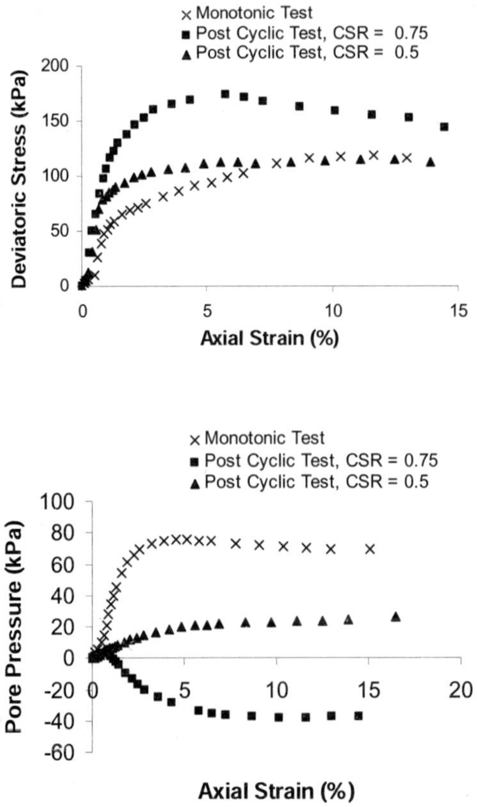

Fig. 3 Variation of deviatoric stress and pore pressure with Strain for Monotonic Tests with and without Prior Cyclic Loading

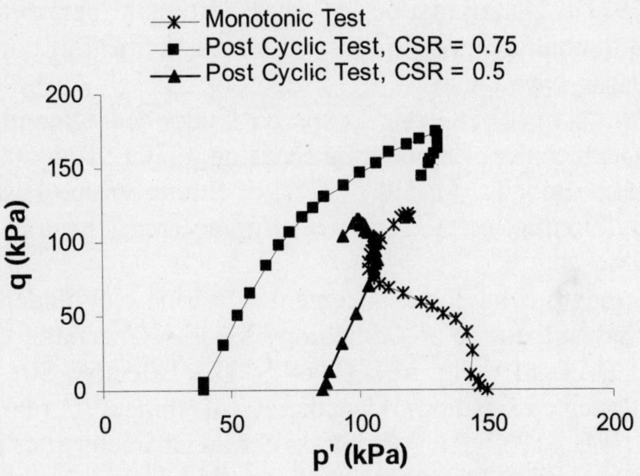

Fig. 4 Stress paths of monotonic loading with and with out cyclic loading

4 CONCLUSIONS

Undrained one way cyclic loading carried out on normally consolidated kaolinite samples showed an increase in the average strain and pore pressure with the number of cycles. The samples subjected to small cyclic stress ratios were found to reach a non failure equilibrium state. But for the tests with higher cyclic stress ratios, the pore pressure and strain was found to increase even after large number of cycles. For higher confining pressures the strains almost reached near 10% for large cyclic stress ratios. It can be expected that for higher cyclic stress ratios the strains may exceed the permissible limits at a less number of cycles which is a concern in earthquake point of view.

The pore pressure developed during cyclic loading leads to a decrease in mean effective confining pressure. From the post cyclic tests it was found that the clay gets apparently over consolidated as a result of reduction of mean effective confining pressure. The post cyclic undrained strengths were found to be higher than the undrained strengths with out cyclic loading. The apparent over consolidation might have made the clay samples to behave in a stiff manner. The samples subjected to cyclic loading with high cyclic stress ratios developed negative pore pressure during post cyclic tests, which shows that they are highly over consolidated. The over consolidated behavior of the samples during post cyclic behavior is clear from the stress paths. It can be seen that the degree of over consolidation increases with the increase in cyclic stress ratio.

REFERENCES

1. Andersen, K.H., Pool, S.F., Brown, S.F. and Rosenbrand, W.F. (1980), "Cyclic and Static Laboratory Tests on Drammen Clay," Journal of the Geotechnical Engineering Division, ASCE, 106, 449-529.
2. Andersen, K.H. (1988), "Properties of Soft Clays under Static and Cyclic Loading," Norwegian Geotechnical Institute Publication, N.R.176, 1-20.
3. Boulanger, R. W. and Idriss, I. M. (2006), "Liquefaction Susceptibility Criteria for Silts and Clays," Journal of Geotechnical and Geoenvironmental Engineering, ASCE, 132, 1413-1426.
4. Castro, G. and Christian, J.T. (1976), "Shear Strength of Soils and Cyclic Loading," Journal of the Geotechnical Engineering Division, ASCE, 102, 887-894.
5. Hyde, A.F.L. and Ward, S.J. (1985), "The Effect of Cyclic Loading on the Undrained Shear Strength of a Silty Clay," Marine Geotechnology, 6(3), 299-314.

6. IS: 2720, part 12 (1971), Determination of shear strength parameter of soil from consolidated undrained triaxial compression test with measurement of pore water pressure, Bureau of Indian Standards, New Delhi.
7. Koutsoftas, D.C. (1978), "Effect of Cyclic Loads on Undrained Strength of Two Marine Clays," Journal of the Geotechnical Engineering Division, ASCE, 104, 609-619.
8. Matsui, T., Ohara, H., Ito, T. (1980), "Cyclic Stress-Strain History and Shear Characteristics of Clay," Journal of Geotechnical Engineering Division, ASCE, 106(10), 1101-1120.
9. Moses, G.G. (2003), Strength behavior of a cemented marine clay under static and cyclic loading, Ph.D. Thesis, Indian Institute of Technology Madras, Chennai.
10. Sangrey, D.A., Henkel D.J. and Esrig, M.I. (1969), "The Effective Stress Response of a Saturated Clay Soil to Repeated Loading," Canadian Geotechnical Journal, 6, 241-252.
11. Thiers, G. R. and Seed, H.B. (1968), "Cyclic Stress-Strain Characteristics of Clay," Journal of Soil Mechanics and Foundation Division, ASCE, 94, 555-569.
12. Toglor, E. and Guler, E. (1984), "Effect of Repeated Loading on the Strength of Clay," Soil Dynamics and Earthquake Engineering, 3(4), 184-190.
13. Wood, D.M. (1982), "Laboratory Investigations of the Behavior of Soils under Cyclic Loading: A Review," In Soil Mechanics, Transient and Cyclic Loads, John Wiley and Sons, England, 513-582.
14. Yasuhara, K., Yamanouchi, T. and Hirao, K. (1982), "Cyclic Strength and Deformation of Normally Consolidated Clay," Soils and Foundations, 22(3), 1-18.
15. Yasuhara, K., Hirao, K. and Hyde, F.H. (1992), "Effects of Cyclic Loading on Undrained Strength and Compressibility of Clay," Soils and Foundations, 32(1), 100-116.
16. Yilmaz, M.T., Pekcan, O. and Bakir B.S. (2004), "Undrained Cyclic Shear and Deformation Behavior of Silt-Clay Mixtures of Adapazari, Turkey," Soil Dynamics and Earthquake Engineering, 24, 497-507.
17. Zergoun, M. and Vaid, Y.P. (1994), "Effective Stress Response of Clay to Undrained Cyclic Loading," Canadian Geotechnical Journal, 31, 714-727.

LOW STRAIN SHEAR MODULUS FROM FIELD AND LABORATORY TESTS

R. Uma Maheswari[1], A. Boominathan[2] and G.R. Dodagoudar[3]
[1]Research Scholar, Indian Institute of Technology Madras, umacivil2003@yahoo.com
[2]Professor, Indian Institute of Technology Madras, boomi@iitm.ac.in
[3]Assistant Professor, Indian Institute of Technology Madras, goudar@iitm.ac.in

ABSTRACT

The shear modulus of soil at a small strain level is normally a required parameter in many earthquake geotechnical problems such as dynamic soil response analysis, modeling the non-linear stress strain relationship and Seismic soil structure interaction analysis. This parameter can be found out from either laboratory or field tests. To develop a greater confidence of in-situ test results, it is helpful to compare field results to conventional laboratory tests. This paper presents the results of shear wave velocity measurements carried out using the Multichannel analysis of surface wave (MASW) tests in the field and the bender element tests conducted on undisturbed soil specimens in the laboratory. The MASW tests are carried out using Geometrics make 24 channels Geode seismic recorder with SGOS operating software. In this test 24 numbers of vertical geophones with natural frequency of 4.5 Hz are used to receive the wavefield generated by the active source of 8 kg sledgehammer. The bender element tests incorporated in cyclic triaxial apparatus of Wykeham Farrance, UK make is used to measure the shear wave velocity of soil specimens. The maximum shear modulus (G_{max}) obtained from the field MASW test is compared with the laboratory bender element tests. It is found that the low strain shear modulus obtained from the field MASW test is higher than the low strain modulus obtained from the laboratory bender element tests. The lower value of G_{max} from bender element tests is due to the disturbance in the sampling processes and boundary conditions involved in the bender element testing procedure.

Keywords: Maximum shear modulus, Shear Wave Velocity, MASW Test, Bender Element Test

1 INTRODUCTION

The evaluation of the dynamic response of a soil or a structure founded on soil requires knowledge of the stress strain properties of the foundation material. The low strain shear modulus G_{max} is an important parameter for the development of modulus reduction curve. The degradation of soil stiffness is represented by modulus reduction curve which is the key parameter for ground response and soil structure interaction analyses during earthquake. The estimation of G_{max} can be made either by carrying out field tests or by laboratory tests, in which the deformation properties of soil are related to elastic shear wave velocity.

Low strain field tests typically induce seismic waves in the soil and measure the velocities at which these waves propagates. The advantage of in-situ techniques includes the test in an undisturbed state and allows larger volume of soil to be tested. Although correlations can be used to estimate the in-situ shear modulus of the soil, a direct measurement of the soil property is always desirable. Field measurements of shear wave velocity include crosshole tests, downhole tests, suspension logging, seismic reflection, seismic refraction and surface wave tests. Surface wave test is the simple and efficient technique compared to other in-situ tests.
Laboratory measurements have been the reference standard for determining the properties of geomaterials. It includes cyclic triaxial, bender element test and resonant column tests. These tests can be conducted in the laboratory on soil specimens to obtain moduli at different strain level. To

develop a greater confidence of the results of in-situ tests, it is helpful to compare field results to conventional laboratory tests.

In this study, an attempt has been made to compare the low strain modulus of soil obtained from bender element tests carried out on undisturbed samples collected from typical sites in Chennai city with the same obtained from the Multichannel Analysis of Surface Wave (MASW) tests carried in the field.

2 SITE DESCRIPTION

Two typical sites in Chennai city namely Tondiarpet and Siruseri are considered for the present study. The subsoil profile and the variation of SPT-N value with depth for Tondiarpet and Siruseri is shown in Figure 1. The soil profile at Tondiarpet site consists of fine to medium sand upto a depth of 9.5 m from ground level. In this layer, the SPT-N value increases from 20 to 60. This layer is followed by soft clay with a low SPT-N value of 2. The ground water table (GWT) occurs at a depth of about 1 m from GL. The soil profile at Siruseri site consists of top silty clay layer with thickness of 0.6 m. This layer is followed by stiff clay upto a depth of 4.4 m with SPT-N value varies from 6 to 19. Subsequently, this layer is followed by fine to medium sand upto a depth of 6.5 m with SPT-N value varies from 20 to 30. The very dense coarse sand with stones occurs at 6.7 m with SPT-N value of above 100. The GWT encounters at a depth of 2.4 m from GL.

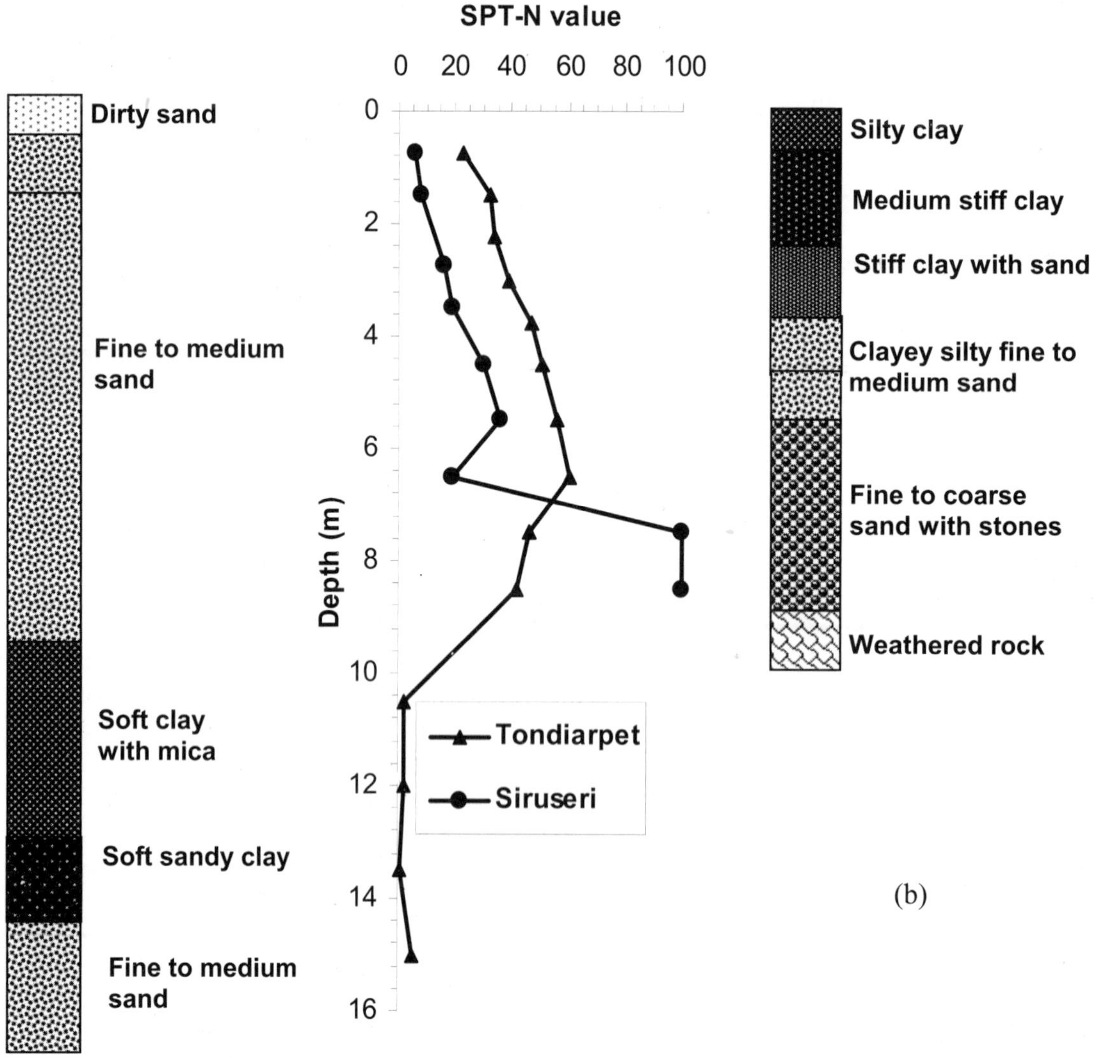

Figure1. Subsoil profile and the variation of SPT-N value variation of (a) Tondiarpet (b) Siruser

3 IN-SITU SEISMIC TESTING PROGRAM

Measurement of shear wave velocity can be grouped into two categories, invasive and noninvasive (surface) procedures. Surface based techniques use surface receivers to measure the travel times of seismic waves with distance along the surface. Surface wave methods offer advantages over other surface-based in-situ seismic techniques which include the ability to measure both shear wave velocity and material damping profiles with depth and the ability to detect low velocity features underneath higher velocity layers, allowing for more accurate site characterization and the detection of susceptible liquefaction features. Rayleigh waves are considered for site characterization and it is characterized by relatively low velocity, low frequency and high amplitude. Particle motion in the fundamental mode of Rayleigh waves moving from left to right is elliptical in a counter-clockwise (retrograde) direction. For a given mode, longer wavelength surface waves will penetrate deeper into the earth than shorter wavelengths.

Nazarian et al (1983) introduced a surface wave method called spectral analysis of surface waves (SASW) to produce near surface shear wave velocity profiles. The main disadvantage of SASW test includes the time consumption for field survey due to single pair of receivers. Later on, a four-phase research project team from Kansas Geological Survey developed an efficient and accurate method to estimate near surface shear wave velocity from ground roll using multichannel seismic data from Multichannel analysis of surface waves (MASW) test. The surface wave method, used to produce a shear wave velocity profile, involves three steps: acquisition of ground roll, construction of dispersion curve (phase velocity vs. frequency (f)), and back calculation (inversion) of the V_s profile from the calculated dispersion curve (Park et al., 1999). The multichannel surface wave data are acquired using multiple numbers of receivers deployed in a linear pattern of equal receiver spacing with each receiver connected to an individual recording channel as shown in Figure 2.

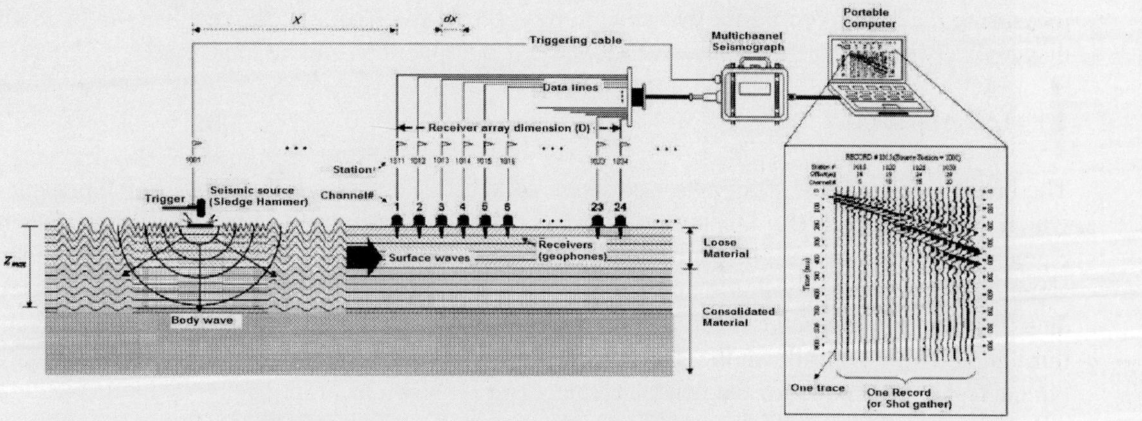

Figure2. The overall test setup for MASW test

In this case the motion generated by an impact source is detected simultaneously at several receiver locations and the corresponding signals are analyzed as a whole using double Fourier transform. Raw field data are transformed into the f-k domain where phase velocities of Rayleigh waves can be calculated to produce a dispersion curve. Then the calculated dispersion curve can be inverted to estimate the V_s profile.

In this study, Geometrics make 24 channels Geode seismic recorder with SGOS operating software is used to carry out MASW tests at Todiarpet and Siruseri site. The vertical geophones of 4.5 Hz (24 Nos.) are used to receive the wavefields generated by the active source of 8 kg sledgehammer. Twenty four geophones were deployed on 1m interval with the nearest source to geophone offset in the range 5 to 15 m to meet the requirement of different type of soil (Xu et al., 2006). The acquired surface wave data were processed using the SURFSEIS software package to develop shear wave velocity profile.

4 LABORATORY TESTING PROGRAM

The measurement of shear wave velocity in soils using piezo electric transducers are not satisfactory due to weak s –wave directivity, poor coupling with the soil, high operating frequency and high impedance mismatch. To avoid these above said problems, Shirley and Hampton (1977) introduced bender elements to soil testing. Later on, Dyvik and Madshus (1985) showed the agreement between G_{max} with bender elements and resonant column. The piezo electric bender element technique is relatively simple and non-destructive test for determining shear wave velocity and hence maximum shear modulus corresponding to very small strain (10^{-5} %).

The bender element consists of two sheets of piezoelectric ceramic materials such as lead zirconate titanate, barium titanate or lead titanate sandwiching a center shim of brass, stainless steel or other ferrous nickel alloys to add strength to it. The bender element is an electromechanical transducer capable of converting mechanical energy into electrical energy and vice versa. When a driving voltage is applied to the bender element, the polarization will cause a bending displacement and, thus, the bender element acts as a signal generator. When the element is forced to bend, a voltage is generated and, thus, the bender element can act as a signal receiver. Benders are available in two versions that differ in the electrical connection to the two polarized plates: a series or parallel connection. The series version is a better receiver because of higher output for a given distortion. The parallel version is better used as a transmitter because it gives largest distortion for a given input. With this connection the available voltage is applied to each ceramic plate and is not divided between the two plates as in the series connection. Bender elements can be inserted into the top and bottom of a soil specimen. The measurement of shear wave velocity is based on transmission of a mechanical signal through a soil sample. A sharp voltage pulse applied to one element causes a shear wave that travels through the soil. When that shear wave reaches the other bender element, the deflection of the receiving bender element produces a voltage that can be measured. By measuring the time required for the wave to travel from the source to the receiver, and knowing the distance between each, the shear wave velocity of the specimen can be measured.

4.1 Test Apparatus

The bender element test setup procured from Wykeham Farrance, UK is used in the present study, which is available in the Department of Civil Engineering, IIT Madras, Chennai. The bender elements are fixed in a servo controlled cyclic triaxial system. The function generator generates a signal with specified amplitude and frequency as input signal and PC based oscilloscope records data from transmitter and receiver bender elements. The PC based oscilloscope is controlled through software that allows the user to quickly and easily calculate the shear wave velocity. The typical test configuration of the bender element test is shown in Figure 3.

4.2 Material Properties and Test procedure

The undisturbed soil samples collected from the sites described in Section 2 are used in the bender element tests. The index properties of these soil samples determined from the laboratory tests are summarized in Table 1.

Table 1. The index properties of test materials

Location	Depth (m)	Soil description (soil classification)	G_s	NMC (%)	Atterberg limits (%)		
					LL	PL	PI
Tondiarpet	10.00	Soft clay with mica (CH)	2.7	60.0	76.8	28.0	48.8
Siruseri	2.25	stiff clay (CH)	2.7	33.9	84.9	25.6	59.3

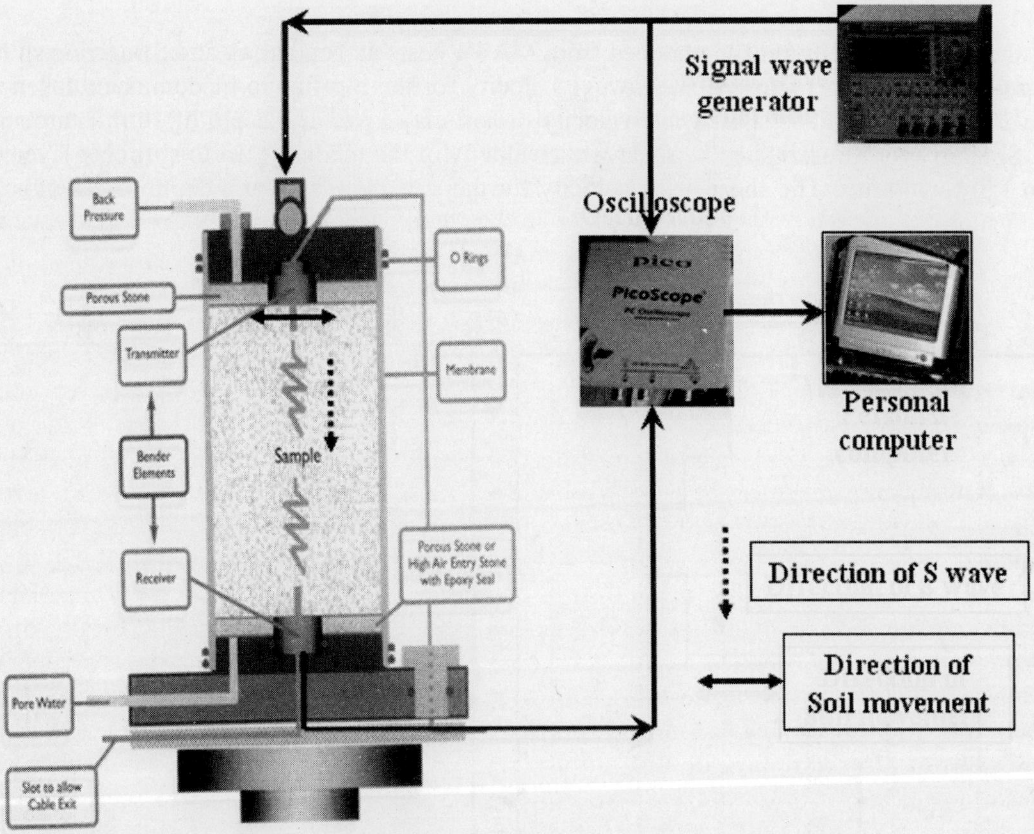

Figure 3. The bender element test configuration

The undisturbed specimens are trimmed to the required size of 50 x 100 mm^2 with minimum sample disturbance to maintain the density and initial water content. The specimen is mounted using the membrane stretcher. After assembling and filling the triaxial chamber with water, back pressure is applied in the specimen in steps and evaluation of degree of saturation is done at appropriate intervals by measuring Skempton's porewater pressure parameter B and a value of more than 0.95 is ensured. Then the saturated soil sample was isotropically consolidated to the required in-situ effective confining stress. The soft clay and stiff clay specimens collected from Tondiarpet site and Siruseri sites were consolidated under an effective consolidation stress of 120 kPa and 34 kPa respectively. The consolidated height and volume of the sample were noted by using the data acquisition system incorporated with the cyclic triaxial apparatus.

After consolidation, wave signal was generated using signal wave generator to transmitter bender element. The simplest way to obtain a bender element trace that may be interpreted objectively is to use a sinusoidal wave rather than a more usual square wave. For better interpretation of arrival time as proposed by Viggiani and Atkinson (1995b) sine wave pulse was adopted in this study. For most interpretable wave forms, Brignoli et al. (1996) suggested frequency range of 3 to 10 kHz for shear wave measurements on clay specimen. In this study, a sine wave form with 5 kHz frequency and input voltage amplitude of 20 V_{pp} were used based on the intensive literature studies. The quality of the receiver signal is improved by applying the following performance criteria: the signal to noise ratio of the receiver signal is at least 4 db and the wave path length to wavelength ratio (L_{tt}/λ) is atleast 3.33. The arrival of first S wave was identified based on the period between the initial/peak /and trough point of the input sine wave and the corresponding output wave.

5 RESULTS AND DISCUSSION

The shear wave velocity profile obtained from MASW tests at Tondiarpet and Siruseri is shown in Figure 4. In Tondiarpet site, the shear wave velocity for the top fine to medium sand layer varies from 200 to 275 m/s. The shear wave velocity of soft clay layer at a depth of 10 m is around 150 m/s. Similarly for Siruseri site, the shear wave velocity of the medium stiff to stiff clay layer varies from 150 to 300 m/s. The shear wave velocity for the stiff clay layer at a depth of 2.3 m is about 220 m/s.

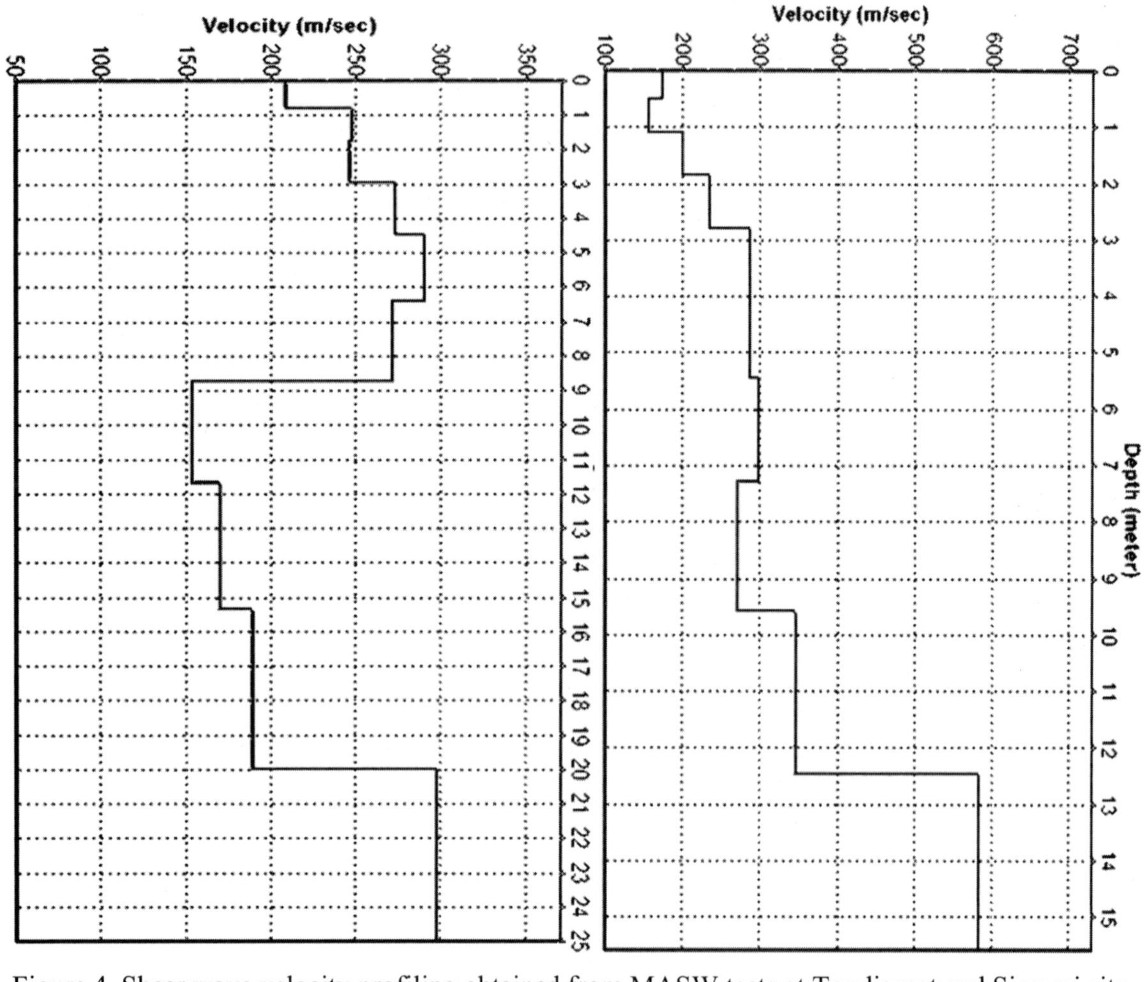

Figure 4. Shear wave velocity profiling obtained from MASW tests at Tondiarpet and Siruseri site

A typical receiver signal (red line) obtained from the bender element tests conducted on Tondiarpet and Siruseri samples at a depth of 10 m and 2.25 m respectively are shown in Figure 5. The shear wave velocity of soft and stiff clay obtained from the bender element tests is found to be 120 m/s and 190 m/s respectively.

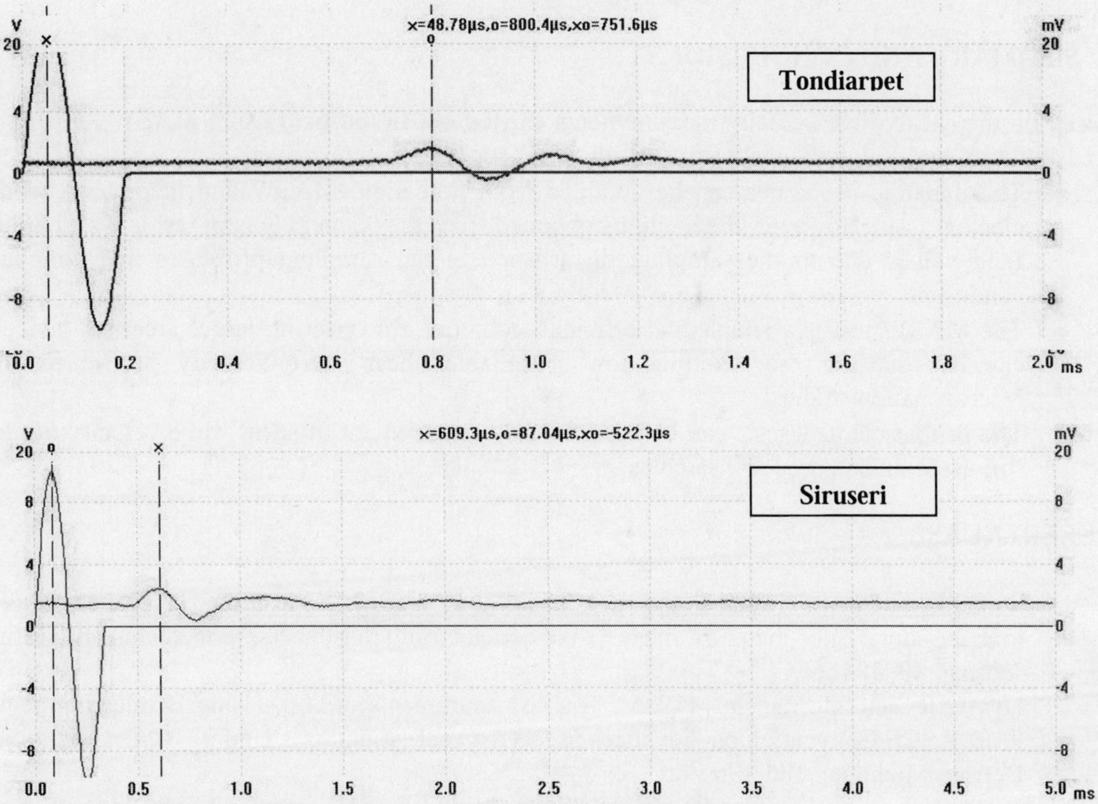

Figure 5. The typical receiver signals obtained from bender element test

In this study, low strain shear modulus (G_{max}) was obtained using the following equation:

$$G_{max} = \rho V_s^2 \qquad (1)$$

where ρ is the unit weight of the soil and V_s is the shear wave velocity. The shear modulus at low strain amplitude has been determined from the laboratory and field tests using the above formulae are listed in the Table 2.

Table 2: Low strain modulus from MASW and Bender Element Tests

Test location	Soil description	MASW tests		Bender element tests	
		V_s (m/s)	G_{max} (kPa)	V_s (m/s)	G_{max} (kPa)
Tondiarpet	Soft clay (CH)	150	36	120	23
Siruseri	Stiff clay (CH)	220	74	190	55

It can be easily noticed from Table 2, that the low strain modulus G_{max} from field MASW test is about 30 % higher than the laboratory bender element tests for both soft and stiff clay. The lower value of G_{max} from bender element tests is due to the disturbance in the sampling processes and boundary conditions involved in the bender element testing procedure. The similar difference between field and laboratory shear modulus is reported by Larsson and Mulabic (1991). It is also reported that the value obtained for laboratory tests may be as low as one half the field values (Marcuson and Joseph 1981).

6 SUMMARY AND CONCLUSIONS

Based on the shear wave velocity measurements carried out in the field using MASW test and in the laboratory by bender element tests, the following conclusions are arrived:

- The dynamic properties can be obtained from both field MASW and laboratory bender element tests. However, the modulus obtained from the bender element test is less than the field values due to the sampling disturbance in the sampling processes and boundary conditions.
- The MASW test is a simple and efficient technique for covering larger area and hence it can be used for the determination of reliable shear wave velocity profile for the Microzonation studies.
- The bender element tests can be used for the measurement of shear wave velocity of clay soil using undisturbed soil samples.

REFERENCES

1. Brignoli, E. G. M., Gotti, M. and Stokoe, K. H. II. (1996). "Measurement of Shear Waves in Laboratory Specimens by means of Piezoelectric Transducers," Geotechnical Testing Journal, **19**, 384-397.
2. Dyvik, R., and C. Madshus (1985). "Lab Measurements of G_{max} Using Bender Element," Proc. ASCE Convention on Advances in the Art of Testing Soils Under Cyclic Conditions, Detroit, Michigan, 186-196.
3. Larsson, R., and M. Mulabdic (1991). "Shear Moduli in Scandinavian Clays: Measurement of Initial Shear Modulus with Seismic Cones," Report No. 40, Swedish Geotechnical Institute, Linkoping.
4. Nazarian, S., Stokoe II, K.H., and W.R. Hudson (1983). Use of Spectral Analysis Surface Waves Method for Determination of Moduli and Thickness of Pavement Systems: Transportation Research Record No. 930, 38-45.
5. Park C.B., Miller, R.D., and J. Xia (1999). "Multichannel Analysis of Surface Waves," Geophysics, **64** (3), 800-808
6. Shirley, D. J., and L.D. Hampton (1977). "Shear Wave Measurements in Laboratory Sediments," Jl. of the Acoustic Society of America, **63**, 601-613.
7. Viggiani, G. and J.H. Atkinson (1995b). "Stiffness of Fine-Grained Soil at Very Small Strains," Geotechnique, **45**, 249-265.
8. Marcuson, W.F., and J. R. Curro (1981). "Field and Laboratory Determination of Soil Moduli," Jl. of the Geotechnical Engineering Division, 107, **GT10**, 1269-1291.
9. Xu, Y., Xia, J., and R.D. Miller (2006). "Quantitative Estimation of Minimum Offset for Multichannel Surface Wave Survey With Actively Exciting Source", Jl. of Applied Geophysics, **59**(2), 117-125.

GRAIN AND SPECIMEN SIZE EFFECTS ON LIQUEFACTION OF FINE SANDS TO UNDRAINED CYCLIC TRIAXIAL LOADING

K. Rangaswamy, [1] A. Boominathan, [2] and K. Rajagopal [3]
[1]Research Scholar, Indian Institute of Technology Madras, rangadr2002@yahoo.co.in
[2]Professor, Indian Institute of Technology Madras, boomi@iitm.ac.in
[3]Professor, Indian Institute of Technology Madras, gopalkr@iitm.ac.in

ABSTRACT

A series of cyclic triaxial tests are performed on fine sand in order to determine the effect of grain and specimen size on liquefaction behavior. There are three types of uniformly graded sands namely DHS, CHS and MBS are chosen with different mean grain diameters of 0.11 mm, 0.26 mm and 0.32 mm respectively. DHS specimens were prepared with different sizes of 50.4 mm, 70 mm and 100 mm ϕ in order to determine the size effect on liquefaction behavior. Moist tamping & under compaction procedure was used to make samples around 40% relative density. Samples were consolidated to 100 kPa and 200 kPa. Two-way cyclic loads were applied sinusoidally to the saturated and consolidated sample with predetermined CSR in the range of 0.10 to 0.26 under undrained triaxial conditions. Liquefaction behavior is examined in terms of buildup pore water pressure ratio or cyclic stress ratio caused to liquefaction at a given number of loading cycles. Axial strain responses with the number of cycles are presented and it shows that extensional failures were observed in specimens of 50.4 mm ϕ size due to dilation. Results are showing that the samples with grain size, D_{50} = 0.32 mm were more resistance to liquefaction about 30%-45% than the sample with average grain sizes of D_{50}=0.112 mm and D50=0.26 mm. It also found that the samples with specimen size of 50.4 mm ϕ were more resistance to liquefaction about 15%-30% than the samples of 70mm ϕ or 100 mm ϕ in size.

Keywords: Sand samples, Pore water pressure ratio, cyclic stress ratio, Liquefaction

1 INTRODUCTION

Liquefaction is a phenomenon wherein the shear resistance of a mass of soil decreases when subjected to monotonic, cyclic, or dynamic loading at constant volume. Liquefaction of soils is a complex problem and caused to huge damages to soil and structures like sand boils, sliding of embankment slopes, failures of dams and buildings. Numerous factors affect the resistance to liquefaction measured in a cyclic triaxial specimen. These factors include the gradation curve and mean grain size of the soil, the void ratio and the relative density of the specimen, the method of specimen preparation, the size of the specimen, the shape of the loading function, the frequency of loading, and the degree of saturation in the specimen. Silver et al. (1976), Townsend (1978) and Tatsuoka et al. (1986) have provided excellent reviews of many of these factors. Among all factors, initially it is essential to study the grain size, gradation of soil and size of specimen on liquefaction behavior to proceed in further parametric investigations in cyclic triaxial testing. In present, there are very limited research publications on these parametric studies. It has been shown that the cyclic undrained triaxial strength increases either with the decrease in specimen diameter for a constant particle size or with the increase in particle size for a constant specimen diameter (Lade and Hernandez, 1977; Martin et al., 1978 and Tatsuoka et al, 1986).

2 LITERATURE REVIEW

Lee and Fitton (1969) tested on prepared samples of course and fine sands with different specimen diameters of 1.4 inches (36 mm) and 2.8 inches (71 mm) in order to determine the effects of specimen size on cyclic strength. The samples were prepared to a constant relative density and tested at a constant confining pressure. Results are found that less than a 5% variation in liquefaction resistance between the two specimen diameters with the larger specimen being slightly weaker. It was also observed that sands with D_{50} in the range of 0.1 mm have the lowest resistance to cyclic loading.

Wong et al. (1975) conducted the cyclic loading tests on various samples of Monterey sand and shell material from Oroville Dam with different Grain sizes. It was seen that as the mean grain size increases from about 0.1 mm to 30 mm, the stress required causing ±2.5% strain increases by about 30% while that required causing ±10% strain increases about 60%. In order to determine the effect of sample size on cyclic stresses causing liquefaction, cyclic loading tests were performed on Monterey Sand. It was found that the cyclic deviator stresses required to cause initial liquefaction in any given number of stress cycles is about 10% less for the 12 inch (300 mm) diameter samples than for the 2.8 inch (70 mm) diameter samples.

Castro et al. (1977) presented the results on cyclic strength of uniform sands having different particle mean size (0.33 mm – 0.74 mm) and showed that the sands with smaller D_{10} have lower resistance to cyclic deformations.

Hosri et al. (1982) carried out undrained cyclic triaxial tests on undisturbed silty clay samples and shown that the cyclic strength increases with the increase of D_{50} in the range of 0.009 mm to 0.132 mm.

There is lot of confusion from review how much these parameters were affected to soil resistance. Therefore, an attempt has been made to examine the grain and specimen size of very fine sand soil on liquefaction behavior by carrying out undrained stress controlled cyclic triaxial testing.

3 EXPERIMENTAL PROGRAM

A series of undrained stress controlled cyclic triaxial tests were conducted on the reconstituted clean sandy specimens at the Geotechnical Engineering Laboratory of IIT Madras. A detailed description of the equipment, material, sample preparation procedure, and testing parameters used is given below

3.1 Test Equipment

Laboratory testing was performed using a servo-pneumatic cyclic triaxial testing device (make: Wykeham Farrance International Ltd, England, United Kingdom), which is fully automated and controlled by software on a personal computer. The loading system having 100 kN load frame and ± 5 kN double acting actuator can able to generate frequencies up to ±70 Hz. A special type of submersible load cell of capacity ± 5 kN is fitted inside of triaxial cell to measure the applied cyclic axial load. A strain control displacement transducer of ± 25 mm capacity is used for the measurement of axial displacement. Three pressure transducers of a diaphragm type with ± 1000 kPa capacity are used to monitor the pore water pressure and measure the cell and back water pressures. A 100 cc volume change transducer is used to monitor the volume of water entering or leaving the sample during saturation and consolidation stages. An internal actuator LVDT of ± 15 mm span is used to detect the peak to peak deformations in a specimen during cyclic loading. All these transducers are connected to the data acquisition system and the data is transfer to the system through RS232 communication link for the processing, display and storage in integrated software called UTM_27.

3.2 Material Tested

Three types of sands were used in this study to investigate the effect of grain size on liquefaction behavior of soils from undrained cyclic triaxial testing. Sand soils collected from three regions, named as Dhuvaran Sand (DHS), Chirala sand (CHS) and Mahabalipuram sand (MBS) with different size of grains are used in the present study. These sands are classified as poorly graded, medium to fine sand (SP) according to the Indian Soil Classification System. The grains were mostly rounded to sub-rounded in shape. Dhuvaran sand is brownish, Chirala sand is white and Mahabalipuram sand is reddish in colors. In case of Dhuvaran sand, 92% of soil is passing through 0.150 mm sieve and retained on 0.075 mm sieve. MBS and CHS samples are more courser than DHS samples and the fines passing through 0.150 mm sieve are only 15 to 20%. The grain size distribution curves of tested soils are shown in Figure 1 and corresponding granular properties are given in Table 1. In order to study the sample size effect on liquefaction, Dhuvaran sand specimens were prepared with 50 mmϕ, 70 mm ϕ and 100 mmϕ sizes.

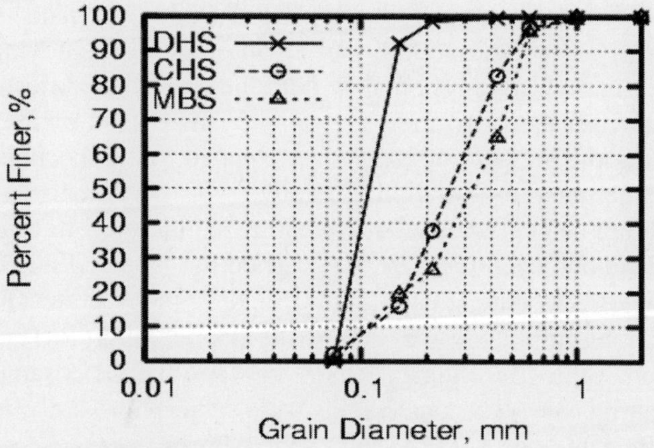

Fig. 1 Grain Size Distribution of the tested Soils

Table 1. Properties of Sands used for testing

Soil type	Gs	D_{10} (mm)	D_{30} (mm)	D_{50} (mm)	D_{60} (mm)	D_{60}/D_{10} (C_u)	$D_{30}^2/(D_{60}*D_{10})$ (C_c)	e_{max}	e_{min}
DHS	2.69	0.081	0.095	0.112	0.125	1.54	0.893	0.975	0.675
CHS	2.61	0.13	0.19	0.26	0.30	2.31	0.925	0.781	0.535
MBS	2.72	0.11	0.23	0.32	0.40	3.64	1.202	0.642	0.457

3.3 Sample preparation

Reconstitute triaxial samples are prepared by moist tamping and undercompaction method as proposed by Ladd (1978). It avoids the particle segregation in case of fine sands. To study the effect of specimen size on liquefaction behavior, DHS samples with D_{50}= 0.112 mm were prepared in various sizes: 50.4 mm, 70 mm and 100 mm in diameter. In addition to that DHS, CHS and MBS samples were prepared to a constant size of 50.4 mm diameter to examine the grain size effect on soil liquefaction. H/D ratio was maintained above or equal to two as per ASTM-D5311. All specimens are reconstituted to around 40% relative density. As per the undercompaction procedure air dried soil sample is divided in to five parts, with each part determined by the weight and relative density of soil with the same volume. To eliminate the consolidation effects and to attain a uniform density throughout the specimen, the relative density at which each part of layer placed in the sand former was different. The relative density of each part of the layer, from top to the bottom of the soil sample, is R_d +2, R_d +1, R_d, R_d −1 and R_d −2% so that the compaction of each succeeding layer could further densify the soil below it and therefore a relatively uniform specimen was obtained. Each part of the sample is placed in a separate sealed container and mixed with 6 % water content which gives 30% degree of saturation initially. In order to obtain very loose specimens, small percentages of water are added to the sand. The capillary tension due to this moisture enables to obtain loosest possible dense condition till the maximum void ratio. For getting the uniform distribution of water content, the water mixed samples are kept in sealed containers for at least eight hours before placing and making the sample in the internal split mould. Each of the five parts of the sample was spooned in layers one by one into the membrane-lined split former and was compacted to the required height with a help of tamping rod having rubber tamping disc of 25 mm diameter. Each layer of 20 mm height is marked on tamping rod for the five layers and the tamping depth for every layer is controlled by horizontal metallic strip which is located at the top of split mould. The top of each compacted layer was slightly scarified before placing the next layer to promote proper bonding.

3.4 Testing procedure

Undrained stress controlled cyclic triaxial tests were performed on reconstituted saturated specimens prepared at 40% relative density as per ASTM-D5311. After the specimen was formed, the specimen top cap was placed and sealed with O-rings. Before removing the former from the sample, internal vacuum of 8 kPa was applied to hold the specimen. After removal of the mould, sample dimensions were measured and then the sample was placed in the triaxial cell and later filled with water. An isotropic pressure of 20 kPa was applied to the specimen and internal vacuum removed from the specimen before saturation. Saturation was performed by purging the specimen with CO_2 before circulating deaired water. Back pressure saturation followed by the application of increments of cell pressure and back pressure was carried out as per BS 1377: Part 8: 1990. After saturation, with a Skempton's B value greater than 0.96, specimens were consolidated isotropically under a desired effective confining stress of σ'_c = 100 kPa or 200 kPa. The samples having D_{50}= 0.112, 0.26 and 0.32 mm were consolidated to 200 kPa and the DHS samples (D_{50}=0.112) were subjected to 100 kPa confinement. Then specimens were subjected to a cyclic deviator stress, σ_{dc} in axial direction with a frequency of 1.0 Hz. The cyclic triaxial tests were performed with various cyclic stress ratio (CSR) = ($\pm\sigma_{dc}/2\sigma'_c$) until the occurrence of initial liquefaction (excess pore pressure, Δu is equal to the applied effective confining pressure, σ'_c) or 5% axial strain. During the cyclic loading, continuous records of the excess pore water pressure, Δu, and cyclic axial strain, ε_c, as well as of the cyclic axial stress, σ_c, applied to the specimen were recorded automatically using the integrated software.

4 RESULTS AND DISCUSSION

A typical result of cyclic triaxial tests carried out on reconstituted DHS sample (D_{50}=0.112) of 100 mm ϕ specimen, isotropically consolidated at a stress of σ'_c =100 kPa and subjected to cyclic loading with the higher cyclic stress ratio (CSR) of 0.15 is shown in Fig. 2. The applied cyclic deviator stress to the sample with the number of cycles is presented in Fig. 2(a). Two-way cyclic loading either in compression and extension was applied to simulate the field earthquake conditions. The variation of pore-water pressure ratio i.e. the excess pore water pressure normalized to effective confining stress ($\Delta u/\sigma'_c$) with number of cycles of loading is shown in Figure 2(b). Figure 2(c) shows the variation of corresponding axial strain with the number of cycles of loading. The plot of deviator stress with axial strain and mean effective confining pressure is shown in Figures 2(d) and 2(e) respectively.

It can be easily noticed from Figure 2(b) that the pore water pressure ratio increases rapidly with an increase of number of cycles and liquefaction failure occurs at about 21 cycles. Figure 2(c) indicates an occurrence of little axial strain before the initial liquefaction and nearly 3.5% of double amplitude strain at 21 cycles of loading. However, once initial liquefaction was triggered, sample experienced large strains rapidly within about 1 loading cycle and then it was failed due to softening of soil. It can be found from Figure 2(d) that at very low strain, the soil specimen experiences the applied deviator stress but at larger strain, the soil is unable to take the applied deviator stress and hence it behaves as a liquid. From Figure 2(c) and Figure 2(d), it was clearly observed that the sample fails in compression side with the 7% of strain. It is the characteristic of loose behavior (contractive) of sample and so that the sample was liquefied within 21 cycles due to the rapid pore-water pressure buildup in the soil. Figure 2(e) shows the effective stress path followed during the test in terms of q' and $p' = \sigma'_1 + 2\sigma'_c/3$. It can be noticed from the figure that the mean effective confining pressure decreases as the increase of pore pressure build-up in the soil.

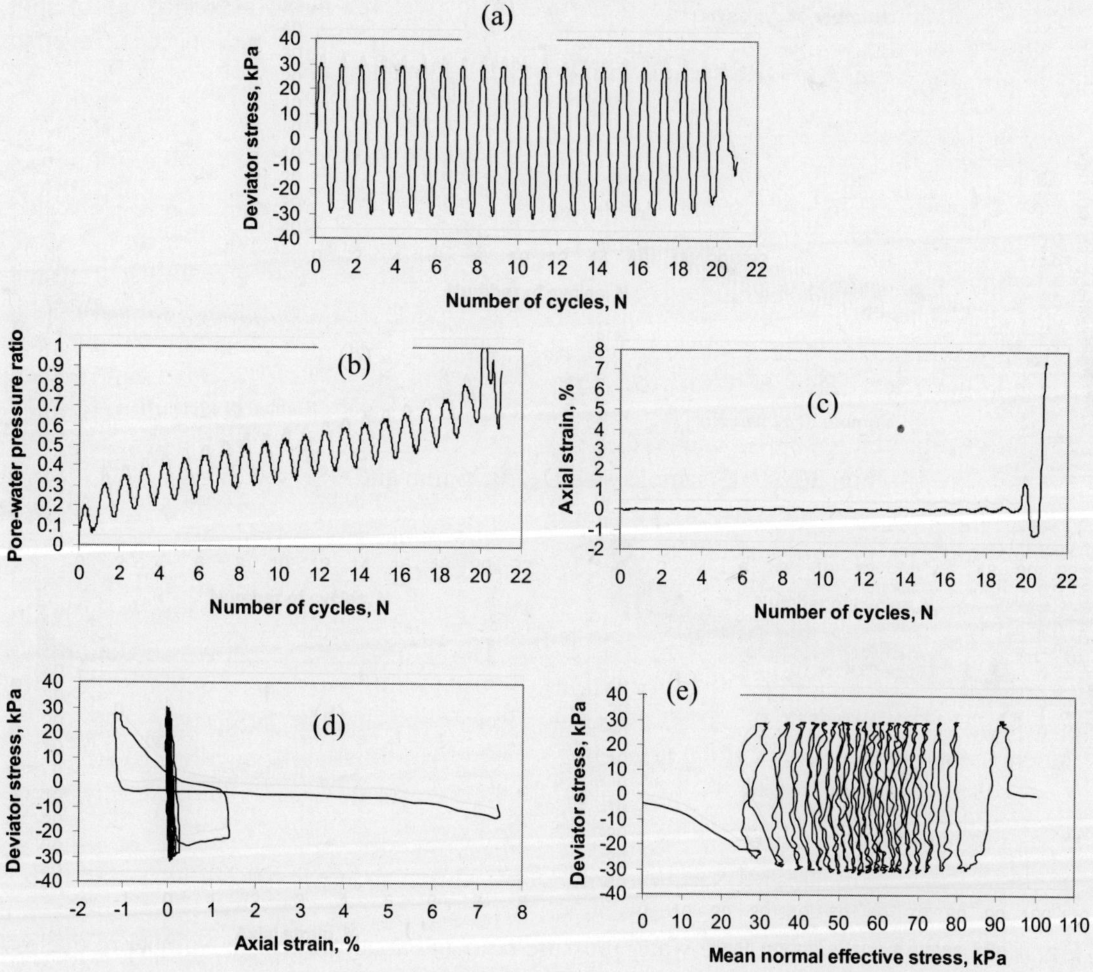

Fig. 2 Typical cyclic response of DHS sample (D_{50}=0.112) under undrained triaxial loading with applied deviator stress of ± 30 kPa (CSR=0.15).

The influence of grain size on pore-water pressure generation and corresponding axial strain responses with number of loading cycles at a constant cyclic stress ratio of 0.15 to 0.2 were shown in Figure 3. The rate of pore-water pressure ratio during cyclic loading with the number of cycles was increases with decrease in grain size, D_{50} from 0.32 mm to 0.112 mm. The samples with D_{50}=0.112 mm were liquefied at about 42 cycles, while the samples with D_{50}=0.26 mm and D_{50}=0.32 mm were liquefied at about 65 and 112 cycles respectively. From the stress-strain plots, it was found that all samples were subjected to either compression or extensional strains and failed in compression side.

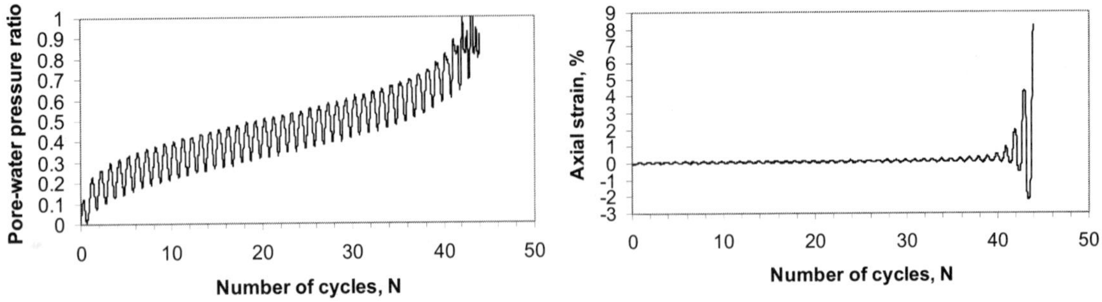

Fig. 3(a) DHS sample with D_{50}=0.112 mm and CSR = 0.15

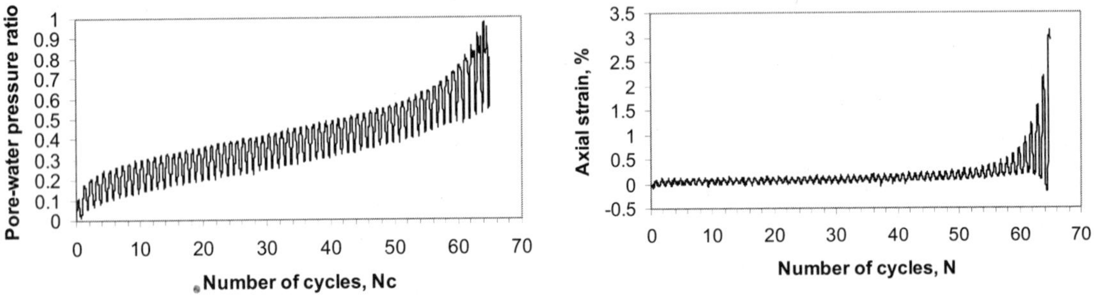

Fig. 3(b) CHS sample with D_{50}=0.26 mm and CSR = 0.2

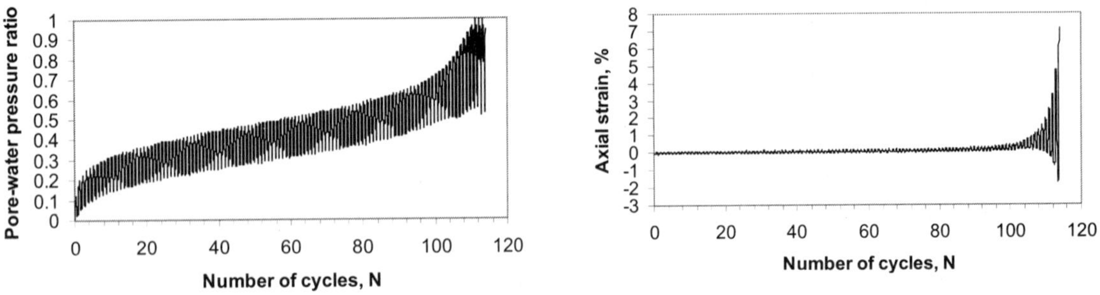

Fig. 3(c) MBS sample with D_{50}=0.32 mm and CSR = 0.2

Fig. 3 Effect of grain size on Pore-water pressure ratio and axial strain with Number of cycles

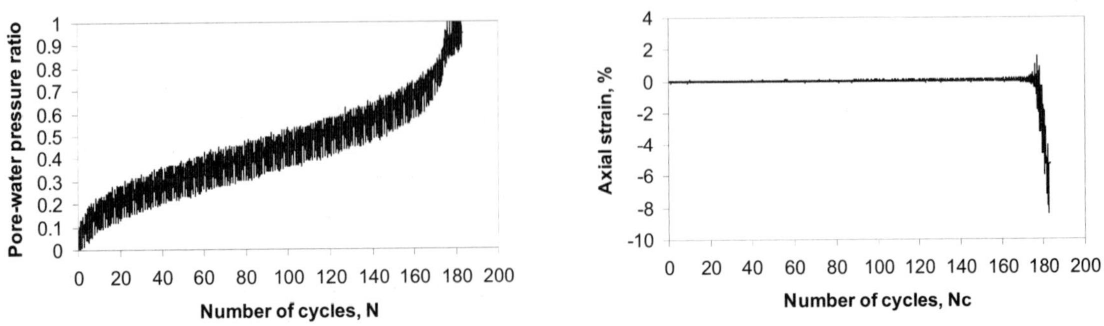

Fig. 4(a) DHS sample specimen with 50.4 mm ϕ and CSR = 0.125

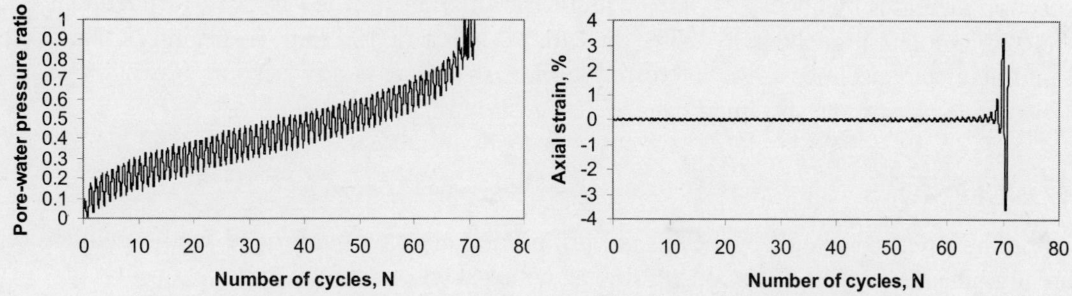

Fig. 4(b) DHS sample specimen with 70 mm ϕ and CSR = 0.125

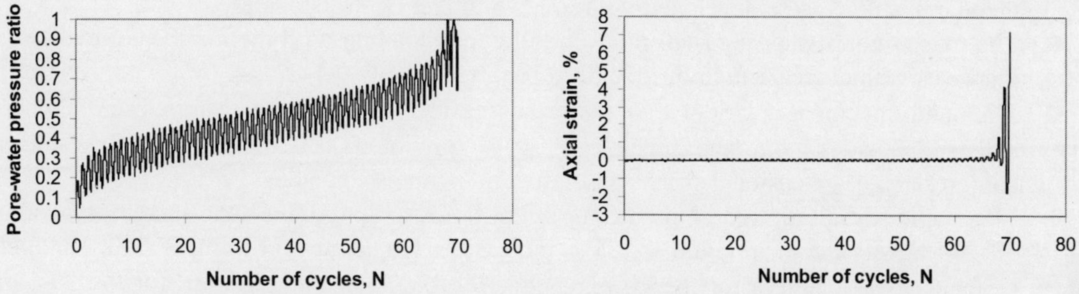

Fig. 4(c) DHS sample specimen with 100 mm ϕ and CSR = 0.125

Fig. 4 Effect of specimen size on Pore-water pressure ratio and axial strain with No. of cycles

The influence of specimen size on pore-water pressure generation and corresponding axial strain responses with number of loading cycles at a constant cyclic stress ratio of 0.125 were shown in Figure 4. Rapid development of pore-water pressure is observed for the specimens of 70 mm ϕ or 100 mm ϕ than the smaller specimens of 50.4 mm ϕ. From the plots of axial strain with number of stress cycles, it can be seen that the specimens of 50.4 mm ϕ were susceptible to extensional strains and failed in extensional side. It indicates that the sample behaves as dense during cyclic loading, so that there is a gradual buildup of pore-water pressure with the stress cycles and the specimen was more resistance to liquefaction. While the specimens with 70 mm ϕ or 100 mm ϕ were taken either compression or extensional strains and failed in compression side.

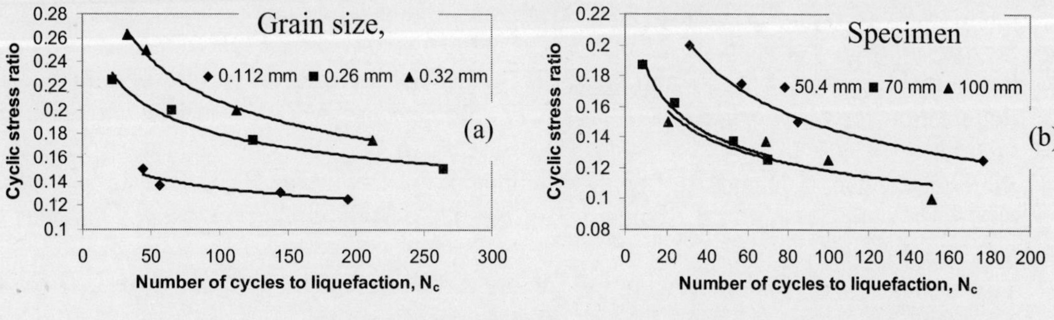

Fig. 5 Influence of grain size and specimen size on liquefaction resistance of sands

The influence of grain size and specimen size on liquefaction resistance of sands in terms of cyclic stress ratio with the number of cycles to cause initial liquefaction was shown in Figure 5. It can be found from Figure 5(a) and 5(b) that the cyclic stress ratio required to initial liquefaction increases either with the increase in grain size, D_{50} or with the decrease in specimen size up to 50.4 mm ϕ at any constant loading cycle. Similar findings are also reported by Lade and Hernandez, 1977; Martin et al., 1978 and Tatsuoka et al, 1986. The cyclic stress ratio required causing initial liquefaction increases by about 30 – 45% with the mean grain size, D_{50} increases from about 0.112 mm to 0.32 mm and it is about 15%-30% less for 70 mm ϕ or 100 mm ϕ specimens than for the 50.4 mm diameter specimens. It is interesting to note that there is no variation in cyclic resistance with loading cycles for the specimen sizes of 70 mm ϕ and 100 mm ϕ samples.

5 CONCLUSIONS

Based on the stress controlled cyclic triaxial test carried out on reconstituted sand specimens with different grain and specimen sizes, the following conclusions are drawn:

- All samples were experienced to little axial strains up to liquefaction was triggered and then large strains were developed with in few number of stress cycles.
- The rate of pore-water pressure ratio during cyclic loading with the number of cycles was increases with decrease in grain size, D_{50} from 0.32 mm to 0.112 mm.
- The sand specimens of 50.4 mm ϕ are susceptible to extensional strains and failed in extensional side, while the specimens of 70 mm ϕ or 100 mm ϕ were taken either compression or extensional strains and failed in compression side.
- The cyclic stress ratio required causing initial liquefaction of fine sand increases by about 30 – 45% with the mean grain size, D_{50} increases from about 0.112 mm to 0.32 mm and it is about 15%-30% less for the 70 mm ϕ or 100 mm ϕ specimens than for the 50.4 mm diameter specimens.

REFERENCES

1. Castro, G., and Poulos, S.J., (1977) "Factors Affecting Liquefaction And Cyclic mobility", *Journal of Geotechnical Engineering*, ASCE, Volume.103, No.GT6, pp. 501-515.
2. EI Hosri, M.S., Biarez, J., and Hicher, P.Y., (1982). "Liquefaction Characteristics Of Silty Clay", *Proceedings of the Eighth World Conference on Earthquake Engineering*, San Francisco, USA, pp.277-284.
3. Lee, K.L., and Fitton, J.A., (1969). "Factors Affecting The Cyclic Loading Strength Of Soil," *Vibration Effects of Earthquakes on Soils and Foundations*, ASTM STP 450, American Society for Testing and Materials, pp. 71-95.
4. Lade, P.V. and Hernandez, S.B. (1977). "Membrane penetration effects in undrained tests," *Journal of Geotechnical Engineering*, ASCE, Vol.103, No.GT2, pp. 109-125
5. Ladd, R.S., (1978). "Preparing Test Specimens Using Undercompaction", *Geotechnical Testing Journal*, Vol.1, pp. 16-23
6. Martin, G.R., Finn, W.D.L. and Seed, H.B. (1978) "Effects of system compliance on liquefaction tests," *Journal of Geotechnical Engineering*, ASCE, Vol.104, No.GT4, pp. 463-479
7. Silver, M., Chan, C., Ladd, R., Lee, K., Tiedmann, D., Townsend, F., Valera, J., Wilson, J., (1976). "Cyclic Strength Of Standard Test Sand" *Journal of Geotechnical Engineering*, Vol.102 (5), pp. 511-523.
8. Townsend, F., (1978). "A Review Of Factors Affecting Cyclic Triaxial Tests", *Dynamic Geotechnical Testing*, ASTM STP 654, ASTM, pp. 356 – 383.
9. Tatsuoka F., Shosuke T., Miura S., Kato H., Okamoto M., Yamada S.I., Yasuda S., and Tanizawa F. (1986) " Some factors affecting cyclic undrained triaxial strength of sand," *Soils and Foundations*, Vol.26, No.3, pp. 99-116.
10. Wong, R.T., Seed, H.B., and Chan C.K., (1975). "Cyclic Loading Liquefaction of Gravelly Soils", *Journal of Geotechnical Engineering*, ASCE, Vol.101, No. GT6, pp. 571 – 583.

STEADY STATE OF SAND IN TRIAXIAL EXTENSION TEST

M. Yoshimine and M. Kataoka
Department of Civil and Environmental Engineering,
Tokyo Metropolitan University
Hachioji, Tokyo, Japan
yoshimine-mitsutoshi@tmu.ac.jp

ABSTRACT

Steady state is an important index for evaluating liquefaction and flow deformation of sandy deposits. There have been extensive arguments on the influence of shear modes on steady state of soils at large deformation, but the problem is unsolved up to date. Triaxial compression and extension are extreme modes of shear in relation to the direction of principal stress and the magnitude of intermediate principal stress; hence comparison of steady states in these two shear modes is efficient for solving the problem. However, it is difficult to attain steady state in triaxial extension tests, because strain non-uniformity (necking of specimen) commonly occurs at large deformation especially in drained condition. Another difficulty arises from highly compressive nature of samples due to its anisotropy of the material in triaxial extension shear. If loose soil is tested in undrained triaxial extension, it is easily liquefied with zero effective stress at smaller deformation, and this makes it impossible to apply larger deformation to bring the specimen to steady state. To solve these problems, a series of triaxial extension tests on sand with partially drained condition under constant shear stress was conducted in this study. It was shown that all of the test results plotted on e (void ratio) – p' (effective mean stress) plane approached toward the steady state line previously obtained by undrained triaxial compression tests. This result implies that the steady state of sand is unique regardless of the magnitude of intermediate principal stress.

Keywords: Steady state, Sand, Triaxial extension, Liquefaction

1 INTRODUCTION

Steady state is a very convenient reference for predicting flow failure of the liquefied sand, if the steady state of specific sand could be uniquely related to its density. However there have been extensive arguments on the influence of initial packing (micro structure) of sand particles and mode of shear on steady state of soils at large deformation. As to the influence of initial packing of sand particles, though some researchers (e.g. Alarcon and Leonards, 1988, Dennis, 1988 and DeGregorio, 1990) reported different steady state depending on the sample preparation method, it is not sure if real steady state was achieved in their testing, and their conclusions could be derived form the observation of quasi steady state at smaller strain levels. We can expect that the initial packing of particles is rearranged in large deformation, thus it could be reasonably assumed that the effects of the initial packing are diminished at steady state.

The source of the effects of shear modes can be divided into two categories. One of them is the direction of principal stress to the initial structure of sand deposit such as bedding plane. This effect is attributed to the initial anisotropy of the material that is developed by the arrangement of particles, and thus we may be able to conclude that the effect is not important in the same reason as the effects of sample preparation on steady state at large deformation.

Another source of the effects of shear mode is the magnitude of the intermediate principal stress (σ_2) compared to the maximum and minimum principal stress (σ_1 and σ_3, respectively). There is much evidence that shows contractive and softer behavior of sand at smaller levels of deformation prior to steady state when σ_2 is larger even if the direction of principal stress is the same (e.g. Nakata et al., 1995, Uthayakumar, and Vaid, 1998, Yoshimine et al., 1998). Nevertheless, the direction of σ_2 was fixed to horizontal (parallel to the bedding plane of the deposits) in these tests, and this condition yields more compressive deformation in horizontal direction and expansive in vertical. Since sand deposited under gravity has anisotropic nature that is dilative and stiffer in vertical compression whereas contractive and softer in horizontal compression, the effects of σ_2 at smaller strain levels are also attributed to the initial fabric of the sand deposit. In fact, Yamada and Ishihara (1979, 1981) reported the opposite effect of σ_2 (larger volume expansion at peak shear stress in drained tests and smaller excess pore pressure at phase transformation in undrained tests for larger σ_2) with the condition in which σ_2 was fixed to the vertical to the bedding plane of the specimens of sand.

These considerations suggest that the differences of shear behavior of sands at relatively smaller strain levels are mainly attributed to the initial anisotropy of the sands, but the problem is unsolved yet if the behavior is affected by σ_2 at very large strain levels, such as steady state where the effect of the initial arrangement of the fabric should cease.

2 DIFFICULTIES OF TRIAXIAL EXTENSION TEST AND ITS SOLUTIONS

Triaxial compression and extension are the extreme modes of shear in the sense of the magnitude of intermediate principal stress; hence comparison of steady states in these two shear modes is efficient for solving the problem. However, it is difficult to attain very large strain levels to observe steady state in triaxial extension tests mainly due to two reasons.

One of the reasons is strain non-uniformity of specimen. Necking of specimen commonly occurs at large deformation in triaxial extension test, because it has flexible boundary in the direction of σ_1 and this condition accelerates non-uniform concentration of stress and deformation in the specimen. Another source of non-uniform deformation is shear banding. Strain concentration due to shear banding is common also in triaxial compression test in drained conditions but it is minor or negligible in undrained conditions for looser specimens. Thus it is desirable to adopt undrained or nearly undrained condition for testing. It is also effective to use looser specimens because contractive sand is more resistant to shear banding. In addition, if the initial state is very near to steady state, the strain levels to attain steady state are smaller, therefore necking of specimen is more unlikely.

Another difficulty arises from the highly contractive nature of samples due to its anisotropy of the material in triaxial extension shear, as described in the former section. If loose soil is tested in undrained triaxial extension, it is easily liquefied with zero effective stress at smaller deformation, and this makes it impossible to apply larger deformation to bring the specimen to steady state. Yoshimine et al. (1999) concluded that steady state line from triaxial extension tests positioned much lower than the steady state line from triaxial compression tests as shown in Fig. 1, by assuming that the zero effective stress state is the steady state of the sample. This conclusion could be misleading, because, if the liquefied specimen could maintain its homogeneity and if further loading was possible, the specimen could dilate with recovering its effective stress and the resultant steady state could be at higher stress levels thus the steady state line could positioned upper on e – p' plane.

To solve these problems, a series of shear tests on loose sand with partially drained condition under constant shear stress was conducted in this study. Fig. 2 shows the e – p' trace of triaxial compression tests ($\sigma_2 = \sigma_3$) of Toyoura sand under constant shear stress conducted by Yoshimine et al. (2006). The test procedure was the same as described in the next section. The steady state line of the same sand obtained by undrained triaxial compression test is also plotted on the same figure. It may be seen that the state of the specimens approached to the steady state line from upper and lower side of the line, and thus we could reasonably specify the position of the steady state line from the test results in the condition of constant stress constant. The sand has some friction ratio in terms of effective stress that reserved non-zero effective confining stress during the shear process under constant shear stress conditions. Volume compression or contraction occurs during the shear process under this shear conditions but it can be minimized if the initial state of the samples were near to the steady state, and it is expected that triaxial extension tests in the constant shear stress could detect the position of the steady state line in the $\sigma_2 = \sigma_1$ conditions.

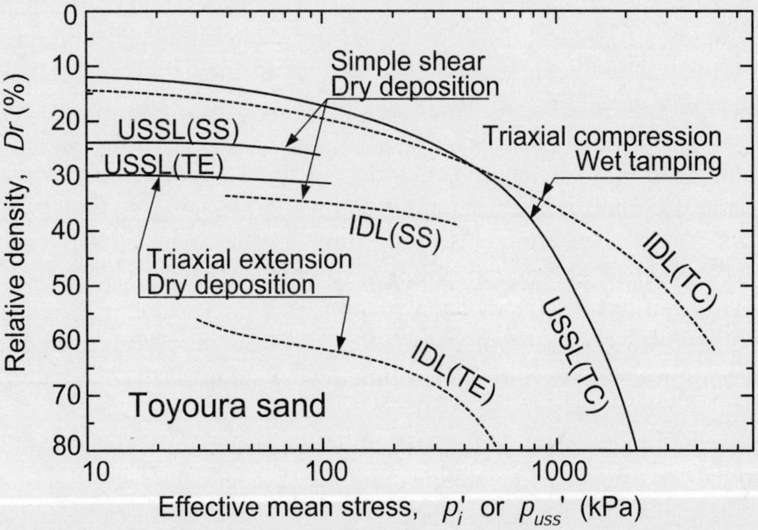

Figure 1. Previously suggested steady state lines (USSL) for different shear modes (Yoshimine et al. 1999)

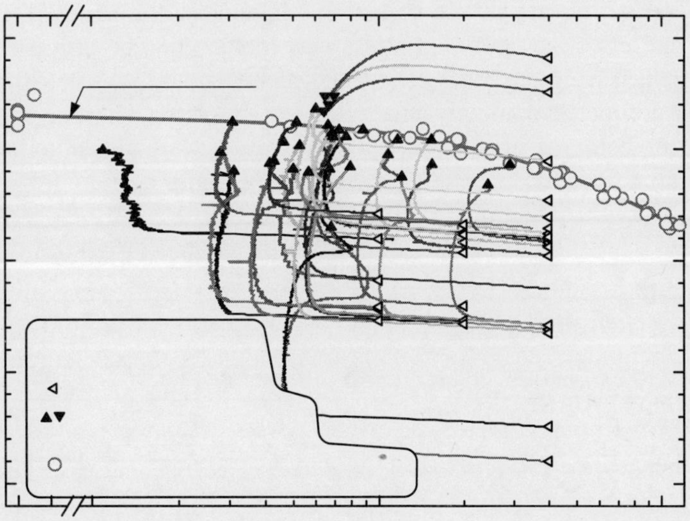

Figure 2. Trace of *e-p* path in triaxial compression tests under constant shear load (Yoshimine et al. 2006)

3 TEST PROCEDURE

The specimens of sand had a diameter of about 50mm and a height of about 100mm. To minimize the end friction of the samples, tiny porous stones those have diameter of 10mm were embedded at the center of top and bottom platens. The platens were rubricated by layers of grease and rubber membranes. Toyoura sand, that has mean diameter of D_{50}=0.2mm, maximum and minimum void ratios of e_{max}=0.977 and e_{min}=0.597 respectively, was used as tested material for all of the tests in this study. Wet tamping method (Verdugo, 1992 and Verdugo and Ishihara, 1996) was adopted to create loose samples. In this method, moist sand with water content of about 5% was deposited in the mold in five layers then each layer was compacted by a tamper that had a diameter of 20mm. After constituting specimen, CO_2 and then de-aired water were circulated through specimen from the bottom to the top to saturate the material. It was confirmed that all of the specimens had Skempton's B-value higher than 0.96.

The saturated samples were firstly consolidated under isotropic effective confining stress, and then lateral confining stress was increased to induce initial static shear stress under drained condition with constant vertical stress. The effective mean stress levels at the end of consolidation were around p'_{ini} = 100kPa whereas the deviator stress ratios were q/p'_{ini} = 0.1, 0.2, 0.3 and 0.55. Consequently we examined the behavior of ten specimens with various initial shear stresses, and the shear stress (deviator stress) was kept constant in the following undrained triaxial extension shear loading. The testing conditions for each specimen were summarized and listed in Table 1.

After waiting about one hour for completing consolidation, triaxial extension test was performed under strain controlled and monotonic loading condition. The lateral confining stress (the cell pressure) $\sigma_1 = \sigma_h$ was constant. The deviator stress $q = \sigma_1 - \sigma_3 = \sigma_h - \sigma_v$ was also kept constant during axial extension process by adjusting pore water pressure by hand. If axial stress $\sigma_3 = \sigma_v$ tended to decreased due to axial extension, then pore water pressure was increased and the effective confining stress was decreased to maintain the total axial stress and the deviator stress constant. If axial stress tended to increase due to softening behavior of the material, then pore water pressure was decreased and effective confining stress was increased. The volume change of the specimen was captured by measuring the water head of a water column in the backpressure line using electric differential pressure meter, amplifier, analog-digital converter and personal computer. Axial load, axial displacement, cell pressure and pore water pressure were also automatically and continuously recorded via electric transducers and the same personal computer.

Table 1. List of the triaxial extension tests

test No.	Initial state			Deviator stress	at $\varepsilon_3 = -10\%$	
	void ratio	D_r (%)	p' (kPa)	q (kPa)	void ratio	p' (kPa)
09	0.936	10.8	90	55	0.922	48.7
10	0.963	3.7	100	20	0.953	26.3
11	0.918	15.5	100	20	0.925	23.6
12	0.968	2.4	100	20	0.947	23.3
13	0.964	3.4	100	30	0.939	33.4
14	0.936	10.8	87	10	0.925	12.0
15	0.956	5.5	100	20	0.948	22.1
16	0.945	8.4	100	20	0.939	21.5
17	0.931	12.1	100	10	0.930	17.2
18	0.905	19.0	92	30	0.911	29.9

4 TEST RESULTS

Fig. 3 shows the deviator stress q - axial strain $\varepsilon_a = \varepsilon_3$ relationships from the triaxial extension tests with constant various deviator stress and various relative densities at initial state from D_r = 2% to 19%. It can be seen from Fig. 3 that deviator stress was kept constant with reasonable accuracy with small errors about $\Delta q = \pm 2$kPa throughout the shear process except just after starting deformation.

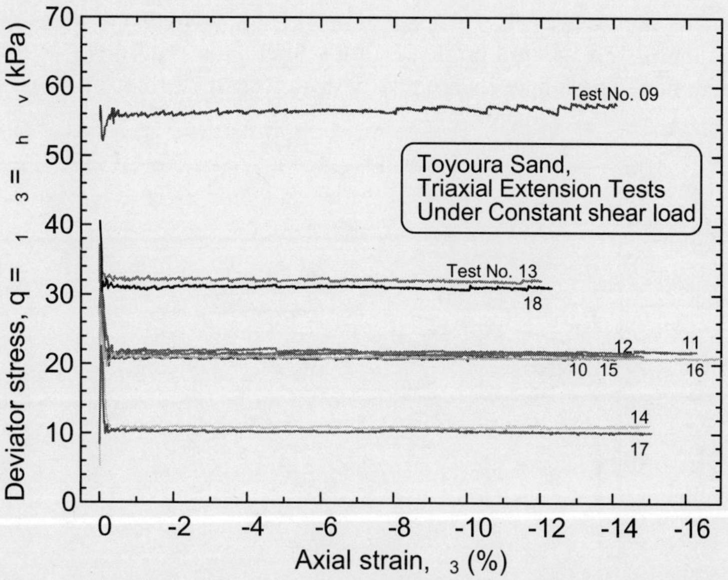

Figure 3. Stress-strain plot of the triaxial extension tests under constant shear load

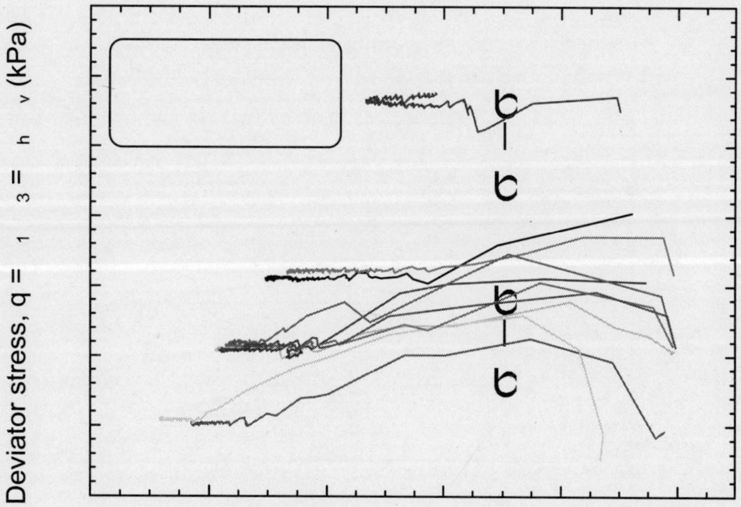

Figure 4. Stress path of the triaxial extension tests under constant shear load

To see the detail of the increase of deviator stress just after the starting of shearing, effective stress path was plotted in Fig. 4. This plot indicates that it was required to apply some amount of shear stress on the specimen to bring the stress to the failure condition at the beginning to develop further deformation with constant shear load. The increment of shear stress was larger for denser specimen. Nevertheless the axial deformation of specimens during the application of this shear stress increment was less than 2 mm and the resultant axial strain was less than 0.2%. Therefore we can say that the deviator stress was kept constant through the following plastic flow deformation process. In reality, earthquake shaking triggers flow deformation, and the cyclic loading brings the stress state to the failure condition. This process is different from the monotonic loading condition adopted in this study, but that the effects of the cyclic loading on the characteristics of the following flow deformation could be minor as reported by Yoshimine et al. (2005).

Measurement of pore water pressure was shown in Fig. 5 for the same tests plotted in Figs. 3 and 4. Fig. 5 indicates that the most part of the excess pore water pressure was applied at the beginning of shearing. Samples subjected to larger shear stress exhibited smaller pore water pressure during flow deformation whereas pore water pressure was larger if the shear stress level was smaller. This fact indicates that the mobilized friction angle is almost the same regardless of shear stress levels and density of the specimen. These aspects are also reflected on the effective stress path shown in Fig. 4.

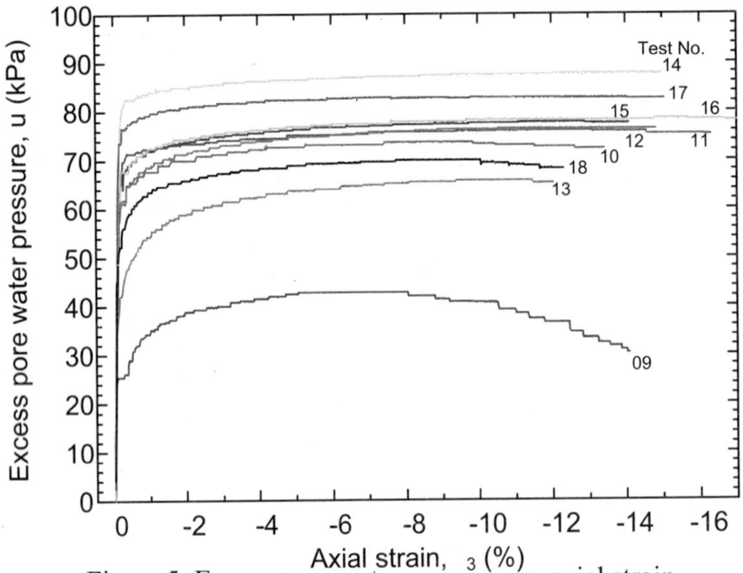

Figure 5. Excess pore water pressure vs. axial strain

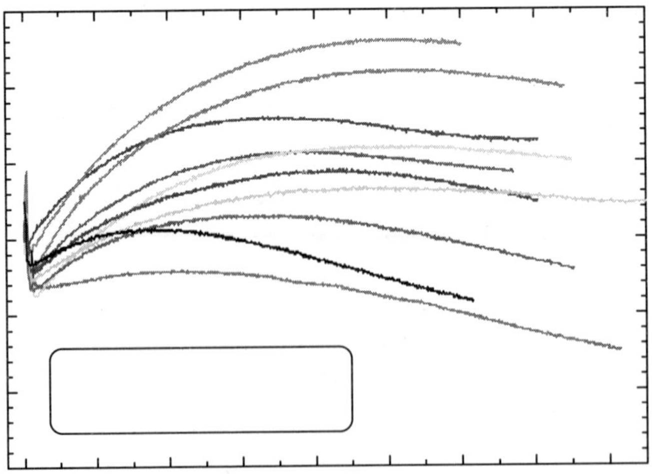

Figure 6. Volumetric strain vs. axial strain

Fig. 6 is the plot of volumetric strain associated with flow deformation with constant shear stress. It can be seen that the volume change of sand during deformation is affected by the initial density of the material. In general, denser sand should swell more water and dilate to exhibit large shear deformation. When sand is looser, smaller volume expansion is enough to create flow deformation, and volume contraction occurred if the initial relative density was less then around 10%.

5 DISCUSSIONS

Fig. 6 indicates that volume change of the specimens continues during the shear deformation and real steady state was not achieved even at the end of shear in these tests. In addition, non-uniform deformation of the samples was generally visible when the axial strain exceeded $\varepsilon_3 = -10\%$. Although it is difficult to quantitatively detect the degree of non-uniformity in the specimen, we judged that the data up to at the strain of $\varepsilon_3 = -10\%$ is reliable, and void ratio (e) and effective mean principal stress (p') at $\varepsilon_3 = -10\%$ for each test (Table 1) are taken as near-steady state. The traces of the $e - p'$ relationships for all of the triaxial extension tests up to $\varepsilon_3 = -10\%$ were plotted in Fig. 7.

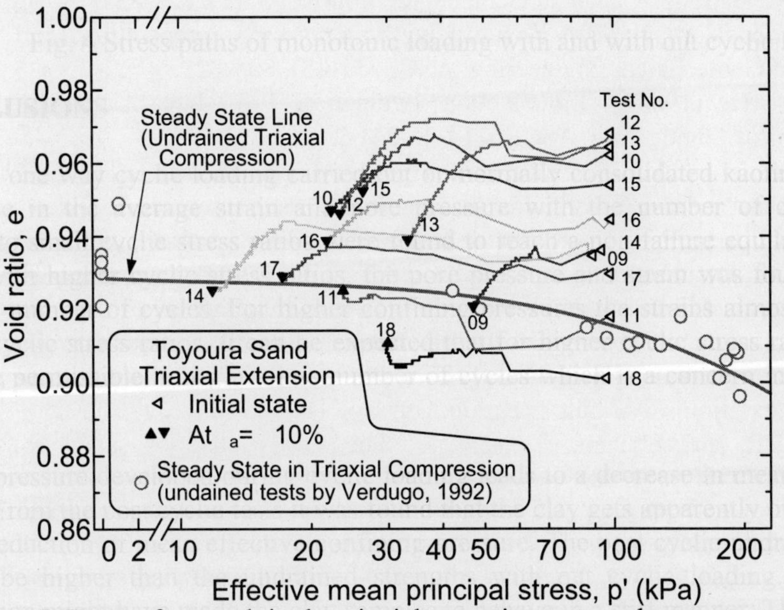

Figure 7. Trace of *e-p* path in the triaxial extension tests under constant shear load

The steady state line of Toyoura sand measured by undrained triaxial compression tests is also indicated on the same figure in the same manner as Fig. 2. It can be seen from Fig. 6 that the traces approach toward the steady state line from both upper and lower sides of the line depending on the initial density of the specimens. This fact strongly suggests that the steady state in triaxial extension condition is the same as the steady state in triaxial compression conditions. The steady state line of Toyoura sand is fairly flat within the target effective stress levels of this study (10 to 50kPa) and it may be able to assumed that the void ratio of all of the specimens should approached to approximately 0.925 irrespective to the initial density if they were sheared further without strain non-uniformity and if the real steady state was attained.

6 CONCLUSIONS

A series of triaxial extension tests on sand with partially drained condition under constant shear stress was conducted in this study. It was shown that all of the test results plotted on e (void ratio) – p' (effective mean stress) plane approached toward the steady state line previously obtained by undrained triaxial compression tests. This result implies that the steady state of sand is unique regardless of the magnitude of intermediate principal stress.

Density of sandy materials at steady state is generally very low at lower confining stress levels. For example, the void ratio of Toyoura sand at steady state observed in this study was around $e = 0.925$

which is corresponding to the relative density of only D_r = 14%. Such loose deposit can not exist in natural conditions, but temporal loosening of deposits to the levels of this density could easily take place due to the re-distribution of void ratio during the process of re-consolidation of liquefied ground as pointed out by Boulanger and Truman (1996), Yoshimine et al. (2006) and others. In this scenario, the knowledge obtained in this study may be important in the analysis of flow failure and deformation of liquefied sandy ground after earthquakes.

REFERENCES

1. Alarcon, A. and Leonards, G.A. (1988), "Liquefaction Evaluation Procedure: Discussion," Journal of Geotechnical Engineering, ASCE, **114**(2), 232-236.
2. Boulanger, R.W. and Truman, S.P. (1996), "Void redistribution in sand under post-earthquake loading," Canadian Geotechnical Journal, **33**(5), 829-834.
3. DeGregorio, V.B. (1990), "Loading systems, sample preparation, and liquefaction," Journal of Geotechnical Engineering, ASCE, **116**(5), 805-821.
4. Dennis, N.D. (1988), "Liquefaction Evaluation Procedure: Discussion," Journal of Geotechnical Engineering, ASCE, **114**(2), 241-243.
5. Nakata, Y., Yasufuku, N., Murata, H., Hyodo, M. Nagashima, T. (1995), "Undrained and drained shear behaviour of inherent anisotropic sand reference to fixed principal stress directions," Journal of Geotechnical Engineering, Japan Society of Civil Engineers, **511**, 209-220. (in Japanese)
6. Uthayakumar, M. and Vaid, Y.P. (1988), "Static liquefaction of sands under multiaxial loading," Canadian Geotechnical Journal, **35**(2), 273-283.
7. Verdugo, R. (1992). "Characterization of sandy soil behavior under large deformation," Ph.D. dissertation, Civil Eng. Program, Univ. of Tokyo, Tokyo, Japan.
8. Verdugo, R. and Ishihara, K. (1996). "The steady state of sandy soils," Soils and Foundations, **36**(2), 81-92.
9. Yamada, Y. and Ishihara, K. (1979), "Anisotropic deformation characteristics of sand under three dimensional stress conditions," Soils and Foundations, **19**(2), 79-94.
10. Yamada, Y. and Ishihara, K. (1981), "Undrained deformation characteristics of loose sand under three-dimensional stress conditions," Soils and Foundations, **21**(1), 97-107.
11. Yoshimine, M., Ishihara, K. and Vargas, W. (1998), "Effects of principal stress direction and intermediate principal stress on undrained shear behavior of sand," Soils and Foundations, **38**(3), 179-188.
12. Yoshimine, M., Robertson, P. K. and Wride, C. E. (1999), "Undrained shear strength of clean sands to trigger flow liquefaction," Canadian Geotechnical Journal, **36**(5), 891-906.
13. Yoshimine, M., Nagashima, N., Amano, K. and Hosono, Y. (2005), "Dilatancy of sand subjected to flow deformation following undrained cyclic loading," The 40th Japan National Conference on Geotechnical Engineering, 523-524. (in Japanese)
14. Yoshimine, M., Nishizaki, H., Amano, K. and Hosono, Y. (2006), "Flow deformation of liquefied sand under constant shear load and its application to analysis of flow slide of infinite slope," Soil Dynamics and Earthquake Engineering, **26**(2-4), 253-264.

MULTI CHANNEL ANALYSIS OF SURFACE WAVE (MASW) TESTING FOR SITE CHARACTERIZATION OF DELHI REGION

D. Neelima Satyam [1] *and K.S.Rao* [2]
[1] *Assistant professor, VNR VJIET, Hyderabad-72, e-mail: neelima.satyam@gmail.com*
[2] *Professor, Indian Institute of Technology Delhi, New Delhi, India*

ABSTRACT

After the Bhuj earthquake the national capital region of Delhi attracted major attention of several scientific studies in the recent times. Since Delhi falls in zone IV (IS 1893: 2002) with high seismic activity, there is a great need for site characterization and seismic hazard mapping of the area. Multi Channel Analysis of Surface Wave (MASW) tests were done in Delhi at 118 sites in predefined grids of 2kmX3km each. Shear-wave velocity, V_S, is an important parameter for evaluating dynamic behavior of soil. This test carried outdone using 48 channel digital seismograph with 4.5 Hz geophones. Data was analyzed using SeisImager/SW software and two dimensional shear wave velocity models at every 5m depth from ground surface was developed. It is observed that the value of Vs is ranging from 400 to 480 m/s in the rocky sites, it is 120 to 250m/s in trans Yamuna region and 250 to 370m/s in western side of the area. The shear wave velocity values are exactly matching with the geological and geotechnical characteristics of the subsoil strata. Also, The field shear wave velocity from MASW test is compared with the estimated values using the empirical formulae developed. It is observed that the experimental shear wave velocity and the empirically estimated values are almost equal up to 20m depth. But at depths greater 20m, Vs from MASW test are slightly higher than the calculated value using empirical formulae developed. Average shear wave velocity at 30m depth i.e., V_{S30} is also calculated and is ranging from 185 to 495 m/s for Delhi region. Also, correlations among Vs, SPT 'N', depth 'D' are developed using the available borehole data. These correlations are useful to estimate the shear wave velocity at any depth with out performing the actual test in Delhi region.

Keywords: Site Characterization, Seismic Microzonation, Multi channel Analysis of Surface Wave Testing, Shear Wave Velocity

1 INTRODUCTION

Geophysical tests based on the generation and propagation of seismic waves are widely used in Earthquake Geotechnical Engineering. Seismic in-situ tests are often the only way to determine soil stiffness in undisturbed conditions, especially for coarse soils, in which undisturbed sampling is problematic. The shear modulus can easily be derived if the velocity of propagation of shear waves and the soil density are known. The shear wave disturbances, as they travel through the soil media, map out dilational and rotational strains. Hence, Vs is currently used as one of the important factors in site characterization. Shear wave velocity structure could be ascertained through invasive techniques like down hole, up hole, cross hole and non invasive methods such as Spectral Analysis of Surface Wave (SASW) method and Multichannel Analysis of Surface Wave (MASW) method (park et al., 1999 and Xia et al., 1999), utilizing theoretically proven relation between Rayleigh and Shear wave velocity as

$$V_R = 0.9 V_S \tag{1}$$

Multi channel analysis of surface wave method is a non-invasive method developed to estimate shear wave velocity profile from surface wave energy. Measurements of phase velocity of Rayleigh waves of different frequencies can be used to determine a velocity depth profile. In the SASW method, the dispersion curve is obtained by using a two-receiver test configuration. But in MASW multi -station recording permits a single survey of a broad depth range and high levels of redundancy with a single field configuration. Multi channel analysis of surface waves overcomes the drawbacks associated with SASW method. Multistation technique increases the reliability of the results and shortens the execution time both in the field and during interpretation for surface wave tests. In the recent years, MASW method attracted researchers all over the world. MASW method is powerful, rapid and cost effective tool for constraining shallow wave velocity structures. This tests were done in Delhi at 118 locations by spreading multiple geophones for estimating 2D velocity profiles.

In this survey spacing between the geophones and source increment had been made equal to 6 meters. The McSeis-SX 48 digital seismograph of 48 channels for the first time in the country is used in this study. Frequency of the equipment ranges from 4.5 to 4600Hz. Geophones of 4.5 Hz are used in the survey and they are connected to the seismograph with connecting cables. A wooden hammer of 11 kg weight is used as a source for creating the energy. Fixed receiver configuration is used and source is placed between each receiver and at both ends of the survey line. Data for each site is recorded and saved in the equipment. The acquired data is then transferred from the seismograph for the analysis using SeisImager/SW software. The two dimensional velocity model is generated for all test locations and the average shear wave velocity up to 30m (V_{S30}) is also calculated.

2. GEOLOGICAL AND GEOTECHNICAL DETAILS OF DELHI

Delhi has interesting geology on account of its being the end of exposed ancient Aravali mountain ranges extending NE in this area. Delhi and its adjoing region is surrounded in the north and east by Indo-Gangetic plaines, in the west by the extension of the great Indian Thar desert and in the south by the Aravali ranges. The rocks of Delhi have undergone multiple folding and different phases of metamorphism with some transverse features. The quarzites are bedded and highly jointed with pegmatite intrusives. The Alwar series and the post Delhi intrusives are covered by the quaternary deposits in the form of aeolian and alluvial deposits. The alluvial deposits belong to the Pleistocene period, i.e., older alluvial deposits and of recent age i.e., newer alluvium. Older alluvium deposits consists of mostly inter bedded lenticular and inter fingering deposits of clay, silt and sand along with kankar (Rao, 2003).

Based on the collected borehole data, soil profiles are made covering almost entire region to study the sub soil heterogeneity. Generally in trans-Yamuna, silt is very predominant. Grain size distribution curves are also drawn at 4 different depths for north, south, east, west and central blocks in Delhi. In the eastern block the soils are sandy silts/silty sands with high percentage of medium to fine sand. The areas like Noida, Mayur Vihar, Yamuna Vihar, Abdul Fazal Enclave, Geeta Colony which falls in the eastern block has very soft soil deposits and high water table. Northern and Western blocks have silty sands with reasonable percentage of clay. The areas like Rohini, Punjabi Bagh, Paschim Vihar, Janakpuri, Dwarka which falls in these blocks. The south and central blocks have gravelly sands with percentage of gravel (Ghitorni, Maidan Garh, Sat Bari,). Rao & Neelima (2004) explained the geotechnical properties of Delhi soils in detail.

3. MULTICHANNEL ANALYSIS OF SURFACE WAVE (MASW) METHOD

MASW technique has been developed in response to the shortcomings of SASW. The simultaneous recording of 12 or more receivers at short (1-2m) to long (50-100m) distances from an impulsive or vibratory source gives statistical redundancy to measurements of phase velocities. Multichannel data displays in a time variable frequency format also allow identification and elimination of non fundamental mode Rayleigh waves and other coherent noise during the analysis process. The field setup, acquisition process and analysis procedure are explained below in detail.

3.1 Field Study

This MASW test is done in Delhi at various locations by spreading multiple geophones for estimating 2D Velocity profiles. Figure 1 shows the field set up of the MASW test. In this survey spacing between the geophones had been made equal to 6 meters. All the geophones are connected to the seismograph and the source is shifted with 6 m increment. An additional geophone is used as a trigger geophone to initialize the recording. The generated Rayleigh wave data are recorded at all shot points.

3.2 Data Acquisition

The 48 channel signal enhancement seismograph is used for acquisition of the seismic data. Geophones of 4.5 Hz are used and they are connected to the seismograph with connecting cables as shown in Fig. 1.

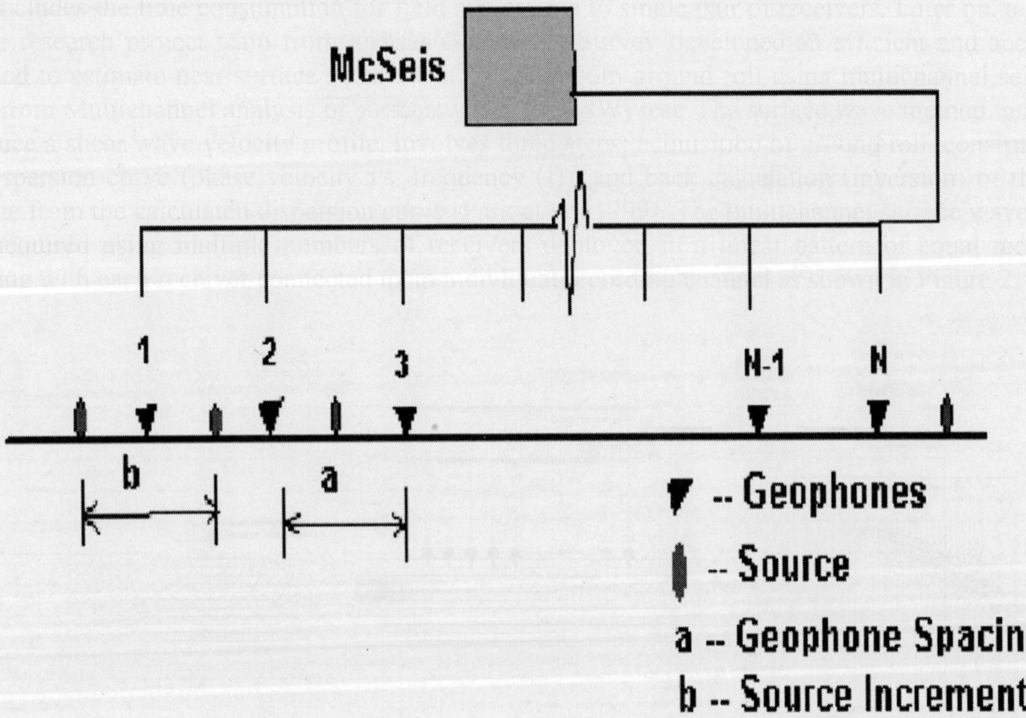

Figure 1. Sketch of MASW Test Spread

A wooden hammer of 11 kg weight manually impacted on 165cm^2 aluminum plate is used for creating the energy. Fixed receiver configuration is used and the source is placed between each receiver and at both ends of the survey line. Around 13 shots are created at each test location in the present study. Data for each shot is digitally recorded and saved in the equipment. Figure 2 shows the multichannel records (shot gathers) of the geophones at JNU campus with source at mid point of the survey line. The acquired data is then transferred for the analysis. The following are the recording parameters (Table 1) during the acquisition process.

SeisImager/SW software has three more softwares i.e., Pickwin95, WaveEq and GeoPlot. Figure 3 gives the flowchart with the detailed steps involved in the analysis of surface wave data for getting two - dimensional shear wave velocity model.

Table 1. Seismic data recording parameters

Recording system:	McSEIS-SX 48
Sampling Interval:	500 μs
Memory:	2kb
Recording Format:	SEG-2
Pre trigger:	ON
Stack mode:	Summation
Geophones:	12 geophones of 4.5 Hz frequency
Geophones array:	Linear with geophone spacing of 6m
Source:	11kg sledge hammer on $165cm^2$ metal plate
Source array:	Source is shifted with 6m interval
Offset distance :	3m

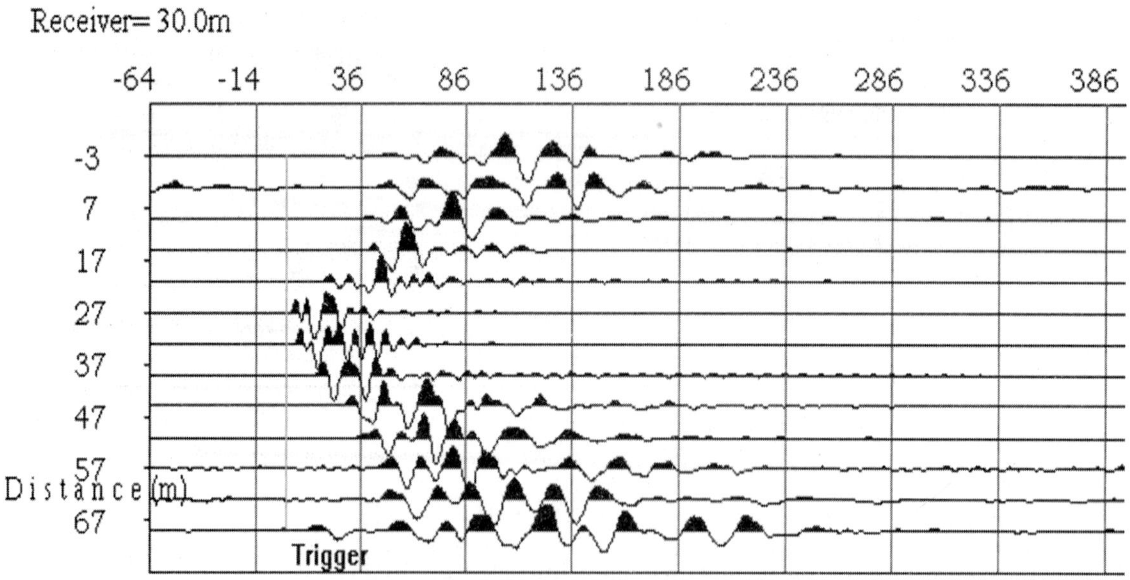

Figure. 2. Multichannel Records (Shot Gathers) at JNU Campus with Source at Mid Point of the Survey Length

The first step in the analysis is making the file list in which all waveform files and source receiver configuration are mentioned and then cross correlation CMP gather is calculated. Dispersion curves are calculated by converting it into frequency domain and then checked. Generation of a dispersion curve is one of the most critical steps for generating an accurate shear wave velocity profile. Dispersion curves are generally displayed as phase velocity versus frequency. This relation can be established by calculating the phase velocity from the linear slope of each component of the swept frequency record. Figure 4 shows the phase velocity curves of JNU campus. The 1D shear wave velocity profiles are calculated using non linear least square method using the dispersion data.

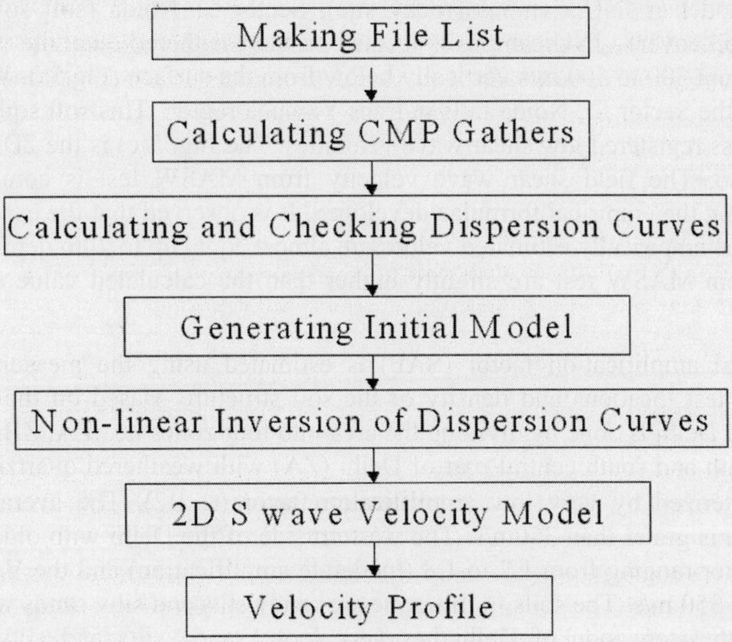

Figure. 3. Flowchart for the Analysis of Surface Wave Data Using SeisImager/SW

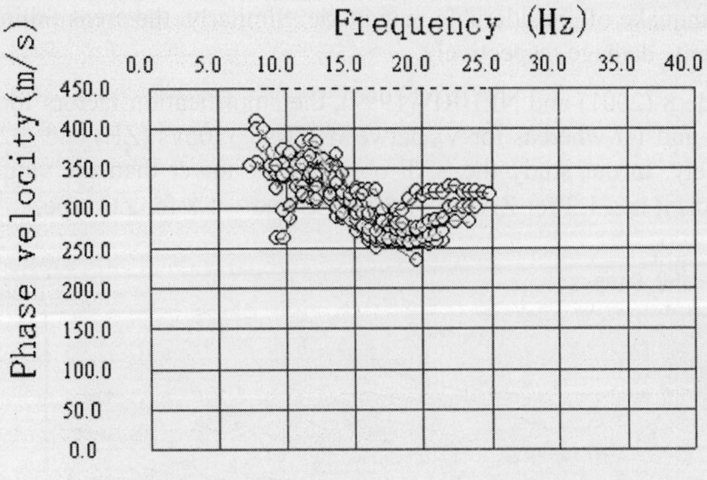

Figure. 4. Phase Velocity Curves for JNU Site

An intial velocity model should be defined such that Vs at a depth Z_f is 1.09 times the measured phase velocity C_f at the frequency where wavelength λ_f satisfies the relationship

$$Z_f = a \ \lambda_f \qquad (2)$$

Where a is a coefficient that changes with frequency slightly and is based on extensive modeling. Finally 2D velocity model can be generated in Geoplot. According to NEHRP guidelines, Vs_{30} for a layered structure is determined for site classification.

4. RESULTS AND DISCUSSION

Surface wave profiling using this MASW method will provide a non destructive, quick (data from 2 sites can be acquired in one day) and accurate shallow subsurface image of the shear wave velocities. Here the velocity model for three sites with varying soil thickness is explained. Figure 5 gives the 2D V_S model at JNU Campus (rocky site), Sector 51 Noida (soft soil site) and IIT Campus (medium soil cover). JNU campus is located on the weathered quartzite with shear wave velocities ranging from 300 to 500 m/s vertically below from the surface (Fig.5a). Where as Figure 5(b) corresponds to the Sector 51, Noida falls in trans Yamuna region. This soft soil sites with high sedimentary thickness registered low shear wave velocities. The Fig. 5(c) is the 2D velocity model at IIT campus. Also, The field shear wave velocity from MASW test is compared with the estimated values using the empirical formulae developed. It is observed that the experimental shear wave velocity and the empirically estimated values are almost equal up to 20m depth. But at depths greater 20m, Vs from MASW test are slightly higher than the calculated value using empirical formulae developed.

The average spectral amplification factor (SAF) is estimated using the measured shear wave velocities at all the test locations and density of the soil structure. Based on the SAF and $V_{S30,}$ microzonation of the Delhi is done by dividing the area into four zones i.e., ZA, ZB, ZC1 and ZC2. The areas in the south and south central part of Delhi (ZA) with weathered quartzite and gravelly deposits are characterized by very low amplification factor ($\leq$ 1.2). The average shear wave velocity in this zone is grater than 350m/s. The western side of the Delhi with older alluvium has an amplification factor ranging from 1.2 to 1.8 (moderate amplification) and the V_{S30} for this zone falls between 250 to 350 m/s. The soils in this zone are sandy silts and silty sands with clay seams. The north and northeastern side of Delhi has very loose sandy silts and silty sands (newer alluvium) and high water table depths. The amplification factor in these zones (ZC1 and ZC2) is quiet high due to high sedimentary thickness also the V_{S30} is less than 250m/s for these zones. It is clear that the amplification factor is high in the trans Yamuna region which can experience severe damage due to an earthquake of considerable magnitude. Similarly, the areas falling in ZA and ZB can have low to moderate damage respectively.

According to Eurocode 8 (2001) and NEHRP (1997), the amplification factors for V_S greater than 350m/s (ZA) are 1.35 and 1.7 whereas for V_S between 250 to 350m/s (ZB) and < 250m/s (ZC) it is 1.5 and 2.4 respectively. In our study the SAF obtained are lower than the values mentioned in these codes. That means it is $\leq$ 1.2 for ZA, 1.2-1.8 for ZB and > 1.8 for ZC type.

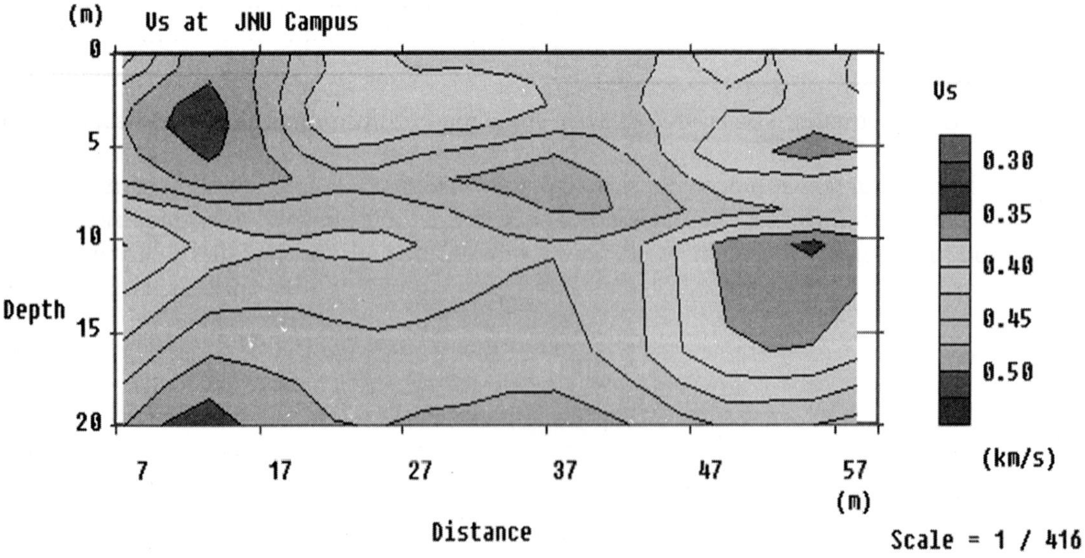

(a)

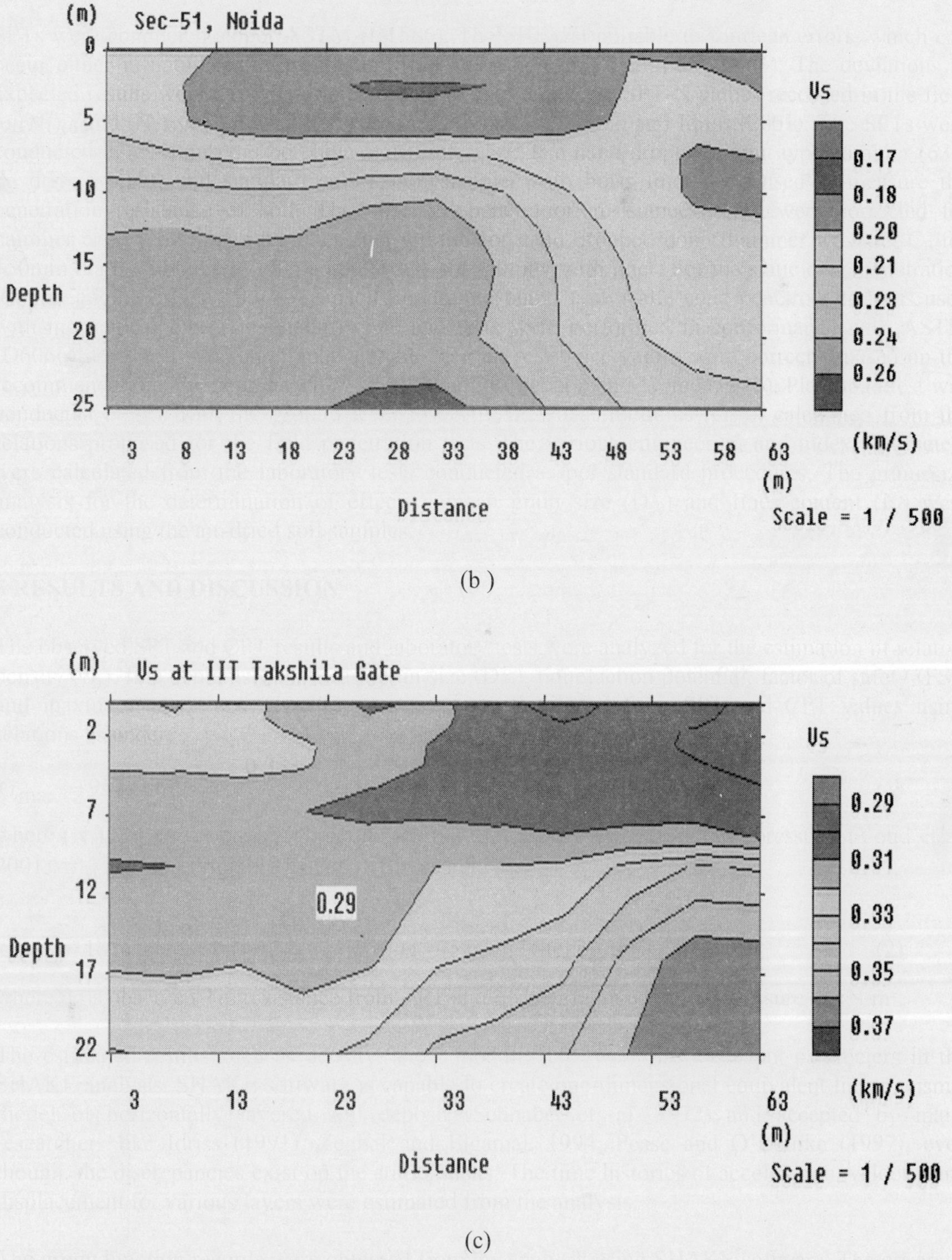

Figure 5. Two Dimensional Velocity Model (a) JNU Campus, (b) Sec-51, Noida and (c) IIT Campus

It is clear that this method is useful for geotechnical engineers/seismologists who need a quick assessment of site characterization or a reliable velocity structure, which is used in earthquake ground motions. Empirically this site classification can be used for estimating site amplifications.

REFERENCES

1. IS: 1893-2002. Criteria for Earthquake Resistant Design of Structures
2. Park, C.B., R.D.Miller, and J.Xia, (1999). "Multi-channel Analysis of Surface Waves", *Geophysics,* **64**, No.3, 800-808.
3. Xia, J., R.D.Miller, and C.B.Park, (1999). "Estimation of Near surface Shear Wave Velocity by Inversion of Rayleigh Waves", *Geophysics*, **64**, 691-700.
4. Rao, K.S. (2003). "Seismic Microzonation of Delhi Region". *12th ARC Soil Mechanics and Geotechnical Engineering*, Singapore, GeoAsia, I,327-330.
5. Rao, K.S. and D.Neelima Satyam (2004). "Liquefaction Mapping for Delhi Region", *Proceedings of Indian Geotechnical Conference*, Warangal, 219-222.
6. EUROCODE-8 - prEN1998-1-3 (2001) - Design Provisions for Earthquake Resistance of Structures – Part 1-3: Specific Rules for Various Materials and Elements, *European Committee for Standardization.*
7. Building Seismic Safety Council (1997). NEHRP Recommended Provisions for Seismic Regulations for new Buildings and Other Structures, Part 1: Provisions, *FEMA 38*, Washington, D.C.

SOIL-STRUCTURE INTERACTION IN THE ANALYSIS OF BRIDGE WITH DEEP WELL FOUNDATION

Supratic Gupta[1], Sumant Gupta[2], Pallavi Agarwal[3], Shailesh Chandra[3] and K. S. Rao[4]

1. *Assistant Professor, Department of Civil Engineering, Indian Institute of Technology, Delhi; New Delhi, India.*
2. *Doctoral Student, Department of Civil Engineering, Indian Institute of Technology, Delhi; New Delhi, India.*
3. *Former UG Student, Department of Civil Engineering, Indian Institute of Technology, Delhi; New Delhi, India.*
4. *Professor, Department of Civil Engineering, Indian Institute of Technology, Delhi; New Delhi, India.*

ABSTRACT

Well Foundation is a popular form of foundation for bridges in India. Not enough research is done to high light the importance of soil structure interaction in case of well foundation. This is because most well foundations have large cross-sections and can be classified as shallow foundation. This phenomenon is not necessarily true when the well foundation has large depth. In this paper, a typical bridge with large well depth is taken to show the importance of adopting appropriate soil-structure interaction. Various conditions for analysis are adopted to highlight the benefit of adopting appropriate modeling.

Key Words: Well Foundation; Dynamic Analysis, Soil Structure Interaction

1. INTRODUCTION

Most of the major bridges in India adopt well foundation. Well foundation has been used in bridges in Japan and other countries also. Analysis of bridges is a challenging subject. The design shear and bending forces often depend on the boundary condition assumed for the foundation. Since well has large cross-sectional dimension, it is generally assumed to be shallow. Most designers in India assume simplistic boundary conditions to analyze the bridge.

This paper presents a typical long span bridge. It is about 1.5 km long and has 10 spans. It is a balanced cantilever bridge made continuous at final stage. Under seismic condition, it can be considered fixed at all piers. Any resemblance to any real bridge is accidental. Relevant details are provided later. Designers rarely adopt three-dimensional analysis incorporating the soil-structure interaction. Such extensive research has been done by both analytical and experimental means for Akashi Keiko Bridge in Japan, which also used well foundation. It was a Cable Stayed Bridge. Hence the structural behaviour is different from this bridge. Designers can surely be expected to use beam elements with soil simulated as spring elements. The main aim of the paper is to highlight the importance of adopting proper boundary condition implementing soil structure interaction. This paper presents a series of logical yet adaptable models high lighting the importance of adopting soil structure interaction. The importance of use High Flood Level Condition for well steining design and Low Flood Level for well pier design is high lighted. The construction stage details are not presented as it has no relevance with the soil-structure interaction.

2. DESCRIPTION OF THE BRIDGE ADOPTED AND MODELLING

The superstructure is 1.5 km long bridge with 11 piers as shown in Fig 1 which also shows the Hydraulic Conditions in LFL and HFL conditions. The superstructure has varying cross-section. This bridge is built by balanced cantilever technique and made continuous at later stage. Hence, here the bridge is considered continuous. The central pier is fixed. STU or Shock Transmission Unit is attached to all other piers so they behave fixed along the bridge axis under seismic load.

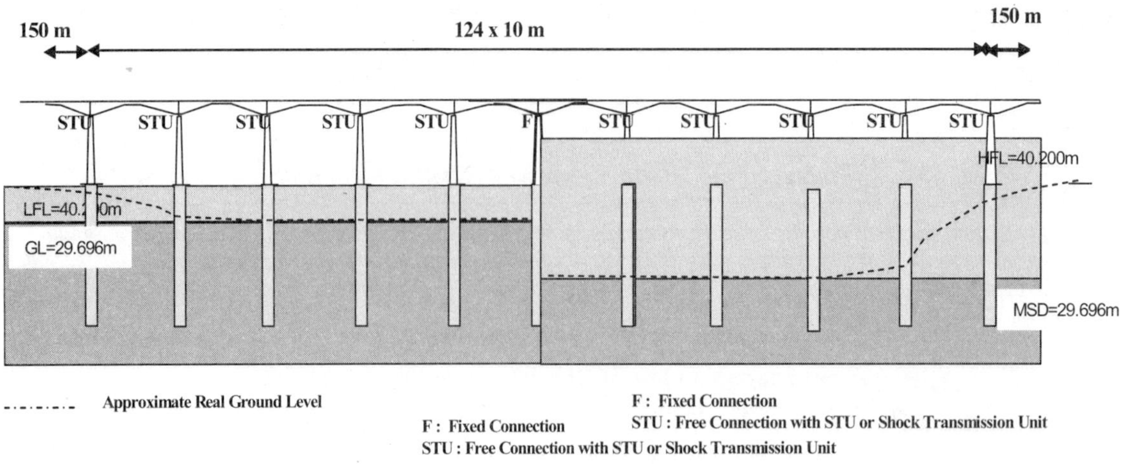

Fig. 1 Bridge dimensions showing both LFL and HFL conditions

3. METHOD OF ANALYSIS

As per IS 1893 (Part 1): 2002, dynamic analysis shall be performed to obtain the design seismic force. Dynamic analysis may be performed using time history method or response spectrum method. The time history method is based on an accepted ground motion and shall be performed using accepted dynamics principles.

The Response Spectrum Analysis of a multiple degree of freedom system is performed. Undamped free vibration analysis of the structure shall be performed as per the established methods of mechanics using appropriate masses and elastic stiffness of the structural system to obtain natural frequencies (T) and mode shapes ( ) of the mode shapes that need to be considered as per the code.

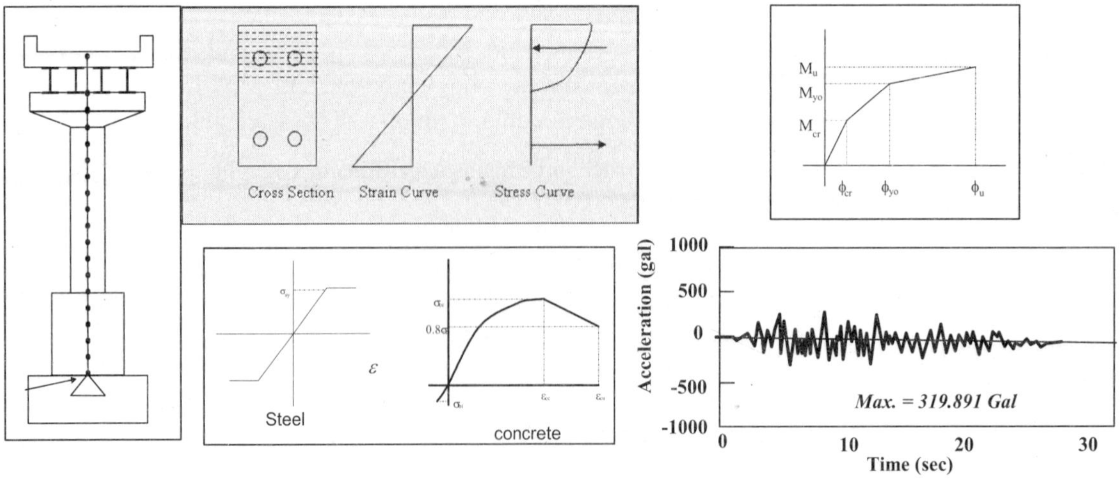

Fig. 2. Typical Non-linear Method of Analysis as adopted in Japan

From the critical Natural Frequency T, we obtain the average response acceleration coefficient (Sa/g), depending on the soil site type. We calculate the

For Rocky or Hard soil Sites, Medium soil types Soft Soil Sites

$$\frac{Sa}{g}=\begin{cases}1+15T & 0.00<T<0.10\\ 2.5 & 0.10<T<0.40\\ 1.00/T & 0.40\leq 4.00\end{cases}, \quad \frac{Sa}{g}=\begin{cases}1+15T & 0.00<T<0.10\\ 2.5 & 0.10<T<0.40\\ 1.00/T & 0.40\leq 4.00\end{cases} \quad \frac{Sa}{g}=\begin{cases}1+15T & 0.00<T<0.10\\ 2.5 & 0.10<T<0.40\\ 1.00/T & 0.40\leq 4.00\end{cases}$$

In other countries, especially Japan, designers indeed design the column and the well assuming non-linear behavior of concrete and steel. After assuming some appropriate stress-strain relationship for concrete and reinforcement, they calculate the M- relationship as a bilinear or tri-linear approximation for the pier or well element, taking care of the increase axial force along the depth. The damping coefficient is calculated. A Site specific acceleration is applied to check the safety of the members.

4. MODELLING, ANALYSIS AND RESULTS

Some designers in India assume that the well is sufficiently stiff and analyse the structure assuming that the structure is fixed on the well cap. This would have been a reasonably correct assumption in the wells were short acting as a deep beam. Even though the bridge can have large dimensions with cross-section of 12m x 8m double D with 3 m wall thickness, it has reasonable flexibility as offered by two reasons a) The well is not necessarily a deep beam as the well has sufficiently large depth. b) The well is restrained by soil which offers sufficient softness.

In this paper, the following 7 conditions are analyzed to exhibit the importance of soil-structure interaction. Even though we can adopt even more sophisticated methods of analysis, designers should at least model the soil as spring elements and check both LFL and HFL conditions

In this case, the bridge is assumed to have fixed at the center and has STU in rest of the piers. Hence, under seismic condition, all pier joints should be assumed to be fixed. This would provide much higher stiffness to the system compared to the flexible stiffness of one pier fixed condition. As a result the seismic force will be higher, and the force will be shared by all the piers. Though one can calculate the seismic load by equally dividing the total value by the number of piers, it is better to reanalyze the structure with equivalent seismic load applied horizontally, in the respective boundary conditions.

It is obvious that the LFL condition would provide higher stiffness and higher stiffness coefficients compared to the HFL conditions. For the Pier base, hence LFL condition is expected to be governing. However, for the Well, HFL provides lower grip, has higher lever arm. Hence, it is more likely that the well would be governed by HFL condition.

STU or Shock transmission unit is like the seat belt we use in our cars. They are actuators with some Fluids that freeze under high velocity. However, it only mobilizes after some displacement. For this bridge, it is assume to freeze after 12 mm displacement. Hence another case of 12 mm static displacement with only central Pier fixed is also considered.

To clarify the effect of soil spring, the conditions of fixed as GL/ MSD levels are also carried out. It is found that the case with soil modeled as spring provides much lower coefficient. To understand the effect of spring, we also compare the case of spring with four times higher stiffness. Even though the coefficient increases, the values are far lower compared to the GL/MSD fixed.

The spring element of soil is determined as per IS2911 Part1.

Table 1: Seismic Coefficient of Different modes of Calculations

	Freq	Time	S_a/g	Coef
Central Pier Considered Fixed	0.44	2.26	0.60177	~~0.065~~
Corrected with GL Fixed	1.33	0.75	1.81	0.195
LWL and GL Cases				
WCF	1.81	0.55	2.46	0.266
GLF	1.651	0.61	2.245	0.2425
GL SP	**1.12**	**0.893**	**1.523**	**0.165**
GLSP1	1.3	0.769	1.768	0.191
HFL and Seismic MSD Cases				
MSDF	1.25	0.8	1.7	0.1836
MSD SP	**0.6**	**1.667**	**0.816**	**0.088**
MSD SP2	0.84	1.19	1.142	0.123

Table 2: Bending Moment for Different cases

Units : Shear : KN BM: KNm

	COEFF.	PIER BOTTOM				WELL			
		Max Mz	Correct Mz	Max Fy	Correct Fy	Max Mz	Correct Mz	Max Fy	Correct Fy
GL-SPRING	0.164	2.45E+06	**4.03E+05**	119230	**19609.561**	3.22E+06	5.30E+05	144157.1	23714
MSD-SPRING	0.088	3.37E+06	2.97E+05	5530.1	487.35665	7.05E+06	**6.22E+05**	377032.9	**33227**
PIER-12	1	0	0.00E+00	900	900	41175.62	4.12E+04	900	900
GL-FIXED	0.24	2.29E+06	5.54E+05	177.	42.9	2.99E+06	7.26E+05	672	163.1
MSD-FIXED	0.184	2.35E+06	4.32E+05	120.	22.1	5.53E+06	1.02E+06	495	90.9
GL-SPRING1	0.19	2.33E+06	4.42E+05	144.5	27.44702	3.09E+06	5.88E+05	14474	2750
MSD-SPRING1	0.123379	2.75E+06	3.39E+05	84.4	10.41	5.84E+06	7.21E+05	36559	4510
WELL CAP TOP-FIXED	0.265853	2224590	5.91E+05	204.602	54.39	693791.3	1.84E+05	6548	1740

Type of Soil	Value	
	Dry	Submerged
Loose sand	0·260	0·146
Medium sand	0·775	0·525
Dense sand	2·075	1·245
Very loose sand under repeated loading or normally loading clays	–	0·040

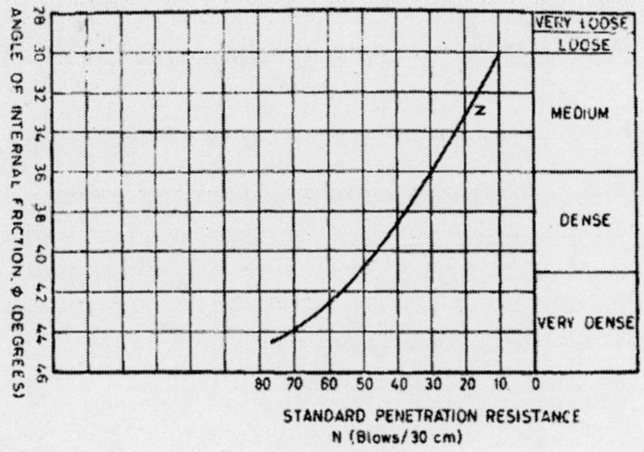

Fig. 3 Determination of the Soil Stiffness as per IS 2911 Part 1

The cases considered are as follow

a) Fixed only in the central Pier Fixed at Pier Cap
b) Fixed in all Piers and Fixed at Pier Cap

Low Flood Level Condition

c) Fixed at Ground Level
d) Ground Considered as Springs
e) Ground Considered with spring with 4 time stiffness

High Flood Level Conditions

f) Fixed at MSD (seismic at 90% of MSD)
g) Ground Considered as Springs
h) Ground Considered with spring with 4 time stiffness
i) Fixed only at the central Pier and 12 mm lateral displacement

5. CONCLUSIONS

A detailed presentation was presented on a series of analysis carried out to understand the importance of seismic analysis. The following conclusions can be drawn.

a) It can be realized that dynamic analysis taking care of soil is important
b) One just cannot assume the Well foundation to be rigid.
c) Analyzing both Low water level condition and High Flood Level condition was found to be important
d) Pier was governed by Low water level condition with a seismic coefficient of 0.164
e) Well foundation was found to be critical under High Flood Level Condition with a seismic coefficient of 0.088
f) One must apply seismic coefficient for the appropriate boundary condition

REFERENCES

1. IS 1893 (Part 1), 2002: Criteria for Earthquake Resistant Design of Structures - Part 1: General Provisions and Buildings.
2. IS 2911 (Part1): Sec 1: 1979 Code of practice for design and construction of pile foundations: Part 1 Concrete piles, Section 1 Driven cast in-situ concrete piles.

3. IS 6403: 1981 Code of practice for determination of bearing capacity of shallow foundations.
4. Colloquium thesis, 2006: Seismic Analysis of Narutohashi Bridge.
5. H. Tajimi. Recent tendency of the practice of soil-structure interaction design analysis in Japan and its theoretical background. College of Science and Technology, Nihon University, Japan, 1983.
6. Do Hyung Lee, Eunsoo Choi, Goangseup Zi. Evaluation of earthquake deformation and performance for RD bridge piers. Department of Civil and Geotechnical Engineering, Paichai University, South Korea, 2005.

EARTHQUAKE HAZARD MITIGATION MEASURES USING TIRE DERIVED RECYCLED GEOMATERIALS

H. Hazarika
Department of Architectural and Environment System
Akita Prefectural University
Yurihonjo, Akita, Japan
hazarika@akita-pu.ac.jp

ABSTRACT

In today's era of performance based design, cost, environment and structural performance are the three key factors that dictate the final design. In this paper, an earthquake disaster mitigation technique for geotechnical structures is described, in which scrap tire derived recycled wastes are utilized. The innovative cost-effective disaster mitigation technique that has been developed is called **SAFETY** (**S**tability **A**nd **F**lexibility of structures during **E**arthquake using **TY**res). The objective of this research is to examine how **SAFETY** technique can reduce the earthquake related damages to structures. To that end, a series of model 1g shaking table experiments were conducted to validate the performance of the technique. The test results have demonstrated that such technique not only reduces the seismic load on structures but also the earthquake induced permanent displacement of structures. Such mitigation measures also leads to cost-effective design and retrofitting of geotechnical structures by reducing the material (both structural and geotechnical) costs.

Keywords: Cost-performance, Disaster Mitigation, Earthquake, Liquefaction, Sustainability

1 INTRODUCTION

Seismologists in Japan have predicted that large-scale devastating earthquakes (Strong metropolitan earthquake, Tokai Earthquake and Tonankai-Nankai Earthquake) may strike the Kanto, Tokai and Nankai areas of Japan any time in the near future. Concerted efforts have been undergoing to mitigate the disasters and minimize the economic implications from these earthquakes (http://www.bousai.go.jp, 2006). Especially after her bitter experiences of the 1995 Hyogoken-Nanbu earthquake, Kobe the seismic stability of the existing and newly built infrastructural facilities continues to remain the top priority of the planners and designers in Japan.

More than 90% of the waterfront structures were severely damaged during the 1995 Hyogoken-Nanbu earthquake Figure 1 shows the damages that the waterfront structures typically suffer, resulting in the loss of serviceability of a port immediately after an earthquake. Following the 1995 Hyogoken-Nanbu earthquake, significant theoretical and experimental works have been done on the subject (Dickenson et al., 1998; Inagaki et al., 1996; Ishihara, 1997; JGS/JSCE, 1996; Kamon et al., 1996; Towhata et al., 1996). Typical waterfront structure such as gravity type quay wall has rubble backfill immediately behind the wall. However, such granular material is vulnerable to deformation under seismic load, and hence can cause large permanent deformations to the structures. Also, liquefaction related damages such as subsidence and lateral spreading are evident in almost every earthquakes affecting the performance of geotechnical structures (Hamada et al., 1996; Yasuda et al., 1996). Various remediation methods against liquefaction prevention for existing structures have been reported in Yasuda (2007). For waterfront structures such as quay walls, if we can substitute the conventional materials (rubble backfill and the landfill beyond it) with some other lightweight granular materials with other beneficial characteristics, then the liquefaction during earthquake can be reduced to a greater extent along with the curtailment of earthquake induced permanent deformation of the structures.

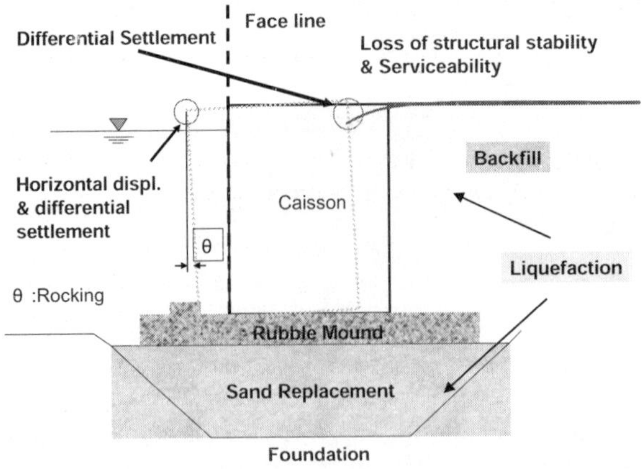

Figure 1. Typical damage pattern of waterfront structures

Since, cost reduction, structural performance and sustainability are the three major players that dictate the final design of geotechnical structures, development of environmentally friendly and economic disaster mitigation technique remains the top priority of the engineers. Recent years have seen the emergence of a new geomaterial called tire chips derived from scrap tires. Tire chips is a recycled material that is lightweight, elastic, compressible, highly permeable, earthquake resistant, thermally insulating and durable. Due to such myriad of beneficial properties, this material has been coined a *smart geomaterial* by Hazarika et al. (2007). A compilation of technical information relating to the applications of scrap tire derived recycle materials can be found in Hazarika and Yasuhara (2007). This material can be used in the backfill of waterfront structures instead of conventionally used granular material (Hazarika, 2007). Various other applications of tire derived geo-materials can be found in Edil (2006), Humphrey (1998), Tweedie et al. (1998) and Yasuhara et al. (2006). In this research, a cost-effective backfilling technique is developed using tire chips that can protect and reduce the damages of structures during devastating earthquakes.

The objective of this research is to evaluate the effect of tire chips on preventing earthquake disaster mitigation such as large permanent displacement and liquefaction, which may lead to disastrous performance of geotechnical structures. An innovative cost-effective disaster mitigation technique is developed using tire chips, in order to enhance the seismic performance of geotechnical structures against liquefaction and liquefaction induced lateral flow. The technique is called **SAFETY** (**S**tability **A**nd **F**lexibility of structures during **E**arthquake using **TY**res). A series of large-scale 1G underwater shaking table test was conducted to validate the performances **SAFETY** when structures are subjected to seismic loadings of high intensity.

2 HOW SAFETY WORKS

SAFETY is a novel technique that involves placing a cushion layer made out of tire chips as a vibration absorber immediately behind the structures. In addition, vertical drains made out of tire chips are installed in the backfill as a preventive measure against soil liquefaction. Yasuhara et al. (2004) used such tire chips as vertical drains for reducing liquefaction induced deformation. Figure 2 shows typical cross section of the earthquake resistant **SAFETY** technique. One function of the cushion is to reduce the load against the structure, due to energy absorption capacity of the cushion material. Another function is to curtail permanent displacement of the structure (either due to incremental dynamics earth pressure or due to lateral spreading of liquefied soils) by making use of the inherited flexibilities of such elastic and compressible material. The principle of **SAFETY**, thus, lies in imparting flexibilities to the structure due to cushioning action of the tire chips layer, which in turn ultimately provides stability to the structure by absorbing the energy and restraining excessive structural displacement during earthquakes.

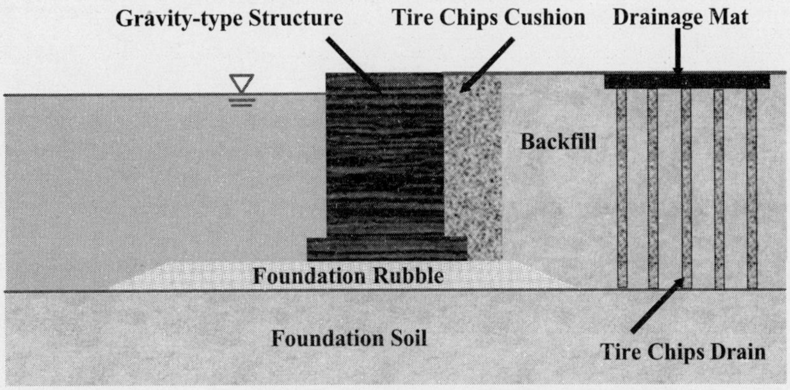

Figure 2. Underlying principle of **SAFETY** technique

3 UNDER WATER 1G SHAKING TABLE TEST

3.1 Experimental Setup

The large three dimensional underwater shaking table assemblies of Port and Airport Research Institute (PARI), Japan were used in the testing program. The shaking table is mounted on a 15 m long by 15 m wide and 2.0 m deep water pool, and thus can simulate the actual state of waterfront structure. A caisson type quay wall (model to prototype ratio of 1/10) was used in the testing. Figure 3 shows the cross section of the soil box, the model caisson and the locations of the various measuring devices (load cells, earth pressure cells, pore water pressure cells, accelerometers and displacement gauges). The model caisson (425 mm in breadth) was made of steel plates filled with dry sand and sinker to bring its center of gravity to a stable position. The caisson consists of three parts; the central part (width 500 mm) and two dummy parts (width 350 mm each). All the monitoring devices were installed at the central caisson to eliminate the effect of sidewall friction on the measurements.

The soil box was made of a steel container 4.0 m long, 1.25 m wide and 1.5 m deep. The foundation rubble beneath the caisson was prepared using Grade 4 crushed stone with particle size of 13 mm ~ 20 mm. The backfill and the seabed layer were prepared using Sohma sand (No. 5). Backfill was prepared in stages using free falling technique, and then compacting using a manually operated vibrator to achieve the target relative density of 50% to 60%. After constructing the foundation and the backfill, and setting up of all the devices, the pool was filled with water gradually elevating the water depth to 1.3 m to saturate the backfill. This submerged condition was maintained for two days so that the backfill attains a complete saturation conditions.

3.2 Testing Procedures

Two test cases were examined (Figure 4) using the same soil box by mounting two gravity type model quay walls with different backfill conditions. In one case (Case A), a caisson with a rubble backfill with conventional sandy backfill was used. In another case (Case B), behind the caisson, a cushion layer of tire chips (average grain size 20 mm) was placed vertically down whose thickness was 0.4 times of the wall height. In addition, vertical drains made out of tire chip (average grain size 7.0 mm), were installed in the backfill. Geotextile bags with the specific drain size were first prepared, which then were filled with the tire chips with a pre-determined density. They were then installed with a spacing of 150 mm in triangular pattern. The drain diameter was chosen to be 50 mm. The top of the entire drains were covered with a 50 mm thick gravel layer underlying a 50 mm thick soil cover. The purpose of such cover layer is twofold: one is to allow the free drainage of the water and other is to prevent the likely uplifting of the tire chips during shaking due to its lightweight nature. The average dry density of the tire chips cushion achieved after filling and tamping was 0.675 t/m^3.

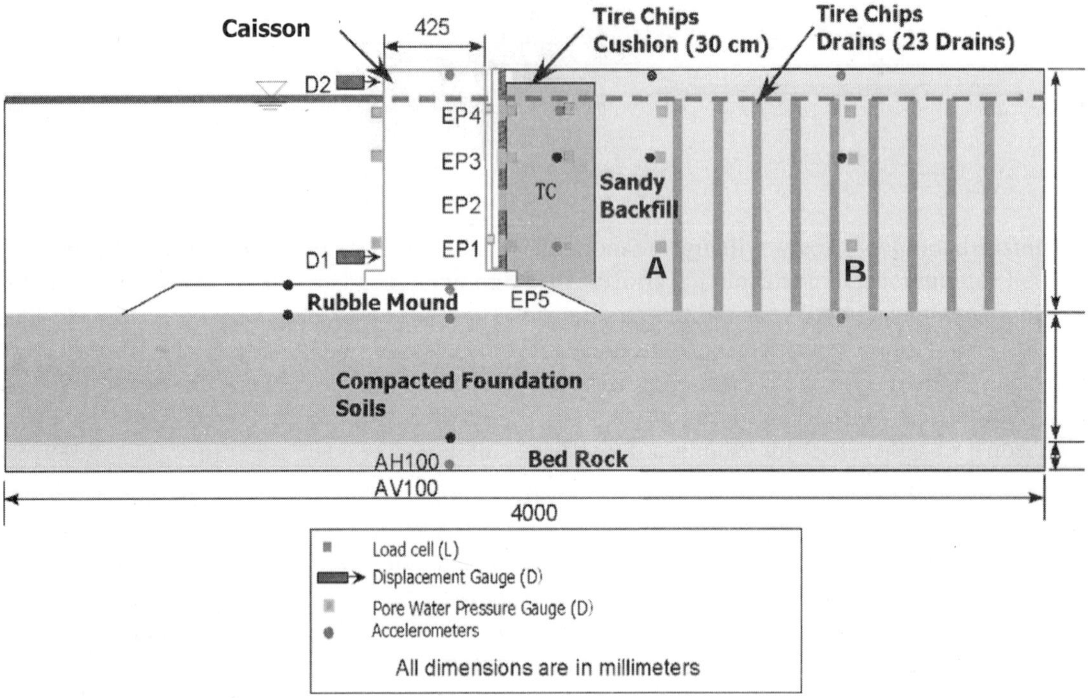

Figure 3. Cross section of the model used in underwater shaking table testing

(a) Case A (Conventional type backfill) (b) Case B (Backfill improved by **SAFETY**)

Figure 4. Test conditions

The relative densities that were achieved after the preparation of the backfill were between 50% to 60%. This implies that the backfill soil is partly liquefiable. On the other hand, the foundation soils were compacted with mechanical vibrator to achieve a relative density of about 80%, implying a non-liquefiable foundation deposit. The input motion used was the N-S component of the strong motion acceleration recorded at the Port Island, Kobe, Japan during the 1995 Hyogo-ken Nanbu earthquake (M 7.2). The wave record of the motion is shown in Figure 5. The similitudes of various parameters in 1g gravitational field for the soil-structure-fluid system were calculated using the relationship given in Iai (1989) for a model to prototype ratio of 1/10. Durations of the shaking in the model testing were based on the time axes of these accelerograms, which were reduced by a factor of 5.62 according to the similitude relationship.

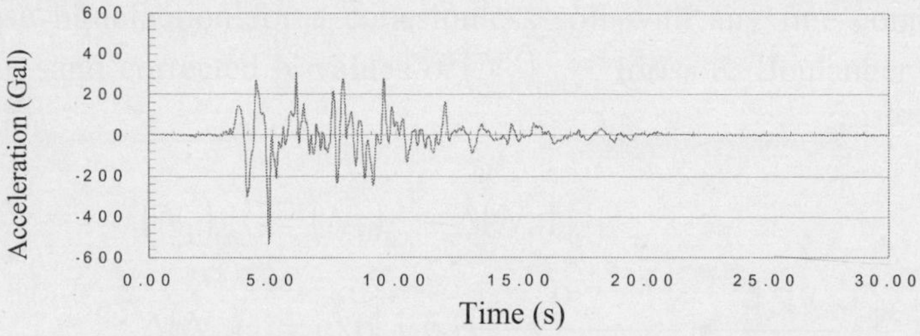

Figure 5. Input earthquake motion (PI record, Kobe)

4 TEST RESULTS

4.1 Seismically Induced Lateral Thrust

Figures 6(a) and 6(b) show the time history of the increment of the seismic earth pressure acting on the quay wall at the lower middle and the upper middle part of the caisson. It can be observed that, compared to conventional backfill condition, the seismic increment is decreased to a considerable extent in tire chips reinforced backfill using SAFETY technique. Considering the fact that the static earth pressure itself will also be reduced (Hazarika et al., 2007) due to low weight and compressible nature of the cushion material, the total seismic thrust acting on the structure can, thus, be reduced to a considerable extent. The end result, thus, is the reduction of the total earth pressure, which will contribute towards the stability of the structure during earthquakes.

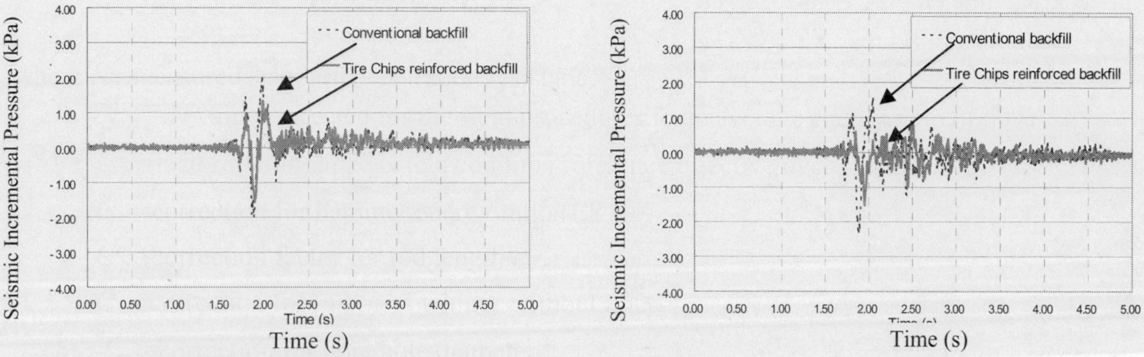

(a) At lower middle part (EP2 of Fig. 3)

(b) At upper middle part (EP2 of Fig. 3)

Figure 6. Incremental seismic load on the quay wall

4.2 Liquefaction Behavior of the Backfill

In order to evaluate, whether the **SAFETY** technique can minimize the liquefaction related damages, the time histories of the excess pore water pressure during the loading at a particular location (B in Figure 3) for the two test cases are compared in Figures 7(a) and 7(b). Comparisons reveal that the building up of pore water pressure is restricted due to dissipation by the permeable backfill condition resulted from the use of tire chips cushion and the tire chips drains. In the case of conventional backfill, as expected, the pore water pressure suddenly rises, and it takes considerable time (about 25 second in the test scale) to dissipate. However, in the case of backfill with SAFETY technique, the built up pore water pressure dissipates within a very short interval (2.5 second), preventing any chance for the backfill to liquefy.

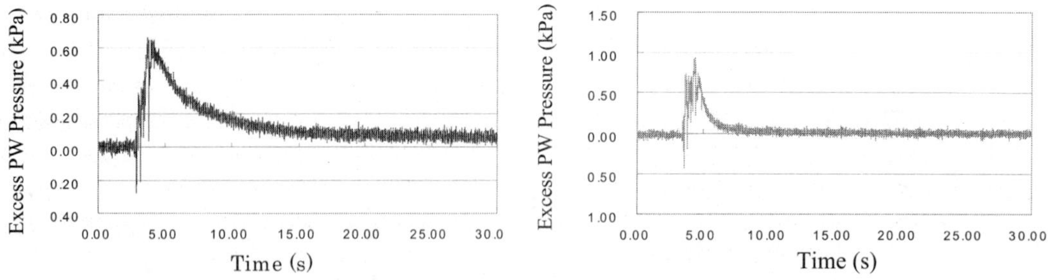

(a) Conventional backfill (b) SAFETY Technique

Figure 7 Prevention of Liquefaction

4.3 Seismically Induced Horizontal Displacement

The time histories of the horizontal displacements (D1 and D2 in Figure 3) during the earthquake loading for the two test cases are compared in Figures 8. Comparisons reveal that the maximum displacement experienced by the quay wall with SAFETY technique (thick continuous line) is toward the backfill in contrast to the quay wall without any earthquake disaster prevention measures (shown in dotted line), in which case it is seaward. The compressibility of the tire chips renders flexibility to the soil-structure system, which allows the quay wall to bounce back under its inertia force, and this tendency ultimately (at the end of the loading cycles) aids in preventing the excessive seaward deformation of the wall. However, the wall with conventional backfill experiences very high seaward displacements right from the beginning due to its inertia. As a consequence, the structure can not move back to the opposite side and ultimately suffers from a huge permanent seaward displacement.

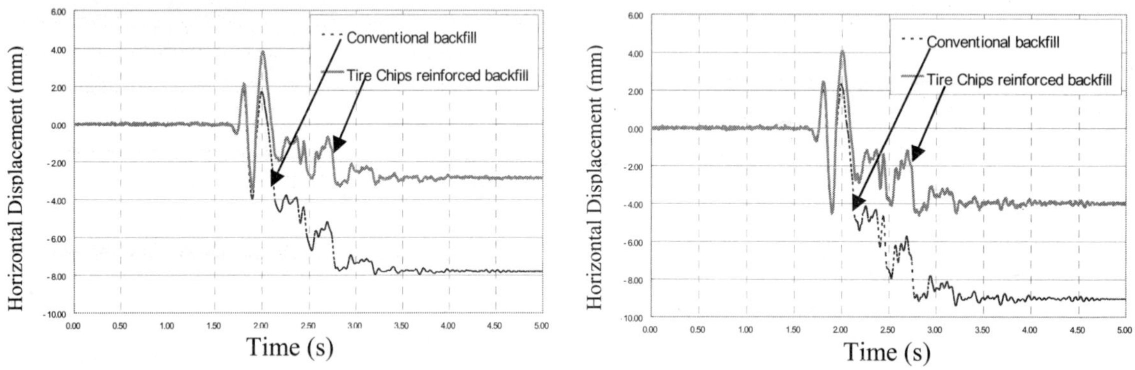

(a) Bottom of the quay wall (D1) (b) Top of the quay wall (D2)

Figure 8. Measured Structural Performance

4.4 Performance Comparison during Intense Earthquakes

In order to examine whether the developed technique can contribute towards the safe operation of port facilities after any devastating earthquakes, differential settlements in the backfill with and without earthquake hazard mitigation measures need to be examined. Figure 9 shows such comparison, in which the states of the ground surface settlement of the backfill at the end of the earthquake loading (maximum acceleration 1.5 times that of the 1995 Hyogoken Nanbu earthquake shown in Figure 5) are compared. It can be observed that, the structure with conventional backfill undergoes a very severe differential settlement. However, the structure protected by **SAFETY** technique does not undergo appreciable differential settlement, confirming that even in such high intensity earthquake conditions, the port facilities can remain operational, if the structure is reinforced with the earthquake resistant technique (**SAFETY**) described here.

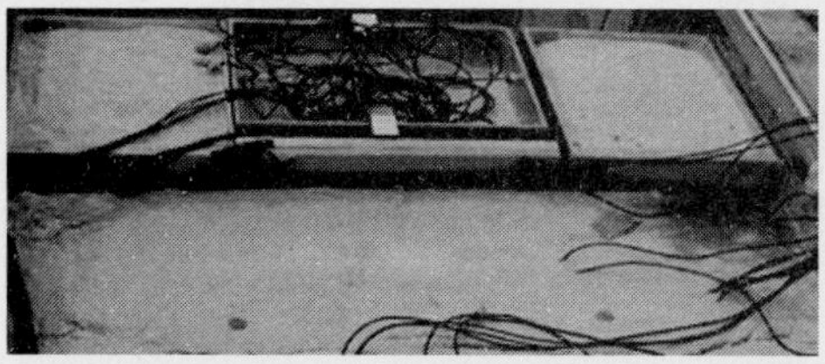

(a) Conventional backfill case

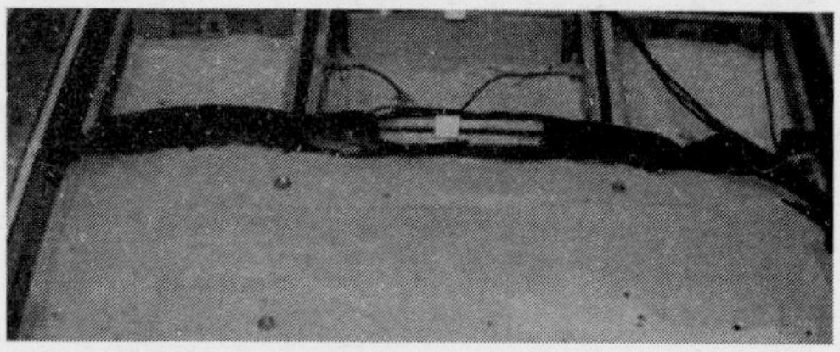

(b) Backfill improved by **SAFETY**

Figure 9. Ground surface settlements

5 CONCLUDING REMARKS

A series of 1g underwater shaking table tests were conducted to examine the performance of a newly developed technique, called **SAFETY**, intended for earthquake hazard mitigation. A substantial reduction of the dynamically induced displacement of structures as well as ground settlement could be achieved using the liquefaction preventive measures embodied in the technique, implying a better structural performance of structures during earthquakes. Also, the technique yields more than 50% of reduction of the dynamics load exerted on structures, implying a designer can opt for a slimmer structure with reduced material costs. Such application of scrap tire derived material, thus, not only contributes towards a sustainable environment by recycling of the scrap tires as materials, but also reduces considerably the construction cost of a project. The earthquake hazard mitigation technique (**SAFETY**), that has been developed, described and evaluated here, thus, satisfy the requirements of emerging and popular design methodology called the performance based design. The technique can also be applied for retrofitting of existing structures, which run the risk of damages during devastating future earthquakes.

ACKNOWLEDGEMENTS

The author gratefully acknowledges the financial support for this research provided by the Japan Society of Promotion of Science (JSPS) and Ministry of Education, Culture, Sports, Science and Technology (MEXT), Japan under the Grant-in-Aid for scientific research (Grant No. 18206052, Principal Investigator: Hemanta Hazarika). The author also would like to acknowledge the supports of Port and Airport Research Institute (PARI), Japan and his former colleagues in PARI, whose assistance in the test execution were extremely valuable. The technical supports from Bridgestone Corporation, Tokyo and Toa Corporation, Yokohama, Japan are also gratefully acknowledged.

REFERENCES

1. Dickenson, S., and Yang, D.S. (1998), "Seismically-induced Deformations of Caisson Retaining Walls in Improved Soils", Geotechnical Earthquake Engineering and Soil Dynamics III, Geotechnical Special Publication, **2**(75), 1071-1082.
2. Edil, T.B. (2006), "Green Highways: Strategy for Recycling Materials for Sustainable Construction Practices", Seventh International Congress on Advances in Civil Engineering, Istanbul, Turkey, 1-20.
3. Hazarika, H. (2007), "Structural Stability and Flexibility during Earthquakes using Tyres (SAFETY) - A Novel Application for Seismic Disaster Mitigation –", In Scrap Tire

Derived Geomaterials – Opportunities and Challenges, Hazarika and Yasuhara Eds. 115-126.

4. Hazarika, H., and Yasuhara, K. (2007), Scrap Tire Derived Geomaterials – Opportunities and Challenges, Taylor and Francis, UK.
5. Hazarika, H., Kohama, E., and Sugano, T. (2007), "Underwater Shake Table Tests on Seismic Performance of Waterfront Structures Protected with Tire Chips Cushion", Journal of Geotechnical and Environmental Engineering, ASCE (In Press).
6. Hamada, M., Isoyama, R., and Wakamatsu, K. (1996), "Liquefaction Induced Ground Displacement and Its Related Damage to Lifeline Facilities", Special Issue of Soils and Foundations, **1**, 81-98.
7. Http://www.bousai.go.jp (2006) Cabinet Office, Government of Japan: Central Disaster Management Council, Available: http://www.bousai.go.jp/jishin/chubou, 2006 (in Japanese)
8. Humphrey D.N. (1998), "Civil Engineering Applications of Tire Shreds", Manuscript Prepared for Asphalt Rubber Technology Service, SC, USA.
9. Iai, S. (1989), "Similitude for Shaking Table Tests on Soil-structure-fluid Model in 1g Gravitational Field", Soils and Foundations, Japanese Geotechnical Society, **29**(1), 105-118.
10. Inagaki, H., Iai, S., Sugano, T., Yamazaki, H., and Inatomi, T. (1996), "Performance of Caisson Type Quay Walls at Kobe Port", Special Issue of Soils and Foundations, Japanese Geotechnical Society, **1**, 119-136.
11. Ishihara, K. (1997), "Geotechnical aspects of the 1995 Kobe earthquake, Terzaghi Orientation", 14th International Conference of International Society of Soil Mechanics and Geotechnical Engineering, Hamburg, Germany, **4**, 2047-2073.
12. Japanese Geotechnical Society (JGS) and Japan Society for Civil Engineers (JSCE) (1996), Joint Report on the Hanshin-Awaji Earthquake Disaster (In Japanese).
13. Kamon, M., Wako, T., Isemura, K., Sawa, K., Mimura, M., Tateyama, K., and Kobayashi, S. (1996), "Geotechnical Disasters on the Waterfront, Special Issue of Soils and Foundations", Japanese Geotechnical Society, **1**, 137-147.
14. Towhata, I., Ghalandarzadeh, A, Sundarraj, K.P., and Vargas-Monge, W. (1996), "Dynamic Failures of Subsoils Observed in Waterfront Area", Soils and Foundations, Special Issue, Japanese Geotechnical Society, **1**, 149-160.
15. Tweedie, J.J., Humphrey, D.N., and Standford, T.C. (1998), "Tire Shreds as Retaining Wall backfill, Active Conditions", Journal of Geotechnical and Geoenvironmental Engineering, ASCE, **124**(11), 1061-1070.
16. Yasuda, S., Ishihara, K., Harada, K., and Shinkawa, N. (1996), "Effect of Soil Improvement on Ground Subsidence Due to Liquefaction", Special Issue of Soils and Foundations, **1**, 99-108.
17. Yasuda, S. (2007), "Remediation methods against liquefaction which can be applied to existing structures", Earthquake Geotechnical Engineering, Pitilakis K.D. Eds, 385-405.
18. Yasuhara, K., Hazarika, H., and Fukutake, K. (2006), "Applications of Scrap Tires in Civil Engineering", Doboku Gijutsu, **61**(10), 79-86 (in Japanese).
19. Yasuhara, K., Unno, T., Komine, H., and Murakami, S. (2004), "Gravel Drain Mitigation of Earthquake-induced Lateral Flow of Sand", 13th WCEE, **146**(CD-ROM).

Part-5

Structural Earthquake Engineering

Part-5

Structural Earthquake Engineering

ELASTO PLASTIC ENERGY DISSIPATION DEVICE FOR PASSIVE SEISMIC RESPONSE CONTROL

K. Muthumani[1], K. Sathish kumar[1], N. Gopalakrishnan[1], R. Sreekala[1], G. R. Reddy[2] and Y. M. Parulekar[2]

[1]Scientist, Advanced Seismic Testing and Research Laboratory, Structural Engineering Research Centre, Chennai, kmm@sercm.org, ksk@sercm.org, ng@sercm.org & kala@sercm.org

[2]Scientific Officer, Reactor Safety Division, Babha Atomic Research Centre, Mumbai, rssed@barc.gov.in

ABSTRACT

The strength based design philosophy depending on the inherent ductility of structures is not suitable for lifeline structures like nuclear power plants where even non-structural damage is a severe safety problem. Failure free design of supporting systems for pipe lines carrying highly toxic or radioactive liquids at very high temperature is an important issue in the safety aspect for a nuclear power installation which is a key topic for researchers all around the world. Generally, these pipeline systems are designed to be held rigid by conventional snubber supports for protection from earthquakes. The piping design must balance seismic deformations and other deformations due to thermal effect. A rigid pipeline system using conventional snubber supports always leads to an increase in thermal stresses; hence a rational seismic design for pipeline supporting systems becomes essential. Contrary to this rigid design, it is possible to design a flexible pipeline system and to decrease the seismic response by increasing the damping through the use of passive energy dissipating elements, which dissipate vibration energy. This paper presents the analytical and experimental studies carried out on a yielding type elasto-plastic passive energy dissipating device for the control of large seismic deformations of pipelines subjected to earthquake loading.

Key words: Passive control, passive energy dissipation, elasto plastic device, seismic support

1 INTRODUCTION

Earthquakes are potentially devastating events which threaten lives, destroy property, and disrupt life sustaining services and societal functions. Historically, seismic resistant design of structures has been based upon a combination of strength and ductility. For small, frequent earthquakes characterized by small deformations, the structure can be expected to remain in the elastic range, with all stresses well below yield levels. However, it is not reasonable to expect that a traditional structure to respond elastically when subjected to a major earthquake characterized by large deformations. Instead, the design engineer relies upon the inherent ductility of structure to prevent catastrophic failure, while accepting a certain level of structural and non-structural damage. However this strength based design philosophy is not suitable for lifeline structures like nuclear power plants where even non-structural damage is a severe safety problem. For example failure free design of supporting system for pipe lines carrying highly toxic or radioactive liquids is an important problem in the safety aspect for a nuclear power installation.

Presently, seismic design of pipeline systems in a nuclear plant is conducted under the rigid design philosophy, where large sized rigid supports like snubbers are used as seismic supports. In the seismic support design, it is necessary to consider the thermal load in addition to the seismic load and the snubbers are used as devices to resist seismic load and allow thermal movement of piping. As these devices are relatively expensive and complex mechanisms, more inexpensive and reliable devices are desirable as alternative to snubbers. Many studies (Chiba 1992 and Namita 1991) have been performed to find more reliable and economical support concepts, with various kinds of high damping supports of elasto-plastic, visco-elastic, friction, and lead extrusion type, aiming to reduce seismic response. To evaluate an alternative seismic support for pipeline systems a cooperative study by SERC, Chennai and BARC, Mumbai has been performed on elasto-plastic type passive energy dissipating (PED) elements/devices applied to pipe segment and piping systems. An X plate metal yielding device is one of the devices considered in the study, because it greatly increases the system damping and there is no fatigue problem in the damper structure. In the present experimental study, X shape metal PED elements having different plate thickness were subjected to tests of static, dynamic and seismic type. The test results were correlated against analytical results obtained using a standard mathematical model.

2 ANALYTICAL STUDIES

2.1 Static loading

Fig.1 shows the basic shape of the PED element along with deformed shape, bending moment and curvature variation along the depth of the element. The material characteristics of the X plate PED element, the elastic modulus (E), and the stress hardening rate (H), are expressed by the bilinear model. The skeleton curve of the force-displacement curve (vide **Fig. 2**) of the X plate for each stress level was investigated by beam theory.

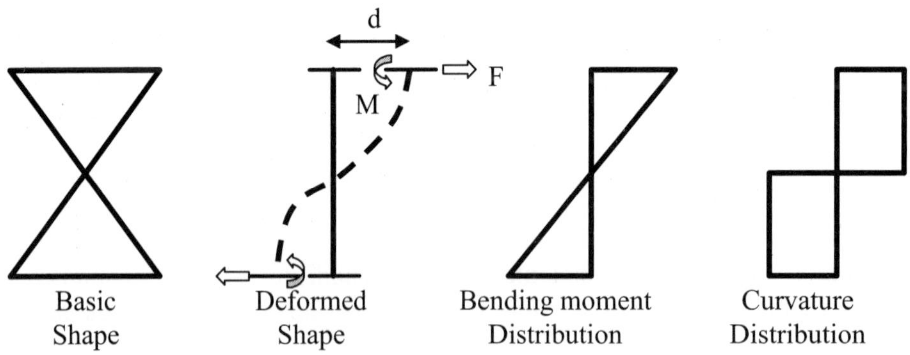

Fig.1 X plate PED element

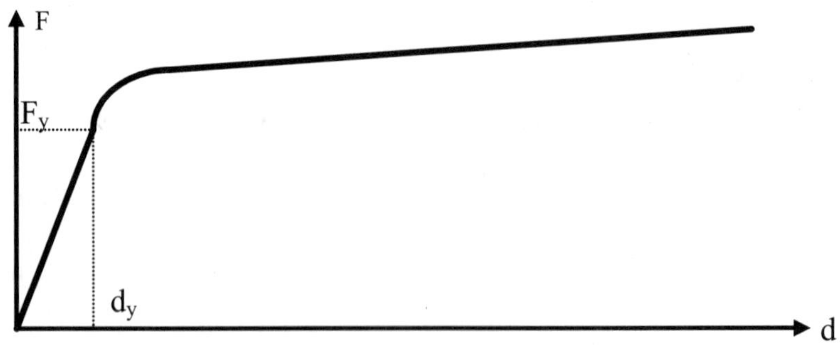

Fig. 2 Static force - displacement curve

In the elastic domain, the force (F) and the displacement (d) relation and the equivalent stiffness (K) are given as follows:

$$F = \frac{Ebt^3 n}{12a^3} d \; ; \; K = \frac{Ebt^3 n}{12a^3} \qquad (1)$$

Where, (a) is the height of the triangular plate, (b) is the maximum breadth, (t) is the plate thickness, and (n) is the number of X plate PED elements. In the plastic domain at the yield point, the X plate surface is plastic and the plate surface stress is (σ_y). Then the yield force (F_y) and corresponding yield displacement (d_y) is given as follows:

$$F_y = \frac{\sigma_y t^2}{6}\frac{b}{a}n; \quad d_y = \frac{2\sigma_y a^2}{Et} \quad (2)$$

Similarly in the plastic region (vide **Fig. 3**), considering the effect due to strain hardening rate of (H), the force-displacement relation is given as follows:

$$F = \frac{nb\sigma_y}{12Ea}\left\{\left(4{y_o}^2 - 3t^2\right)\left(H - E\right) + H\frac{t^3}{y_o}\right\}; \quad (3)$$

Where, (y_o) is the elastic depth, which can be evaluated using,

$$y_o = \frac{\sigma_y a^2}{Ed} \quad (4)$$

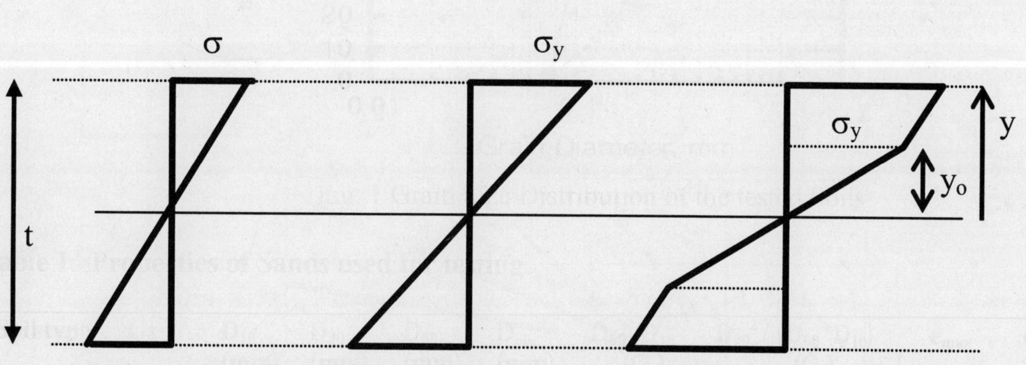

Fig. 3 Variation of stress across thickness at various stages of loading

The hysteretic loop can be obtained by giving the strain-stress characteristics of the very small part as bilinear in each stress condition. The equivalent stiffness (K) and the element damping ratio (h) can show the dynamic characteristics of PED. The former is obtained from the tangential stiffness at the reversal point of the skeleton curve. The latter is obtained as the ratio between the element strain energy and the dissipating energy per cycle (ΔE).

$$\Delta E = 3F_y\left(d + d_1\right) + \frac{{d_y}^2 F_y}{\left(d + d_1\right)} - \left(3d_y F_y\right) - \left[Ft^2 / K\right] \quad (5)$$

$$h = \frac{\Delta E}{2\pi K d^2} \quad (6)$$

2.2 Dynamic / Cyclic loading

In the elasto-plastic analysis of these X plate PED elements under dynamic or cyclic loading, standard hysteretic models are usually used, such as ideal elasto-plastic, bilinear, Bauschinger, Ramberg–Osgood, Takeda, Wen & Polygonal model. Etc. The basic form of the force-displacement model is selected, usually based upon an analogy with plasticity theory, and then the model parameters are determined via a curve fitting procedure. Of all the hysteretic models stated the Ramberg–Osgood model suits closely to such X plate PED elements.

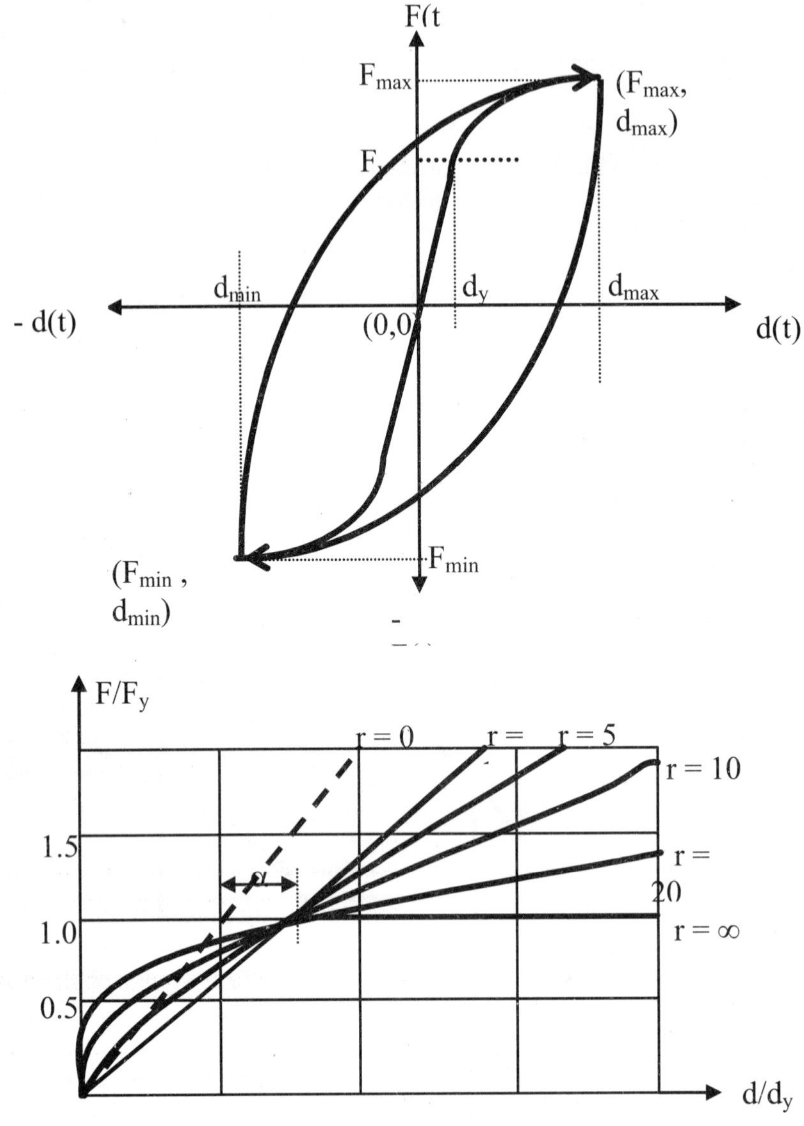

Fig. 4 Ramberg-Osgood model adopted for modeling X plate PED element under dynamic/cyclic loading

The Ramberg-Osgood model (vide **Fig. 4**) essentially establishes a power law relationship between the force and inelastic displacement, and consequently is effective in modeling the response of a variety of metals under cyclic/dynamic loading. This model approximates the shape of the force-displacement curve by appropriate selection of the auxiliary parameters called Ramberg-Osgood exponent (r) and Jennings constant (α), according to the shape of the actual hysteretic curve. In the present study the values of these auxiliary parameters were realistically evaluated from the static test results using curve-fitting procedure. The force-displacement relations at various stages in the hysteretic loop namely the basic loading branch between the coordinates (0,0) and (F_{max},d_{max}), unloading branch between the coordinates (F_{max},d_{max}) and (F_{min},d_{min}) and the reloading branch between the coordinates (F_{min},d_{min}) and (F_{max},d_{max}) can be obtained using the following expressions:

$$\frac{d(t)}{d_y} = \frac{F(t)}{F_y}\left(1 + \alpha\left(abs\frac{F(t)}{F_y}\right)^{r-1}\right) \quad (7)$$

$$\frac{d(t) - d_{\max}}{d_y} = \frac{F(t) - F_{\max}}{F_y}\left(1 + \alpha\left(abs\left(\frac{F(t) - F_{\max}}{2F_y}\right)\right)^{r-1}\right). \quad (8)$$

$$\frac{d(t) - d_{\min}}{d_y} = \frac{F(t) - F_{\min}}{F_y}\left(1 + \alpha\left(abs\left(\frac{F(t) - F_{\min}}{2F_y}\right)\right)^{r-1}\right). \quad (9)$$

The hysteretic energy dissipation index (h) of the X plate PED element using the Ramberg-Osgood model is expressed as follows:

$$h = \frac{2}{\pi}\left(1 - \frac{2\alpha}{r+1}\right)\left(1 - \frac{d_y F_{\max}}{F_y d_{\max}}\right) \quad (10)$$

3 EXPERIMENTAL STUDIES

3.1 PED elements / device

X plate PED elements made of mild steel having three different thickness values of say 3, 4 and 6 mm were adopted in the present study. **Fig. 5** shows the construction of the yielding type elasto-plastic PED device and the individual X plate PED elements used inside the device. The PED element is basically made of mild steel plate having an X shape in plan. The energy dissipation is achieved by deforming the X plate to plastic region. The force-displacement curve shows nonlinear hysteretic characteristics. The seismic response of the piping system is reduced by the energy dissipation of the hysteretic loop. Because mild steel is used in the plastic region, this region must be generated to prevent plastic hinges from developing. Hence a strong uniformly thick X plate is used as a plate spring. When this X plate has large deformation, the plastic domain progresses in the thickness direction of the plate leading to simultaneous yielding over the triangular region. In the present study the height and base of the triangle portion in an individual X plate was kept as 40mm and 40mm. The overall height and width of the single element was kept as 160mm and 160mm respectively. The properties like yield stress (σ_y), Elastic modulus (E) and Hardening rate (H) for the element material were evaluated as 235 MPa, $1.94\text{x}10^5$ MPa and $5.00\text{x}10^3$Mpa respectively.

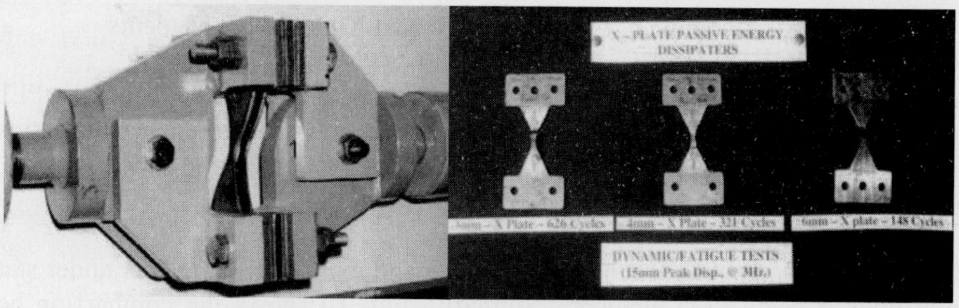

Fig. 5 Yielding type elasto plastic PED device and X plate PED element

3.2 Tests conducted on PED elements

The PED element was initially tested using a displacement controlled 50kN actuator (vide **Fig. 6)** under static conditions for evaluating the yield and post yield behaviour. Then using the same facility dynamic, fatigue and seismic loading tests were also carried out on similar devices up to failure for evaluating the performance in a seismic environment. For the dynamic and fatigue loading the rate of loading was kept as 3.0Hz and with different increasing PED displacement amplitudes of say 5, 10 and 15mm. From the force-displacement (Hysteretic) loops the dynamic stiffness and damping of the element were evaluated. For the seismic loading tests a spectrum compatible displacement time history for a specific site was generated and applied to the element in succession to evaluate their design life.

Fig. 6 X plate PED element under testing

Fig. 7 shows the dynamic test results conducted on X plate PED elements of three different thicknesses at constant displacement amplitude of 15mm at 3.0 Hz and on 6mm thick PED elements at different displacement amplitudes at 3.0 Hz.

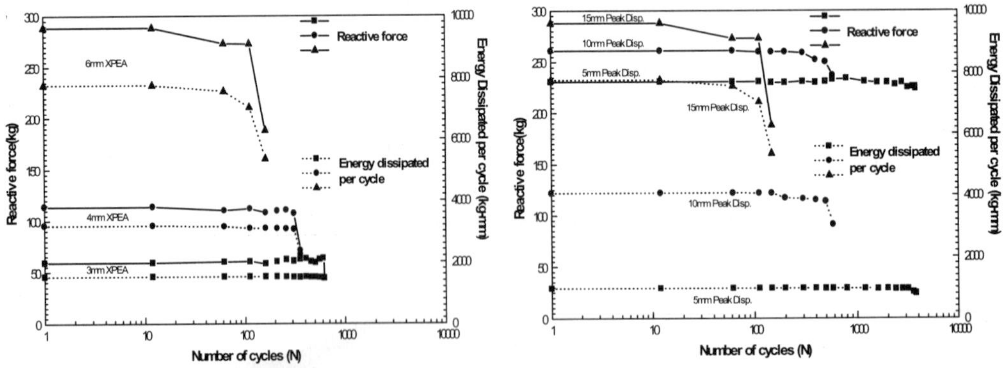

Fig. 7 Dynamic test results on different X plate PED elements

From the reactive forces and force-displacement hysteretic curves obtained from the PED element, the design stiffness and damping of the individual PED elements were evaluated. A typical force-displacement response obtained on a 4mm thick X plate PED element under seismic loading is shown in **Fig. 8**.

Fig. 9 shows the comparison between the theoretical and experimental results under static loading on a 4 mm thick X plate PED element. Similarly, **Fig. 10** shows the comparison between the theoretical and experimental results under dynamic loading at 3.0 Hz. on a 4 mm thick X plate PED element.

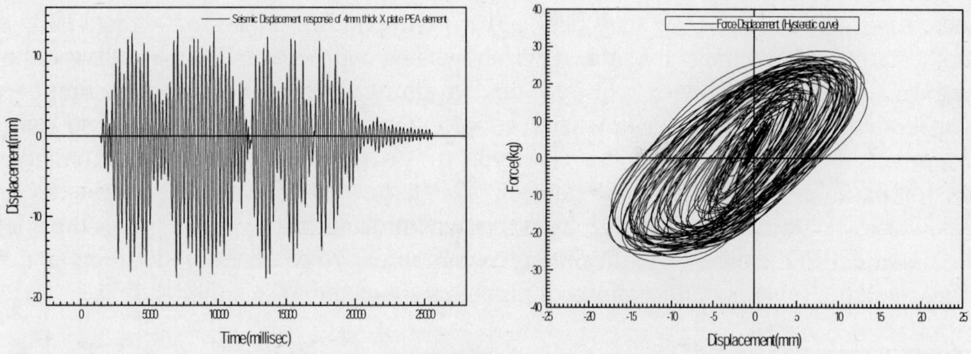

Fig. 8 Typical seismic responses measured on the X plate PED element under seismic loading

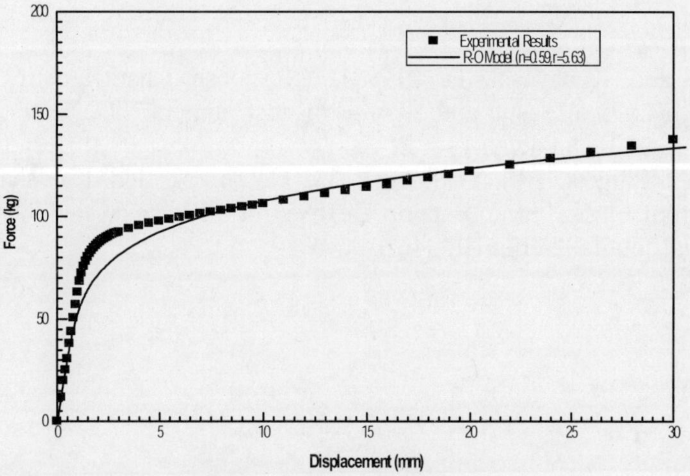

Fig. 9 Typical comparison of analytical and experimental results for static loading on a 4 mm thick X plate PED element

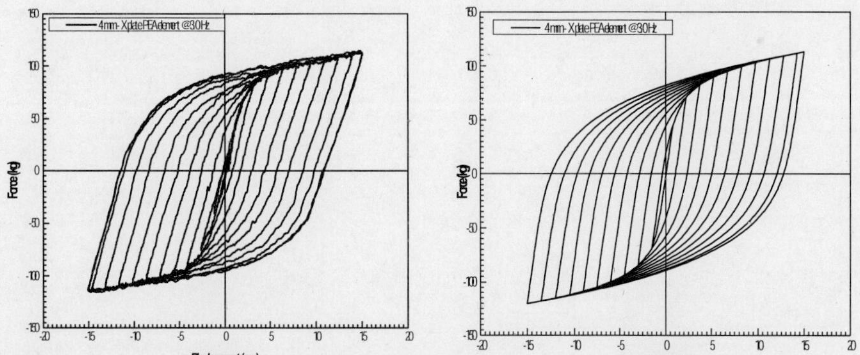

Fig. 10 Typical comparison of analytical and experimental results for dynamic loading at 3.0 Hz. on a 4 mm thick X plate PED element

4 OBSERVATIONS AND CONCLUSIONS

Close match is found between the analytical and experimental responses of X plate PED elements under both static and dynamic loading. This shows the validity of Ramberg-Osgood model in predicting the design parameters required for designing PED devices of yielding type. The analytical model so generated can be easily adopted to standard FEM software so that detailed seismic analysis can be carried out for any type of structural system incorporating these PED elements for passive seismic response control. The large and stable force-displacement curves (hysteretic curves) obtained up to failure from the dynamic and seismic tests shows the adaptability of such X plate PED elements in designing a suitable passive energy dissipating device for controlling the large seismic deformations of pipelines subjected to seismic loading.

5 ACKNOWLEDGEMENTS

The authors wish to express their deep appreciation to Director, SERC, Chennai and Head, Reactor safety division, BARC, Mumbai, for their precious advice and kind permission to publish the combined research work carried out at SERC.

REFERENCES

1. Chiba, T. and Kobayashi, H. (1992), "Response Characteristics of Piping System Supported by Visco-Elastic and Elasto-Plastic Dampers," Transactions of the ASME on Pressure Vessel Technology, 112, 34-38.
2. Namita, Y., Shigeta, M., Yoshinaga, T., Gotoh, N. and Kawahata, J. (1991), "The Application of Elasto Plastic Support Devices for a Piping System," Proceedings of JSME International Journal, Series III, 34(1), 42-47.

MODELLING POST PEAK BEHAVIOUR OF CONCRETE FOR SEISMIC APPLICATIONS

R.Sreekala, N.Lakshmanan, K.Muthumani, N.Gopalakrishnan and K.Sathish Kumar
Structural Engineering Research Centre, CSIR Campus, Taramani, Chennai 600 113, India
kala@sercm.org, nlaxman@sercm.org kmm@sercm.org, , ng@sercm.org & ksk@sercm.org

ABSTRACT

Considerable research has been carried out in the recent years in the development of models to simulate the inelastic responses of reinforced concrete elements. Post-peak behavior and enhancement of ductility are of special interest to the seismic design of concrete structures. Ductility analysis of reinforced concrete is seldom carried out in normal design practice, because no direct methods are available. To evaluate structural ductility of a beam section, it is necessary to analyze the complete moment curvature relation of the section covering the post peak ranges and then calculate the inelastic curvature that the section can sustain before failure. Here an attempt has been made to study the post peak behavior of two types of reinforced concrete beams, namely normal, and ductile concrete tested monotonically till failure. An experimental investigation was carried out in a displacement-controlled test system. The beams were modeled in a finite element software-ANSYS and excellent agreement with the experimental load-deflection response and the strains were obtained. The moment-curvature relationship was evaluated using the strain profile thus obtained using the model. The analytical model can be extended to study the behavior under cyclic loading. The paper also highlights the importance of post-peak behavior in the inelastic response of concrete structural elements and its relevance in seismic resistant design.

Key words: Ductility, concrete strain, flexural behaviour, finite element modeling

1 INTRODUCTION

Inelastic response of concrete structural elements is of high significance to seismic resistant design. Structures expected to be exposed to ground motion are designed according to their ductility and for such structures, the post peak information is required to check the structural performance. Ductility evaluation is essential to arrive at response reduction factors in the design stage calculations. Moment-curvature relationship from non-linear stress strain diagrams is needed for a more accurate and general method of reinforced concrete member analysis. Towards this objective a study was made on the behavior of a flexural member. Traditional flexural design of reinforced concrete (RC) beams concentrates on the provision of adequate strength for resisting applied loads at ultimate limit state and sufficient stiffness for limiting the deflection at serviceability limit state. The post peak behaviour is usually ignored and the nominal ductility is provided by imposing certain empirical rules on reinforcement detailing. From the structural safety point of view, ductility is as important as strength. Possession of good flexural ductility would enable a structure to dissipate excessive energy through inelastic deformations within the potential plastic hinge regions while maintaining sufficient flexural strength to resist applied loads. A relatively high level of flexural ductility would provide the structure an increased chance of survival against accidental impact and seismic attack. Flexural ductility is particularly important when the capacity design philosophy for seismic resistant structures is adopted (Park et al, 2001). According to this philosophy, which is also called the strong column- weak beam approach, the beams should yield before the columns yield and the beams should be required to have sufficient flexural ductility such that the potential plastic hinges in the beams are able to maintain their moment resisting capacities until the columns fail. So a non-linear analysis extended into the post peak range of the beam section is necessary to evaluate the flexural ductility. Ductility of reinforced concrete elements can be taken as an added advantage for the design of structures against seismic loading (Lakshmanan et al, 2003). The philosophy of most modern building codes concentrates essentially on providing ductility in the main structure to prevent its collapse during a major seismic event.

Ductility can be incorporated in a structural element by seismic detailing or at the material level of concrete modification. It is also well known that inclusion of fibers in concrete matrix, and use of natural rubber latex polymer-modified concrete improves the ductility of reinforced concrete elements (SERC Research Report, 2005). In the presence of fibers, void content was found to be minimum when latex cement ratio is 0.15(Pu-woeichen, 1997) along with a reduction in the compressive to tensile strength ratio, an improvement in the extensibility and a relatively larger strain at failure was observed. The paper describes a method by which ductility is enhanced in the material level and the results of the non-linear analysis of the beam section through FEM software. An experimental investigation was carried out in a displacement-controlled set up. Two types of concretes were considered, one normal and the other ductile using the natural rubber latex modified concrete. The steel ratio (tension and compression) and steel yield strength were kept same for the beams. The load deflection behavior including the monotonic moment curvature relationship has been found out using the non linear FEM analysis and excellent agreement is observed in the moment-curvature, failure pattern and the force displacement behavior, with the experimental results. The moment curvature relationship under monotonic loading is analyzed for RC sections based on the different stress strain relationships. The increased ductility and curvature for ductile concrete is of significance for seismic design of structures, especially for flexural members of statically indeterminate structures.

2 PRESENT INVESTIGATIONS

The materials chosen for the stress-strain studies are one normal strength concrete designated as RC, and natural rubber latex modified concrete with fibers (Dry Rubber Content 0.5) designated as LMFC 0.5. Volume fraction of fibers was one percent with the aspect ratio of 60. The ductile concrete was prepared as per ACI mix design procedure conforming to ASTM standards for fiber reinforced concrete. The stress-strain characteristics were obtained using 200mm×100mm cylinder specimen subjected to compressive loading (rate of displacement 0.2 mm/min). These were obtained using the MTS servo controlled 2500KN universal testing machine. The closed loop control was achieved using lateral strain. Fig.1. shows the stress-strain characteristics of the selected concrete matrices based on average of three specimens. The photograph of the cylinder testing is shown in fig.2(a). Fig.2b.shows the observed ultimate failure pattern in the specimens. The failure pattern as observed in the cylinders shows the ductile mode of failure in case of polymer modified concrete, namely LMFC0.5.

Further flexural tests were carried out on the concrete beams RC and LMFC, which are of approximately the same concrete-grade as observed from the cylinder tests. The selected RC beam serves the purpose of comparison. Three point bending tests were carried out with a central concentrated load. The schematic sketch of the test set up is shown in Fig. 3. The beam specimens had breadth of 100mm, depth of 150mm, and were tested over an effective span of 1.4m. The beams were reinforced using 2-10# bars as tension reinforcement. The stirrups hangers were also 2-10# bars provided with an effective cover of 30mm for all beams. 6# stirrups were used throughout at 90mm spacing as the shear reinforcement. Analytical model of the beams were developed using user-friendly finite element software.

3 OBSERVATIONS

3.1 Stress- Strain Behavior of the Concrete

In Figure 1 stress-strain curves for the concretes are shown. It is interesting to observe that, in LMFC behaves more ductile and almost equal values of initial tangent modulus with RC were noticed. The stress-strain curve for RC has a sharp peak and a steeper descending branch after the peak load. It is quite evident that the natural rubber latex polymer with fiber modification helps to improve the post-peak response of the beams. The research here mentioned also focuses on how to improve the compressive ductility of plain high strength concrete. During cylinder testing it was observed that the failure of the rubberized concrete specimens under compression loading did not demonstrate a brittle failure, but rather a ductile, plastic failure. The typical photograph of the failure pattern is shown in fig. 2(b). It is clear that a tri axial state of stress gets developed in the concrete matrix in the ductile composites with fibers in the post peak region, which is responsible for the above phenomena (Van Mier-1986).

3.2 Flexure Behavior of the Concrete Beams

The observed failures in the beams were flexural failures. For evaluating the ductility factor, the theoretical yield moment was computed. Theoretical capacities were computed using the stress-strain curves of steel and concrete, and assuming linearity of strain across depth. The peak loads were computed as 25KN for both the beams. For evaluating the ductility factor, the theoretical yield moment was computed using the relation.

$$My = As.fy.\ (0.87d) \quad (1)$$

The lever arm at yield has been assumed at 0.87d, and quite acceptable. After evaluating the yield moment, the deflection at yield was taken from the load-deflection behavior of test beams. All the beams failed in flexure at the loading point. Typical photographs of the failed specimens are shown in fig.4. Table.1. gives the significant data taken from load deflection curves. The displacement ductility was computed, $\mu_{\Delta} = \Delta_u/\Delta_y$ and are included in the table.1. The ultimate deformation was computed for 15 percent load drop (Rao P.S., 1998). A displacement ductility factor of 5.5 has been arrived with LMFC instead of 3.10 in RC beams. The studies revealed that the LMFC-ductile composite maintains composite integrity and a flexural failure mode are observed. A compatible inelastic deformation between reinforcement and the composite reduces the interfacial bond stress. The analysis of the flexural member has been carried out by suitably modeling in a Finite element software and excellent agreement observed in the failure pattern and the monotonic load-deflection behavior. The strain values obtained from the modeling matches well with the theoretical results.

4 MODELLING THE BEHAVIOUR OF BEAMS

The model developed takes into account the gradual spread of inelastic deformations in the case of a simply supported beam loaded by a concentrated load at the center.

The concrete element is a 3-D structural solid element, with special cracking and crushing capabilities. 'Solid 65' an eight node brick element used for the 3-D modeling of solids with rebar capabilities have been used for the purpose. The treatment of non-linear material properties is the most important aspect of this element. The concrete is capable of cracking (in three orthogonal directions), crushing, plastic deformation, and creep. Rebar specifications can be input as real constants, include the material number, the volume ratio and the orientation angles. Rate independent multi linear kinematic hardening option with Von-Mises yield criterion has been used to define the material property of steel rebar. For capturing the constitutive behavior of concrete the element, a low-tension material model is chosen which provides a smeared crack approach accounting the cracking contribution in RC members. The model is as shown in fig.5.

The compressive stress-strain values of the two concretes namely RC and LMFC were given as input. In order to capture the failure point two analyses were carried out by changing the shear transfer coefficients in opening and closing for the two types of concretes considered. One in which the failure strain of concrete has been defined and another in which the failure deflections were arrived. It was interesting to observe that the analysis converges to a failure point when the ultimate compressive strain in the concrete material reaches in the first case. To arrive at the deflections beyond the ultimate strain capacity of concrete, the stress-path dependence of compressive steel beyond bond-slip has to be considered in this case. A linear approximation has been used to fit the FEM data (like the Kent and Park Model) in order to arrive the full theoretical curve in this region (Kent et al, 1971). The strain profile at the ultimate stage (strain capacity of concrete) using the model developed is shown in fig.6. Fig.7. shows the comparison of the experimental and theoretical curves in the case of RC. The load-deflection response comparison for RC and LMFC is shown in the fig.7 and fig.9 respectively. Table.2. gives the extreme compressive fiber strain and the tensile strain developed in the tensile steel during successive stages of loading. It gives a clear indication regarding the increased ductility capacity of the Latex Modified Fibre Concrete while capturing the initial cracking and the subsequent hardening features. It is observed from the values that the model well predicts the behavior of the LMFC as shown in figure.9.

The neutral axis depth was arrived from the strain values obtained from the analysis. The moment curvature relationship was established using the compression stress block approach. The compressive strain on the concrete extreme fiber was limited to the ultimate values given in the input and the variation in the calculated moment-curvature reveals the effect of this. The moment-curvature relationships for the two types of concretes are plotted in fig 10. The increased curvatures in the case of ductile concrete can be noted and is of special advantage for seismically loaded structures. This shows the moment redistribution in the plastic hinge zone. The displacement and curvature ductility of 3 and 5.86 in normal concrete is replaced by 6.97 and 8.68 in the ductile concrete. The study can be extended for different structural elements and assemblages.

5 CONCLUSIONS

Presence of ductile elements enables the design of members with lesser strength than the elastic strength demand imposed by ground motions. The experimental program indicated that addition of natural rubber latex in concrete along with fibers is an effective measure to enhance the flexural ductility. Improvement of our modeling tools is very much essential for very important, yet un-quantified, properties like ductility, failure mode etc. The analysis of a flexural member using a ductile material has been carried out using a user friendly Finite Element Software and excellent agreement observed in the strains developed, the failure pattern and the force displacement behavior, with the theoretical results. A good correlation between experimental and non-linear finite element results can be achieved. Test results reveal the advantage of using ductile composites in seismic resistant construction. As the progressive damage occurring to the beams could be found from the finite element analysis, design of members using these ductile concretes can be made easier based on performance criteria.

6 ACKNOWLEDGEMENTS

The authors wish to express their deep appreciation to Director, SERC, Chennai, for the constant encouragement, and support in carrying out the research work and publishing the paper. The kind help and support provided by the laboratory staff during the experiment is gratefully acknowledged.

REFERENCES

1. Kent D.C and Park R.(1971)'*Flexural members with confined concrete,",Journal of structural division,ASCE*,vol.97,ST7,pp 1969-1990.
2. Lakshmanan.N(2003), "Ductility, energy absorption, and damage modeling of concrete structural elements", keynote paper, *International conference on recent trends in concrete technology and structures, INCONTEST-2003*, sept 2003, pp 69-85.
3. Pu-woei chen, Xuli Fu, and Chung, D.D.L., "Micro Structural and Mechanical Effects of Latex-Methyl Cellulose and Silica Fume on Carbon Fibre Reinforced Cement", *ACI Materials journal, pp147-155, Mar-April,1997*
4. Park.R (2001), "Improving the resistance of structures to earthquakes", Bulletin of the New Zealand National Society for Earthquake Engineering, 34, No. 1.,1-39.
5. Rao.P.S, Sharma.B.S, Lakshamanan.N, and Strangenberg.F(1998), "Damage model for reinforced concrete elements under combined loading", *ACI material journal, Nov-Dec, 1998, pp682-690.*
6. SERC Research Report –Ductility Improvement using natural Rubber Latex Modified Concrete Beams- MLP 9641.
7. Van Mier J G M.'' Influence of Damage Orientation distribution on the multi axial stress-strain behaviour of concrete",*Cement and Concrete Research ,1985;Vo.l 15:pp849-862*

Table .1. Characteristics Of Tested Beams

Sl. No	Type of beams	Yield		Peak		Ultimate		Disp Duct-ility
		Load *	Def *	Load	Def	Load	Def	
1	RC	20	8.2	25	15.35	18.75	25.36	3.10
2	LMFC0.5	23	7.8	24.4	10.65	20	42.6	5.46

Load in kN, Deflection in mm

Table. 2. Strains developed during displacement controlled loading. FEM results of non-linear Static analysis

Displacement (mm)	RC		LMFC	
	Extr.fibre Compressive Strain	Extr.fibre Tensile strain	Extr.fibre Compressive Strain	Extr.fibre Tensile strain
3	0.00065	0.001001	0.000619	0.001028
6	0.001143	0.001851	0.00157	0.002801
9	0.002049	0.004073	0.001967	0.004004
12	0.002367	0.005335	0.002528	0.004907
15	0.003124	0.006566	0.003995	0.005879
18			0.004896	0.006417
21			0.005244	0.007787
24			0.005708	0.008709
27			0.006493	0.009698
30			0.007356	0.010492

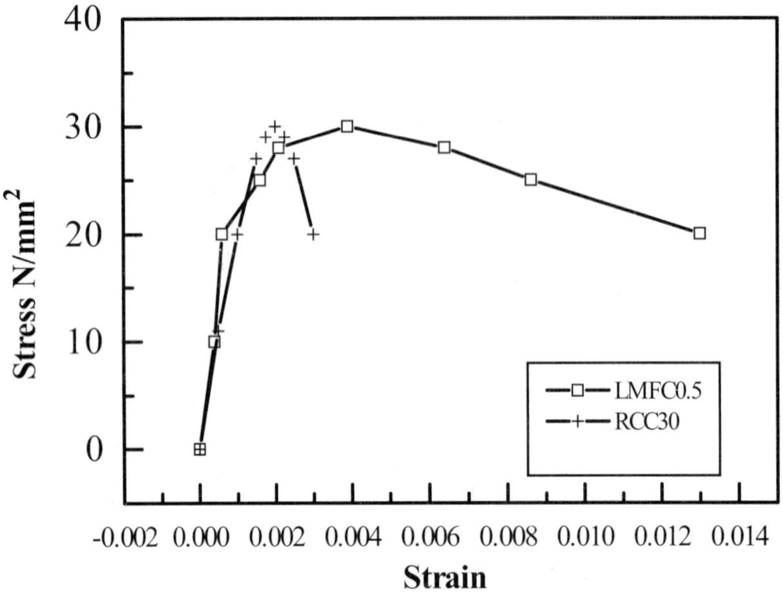

Fig. 1 Average stress- strain curves of the concretes selected

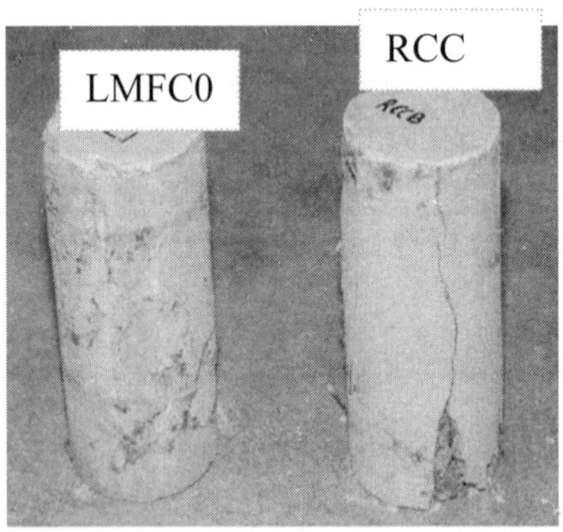

Fig.2(a) **Fig.2(b)**

Fig. 2(a) Cylinder testing using MTS test facility, 2(b) failure pattern observed

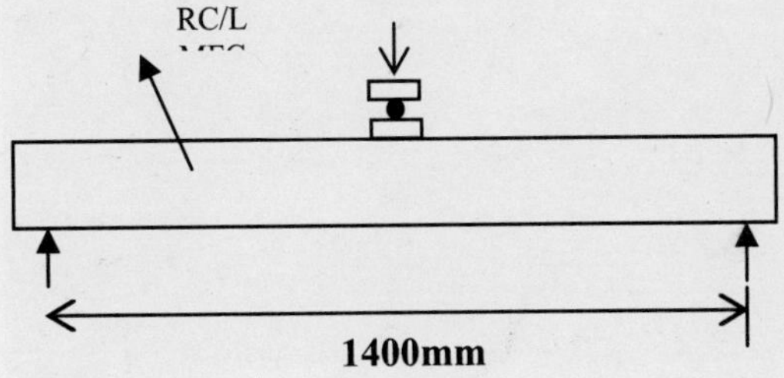

Fig. 3. Schematic sketch of the Test Set-up

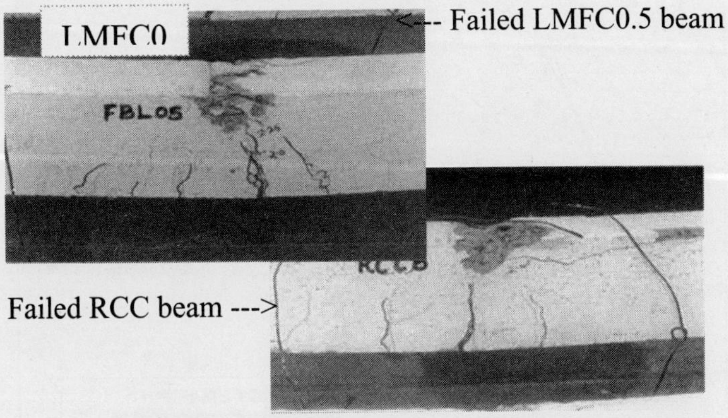

Fig. 4 . Typical Failure of beams

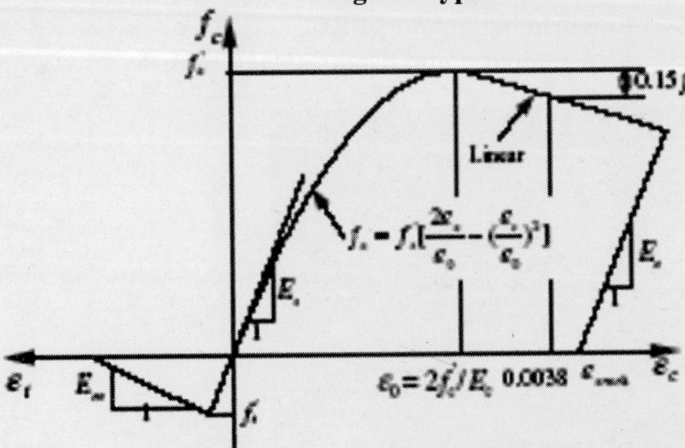

Fig. 5. Non linear material model of Concrete

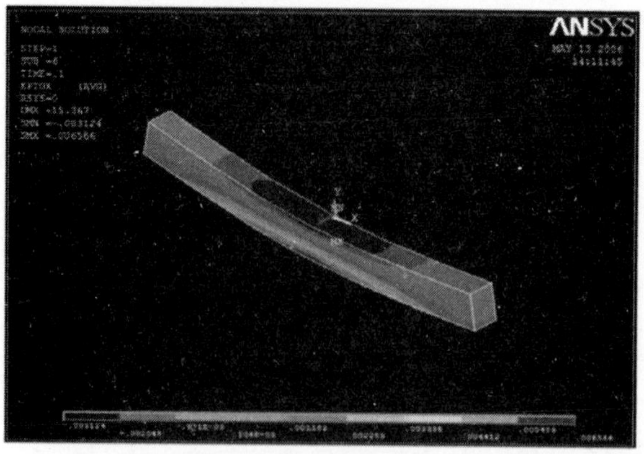

Fig. 6. Strain Profile at the ultimate stage using the model developed-RC

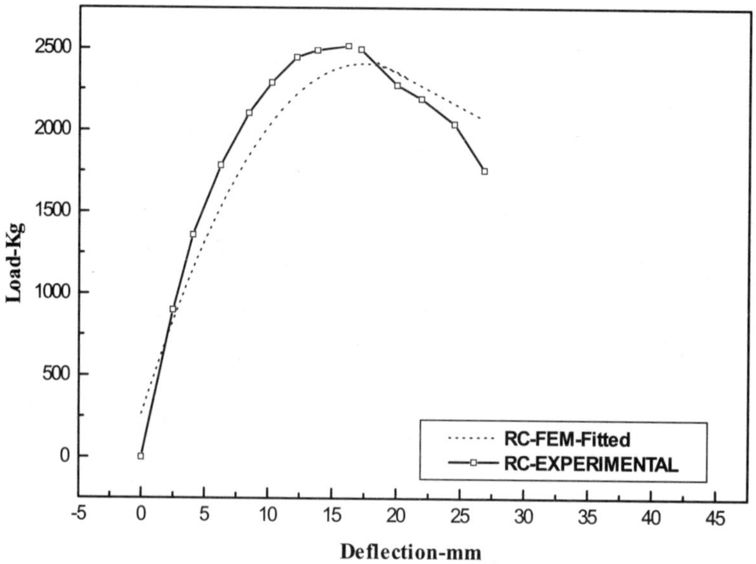

Fig. 7. Flexural behaviour of RC, Experimental and Theoretical Curves

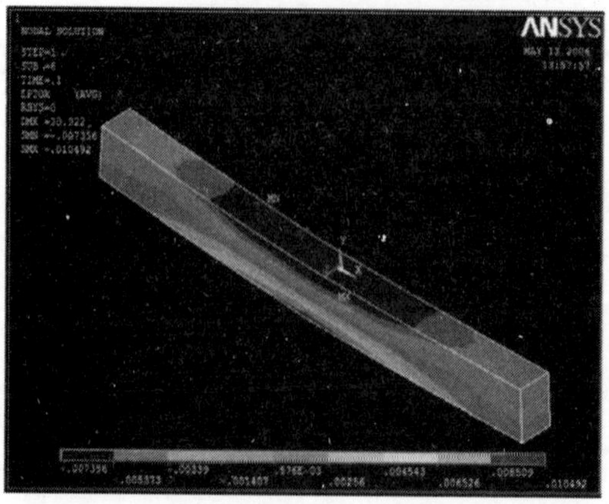

Fig. 8. Strain Profile at the ultimate stage using the model developed-LMFC

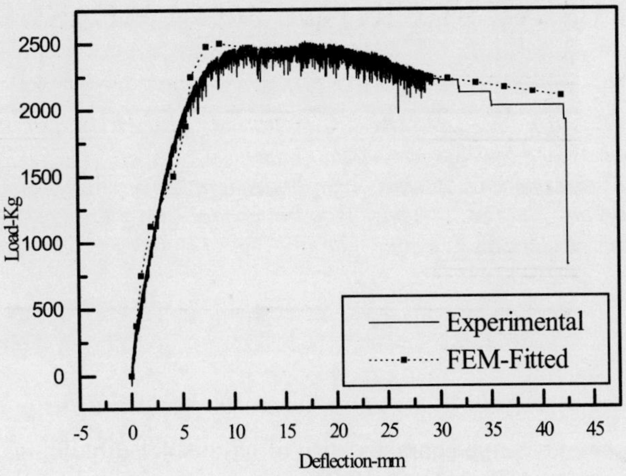

Fig. 9. Flexural behaviour of LMFC0.5- Experimental and Theoretical Curves

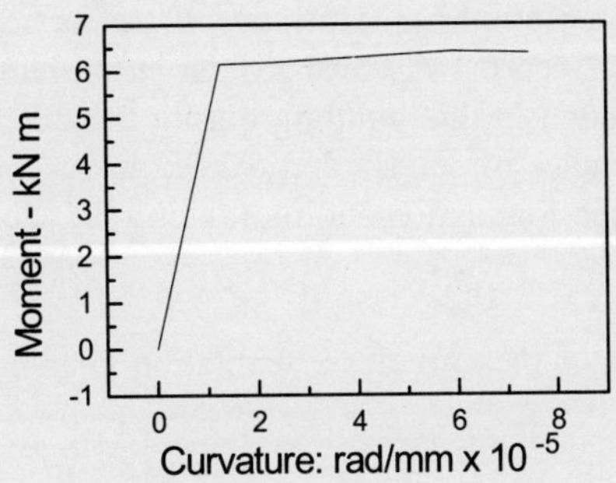

10(a) RC

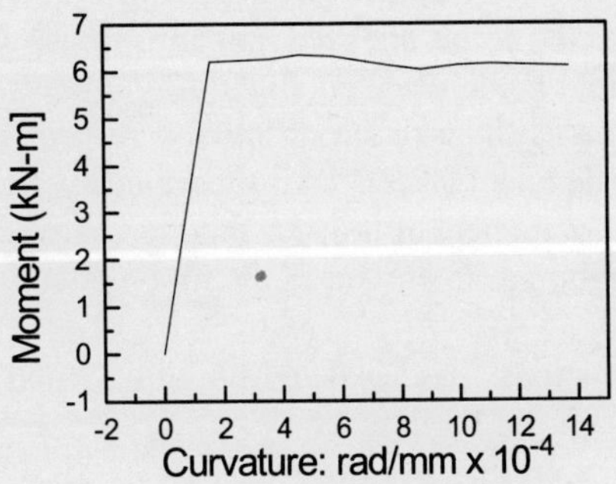

10(b) LMFC 0.5

Fig. 10. Moment Curvature relation ship developed using the model

VIBRATION CHARACTERISTICS OF BASE ISOLATED LOW RISE BUILDING USING MICROTREMOR MEASUREMENTS

***T. Sato*[1], *K. Tokeshi*[2], *C. Cuadra*[3] and *H. Hazarika*[4]**

[1]*Dept. Architecture and Env. Systems, Akita Prefectural University, m09c004@akita-pu.ac.jp*
[2]*Dept. Architecture and Env. Systems, Akita Prefectural University, tokeshi@akita-pu.ac.jp*
[3]*Dept. Architecture and Env. Systems, Akita Prefectural University, carlos@akita-pu.ac.jp*
[4]*Dept. Architecture and Env. Systems, Akita Prefectural University, hazarika@akita-pu.ac.jp*

ABSTRACT

An attempt to infer some dynamic characteristics of base isolated buildings using microtremor measurements is shown here. The target building is this research is a base isolated RC 3-story building, which was built by Tohoku University in 1980. Two predominant peaks in transfer functions between slab foundation and roof were obtained in the NS component microtremor measurements. The values of 1.9 Hz and 5.91 Hz for the first and the second predominant frequencies were obtained, which would correspond to the first and the second normal mode of the experimental building. These assumptions were obtained after filtering microtremor records for above mentioned predominant frequencies. In general, the first normal mode of vibration is expected to be the predominant in base isolated buildings. However, the effect of the second normal mode detected slightly by microtremor records was strong in target building, according to the earthquake strong motion records obtained in this building during the 2003 Miyagi-oki Earthquake (M7.2). Two predominant frequencies of 1.1 Hz and 5.2 Hz were obtained in the NS component of transfer function between slab foundation and roof, being the second mode of vibration as important as the first mode.

Keywords: Base isolated building, Microtremor measurements, Vibration characteristics, 2003 Miyagi-oki Earthquake

1 INTRODUCTION

Since the large damages caused by the 1995 Hyogo-ken Nambu Earthquake, several techniques to reduce the impact of seismic force on structures have been developed in Japan. One of these techniques is the installation of vibration control devices in buildings. Among these kinds of devices, seismic isolators are popular, and buildings using them are called "base isolated buildings". It is known that the elasto-plastic behavior of seismic isolators switch former vibration characteristics of buildings to lower frequencies during earthquake motions.

In general, the first normal mode of vibration is expected to be the predominant in base isolated buildings. However, the use of seismic isolators does not switch only the first normal mode of former upper structure, but others higher frequencies too. The estimation of all preponderant vibration modes is essential to prevent the so-called resonance phenomenon. In recent years, several papers have reported successful results in using microtremor measurements for estimating the vibration characteristics of buildings [e.g., Nishikawa *et al.* (2003), Kanai *et al.* (2006)]. In this sense, this paper reports a case study on the application of microtremor measurements in a base isolated building to infer preponderant modes. Also, the vibration characteristics estimated from microtremor measurements are compared with the predominant frequencies of earthquake strong motion records obtained at target buildings during the 2003 Miyagi-oki Earthquake (M7.2).

EXPERIMENTAL BUILDING

The target building used as case study in this research is an experimental RC 3-story building, which was built by Tohoku University in 1980 (Photo 1). This base isolated building has 6 high damping laminated rubber bearings. The outer walls are autoclaved lightweight concrete (ALC) timber panels.

2 MICROTREMOR MEASUREMENTS

Simultaneous synchronized time microtremor measurements at 7 observation points (① ~ ⑦) in experimental building were recorded during 600 s with 100 Hz sampling frequency, respectively (Fig. 1). Two horizontal sensors were set in each observation point. Simultaneous frames of 40.96 s containing stationary vibrations were selected in all records.

Photo 1. A view of experimental base isolated building

Units: mm

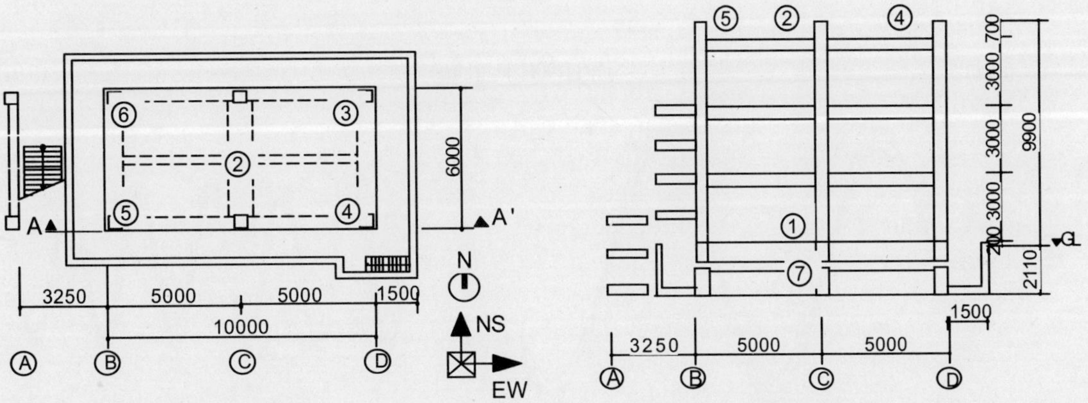

Floor plan

Section A-A'

Figure 1. Location of observation points in experimental base isolated building. Microtremor measurement points are ① ~ ⑦, and earthquake strong motion observation points are ①, ② and ⑦.

Two predominant peaks were obtained from transfer functions between the ground and the roof of NS component microtremor measurements (Fig. 2). The values of 1.9 Hz and 5.91 Hz for the first and the second predominant frequencies of the transfer functions were used for filtering all former microtremor measurements recorded at the roof (② ~ ⑥). Figure 3 shows corresponding waveforms with same phase for 1.9 Hz and 5.91 Hz, which could be interpreted as the first and the second normal mode of the base isolated building. These assumptions were verified when comparing waveforms of observation points at slab foundation ⑦, at first floor ①, and at roof ② filtered for 1.9 Hz and 5.91 Hz (Fig. 4).

The waveforms of 1.9 Hz show relative movements in phase for observation points at first floor and at roof (① and ②), while the slab foundation movements is very tiny (⑦) (Fig. 5a), meaning this frequency would represent the first normal vibration mode of the system. The waveforms obtained for 5.91 Hz show inverse phase for observation points at first floor and at roof (① and ②), and very small movements for slab foundation observation point ⑦ (Fig. 5b), meaning that 5.91 Hz could represent the frequency of the second normal vibration mode.

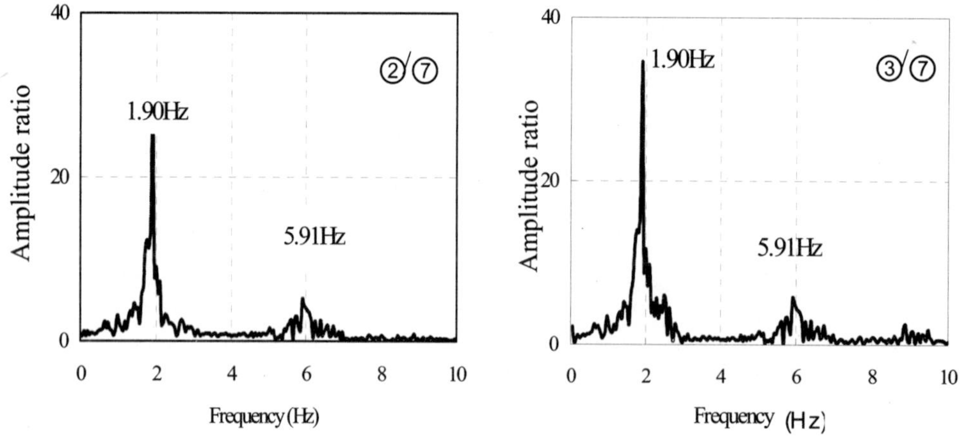

Figure 2. Transfer functions of NS comp. microtremor records

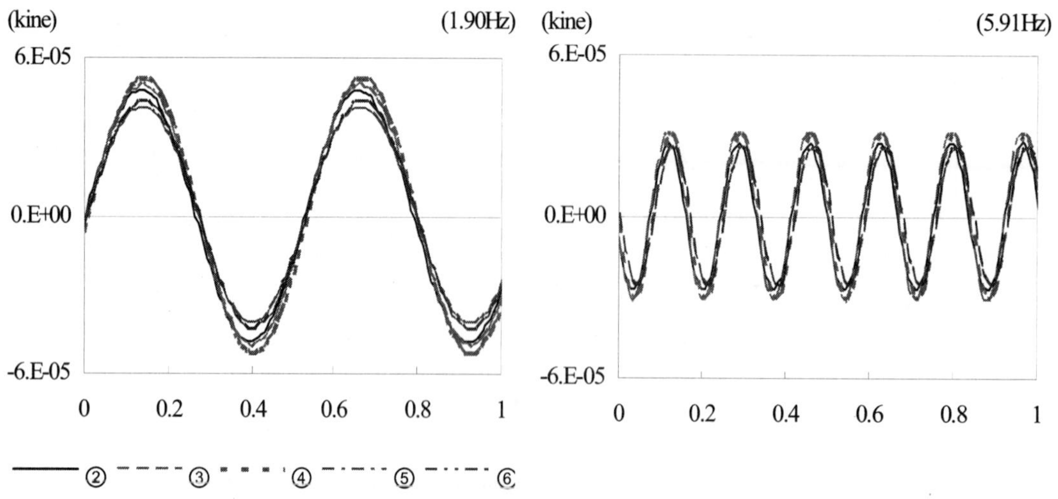

Figure 3. Waveforms obtained after filtering for1.90 Hz and 5.91 Hz

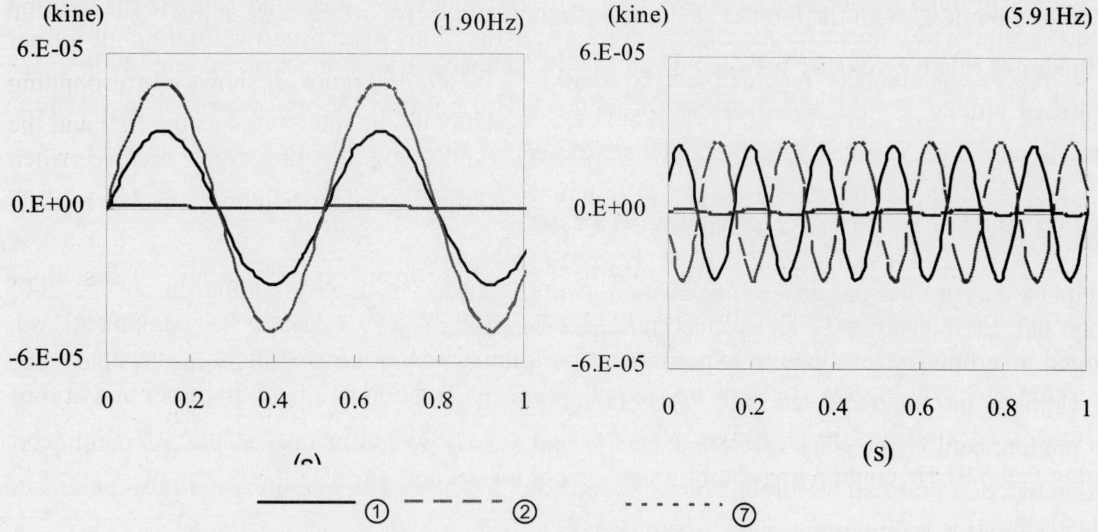

Figure 4. Waveforms obtained for slab foundation ⑦, first floor ① and roof ②, filtered for 1.90 Hz and 5.91 Hz, respectively.

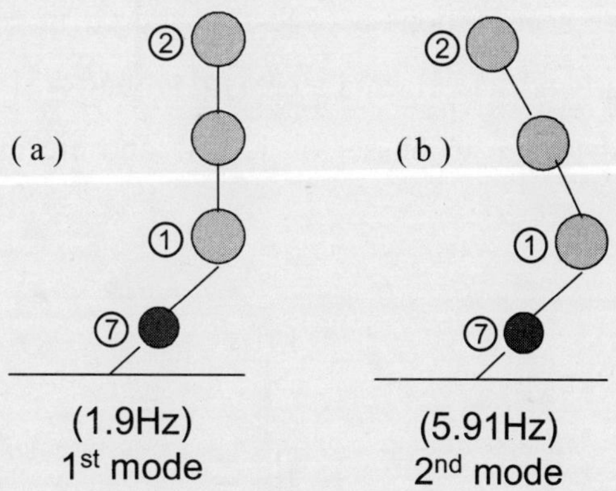

Figure 5. First and second normal vibration modes estimated from microtremor records

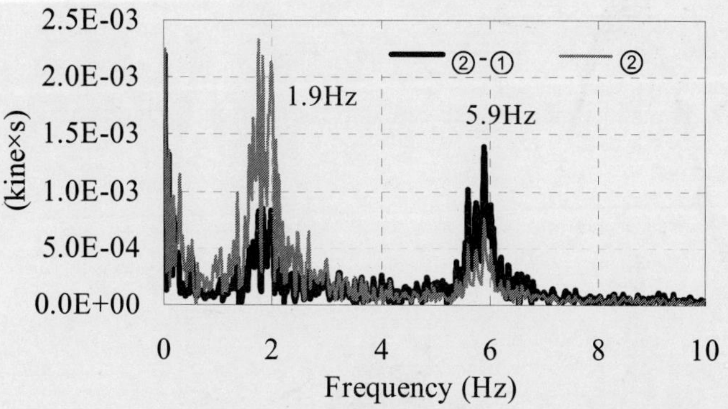

Figure 6. Fourier spectra of microtremor observation point ② and relative velocities between ① and ②

Figure 6 shows another verification by comparing the Fourier spectrum of microtremor records at observation point ② with the Fourier spectrum of relative velocity between ① and ②. The spectral amplitudes of relative velocity between ① and ② in comparison with the spectral amplitude of ②, diminished around 1.9 Hz and increased around 5.9 Hz, because the movements at observation points ① and ② are in phase for 1.9 Hz and are in inverse phase for 5.9 Hz, respectively.

3 EARTHQUAKE STRONG MOTION RECORDS

Earthquake strong motions at several locations were recorded in target building during the 2003 Miyagi-oki Earthquake (M7.1). A maximum acceleration of 159 Gals in NS component was recorded at ground level close to experimental building. Locations of earthquake strong motion observation points are displayed as ①, ② and ⑦ in Fig. 1.

Two predominant frequencies of around 1.1 Hz and 5.2 Hz were obtained in the NS component transfer function between slab foundation ⑦ and roof ②, being the amplification ratio of second mode of vibration as important as the first mode (Fig. 7).

Figure 8 shows similar verification shown in Fig. 6 but using earthquake strong motion records. Fourier spectrum of NS component acceleration strong motion recorded at observation point ② is compared with the Fourier spectrum of relative accelerations obtained between ① and ②. The spectral amplitudes of relative accelerations between ① and ② in comparison with the spectral amplitude of ②, diminished around 1.17 Hz and increased around 5.21 Hz, because the movements at observation points ① and ② are in phase for 1.9 Hz and are in inverse phase for 5.9 Hz, respectively.

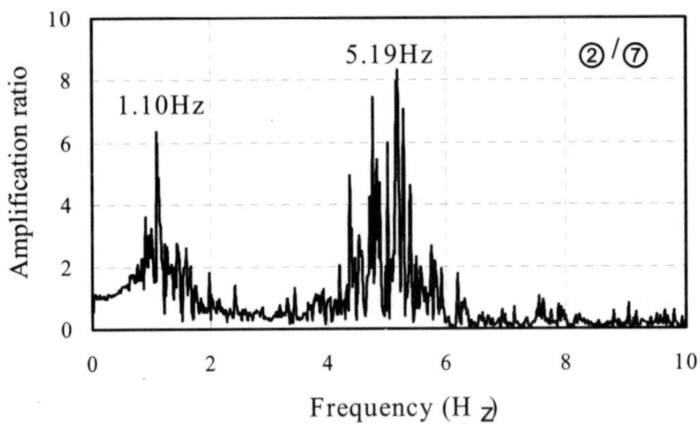

Figure 7. Transfer function between slab foundation ⑦ and roof ②.

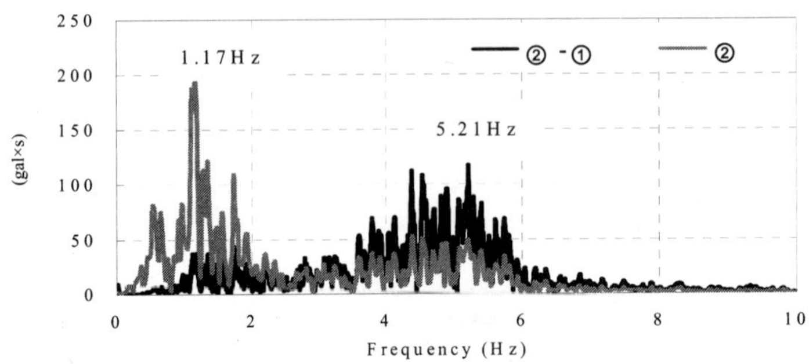

Figure 8. Fourier spectra of NS component accelerogram recorded at observation point ②, and for relative accelerations between ① and ②

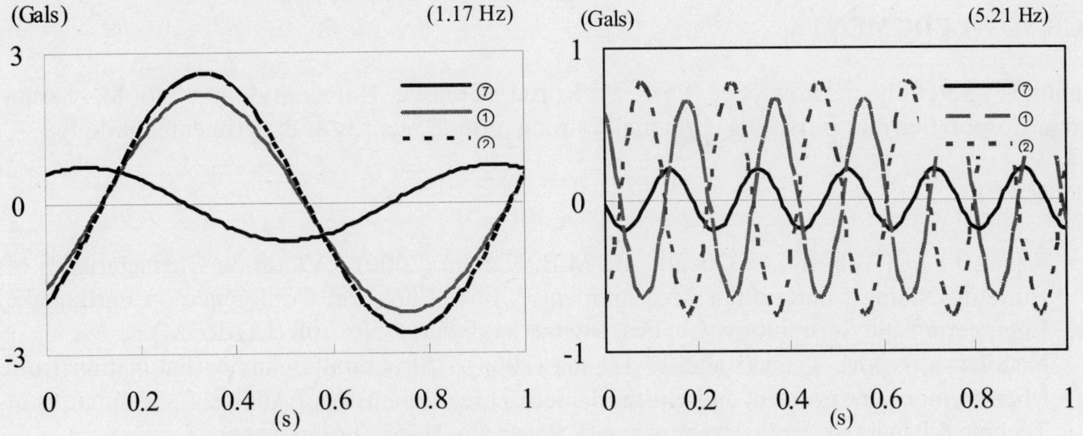

Figure 9. Waveforms obtained for slab foundation ⑦, first floor ① and roof ②, filtered for 1.17 Hz and 5.21 Hz, respectively.

Table 1. Comparison of NS component predominant frequencies obtained at base isolated building (Hz)

Vibration mode	Microtremor measurements	2003 Miyagi-oki Eathquake strong motion records
1st normal mode	1.90	1.17
2nd normal mode	5.91	5.21

Waveforms of earthquake strong motions at first floor ①, and at roof ② are in phase for 1.17 Hz, and in inverse phase for 5.21 Hz, respectively. The waveforms at slab foundation ⑦ for 5.21 Hz have a phase angle of 90° in comparison with waveforms at first floor ①, and a phase angle of 270° in comparison with waveforms at roof ② (Fig. 9).

In consequence, the presence of the first and the second normal mode of vibration estimated from microtremor measurements was confirmed by the predominant frequencies of transfer functions of earthquake strong motions recorded in base isolated building.

Finally, Table 1 shows a comparison of predominant frequencies for experimental building obtained from microtremor measurements and earthquake strong motion records, as well as, the vibration characteristic modes concluded from these data analysis. As expected, lower predominant frequencies for both modes for earthquake strong motions were obtained, when comparing with the corresponding of microtremor measurements.

4 CONCLUSIONS

Microtremor measurements were performed in a base isolated building to infer some dynamic characteristics of this kind of building.

The conclusions for the NS component are as follows:

1. The predominant frequencies around 1.90 Hz and 5.91 Hz were estimated for the first and the second normal modes.
2. Lower predominant frequencies of 1.17 Hz and 5.21 Hz were obtained from the 2003 Miyagi-oki Earthquake motions recorded in this experimental building.

3. Amplification ratio for the second predominant frequency of transfer function was as preponderant as the first predominant one, meaning that higher vibration characteristics would be taken into account in seismic resistant analysis of this kind of building.

5 ACKNOWLEDGMENTS

The authors gratefully acknowledge Prof. S. Kurita (Tohoku University) and Dr. M. Saruta (Shimizu Corporation) for providing earthquake strong motion records of experimental building.

REFERENCES

1. Kanai, J., J.C. Tokeshi, C. Cuadra and M.B. Karkee (2006). "Vibration Characteristics of Buildings Using Microtremor Measurements", First European Conference on Earthquake Engineering and Seismology, Geneva, Switzerland, Paper No. 708 (CD-ROM).
2. Nishikawa, N., M. Inagaki and T. Takada (2003), "Structural Behavior Estimation from Microtremor Measurement and Finite Element Three-dimensional Analysis", Summaries of Technical Papers of Annual Meeting, AIJ, Paper No.21467 (in Japanese).

EXPERIMENTAL INVESTIGATION ON RETROFITTING AND REHABILITATION OF REINFORCED CONCRETE OPENING CORNERS

Asha Patel[1], More Rajkumar Champatrao[2]

1. Lecturer ,Department of Civil Engineering
2. Post graduate student (Structural Engineering)
Department of Civil Engineering, National Institute of Technology, Rourkela

ABSTRACT

A beam column joint is a critical portion in a reinforced concrete moment resisting frames. The function of beam-column joint in a frame is to transmit the loads and moments from beams to columns hence they should maintain the integrity and must be designed stronger than the members framing into it.Different studies on beam-column joints have demonstrated that opening corners in reinforced concrete structures have significantly quashed the efficiency and ductility which tends to form cracks at service loads. During earthquake reversal of loading takes place which causes the closing moments as well as opening moments at the corner joint. It has been observed that the closing joints show better efficiency than the opening joints hence opening corners are more vulnerable to seismic loading. This investigation has motivated the present study. In the present study the parameters studied are effect of GFRP reinforcement bonded in different resin matrix on rehabilitation and retrofitting works in terms of joint efficiency, ductility and crack pattern. Results show that the application of GFRP sheets is greatly effective in improving the joint strength and load carrying capacity of the rehabilitated structures. It also helps to improve the ductility of retrofitted structures.

1 INTRODUCTION

In various parts of the world, including India, many structures deteriorate due to the problems associated with reinforced concrete. Haphazard like earthquakes have repeatedly demonstrated the susceptibility of existing structures to seismic effect. Like many earthquakes in the past, the recent earthquakes in India of Killari (1993), Bhuj (2001) and Kashmir (2005) have shown the seismic vulnerability of structures due to faulty construction practices being followed in the country. Many structures situated in high seismic regions of India are found to be designed without consideration of seismic codal provisions. In these regions, implements like retrofitting and rehabilitation of deteriorated concrete structures are important. Since the cost of replacement of structures that do not have the adequate earthquake resistance is very high these structures have to be retrofitted by adopting suitable technique. The vulnerability of inadequately designed structures represents seismic risk to occupants and this fact justifies a strong social need to retrofit the existing buildings and rehabilitate the damaged one.

Seismic resistance of structures can be enhanced by strengthening the structural elements through local retrofitting technique. Local retrofitting allows strengthening of elements to resist the strength demands predicted by the analysis, without significantly affecting the overall response of the structure. Thus the ductile global behavior of structures can be achieved with the local strengthening of members by RC jacketing, steel jacketing and by bonding fiber reinforced polymer composites.

Post earthquake damages of reinforced concrete structures revealed that one of the weakest links in the lateral load resisting system is the beam column joint It is a critical portion in reinforced construction. The function of beam-column joint in a frame is to transmit the loads and moments from beams to columns safely while maintaining the integrity of the structure. In design, the reinforcement at the joint is not taken care of completely. Post earthquake damage studies showed that, the bending moment and shear force are high at the beam-column joints during seismic action Under such conditions, seismic qualification of joints has become extremely important.

Exterior Beam-column joints are more vulnerable than interior joints owing to partial confinement. The shear force in the exterior joint gives rise to diagonal cracks. Opening corners displays parturient cracks and failure is noted by development of diagonal tensile crack. The force built up in opening joint causes the major crack along the corner diagonal. In the absence of rational reinforcement system, the corner concrete is not able to sustain the resultant diagonal tensile force and a premature brittle failure characterized by low toughness and ductility is imminent. During seismic action, the joint is subjected to closing moments as well as opening moments due to cyclic nature of forces. Hence the corner joints have to be conservatively designed as opening joints with sufficient reinforcement..

Enough information about behavior of beam-column corner joint retrofitted and rehabilitated with Glass Fiber Reinforced Polymer is not available. During earthquake reversal of loading takes place which induces both the closing moments as well as opening moments at the corner joint. It has been observed that the closing joints show better efficiency than the opening joints hence opening corners are more vulnerable to seismic loading. This investigation has motivated the present study. This study will be very important in case of structures like cantilever retaining walls, bridge abutments, rectangular liquid retaining structures and portal frames

The main objective of this investigation is to study the behavior of reinforced concrete opening corners retrofitted and rehabilitated by using glass fiber fabric in different matrix under statically applied horizontal loading. The parameters of investigation are the strength measured in terms of joint efficiency, ductility, energy loss, and failure pattern.

2. EXPERIMENTAL PROGRAMME

The experiment consists of four portal frames of identical size. The outer dimensions of the frame were 1250 mm x 1300 mm. The cross sections of both the legs were 200 mm x 250 mm as shown in Fig. 3.1. All legs were reinforced with 3 bars of 12 mm diameter in tension side and 2 bars of 12 mm diameter in compression side. Two legged 8 mm dia. stirrups were provided @ 140 mm c/c throughout the specimen except at corners. No lateral reinforcements were provided in the joint region. The detail of reinforcements is shown in Fig. 3.2. The reinforcement used were HYSD bar of characteristic strength 415 Mpa . Nominal mix of concrete of M15 grade having mix proportion 1:2.2:4.4 by weight with 0.64 water cement ratio was used. Portland slag cement conforming to IS 455-1976 and naturally available river sand as fine aggregate and crushed stone of nominal size 12.5 mm as course aggregate were used in making the concrete mix. Six cubes of 150 mm size were cast for each batch of concrete for evaluating 28 days compressive strength of concrete.

For casting the form work was thoroughly cleaned .Before oiling the inner faces the edges and corners were coated with plaster of Paris to have round edges. The reinforcement cage was then placed in position inside the form work carefully keeping in view a clear cover of 20 mm for bars. Freshly mixed concrete was poured to the mould and compacted properly using needle vibrator. Specimens were removed after 24 hours and kept submerged in water for curing for 28 days. Each specimen was kept in air for 24 hours before test so as to attain surface dry condition. Thereafter it was white washed for ease in studying the crack pattern formed during different stages of loading.

2.2 Rehabilitation Scheme

Specimens A, B and C were controlled specimens tested under similar conditions till failure. After this loose concrete were removed from the cracked portion and repaired by applying cement paste. Then the cracked specimen A is rehabbed with 2-layers of GFRP sheet glued only in vertical edge of the corners (Ref.Figure3a). The rehabbed specimen was named AR and further tested till complete failure. Similarly cracked specimens B and C were rehabilitated by sticking 2-layers of GFRP sheets in the vertical inner face and top and bottom joint region (Ref. Fig.3b) in different resin matrix.

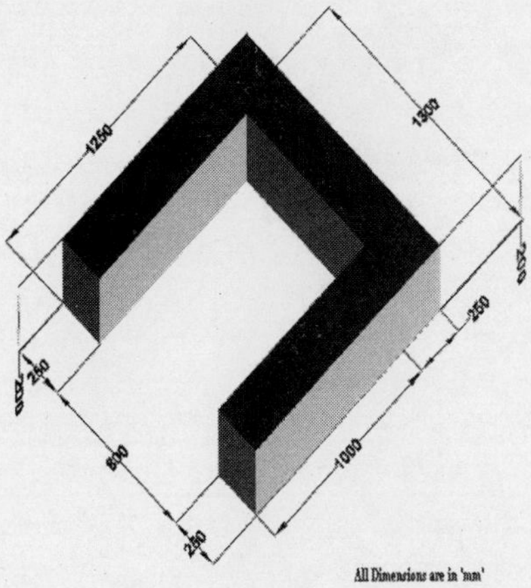

Fig. 1 Dimensions of specimen

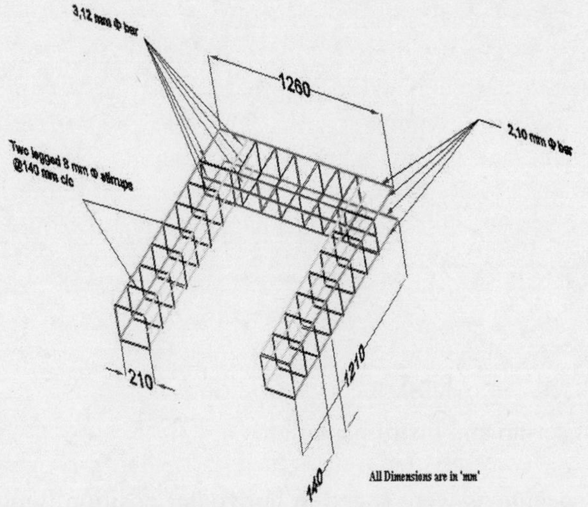

Fig. 2 Reinforcement detailing of specimens

Table 1

Sln o	Speci-men design-ation	Description of specimen	Purpose Of wrapping
1	**A**	Control specimen.	
2	**AR**	Rehabilitated specimen A with 2- layers of bidirectional woven roving mat at inner vertical surface of corners in LY556 epoxy resin.	Rehabilitation
3	**B**	Control Specimen.	
4	**BR**	Rehabilitated specimen B with 2-layers of bidirectional woven roving mat at inner vertical surface and top & bottom faces of corners in LY556 epoxy resin.	Rehabilitation
5	**C**	Control Specimen.	
6	**CR**	Rehabilitated specimen c with 2-layers of bidirectional woven roving mat at inner vertical surface and top & bottom faces of corners in Vinyl ester resin.	Rehabilitation
7	**D**	Retrofitted with 2-layers of bidirectional woven roving mat at inner vertical surface and top & bottom faces of corners in LY556 epoxy resin.	Retrofitting

They were named as BR and CR respectively and further tested till failure occurred. The specimen D was retrofitted with 2-layers of GFRP sheets adopting second scheme of application of sheets. Fibre sheet used was E-Glass, bi-directional woven roving mat. It is chemically resistive and anticorrosive. Two kinds of resins were used. One was LY-556 from Geigy-Ciba with Hardener HY-951. The chemical name of this resin used for bonding between fibre mat and concrete is Diglycidyl Ether of Bisphenol – A. Another was Vinyl ester. Both were of thermosetting type resin. The Epoxy resin was mixed with 9% hardener by weight and applied on the specified surfaces of corners as well as the surface of fiber mat. Then this sheet was glued to the surface and left for 48 hours for hardening. Special care was taken by using hand gloves during application of glass fiber mat. The proposed rehabilitation and retrofitting schemes are summarized in Table 1.

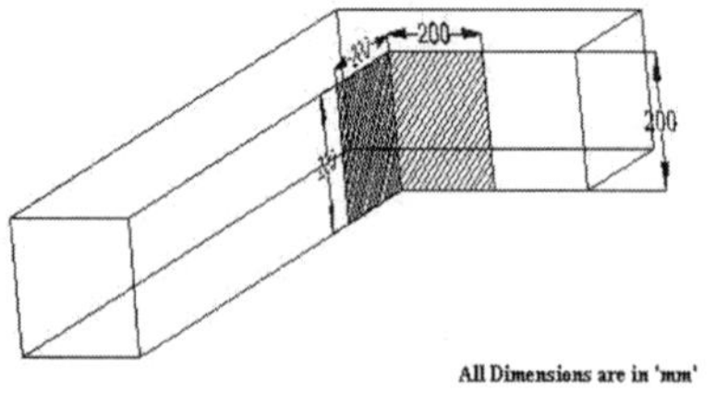

Fig.3 -a

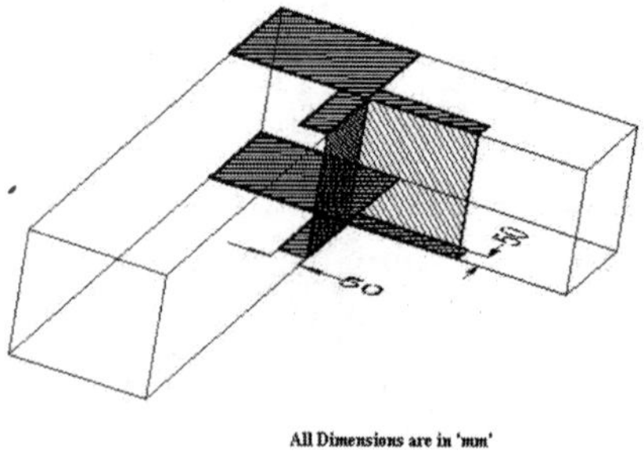

Fig 3-b

2.3 Test set-up and Instrumentation

All the specimens were tested in horizontal position lying on roller supports on the ground. Static loading was applied by means of two hydraulic jacks each of capacity 100 KN. The loads were gradually increased till the failure occurs. The deflections were observed at center of horizontal leg and at two points on the vertical legs at a distance of 600 mm and 950 mm from the top end respectively by means of dial gauges . At the corner strain were measured by strain gauge. The least count of strain gauge is 0.0001. Dummies were used in the corners to observe the strain by strain gauge. All the specimens were tested under pure opening moment using the loading arrangements shown in Fig. 4.

The monotonic increasing load was applied in increments using manually operated hydraulic jack. Readings of dial gauges were recorded at each load increments. All the specimens were tested till failure.

As the loading progressed the specimen surface specially the corners were keenly observed for formation development of cracks. The cracks propagation were marked on the surface. Load at failure was taken as the ultimate load for the specimen.

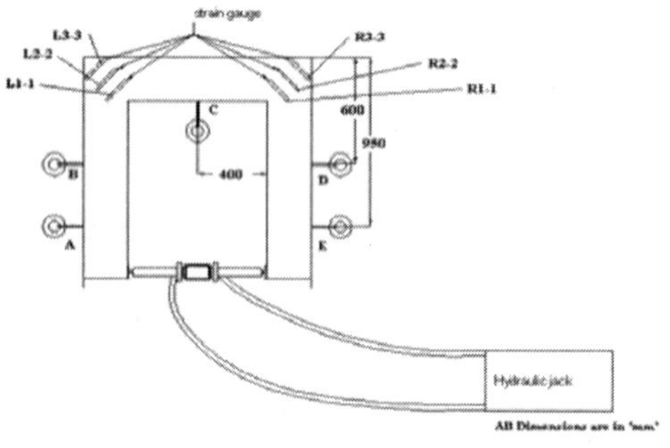

Fig.4

3. RESULT AND DISCUSSION:

The result is compared with retrofitted specimen 'D'. Specimen were placed horizontally and lateral horizontal load was applied at the tip with the help of two hydraulic jack as shown in Fig 3.3.The Load was applied at an increment of 4 KN. The deflections at different points was measured by dial gauges.

3.1 Specimens A,B and C

The initial crack was observed at the centre of the inner vertical face of the corner on both sides simultaneously at 12 KN Load. The cracks propagated vertically and with further increment of loading many cracks started forming within the joint and ultimately the right corner failed at 20 KN. At this instant some cracks of similar pattern were observed on left corner also. The failure pattern shown a combination of flexural and shear failure in joint region causing collapse of the corner.Almost identical crack patterns were observed for specimen 'B' and 'C'. In case of specimen 'B' the initial crack was observed at 12 KN and for specimen 'C' it was observed at 14 KN. The failure load of specimen 'B' was 20 KN and specimen 'C' was 22 KN. The Load Deflection curves for specimens A, B and C at bottom of left leg are presented in Fig 5, Fig 6, and Fig 7 respectively.

3.2 Specimen AR

The specimen 'AR' was rehabbed with two layers of L-Shaped bidirectional woven roving Glass fibre mat with matrix of LY-556 resin adopting the first scheme shown in Fig 3a and tested in the similar manner. The initial cracks were observed at 8KN Load. Further increase in the load caused cracks followed the same pattern and right corner failed at 11 KN. The load deflection curve for 'AR' is presented in Fig 5.

Testing of this specimen indicated that this scheme of rehabilitation is not sufficient to strengthen the repaired specimen as the strength gain is 55 % (Ref: Table 2) of the original strength. This is because the joint area was not confined. While sticking the mat some air packets were trapped inside which might have caused this type of failure. This failure directed us to follow second scheme of jacketing as shown in Fig.3.b.

3.3 Specimen BR

The specimen 'BR' was rehabilitated with two layers of bidirectional woven roving Glass fibre mat with LY-556 resin matrix and tested under the same condition. First crack was observed at 38 KN load. At this moment some small cracks were observed on opposite vertical (outer) face of the corner indicating crushing of concrete at that face. The load deflection curves are shown in Fig 6.

3.4 Specimen CR

The specimen 'CR' was rehabilitated with two layers of bidirectional woven roving Glass fibre mat with vinyl ester matrix and tested in the same manner. Initial cracks were observes at 20 KN. Compared with the specimen 'BR' this specimen has failed earlier because the matrix used here has possessed less bond strength. This indicates that the matrix LY-556 is more suitable for rehabilitation purpose in comparison with vinyl ester. The load deflection curves are shown in Fig 7.

3.5 Specimen D

The specimen 'D' was retrofitted with two layers of bidirectional woven roving Glass Fiber mat with LY-556 Resin matrix. In this case due to jacketing the cracks could not be observed and the failure occurred at 40 KN. The failure is bondage failure at the interface. The load-deflection curve is shown in Fig.8.

Comparison of load-deflection curves for 'A', 'AR' and 'D' (Ref: Fig 5) shows that the behavior of the retrofitted specimen 'D' is more ductile than the rehabilitated and the control specimen. Again it shows that the loss of energy in the specimen 'AR' is more in comparison with specimen 'A'. This indicates the failure of first proposed scheme of rehabilitation.

Table 2. Joint Efficiency

Specimen No.	Ultimate moment of specimen without GFRP application, $M_{ul.}$ (KN-m)	Failure moment after GFRP application, $M_{ult..}$ (KN-m)	Corner Efficiency ($M_{ult.}$/ $M_{ul.}$) × 100
A	22	12.1	55 %
B	22	41.8	190 %
C	24.2	25.3	104.545 %
D	21.279*	45.1	211.946 %

*** indicates Theoretical Moment of resistance calculated by testing cubes specimens.**

Comparison between specimens 'B', 'BR' and 'D' shows that the behavior of the retrofitted specimen 'D' is much stiffer than the other two specimens. Specimen showed stiffer behavior because deflections are observed at point D which is nearer to the joint. Jacketing makes the joint stiffer. But rehabbed BR showed ductile behavior compared with control specimen B. The loss of energy in the rehabilitated specimen is more compared retrofitted specimen 'D' and is more ductile compared with the specimen 'B' with less energy loss.

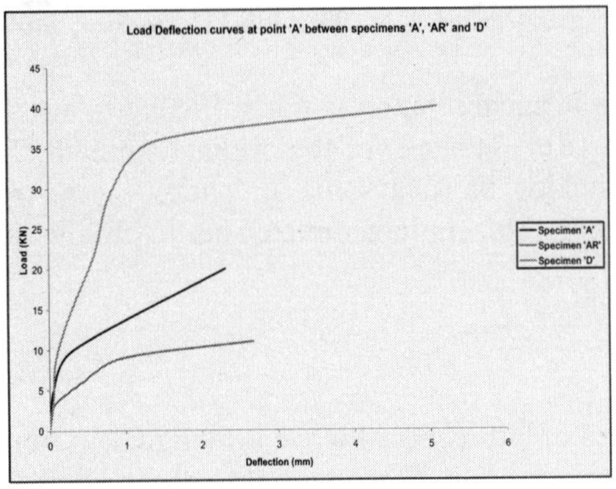

Fig. 5 Load deflection curves for 'A', 'AR' and 'D'at pt.E

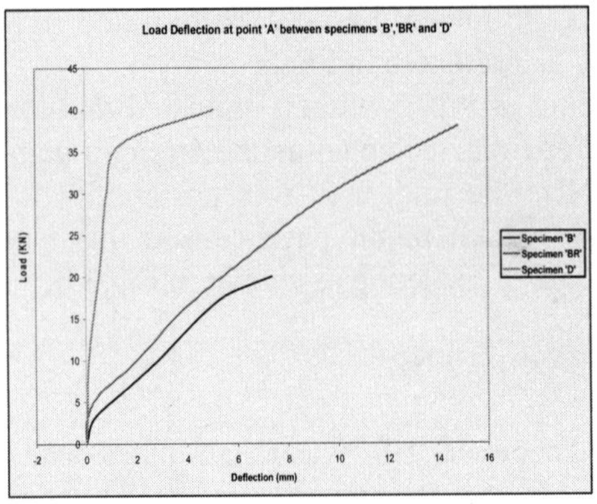

Fig. 6 Load deflection for specimens 'B','BR' and 'D'at pt

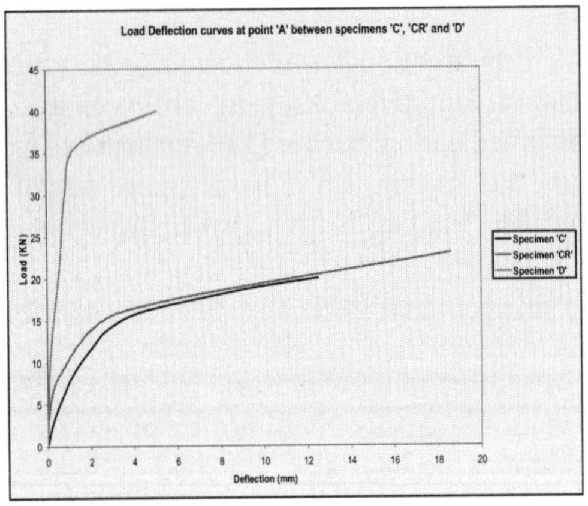

Fig. 7 Load deflection curves for 'C', 'CR' and 'D'at pt.D

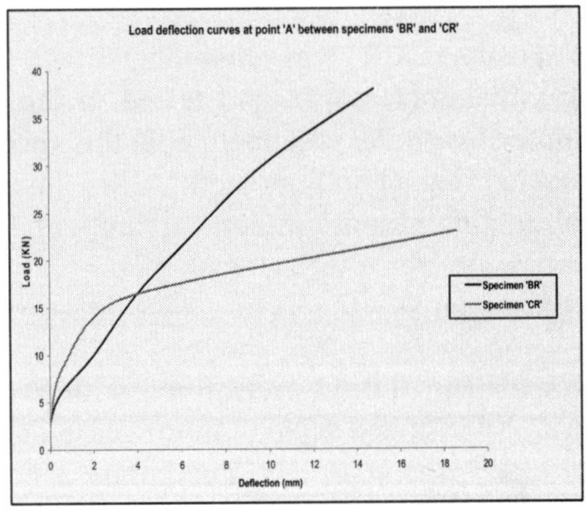

Fig. 8. Load deflection for specimens 'BR' and 'CR'at pt D

The same behavior was observed for set 'C','CR' and 'D'.

Comparison between 'BR' and 'CR' shows that the behavior of the specimen 'BR' is stiffer than specimen 'CR'. Also the loss of energy in the specimen with vinyl ester was observed to be more than that of the specimen bonded with LY-556 epoxy resin. This indicates that bond strength of LY-556 resin is more than vinyl ester.

Mainly bondage failure at interface occurred in all rehabbed and retrofitted specimens. After removing the jacketing it was observed that this failure is followed by crushing of concrete in compression side as the jacketing made the tension side stronger. The cracks are not very prominent in the joint region. (Ref. Fig.9).

The load carrying capacity in terms of joint efficiency is calculated for all specimens and given in Table 2.

CRACK PATTERN ON SPECIMENs

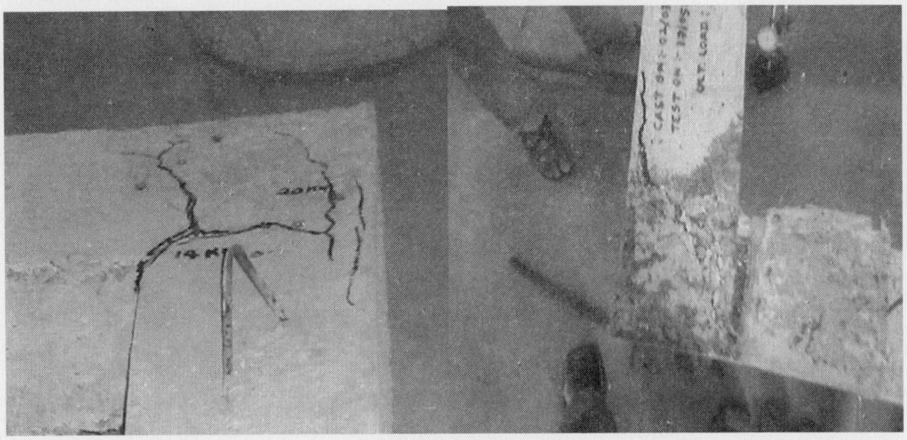

Failure pattern at corner of control specimen

Failure pattern after removing fiber jacket. of rehabbeb specimen

Fig 9

4. CONCLUSIONS

The following conclusions can be drawn on the basis of the experimental studies on rehabilitated and retrofitted opening corners with FRP reinforcement.

1. Comprehensive static tests conducted on failed specimen rehabbed with two layers of GFRP mats in LY-556 demonstrated 190% increase in joint efficiency.
2. Similar test conducted on failed specimen with two layers of GFRP mats with Vinyl Ester verified 104% increase in joint efficiency.
3. The investigation suggests that LY-556 resin matrix provides more bondage as well as joint efficiency and less curing time in comparison with matrix of Vinyl Ester for concrete works.
4. Due to the more number of cross linking sites in Epoxy cyclic structure, it forms 3D network with glass fiber at the interface of fiber and matrix which is not so prominent in vinyl ester structure.
5. GFRP retrofitting with two layers of sheet in LY-556 offers the promise of a significant improvement in stiffness as well as about 210% increase in joint efficiency.
6. Second scheme of application of Glass Fiber sheets is more effective than the first scheme adapted in the experiment. This is because gluing the sheets on top and bottom faces confined the concrete in the corner region causing significant improvement in joint efficiency and stiffness for both rehabilitated and retrofitted specimens.
7. Local strains reduce due to the presence of GFRP.
8. Specimens with jacketing are able to sustain significantly large tensile strains hence structural performance and integrity in presumed over a long range.
9. It may be practically feasible to provide GFRP jackets on corners like junction between abutments, retaining walls or the junction between stem slab and base slab in cantilever and retaining wall. However difficulty in handling the materials, their applications and cost may discourage the use of this method.
10. The experimental work confirmed that the use of Glass Fiber sheet in suitable resin matrix could be a promising technique for rehabilitation & retrofitting of joint

5. REFRENCES

1. El-Amoury T. and Ghobarah A., (2002)."Seismic rehabilitation of beam-column joint using GFRP sheets", Jou. of Engineering Structures, **24**,1397-1407.
2. Ghobarah A., Said A.,(2002)."Shear Strengthening of beam-column joints". Journal of Engineering Structures, **21**, 881-888.
3. Singh B., Prof. Kaushik S. K.,(2003),"Investigations on Fibre Reinforced Concrete Opening Corners" IE (I) Journal. CV,_ **84**, 201–209.
4. Li Jianchun, Bakoss Steve L., Samali Bijan and Ye Lin,(1999),"Reinforcement of concrete beam-column connections with hybrid FRP Sheets", Jou. of Composite Structures , **47**, 805-812.
5. Li Jianchun, Samali Bijan Ye Lin and Bakoss Steve,(2002),"Behaviour of concrete beam-column connections reinforced with hybrid FRP sheet", Journal. of Composite Structures ,**57** ,357-365.
6. Mukherjee Abhijit, Joshi Mangesh,(2005)," 'FPRC reinforced concrete beam-column joints under cyclic excitation",Journal of Composite Structures,**57,** 185-199.
7. Stanton J. F., Lehman D. E., Walker S. G. and Yeargin C. M.,(2002),"Shear in reinforced concrete beam-column joint without transverse reinforcement", Publication of pacific earthquake research center- Annual meeting research digests no **2002-9**.

RESEARCH ON THE EVALUATION OF THE RESISTANCE TO EARTHQUAKE FOR THE WOODEN HOUSE STRUCTURE BY USING MICROTREMOR MEASUREMENT

Koushige,N[1]., Hayakawa,K[2]., Honda,S[1]., Edane,K[2].

[1]Nikken Sekkei Civil Engineering Ltd. Japan.

[2]Ritsumeikan University. Japan.

ABSTRACT

To compensate with present diagnosis of seismic performance, the method of evaluating the earthquake performance of the building to have used the microtremor measurement was examined. It can measure with simple vibrometer, and doesn't use a special seismic source. The natural period of structure from a wall and diagonal direction was obtained by measuring wooden houses and temple with typical wooden frame structure. Then, the earthquake resistant capacity index C-value to wooden houses was calculated, and the comparison with each building was inspected. As a result, it turned out that the difference at the structure and years since construction was shown. The high applicability of this method was understood. According to the characteristic of diagonal natural period, it's suggested the degree of binding between floor and wall control earthquake performance of whole structure.

Keywords: Microtremor measurement, Wooden structure, Natural period

1 INTRODUCTION

Record-breaking damage appeared by Kobe Earthquake on January 17, 1995, and about 80% of the victims were compression death due to the collapse of the residence. Specially, much collapse of the wooden house of long years occurs, and improvement in the residence to earthquake play of the wooden house is being asked. It is hoping for the development of the way of evaluating the standard of the residence to earthquake play of the building as the way of complementing present earthquake resistance diagnosis of such.

Microtremor measurement is always shown as a tool that the vibration character of the wooden building structure thing is grasped in easily target. If a vibrometer is always set, a microtremor can be always measured easily, and it is economically cheap, and has a very excellent character as simple measurement tool. Though even the research of previous report by Maiguma is pointed out, a few detailed measurement cases seem that still there is it in the possibility to make use of a microtremor for the estimation of the shock character of the building always.

So, when the shock character of the wooden building thing was grasped, the measurement of the parallel direction and the oblique direction was carried out about the temple which was the building of the typical wooden structure shaft party method of construction which actually has a wooden structure door house and a big opening for the purpose of verifying whether it is always the way that microtremor measurement is effective by this paper in many. In comparing with each measurement element, the evaluation technique of the many-sided resistance to earthquake play in accordance with the nature of the characteristic period is examined.

2 OUTLINE OF MEASUREMENT

The microtremor measurement was carried out in Kusatsu city, Shiga. Shiga prefecture is located in center of Japan. Shiga has Biwa-ko lake which is the lake of the biggest in Japan, and it is formed in hilly terrain and alluvial fan. Then, It's a position of Kusatsu city in Fig.1. There is much residential area on the low land in Kusatsu city. Kusatsu city develops as a commuter town in Osaka and Kyoto. In addition, people live for a long time historically, and it is characteristic that a lot of Buddhist temples are distributed. The servo velocity-meter which is produced by Tokyo Sokushin ltd is used in microtremor measurement, A photograph of a measuring equipment is as shown in Fig.2 and a measuring equipment specifications is as shown in Tab.1. The amplitude of velocity two horizontal elements and one vertical element can be measured at the same time. The measurement was carried out with this one set of vibrometer. Measurement frequency is gathered data of 10 minutes when we continued at 50Hz and offered it in indoor analysis. Wave analysis is extracted a part with a few noises based on provided digital data. The Fourier spectrum by Fast Fourier Transform was made from each wave pattern. It performed wave pattern handling of 4 for each measurement elements to decline unevenness of a wave pattern and it was regarded as a representative spectrum of an investigation spot by overlapping them.

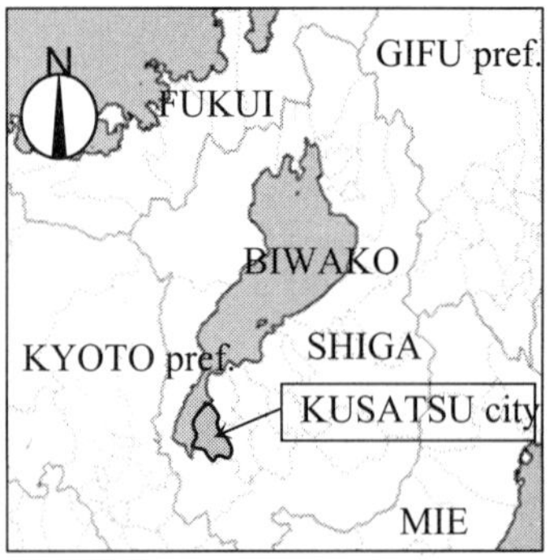

Fig.1 Guide chart

Fig.2 Vibrometer externals

3 MICROTREMORE MEASUREMENT OF WOODEN HOUSES

Tab.1 Measuring equipment specification

Measurement unit	Tokyo Sokushin co., ltd. SPC-35N
Vibrometer	Tokyo Sokushin co., ltd. VSE-15D
Measurement system	Servo velocity-meter
Range of frequency	0.25 ~ 70Hz
Range of velocity	±10kine

Fig.3 show the photograph of the representative investigation house. The Japanese wooden house generally has a square and rectangular plane shape, and plus faces the North, South, East, West, and there are almost many cases that a big opening (window and door)is posted in a Southern aspect by convenience of the lighting. All the measurement houses are more than 2 stories. It carried out microtremor measurement on floor (1F,2F,3F) of the building and the ground side of the building and measured each North and South (or wall where opening is few) (NS), East and West (or wall where a lot of opening exist) (EW), vertical (UD) and the element of the diagonal (DIG) direction. Fig.4 show a outline of the measurement. The number of years of the building old is spread from 2 to 30 years and more. The Building Standards Law is revised in Japan in 1981, and high-level earthquake resistant performance is demanded. This measurement executes three buildings after the Building Standards Law revision and before the Building Standards Law revision (cf. Tab.2). Fig.5-Fig.6. shows result of measurement / analysis of representative house ([A] house) . Fig.5 was the Fourier spectrum on the third floor and first floor

Fig.3 Wooden house externals

Tab.2 Table of wooden house measurement

Sign	Years since construction	Building Standards Law revision	Remarks
A	2	after	Three stories. 2×4 method.
B	20	after	2×4 method. There are openings in the East and South .
C	24	after	There is opening in the South.
D	26	before	There are openings in the North and South.
E	30	before	Extension history having.
F	30	before	-

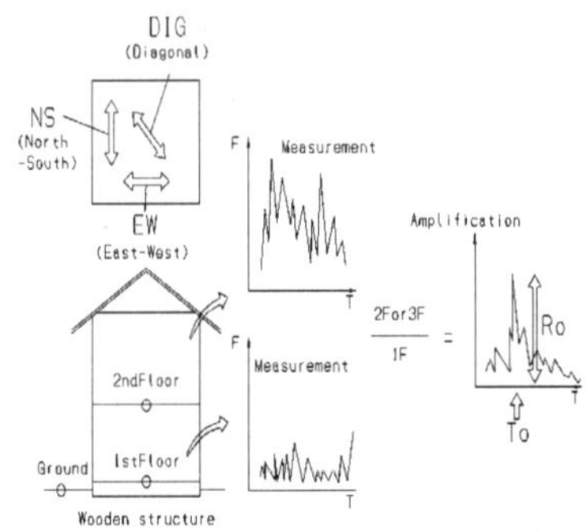

Fig.4 Outline of measurement

Fig.6 show amplifying characteristic to exclude the Fourier spectrum of the third floor by the Fourier spectrum of the first floor. The peak of the spectrum is accepted 0.14sec and 0.20sec in NS direction, 0.20sec in EW direction , 0.20sec in DIG direction. This spectrum is consisted of input from the underground and the reply of the building, and put them together. The predominant period is seen in the vicinity of 0.2sec like Fourier spectrum when I watch an amplification characteristic to show in Fig.6. This peak can be judged the amplification of the building.

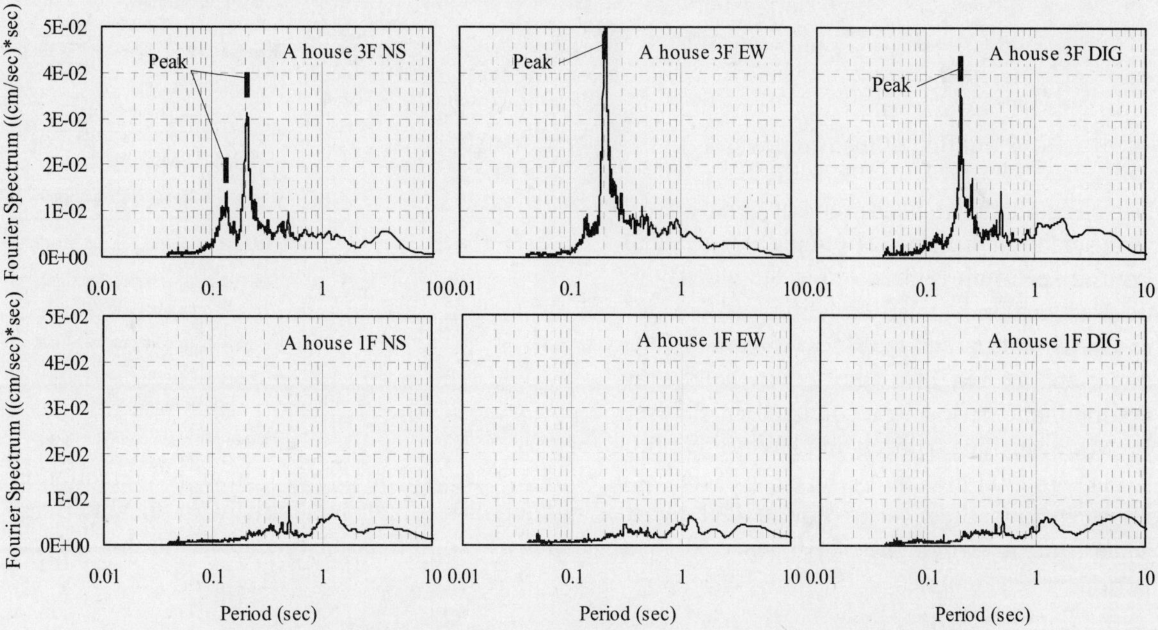

Fig.5 Fourier spectrum of Wooden house measurement ([A] house)

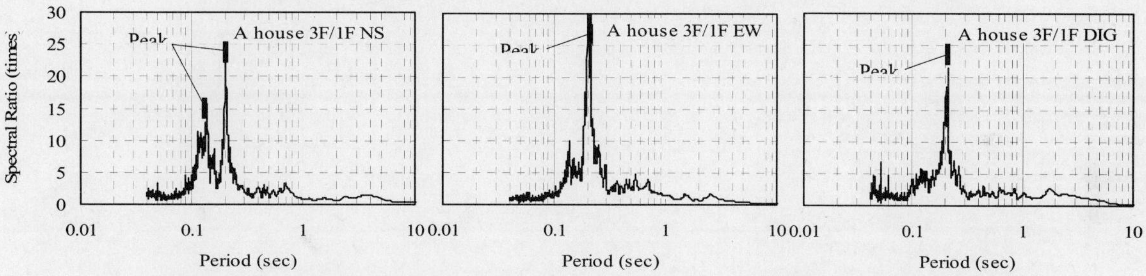

Fig.6 Amplifying characteristic of Wooden house measurement ([A] house)

4 MICROTREMORE MEASUREMENT OF TEMPLE

Fig.7 shows the photograph of the representative investigation temple. The room of temple is taken widely for many visitors, and compare with a wooden house, there is little amount of wall, and the opening of the entrance is very big, and Roof tile is stately and heavy. That is the structural characteristics of temple. Furthermore, the constructed years of a temple are very long, and it goes back in the oldest thing till A.D. 740. The repair of several degrees is performed, but the number of years old is very long in comparison with the wooden house. The measurement was carried out this time on floor (1F) of the main hall of temple and the ground side of the building in 11 temples to show in Tab.3. The outline of measurement is as shown in Fig.8.

Fig.7 Temple externals (Denkyu-ji)

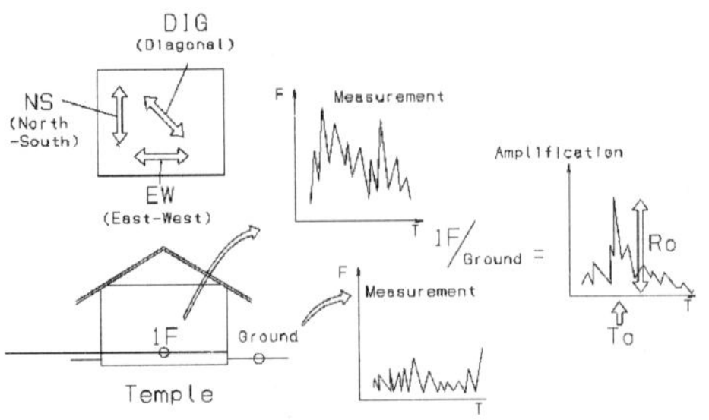

Fig.8 Outline of measurement

Tab.3 Outline of temple measurement

Temple name	Sign	Construction year	Remarks
Denkyu-ji	a	-	There is an opening in the South.
Entsu-ji	b	-	There is an opening in the South.
Korin-ji	c	1924	There is an opening in the South.
Ganshin-ji	d	1807	There are openings in the South and West
Ganrin-ji	e	1512	There is are openings in the South and East
Tyozen-ji	f	1743	There is an opening in the South
Jyosyo-ji	g	1700	There is an opening in the South
Muryojyu-ji	h	740	There is an opening in the South. Rebuilding in 1989.
Saihou-ji	i	-	There is an opening in the South.
Kouen-ji	j	500	There is an opening in the North. Rebuilding in 1990.
Kyozen-ji	k	-	There is an opening in the West.

Fig.9-Fig.10 shows measurement / analysis of representative temple ([a] temple). A peak of Fourier spectrum is observed in the vicinity of 0.39sec, 0.58sec in NS direction and 0.45sec in EW direction and 0.38sec, 0.58sec in DIG direction in Fig.9.On the other hand, the predominant period is recognized in 0.38sec in NS direction 0.45sec in EW direction, 0.38sec in DIG direction. The peak of the vicinity of 0.58sec which appeared in an NS direction is not seen, and there is possibility that there is not amplification characteristic of the building. Therefore, it is thought that 0.38sec, 0.45sec seen with amplification characteristic are natural periods in this temple.

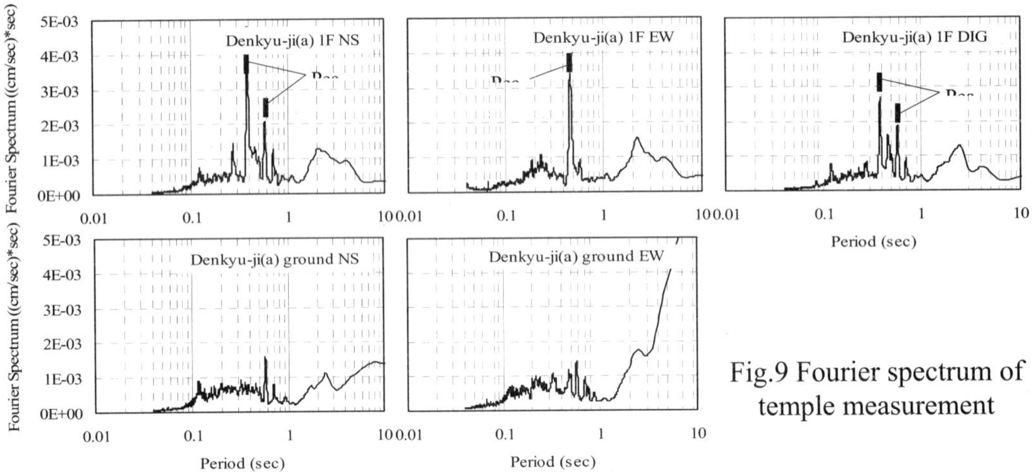

Fig.9 Fourier spectrum of temple measurement

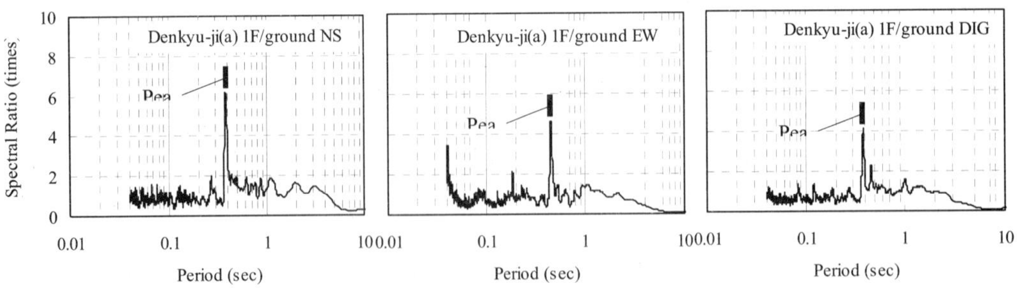

Fig.10 Amplifying characteristic of temple measurement (Denkyu-ji[a])

4 CHARACTERISTIC OF NATURAL PERIOD OF BUILDING

Tab.4 and Tab.5 shows the natural period of each direction (NS,EW,DIG) provided from microtremor measurement result of 6 wooden houses and 11 temples. Fig.11 expressed relations between ToNS (the natural period in NS direction) and ToEW (the natural period in EW direction). The measurement result pointed on solid lines (ToNS=ToEW) in a figure shows that the natural period of the building does not have directionality. Furthermore, in the building pointed by the top right corner of the figure, the natural period of building in itself is long; in other words, Rigidity is small. As for wooden house , the natural period of building spread from 0.2 to 0.4sec, mainly 0.2 around. There is little number of the samples about the wooden house, but the differences of the natural period are big before the Building Standards Law revision as a tendency. And the natural period almost concentrates on appearing in the vicinity of 0.2sec after Building Standards Law revision, and the quality of the building are seemed to be constant. In paying attention to the difference of the natural period in EW direction and NS direction, the natural period of EW direction is existed depending on a building, however there is little difference. This phenomenon seems to be caused by that roof weight is small, and, directionality being hard to appear and the rigidity of building in itself become higher by partition wall. On the other hand, there is the natural period of the building in the vicinity of 0.3sec - 0.4sec by the measurement result of the temple. In spite of one-storied

Tab.4 Table of natural period of wooden house.

Sign	Natural period (sec)		
	NS	EW	DIG
A	0.20	0.20	0.21
B	0.22	0.20	0.22
C	0.16	0.19	0.20
D	0.24	0.25	0.25
E	0.16	0.19	0.16
F	0.40	0.41	-

Tab.5 Table of natural period of temple.

Sign	Natural period (sec)		
	NS	EW	DIG
Denkyu-ji[a]	0.38	0.43	0.38
Entsu-ji[b]	0.31	0.42	0.31
Korin-ji[c]	0.40	0.45	0.45
Ganshin-ji[d]	0.32	0.41	0.43
Ganrin-ji[e]	0.35	0.40	0.40
Tyozen-ji[f]	0.30	0.42	0.30
Jyosyo-ji[g]	0.36	0.42	0.37
Muryojyu-ji[h]	0.30	0.39	0.39
Saihou-ji[i]	0.30	0.42	0.42
Koen-ji[j]	0.29	0.29	0.29
Kyozen-ji[k]	0.35	0.41	0.34

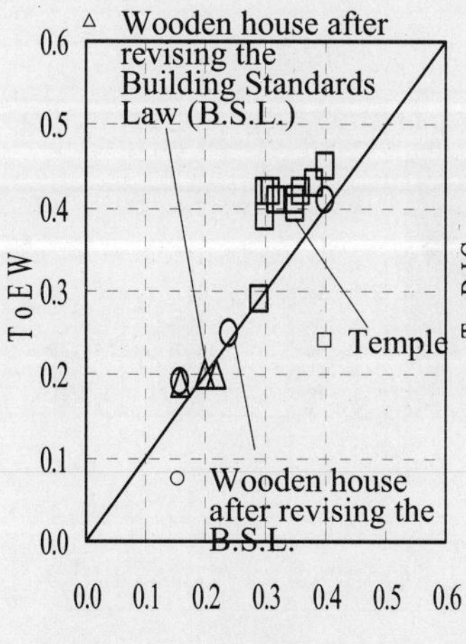

Fig.11 Comparison of natural period (ToNS vs ToEW)

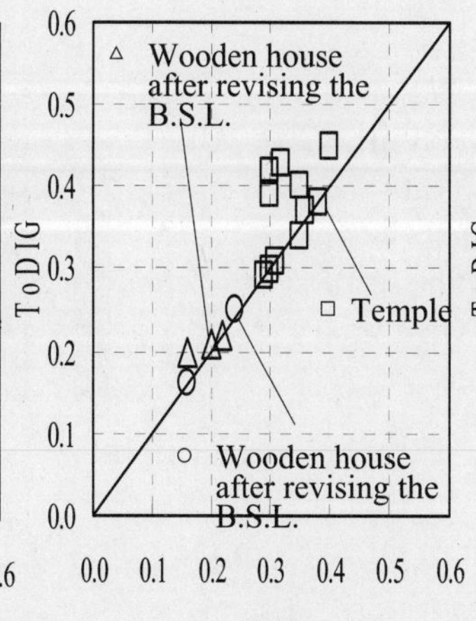

Fig.12 Comparison of natural period (ToNS vs ToDIG)

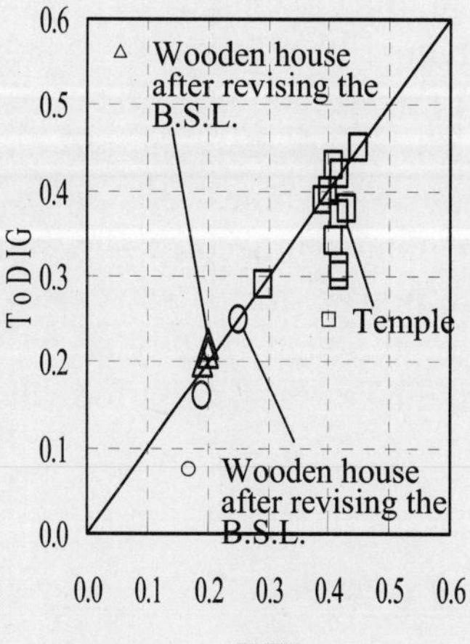

Fig.13 Comparison of natural period (ToEW vs ToDIG)

house, the natural period of the temple is bigger than a wooden house of two or three stories. In comparison with the wooden house, the cause is a little amount of wall and heavy weight of roof tile in comparison with wooden house. Furthermore, when the difference of the natural period in the NS direction and EW direction is seen, the natural period in EW direction of most temples are longer than NS direction. It is suggested that a EW direction rigidity was few. Fig.12-Fig.13 shows the comparison that the natural period of the diagonal (ToDIG) direction for ToNS and ToEW. At half temples, the ToDIG was biger than the ToNS, and it followed that the ToDIG was smaller than the ToEW. This shows that the natural period of the diagonal direction has relations of ToEW < ToDIG < ToNS. In addition,it becomes ToNS=ToDIG in other temples and the wooden house, and this difference is guessed by degree of binding between floor and wall. Therefore, there is a possibility that the wall rigidity of NS direction is not demonstrated enough when there was the input such as the earthquake vibration in a diagonal direction in the building of ToNS < ToDIG.

5 EVALUATION OF RESISTANCE TO EARTHQUAKE

The earthquake resistant capacity index (C-value) that Dr. Maiguma had proposed was calculated, and compared earthquake performance as for the index of earthquake-resistant of wooden house where the measurement was carried out at the second floor and the ground. C-value is expressed in an equation(1), and a criterion for the earthquake is provided by C-value. The earthquake vibration assumes Kobe Earthquake (M7.2 in 1995.). Ro in the equation is used the amplitude width of the amplification characteristic curve and Q is calculated by damping constant by half power method from an amplification characteristic curve. Fig.14 show C-value of each house. C-value shows width from 20 to 80, but, As for [A] , [I]'s house, C-value exceed 40 in the EW direction. As for [B] , [C]'s house, C-value exceed 40 in the NS and EW directions. It is judged that the possibility of the house collapse at the time of a large scale earthquake. The [A],[B],[C] residences in a figure are houses before the Building Standards Law revision, and [E],[G],[I] residences is a house after the Building Standards Law revision. It is understood that C-value after the Building Standards Law revision is less than C-value before the Building Standard Law revision. Furthermore, the difference between C-value of EW direction and NS direction in each building becomes 30 around at the maximum, and it become important how earthquake power stream down into the wall direction of the high rigidity when we think about earthquake-resistant performance.

$$C = 100 \cdot T_o^{\ 2} \cdot \sqrt{Q \cdot R_o} \quad \text{—— Eq. (1)}$$

here、

To : Natural period of building (sec)

Q : Index that shows resonance sharpness. There is a relation of Q=1/2h up to damping constant h.

<Criterion>

$C < 20$: Safety

$20 \leqq C < 30$: The fear of the collapse is a little.

$30 \leqq C < 40$: The fear of the collapse is middle.

B.S.L = Building Standards Law

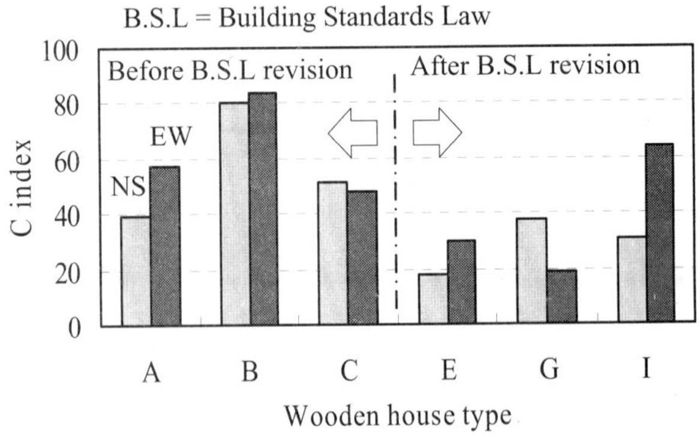

Fig.14 Comparison of C index

6 CONCLUSIONS

When the vibratory behavior in the building and the earthquake performance were evaluated from the natural period of wooden house and temple by the microtremor measurement, it turned out that the difference at the structure and years since construction was shown. The high applicability of this method has been understood. Furthermore, it is thought that the direction of the high rigidity wall is transmitted the seismic force by improving the binding degree of the floor and the wall as an idea of the seismic retrofit. Therefore the earthquake performance can be improved. Experienced judgment with a spectral analysis is needed to evaluate the natural period in the building because the microtremor measurement cannot specify the source of the vibration. In future, the method of judging the accurate natural period in the building is improved, and it is thought that the accumulation of the measurement data and experimental verification is necessary.

REFERENCES

1. Tsuji, Maiguma, Kobayashi, Yamada(2005). A method for evaluating the earthquake resistant capacity of small-scale buildings by means of microtremors analysis., Butsuri-Tansa vol.58.
2. Hasegawa(2006). Research on evaluation of earthquake performance of wooden house by microtremor. Master's thesis, Ritsumeikan University.
3. Koushige, Honda (2005). Study of ground vibration characteristics and factors of disturbance included in microtremor , IW-SHIGA.

EFFECT OF CHANGE IN FOUNDING DEPTH ON THE STRUCTURAL RESPONSE OF SAFETY RELATED NUCLEAR POWER PLANT STRUCTURE

Patil G.R[1]. Ranjan Rajiv[2], Giridhar K[3], Prabhakar G.[4], Jain D.K.[5], Verma U S P[6]

Civil Group, Nuclear Power Corporation of India Ltd., Mumbai.
[1]Engineer(Civil), [2] Sr.Executive Engineer(Civil), [3] Dy. Chief Engineer(Civil), [4] Addl. Chief Engineer (Civil), [5] Chief Engineer (Civil), [6]Associate Director(Civil)

ABSTRACT

A parametric study has been carried with the overall objective of containing the effect of different founding levels on the structural response in terms of floor response spectra for Control Building (CB) of a typical Nuclear Power Plant. CB is a multi-storeyed RCC structure with a basement that houses various safety related systems. Three different parametric models of the building with Baffle-Wall sub-foundation system of depth 3m, 5m and 10m respectively, have been used to evaluate the structural response. Structural engineering software ETABS V9.0.3 has been used to mathematically model the structure in terms of stiffness, mass and damping and carry out Response Spectrum and Time History Analysis. The comparison of Floor Response Spectra shows a close match in terms of peak acceleration and Zero Period Acceleration (ZPA) and hence, the systems and equipments need not be re-qualified. However, with an increase in the founding level of the structure, there is an increase in the base shear, which is attributed to the additional mass of the baffle wall system, which may call for a design adequacy check of foundation system.

Keywords: Baffle-Wall Sub-foundation system, Floor Response Spectra, Equipment requalification, Control Building

1. INTRODUCTION

The structures, systems and components of a Nuclear Power Plant (NPP) are designed considering very low probability of failure and sufficient structural integrity against different loads and hazards. Various safety related systems, equipments and components housed in safety related structures are qualified for the corresponding floor response spectra under site specific ground motions. In order to provide the possibility of NPP allocation to any suitable site that meets the regulatory criteria, the layout and design of safety related structures has to be standardized such that the re-qualification of structures, systems and components are kept to a minimum. The foundation system for various main plant buildings has to be chosen in such a way that the design of superstructures and associated system, equipments and components does not change and are within the range for which they are qualified. The foundation system needs to have sufficient flexibility, such that, it is possible to alter the foundation depth depending upon the site conditions. Further, provision should be available in the foundation such that the required floor spectra of various buildings are contained and within those for which the system, equipments and components of NPP are designed. Normally adopted foundation types for regular framed RCC structures are:

- Combined footings, where eccentric loads or locally large loads are coming.
- Raft foundation where the loads to be supported are comparatively large
- Baffle walled foundation, to transfer heavy loads to a deeper founding media
- Pile foundations with Rigid pile cap, to transfer loads to hard strata

Baffle wall sub-foundation system having high rigidity has been adopted in the study, so that, it does not magnify the seismic response under different founding levels and the design of the structures, systems and components located at various floor levels does not require re-qualification.

2. CONTROL BUILDING DESCRIPTION

The Control Building of 220MWe NPP is a multi-storied reinforced concrete framed structure having lengthwise orientation along the N-S (Construction) direction. Typically the Finished Ground Level (FGL) is assumed at EL 100m. The structural configuration consists of raft foundation at EL91.0 and beam-column framework supported on the raft foundation. The general arrangements of the Control Building are shown in Fig-1. The framed structure is stiffened by shear walls along the periphery of the structure (Fig.2) and has floors at EL.96.85, EL100.0, EL106.5, EL111.0, EL116.8, and EL121.5.

The building has a dimension of about 86m length, 29m wide and 25.5m height above finished grade and has a 9m deep basement of the same plan size as the building itself. R.C.C. retaining walls with buttresses at short intervals are provided on the east and west side of the basement, while the north and south ends are provided with tapered R.C.C. walls. The retaining walls have been kept structurally isolated from the main structural system, to ensure forces due to soil pressure, ground water and earthquake on the retaining walls are transmitted directly to the base raft and not to the main building structural frame. The basement and foundation raft has been kept structurally separated from the adjacent Service Building, Turbine Building and Reactor Building and Diesel Generator Building foundations and structures. Structural seismic gap of 100mm has been provided to take care of out of phase seismic deformations between two adjacent structures.

Fig.1 : 3-D Model of Control Building

Fig.2: Isometric View of Shear wall of CB

3. STRUCTURAL MODELING

3.1. Stiffness Modeling

A detailed 3-D finite element model [4], using advanced structural engineering software ETABS, has been developed for the Control Building to study the behavior of the structure under seismic ground motions. The mass, stiffness and material damping has been adequately represented in the finite element model. The 3-D finite element model for the entire nuclear building consists of the following elements [3]:

- 3-D shell element has been used for modeling shear walls and floors.
- 3-D frame element has been used for modeling beams and columns.

The 3-D finite element model of the Control Building consists of stiffness components of the entire structural configuration so as to arrive at the dynamic response of total structure as a whole. All concrete sections are modeled as uncracked sections.
The peripheral retaining walls have not been modeled as it is structurally independent from the framed structure and will not affect the structural response of the Control Building under seismic excitation. However, the mass of the peripheral walls have been accounted by mass lumping at the raft level along the peripheral nodes.

3.2. Inertial Modeling

The 3-D Finite element model has been developed considering the inertial effects of all the concrete structural elements. The mass of various equipments housed in the Control Building, dead weight of various non-structural elements such as outer masonry walls, partition walls etc. has been considered with lumping the mass at appropriate locations. 100% contributions of such equipment and non-structural loads towards inertial effect have been considered for dynamic analysis [1].In addition, 50% contribution of live loads on floors and 25% contribution of live load on roof have been considered for inertial effect during dynamic analysis. The entire finite element model of the Control Building consists of beams and columns as 3-D frame element and slabs and shear walls as 3-D plate shell element. The shell element has been used with auto-meshing options having a maximum size of 1.0m.

In a detailed 3-D model of such a large structure, large number of dynamic degree of freedom (DDOF) are present and are generally not required for evaluating the dynamic response of the structure as the contribution to higher modes are generally negligible. Further, truncation of DDOF reduces the computational time in step-by-step time history analysis. Structural idealization in the form of Kinematic constraints are imposed for such systems with large DDOF so that the mathematical model with constraint degree of freedom represents actually the true behaviour of the structure without much significant error*[3]*. Kinematic constraints in the form of rigid diaphragm for each floor have been used in the present analysis. The software ETABS has an in-built feature of assigning rigid diaphragm action to node points at a particular storey level and this has been used to constrain the DDOF *[4]*.

3.3. Material Properties and FE mesh details:

The grade of concrete for sub-base foundation and raft is M20 and for all other structural component is M25. Details of material properties and FE mesh are shown in Table 1, 2

<table>
<tr><th colspan="5">Table 1: Material Properties</th><th colspan="2">Table 2: FE Mesh Details</th></tr>
<tr><th>Materials</th><th>Density (kN/m^3)</th><th>f_{ck} (MPa)</th><th>Poisson's ratio</th><th>E_c (MPa)</th><th>Parameter</th><th>Total Numbers</th></tr>
<tr><td rowspan="2">Concrete, M25</td><td rowspan="2">25</td><td rowspan="2">25</td><td rowspan="2">0.2</td><td rowspan="2">25000</td><td>Nodes</td><td>4552</td></tr>
<tr><td>Shell Elements</td><td>4132</td></tr>
<tr><td>Concrete, M20</td><td>25</td><td>20</td><td>0.2</td><td>22360</td><td>Frame Elements</td><td>1351</td></tr>
</table>

4. ANALYSIS METHODOLOGY

Modal analysis has been carried out to study the dynamic behaviour of the Control Building structure. Modal Superposition Method is used to decouple the equation of motion in nodal coordinates into 'N' number of SDF generalized modal coordinates system. Modal frequency, mass participation factor, mode shape vector and spectral acceleration are calculated for each of these SDF systems in generalized modal coordinate. The generalized response of each mode is then used in Time History and Response Spectrum analysis for obtaining the total response of the structure *[2]*. For such MDF system being uncoupled to SDF systems the number of modes to be considered for analysis depends upon three parameters, viz., response type, cut-off frequency and cumulative modal mass participations. The numbers of modes included in the analysis has been decided on the basis such that 90% mass participation in each of the three orthogonal directions. For masses that do not participate in the dynamic analysis within 33 Hz, missing mass correction has been performed.

5. SEISMIC INPUT

Site specific response spectrum and spectrum compatible time history for a magnitude of 6.5 (Richter Scale) of Kaiga site shown in Fig.3, corresponding to 5 % of damping, has been used seismic analysis. The horizontal response spectra, normalized to 0.2g for Safe Shutdown Earthquake (SSE) and 0.1g for Operating Basis Earthquake (OBE) are applied at the sub-foundation level. The response spectra and time history for vertical direction is taken as 2/3 that of horizontal design response spectra and time history respectively [1].

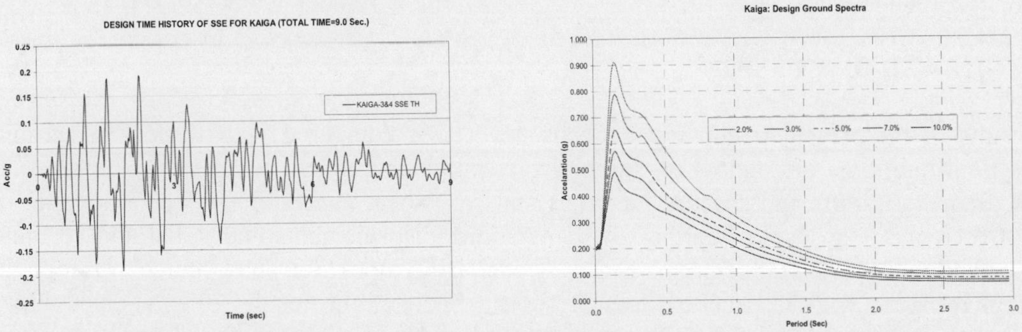

Fig.3:Compatible time-history and response spectra for KAIGA- 3&4

6. FLOOR RESPONSE SPECTRA GENERATION

The time history acceleration response at the floors of the structure is used as input for the analysis of single-degree-of-freedom oscillators, with varying natural frequencies over the range of interests (0.5 Hz to 33 Hz) and for specified damping ratios. ETABS adopts an approach for the calculation of response at discrete time intervals, based on the exact analytical solution of the governing differential equation for the successive linear segments of the excitation [4]. For generating the response spectrum curves from the time-history results, the absolute accelerations at the joints are obtained from the time-history case. For each frequency plotted in the response spectrum curve, the modal equations are analytically integrated. The maxima of the acceleration response obtained within the total duration of motion for the simple oscillators for each damping ratio is then plotted in the frequency domain to obtain the response spectra. In generation of in-structure response spectra, the spectrum ordinates have to be computed at sufficiently small frequency intervals to produce accurate response spectra. The response spectra have been calculated at the frequencies intervals given in ASCE [1*]*

7. PARAMETRIC STUDY

Model-BW3 is the base model and other two models are modification in base model in terms of founding level. The raft has been kept at EL 91m and the baffle-wall sub-foundation system is having a depth of 3m, 5m and 10m, respectively. The numeric nomenclature of the model name gives the height of the baffle-wall sub-foundation system. A total of three different foundation cases are considered in the present study (Fig. 4, 5), typically,

Model BW3: Fixed Model with 3.0m height baffle wall below El.91m
Model BW5: Fixed Model with 5.0m height baffle wall below El.91m
Model BW10: Fixed Model with 10.0m height baffle wall below El.91m

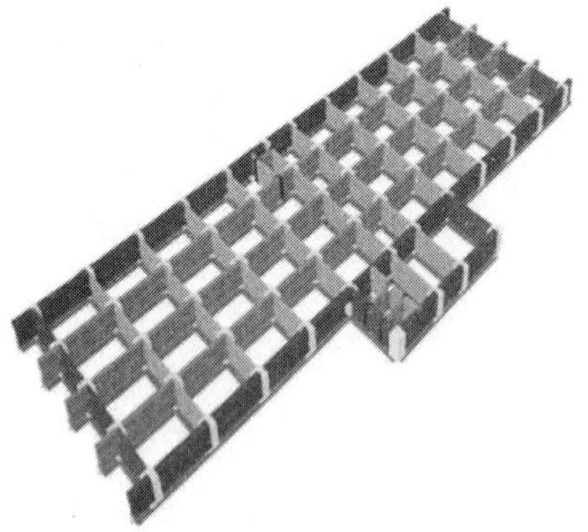

Fig.4: Sub-foundation System

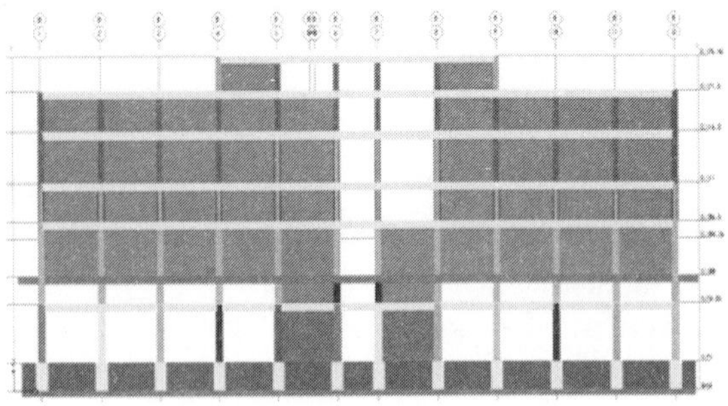

Fig.5: Sectional View of BW3

8. RESULT AND DISCUSSION

The results of the present parametric analysis have been compared with respect to seismic masses in all the three orthogonal directions, corresponding frequency and mass participation factors, frequency shifts and spectral locations and base shear and moments corresponding to PGA of 0.1g and 0.2g level of earthquakes, maximum displacements and nodal and absolute enveloped Floor Response Spectrum. Since, Model BW5 does not show much variation in structural response with respect to Floor Response Spectra, the results of enveloped Floor Response Spectra of Model BW5 are not included in the paper. However, other dynamic characteristic are compared.

8.1. FREQUENCY AND MPF COMPARISION

The frequency and mass participation factors as obtained from Ritz modal analysis for Model BW3, Model BW5 and Model BW10 for all the three orthogonal directions are given in Table 3.Fig. 6, 7, 8 shows the variation of the modal mass participation factors in all the three directions with respect to frequency.

Table-3: Dynamic Characteristics and Structural Response of Parametric Models

	Model: BW3		Model: BW5		Model: BW10	
Direction	X-Hor.	Y-Hor.	X-Hor.	Y-Hor.	X-Hor.	Y-Hor.
Fundamental Frequency(Hz)	2.15	2.03	2.13	2.00	2.10	1.95
Predominant Mode MPF (%)	55.77	55.56	55.16	55.30	53.33	54.9
Maximum Displacement @ EL121.5 (mm)	26.2	43.6	26.6	44.9	27.3	48.0
Enveloped FRS ZPA @ EL121.5	**0.457*g***	**0.797*g***	**0.478*g***	**0.805*g***	**0.50*g***	**0.814*g***

8.2. SEISMIC MASS COMPARISION

The seismic masses in all the three translational orthogonal directions and vertical rotational direction for the parametric models, Model BW3, Model BW5 and Model BW10 are given in

Table 4. The rotational masses in the horizontal X- and Y-Directions are zero due to the kinematic constraints imposed at all the storey slabs and roof slab wherein the rotational inertias have been neglected as it has negligible impact on the dynamic behaviour of the structure.

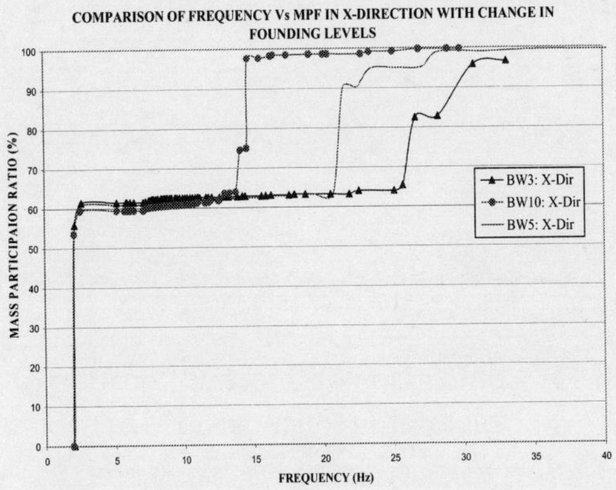

Fig.6: Comparison of Freq vs. MPF (X-Dir)

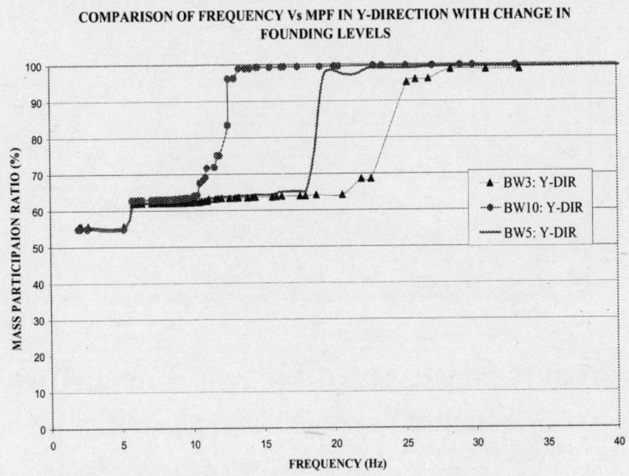

Fig.7: Comparison of Freq vs. MPF (Y- Dir)

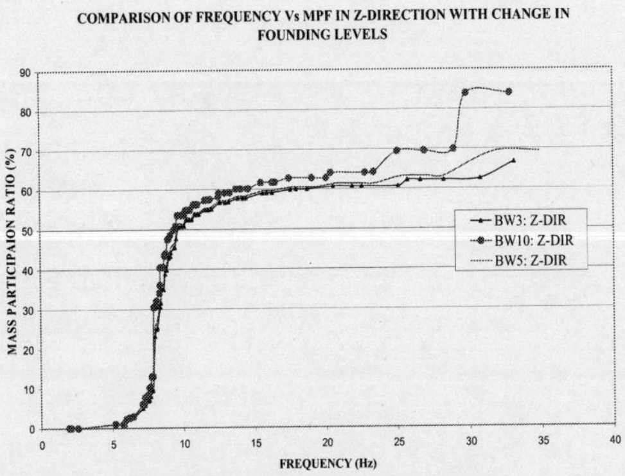

Fig.8: Comparison of Freq vs. MPF (Z- Dir)

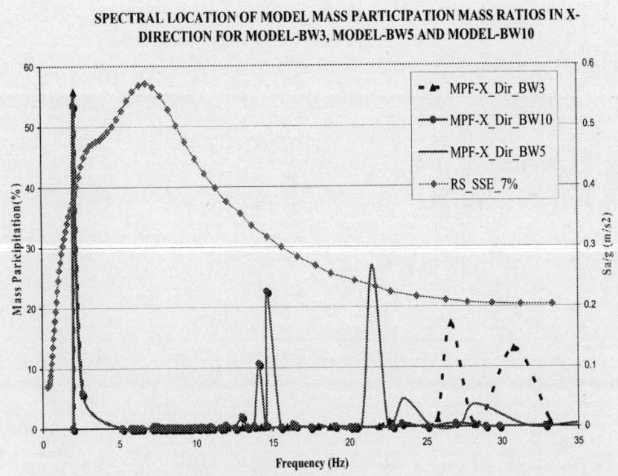

Fig. 9: Comparison of Spectral Location of Structural Frequencies (X-Dir)

Table 4: Seismic mass comparison for Model BW3, BW5 & BW10						
Model	UX	UY	UZ	RX	RY	RZ
BW3	61170.06	61170.06	61170.06	0	0	25933251
BW5	63250.43	63250.43	63250.43	0	0	25933251
BW10	69460.47	69460.47	69460.47	0	0	25933251
Note: All units are in kN and m.						

8.3. FREQUENCY SHIFT AND SPECTRAL LOCATIONS

The spectral locations of modal mass participation ratios for Model BW3, Model BW5 and Model BW10 in all the three directions are shown in Fig. 9, 10, 11, respectively. It is observed that with an increase in the founding level depth, there is not only a shift in the frequency towards the left side of ground response spectrum; there is an increase in the structural response in terms of peak displacements and base shear. However, it is observed that the shift is more at higher frequency modes where the mass participation is less. Hence, this is not contributing to the magnification in the structural response in terms of floor response spectra.

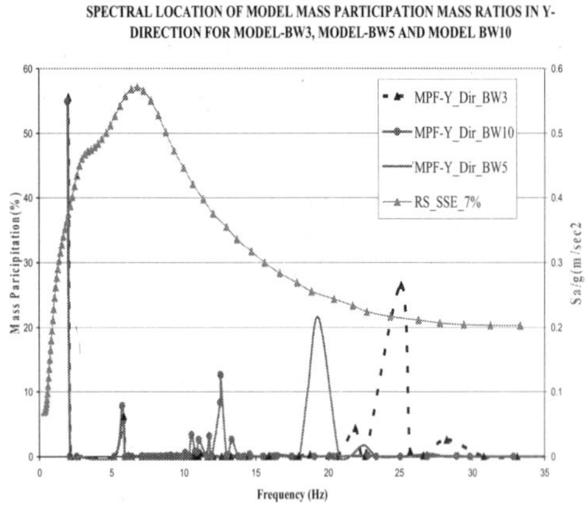

Fig. 10: Comparison of Spectral Location of Structural Frequencies (Y-Dir)

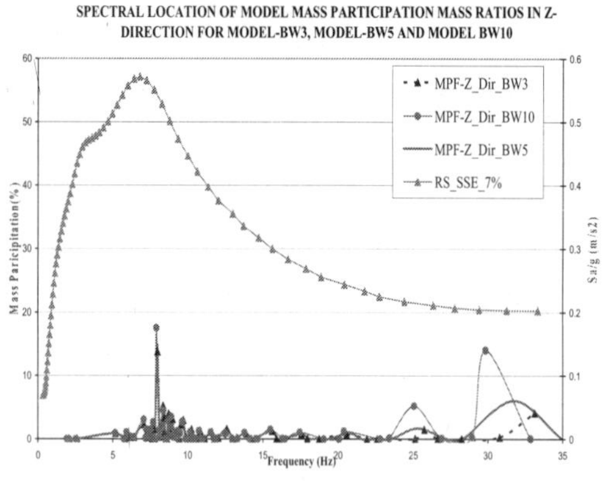

Fig. 11: Comparison of Spectral Location of Structural Frequencies (Z-Dir)

8.4. BASE SHEAR COMPARISON

The base shear and base moments as obtained from Response Spectrum Analysis (RSA) expressed as percentage increase w.r.t Model BW3 for Model BW5 and Model BW10 for all the three translational and rotational directions are given in Table 5.

Table 5: Base shear and moments comparison for Model BW5 and Model BW10

Percentage increase for Model BW5 (w.r.t Mode BW3)							
Sr No.	Case	F1(KN)	F2(KN)	F3(KN)	M1 (KN-m)	M2 (KN-m)	M3 (KN-m)
1	RS OBE*	2.7	3.6	3.4	6.2	5.3	3.4
2	RS SSE**	2.3	3.5	3.4	6.1	5.0	3.2
Percentage increase for Model-BW10 (w.r.t Model BW3)							
Sr No.	Case	F1(KN)	F2(KN)	F3(KN)	M1 (KN-m)	M2 (KN-m)	M3 (KN-m)
1	RS OBE*	11.5	12.2	16.3	27.3	23.7	11.7
2	RS SSE**	12.1	15.2	15.0	27.0	22.2	14.3

* RS OBE: Base Shear using Response Spectrum for OBE of Kaiga site with 4% Damping
**RS SSE : Base Shear using Response Spectrum for SSE of Kaiga site with 7% Damping
Note: F1, F2 and F3 are bases shears in X, Y and Z directions, M1, M2 and M3 are base moments in X, Y and Z directions respectively.

8.5. FLOOR RESPONSE SPECTRA COMPARISON

The floor response spectra have been compared with respect to nodal floor response spectra and enveloped floor response spectra. Nodal floor response spectra results are not covered in the paper. For presentation purpose, since the ground motion parameter in terms of peak spectral acceleration and frequency content for OBE level of EQ is only scaled to SSE level having a PGA of 0.2g, the comparison of floor response spectra presented is for different parametric model subjected to SSE time history only.

The comparison shows a close match of enveloped floor response spectra in terms of peak acceleration and the corresponding frequency. The Zero Period Acceleration (ZPA) in the enveloped FRS for EL 121.5 in X-direction for Model BW10 is 0.50g compared to 0.457g for Model-BW3, which can be considered as a close match.

The structural behaviour at lower elevation i.e., floor @ EL 121.5, EL116.8, EL 111.0, EL106.5, EL104.26, EL 100 and EL 96.85 is similar to that of floor @ EL121.5. Less difference in the spectral acceleration in the floor response spectra is observed for all the three directions, hence, results of floors other than EL 121.5, EL 111.0 are not presented. Fig.12 to 17 shows the comparison of enveloped floor response spectrum for floors at EL 121.5, EL 111.0.

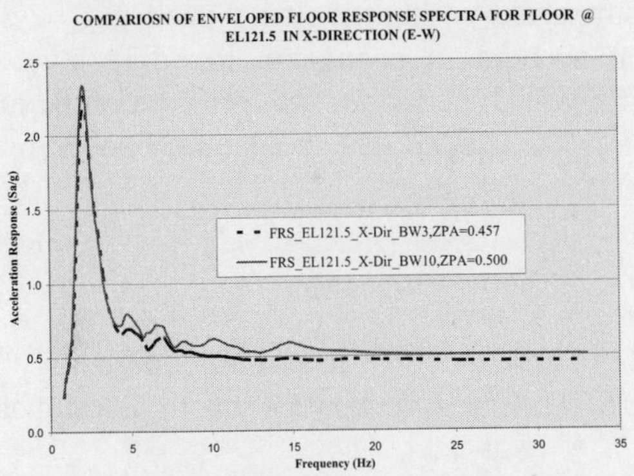

Fig.12: Enveloped FRS @ EL121.5 for Model BW3 and BW10 (X-Dir)

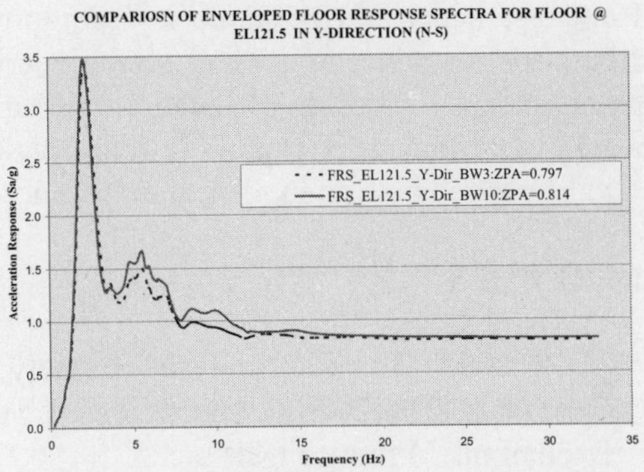

Fig.13: Enveloped FRS @ EL121.5 for Model BW3 and BW10 (Y-Dir)

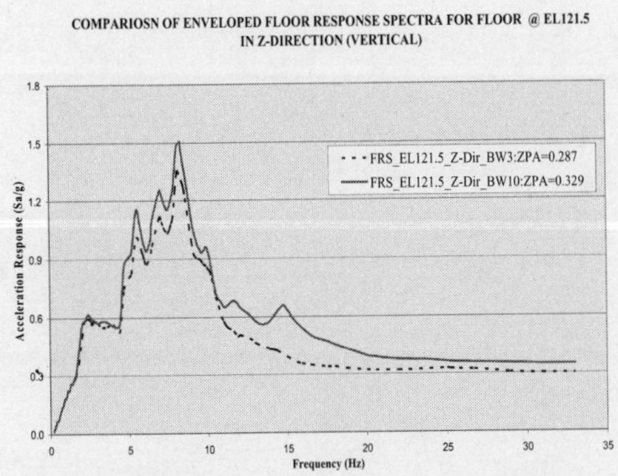

Fig.14: Enveloped FRS @ EL121.5 for Model BW3 and BW10 (Z-Dir)

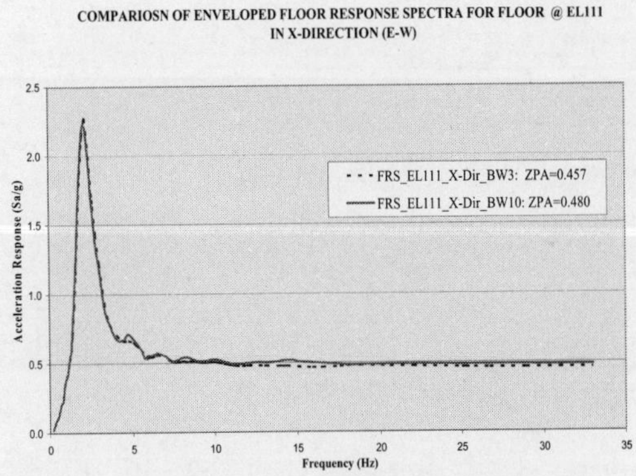

Fig.15: Enveloped FRS @ EL111.0 for Model BW3 and BW10 (X-Dir)

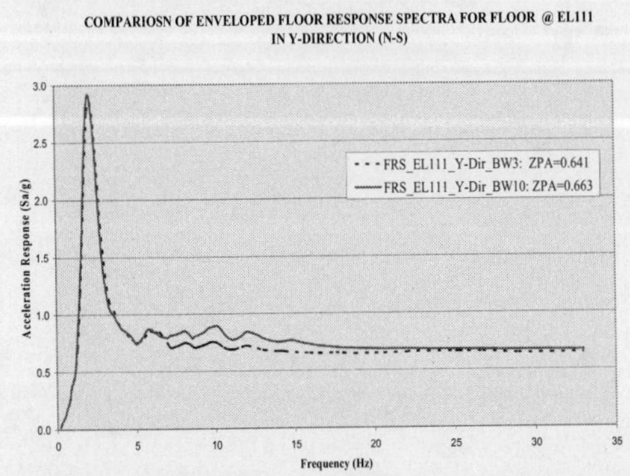

Fig.16: Enveloped FRS @ EL111.0 for Model BW3 and BW10 (Y-Dir)

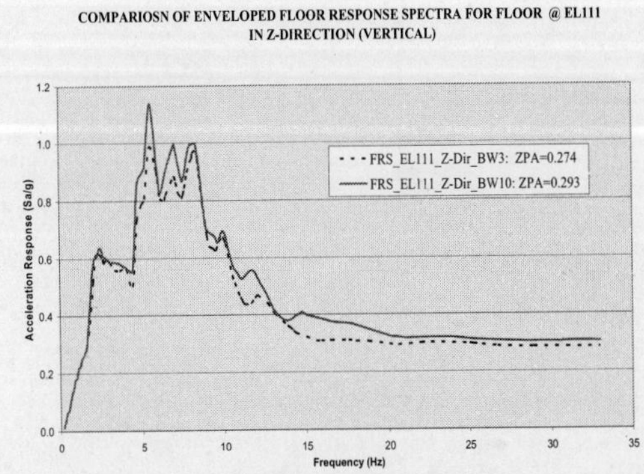

Fig.17: Enveloped FRS @ EL111.0 for Model BW3 and BW10 (Z-Dir)

9. CONCLUSION

From the present study, it is evident that foundation system with baffle walls seems to be acceptable for site conditions where stiff soil / hard rock is not available at shallow depths. The floor response for such system will not alter significantly, with increased depth of foundation up to 10meters, hence, the systems and equipment supported at higher floors need not be re-qualified for revised FRS. Thus re-engineering on account of change in founding level is minimal, except for design of the sub-structure baffle wall foundations and preparation of few drawings with foundation details of relevant structures. A detailed design adequacy check may need to be carried out for sub-structure for increase in the base shear.

REFERENCES

1. ASCE 4-98, 'Seismic Analysis of Safety Related Nuclear Structure and its Commentary,'
2. Chopra A. K. (2006). 'Dynamics of Structures-Theory and Application to Earthquake Excitations,' Prentice Hall Inc
3. Cook Robert D., Malkus David S.,_Plesha Michael E.,_Witt Robert J.(2005).'Concepts and Application of Finite Element Analysis,' Wiley Student Edition
4. ETABS and SAP. 'CSI Analysis Reference Manual,'

SEISMIC RESPONSE CONTROL OF COMPLEX PIPING SYSTEMS USING ELASTO-PLASTIC DAMPERS- EXPERIMENTS AND ANALYSIS

Y. M. Parulekar[1], G.R. Reddy[1], K.K. Vaze[2], A.K. Ghosh[3], H.S. Kushwaha[4], Ramesh Babu[5]
[1-4]Bhabha Atomic Research Centre, Mumbai-400085, India
[5]Central Power Research Institute, Bangalore, India

ABSTRACT

In Nuclear Power plants, snubbers that are expensive, complex and require frequent maintenance have often been used as seismic supports in piping systems. Passive dampers such as Elasto-Plastic Damper (EPD) and Shape Memory Alloy Damper (SMAD) are cheap, simple, cost effective and reliable. They can be used for replacing snubbers in piping systems as they maintain the flexibility of piping and also increase the damping. In this paper, in order to quantify the effectiveness of multiple elasto-plastic dampers on realistic piping systems with 3D earthquake excitation, a complex piping model with internal pressure is tested with and without two EPDs on shake table. Tests are also carried out for piping with only one EPD attached in vertical direction and only one EPD attached in horizontal direction. Linear and nonlinear dynamic analysis is carried out for the piping system using Newmark's time integration technique.

1 INTRODUCTION

It is very essential that important equipment / piping systems of a nuclear power plant need to be safeguarded against natural disaster such as earthquake. In NPP, snubbers that are expensive, complex and require frequent maintenance have often been used as seismic supports. Hence recently, world wide there is a trend to use passive control devices to reduce the response of the piping systems subjected to seismic loads [1-3]. The objective of these devices is to absorb a significant amount of seismic input energy, thus reducing the demand on the piping system.

Elasto-plastic dampers [EPDs] consists of X shaped plates which can sustain many cycles of stable yielding deformation. The damper as shown in Fig. 1 is designed and manufactured. The characteristics of the damper were studied by carrying out, static, dynamic, and fatigue tests on the X plate [4, 5] and the test results showed good agreement with theoretical calculations. Tests were also initially carried out on simple cantilever piping systems with EPDs [6] for unidirectional excitation of the shake table. It was found that the dampers are very effective is reducing the response of simple piping subjected to unidirectional excitation by 30 to 40 %. Many researchers have also done significant contribution in evaluating response of piping systems with Elasto-Plastic Dampers [1-3]. They have carried out tests on piping with simple piping configurations with unidirectional excitations. In this paper an attempt has been made to obtain the effectiveness of multiple elasto-plastic dampers on complex piping system under three directional earthquake excitation by carrying out experiments and proving by analysis. This study is carried out to make certain that the damper performance in one direction is not affected by the hinges acing in the other two perpendicular directions during the three directional motion. It also gives the confidence that using multiple dampers in the piping system does not alter the free vibration characteristics of the piping system. This is the most essential factor for NPP piping systems as they have to cater thermal load along with seismic load.

The test response was compared with the analysis results. Nonlinear dynamic analysis of the piping system is carried out using Newmark –Beta time integration method. The non-linearity of damper plates was addressed using Von-mises kinematic hardening model. The reduction in response of piping with multiple dampers is more than with a single damper. It is observed that there is 40-50 % reduction in the response of the piping system attached with two dampers.

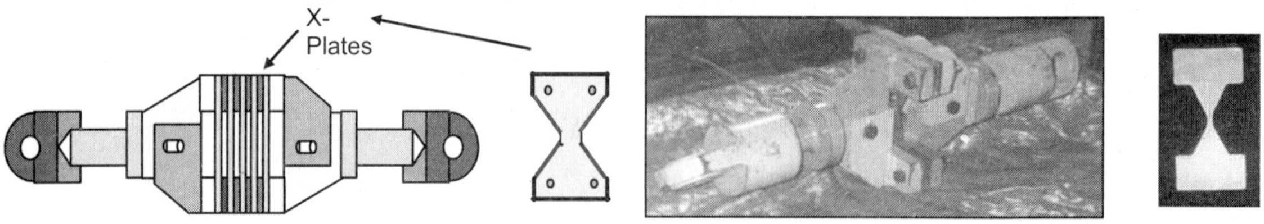

Fig. 1(a) Elasto-plastic Damper

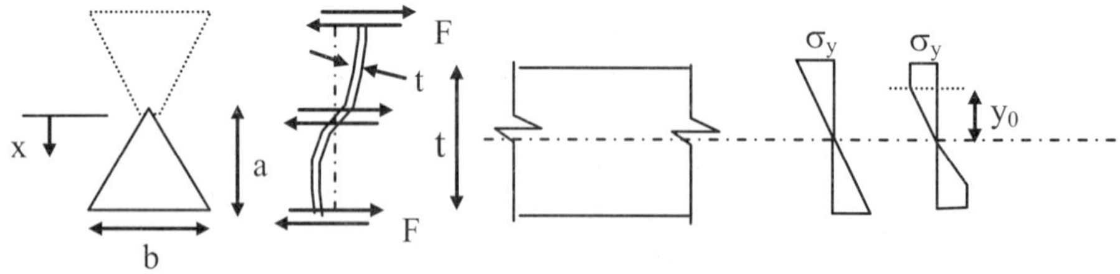

Fig. 1(b) Characteristics of X shaped plate

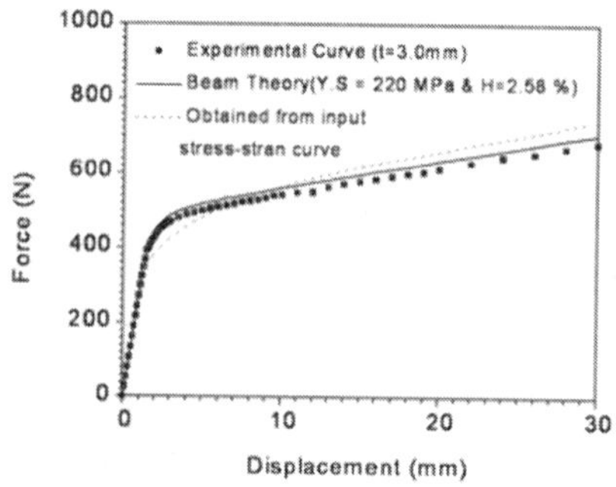

Fig. 1(c) Comparison of Experimental and theoretical force displacement curve for 3mm thick plate under static load.

2 FORCE DISPLACEMENT CHARACTERISTICS OF X PLATE

The force displacement characteristics of X plate were obtained by analytical formulations and also by conducting static and cyclic tests on X plates of different thickness.
Considering bending stress in X- plate shown in Fig. 1(b) is elastic the X shaped plate connects the top of two triangular plates. Thus considering only triangular plate

$$EI\frac{d^2y}{dx^2} = -Fx$$

$$F = \frac{Ebt^3d}{6a^3} \tag{1}$$

For X plate, the stiffness given by Eq. 2, where d is the displacement, b is the width, t is the thickness, a is the height of the triangular plate and E is the Young's Modulus of Elasticity.

$$K = \frac{F}{2d} = \frac{Ebt^3}{12a^3} \tag{2}$$

Considering bending stress in X-plate just reaching yield stress,

$$F_y = \frac{\sigma_y bt^2}{6a} \tag{3}$$

$$d_y = 2\frac{\sigma_y a^2}{Et} \tag{4}$$

Considering the bending stress in which elastic depth of the X plate reaches $2y_o$ and H is the strain hardening rate,

$$F = \frac{b\sigma_y}{12Ea}\left\{(4y_0^{\ 2} - 3t^2)(H - E) + \frac{Ht^3}{y_0}\right\} \tag{5}$$

$$d = \frac{\sigma_y a^2}{Ey_0} \tag{6}$$

Using the above results the force displacement curve is obtained for a 3mm thick X-plate of carbon steel having the following properties E= 2.1 x 10^5 N/mm^2, H=2.58% of E=5418 N/mm^2 and σ_y= 220 N/mm^2. Load deflection curve of the X-plate is also obtained from input stress strain curve and is compared with test results and beam theory results as shown in Fig.1(c)

3 SHAKE TABLE TESTS ON COMPLEX PIPING MODEL

Shake table tests were carried out on the complex piping model shown in Fig.2(a). The piping model of material SA 106 Gr B was made of 10 Nos. of 3 in. NB sch. 40 straight pipes and 9 Nos. of elbows of radius 197 mm. One end of the pipe element was rigidly connected to the shake table through a horizontal flange type fixture, and the other end was fixed to the shake table through a vertical flange type fixture. In the middle of the horizontal straight pipes as shown in Fig. 2(a), masses of 52kg, 36 kg and 36 kg respectively were attached in the form concentric elements made of lead. The piping system is filled with water and the total mass of the piping system with masses attached is obtained as 260 kgs. Four cases of test setup were considered. The first case was with no damper as shown in Fig. 2(a). In the second case, vertical damper consisting of one X plate was attached to the piping as shown in Fig. 2(b). In this case, the damper attached was effective only in vertical direction and in other two directions it was designed to perform like hinge. In the third case, horizontal damper housing two plates was attached to the piping system as shown in Fig. 2(c). Similarly in this case the damper was effective only in horizontal direction (X- direction) and for other two directional excitation the damper was designed to perform like hinge. In the forth case as shown in Fig. 2(d) both the dampers were simultaneously attached to the piping system.

Fig. 2(a) Piping system without Damper (Case I)

Fig. 2(b) Piping system with one vertical Damper (Case II)

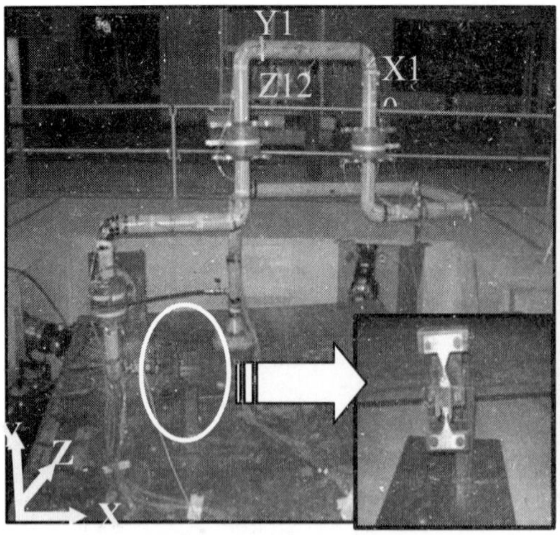

Fig. 2(c) Piping system with one horizontal Damper (Case III)

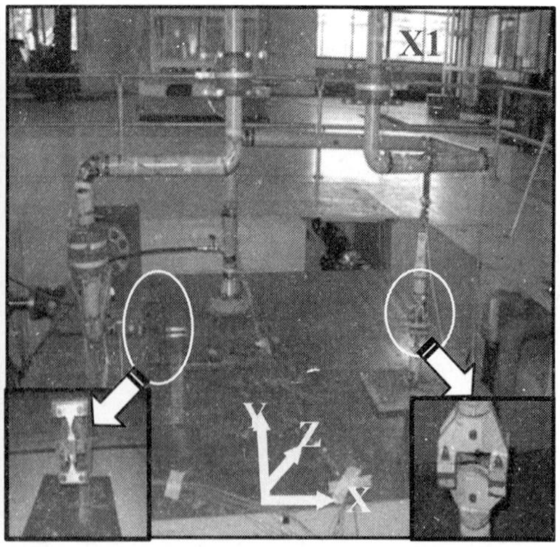

Fig. 2(d) Piping system with a horizontal and a vertical damper (Case IV)

4 SINE SWEEP WAVE TESTS

In order to obtain the frequencies and damping ratios, sine sweep tests were conducted on the piping system for all the four cases of piping system. Sweep since excitation tests were conducted with a constant base acceleration of 0.1g peak in the frequency band of 1 to 40 Hz with a constant linear sweep rate one octave/min. During this process acceleration responses were collected at various locations in X, Y and Z directions individually at accelerometers X10 Y11 and Z12. The location of the accelerometers is shown in Fig. 2. These tests were carried out for all the four cases of piping system.

The frequency response function (FRF) of the pipe acceleration (at accelerometers X10 and Y11) to the table acceleration was obtained by the sweep sine tests. The resonant frequency, which is the peak frequency of the FRF is obtained. Also the damping ratio is obtained using the half power method and response amplitude method. Harmonic analysis [7] was also carried out for the piping model and the FRF was obtained. Table 1 shows the comparison of resonant frequencies obtained from tests and analysis for the piping system without dampers. Table 1 also shows the damping ratios obtained from tests using half-power method for different frequencies of the piping without EPDs. It is observed that the analysis and test frequencies are in good agreement. The damping ratio at all frequencies obtained from test data is around 1.0 %.

Table 1. Comparison of test and analysis frequencies and damping ratios

Accelerometer No (Direction)	Mode No.	Frequency Hz (Analysis)	Frequency Hz (Tests)	% Damping (Tests)
X10 (X direction)	2	5.13	4.45	1.77
X10 (X direction)	3	10.13	8.9	0.96
Y11(Y direction)	1	4.54	4.06	1.04
Y11(Y direction)	3	10.13	8.9	1.1
Z12(Z direction)	1	4.54	4.06	1.02
Z12(Z direction)	2	5.13	4.45	2.0
Z12(Z direction)	3	10.13	8.9	1.23

For other three cases of the piping systems with EPDs the fundamental frequency of the piping remains almost same. For the piping system with both dampers the frequency of first mode is obtained from tests as 4.125 Hz (Frequency of the piping without damper is 4.06 Hz.). EPDs are attached to the piping to allow free thermal expansion in case of thermal loads and reduce the earthquake loads by absorbing energy through hysterisis deformation. The thermal loads can be minimized by using EPDs, if the frequency of the piping with EPD, do not change much. It is observed that the difference between the frequencies of the piping with and without EPDs is small (maximum upto 10%).

5 DESIGN SEISMIC WAVE TESTS

The spectrum compatible acceleration time history was obtained from the 2% damped required response spectrum shown in Fig. 3. This random seismic wave shown in Fig. 4 was applied to the piping system in X, Y and Z directions simultaneously. Tests were carried out for the piping system considering all the four cases shown in Fig. 2. The random seismic wave in three directions with peak RMS acceleration 0.35g, 0.7g, 1.05g, 1.45g was applied for each case. The acceleration applied in vertical direction was 2/3rd of the acceleration in two horizontal directions. The acceleration responses were collected at points X10(X Direction), Y11(Y direction) and Z12 (Z Direction) using accelerometers. The locations of the accelerometers on the piping system are shown in Fig.2.

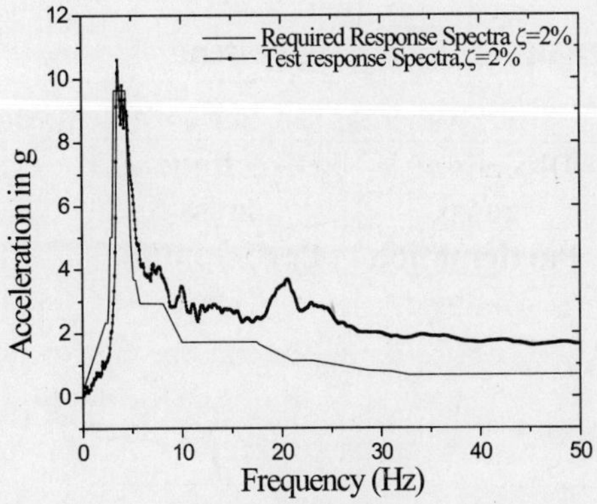

Fig. 3 Acceleration Response Spectra

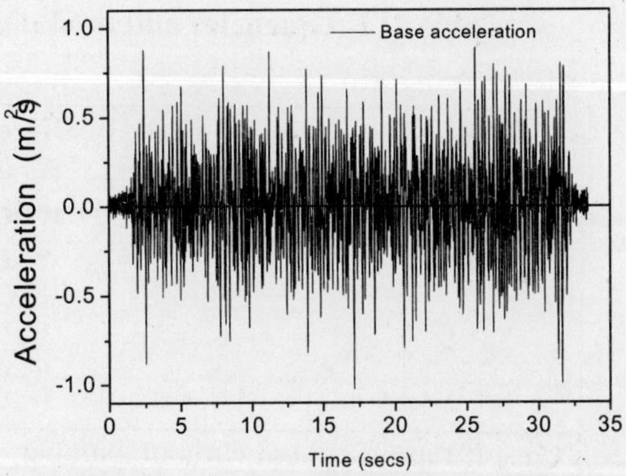

Fig. 4 Table Acceleration time History with 0.1g

6 DYNAMIC ANALYSIS OF PIPING SYSTEM

Finite element models of the piping system, as shown in Fig. 5 are made. The straight pipe portion was modeled using 3D beam element, elbow element was used for elbows, nonlinear truss element was used for EPDs and mass element was used to model the attached masses. The material characteristics of the EPD were given in the form of stress-strain relationship.

6.1 Free Vibration Analysis

Free vibration analysis is carried out and the frequencies, mode shapes and the modal mass participations of the piping system for all the four cases are obtained. The first mode shape of the piping without damper is shown in Fig. 6. The frequencies and the modal mass participations of the piping system without dampers for the dominant vibration modes are listed in Table 2. It is observed that the first two modes are coupled modes in Y, Z and X, Z directions respectively and the third mode is coupled mode in all the three directions. The results of free vibration analysis are in good comparison with sine sweep test results.

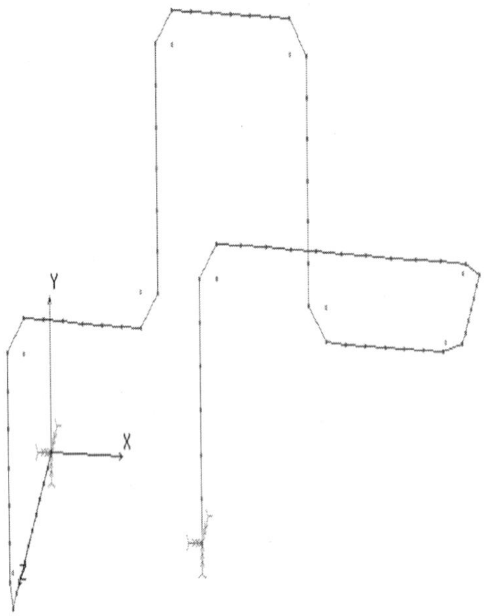

Fig. 5 FEM Model of piping system without damper

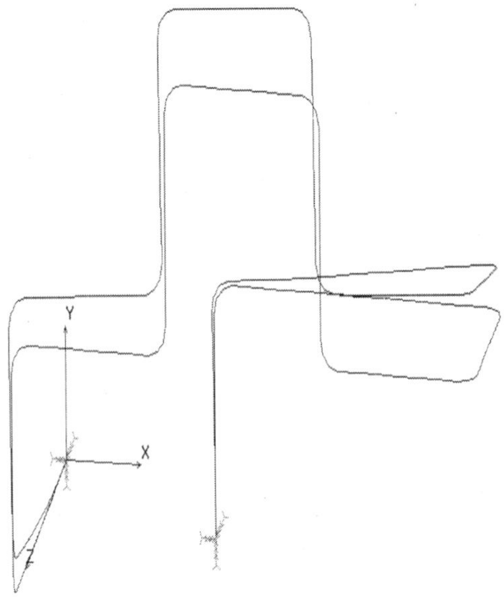

Fig. 6 First mode shape of the piping system without Damper (Freq=4.54Hz)

Table 2: Frequencies and modal participations of the piping system

Pre-dominant Mode	**Freq Analysis (Hz)**	**Freq Test (Hz)**	**Direction: X mass Participation %age**	**Direction: Y mass Participation %age**	**Direction: Z mass Participation %age**
Case I: Piping system without Damper					
1 Y,Z	4.54	4.06	3.2	48.0	23.3
2 X, Z	5.13	4.45	41.2	2.0	23.1
3 X, Y, Z	10.31	8.9	27.7	27.5	17.1

6.2 Time History Analysis

Linear and nonlinear time history analysis is performed for the piping system without and with dampers subjected to the spectrum compatible time history in all the three directions using Newmark- β time integration technique[7] with factors,γ=1/2 and β=1/6. The damping ratio considered for the time history analysis as obtained from tests is 1.0 %. The test and analysis results showing Y direction acceleration time history obtained at Y11 (accelerometer) location for the piping system without dampers (Case I) subjected to with 0.7g peak base acceleration is shown in Fig. 7(a). The same for piping system with two dampers is shown in Fig. 7(b). It is observed that the response of the piping system of 25 m/s^2 at the top of the piping is reduced to 13 m/s^2 at the same location with two dampers. Thus it is observed that for 0.7g peak base excitation the response reduction is about 40 %.

The force v/s displacement graphs at the damper locations for the piping with both dampers (case IV, Fig. 2(d)) for 0.35g and 1.05g base acceleration obtained from nonlinear time history analysis is shown in Fig. 8(a) and 8(b). It is observed that the displacement at the damper location of horizontal damper is about one third the displacement at the damper location of vertical damper. Hence large amount of energy is absorbed in deformation of vertical damper (one X plate) than the horizontal damper (housing two X plates). Thus vertical damper is found to be more effective than the horizontal damper in reducing the response of piping system. The energy dissipated in the hysteisis loops increases as the base RMS acceleration increases from 0.35g to 1.05g as shown in Figs. 8(a) and 8(b)

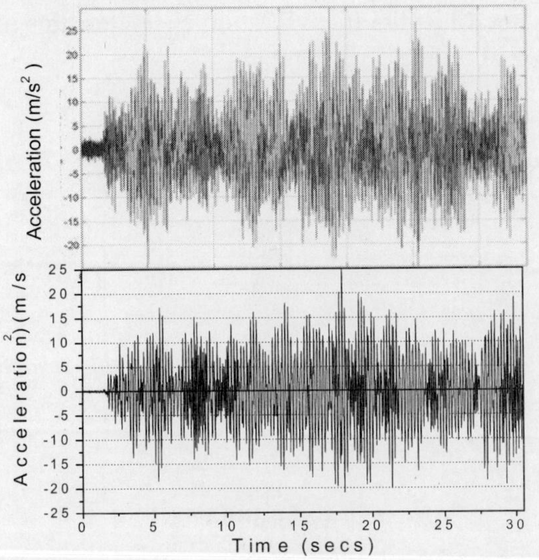

Fig. 7(a) Tests and analysis acceleration time history at Y11 location for piping system without damper (Case I)

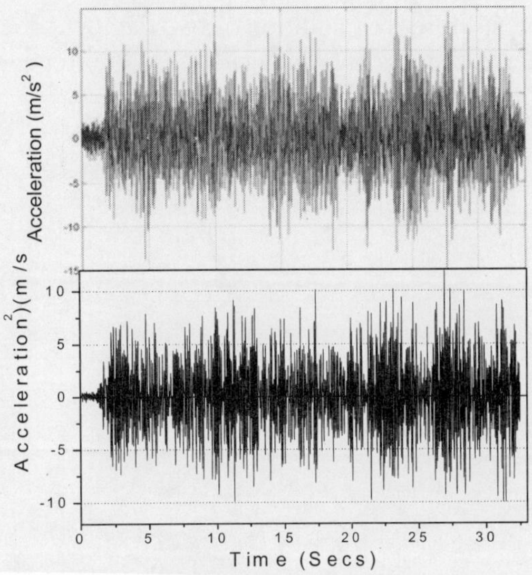

Fig. 7(b) Tests and analysis acceleration time history at Y11 location for piping system with two dampers (Case IV)

7 DISCUSSION OF TEST AND ANALYSIS RESULTS

The comparison of maximum values test and analysis response for all the four cases of piping without and with dampers for increasing base excitation from 0.35g to 1.45g is shown in Fig. 9. The result obtained at the location of X10, Y11 and Z12 accelerometers is shown in Fig. 9(a), Fig. 9(b) and Fig. 9(c) respectively. It is observed that the test and the analysis results are in good agreement. It is observed that for piping with vertical damper alone (case II) the response reduction in all the three directions is more than that of piping with horizontal damper alone (case III). This is because the piping is more flexible in vertical direction and its frequency in vertical direction is 4Hz, (Table I) which is lesser than that in horizontal direction. Hence it results in more energy dissipation in Y (vertical) direction than X direction. For the piping with horizontal damper alone (case III) the response reduction in X direction accelerometer, X10 is more than that of Y (Y11) and Z (Z12) directions. Moreover the response reduction in Z direction (Z12 accelerometer) for all the cases is lesser than that in other two directions. This is because the dampers are attached in X and Y directions only. If the damper had been attached in Z direction then the reduction in Z direction would have been increased. For the piping with both dampers (Case IV), it is observed that net reduction in response in Z direction (Z12) is 25 %, in Y direction it is 42 % and in X direction (X10) it 44%. The response reduction in all directions due to both dampers simultaneously acting together (case IV) is more than when anyone damper is attached but the increase is not the same as the addition of responses of two dampers acting separately. This is because when both dampers are attached simultaneously there is reduction in displacement due to effect of each damper and thus the energy absorbed is shared by two dampers. It is also observed that as the table RMS excitation increases from 0.35g to 1.45g, the percentage reduction in the response due to dampers increases initially but at 1.45g the % reduction in response remains same as the % reduction in response at 0.7g table excitation. The damping ratio of EPD thus does not increase with increasing displacements for large excitations. This can be explained henceforth. The damping ratio of the EPD depends on the ratio of hysterisis energy to the total strain energy [8]. The hysterisis energy is proportional to the displacement of the damper while the strain energy is proportional to the square of displacement. At small values of displacement when the damper just enters nonlinear range, the hysterisis energy to strain energy ratio is larger. As the excitation level increases, the increase in strain energy is more than that in hysterisis energy thus not increasing the damping value.

It can be inferred from Fig. 9 that there is about 25-50 % reduction in seismic response of the complex piping at all locations in all directions of the piping due to the addition of Elasto-plastic dampers. Moreover, addition of two dampers does not affect the free vibration characteristics of the damper as there is not much change is the frequency of piping with dampers.

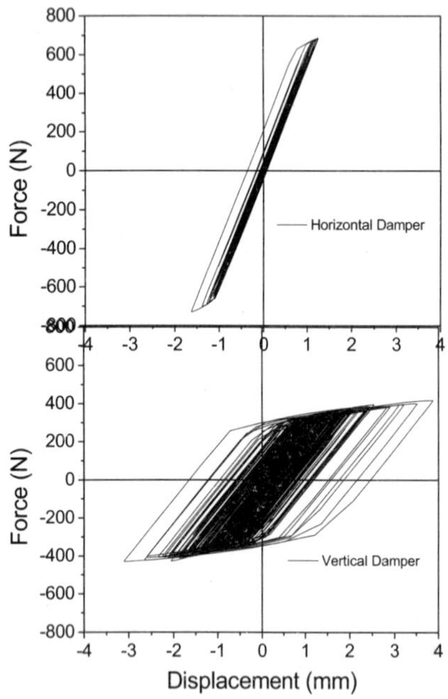

Fig. 8(a) Force v/s displacement plots at damper locations for piping with two dampers (case IV) with 0.35g RMS base

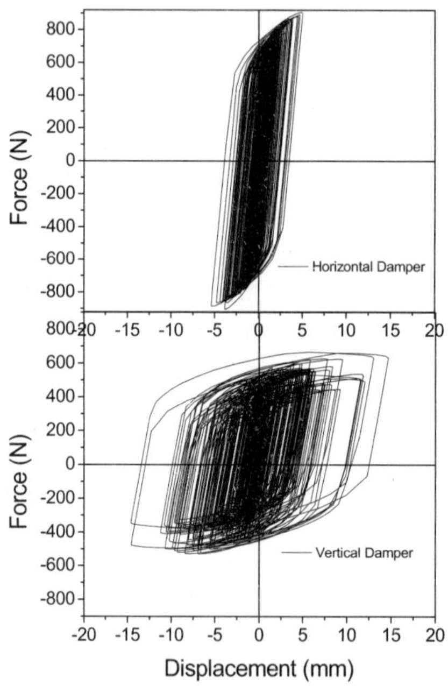

Fig. 8(b) Force v/s displacement plots at damper locations for piping with two dampers (case IV) with 1.05g RMS base

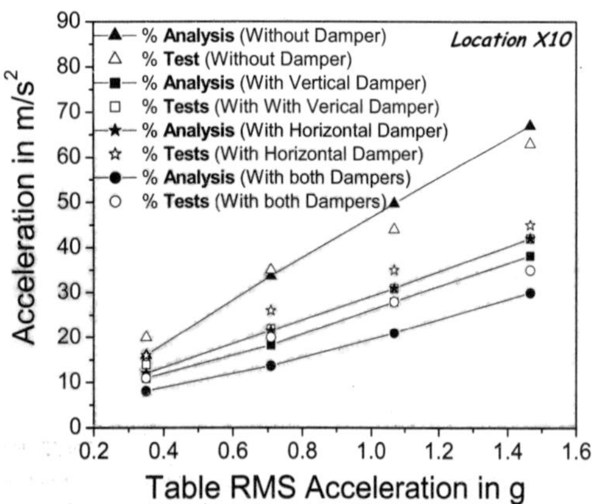

Fig.9(a) Maximum response acceleration of pipe at location X10

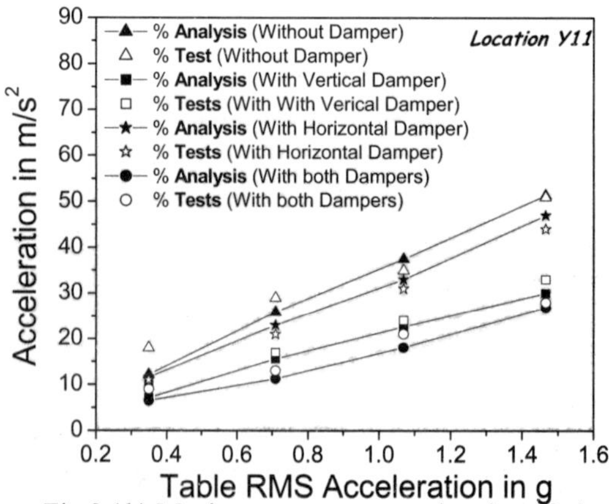

Fig.9 (b) Maximum response acceleration of pipe at location Y11

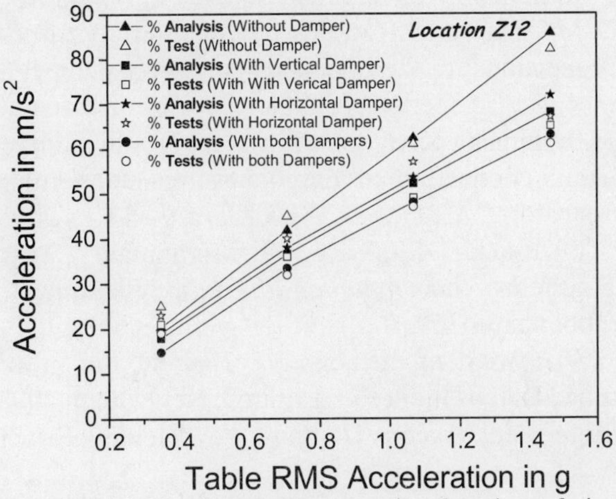

Fig.9(c) Maximum response acceleration of pipe at location Z12

8 CONCLUSIONS

Seismic proving tests were carried out on 3 dimensional complex piping system with and without Elasto-plastic dampers at multiple supports. It was observed that there is not much change in frequency of the piping system when dampers are attached to it. It was also observed that there is about 25-50 % reduction in response of the piping due to effect of two dampers. Time history analysis has been carried out for piping with the nonlinear supports and it is observed that analysis results are in good agreement with test results. The vertical damper is more effective in reducing the response of the piping than the horizontal damper due to flexibility of the piping in vertical direction. The response reduction with two dampers attached simultaneously to the piping is more than only one damper attached. Damper should be attached to the piping at the location of maximum displacement for optimum effectiveness of the damper. In order to obtain effective response reduction of piping subjected to 3- directional excitation, the dampers should be attached at different locations in all the three directions. In general multiple Elasto-plastic dampers can be effectively used for reducing response of complex piping systems subjected to multi directional earthquake excitation.

ACKNOWLEDGEMENT

The authors like to the acknowledge the contribution of Mr. Srinivas Murthy, Mr. Hegde, Mr. Paneer Selvam and Mr. Chetan from CPRI, Bangalore in conducting the experiments.

REFERENCES

1. Namita, Y., Yoshinaga, T., Shibata, H., Kunieda, M., Hara, F., Suzuki, K.,Ichihashi, I., Iiyama, K., Matsuda, T., and Murota, M., "Development of Energy Absorber and its Application to Piping System in Nuclear Power Plants," *ASME PVP* 211, pp. 51–57, 1991.
2. Namita, Y., Ono, S., Suzuki, K., Sasaki, Y., Abe, H., and Kuroda, K., "Proving Test on the Seismic Reliability of the Main Steam Piping System: Part 3—Analytical Evaluation of BWR Feed Water Piping Supported by Energy Absorbing Support," *Seismic Engineering Conference Proceedings, ASME PVP* - Vol. 345, pp. 125–131, 1997.
3. Shibata, H., Kunieda, M., Hara, F., Suzuki, K., Ichihashi, I., Fukuda, T., Satoh, A., Takata, K., Furukawa, S., and Kobayashi, H., "Development of Elastoplastic Damper as a Seismic Support for Piping System in Nuclear Power Plants," *ASME PVP* - Vol. 211, pp. 63–69, 1991.

4. Satish Kumar, K., Muthumani, K., Gopalakrishnan, N., Sivarama Sarma, B., Reddy, G. R., and Parulekar, Y. M., "Seismic Response Reduction of Structure Using Elastoplastic Passive Energy Dissipation," *Indian Society for Earthquake Technology Journal*, 39(3), 2002, p.p. 85-99.
5. Satish Kumar, K., Muthumani, K., Gopalakrishnan, N., Sivarama Sarma, B., Reddy, G. R., and Parulekar, Y. M., "Reduction of Large Seismic De- formations using Elastoplastic Passive Energy Dissipators," *Science Defense Journal.*, **53**-1, 2003, pp. 95–103.
6. Y.M. Parulekar, G.R.Reddy, K.K.Vaze, K. Muthumani, "Passive control of Seismic response of piping systems." *International Journal of Pressure Vessels Technology* , Vol. 128, No.3, Aug- 2006, pp364-369.
7. Chopra, A. K., "*Dynamics of Structures, Theory and Application to Earthquake Engineering*," Prentice Hall of India Pvt. Limited, New Delhi, 1998.
8. Jennings, P. C., "Equivalent Viscous Damping For Yielding Structures,"*J. Eng. Mech.*, **94**, 1968, pp. 103–116.

PERFORMANCE OF BASE ISOLATED RCC FRAMED BUILDING UNDER ACTUAL EARTHQUAKE

P.N. Dubey[1], G.R. Reddy[1], S. K. Deb[2], K.K. Vaze[1], A.K.Ghosh[1], H.S. Kushwaha[1]

[1]*Reactor Safety Division, Bhabha Atomic Research Centre, Mumbai, India.*
e-mail: pndubey@barc.gov.in
[2]*Department of Civil Engineering, Indian Institute of Technology, Guwahati, India*
e-mail: skdeb@iitg.ernet.in

ABSTRACT

The concept of base isolation for equipments is well known. Its application in buildings and structures is rather challenging. To study the behaviour base isolated buildings under actual earthquakes, two, three-stored RCC framed buildings have been constructed at IIT, Guwahati campus, one with base isolation using lead plug bearings (LPB) and other with conventional foundation. Accelerometers are installed at first floor and third floor of both the building in order to record and compare the seismic responses of base isolated and conventional foundation buildings. The experimental buildings experienced earthquake of 5.2 Ritcher magnitude on November 06, 2006. A study has been made on data recorded on both the buildings and a comparison has also been made on recorded and analysis data of the buildings. Improvements required in the conventional analysis procedures are clearly brought out to obtain realistic structure seismic response in this paper.

Keywords: Isolator, LPB, Ritcher Magnitude, Dual slope

1 INTRODUCTION

The base isolation is aimed to attenuate the horizontal acceleration transmitted to the superstructure. The isolation system attempts to decouple the building or structure from the horizontal components of the ground motion by interposing structural elements with low horizontal stiffness between the structure and the foundation, which results the shifting of fundamental frequency of the structure out of the range of dominant earthquake energy and increasing the energy absorbing capacity. Thus the high energy in the ground motion at the higher mode frequency is filtered out.

To study, the behavior of isolated building under actual earthquake scenario, two, three-storeyed framed RCC buildings with similar construction, one building with conventional foundation (here onward called as normal building) and other with base isolation has been built at IIT, Guwahati campus. Guwahati is situated in the most severe seismic zone (Zone-V) of Northeast India. Northeast India is lying at the juncture of Himalayan Arc to the north and Burmese Arc to the east, and is one of the most active regions of the world. Eighteen large earthquakes ($M \geq 7.0$) including two great earthquakes ($M \geq 8.0$) occurred in this region during the last 100 years [1]. The normal building and base isolated buildings are shown in Fig. 1. Plan view and the orientation of X & Z axes are shown in Fig. 2, whereas Y-axis is aligned with the height of the building. For base isolation four Lead Plug Bearing (LPB), 50T each vertical load capacity have been used under each column of the building. A twelve channel dynamic structural recording system with GPS and 32 MB PCMCIA card has been employed for recording the seismic ground motion and structural response of normal and isolated buildings [2]. One surface tri-axial force balance accelerometer has been installed on the ground to capture earthquake induced ground motion and eight numbers of surface uni-axial force balance accelerometers have been installed at 1st floor and 3rd floor of the buildings to record the bi-directional accelerations. The responses of both buildings have been recorded for an earthquake occurred on November 06, 2006. Its magnitude is 5.2 on Ritcher scale and focal depth is 122.6 km (24.736° N, 95.223° E) and epicenter is in the Southeast of the site of buildings at Guwahati. Recorded responses of both the buildings have been compared and a comparison has also been made on recorded and analysis data of the buildings. Improvements required in the conventional analysis procedures are clearly brought out to obtain realistic structure seismic response under following headings.

2 ISOLATOR CHARACTERISTICS

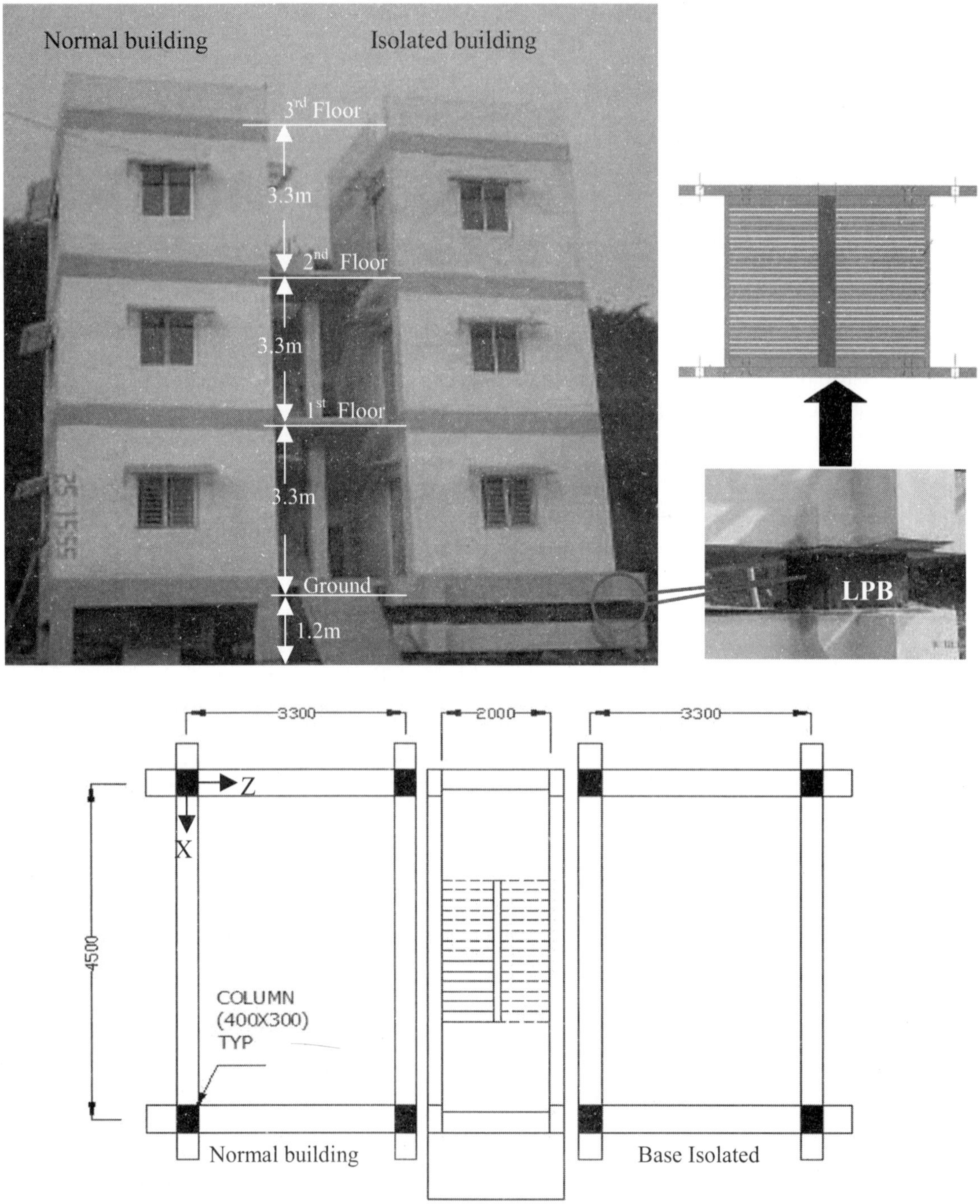

Fig. 2 Plan view of buildings

Isolators used for base isolation of the building are lead plug bearings, having alternate layers of rubber and steel with a central lead core. Each isolator has 50T vertical load capacity. Same isolator has been tested for cyclic characteristic and the hysteretic curve obtained is shown in Fig. 3.

Vertical stiffness of isolator is 188.96 kN/mm and horizontal stiffness of isolator considering the first slope (K_1) is 17.18 kN/mm and 0.7968kN/mm for second slope (K_2) as shown in Fig. 3.

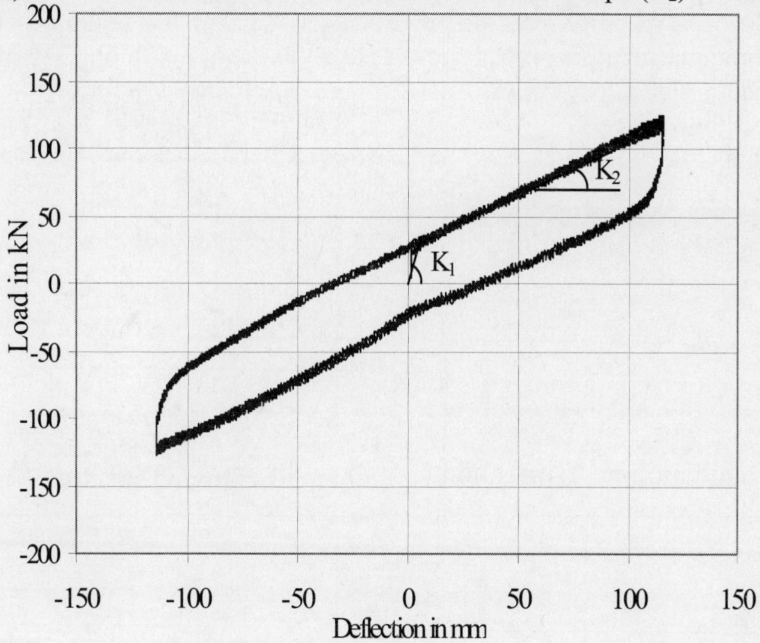

Fig. 3 Cyclic load deflection characteristic of isolator

3 GROUND MOTION RECORDS

Recorded ground time histories along X, Y and Z directions have been shown in Figs. 4 (a), 5 (a), & 6(a) and corresponding response spectra at 5% damping have been shown in Figs. 4(b), 5(b), & 6(b) respectively. From Figs. 4(b), 5(b), & 6(b), it can be observed that the earthquake ground motion signal consists of mainly higher frequency excitations and spectral peaks are occurring at frequencies 24.69Hz and 24.35Hz in horizontal directions (X & Z). In vertical direction very high spectral peaks can be observed at 40 Hz frequency, which is unusual. The ground motion data have been recorded from the instrument that is placed just below the convention building and because of soil-structure interaction, spectral peaks are occurring at very high frequency.

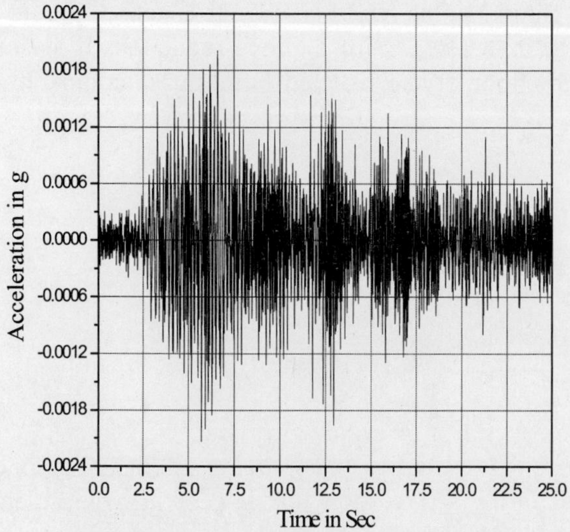

Fig. 4(a) Ground motion, X-direction

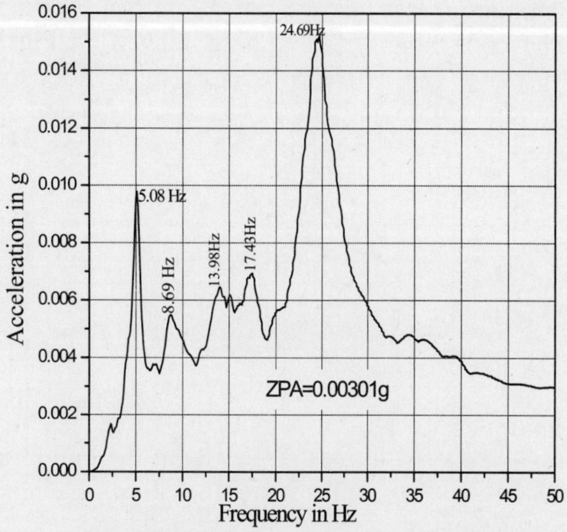

Fig. 4(b) Ground spectrum, X-direction

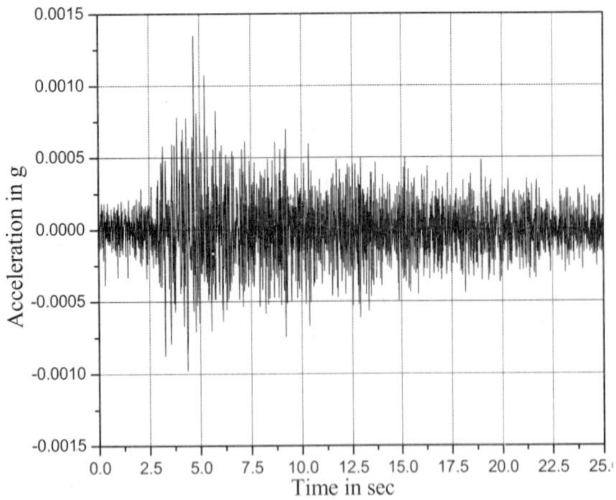

Fig. 5(a) Ground motion, Y-direction

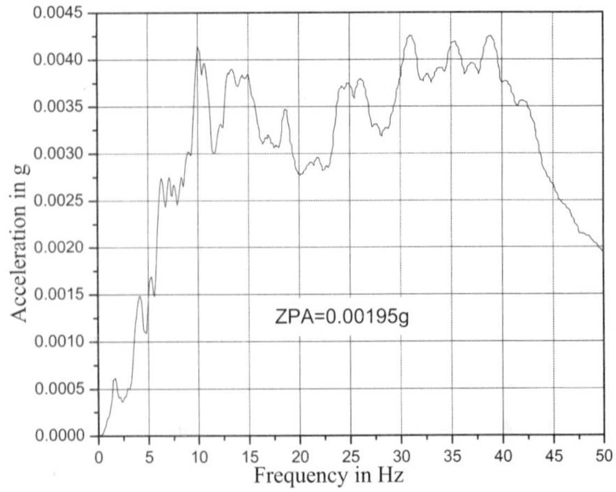

Fig. 5(b) Ground spectrum, Y-direction

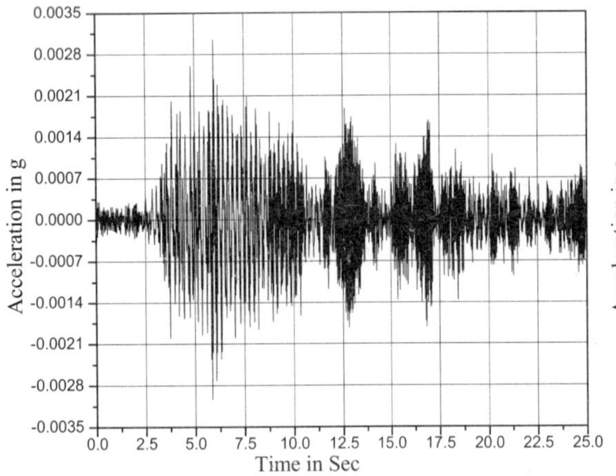

Fig. 6(a) Ground motion, Z-direction

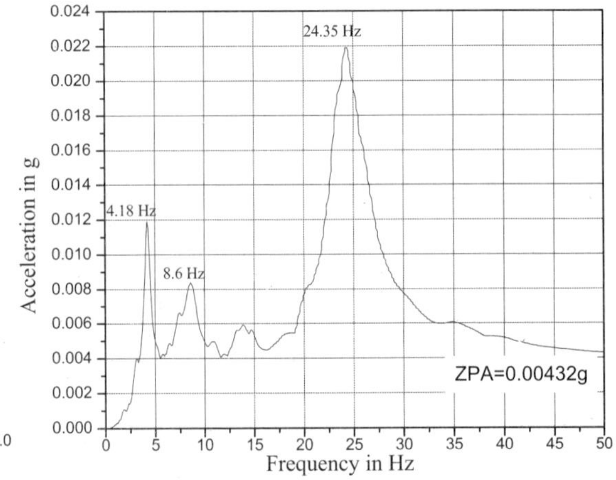

Fig. 6(b) Ground spectrum, Z-direction

4 COMPARISON OF RECORDED RESPONSE OF BUILDINGS

The recorded time histories at 3rd floor of the normal and isolated buildings in X-direction have been shown in Fig. 7, where maximum recorded response of normal building is 0.0094g and maximum recoreded response of isolated building is only 0.0017g. The response of isolated building is less than 1/5th of the normal buoding. Similarly the time histories at 1st and 3rd floors of the both buildings have been shown in Figs. 8, 9, where 2.5 to 5 times reduction of seismic response can be clearly seen from 1st floor to 3rd floor of the isolated building as compared to normal building respectively.

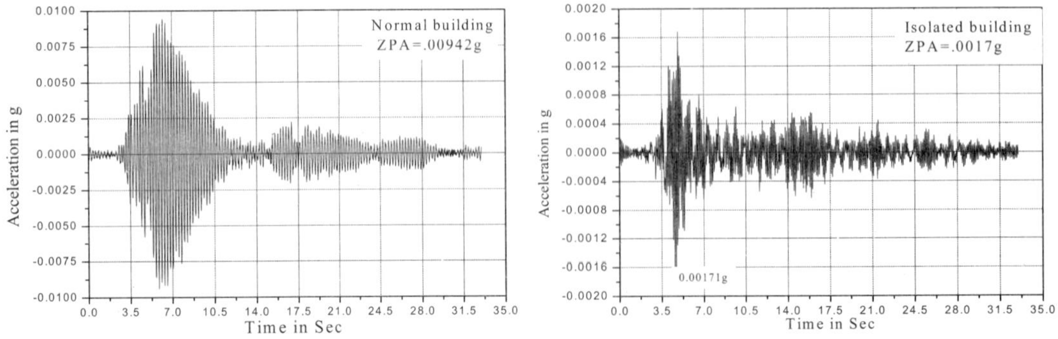

Fig 7. Recorded response at 3rd floor in X-direction

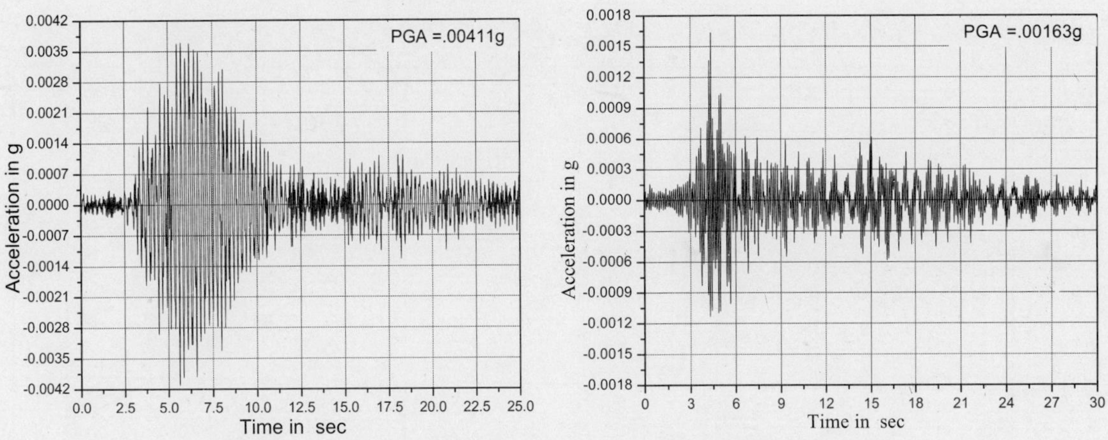

Fig 8. Recorded response at 1st floor in Z-direction

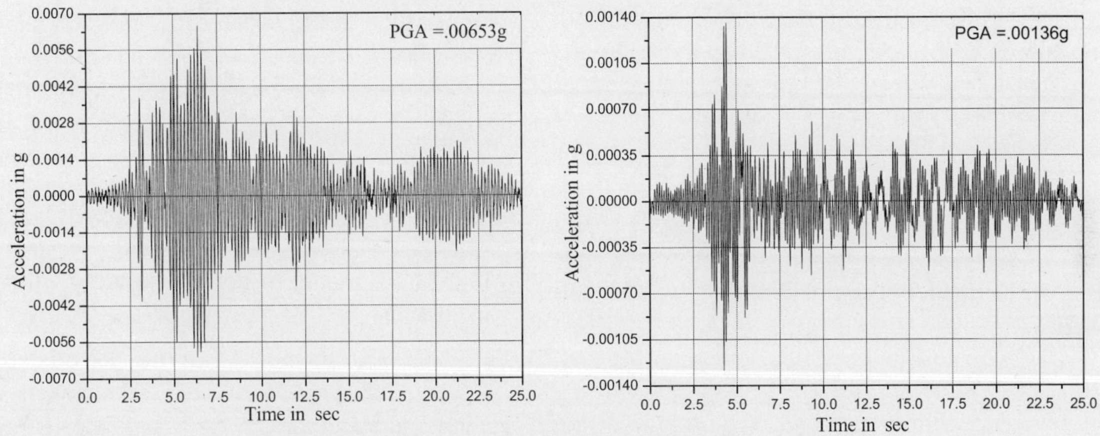

Fig 9. Recorded response at 3rd floor in Z-direction

5 ANALYSIS OF NORMAL BUILDING WITHOUT INFILL WALLS

Using designer's approach Finite Element (FE) model of normal building has been made without considering the stiffness of the infill walls and lumping the infill wall masses on the nodes of respective floor beams as shown in Fig 10. The assumption made in this approach is that the brick walls will not affect the characteristics of the building. The total mass of normal building is 116.70 T. Seismic analysis of the same was performed applying recorded ground time histories as shown in Figs. 4(a), 5(a) & 6(a), as input excitation at the base in X, Y, &Z directions respectively. Modal frequencies, and modal masses for significant modes have been listed in Table 1.

6 ANALYSIS OF NORMAL BUILDING WITH INFILL WALLS

Infill walls have been modeled as equivalent struts [3] as shown in Fig.11. Calculations of the studs have been performed as follows:

The following Eq. (1), is recommended to calculate the effective width (a) of diagonal compression strut. Geometric properties of strut have been derived from the geometry of the brick wall using the procedures given in reference [4].

$$a = 0.175\left(\lambda_1 h_{col}\right)^{-0.4} r_{inf} \qquad (1)$$

Where, h_{col} is the column height (inch), r_{inf} is the diagonal length of the masonry infill (MI) panel (inch), as shown in Fig 12, and λ_1 is given by

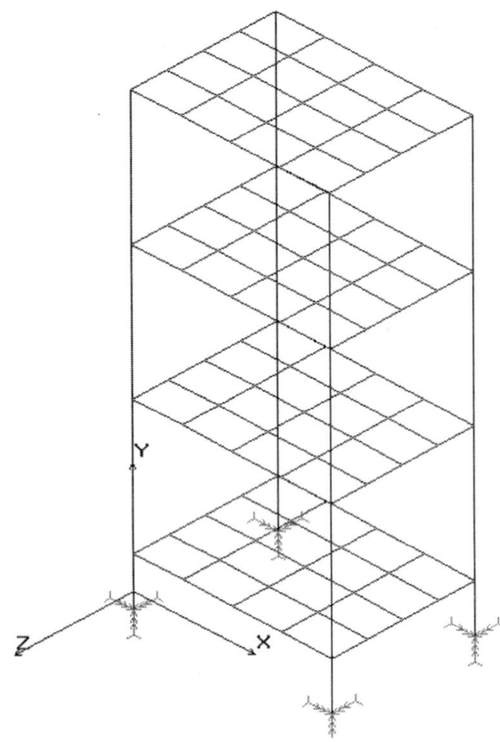

Fig. 10 FE model of normal building without infill wall

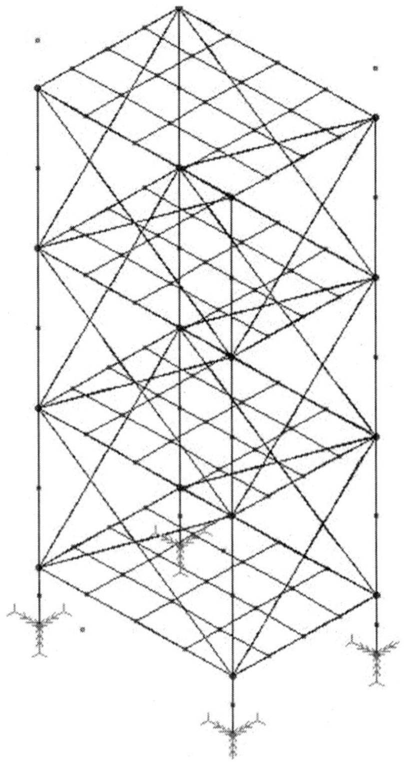

Fig. 11 FE model of normal building with infill wall

$$\lambda_1 = \left[\frac{E_{me} t_{\text{inf}} \sin 2\theta}{4 E_{fe} I_{col} h_{\text{inf}}} \right]^{0.25} \qquad (2)$$

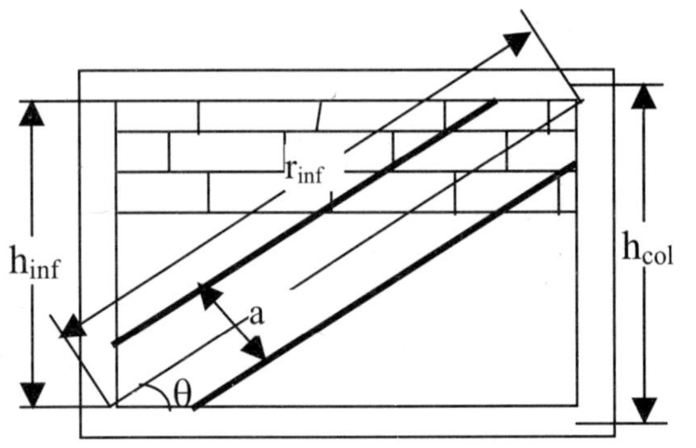

Fig. 12 Strut model of MI

Where E_{me} and E_{fe} are expected modulus of elasticity of masonry (secant modulus of elasticity between 5% and 33% of masonry prism strength) and frame material, respectively. t_{inf} is the actual thickness of MI in contact with frame, θ is the inclination of diagonal strut with horizontal, I_{col} is the moment of inertia of column, and h_{inf} is the height of MI panel. Thickness of equivalent strut is taken to be equal to actual thickness of the wall. Various parameters for the building being analysed are given below:

h_{inf}= 130 in, h_{col}=130 in, r_{if}= 219.698 in (longer side), r_{if}= 183.73 in (shorter side), $\theta = 36^0$(longer side), $\theta = 45^0$ (shorter side), $I_{col} = 3.85 \times 10^3$ in^4 (longer side), $I_{col} = 2.162 \times 10^3$ in^4 (shorter side). After subsisting the values above in the Eqs. (1 & 2), following values of strut parameters were obtained:

λ_1 = 0.08238 and a = 0.3783 m for longer side and similarly λ_1= 0.09637 and a=0.2971m for shorter of the building are obtained. Struts of area ($a \times t_{inf}$) applied along the diagonal at each floor of the building as shown in Fig.11.

From the frequencies listed in Table 1, for above two cases it can be clearly observed that there is three fold hike in fundamental frequency of normal building when stiffness of infill walls are considered and mass participation in higher modes are insignificant.

Table. 1 Modal frequencies, mass participation of normal building

Without infill wall					With infill walls				
Mode No	Freq (Hz)	Mass Participation (kg)			Mode No	Freq (Hz)	Mass Participation (kg)		
		M_x	M_y	M_z			M_x	M_y	M_z
1 (Z-dir)	1.25	0	0	88853	1 (Z-	4.42	0	0	94182
2 (X-dir)	1.55	88680	0	0	2 (X-	5.65	95275	0	0
4 (Z-dir)	4.25	0	0	10308	4 (Z-	13.7	0	18.478	19390
5 (X-dir)	5.35	10579	0	0	5 (Y-	14.4	0	47350	4.5258
10 (Y-	14.03	0	57741	0	6 (X-	16.9	17763	0	0
11(Y-	16.66	0	22062	0	7 (Y-	16.92	0	27516	0

From the floor time histories, floor response spectra have been generated at 5% damping and shown in Figs. 13, 14 &15, at 3rd floor X-direction, 1st & 3rd floor in Z- direction respectively. From Figs. 13, 14 & 15, it can be observed that spectral peaks of recorded response and analysis response are better matching i.e. occurring at almost same frequency, when stiffness of infill walls are considered in the analysis.

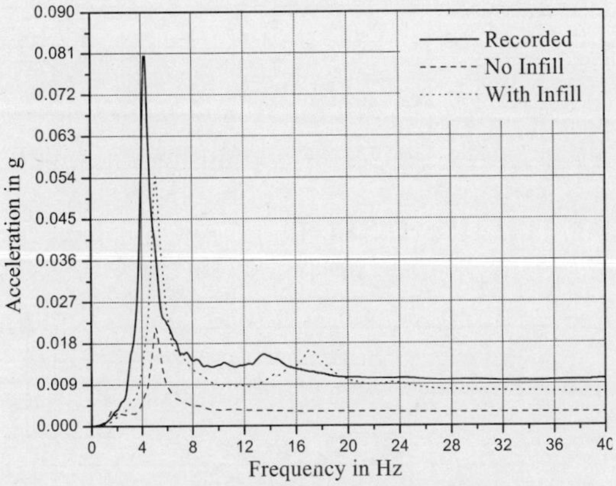

Fig. 13 Response spectrum at 3rd floor, X-dir

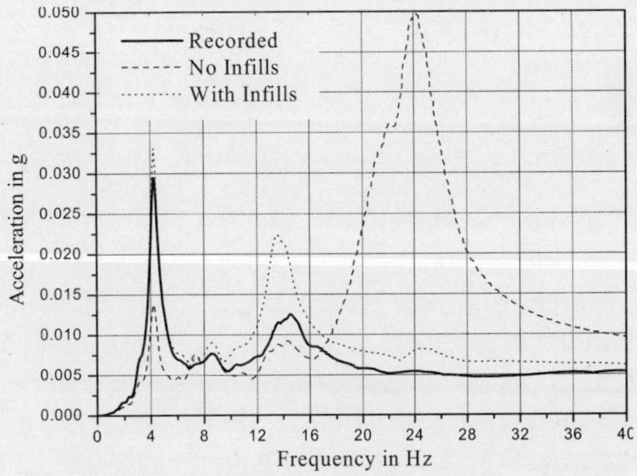

Fig. 14 Response spectrum at 1st floor, Z-dir

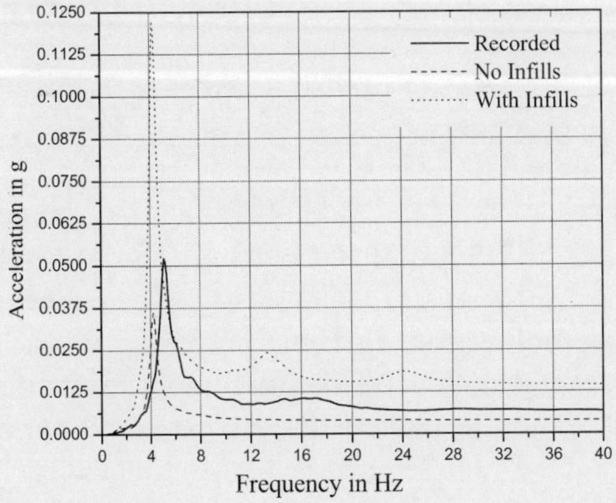

Fig. 15 Response spectrum at 3rd floor, Z-dir

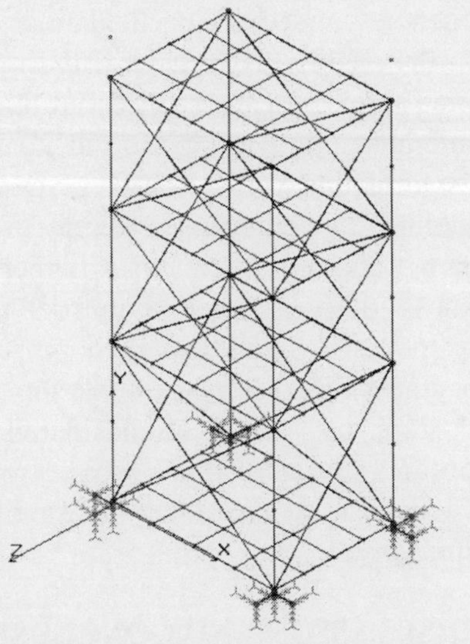

Fig. 16 FE model of base isolated of building

7 ANALYSIS OF BASE ISOLATED BUILDING WITH 2ND SLOPE

Isolated building has been modeled same as normal building except base isolators at the bottom as shown in Fig. 16. They have been modeled as linear truss element and real constants (area) of truss members have been calculated corresponding to the 2nd slope (K_2) of the base isolator as shown in Fig. 3. Modal frequencies and mass participations have been listed in Table 2. From Table 2, it can be observed that 1st and 2nd modes are in X & Z direction with 100% mass participation, whereas 6th mode is in Y-direction with 96% mass participation. In case of isolated building all mass is getting participated in 1st mode in each direction, whereas in case of normal building mass participation is distributed in higher modes also.

Table. 2 Modal frequencies, mass participation of isolated building

With 2nd slope					With Dual Slope				
Mode No	Freq (Hz)	Mass Participation in kg			Mode No	Freq (Hz)	Mass Participation in kg		
		M_x					M_x	M_y	M_z
1 (X-	0.796	0	0	116000	1 (Z-dir)	2.07	0	0	98718
2 (Z-dir)	0.812	116000	0	0	2 (X-dir)	2.53	1.04E+05	0	0
6 (Y-	10.10	0	112000	0	4 (Z-dir)	7.58	0	0	17367
					5 (X-dir)	8.42	12211	0	0
					6 (Y-dir)	10.1	0	1.11E+05	0

8 ANALYSIS OF BASE ISOLATED BUILDING WITH DUAL SLOPE

In present analysis isolators are modeled as nonlinear truss element, taking real constant (area) corresponding to the first slope of load deflection characteristic of the isolator as shown in Fig. 3. Nonlinear analysis has been performed considering the dual slope characteristic of the base isolators. The modal frequencies and mass participations obtained from the analysis have been listed in Table 2. From Table 2, it can be observed that there is significant mass participation in 2nd modes in each direction unlike above case. Input excitation is very low therefore isolators remain elastic, imposing higher elastic stiffness at the base, therefore three-fold frequency hike can be observed in present case also. From Fig. 17, it can be observed that if dual slope of isolator is considered the analytical spectral peaks are better matching with the recorded spectral peaks in comparison to previous case. It depicts that for small input excitations the deflection of the isolator is small hence natural vibration characteristics are governed by the elastic slope.

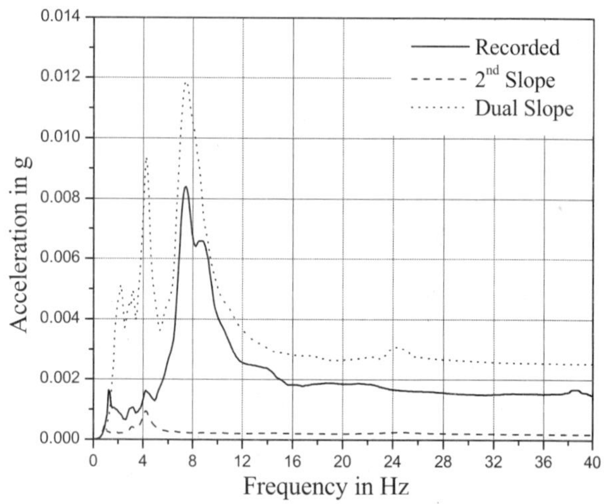

Fig. 17 Response spectrum at 3rd floor in Z-dir

9 DISCUSSION AND CONCLUSION

In ground response spectrum significant peak have been observed at 24 Hz which generally not seen. This may be because of soil-structure interaction. The response at the 3rd floor of isolated building is less than 1/5th response of normal building at the same elevation. There is significant contribution of infill walls in the dynamic characteristic of the buildings. The frequency shift is of the more than three times that of normal building without infill walls. If stiffness of the infill walls is included in the analysis, the analytical spectral peaks are matching with the recorded data.

At low magnitude earthquakes, infill walls will remain intact and increase the frequency. The accelerations corresponding to that frequency may go beyond the calculated value at that level of earthquake. This is a concern only if there is a different allowable value at lower level of earthquake as shown in Fig. 18. Therefore in the design the normal buildings smaller as well as higher magnitude i.e. two levels of earthquakes should be considered. Because of weak input excitations, base isolators remain elastic and impose higher stiffness. Recorded response of the isolated building matches with calculated response, if the isolator is considered elastic for small earthquakes. The recorded response of the base isolated building is almost constant along the height of the building, which also depicts the significant effect of base isolation.

For low magnitude earthquakes, isolators remain elastic and the actual frequency (f_2) of the isolated structure will be more than the design frequency (f_1=0.5Hz) and acceleration corresponding to f_2 may be higher than the design value, as shown in Fig. 19. Therefore it can be concluded that there is a chance that even for small excitations, isolated structure will experience large acceleration because of ineffectiveness base isolation and it may have large deformation, which may cross the design limits. Therefore isolators should be designed so that at a level of earthquake, which the un-isolated buildings can withstand and the isolator should become effective. Finally the level, which an un-isolated building can withstand, does require a check.

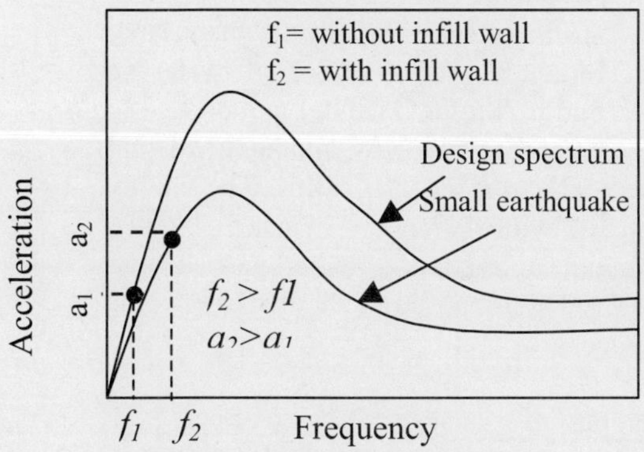

Fig. 18 Normal Building

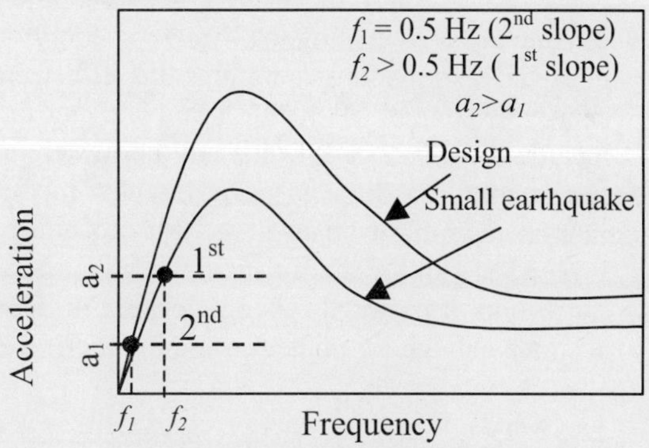

Fig. 19 Isolated Building

REFERENCES

1. J.R. Kayal "Seismicity of Northeast India and Surroundings- Development Over the Past 100 Years" Journal of Geographics, Jan 1998, vol. XIX No.1 pp 9 to 34.
2 S.K. Deb, A. Dutta, " Verification Test of a Prototype Two Storeyed RCC Framed Building Subjected to Actual Earthquakes" Project report for 8th Technical Programe Discussi0n Meeting.
3 Hemant B. Kaushik, Durgesh C. Rai, and Sudhir K. Jain, "Code Approaches to Seismic Design of Masonry-Infilled Reinforced Concrete Frames: A State-of-the-Art Review" Earthquake Spectra, Volume 22, No. 4, pages 961–983, November 2006; © 2006, Earthquake Engineering Research Institute.
4 Federal Emergency Management agency (FEMA)-306, " Evaluation of earthquake Damaged concrete and masonry walls buildings" 1998.

ESTIMATION OF HYSTERETIC ENERGY DEMAND INCLUDING P-DELTA EFFECT USING EQUIVALENT SYSTEMS

Amarnath Roy Chowdhury [1] ***and Siddhartha Ghosh*** [2]

[1]*Post graduate student, Department of Civil Engineering*
Indian Institute of Technology Bombay
Mumbai, 400076 India
amar.roychowdhury@iitb.ac.in
[2]*Assistant Professor, Department of Civil Engineering*
Indian Institute of Technology Bombay
Mumbai, 400076, India
sghosh@civil.iitb.ac.in

ABSTRACT

Hysteretic energy demand on a structure gives a good estimation of seismic damage. Hysteretic energy is the energy associated with inelastic deformations. Hysteretic energy demand can be estimated exactly by non-linear response history analysis (NRHA). Approximate procedures of estimating hysteretic energy demand based on equivalent systems were proposed by Ghosh (2003) and Prasanth (2006). Ghosh estimated the hysteretic energy demand on a structure based on a single equivalent single degree of freedom (ESDOF) system. Prasanth proposed a modified method of estimating hysteretic demand using multiple modal ESDOF systems based on the concepts of modal pushover analysis (MPA).

However, in both of these procedures the P-Delta (secondary moment effect) was not considered in estimating the hysteretic energy demand. P-Delta effect is significant under large inelastic deformations therefore, this effect cannot be ignored in the estimation of hysteretic energy demand on a structure. In this paper the method used by Prasanth (2006) is modified to include the effect of P-Delta and the hysteretic energy demand on two SAC moment resisting steel frames is estimated using the new method for several earthquake scenarios. The energy estimated by the equivalent system procedure is compared with the energy demand obtained from NRHA of the multi-degree model. It is observed that the approximate method gives a good estimation of hysteretic energy demand in these test cases. The proposed method is simple and accurate enough for implementing in practical design procedures.

Keywords: Hysteretic energy, equivalent system, modal pushover analysis, P-Delta effect.

1 INTRODUCTION

Structural designers are familiar with elastic force based seismic design procedure which is based on strength principles (based on acceleration spectra). However, this static elastic approach cannot take into account the dynamic nature of seismic damage. In the past researchers tried to develop more rational seismic design procedure through better characterization of the damage potential. Many researchers identified that the energy associated with inelastic deformation occurring in a structure gives a better estimation of seismic damage. This energy is known as the hysteretic energy. This demand parameter can take into account the dynamic nature of earthquake forces i.e. the duration of earthquake, and the number of cyclic elastic plastic deformation.

The necessity of seismic design based on energy demand was first proposed by Housner (1956). Energy based design gained recognition as one of the prescriptive design procedures proposed by Vision 2000 document (1995). Estimation of the hysteretic energy demand on a structure is one of the major aspects of energy based seismic design. This demand can be estimated exactly by nonlinear response history analysis (NRHA). But this method is computation heavy and hence it is not suitable for design implementation. Therefore, alternate simple procedures of estimating hysteretic energy demand were proposed.

Ghosh (2003) proposed estimation of hysteretic energy demand on a structure based on an equivalent single degree of freedom (ESDOF) system. The ESDOF system was developed by conducting nonlinear static pushover analysis on the multi degree of freedom system (MDOF) system. The pushover analysis was carried out with a time-invariant lateral load distribution as proposed in uniform building code (UBC). Ghosh and Collins (2006) tested this method on three moment resisting steel frames. It was found that the method proposed by Ghosh gave good estimations of hysteretic energy demand on low rise structures. However the level of accuracy reduced for high rise structures.

Later, Prasanth (2006) developed a method based on the modal pushover analysis concept proposed by Goel & Chopra (2002). Instead of using a single ESDOF system multiple modal ESDOF systems were used to evaluate the hysteretic energy demand on a structure. On the basis of case studies on three SAC steel moment resisting frames Prasanth found that this method gave a good estimation of hysteretic energy demand for medium and high rise frames as well.

Secondary effect coming in a structure due to its lateral deformation is known as P-Delta effect. This effect causes an increase in story shear and moment demand in the structure. The magnitude of this effect depends on gravity load intensity and the amount of lateral displacement. Under large inelastic deformation the P-Delta effect becomes significant. Static P-Delta effect reduces the post yield stiffness, increases lateral displacement and reduces the lateral load resisting capacity of a building frame therefore inclusion of the P-Delta effect in estimation of the hysteretic energy demand is necessary. In this paper the method proposed by Prasanth is modified to include the effect of P-Delta. The method proposed in this paper is tested on two steel moment resisting frames (9-story and 20-story) under various earthquake scenarios.

2 MODAL PUSHOVER ANALYSIS

Modal pushover analysis is a modification over traditional pushover analysis and it can take into account the contribution of higher modes in estimation of seismic demands. This method was developed by Chopra and Goel (2002) where the MDOF system was divided into multiple ESDOF systems. The force deformation parameters of the nth mode ESDOF system were determined by conducting pushover analysis on the MDOF system with lateral load distribution $\mathbf{s}_n^* = \mathbf{M}\boldsymbol{\varphi}_n$, where $\boldsymbol{\varphi}_n$ is the nth mode shape. The governing equation of motion of the nth mode inelastic ESDOF system was given as

$$\ddot{q}_n + 2\xi_n(\omega_n^*)\dot{q}_n + \frac{F_{sn}}{L_n} = -\ddot{u}_g(t) \tag{1}$$

The nth mode base shear (V_{bn}) versus roof displacement (u_{rn}) curve was idealized as a bilinear curve and the relation between the resisting force (F_{sn}) and modal displacement (q_n) was obtained for the nth mode ESDOF system as

$$F_{sn} = \frac{V_{bn}}{\Gamma_n} \text{ and } q_n = \frac{u_{rn}}{\phi_{rn}\Gamma_n} \tag{2}$$

For each mode, Eq. (1) was solved to get the peak value of modal displacement q_n. Based on this peak modal displacement, peak modal seismic responses (forces, displacements) were calculated and these modal responses calculated for first few modes were combined by SRSS or CQC rule to get peak seismic response of the MDOF system. This method was used by Chopra to calculate the force and displacement demand on three SAC steel moment resisting frames for several earthquake scenarios. It was found that this method gave a good estimation of seismic demand for low rise as well as for high rise steel frames.

This procedure was extended later by Goel and Chopra to estimate seismic demand including the P-Delta effect (Goel and Chopra, 2004). For this gravity load analysis was carried out. During the pushover analysis the effect of P-Delta was taken into consideration. This method was tested on the 9-story and 20-story SAC steel frames. It was found that the inclusion of P-Delta effect increases the error in estimating the seismic demand by the MPA procedure. The additional error introduced by the P-delta effect due to gravity loads was found to be small unless the structure is deformed far into inelastic range with significant degradation in lateral load resisting capacity.

3 ESTIMATION OF HYSTERETIC ENERGY DEMAND USING MULTIPLE EQUIALENT SYSTEMS

The method of estimating hysteretic energy demand based on single equivalent system proposed by Ghosh (2003) was developed based on traditional pushover analysis. Since traditional pushover analysis cannot take into account the contribution of higher modes in seismic demand this method was unable to give good estimation of hysteretic energy demand for high rise structures. Prasanth modified that method and developed a procedure based on the concept of modal pushover analysis. The MDOF system was converted into multiple modal ESDOF systems. And for each mode the equation of motion of the ESDOF system was written as follows

$$\ddot{D}_n(t) + 2\xi_n\omega_n^*\dot{D}_n(t) + (\omega_n^*)^2 G_n(D_n) = -\Gamma_n^*\ddot{u}_g \tag{3}$$

For each mode a pushover analysis was carried out on the MDOF system with a lateral load distribution $\mathbf{s}_\mathbf{n}^* = \mathbf{M}\boldsymbol{\varphi}_\mathbf{n}$ up to a 2.5% maximum story drift. The base shear v/s roof displacement plot obtained from pushover analysis for each mode was converted into a bilinear plot. The elastic stiffness (K_n) yield displacement (D_{ny}) post yield stiffness ratio (α_n) were obtained from this bilinear plot for a particular mode as shown in Fig. 1.

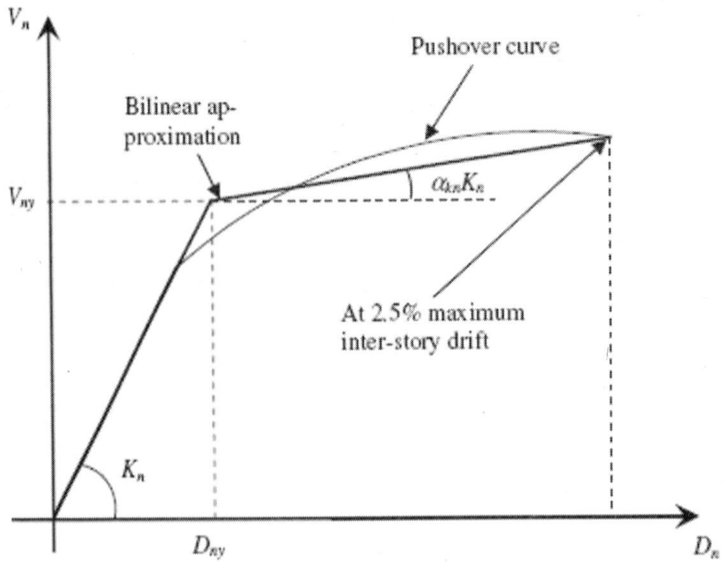

Figure 1 Obtaining the nth mode force deformation parameters from bilinear approximation of pushover curve.

The stiffness of the nth mode ESDOF system was calculated as $K_n^* = K_n\boldsymbol{\varphi}_\mathbf{n}^\mathbf{T}\mathbf{f}_\mathbf{n}$ and the yield displacement and yield base shear of nth mode ESDOF system were evaluated as $D_{ny}^* = D_{ny}$ and $V_{ny}^* = K_n^* D_{ny}^*$ respectively. The force deformation parameters ($K_n^*, D_{ny}^*, V_{ny}^*, \alpha_n$) estimated for n^th^

mode ESDOF system defined the function $G_n(D_n)$. Modal participation factor (Γ_n^*) was calculated as follows:

$$\Gamma_n^* = \frac{\boldsymbol{\varphi}_n^T \mathbf{M} \boldsymbol{\iota}}{\boldsymbol{\varphi}_n^T \mathbf{M} \boldsymbol{\varphi}_n} \tag{4}$$

where, ι is the influence vector.

For each ESDOF system, Eq. (3) was solved by NRHA and the modal hysteretic energy demand (E_{hn}) was calculated. Since E_{hn} is cumulative in nature (with time) the hysteretic energy demand for the first few modes were directly added to get the total hysteretic energy demand on the MDOF system. Prasanth estimated hysteretic energy demand on 3-story, 9-story, 20-story steel moment resisting frames for a set of 18 earthquakes and compared the results with the exact hysteretic energy demand obtained from NRHA on MDOF system. It was observed that the multiple modal ESDOF based approximate method gave a good estimation of hysteretic energy demand with less computational effort.

4 ESTIMATION OF HYSTERETIC ENERGY DEMAND USING MULTIPLE EQUIALENT SYSTEMS INCLUDING P-DELTA EFFECT

The approximate methods proposed by Ghosh (2003) and Prasanth (2006) did not consider the effect of P-Delta in the estimation of hysteretic energy demand. However under large inelastic deformations the P-Delta effect becomes significant and therefore this effect cannot be neglected in the estimation of hysteretic energy demand. A method of estimating hysteretic energy demand including P-Delta effect is proposed here. The proposed method is similar to that of Prasanth. Gravity loads and the P-Delta effects are included in the modal pushover analysis for each mode. In this case the bilinear pushover plot provides a set of K_n, D_{ny}, α_n values which are different from those obtained without considering P-Delta effect. The pushover parameters are used for defining the modal ESDOF systems and hysteretic energy is estimated in same way as in Prasanth's method.

5 TWO TEST CASES FOR PROPOSED PROCEDURE

5.1 Description of the test structures

The proposed procedure is tested for two steel moment frame buildings under various earthquake scenarios. These are the two "pre-Northridge" designs of 9- and 20-story buildings in Los Angeles, USA, considered in the SAC Steel Project. These structures meet the seismic code requirements as per UBC-94 (ICBO, 1994) and represent typical medium and high-rise buildings designed for Los Angeles at that time. Details regarding frame dimensions, material properties and loads can be obtained from previous research works (Gupta and Krawinkler, 1999) and are avoided here. For each building, a two-dimensional model of the North-South moment frame representing the building is considered. This frame is modeled for static and dynamic analyses in DRAIN-2DX, based on its centerline dimensions. A simple bilinear inelastic beam column element ("Element 02") with no strain hardening is used for modeling beams and columns. Strength, dimension, and shear distortion of panel zones are neglected. Stiffness contribution of the gravity frames (interior frames) is not considered. At each floor level, the frame is assigned half of the seismic masses of the whole building.

5.2 Estimation of hysteretic energy demand

A stepwise description of estimating hys' :retic energy demand on MDOF system using multiple equivalent systems including P-Delta effect is provided next.

1 From eigen value analysis of the two dimensional frame model the mode shape vectors $\boldsymbol{\varphi}_{\mathbf{n}}$ are obtained. The modal frequency, generalized modal mass M_n^* and the modal participation factor Γ_n^* are calculated for each mode. Three generalized modal mass is estimated as $M_n^* = \boldsymbol{\varphi}_{\mathbf{n}}^{\mathbf{T}} \mathbf{M} \boldsymbol{\varphi}_{\mathbf{n}}$.

2 Gravity load analysis (dead load + live load) is carried out on the frame.

1. Nonlinear pushover analysis is carried out including P-Delta effect for each mode with a lateral force distribution $\mathbf{f}_{\mathbf{n}}$ (that is, $\mathbf{M}\boldsymbol{\varphi}_{\mathbf{n}}$ normalized to a unit base shear) until the maximum of the story drifts reaches 2.5%.
2. The V_n vs. D_n pushover cuve obtained for each mode is approximated by a bilinear curve and the force deformation parameters (K_n , V_{ny}, D_{ny}, α_{kn}) are determined. From these parameters the force deformation parameters of inelastic ESDOF systems are calculated. The stiffness of the nth mode ESDOF system is given by $K_n^* = K_n \boldsymbol{\varphi}_{\mathbf{n}}^{\mathbf{T}} \mathbf{f}_{\mathbf{n}}$ and the yield displacement and yield base shear of the nth mode ESDOF system are given as $D_{ny}^* = D_{ny}$, $V_{ny}^* = K_n^* D_{ny}^*$.For example the force deformation parameters for the first five modes of the 9 –story frame are given in Table 1.

Table 1: ESDOF parameters for the first five modes of the 9-story frame.

Mode	M_n^* (kN-m/sec^2)	Γ_n^*	K_n^* (kN/m)	D_{ny}^* (m)	V_{ny}^* (kN)	α_{kn} (%)
1	2050.004	1.366	14687.248	0.292	4283.536	2.88
2	1773.867	0.529	94561.286	0.076275	7212.662	9.595
3	2952.125	0.238	462508.644	0.036575	16916.254	11.333
4	4228.748	0.1127	1507564.70	0.0238	35880.04	7.2
5	7608.172	0.0531	6134680.15	0.0125	76683.502	1.4

3 Hysteretic energy demand on the nth mode ESDOF system (E_{hn}) is obtained from NRHA.

4 E_{hn} obtained for the first five modes are added directly to obtain the hysteretic energy demand on the MDOF system (E_h).

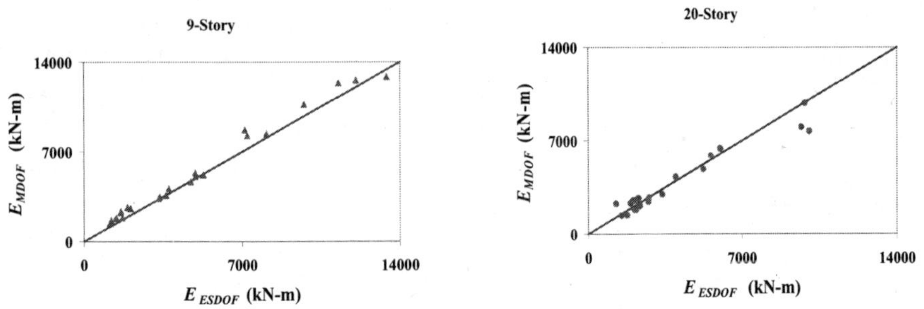

Figure 2 Scatterplots for the 9 and 20-story buildings.

6 DISCUSSIONS

The hysteretic energy demand calculated using equivalent system (E_{ESDOF}) is compared with the actual hysteretic energy demand (E_{MDOF}). For an easy comparison scatterplots are given in Fig.2 The diagonal line accross a scatterplot represents the ideal ESDOF estimate for which E_{ESDOF} = E_{MDOF}. Each point in the scatterplot represents a single earthquake. Points above the diagonal line in the plot represents underestimation of hysteretic energy (E_h) by the approximate method, and vice versa.

Table 2 Bias factor and statistics for the two frames

Ground motion	9-story frame			20-story frame		
	E_{MDOF} (kN-m)	E_{ESDOF} (kN-m)	Bias factor N	E_{MDOF} (kN-m)	E_{ESDOF} (kN-m)	Bias factor N
chy0809	5067.862	4914.544	1.031	2531.446	2033.490	1.245
chy08036	4116.624	3748.182	1.098	2296.080	1265.934	1.814
newh360	2259.918	1612.930	1.401	2118.282	2357.028	0.899
nh	5187.802	5266.146	0.985	1833.621	2113.848	0.867
nr	4658.424	4725.606	0.986	3007.271	3382.929	0.889
ns	5347.886	4914.537	1.088	2704.095	2296.287	1.178
s050	8398.380	8094.053	1.038	5866.078	5598.502	1.048
s065	2350.957	1633.560	1.439	2729.493	2762.104	0.988
s212	10692.047	9769.197	1.094	7711.542	10052.97	0.767
s305	12581.946	12040.308	1.045	8020.411	9703.287	0.827
s503	1801.407	1401.106	1.286	2458.866	2736.731	0.898
s549	12838.040	13407.738	0.958	9816.725	9845.179	0.997
s621	3470.619	3349.605	1.036	***	***	-
s640	1399.852	1152.312	1.215	***	***	-
syl90	3360.164	3310.167	1.015	***	***	-
syl360	5200.179	5280.570	0.985	1855.928	2177.099	0.852
tcu0659	12377.894	11280.622	1.097	13221.57	14445.73	0.915
tcu06536	8695.117	7119.801	1.221	6414.261	6016.761	1.066
cpm00	***	***	-	1405.601	1534.698	0.916
cpm90	***	***	-	0.000	0.000	-
gaz90	***	***	-	1428.731	1779.165	0.803
kjm00	2586.746	2057.972	1.257	1664.975	934.221	1.782
kjm90	1651.943	1204.798	1.371	0.000	0.000	-
lcn275	1859.175	1697.905	1.095	4882.347	5248.043	0.930
lgp00	8259.324	7231.947	1.142	4320.376	3994.249	1.082
lgp90	***	***	-	***	***	-
nis00	***	***	-	***	***	-
pcd164	2644.314	1923.872	1.374	2310.892	1930.638	1.197
tabln	3594.840	3621.773	0.993	2549.371	2134.764	1.194
tar90	1825.759	1432.134	1.275	0.000	0.000	-
gaz00	***	***	-	2300.461	2300.635	1.000
		MEAN	1.141		MEAN	1.050
		SD	0.148		SD	0.271
		COV	0.130		COV	0.258

*** refers to discarded results.

In addition to the scatterplots a bias factor N is used for measuring the degree of accuracy of the proposed method.

$$N = \frac{E_{MDOF}}{E_{ESDOF}} \quad (5)$$

A bias (N) statistics for the chosen ground motion scenarios is presented in Table 2. In calculating the bias statistics the results for which E_{MDOF} is less than 1000 kN-m are discarded. A lower limit of E_{MDOF} eliminates only those cases for which a hysteretic energy-based design is not needed. Low hysteretic energy demand indicates that inelasticity developed in the structure is small, therefore in that case there is no need to go for design based on hysteretic energy demand.

The bias statistics obtained in Table 2 shows that for the 9-story frame the mean bias is 1.141 and that for 20-story frame is 1.05, which are close to their ideal value 1. The standard deviation and coefficient of variation (COV) values obtained for the 9 and 20-story frames are close to 0. Based on these observations it can be concluded that the proposed method gives a good estimation of hysteretic energy demand including P-Delta effect.

7 CONCLUSIONS

From the case study made on the proposed method the following conclusions can be drawn.

a) The proposed procedure gives a good estimation of hysteretic energy demand including the P-Delta effect without going through a computation intensive nonlinear response history analysis of the MDOF model.
b) The proposed method can use energy response spectra, therefore it can be used conveniently in energy based design.
c) This concept needs to be tested for structures with different structural configurations.

REFERENCES

1. Carl, F. and Bruce F. Maison, (1984). "Analysis for $P-\Delta$ effects in seismic response of buildings", Computers & Structures, 19(3), 369-380.
2. Chopra, A.K., and Goel, R.K. (2001). A modal pushover analysis to estimate seismic demands for buildings: Theory and preliminary evaluation. PEER Report 2001/03, University of California Berkeley.
3. Chopra, A.K., and Goel, R.K., (2002). "A modal pushover analysis procedure for estimating seismic demands for buildings", Earthquake Eng. Struct. Dyn, 31(3), 561-582.
4. Chopra, A.K., 2005.Dynamics of sttructures.2nd ed. New Delhi: Prentice Hall of India.
5. Ghosh,S.,2003.Two Alternatives of Implementing Performance-Based Seismic Design of Buildings-Life Cycle Cost and Seismic Energy Demand, Dissertation, (PhD).Department of Civil Engineering, University of Michigan, Ann Arbor, USA.
6. Ghosh, S., and Collins, K.R. (2006). "Merging energy-based design criteria and reliability based methods: Exploring a new concept", Earthquake Eng. Struct. Dyn, 35, 1677-1698.
7. Goel, R.K., and Chopra, A.K. (2004). Evaluation of Modal and FEMA pushover analyses: SAC buildings, Earthquake Spectra, 20(1), 225-254.
8. Goel, R.K., and Chopra, A.K. (2005). "Role of higher-"mode" pushover analyses in seismic analysis of buildings", Earthquake Spectra, 21(4), 1027-1041.
9. Gupta, A., and Krawinkler, H., (1999). Seismic demands for performance evaluation of steel moment resisting frame structures (SAC task 5.4.3); Report No. 132 John A. Blume Earthquake Engineering Center, Department of Civil and Environmental Engineering Stanford University, Stanford, USA.
10. Gupta, A., and Krawinkler, H., (2000). "Dynamic P-Delta effects for flexible inelastic steel structures", J.struct.eng.ASCE, 126(1), 145-154.

11. Powell, G.H. (1993). Drain-2Dx element description and user guide for element type01, type02, type04, type06, type09 and type15, Manual. Department of Civil Engineering, University of California Berkeley, California.

12. Prasanth, T., (2006). Application of Modal Pushover Concept in Estimation of Hysteretic Energy Demand, Dissertation, (M.Tech.). Department of Civil Engineering, Indian Institute of Technology Bombay, Mumbai, India.

13. Tjondro, J.A., Moss, P.J., and Carr, J. (1992). "Seismic $P-\delta$ effects in medium height moment resisting steel frames", Eng. Struct. 14(2), 75-89.

14. ICBO, Uniform building code, (1994), ICBO, Whittier, USA.

NON LINEAR ANALYSIS OF SHEAR WALL – SLAB INTERFACE UNDER LATERAL LOADING

*** S. Greeshma, ** K.P.Jaya, ***S. Asadhullah, *** S. Balakumar**
***** V. Palanisamy and ***C. Prakash**

* S.Greeshma, Lecturer, Structural Engineering Division, AnnaUniversity , greeshmas@annauniv.edu
** K.P.Jaya, Assistant Professor, Structural Engineering Division, AnnaUniversity jayakp@annauniv.edu .
*** S Asadhullah, S Balakumar, V Palanisamy and C Prakash, Graduate Students in Civil Engineering, College of Engineering, Guindy, Anna University, cprakash_be@yahoo.com .

ABSTRACT

The frequent occurrence of the major earthquakes in the Indian subcontinent, and the construction of tall buildings, especially over the last two decades demands for the construction of earthquake resistant buildings. Most of the studies shows that majority of the failures of the earthquake is due to the joint failure. Hence great care must be taken to design and detail connections between the vertical resisting elements and the horizontal resisting elements of the building so that all the elements act together to provide an integral structural system. A positive means of connection will be provided to transfer the diaphragm shear forces into the shear walls. Shear wall is one of the best lateral loading systems. The vulnerable location in such type of buildings is the interface between shear wall and slab. The connections between floor slabs and shear walls- an essential link in the lateral load resisting mechanism of slab- wall systems. The performance of the connection can influence the pattern and distribution of lateral forces among the vertical elements of the structure.

The earthquake-resistant design philosophy anticipates that in the event of a severe ground shaking, the structure will undergo some level of damage, and behaves in elastically. Hence it is important that the structure should respond in a ductile manner beyond its yield point. The connections between the slab and wall must be capable of resisting the stresses induced by external, preferably lateral loadings. The present paper aims to model and analyze the Shear wall-Slab interface in order to study the responses (maximum displacements and critical stresses) at the joint for different load cases.

Keywords: Shear Wall, Interface, Response, Non linear Analysis.

1 INTRODUCTION

Shear walls are vertical elements in the lateral force resisting system. They transmit lateral forces from the diaphragm above to the diaphragm below or the foundation. Within every building, there are multiple elements that are used to transmit and resist lateral forces. These transmitting and resisting elements define the building's lateral-load path. This path extends from the uppermost roof or parapet, through each element and connection, to the foundation. A primary structural load path element is shown in Fig 1. The roof and floor elements are known as diaphragms. Diaphragms serve primarily as force-transmitting or force-distributing elements that take horizontal forces from the stories at and above their level and deliver them to walls or frames in the story immediately below. Diaphragms are classified as either flexible or rigid, and the method of distributing earthquake forces from the diaphragm to the resisting elements depends on that classification.

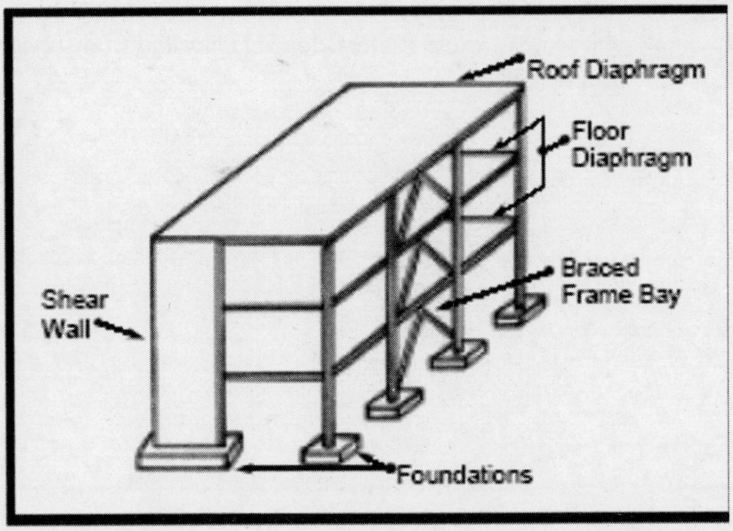

Fig 1 Primary structural load path elements

Within the primary load-path elements, there are individual secondary elements needed to resist specific forces or to provide specific pathways along which lateral forces are transmitted. Particular attention must be given to transmitting forces between horizontal seismic elements (diaphragms) and vertical seismic elements. Two important secondary elements are chords and collectors. A chord is a structural member along the boundary of a diaphragm that resists tension and compression forces. A collector is a structural member that transmits diaphragm forces into shear walls or frames. Fig 1.2 depicts the overall function of chords and collectors. Collectors are needed when an individual shear wall or frame in the story immediately below the diaphragm is not continuous along the diaphragm boundary.

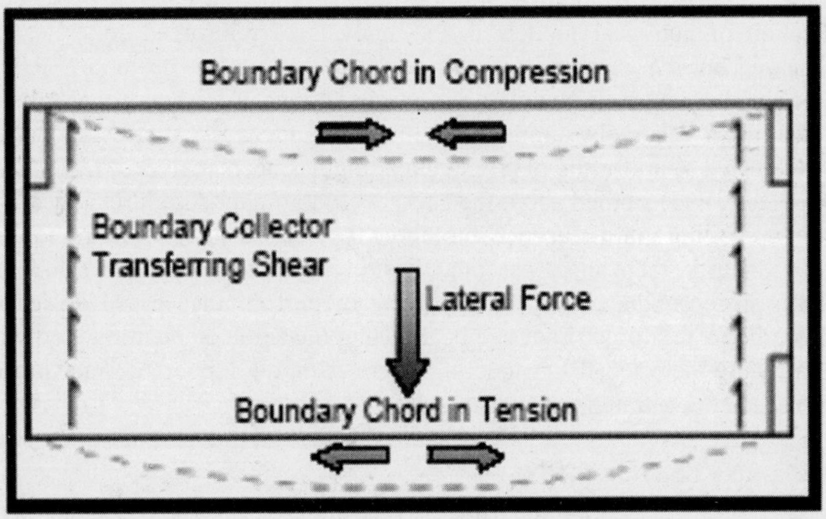

Fig 2 – Function of diaphragm chords and collectors

2 MODES OF FAILURE

Shear wall act as deep cantilever beams to resist lateral loads. They fail in distinct modes, namely flexure, shear, sliding or overturning (Fig 3)

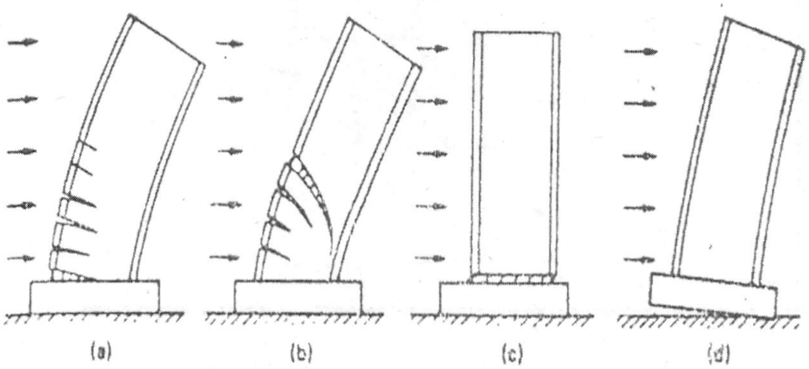

Fig 3 – Failure modes of a cantilever shear wall (a) Flexure failure, (b) shear failure, (c) Sliding failure (d) Rotation of foundation.

3 LITERATURE REVIEW

Joel. M. Barron et al (2004) studied the diaphragm effect in rectangular reinforced concrete building. Under seismic loading, floor and roof systems in RC building acts as diaphragms to transfer lateral earthquake loads to the vertical lateral force resisting system. The impact of in-plane diaphragm deformation on the structural response of RC building is evaluated using a performance based approach. Antonio F. Barbosa et al (2000) presented a paper considering the practical application of nonlinear models in the analysis of reinforced concrete structures. The differences observed in the response of the reinforced concrete beam are emphasized. The consequences of small changes in modeling are discussed and it is shown that satisfactory results may be obtained from relatively simple and limited models. A simply supported reinforced concrete beam subjected to uniformly distributed loading has been analyzed.

Can Balkaya et al (1993) studied about the shear wall dominant structures. Shear-wall dominant buildings are the prevailing multi-story RC buildings type particularly in the regions prone to high seismic risk. To identify their most essential design parameters, dynamic and inelastic static pushover analyses were conducted on the backbone of performance based design methodology. Faherty (1972) studied a reinforced concrete beam using the finite element method of analysis. The two beams that were selected for modeling were simply supported and loaded with two symmetrically placed concentrated transverse loads.

4 ANALYTICAL INVESTIGATION:

The present paper aims to analyse the behavior of shear wall-slab junctions using Structural Analysis Program (SAP) software and find out the critical stresses. The modeling of the shearwall – slab connection has been carried out using "SAP2000 non linear" (Structural Analysis Program) Software (version 7.4).

4.1 Modeling Of Shear Wall

For modeling the shear wall in SAP 2000 , initially a building with lift core was selected (Fig.4) and one of the Shear wall (4 x 3 x 0.2m) and Slab(3m x 4m) was modeled in SAP software.

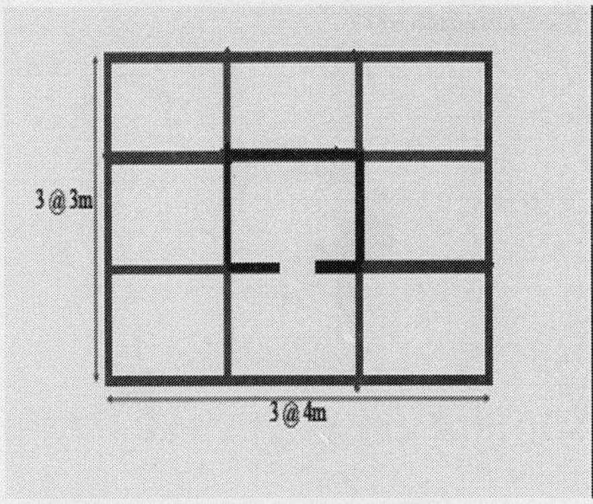

Fig.4 Building with Lift core

Dimensions:

Beam : 230 x 300 mm

Column : 300 x 300 mm

Slab : 150 mm thick.

Wall : 300 mm thick.

Storey Height : 4000 mm (for bottom storey and 3000 mm for other storeys)

No.of Storeys : 7

Elements:

Shear Wall : Plate elements were used to model the shear wall .
Slab : For modeling the slab, shell elements were used.
Beam and Column: Frame elements were used for modeling the beams and columns.

Loading:

The loads considered for the analysis were

1. Dead load
2. Live load (as per IS 875)
3. Earthquake Load (as per IS 1893-2002)-Earthquake loads are applied as equivalent static load as well as dynamic load (Time history data)

The Earthquake Load can be assigned as either response spectra or Time History data. In the Present work, the seismic load is assigned as time history functions (El centro data is used – shown in Fig.5).

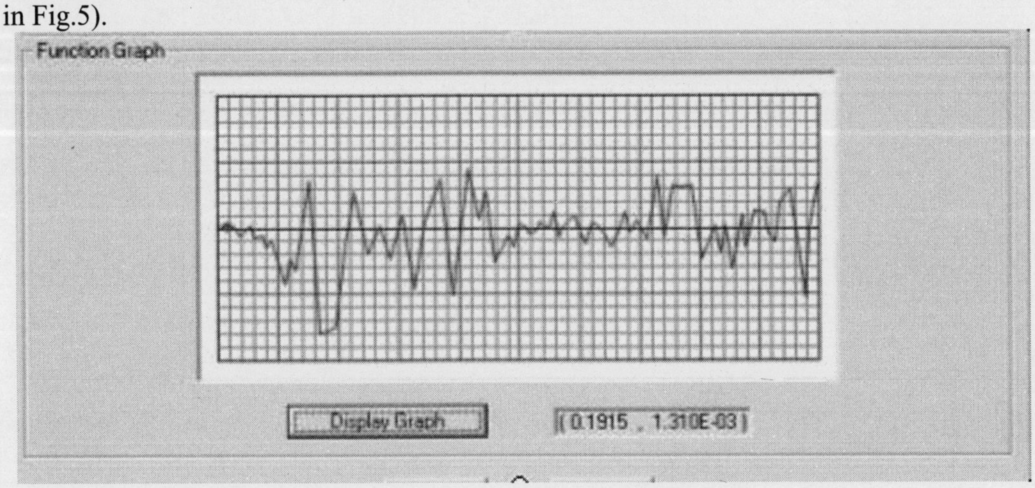

Fig 5. El Centro Data- Acceleration-time

Load Combinations

In the limit state design of reinforced concrete structures, the following load combinations shall be accounted for the analysis (IS 1893, 2002).

1. 1.5 (DL + IL)
2. 1.2 (DL + IL ± EL)
3. 1.5 (DL ± EL)
4. 0.9 DL ±1.5 EL

5 NUMERICAL RESULTS AND DISCUSSIONS

The modeling and analysis of shear wall –slab connection has been carried out with the following conditions.

1. Shear wall –slab connection with inplane static loading.(Equivalent static analysis).
2. Shear wall –slab connection with seismic loading.(Time History analysis).

5.1 Equivalent Static Analysis

For static analysis, earthquake loads, are assigned as equivalent static load applied at the corresponding joints for each floor. The following critical load combinations were considered for the analysis (IS 456 and IS 1893).

1. Comb 1 = 1.5 (DL + IL)
2. Comb 2 = 1.2 (DL + IL + EL)
3. Comb 3 = 1.5 (DL - EL)
4. Comb 4 = 0.9 DL + 1.5 EL
5. Comb 5 = 0.9 DL – 1.5 EL

Static analysis has been carried out to find the maximum stresses and displacements. The maximum principal stresses in shear wall obtained from SAP is shown in Table 1 and Fig 6 .

Table 1 Principle stresses in shear wall kN/m^2

Storey	Comb1	Comb2	Comb3	Comb4	Comb5
1	59.403	1073.4	1405.5	27.55	67.99
2	41.76	779.71	991.34	17.97	53.32
3	35.88	536.166	674.11	14.9	47.96
4	30.32	331.5	417.05	11.85	43.43
5	24.3	171.72	219.73	8.4	39.135
6	19.85	61.15	88.04	5.6	37.33
7	259.33	361.08	493.1	138.86	222.64

The maximum principal stresses in slab is obtained from the analysis using SAP 2000 and is shown in Table 2 and Fig 7.

Table 2 Maximum Principal Stresses in Slab in kN/m^2

Storey	Comb1	Comb2	Comb3	Comb4	Comb5
1	1066.00	1198.70	2232.40	340.38	1871.86
2	919.06	856.77	1703.21	279.10	1641.90
3	852.25	873.33	1717.79	255.77	1534.14
4	810.94	1002.78	1866.15	237.79	1482.06
5	779.12	1128.27	2026.35	219.03	1461.53
6	737.73	1283.23	22389.32	187.87	1461.74
7	1192.42	1536.39	2117.70	636.94	1030.04

The maximum displacement in X-Direction is shown in Table.3 and Fig. 8 and the displacement in Y-direction is shown in Table. 4 and Fig. 9.

Table.3 Maximum Displacement in X-Direction in m

Storey	Comb1	Comb2	Comb3	Comb4	Comb5
1	0.000293	0.000521	0.000117	0.000102	0.00047
2	0.000895	0.00136	0.00226	0.000311	0.00144
3	0.00165	0.00226	0.00151	0.000576	0.00266
5	0.0025	0.00318	0.00257	0.000877	0.00398
6	0.00437	0.00208	0.00518	0.00159	0.00675
7	0.00543	0.00229	0.01	0.00204	0.00811

Table. 4 Maximum displacements in Y-Direction in m

Storey	Comb1	Comb2	Comb3	Comb4	Comb5
1	0.0001	0.00165	0.002	0.00014	0.00034
2	0.00045	0.00141	0.00176	0.00021	0.00041
3	0.00047	0.00714	0.00356	0.00011	0.00031
4	0.00051	0.001	0.0127	0.00018	0.00038
5	0.00058	0.0131	0.0165	0.00025	0.00045
6	0.00068	0.0159	0.011	0.00058	0.00078
7	0.00071	0.011	0.013	0.00078	0.00098

5.2 Time History Analysis (El-Centro Earthquake Data)

Time History analysis has been carried out for El-centro earthquake acceleration 0.34g. (www.berkeley.com). The maximum principal stresses in slab is obtained from SAP 2000 and is shown in Table 5 and Fig 10.

Table 5 Maximum principal stress in slab (kN/m^2)

STOREY	MAXIMUM PRINCIPAL STRESSES
1	9047.00
2	3179.00
3	2350.00
4	1981.00
5	2054.00
6	2165.00
7	815.70

The stress concentration in wall- slab junction is studied and it reveals that principal stress in bottom most storey shear wall is more than other stories. The maximum principal stress in shear wall is shown in Table 5.2 and Fig 5.2.

Table. 6 Maximum principal stress in shear wall (kN/m^2)

Storey	Maximum Principal Stresses
1	10370
2	6071
3	3225
4	1382
5	1809
7	2848

The maximum displacement along Y Direction (in plane) is shown in Table 7 and Fig 12.

Table 7. Max displacement in Y-Direction in m

Storey	U2
1	0.00042
2	0.00143
3	0.00283
4	0.00449
5	0.00627
6	0.00809
7	0.00991

6 CONCLUSION

It is observed that the displacements in both in plane and out of plane directions increased, when the number of stories increased. For the shear wall, the displacement in plane direction is much lesser than that of out of plane direction. The displacements in both directions in static analysis were noticed to be less than that of dynamic analysis. Therefore for multi storey buildings dynamic analysis should also be considered.

The critical stresses for the shear wall-slab junction and beam slab junction were calculated and tabulated. As the height of the building increases, the stresses at the bottom increases and it reduces towards top. The maximum displacement was noticed to be at the top of the shear wall (at 7th storey) and its value is approximately 10mm.

REFERENCES

1. BALKAYA C. and Schnobrich W.C.(1993) 'Nonlinear 3D Behavior of Shear-Wall Dominant RC Building Structures.' Structural Engineering and Mechanics, 1(1), 1-16.
2. Antonio F Barbosa and Gabriel O Ribeiro(2000),Analysis Of Reinforced Concrete Structures Using ANSYS Non Linear Concrete Model, Federal University of Minas Gerais, Department of Structural Engineering,Avenida Do Contorno, Belo Horizonte-MG-Brazil.
3. Faherty (1972),Reinforced Concrete Beam Using Finite Element Analysis.
4. Indian Standard Code of practice of “Ductile detailing of reinforced concrete structures subjected to seismic forces”, IS 13920,1993, Bureau of Indian Standards, New Delhi.
5. Indian Standard Code of practice of “Criteria for earthquake resistant design of structures”,Part1 , IS 1893,2002, Bureau of Indian Standards, New Delhi.

BIBILIOGRAPHY

1. Seismic Shear design of RC Structural walls- Review of Codal Provisions, Kaustubh Dasgupta, C.V.R.Murthy and Shailesh.K.Agrawal.
2. Technical Committee No.21, Elastic Analysis- Strength of members and connections, Masahide Tomii
3. “Amlan.K.Sengupta,“Constitutive Models For Plain And Reinforced Concrete”-An Overview, Indian Concrete Journal, August 2003.

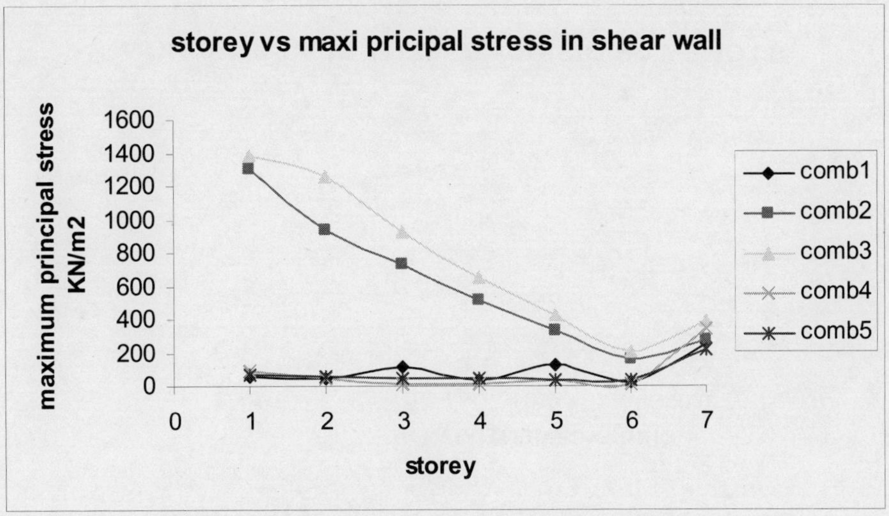

Fig. 6 Principal Stress distribution in shear wall

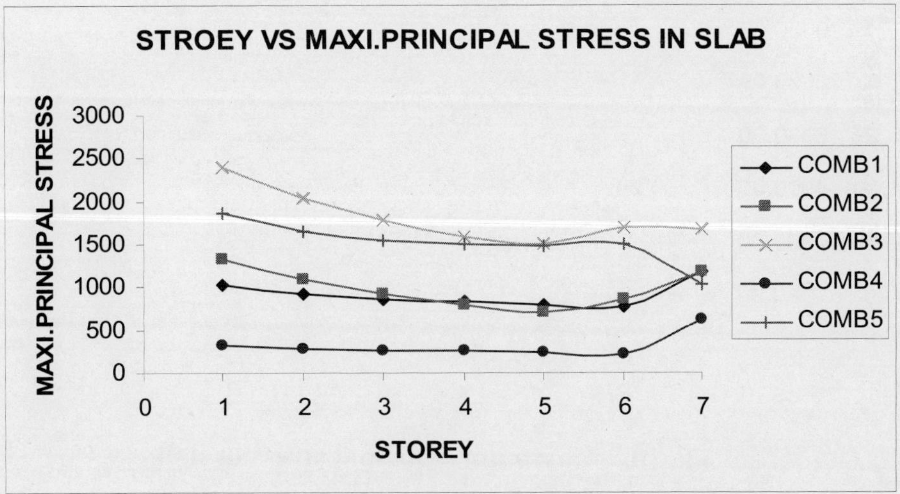

Fig. 7 Maximum Principal Stress Distribution in Slab

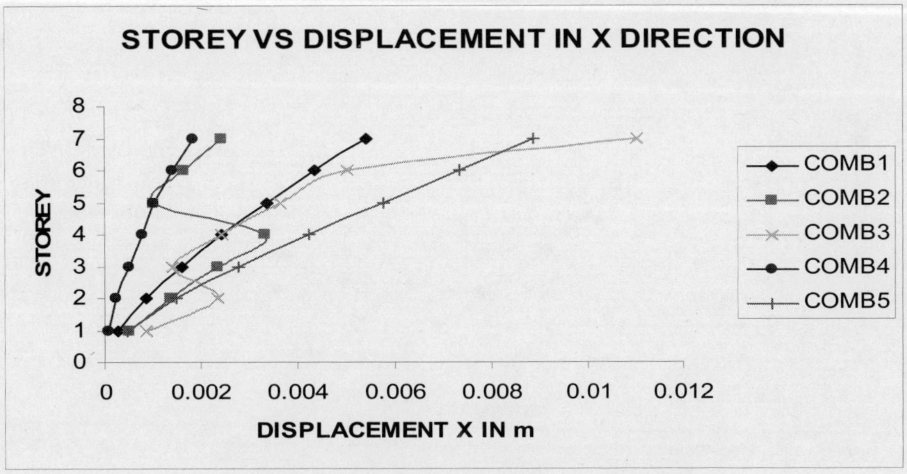

Fig.8 Displacement in X-Direction

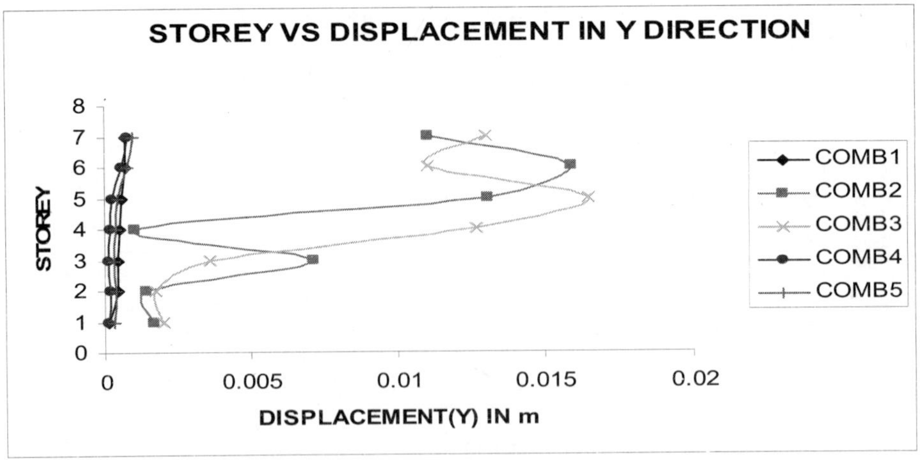

Fig .9 Maximum Displacement in Y-Direction

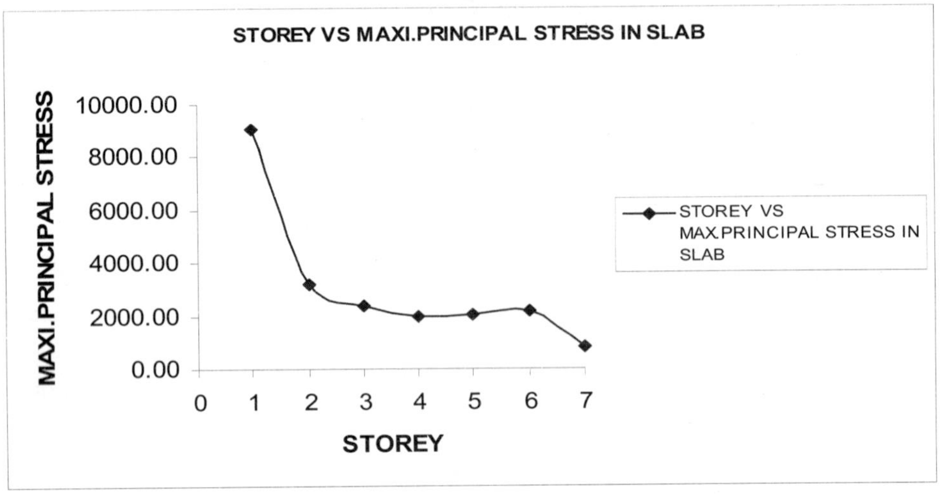

Fig 10. Maximum principal stress in slab

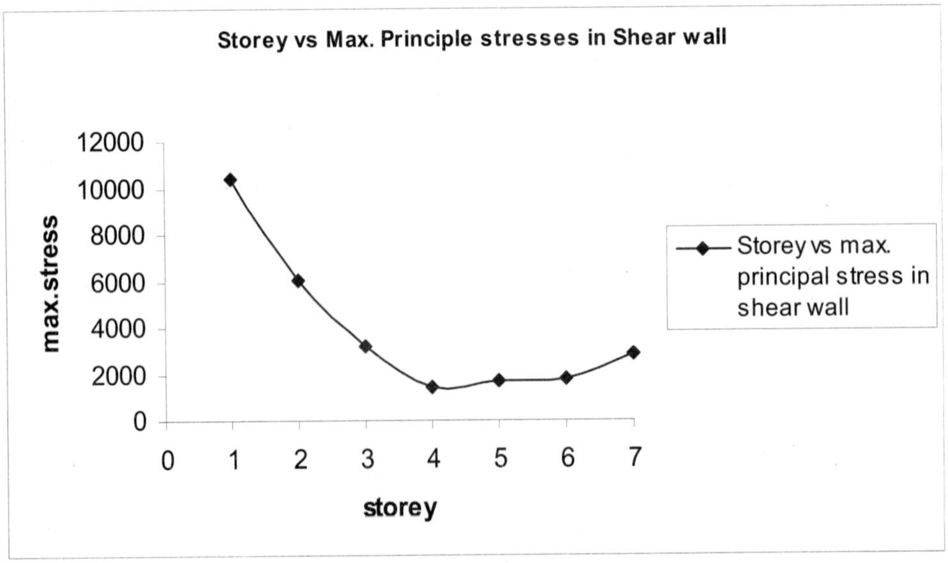

Fig 11. Maximum principal stress in shear wall

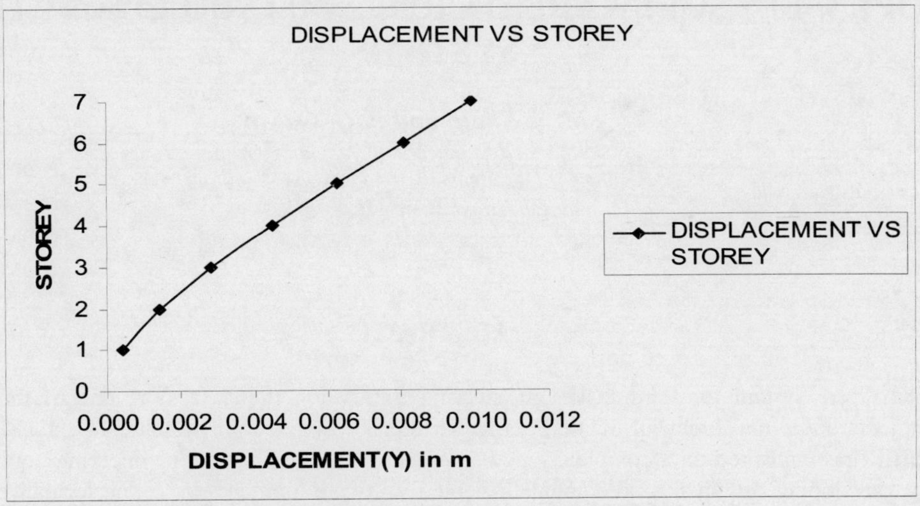

Fig .12 Displacement in Y-Direction

PERFORMANCE EVALUATION OF STRUCTURAL ENGINEERING OPENSEES SOFTWARE ON HPC AND GRID-COMPUTING SYSTEMS

***M. S. Shah*[1] *and P. G. Gavali*[2]**

Scientific and Engineering Computing Group
Center for Development of Advanced Computing
Ganesh Khind, Pune 411 007, INDIA
e-mail:, mahesh@cdac.in, pallavig@cdac.in

ABSTRACT

The Open System for Earthquake Engineering Simulation (OpenSees) is one of the many projects under development, of the Pacific Earthquake Engineering Research Center (PEER). PEER has embarked on a completely new approach in the earthquake engineering community by developing an open-source, object-oriented software framework. In performance-based earthquake engineering methodology the importance is to simulate the performance of structural and geotechnical systems. There are substantial problems and open questions on how to model the highly nonlinear behavior of structural systems with degrading components, or soil undergoing large deformation because of liquefaction, and the interaction between foundations and soils during large deformation. To address these challenges use of high-end computers for solving large-scale problems; databases for searching for new information from experimental data, simulation data, or observed data such as ground motion and field data; and visualization technology for providing engineers, design professionals, understanding about the performance of their systems.

The capabilities of OpenSees currently include Modeling, Analysis and Structural Reliability. In this investigation, OpenSees software has been implemented on our HPC Param Padma in India and its performance has been evaluated using simulation studies. Various simulation studies can be carried out using our Supercomputing facility - Param Padma.

Keywords: Earthquake Engineering, OpenSees Software, Simulation Analysis, Parallel Computing

1 INTRODUCTION

A central aspect of performance-based earthquake engineering (PBEE) involves simulating the performance of facilities (broadly defined to include the foundations, soil, and nonstructural components) during and after an earthquake to assess impacts in terms of direct loss, functional loss, and casualty loss. The current software for simulating the nonlinear behavior of soils and structures is inadequate, with incomplete or outdated models, problematic solution methods, and an inflexible software architecture that inhibits innovative use of modern information technology and high-performance computing. To address this shortcoming, PEER has embarked on a multiyear effort to develop the Open System for Earthquake Engineering Simulation, or OpenSees.

OpenSees is an advanced performance simulation software framework for structural and geotechnical facilities. The software is designed to integrate the implementations of models for structural behavior, soil and foundation behavior, and damage measures. Unlike traditional "codes," OpenSees is designed to be implemented in a modular, object-oriented manner with a clearly defined application program interface (API). The modules for modeling, solution, equation solving, databases, and visualization are independent, which allows great flexibility in combining modules to solve classes of simulation problems.

One of the best features of OpenSees is its suitability to support a multidisciplinary approach to Performance Based Earthquake Engineering simulation problems. The modular design allows researchers from different disciplines to combine their software implementations. As one example, OpenSees is linking new models in structural and soil behavior for affecting a complete soil-foundation-structure interaction analysis. As another, parallel and distributed equation solvers developed by computer scientists and mathematicians are integrated into the framework for the simulation of very large models. The framework has been combined with advanced ground motion simulation software to investigate the earthquake behavior of structures distributed through an urban region. Researchers are developing reliability modules using the framework. In another example, other researchers are using OpenSees to develop computational reliability modules for performance-based earthquake engineering.

Structural models have been developed with a general representation of shear behavior in reinforced concrete. An accurate assessment of reinforced concrete building damage must include shear and shear-flexure modes of behavior in order to capture nonductile failure modes that can occur and to track the degradation of shear strength and stiffness during an earthquake.

One of the major objectives of the OpenSees Processor is to provide a user-friendly mechanism to researchers and engineers so that they can employ the renowned computational technology and handle the complex problems in an efficient manner. Therefore, the functions available in the OpenSees toolbox must be accessible. Firstly OpenSees user can use any available tools, such as GID, SAP2000, to gather information regarding to the coordinates of the nodes and elements. Secondly, by inputting these coordinate data, users can easily create tcl files that accurately represent the shape and the load conditions of the structures by using OpenSees Model Builder. Thirdly, the users can query the results and download the output files in the local machine after submitting the completed OpenSees model to the collaborative framework or High performance Computing Facility. Thereafter, users can also examine the analytical results by plotting the deformation time history, displacement plot, acceleration plot, the stress-strain contour and numerous other types of plots.

2 GENERAL FEATURES OF OPENSEES

2.1 OpenSees Interpreter

OpenSees is an object-oriented framework for finite element analysis. A key feature of OpenSees is the interchangeability of components and the ability to integrate existing libraries and new components into the framework without the need to change the existing code. The basic commands currently available with the OpenSees interpreter are using the Tcl/Tk language.
The Tcl scripting language was chosen to support the OpenSees commands, which are used to define the problem geometry, loading, formulation and solution. These commands are one-line commands, which have specific tasks, as described in this manual. The Tcl language provides useful programming tools, such as variables manipulation, mathematical-expression evaluation and control structures.

Tcl is a string-based scripting language, which allows the following:

- Variables and variable substitution
- Mathematical-expression evaluation
- Basic control structures (if, while, for, foreach)
- Procedures File manipulation

For OpenSees we have added commands to Tcl for finite element analysis:

- Modeling – create nodes, elements, loads and constraints
- Analysis – specify the analysis procedure
- Output specification – specify what it is you want to monitor during the analysis

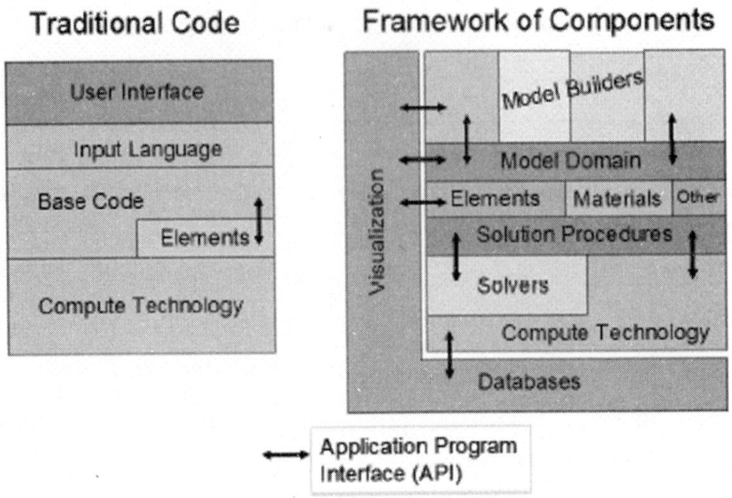

Fig.1 Traditional Code Vs OpenSees

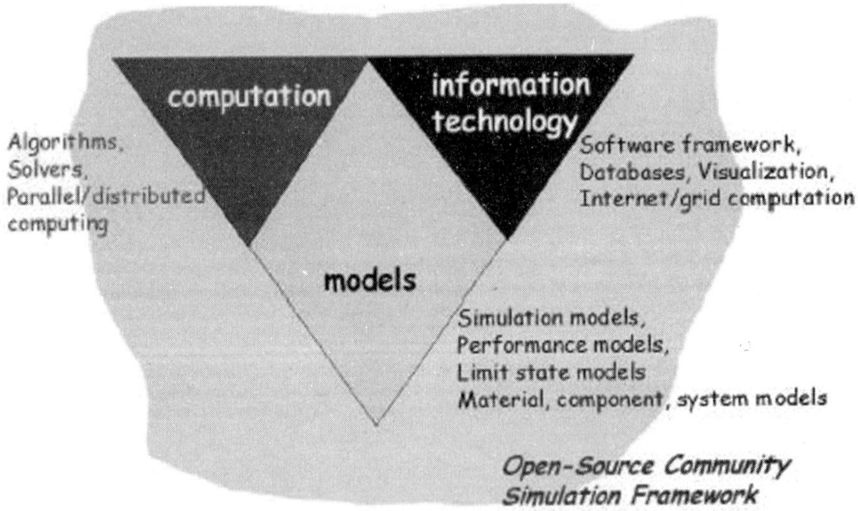

Fig. 2 Simulation Framework Network

2.2 OpenSees Models and Simulations

Linear & nonlinear structural and Geotechnical models can be analyzed using OpenSees. The modeling of any structural or Geotechnical system could be done using OSP or GiD pre and post processing software packages.

The Following types of simulations can be carried out using OpenSees

- Static pushover analyses
- Static reversed-cyclic analyses
- Dynamic time-series analyses
- Uniform-support excitation
- Multi-support excitation

3 INSTALLATION OF OPENSEES ON HPC

3.1 OpenSees on Param Padma

For the Simulation of large scale structural and geotechnical system using the availability of parallel OpenSees, the source code in successfully installed on the C-DAC's Supercomputing facility Param Padma. The user can directly submit their structural and geotechnical simulation problems through access to Param Padma and could get the simulations results.
Using the freely available pre and post processing GUI i.e. OSP and GiD the Graphical representation of the Results can be seen.

3.2 Example Problem

A simulation example problem was solved using OpenSees, the material is elastic pressure dependent material, plain strain, single element, dynamic analysis is carried out (input motion: sinusoidal acceleration at base). The SI units are used.
The input file is as follows

```
#      4     3
#      -------
#      |     |
#      |     |
#      |     |
#      1-------2  (nodes 1 and 2 fixed)
#      ^      ^
#      <--> input motion: sinusoidal acceleration at base
wipe
#some user defined variables
set accMul  9.81    ;
set massDen  2.     ;# mass density
set massProportionalDamping  0.0 ;
set stiffnessProportionalDamping 0.001 ;
set fangle 31.4        ;#friction angle
set ptangle 26.5       ;#phase transformation angle
set E1     90000.0     ;#Young's modulus
set poisson1 0.40 ;
set G [expr $E1/(2*(1+$poisson1))] ;
set B [expr $E1/(3*(1-2*$poisson1))] ;
set press  0 ;# isotropic consolidation pressure on quad element(s)
set period  1     ;# Period of applied sinusoidal load
set deltaT  0.01  ;# time step for analysis
set numSteps 2000   ;# Number of analysis steps
set gamma   0.5   ;# Newmark integration parameter
set unitWeightX 0.0                    ;#unit weight in X direction
set unitWeightY [expr -9.81*($massDen)] ;#unit weight in Y direction
```

```
#create the ModelBuilder
model basic -ndm 2 -ndf 2
# define material and properties
nDMaterial PressureDependMultiYield 2 2 $massDen $G $B  $fangle .1 80 0.5 \
                          $ptangle 0.17 0.4 10  10 0.015 1.0
nDMaterial FluidSolidPorous 1 2 2 2.2e6
updateMaterialStage -material 1 -stage 0
updateMaterialStage -material 2 -stage 0
# define the nodes
node 1   0.0 0.0
node 2   1.0 0.0
node 3   1.0 1.0
node 4   0.0 1.0
# define the element     thick  material     maTag  press   mDensity  gravity
element quad  1  1 2 3 4  1.0  "PlaneStrain"   2  $press    0.     $unitWeightX  $unitWeightY
# fix the base in vertical direction
fix 1 1 1
fix 2 1 1
# GRAVITY APPLICATION (elastic behavior)
# create the SOE, ConstraintHandler, Integrator, Algorithm and Numberer
system ProfileSPD
test NormDispIncr 1.e-12 25 0
constraints Transformation
integrator LoadControl 1 1 1 1
algorithm Newton
numberer RCM
# create the Analysis
analysis Static
#analyze
analyze 2
# switch the material to elastic pressure dependent
updateMaterialStage -material 2 -stage 2
#analyze
analyze 1
# NOW APPLY LOADING SEQUENCE AND ANALYZE (plastic)
# rezero time
setTime 0.0
wipeAnalysis
equalDOF 3 4   1 2   ;#tie nodes 3 and 4
pattern UniformExcitation 1 1 -accel "Sine 0 2000 $period -factor $accMul"
# create the Analysis
constraints Penalty 1.0e18 1.0e18  ;# Transformation;  #
test NormDispIncr 1.e-12 25 0
algorithm Newton
numberer RCM
system ProfileSPD
integrator Newmark $gamma  [expr pow($gamma+0.5, 2)/4]  \
        $massProportionalDamping 0.0 $stiffnessProportionalDamping 0.
analysis VariableTransient
#create the recorder
recorder Node -file disp8.out   -time  -node 1 2 3 4 -dof 1 2 -dT  0.01 disp
recorder Node -file acce8.out  -time  -node 1 2 3 4 -dof 1 2 -dT 0.01 accel
recorder Element 1 -time -file stress1.out material 1 stress -dT 0.01
```

```
recorder Element 1 -time -file strain1.out material 1 strain -dT 0.01
recorder Element 1 -time -file stress3.out material 3 stress -dT 0.01
recorder Element 1 -time -file strain3.out material 3 strain -dT 0.01
#analyze
set startT [clock seconds]
analyze $numSteps $deltaT [expr $deltaT/100] $deltaT 10
set endT [clock seconds]
puts "Execution time: [expr $endT-$startT] seconds."

wipe  #flush ouput stream
```

3.3 Example Results

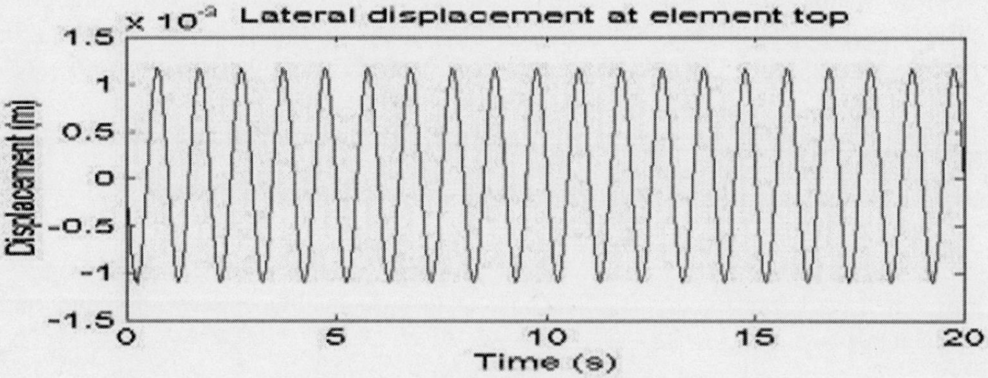

Fig 3. Displacement Output

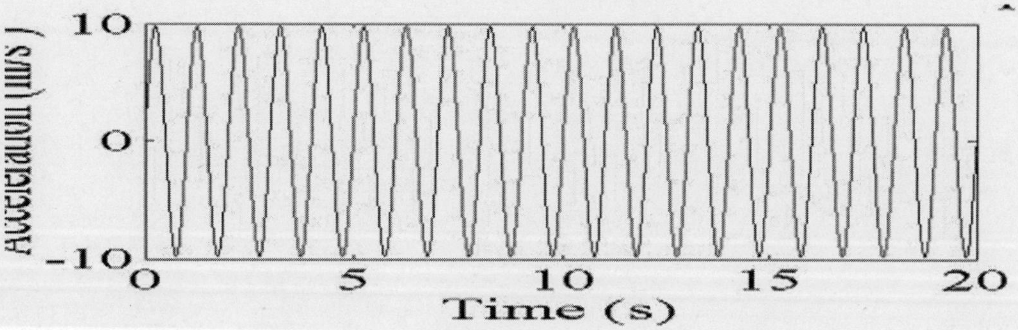

Fig 4 Lateral acceleration Output at element top

4 CONCLUSIONS

The significance of the present work is to create the awareness among the researchers, academicians and structural consultants regarding the simulation of Structural and Geotechnical systems subjected to earthquakes. OpenSees is intended to serve as the computational platform for research in performance-based earthquake engineering.

The successful installation of parallel OpenSees on Param Padma facilitates the large scale simulations of structural and geotechnical systems in India. So that the researchers, academicians and consultants could get the benifits of OpenSees and HPC facility.

The current limitations of OpenSees is the pre and post processors applicability, since more refined software should be available so that the modeling and post processing results could be easily obtained.

REFERENCES

1. Gregory L. Fenves, Frank Mckenna, Silvia Mazzani, OpenSees Computational Simulation Services in NEESit, University of California, Berkeley, Version 1, (2007) (http://it.nees.org/support/workshops/2007/2wfcree/)

2. Ravikumar C. P., Shukla Uday, New Frontiers in High-Performance Computing, Int. Conf. on High Performance Computing (2003).

3. M.S. Shah, S.Chakraborty and R. Rastogi, 'Earthquake Engineering Grid in India – A Concept', Proc. of 41st Annual Convention and Meeting on 'Inter and Intraplate Seismicity in India – Present Knowledge and Future Strategy', NGRI-Hyderabad, 29-31 Dec 2004.

4. Stephen Mahin, Robert Nigbor, Cherri Pancake, Robert Reitherman and Sharon Wood, 'The Development of the Network for Earthquake Engineering Simulation (NEES)', NEESgrid.org, 2001. Earthquake Information Network (http://www.eqnet.org/)

5. Timothy Mark Pinkstone, Viktor K. Prasanna, 'High Performance Computing – HiPC 2003, 10th International Conference, Dec 2003.

6. http://opensees.berkeley.edu/index.php

7. http://opensees.berkeley.edu/OpenSees/manuals/usermanual/OpenSeesCommandLanguageManual.pdf

8. http://cyclic.ucsd.edu/opensees

Part-6

Other Related Issues

DISASTER MANAGEMENT STRATEGIES IN TRANSPORTATION SYSTEM – A RESEARCH REVIEW

Ella Shalparni & Ashish Verma
4th year B-Tech, IIT Guwahati, ella@iitg.ernet.in
Assistant Professor, IIT Guwahati, ashishv@iitg.ernet.in

ABSTRACT

1 INTRODUCTION

Transportation infrastructure acts as backbone for any economy. The development of a system is greatly enhanced by an efficient transportation network. And so does the disruption of the transportation system can cause the collapse of economy, it can lead bring the entire city to a standstill. The destruction of traffic and transportation facility of a city due to unforeseen disaster, natural or unnatural like earthquake, hurricane, flood, gas leak etc is a matter of great concern. It has been clear from past events that the quick response of the public authorities in emergency situations is mostly critical to reduce the magnitude of the effects. The large size of cities extends the problems even more. Several occurrences, including fire, flood, thunderstorm, earthquake, chemical accident etc. may cause loss of life or injury to people and severe damage to structures and facility networks. The impact of a disaster event in a city can be local (e.g. fire event) or spread all over the urban area (e.g. earthquake event). Earthquakes represent one of the most destructive natural hazards known to man. A large magnitude earthquake near an urban area can cause human loss and extensive damage to infrastructure, disrupting the socioeconomic environment for a long period of time. The immediate and serious result is the disruption of transportation systems. It is very usual that the movement of emergency vehicles, such as ambulances, police cars and fire trucks, is blocked. The earthquake damage will interfere with the delivery of response personnel into the disaster area, as well as the evacuation of the injured and homeless. Traffic management is an important tool for disaster management. It holds a wide scope for research work to be done. This paper briefly outlines the various researches being done in this field.

2 DISASTER STRATEGIES

Minhans (2007) presented the understanding of the-state-of-the-art disaster management and relationship between the disaster management and traffic management. He defines traffic management strategies as demand reduction; static and dynamic traffic management; and assessed them for their effectiveness. He used AHP (Analytic Hierarchy Process) for measuring the effectiveness of the strategies. AHP is a mathematical technique for decision making, where there is a limited no of choices but is determined by various parameters. It enables to make decisions involving planning, setting priorities, selecting the best among given options. It also emphasizes the importance of intuitive judgments of decision makers as well as the consistency of alternatives comparisons. It is used to obtain weights of goals and objectives in disasters and implementation constraints.

$$ES(ij) = \Sigma WG_m * (\Sigma WO_{mn} * IR^{ij}_{mn})$$

Where ES(ij): Effective Score of Measure j, category i; WGm: Weight of goals m (m = 1 to 4); WO_{mn} : Weight of the objective n, under goal m; IR^{ij}_{mn} : Impact Rate of measure j, category i, on objective n, under goal m. the goals were divided in the priorities on the basis of ES values: first priority for the ES>2.4 with difficulty score (DS) <1.5; second priority for ES>1.5, 1.5<DS>2.4 and 1.5<ES>2.4, DS<1.5; third priority group for 1.5<ES>2.4, 1.5<DS>2.4; and the fourth priority group for the rest of measures. The disaster management in transportation can be broadly classified as

i. Disaster preparedness: To ensure the preparedness in anticipation of disaster
ii. Disaster response: To ensure the prompt assistance to the affected population
iii. Disaster recovery: To ensure the rapid and effective recovery from disasters

2.1 Disaster mitigation strategy:

It is necessary to build prevention strategies to ensure smooth function of transportation system during emergency disasters. The levels of services of the infrastructures are improved so that they can withstand the disasters and hence the havoc caused by them can be reduced. This prevents the collapse or total failure of the transportation services during emergency. It saves the policy maker from urgent formulation of evacuation strategies at the time of disaster. These strategies are especially helpful in the regions where the frequency of disaster is more, as in the earthquake prone areas (areas lying in zone V region), regions more prone to heavy rainfalls and floods, etc.

Peeta (2006) proposed to increase the level of service of the bridges by investment of funds, called upgrading cost, within the constraint of the budget limitation. The vulnerable links which needed to be improved, the probability of their sustainability with and without investment, were identified. He worked on the assumption that the failures are independent of each other. While selecting the links for improvement, the network reliability and performance factors were considered simultaneously. Two-stage stochastic programming model with decision-dependent probability distribution was used. The derived structural model resulted in better system level benefits of link investments.

The model was tested to strengthen the urban highway system of Istanbul against earthquake risk. Istanbul has high probability of 62.5±15% of having a major earthquake in next 30 years (parsons et al, 2000).

2.2 Disaster response strategy:

The quick response of civil authorities after a major disaster event in an urban area is essential for the reduction of damages and impacts to human lives. Evacuation and logistics support are the two major activities in disaster response. Evacuation activities are during the initial response phase, whereas logistics support operations tend to continue for a longer time for sustaining the basic needs of survivors who remain in the affected area. The timely availability of commodities such as food, shelter and medicine and the effective transportation of the wounded have an effect on the survival rate in the affected areas. This requires the commodity transportation from major supply centers to distribution centers in the affected areas and the transportation of wounded people from those areas to the emergency medical centers. The goal is to minimize the delay in the provision of prioritized commodities to survivors and health care services to injured people.

2.3 Evacuation strategies:

Virginia (2007) used microscopic traffic simulation to assist decision making for regional emergency preparedness through a case study implemented on Birmingham's regional transportation network. He used CORSIM as simulation tool to evaluate different emergency response procedures (CORSIM events are stochastic in nature). The study network included all major interstates in and around the Birmingham, Ala. Area.

This model was used to analyze the impacts of the plan proposed by the Jefferson County Emergency management Agency (EMA) for evacuation due to an emergency, on transportation network operations and recommended potential improvements. The original plans were not feasible due to capacity restrictions of the evacuation routes and modifications were proposed and tested, leading to more reasonable evacuation procedures. The simulation analysis results saved delay and queue lengths on impacted links. Moreover, the proposed adjustments increased the average speed on critical sections of the transportation network considerably. The results reaffirm that transportation and other emergency response agencies need to interact regularly in preparing emergency plans. The study demonstrated how traffic simulation can be used as a tool to support emergency management. But it is limited in its ability to simulate emergencies in real time and

does not model travel behavior at the network level. As a result the scenarios tested have only limited impacts and do not capture the real dynamics of emergency planning at the regional level.
A future extension of this work should involve the integration of the CORSIM microscopic transportation model and a dynamic traffic assignment model in an attempt to develop a comprehensive model for emergency planning at the regional level.

Paraskevas (2002) used vehicles as sensors for monitoring the traffic conditions in an urban area after a major disaster event. The vehicle fleets in the urban area records travel time with the help of GPS system for every road segment, thus providing a digital time database from which traffic parameters can be computed. It can be used for real time monitoring of traffic conditions under disaster or emergency situations where all the previously available data become invalid or unreliable. This is a dynamic system, as it keeps updating and correcting its database continuously, and thus optimizing its effectiveness.

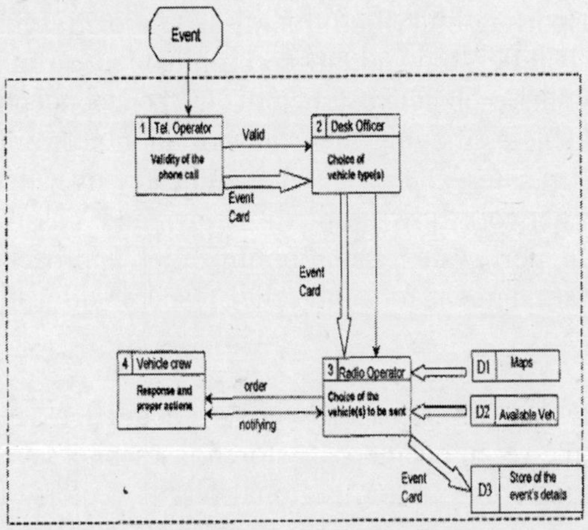

Figure 1: Data flow diagram to represent the response procedure at the moment of an emergency call.

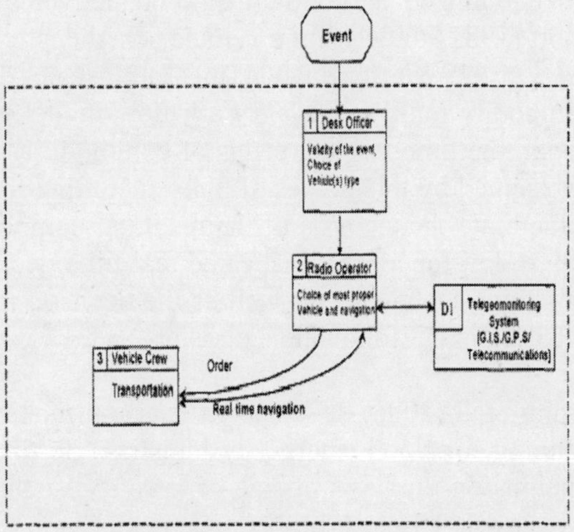

Figure 2: Data Flow Chart of the modified responding procedure with the help of a tele-geomonitoring

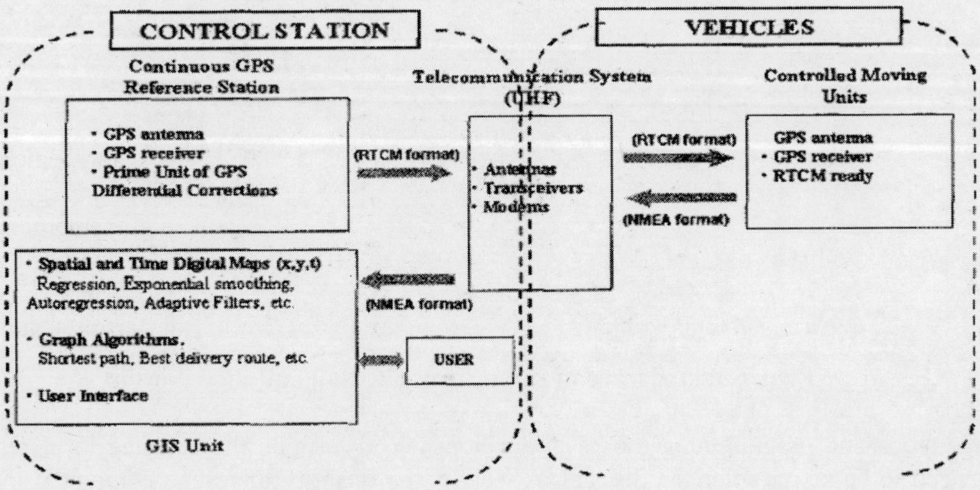

General frameworks of the OontroE Operation Center: The continuous reference GPS station computes and transmits differential corrections in RTCn/S format. The vehicle receiver corrects and transmits. Its position to the control center in HA format. In the G5S unit of the control center, several techniques are used for the computation of the digital spatial maps and travel times per road segment. Based on these, graph algorithms are used for the determination of shortest paths when necessary.

3 LOGISTICS APPROACH

Emergency logistics planning involves dispatching commodities (medical materials, specialized rescue equipment, rescue teams, etc.) to distribution centers in affected areas as soon as possible to accelerate relief operations. It can be obtained by repetitively solving the dynamic time- dependent transportation problem.

Ray (1987) developed a single-commodity, multi-modal network flow model on a capacitated network over a multi-period planning horizon. The sum of incurred costs was minimized for the transportation and storage of food. A linear programming model was proposed for the bulk food transportation problem in which efficient use of the truck fleet minimized the transportation cost and maximized the amount of food delivered. A linear programming model was developed by Knott (1987) to determine vehicle schedules for transporting bulk food to a disaster area. Haghani and Oh (1996) formulated the large-scale disaster relief transportation problem as a multi-commodity multi-modal network flow model with a single objective function. In their conceptual model, the time-varying status of commodities and vehicles moving in a transportation network is represented by three types of links, including routing, transfer, and supply demand carry-over links, to facilitate the analysis of the resulting complicated network flow problem. Brown and Vassiliou (1993) developed a sophisticated real-time decision support system using optimization approaches, simulation techniques as well as the decision maker's judgment for both relief resource allocation and assignment following a disaster.

Chang et al (2006) developed a decision-making tool that can be used by government agencies in planning for flood emergency logistics. He divided the rescue centers or storehouses into several echelons on the basis of size of territory, such as head rescue center, regional rescue centers, and local rescue bases. A head rescue center is in charge of the rescue operations at any demand point within the disaster area. Disaster districts in the same group are classified into three levels indicating the different imperiousness of the rescue demand: first-level disaster district has the highest risk of being attacked by a flood; while the third-level has the lowest.

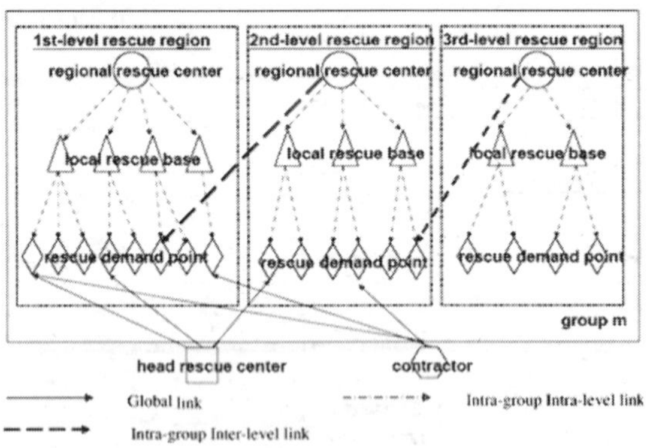

Figure 3: structure of rescue organization and distribution

A two-stage stochastic programming model is developed to determine the selected local rescue bases that need to be set up after the disaster as well as the quantity of rescue equipment in the storehouses of all echelons and the transportation plans of rescue equipment.

The proposed model is applied to Taipei City, in Northern Taiwan, 272 square kilometers wide with a population density of over 10,000 people per square kilometer. The entire disaster rescue operation should be completed within 36 hours. The Taipei City Government is designated as the head disaster rescue center; 12 administrative district offices are set up as regional rescue centers; and 84 offices of the subdivision of the district are chosen as candidate points for the local rescue bases. The GIS package ESRI ArcGIS 9.x was used to perform the spatial data analysis of the testing problem. After the spatial data analysis, the mathematical programming package LINGO was used to solve the proposed models.

4 DISASTER RECOVERY

A disaster on large scale like earthquake can lead to total failure of the transportation services and hence leading to chaos in the city, and requires restoration of the system. Nicholson and Du 1997; Chang and Nojima 2001; Sohn et al. 2003 focused on modeling and analyzing network performance of the post disaster infrastructure. Wakayabashi 1996; Karauchi et al. 2003; Li and Tsukaguchi 2003; Sakakibara et al. 2004; Sohn 2006 had evaluated the significance and reliability of transportation links in a larger network. Nicholson and Du 1997; Bell 2000; Peeta 2006 have discussed the network connectivity and critical route planning following a natural disaster. Wakabayashi and Kameda (1992) worked on network reliability analysis and modeled traffic flow adjustments in Loma Prieta. Basöz and Kiemidjian (1995) in their methodology for bridge retrofit prioritization utilized a bridge importance measure based on network connectivity analysis. Nojima (1997) proposed road traffic capacity as a basic post earthquake performance measure for highway systems. The capacity consists of the aggregate flow capacity of links connecting a specific origin-destination pair of nodes. Werner et al (1997) modeled bridge damage, network traffic flows, and the costs for travel time delay. Shinozuka et al (1998) linked traffic flows, transport costs, and regional production losses.

Matthew et al (2006) proposed a methodology for optimally allocating funds for repairing an urban infrastructure network following natural disasters using a three-stage genetic algorithm based approach. Inspection crews are dispatched shortly after the event to assess post-event bridge condition. Repairs are prioritized so that network operations are restored in the least possible time yet within budget constraints. The methodology is prioritized on the basis of 1.bidge importance rating and 2.bidge condition ratings (evaluated using discrete ratings from 1: very bad to 5: excellent, in accordance with the BRIME Project 1999). The paper proposes a three-stage approach for allocating repair funds in an urban bridge network following a natural disaster. The methodology allocates available funds to repairing bridges up to their lowest acceptable operational or pre-earthquake condition and then estimates necessary funds for repairing the entire bridge network to its pre-event condition. A genetic algorithm was developed for supporting this approach. This paper has shown that the methodology proposed provides useful insights regarding the magnitude of repairs and necessary funds for repairing a bridge network following an unforeseen event. Further, the methodology indicates the priority of bridges to be repaired as well as cost estimates for different importance levels. As far as genetic algorithms are concerned and given the nature of the models, both their application and performance proved to be adequate for the proposed approach. The case study results have demonstrated that the GAs performs well, providing logical results for a problem combinatorial in nature. Seismic events can cause significant disruptions to the operation of a bridge network and, as such, the developed methodology is intended as a tool for estimating bridge repair needs following such events.
The proposed methodology is applied to a set of 400 existing bridges in the Athens area. Two seismic scenarios, where 15 and 35% of the bridges, respectively, are damaged, are examined. A number of bridges correspond to each condition rating for each scenario. The available budget is set to 20 million Euros.

Chang et al (1999) measured the performance of post disaster system at Kobe (earthquake in 1995). Although there has been much attention paid to the understanding and predicting the performance of individual bridge structures, not much interest has been shown in the performance of the transport system as a whole. System performance measures are important for several reasons: fist, they enable comparisons of system conditions across disaster events in different urban areas. (They thus facilitate the development of generalized rather than case study understanding of earthquakes

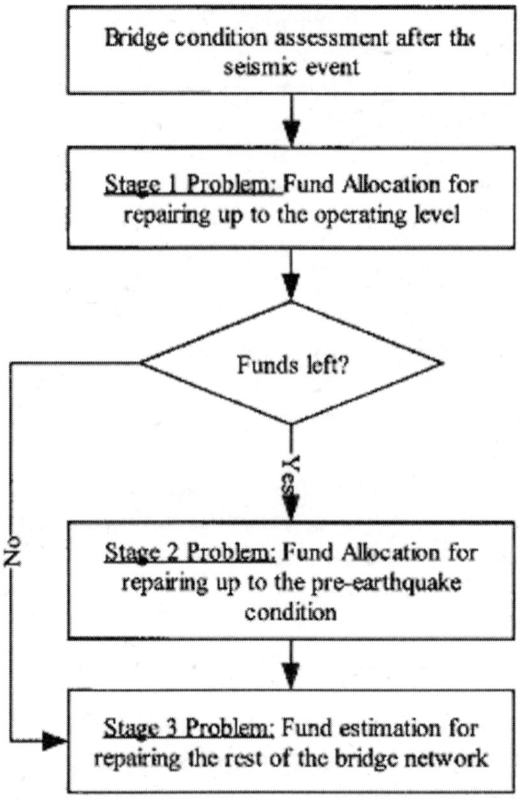

and their impact). Second, they allow comparisons across scenario disaster events for a single study region. Third, the measures can be used to design efficient post disaster restoration strategies by prioritizing damage repair such that overall system performance can be optimized. Fourthly, it can be implemented for estimating economic impacts in the context of real time earthquake loss models for emergency response and recovery planning. Three system performance measures are proposed here: total length of open network; total distance based accessibility; areal distance based accessibility, which requires only readily available data. The post disaster performance measures wee developed and applied to the case of Kobe City after the 1995 earthquake. Further research can improve the accuracy and applicability of the system performance measures developed.

5 SUMMARY

Earthquakes represent one of the most destructive natural hazards known to man. The earthquake damage will interfere with the delivery of response personnel into the disaster area, as well as the evacuation of the injured and homeless. Traffic management is an important tool for disaster management. It holds a wide scope for research work to be done. This paper briefly outlines the various researches being done in this field and their shortcomings along with possibilities of further research work.

REFERENCES

1 Brown, G.G., Vassiliou, A.L., (1993). Optimizing disaster relief: real-time operational and tactical decision support. Naval Research Logistics 40, 1–23.

2 Basöz, n., Kiremidijian, A.S., (1995). A bridge prioritization method based on ransportation system performance using GIS. Proceedings of the Sixth US-Japan Workshop on arthquake Disaster Prevention for Lifeline Systems, Tsukuba Science City, Japan, pp 437-449.

3 Bell, M. G. H. (2000). "A game theory approach to measuring the performance reliability of transportation networks." Transp. Res., Part B: Methodol., 34(6), 533–545.

4 BRIME. (1999). "Deliverable 13: Bridge management systems: Extended review of existing systems and outline framework for a European system." Web document, (http ://www.trl.co.uk/brime/d13.pdf), EU.

5 Chang, S.E., Nojima, Nobuoto,(1999) "Measuring post-disaster transportation system performance: the 1995 Kobe earthquake in comparative perspective.", Transportation research part A 35(2001) 475-494.

6 Chang Mei-Shiang , Tseng Ya-Ling, Chen Jing-Wen, "A scenario planning approach for the flood emergency logistics preparation problem under uncertainty" Transportation Research Part E 43 (2007) 737–754

7 Fiedrich, F., Gehbauer, F., Rickers, U., (2000). Optimized resource allocation for emergency response after earthquake disasters. Safety Science 35, 41–57.

8 Haghani, A., Oh, S., (1996). Formulation and solution of a multi-commodity, multi-modal network flow model for disaster relief operations. Transportation Research A 30, 231–250.

9 Karaouchi, F., Iida, Y., and Shimada, H. (2003). "Evaluation of road network reliability considering traffic regulation after a disaster." The Network Reliability of Transport: Proc., 1st Int. Symp. on Transportation Network Reliability (INSTR), Kyoto, Japan, 2001, M. G. H. Bell and Y. Iida, eds., Elsevier, Oxford, U.K., 289–300.

10 Nott, R., (1987). The logistics of bulk relief suppliers. Disaster 11, 113–115.

11 Knott, R., (1988). Vehicle routing for emergency relief management: a knowledge-based approach. Disaster 12, 285–293.

12 Karlaftis Matthew G.; Kepaptsoglou Konstantinos L; and Sergios Lambropoulos,.(2007), "Fund Allocation for Transportation Network Recovery Following Natural Disasters" Journal of Urban Planning and Development, Vol. 133, No. 1.

13 pofsky, Mark, Abkowitz, Mark, Cheng Paul (1993), "Transportation Hazard Analysis Integrated GIS Environment", Journal of Transportation Engineering, Vol. 119, No. 2

14 Li, Y., and Tsukaguchi, H. (2003). "Improving the reliability of street networks in highly densely populated urban areas." The Network Reliability of Transport: Proc., 1st Int. Symp. On Transportation Network Reliability (INSTR), Kyoto, Japan, 2001, M. G. H. Bell and Y. Iida, eds., Elsevier, Oxford, U.K., 261–272.

15 Minhans Anil, (2007) "Traffic Management Strategies in Cases of Disasters", Ph.D. Thesis, TU-Darmstadt, Germany.

16 Nicholson, A., and Du, Z.-P. (1997). "Degradable transportation systems: An integrated equilibrium model." Transp. Res., Part B: Methodol., 31(3), 209–223.

17 Nojima, N., (1997).Performance-based prioritization in upgrading seismic reliability of road network. Proceedings of the 24th JSCE Earthquake Engineering Symposium (in Japanese)

18 O¨zdamar, L., Ekinci, E., Ku¨c¸u¨kyazaci, B., (2004). Emergency logistics planning in natural disasters. Annals of Operations Research 129, 217–245.

19 Peeta Sreenivas, Viswanath Kanan, Salman Sibel, (2006) "Strengthening the Links of a Network for Disaster Response", 7th International Workshop TPMDC-06, IIT Bombay.

20 Ray, J., (1987). A multi-period linear programming model for optimally scheduling the distribution of food-aid in West Africa. M.S. Thesis, University if Tennessee, Knoxville, TN.

21 Rathi, A.K., Church, R.L., Solanki, R.S., (1992). Allocating resources to support a multicommodity flow with time windows. Logistics and Transportation Review 28 (2), 167–188.

22 Sheu, Jiuh-Biing, (2006), An emergency logistics distribution approach for quick response to urgent relief demand in disasters

23 Sawaidis, Paraskeuas, Lakakis, Konstantinos, Ifadis, Ioannis,(2002), "Organization of Emergency Response After a Major Disaster Event in an Urban Area with the Help of an Automatic Vehicle Location and Control System" GPS Solutions, Vol. 5, No. 4, pp. 70-79 (2002)

24 Shinojuka, M., Moore, J., Gordon, P., Richardson, H.W., Chang, S., Cho., S.-B., (1998). An integrated model of highway networks and the spatial metropolitan economy : towards a general model of how earthquake losses affect the economy. NCEER Bulletin 12(1), 8-15.

25 Sohn, J. (2006). "Evaluating the significance of highway network links under the flood damage: An accessibility approach." Transp. Res., Part A: Policy Pract., 40 (6), 491–506.

26 Sohn, J., Kim, T. J., Hewings, G. J. D., Lee, J. S., and Jang, S.-G. (2003). "Retrofit priority of transport network links under an earthquake." J. Urban Plann. Dev., 129(4), 195–210.

27 Virginia P. Sisiopiku (2007) Application of Traffic Simulation Modeling for Improved Emergency Preparedness Planning Journal of Urban Planning and Development, Vol. 133, No. 1, March 1, 2007.

28 wakabayashi, H.,Kameda,H., (1992). Network performance of highway systems under the earthquake effects: a case study of the 1989 Loma Prieta earthquake. Proceedings of the US-Japan Workshop on earthquake Disaster Prevention for Lifeline Systems. Tsukuba Science City, Japan, pp 215-232.

29 Wei Yi, Arun Kumar (2006), "Ant colony optimization for disaster relief operations", Transportation Research Part E 43 (2007) 660–672